ISBN 978-1-5278-9102-9
PIBN 10926012

# 1 MONTH OF
# FREE
# READING

## at
## www.ForgottenBooks.com

By purchasing this book you are eligible for one month membership to ForgottenBooks.com, giving you unlimited access to our entire collection of over 700,000 titles via our web site and mobile apps.

To claim your free month visit:
www.forgottenbooks.com/free926012

# TRANSACTIONS

(OF THE)

## AMERICAN INSTITUTE OF MINING AND METALLURGICAL ENGINEERS

(INCORPORATED)

WITH WHICH IS CONSOLIDATED THE

### AMERICAN INSTITUTE OF METALS

## VOL. LXI

CONTAINING PAPERS AND DISCUSSIONS ON GEOLOGY, MINING AND
MILLING, COAL, OIL AND GAS, PRESENTED AT THE COLORADO
MEETING, SEPTEMBER, 1918, AT THE MILWAUKEE
MEETING, OCTOBER, 1918, AND AT THE NEW
YORK MEETING, FEBRUARY, 1919.

NEW YORK, N. Y.
PUBLISHED BY THE INSTITUTE
AT THE OFFICE OF THE SECRETARY
29 WEST 39TH STREET
1920

THE MAPLE PRESS YORK PA

# PREFACE

In this volume are the papers and discussions on Geology, Mining and Milling, Coal, Oil and Gas that were presented at the Colorado and Milwaukee meetings in the Fall of 1918 and at the New York meeting in February, 1919.

These papers were printed in *Bulletins* 135 to 147, but because the papers on Iron and Steel that appear in these *Bulletins* are reserved for Vol. LXII, Vols. LX and LXI together supersede only *Bulletins* 135 to 139, inclusive.

An account of the Proceedings of the New York meeting, February, 1919, is printed in this volume, and special mention is made of the men who died in service during the war.

# CONTENTS

# OFFICERS AND DIRECTORS

### For the year ending February, 1920

#### PRESIDENT

HORACE V. WINCHELL,. . . . . . . . . . . . . . . MINNEAPOLIS, MINN.

#### PAST PRESIDENTS

PHILIP N. MOORE, . . . . . . . . . . . . . . . . . . . . . .ST. LOUIS, MO.
SIDNEY J. JENNINGS, . . . . . . . . . . . . . . . . NEW YORK, N. Y.

#### FIRST VICE-PRESIDENT

A. R. LEDOUX, . . . . . . . . . . . . . . . . . . . NEW YORK, N. Y.

#### TREASURER

GEORGE D. BARRON, . . . . . . . . . . . . . . . NEW YORK, N. Y.

#### SECRETARY

BRADLEY STOUGHTON, . . . . . . . . . . . . . . . NEW YORK, N. Y.

#### VICE-PRESIDENTS

CHARLES W. GOODALE,[1] . . . . District 5. . . . . . . . . BUTTE, MONT.
MARK L. REQUA,[1] . . . . . . District 6. . . . . . SAN FRANCISCO, CAL.
HENRY S. DRINKER,[2] . . . . . District 2. . . . . . . .So. BETHLEHEM, PA.
ROBERT M. RAYMOND,[2] . . . . District 0 . . . . . . . NEW YORK, N. Y.
EDWIN LUDLOW,[3] . . . . . . . District 2. . . . . . . . . .LANSFORD, PA.
A. R. LEDOUX,[3] . . . . . . . . District 0 . . . . . . . NEW YORK, N. Y.

#### DIRECTORS

ALLEN H. ROGERS,[1]. . . . . . . District 1. . . . . . . . .BOSTON, MASS.
HOWARD N. EAVENSON,[1] . . . District 9. . . . . . . . . GARY, W. VA.
J. V. N. DORR,[1] . . . . . . . . District 0. . . . . . . NEW YORK, N. Y.
WILLIAM R. WALKER,[1] . . . . District 0. . . . . . . NEW YORK, N. Y.
WILLET G. MILLER,[1] . . . . . District 11 . . . . . TORONTO, CANADA.
FREDERICK G. COTTRELL,[2] . . District 6. . . . . . SAN FRANCISCO, CAL.
HENNEN JENNINGS,[2] . . . . . District 9. . . . . . WASHINGTON, D. C.
GEORGE C. STONE,[2] . . . . . . District 0. . . . . . . NEW YORK, N. Y.
SAMUEL A. TAYLOR,[2] . . . . . District 2. . . . . . . . PITTSBURGH, PA.
ARTHUR THACHER,[2] . . . . . District 3. . . . . . . . .ST. LOUIS, MO.
J. V. W. REYNDERS,[3] . . . . . District 0. . . . . . . NEW YORK, N. Y.
GEORGE D. BARRON,[3] . . . . . District 0. . . . . . . NEW YORK, N. Y.
CHARLES F. RAND,[3] . . . . . . District 0. . . . . . . NEW YORK, N. Y.
LOUIS S. CATES,[3] . . . . . . . District 7. . . . . . . . . . RAY, ARIZ.
STANLY A. EASTON,[3]. . . . . . District 5 . . . . . . . . .KELLOGG, IDA

---

[1] Until Feb., 1920.　　　[2] Until Feb., 1921.　　　[3] Until Feb., 1922.

| Year of Election | | |
|---|---|---|
| | **HONORARY MEMBERS** | |

1876. PROF. RICHARD ÅKERMAN............................Stockholm, Sweden
1913. DR. FRANK DAWSON ADAMS.........................Montreal, Canada.
1888. PROF. HATON DE LA GOUPILLIERE ... ... ............Paris, France.
1906. SIR ROBERT A. HADFIELD   .... ....................London, England.
1917. HERBERT C. HOOVER...... .................. .........San Francisco, Cal.
1919. ROBERT W. HUNT..............................Chicago, Ill.
1915. JAMES FURMAN KEMP ..... .........................New York, N. Y.
1905. PROF. HENRI LOUIS LE CHATELIER ......................Paris, France.
1913. EZEQUIEL ORDOÑEZ ......................Mexico City, Mexico.
1909. ALEXANDRE POURCEL................................Paris, France.
1911. PROF. ROBERT H. RICHARDS............................Boston, Mass.
1906. JOHN E. STEAD..............................Middlesbrough, England.
1902. PROF. DIMITRY CONSTANTIN TSCHERNOFF ...............Petrograd, Russia.
1910. PROF. TSUNASHIRO WADA...........................·....Tokyo, Japan.
1907. CHARLES D. WALCOTT................................Washington, D. C.

| HONORARY MEMBERS (*Deceased*) | | Year of Decease |
|---|---|---|

1872. BELL, SIR LOWTHIAN....................................1904
1905. CARNEGIE, ANDREW....................................1919
1892. CASTILLO, A. DEL....................................1895
1902. CONTRERAS, MANUEL MARIA........................................1902
1888. DAUBRÉE, A....................................1896
1906. DOUGLAS, JAMES........  ............... .....................1918
1884. DROWN, THOMAS M....................................1904
1890. GAETZSCHMANN, MORITZ........................................1895
1873. GRUNER, L....................................1883
1891. KERL, BRUNO....................................1905
1895. LE CONTE, JOSEPH....................................1901
1891. LESLEY, J. P....................................1896
1899. OSMOND, FLORIS....................................1912
1890. PATERA, ADOLPH....................................1890
1886. PERCY, JOHN....................................1889
1888. POSEPNY, FRANZ....................................1895
1911. RAYMOND, ROSSITER W ........................................1918
1884. RICHTER, THEODOR....................................1898
1899. ROBERTS-AUSTEN, W. C....................................1902
1890. SERLO, ALBERT....................................1898
1880. SIEMENS, C. WILLIAM ....................................1883
1909. SWANK, JAMES M....................................1914
1872. THOMAS, DAVID....................................1882
1873. TUNNER, PETER R. VON....................................1897
1885. WEDDING, HERMANN ....................................1908

## PROCEEDINGS OF THE 119TH MEETING

### New York, February, 1919

From the beginning to the end, the attendance and enthusiasm of the 119th meeting, which was held in New York, Feb. 17 to 20, surpassed all expectations. Besides the ten technical sessions, one of which was in conjunction with the American Institute of Electrical Engineers, and one a session of the National Research Council, there were two memorial meetings, one for Dr. Raymond and one for the members who died in service; two joint sessions with the Canadian Mining Institute; and a meeting devoted to pictures of copper mining, milling, and smelting.

Monday's sessions were devoted to the problems of industrial organization, the Institute of Metals Division, and to petroleum and gas. Among the special features of these sessions were the topical discussions on housing and Americanization. Americanization, too, was the topic for discussion by the Woman's Auxiliary Wednesday morning.

On Tuesday, nearly 100 members of the Canadian Mining Institute were present to discuss the possibility of bringing about uniform mining laws for the United States, Canada, and Mexico, to obviate the maintaining separate legal departments and managerial forces for the several countries; and to avoid the confusion which, it is said, has led to a duplication of effort and has sometimes created a barrier to international coöperation in mining. With the Canadian delegation were D. B Dowling, President of the Canadian Institute; John McLeish, Statistician in the Department of Mines; O. S. Finnie, Mining Engineer in the Department of the Interior; H. H. Rowatt, Controller in the Department of the Interior; and Colonel Machin, of the Department of Justice.

Wednesday was devoted to the work of the National Research Council, which occupied the entire day and not only the morning as was planned; to the problems of mining, milling, and geology; and to the study of welding problems, together with the American Institute of Electrical Engineers. In all cases the discussion was most animated.

The committees in charge of the arrangements were as follows:

*Committee on Arrangements and Reception.*—ALLEN H. ROGERS, Chairman; WALTER S. DICKSON, Secretary; S. H. BALL, G. D. BARRON, H. W. HARDINGE, P. W. HENRY, MRS. SIDNEY J. JENNINGS, J. E. JOHNSON, JR., A. C. LUDLUM, E. P. MATHEWSON, P. A. MOSMAN, H. C. PARMELEE, F. T. RUBIDGE, FOREST RUTHERFORD.

*Finance Committee.*—GEORGE D. BARRON, Chairman; J. PARKE CHANNING, WALTER DOUGLAS, JAMES GAYLEY, J. H. JANEWAY, C. F. KELLEY, W. H. NICHOLS, JR., W. W. MEIN, WILLIAM L. SAUNDERS.

*Excursion Committee.*—P. W. HENRY, Chairman; J. E. JOHNSON, JR.

*Banquet Committee.*—A. C. LUDLUM, Chairman; F. T. RUBIDGE, E. B STURGIS.

*Luncheon Committee.*—P. A. MOSMAN, Chairman; E. MALTBY SHIPP, C. A. BOHN.

*Automobile Committee.*—H. W. HARDINGE, Chairman.

*Entertainment Committee.*—E. P. MATHEWSON, Chairman; LAWRENCE ADDICKS, LUCIUS W. MAYER.

*Tellers.—For Officers.*—W. S. DICKSON, H. N. SPICER, WILLIAM YOUNG WESTERVELT. *For Proposed Amendment.*—C. Q. PAYNE, R. C. WARRINER, A. D. BEERS.

### NEWLY ELECTED OFFICERS OF THE INSTITUTE

The following officers were elected: Horace V. Winchell, of Minneapolis, President; Vice-presidents: Edwin Ludlow, Lansford, Pa.; A. R. Ledoux, New York; Directors: J. V. W. Reynders, New York; George D. Barron, Rye, N. Y.; Charles F. Rand, New York; Louis S. Cates, Ray, Ariz.; Stanly A. Easton, Kellogg, Idaho.

## CHANGE IN NAME OF THE INSTITUTE

By a letter ballot of the members of the Institute, as reported by the tellers and announced by the President at the annual business meeting of the Institute on Feb. 18, 1919, the name of the American Institute of Mining Engineers has been changed to the American Institute of Mining and Metallurgical Engineers. The vote for the amendment to the Constitution making this change was a heavy one for such matters, showing the very wide and deep interest taken by the members in this vital change. The actual count was 1274 for and 672 against.

It is the consensus of opinion that this marks an epoch in the history and usefulness of this Institute, and it is expected that the change will presage a large and healthy growth in metallurgical fields. Many metallurgists who knew nothing about mining were loath to join a society of mining engineers, despite the fact that over 35 per cent. of the papers published in our *Transactions* during the past five years have dealt with metallurgical subjects.

The choice of the name to be balloted for had been made by popular canvass and has the double advantage, much to be desired, of retaining the word Engineers and maintaining the former initials of the Institute, A. I. M. E. This change will doubtless make the Institute *the* metallurgical society of the country as it has been *the* mining society.

## CANADIAN MINING INSTITUTE DAY

Four joint discussions formed the program of the sessions with the Canadian Mining Institute on Tuesday. T. W. Gibson, the representative of the Canadian Mining Institute, opened the discussion on the subject of "Principles of Mine Taxation;" the American Institute was represented by Ralph Arnold, of the United States Treasury Department. Later, Alfred G. Heggem, of Tulsa, Okla., cited an instance in which a high rate of taxation decreased the Government's revenue by retarding the transfer of property. The second discussion in the morning was on the subject "Industry Democracy, and Education." C. V. Corless represented the visitors and President Jennings the Americans. During the latter part of this discussion, D. B. Dowling, President of the Canadian Mining Institute, presided.

In the afternoon, T. A. Rickard, of San Francisco, Cal., opened the discussion by a paper entitled "The English-speaking People." He was followed by Dr. A. R. Ledoux, of New York, who gave a brief talk on "International Coöperation." In it he spoke of the necessity of closer coöperation, also, with Mexico. President Jennings later said that he is planning a trip to Mexico, when he expects to meet with the Local Section there, at which time he hopes to initiate some discussion of problems that confront the American engineer. The question of a uniform mining law for North America provoked considerable discussion. While most of the talk was confined to the various Canadian laws, the possibilities of uniform laws with Mexico and Central America were also considered.

## TECHNICAL SESSIONS

### *Institute of Metals Division*

One session of the Institute of Metals Division was held on Monday morning, Feb. 17, Mr. W. M. Corse presiding. The following papers were presented:

Effect of Temperature, Deformation and Grain Size on the Mechanical Properties of Metals. By Zay Jeffries. (Presented by the author; discussed by C. H. Mathewson, P. D. Merica, S. L. Hoyt, and the author.)

Volatilization of Cuprous Chloride on Melting Copper Containing Chlorine. By S. Skowronski and K. W. McComas. (Presented by Mr. Skowronski.)

Automatic Copper Plating. By J. W. Richards. (Presented by the author; discussed by A. Silverman.)

Two I sta s of Mobility of Gold in the Solid State. By E. Keller. (Presented by title.) n nce

First Year of Leaching by the New Cornelia Copper Co. By H. A. Tobelmann and J. A. Potter. (Presented by title; written discussion by C. A. Rose.)

Die Castings and Their Application to the War Program. By Charles Pack. (Presented by title; written discussion by Jesse L. Jones.)

Electric Furnace Problems. By J. L. McK. Yardley. (Presented by title.)

---

The second session was held on Monday afternoon, Feb. 17, Mr. William B. Price presiding. The following papers were presented:

Comparison of Grain-size Measurements and Brinell Hardness of Cartridge Brass. By W. H. Bassett and C. H. Davis. (Presented by Mr. Bassett; written discussions by W. R. Hibbard, Arthur Phillips, C. H. Mathewson, W. B. Price.)

Manganese Bronze. By P. E. McKinney. (Presented by the author; discussed by H. J. Roast, G. H. Clamer, C. R. Spare, W. Campbell, W. Hamilton, H. Traphagen; written discussions by Jesse L. Jones, W. M. Corse.)

Metals and Alloys from a Colloid-chemical Viewpoint. By Jerome Alexander. (Presented by the author; written discussion by W. D. Bancroft.)

Standards for Brass and Bronze Foundries and Metal-finishing Processes. By Lillian Erskine. (Presented by W. M. Corse.)

## Industrial Organization

A session on Industrial Organization was held on Monday morning, Feb. 17, Mr. B. F. Tillson presiding, at which the following papers were presented:

New Spirit in Industrial Relations. By H. M. Wilson. (Presented by the author; discussed by C. W. Goodale, M. D. Cooper, B. F. Tillson, E. A. Holbrook; written discussions by W. D. Brennan, W. L. Clark, and R. R. Goodrich.)

Use of Cripples in Industry. By J. P. Munroe. (Presented by the author.)

Mental Factors in Industrial Organization. By T. T. Read. (Presented by the author.)

Prevention of Illness Among Employees in Mines. By A. J. Lanza. (Presented by the author; discussed by C. W. Goodale, B. F. Tillson, H. M. Wilson, and the author; written discussions by C. E. Calvert and J. J. Carrigan.)

---

A second session on Industrial Organization was held on Monday afternoon, Feb. 17, with Mr. T. T. Read in the chair. The following papers were presented:

Mental Tests in Industry. (Presented by Robert M. Yerkes; discussed by Major Yerkes, Bradley Stoughton, B. F. Tillson.)

Need for Vocational Schools in Mining Communities. By J. C. Wright. (Presented by the author; discussed by Marguerite Walker Jordan.)

Professional Lecture—U. S. Employment Service. By I. W. Litchfield.

Topical Discussion on Housing. The speakers were D. Eppelsheimer and Laurence Veiller.

Topical Discussion on Americanization. The speakers were A. J. Beatty and W. C. Smith.

## Petroleum and Gas

On Monday afternoon, Feb. 17, there was a session on Petroleum and Gas. Capt. A. F. Lucas presided, and the following papers were presented:

Water Troubles in the Mid-continent Oil Fields, and Their Remedies. By Dorsey Hager and G. W. McPherson. (Presented by J. F. Kemp.)

Natural-gas Storage. By L. S. Panyity. (Presented by the author; discussed by C. H. Shaw, David White, Eugene Coste, I. N. Knapp, and the author.)

Economic and Geologic Conditions Pertaining to the Occurrence of Oil in the North Argentine-Bolivian Field of South America. By S. C. Herold. (Presented by M. Bates.)

Petroleum Hydrology Applied to Mid-Continent Field. By R. O. Neal. (Presented by title; written discussion by G. Sherburne Rogers.)

Cement Plugging for Exclusion of Bottom Water in the Augusta Field, Kansas. By H. R. Shidel. (Presented by title.)

## Session with Canadian Mining Institute

On Tuesday, Feb. 18, known as Canadian Mining Institute Day, three discussions took place, as follows:

Principles of Mine Taxation. President Sidney J. Jennings, A. I. M. E., presided. Mr. T. W. Gibson, Canadian Mining Institute, opened the discussion, and was followed by Mr. Ralph Arnold and Mr. A. G. Heggem, A. I. M. E.

Industry, Democracy and Education. President D. B. Dowling, Canadian Mining Institute, presided. Mr. C. V. Corless, Canadian Mining Institute, presented the paper on this subject. Mr. S. J. Jennings, A. I. M. E., made a short address.

International Coöperation in Mining in North America and Uniform Mining Law for North America. President S. J. Jennings, A. I. M. E., presided. The speakers were: For the Canadian Mining Institute, Mr. T. E. Godson, Mr. H. H. Rowatt, Mr. T. W. Gibson, Mr. T. C. Denis, Mr. A. R. Chambers; for the A. I. M. E., Mr. T. A. Rickard, Dr. A. R. Ledoux, Mr. H. V. Winchell.

## Iron and Steel

There were two sessions on Iron and Steel on Tuesday, Feb. 18. At the first session, Dr. J. W. Richards presided, and at the second session, Mr. J. E. Johnson, Jr., was in the chair. The following papers were presented:

Does Forging Increase Specific Density of Steel? By H. E. Doerr. (Presented by title; written discussion by J. S. Unger.)

Flaky and Woody Fractures in Nickel-steel Gun Forgings. By C. Y. Clayton, F. B. Foley and F. B. Laney. (Presented by Mr. Foley; written discussion by O. A. Knight, J. A. Mathews.)

Static, Dynamic and Notch Toughness. By Samuel L. Hoyt. (Presented by the author; written discussion by P. D. Merica, G. Charpy, J. A. Mathews.)

Development of Grain Boundaries in Heat-treated Alloy Steels. By R. S. Archer. (Presented by title; written discussion by J. A. Mathews.)

Water-cooled Equipment for Open-hearth Steel Furnaces. By W. C. Coffin. (Presented by the author; written discussions by L. L. Knox, J. S. Unger.)

Prevention of Columnar Crystallization by Rotation During Solidification. By H. M. Howe and E. C. Groesbeck. (Presented by Dr. Howe.)

Basic Refractories for the Open Hearth. By J. Spotts McDowell and R. M. Howe. (Presented by Mr. McDowell; written discussions by J. S. Unger, F. R. Pyne.)

The Shimer Case-hardening Process. By J. W. Richards. (Presented by the author.)

Metallographic Investigation of Transverse-fissure Rails with Special Reference to High-phosphorus Streaks. By G. F. Comstock. (Presented by the author; written discussions by J. E. Howard, P. H. Dudley, C. B. Bronson.)

Effect of Rate of Temperature Change on Transformations in an Alloy Steel. By H. Scott. (Presented by the author.)

Davidson Process of Casting Formed Tools. By J. E. Johnson, Jr. (Presented by the author; written discussion by J. A. Mathews.)

## Session of National Research Council

The session of the National Research Council was begun on Wednesday morning, Feb. 19, and was continued throughout the afternoon. Mr. A. A. Stevenson and Dr. H. M. Howe presided. The following papers were presented:

Production of Ferromanganese in the Blast Furnace. By P. H. Royster. (Presented by the author.)

Effect of Cold-working and Rest on Resistance of Steel to Fatigue under Reversed Stress. By H. F. Moore and W. J. Putnam. (Presented by Mr. Moore; written discussions by J. E. Howard, J. B. Kommers.)

Use of Manganese Alloys in Open-hearth Practice. By Samuel L. Hoyt. (Presented by the author; discussed by H. M. Howe, F. N. Speller, H. Traphagen, and the author; written discussions by C. L. Kinney, Jr., J. R. Cain.)

A Volute Aging Break. By H. M. Howe and E. C. Groesbeck. (Presented by Dr. Howe.)

Report of Steel Ingot Committee, National Research Council. By H. M. Howe, H. S. Rawdon and F. H. Schoenfuss. (Presented by Dr. Howe.)

Microstructural Features of Flaky Steel. By H. S. Rawdon. (Presented by the author.)

In connection with the last-named paper, there was a general discussion on Flaky Steel, involving also the paper by Messrs. Clayton, Foley and Laney, presented at the Iron and Steel Session on Tuesday morning. This discussion was participated in by: M. H. Wickhorst, L. F. Fry, C. B. Bronson, A. A. Stevenson, D. E. Field, J. A. Mathews, H. C. Boynton, P. McKinney, H. D. Hibbard, H. Traphagen, R. J. Wysor, Leonard Waldo, A. H. Coles, J. E. Johnson, Jr., Boyd Dudley, Jr.

## Mining, Milling and Geology

The session on Mining, Milling and Geology was held on Wednesday morning, Feb. 19, Mr. Charles W. Goodale presiding. The following papers were presented:

Mining Methods of United Verde Extension Mining Co. By C. A. Mitke. (Presented by title.)

Study of Shoveling as Applied to Mining. By G. Townsend Harley. (Presented by title.)

Fine Crushing in Ball-mills. By E. W. Davis. (Presented by title; written discussion by A. L. Blomfield.)

Problems Involved in Concentration and Utilization of Domestic Low-grade Manganese Ore. By Edmund Newton. (Presented by title; discussed by C. W. Goodale, Kirby Thomas.)

Notes on Certain Ore Deposits of the Southwest. By W. Tovote. (Presented by title; discussed by W. G. Mitchell, written discussion by Philip D. Wilson.)

Petrographic Notes on the Ore Deposits of Jerome, Ariz. By Marion Rice. (Presented by title.)

Anthracite Mining Costs. By R. V. Norris. (Presented by the author; discussed by E. W. Parker, S. D. Warriner, Paul Sterling, W. V. de Camp, Edwin Ludlow, and the author.)

Work of National Production Committee, U. S. Fuel Administration. By J. B. Neale. (Presented by the author; discussed by Robert Peele, R. D. Hall.)

Distribution of Coal Under the U. S. Fuel Administration. By J. D. A. Morrow. (Presented by title.)

*Session with American Institute of Electrical Engineers, on Electric Welding*

This session on Electric Welding was held on Wednesday afternoon, Feb. 19, Dr. Comfort A. Adams presiding.   The following papers were presented:

Microstructure of Iron Deposited by Electric Arc Welding.   By G. F. Comstock. (Presented by the author.)
Path of Rupture in Steel Fusion Welds.   By S. W. Miller.   (Presented by the author; discussed by A. M. Candy, G. F. Comstock.)
Welding Mild Steel.   By H. M. Hobart.   (Presented by the author; discussed by A. M. Candy, C. J. Holslag, W. L. Merrill, D. C. Alexander, Jr., W. Spraragen, J. C. Armor, S. V. Goodall, R. P. Jackson.)
*Electric Welding in Shipbuilding.   By S. V. Goodall.   (Presented by the author; discussed by W. H. Hill, R. R. Horner, Capt. Corbett.)
*Fusion in Arc Welding.   By O. H. Eschholz.   (Presented by the author.)

## THE SMOKER

The annual Smoker of the Institute was held in the rooms of the Engineering Societies' Building, Monday evening, Feb. 17, and was attended by about 400 members and guests.   Mr. E. P. Matthewson was Chairman of the Entertainment Committee, and produced one of the most varied and interesting—not to say exciting—programs provided for any of the Smokers.   Patriotic and other songs, comprising largely the favorites in the Army today, were sung.   The members were seated around small tables decorated with college flags, and laden with refreshments.   A cartoonist made rapid-fire charcoal sketches of prominent members of the Institute, which were so cleverly done that the identification was made before each sketch was completed.   Remarks by the cartoonist added to the appreciation of the sketches themselves.

Following this, were various new and novel films: among them one showing the interior mechanism of an automatic adding machine at work, and a most interesting one showing in section the workings of the Lewis machine gun.

It was the largest and most successful smoker ever held by the Institute, and if Mr. Mathewson were not such a good metallurgist, he would have been an equally famous purveyor of amusements.

## LIEUTENANT PAT O'BRIEN

On Tuesday night, Feb. 18, to an auditorium filled with members and lady guests, Lieutenant Pat O'Brien, formerly of the Royal Flying Corps of Canada, the United States Army, and the French Foreign Legion, gave a vivid and intensely interesting description of his flying exploits, which landed him wounded and unconscious inside the German lines; of his life in various German prisons and prison camps; and of his successful escape to England, through Luxemburg, Belgium, and Holland. Lieutenant O'Brien's marvelous escape after a fall of 2000 ft., after which the wreckage of his machine had to be cut away in order to extract him, was even more thrilling when told than as written in his book "Outwitting the Hun," which is his published story of the episode.   His

---

* Published in *Proceedings*, American Institute of Electrical Engineers.

recital of how he executed a daring plunge through the window of a train loaded with prisoners, running at 30 miles an hour; how it took him seventy-two days, or rather nights, to cover 300 miles to the Dutch border; how he lived on raw vegetables, swam rivers and canals, and finally dug under the highly charged barbed wire barrier between Belgium and Holland, as well as the many instances of grim humor with which his speech was interspersed, brought many bursts of applause from a highly appreciative audience. At the conclusion of the speech, many of the audience went to the fifth floor, where an informal dancing party was held, during which ice cream, cake, and punch were served. It was a bit different from the other evening entertainments provided for the members and ladies, and was an unusual success.

## THE ANNUAL BANQUET

The annual banquet of the American Institute of Mining Engineers was held at the Hotel Biltmore on Wednesday evening, Feb. 19. The dinner was preceded by the President's Reception, Mr. and Mrs. Horace V. Winchell and Mr. and Mrs. Sidney J. Jennings receiving. The ballroom was beautifully decorated with flags of the Allies and the electric sign of the A. I. M. E. There were 450 members and guests present, and it was the largest and one of the most enjoyable of the dinners ever held by the Institute.

An innovation in the dinner arrangements was cutting the list of speakers to three, who were the President of the Canadian Mining Institute, and the retiring and incoming Presidents of the American Institute. The speeches were all short and very interesting.

After the dinner, a larger number remained for the dance than heretofore, and dancing was indulged in until the wee small hours of the morning. There was a large sprinkling of French, Canadian, and American uniforms which with the many beautiful and elaborately gowned ladies made it the most brilliant affair of its kind the Institute has ever given.

### MENU

Blue Point Oysters
Velouté Favorite
Celery      Salted Almonds      Olives
Halibut Steak, Italian Style
Risotto Joinville
Roast Philadelphia Capon Chaructière
Salad Parisien
Nelusko Praliné
Cakes
Demi Tasse

### TOASTMASTER'S REMARKS

Each speaker was introduced by the toastmaster, Lawrence Addicks, who said:

In this great year of victory it seems but fitting that we, as members of the *American* Institute of Mining and Metallurgical Engineers should first rise in response to a toast to America. (America was then sung by all.)

My instructions from the Dinner Committee were exceedingly brief. They consisted of three or four typewritten lines reading: "You will have three speakers to introduce—Mr. Dowling, Mr. Jennings and Mr. Winchell. Do not let any of them speak long, give each of them a good rub, especially Winchell, and do not say much yourself." The last line leaves me much in the position of the young husband who looked adoringly at his wife and said: "Dearest, I do worry so when I think what would become of you if I should die.' "Why," she replied, "nothing would happen to me; I should stay right here; the real question is what would become of you." So I shall proceed with some caution.

As the smoke of battle clears away, we begin to see that the war has not been wholly a thing of evil. The ladies, in a new sense, have become our companions in arms; and as peace on earth is reëstablished we find an added measure of goodwill toward men. The Allies have been bound together not by the ties of mere political expediency but by those of a true blood brotherhood. This is especially true of the English-speaking peoples and of the United States of America and her near neighbor Canada. Some of us make frequent trips across the border and we know with what uniform cordiality we are received. Our worthy president here about two weeks ago came near making an unannounced visit to Canada. You know he is very fond of horseback. He has a theory that there is no place where a man can think so clearly as on the back of a horse; and, like ex-president Taft, he does not care what the horse thinks. Well, as I have it, Allen Rogers and Galen Stone were standing outside of Mr. Rogers' house in Brookline discussing flotation of mining stocks. All of a sudden Mr. Jennings comes tearing down the street on a horse, with one foot in the stirrup and both hands buried in the animal's mane. "Hey," Mr. Rogers called out, "where are you going?" From the distance Mr. Jennings cried: "I do not know, but the horse belongs in Montreal."

We in our turn have always been glad to see the delegation from the Canadian Mining Institute who have made a practice of attending our annual meetings for some years past. But I think that it is with a new feeling of fellowship that we welcome them this year and it is in this spirit that I greet Mr. D. B. Dowling, president of the Canadian Mining Institute, and ask him to say a few words to us.

## PRESIDENT DOWLING'S ADDRESS

I feel that it is a great honor to be invited to reply to this toast on behalf of the Canadian Mining Institute. I wish to thank you for your very cordial reception of the toast and for the sentiments expressed in your invitation to the Institute to coöperate in a series of joint meetings. It is an earnest of the mutual regard which the peoples of the two countries have for each other. We have been allies in a very great war and the feelings of good-will engendered by that alliance will survive, I believe, through the strenuous times of peace. The other nations of the alliance had, through many changes in history, fought each other and for each other and had thereby gained a clear conception of the social aims and moral character of their peoples. With this knowledge came mutual trust—the one element necessary for united action or lasting agreement. On this continent the two peoples are drawn together by bonds of a common origin and, largely, a common language. No war has been necessary to force declarations of national aims on which to base terms of amity. And now our peoples who had studied the arts of peace together have together passed through fire—drawn into the conflict by a like horror at the sacrilege of the crucifixion of the rights of humanity. Their courage and manhood has been tested at Ypres and Chateau Thierry by the great assayer war, and, together, with thankful hearts we see the triumph of right over might and better than ever know that we have the same conception of the government of the people for the people.

We are met to celebrate the close of the association of our brethren at the front in the fight for freedom for the world—a freedom that to us seems natural but which to the majority of the allied nations is a heritage derived from the protest made by the American colonies against German ideas of government of which remnants had until then survived in Great Britain.

The common heritage of freedom enjoyed by the many units of the present British Empire proved to be a bond which, although almost invisible to German perception, was of the strongest fiber. The act of offering help to the motherland brought to us the realization that as a people we were assuming the responsibilities of a nation and it is this realization which fills us with a modest pride when as members of a new nation within the Empire we are invited to coöperate in your deliberations. We in return offer the hospitality of the Lady of the Snows at our Montreal meeting and let me assure you that the reception will be one that will indicate the beating of a warm heart. Our people are essentially homemakers, our industry is largely agricultural, but our problems in mining and metallurgy, while not on a grand scale, are interesting. We hope to discuss regions of greatly varying topography and climate. Those of you who can attend will see views of the Arctic coast, of the more southern plains and mountains, and—a theme that has been of common interest since the days of the first crusade—views of the last crusade to the Holy Land under General Allenby. For our own people whose hearts are still with the boys at the front, we shall have the War Records pictures of the last days of the war with the Canadians at the taking of Mons. This city, it will be remembered, was the point at which the first British Army, the "Contemptibles," took their bulldog grip on the German legions.

Again, Mr. Chairman, may I in closing tender on behalf of the Canadian Mining Institute our greetings and our wish for continued cordial relations and a mutual prosperity.

At the close of Mr. Dowling's speech, Mr. Addicks said: It has long been the custom to afford an opportunity to the retiring president to sing his swan song, and it is always hard to break old customs. You know they say whenever you put a tractor on a farm the hired man quits and when you ask him what the trouble is, he says, "Wal, fer 20 years I been tryin' to learn to understan' a mule and at my time of life I don't intend ter take on this new fangled thing." Now for 20 years we members of the Institute have been trying to learn to understand these farewell orations and we do not intend to give up now. Some of you may think that the words and music of this song are to be found in the Secretary's annual report in the current issue of the Bulletin, but I assure you that this is not so. It does grave injustice to the responsibilities of the office of President of this Institute. Let us just consider for a moment his manifold duties.

First, his foremost duty is to bring the membership of the Institute up to the point where the incoming president will have to rely solely upon the birthrate of the population. Second, and this is most important, he must devise some solemn stunt with which to make the annual attempt to coax the Mining and Metallurgical Society into the fold. Then, he must make a semi-annual tour of the provinces to visit all of the Local Sections and deal with incipient cases of heresy and schism, diseases that appear to be endemic in Southern California and Southeastern Missouri. Next, ex officio, he has to try to keep the peace between the Editor of the *Engineering and Mining Journal* and the Editor of the *Mining and Scientific Press.* Finally, this year there was a matter of the attitude of the Board of Directors toward the prohibition amendment—a most serious question.

However, while there may still be a few citizens not yet members of the Institute, who will have twelve dollars after March 15, and while the Mining and Metallurgical Society may still claim to represent the certified, pasteurized, grade A variety of mining engineer, this Institute has never had a more successful year than under the able administration of Mr. Sidney Johnston Jennings, whom I now have the pleasure to introduce.

### PRESIDENT JENNINGS' ADDRESS

It is with a very special sense of pride and gratification that I am able to extend to the Canadian Mining Institute a welcome on behalf of the American Institute of Mining and Metallurgical Engineers. Last October, we were lured to Colorado by the tales of its wonderful climate —where it only rained once in 6 months. We were in Colorado Springs four days and took away with us all their bad weather for the next 2 years; but we also took away with us memories of a warm and enduring hospitality that more than made up for the rainy weather. At that time, I invited all our western members to this meeting but pointed out to them that New York had the second worst climate in the United States. New York, the superb, has extended its all-embracing arms in a most hearty welcome and has shown by the bright sunshine obtaining during this meeting that I had maligned her climate. A New Yorker, when he is not a Southerner, is a New Englander, who has gone west and come back to New York. We can under this definition all claim to be New Yorkers.

I am not going to take up in detail the reports of our Secretary and Treasurer, which give you a full account of the work of our Institute.

I want to draw your attention to the numbers of your membership, which now amounts to, I think, 7300, a greater membership than ever recorded in our history. I estimate, however, that there are at least 14,000 people in the United States who as members, associates, or junior associates, should belong to our Institute. That is a figure that we should set as a mark and not rest content until it is achieved. With the increase in membership, the capacity of the Institute to serve its members will be correspondingly increased.

As far as the financial condition is concerned, I shall only say that we are solvent. This is due, in large measure, to the generosity of one of our Directors who, when it looked as if we would not be able to carry out our plan of remitting the dues of our members in active service, made a large donation, which enabled us to close the year with a balance on the right side.   Our budget for the coming year has been conservatively planned and our activities will be confined to our resources.

The work of our members as individuals and the Institute as a whole has contributed to help win the war.   The service flag opposite me shows that 845 members wear the uniform of the United States and the Allies; twenty-three are known to have paid the supreme sacrifice.   Although exact details are not obtainable, enough is known to be able to state that there are many others.   When details are obtainable their names and records will be published in the *Bulletin*.

During the past year I paid a visit to six of our Local Sections, of which there are now fifteen.   Our Secretary and our First Vice-president, Mr. Goodale, visited several others.   The ferment that is stirring all hearts and minds is strongly operating among the engineers.   This is natural as Major Yerkes has shown us that in the Army, representing all classes of the community, the engineers were shown to be the most mentally alert and intelligent.   I found in many states a movement to found State Engineering Societies, on the order of trade unions, with the idea of obtaining legislation to license engineers, to increase the emoluments of the younger men in the profession and to give the engineer more prominence in the community as a man and as a citizen.   While I am heartily in sympathy with the last two objects, it does not seem to me that the engineers interested are using the right way to achieve these objects.   The Engineering Council, which represents the four national Engineering Societies in matters that concern the relationship of the engineer and the public, is the proper body to take up these questions and has here a great opportunity, which it should utilize to the full.

In this review of the year's accomplishments, I want to give full value to the devoted work of our Secretary.   Presidents come and go and are more or less figureheads, but the brains, initiative, and energy that keep our Institute thriving are supplied by our Secretary.

The coming year is full of promise.   With the change of name, the activities of a large and increasingly important part of your membership has been recognized.   It is hoped that those who previously were unwilling to call themselves mining engineers will find themselves attracted to that scientific society that not only provides them with the class of papers they are interested in discussing but also recognizes them in its name.

Under the strong and skillful guidance of our new President, aided by an interested board of Directors, a devoted Secretary and a loyal membership, the American Institute of Mining and Metallurgical Engineers will increase in numbers and renown.

At these last sessions we have had many papers concerned with what I call "human engineering."   Many visions dependent for their beauty on the perfectibility of human nature, in which I ardently believe, have been described in words.   May at least some of them be realized during the coming year in deeds.

In introducing the last speaker, Mr. Addicks said: We are living in strenuous times.   Governments are being overthrown every day.   In fact, I understand that the President even now is hurrying home from France on account of rumors that a republic is about to be declared in Washington.   So it is not strange that the Institute itself has decided upon a radical, nay a revolutionary, step.   It has placed the seal of its approval upon mining geology.

The geologist has had a hard time of it.   You would think that a man
who had made the study of rock formation and ore deposits his life work
would be the ideal person to send out prospecting.   Great mines have
been discovered by bartenders, by livery stable keepers, by one, or
perhaps two, Columbia graduates, but by a mining geologist, never.
If you already have a mine, he will make a beautiful map of it, all in colors,
for he has an artistic soul, with the quartz in blue and the limestone in
yellow, and the ore as a red streak in between.   He will show you that if
you drill a hole over in the next lot while your neighbor is at the seashore,
you will discover an extension of the vein.   You try it and find the limestone where the quartz ought to be, the quartz where the lime ought to be,
and the ore vanished altogether, and then go to your geologist; he will
coolly meet the ascending waters of your wrath by taking his pencil,
drawing a line across the map and writing "faults"—meaning the fault
is nature's and not his.   "Well," you say, "if a geologist cannot discover a mine and cannot develop a mine, of what good is he?"   I will
tell you.   If you have a real good one, and he really likes you, you take
him into court and he can prove *anything*.

An old gentleman was coming down the street and saw some colored
boys shooting craps.   "Rastus, don't you know that it's just as wicked to
gamble when you win as when you lose?"   "Yessah, yessah, de immo-
rality am jes as great but the inconvenience ain't."   Well, what is the
use of my discussing this ethical question when the Institute has settled
the matter by electing as its next president the very best mining geologist in the whole city of Minneapolis, Mr. Horace V. Winchell.

### President Winchell's Speech

I am told that it is customary on the recurrence of this annual event,
for the incoming President to present for your consideration and that of
the members at large a few thoughts as to the work of our Institute,
and the industry which it is supposed to represent; to sound the key-
note of progress for the coming year and to direct attention to the work
of the hour.

If this be already a custom, it is something for which neither you nor
I can be held responsible, and yet something which neither of us can
escape.   It is therefore my duty and high privilege to prepare the dose,
and yours to take it as gracefully as possible, in the full belief that what
is customary is proper, and what the doctor prescribes is beneficial,
though not always palatable.   In this case I trust it may be both.

It is a trite saying and a true one that we are living in the most wonderful period of history.   We are witnessing the evolution of a new order of
things.   We are participants in movements which are everlasting in
their effect and world wide in their scope.   Forces and activities now
set in motion will have their influence on the human race throughout its
entire period of existence, and the lustrum of time of which 1919 is the
closing year will be forever marked in red and gold on the pages of history.

It is entirely proper for us to reflect with pride upon the service which
our Institute and its members have rendered the cause of humanity in
the recent conflict.   The names of those who have given their money,
their time, their efforts, and in many cases, their lives, will ever be held
in honored memory.   But it is not sufficient to complacently remember
the past.   Our work is still ahead, and it is timely for us to consider

our relation to our profession, to our country and to ourselves in the working out of the problems of the future. What part are we able to play in reaping the benefit which should and must accrue from the sacrifices of the world since 1914? Is there any work which the engineer is better qualified than any other class to undertake? Is he now already preparing in true engineering fashion to get 100 per cent. efficiency out of himself and his organization? As educated and reasoning citizens of the world's greatest democracy, as trained engineers, as members of the American Institute of Mining and Metallurgical Engineers, what can we as individuals and as an organization do, what are we morally and in a sense of reciprocity obligated to do, under the new conditions with which we are confronted, to improve our own situation and that of all mankind?

The answer depends on the nature of the problem and the conditions with which it is surrounded. Do we fully realize the momentous changes which have already taken place? A nation which was taught by Washington to avoid foreign entanglements has in the interests of self-preservation been forced to take an active part in the settlement of a war in Europe, and is now occupying a prominent seat at the council of nations. Within a single twelve months we have abandoned the policy of our forefathers and become an arbiter and co-administrator of the affairs of all countries. Never again can we pretend to stand aloof and unconcerned while war flames are kindled abroad. No longer can we afford to be provincial. The ocean now is narrower than was the English Channel before the construction of telegraph lines, steamships, submarines, and airplanes. The food supplies, the raw materials, the factories, the mines, the schools, the people of foreign lands are now subjects for our careful study and consideration. We are and shall be impelled to it by motives of every variety, materialistic and humanitarian. The better we understand the world and its problems, the more successful we shall be in our commerce, in the development and handling of our internal affairs, our finances and our labor questions, our crops, our mineral production; everything that concerns our prosperity and welfare.

Then, too, from a nation of borrowers, needing foreign capital to develop our resources and owing vast sums upon which we were paying interest to the bankers of Europe, we have suddenly by fortune of war become a creditor nation. The nations of the world owe us inconceivable sums. The interest on our loans abroad would have paid the entire expense of our national government when we men were youths. This change in our status is of tremendous importance to the engineer, whose field of operations has thus suddenly become widely extended. The first and immediate demand for engineering talent probably lies in connection with the problems of restoration and reconstruction in the devastated areas. To what extent American engineers will be used in this work depends largely upon the supply of materials of construction and of engineers in Europe and upon the ability of our trans-Atlantic neighbors to finance their operations without our aid. In many districts, mines and manufacturing plants have been destroyed; in others, worn out; in most cases on the Continent, production has been pushed to the maximum without stopping for repairs; railroads and highways have been worn unceasingly, and must need renewing. And in addition to this plant and material exhaustion, it must be remembered that there has been appalling loss of life among the engineers and those of sufficient

mentality to make army officers.   Increased and intensified demand for
materials and a shortage of designing and constructing engineers would
seem to create unprecedented opportunities for this country of matchless
resources in both.   And here is where the advantage of the United
States as an opulent and creditor nation becomes doubly accentuated.
Now for the first time in our history we have more money than we can
spend at home; now the wide world is our field of operations; now we
are compelled to search for development enterprises in which our
surplus capital may be employed; now we must, in spite of ourselves,
become cosmopolitan.   And since those very countries which are our
debtors are the ones which most need our assistance, it is only by increasing our business with them and our investments there that those debts
will ever be paid.   By enabling them to rebuild, by establishing them
again on a prosperous basis we shall at the same time protect the investments and loans already made under the necessities of war.   Does this
situation not promise work and opportunity for the engineer?

But the war has brought other changes.   The employment of man
power and materials for the sole purpose of destruction has been so general
and widespread as to change living conditions over a large portion of the
inhabited globe.   Economic laws have been arbitrarily set aside, and the
governments have been compelled to establish artificial prices and regulations for the transportation and allocation of labor and materials.   The
inevitable result has been dissatisfaction, uncertainty, and confusion.
The workman who has recently been receiving fifty dollars per week instead of the former twenty-five cannot understand why there should be
any reduction now that the war is over.   The farmer still wants war
price for wheat; the southern planter wants war price for cotton; and
the field hand still desires to work half time and get double wages.   On
the other hand there is a general desire for a reduction of the cost of the
every day necessities of life.   In short, we are faced with a condition of
unrest and uncertainty in all quarters of the globe.   Every thoughtful
person knows that there must be readjustment, but no one can foresee its
precise trend and effect.   Is there not here again a demand for the influence and effort of the intelligent engineer?

Those of us who have watched the spread of Marxian socialism
abroad, who have seen its adoption by the Russian Bolsheviki and have
seen their poisonous propaganda insidiously inoculating the workingmen
of Russia, Austria, Germany and even England, and who have read the
anarchistic publications of their disciples in this country cannot but feel
that it is high time something was done to counteract it.   Bolshevism is
the antithesis of democracy; it is the foe of freedom; it is a rule by a class,
and that class the most ignorant and least civilized in the community.
It matters not whether it be found in the parlor or in the revolutionary
parade, in the poisoned press or on the street platform, bolshevism is an
abomination subversive of order and government, and must be opposed
by every patriot and loyal citizen, by every influence and power which
desires the welfare of mankind.

It has often seemed to me that the engineer is not fully awake to his
duties and privileges as a citizen; that he is too engrossed with the
details and the mechanism of his profession; that his mind dwells too
much on facts and figures and processes; that he is too retiring by training and disposition, too little of a publicist and a humanitarian, and
too much of a materialist.   I believe that individually and through

his organizations he should take an active part in every movement that concerns the good of society; that he should take the initiative in shaping the policies of government; that he should be an aggressive educating and moral force in every community.

On his visit to this country in 1876, at the founding of the Johns Hopkins University at Baltimore, Thomas Huxley addressed us as follows: "Truly America has a great future before her—great in toil, in care and in responsibility, great in true glory if she be guided in wisdom and righteousness, great in shame if she fail. I cannot understand why other nations should envy you or be blind to the fact that it is for the highest interest of mankind that you should succeed; but the one condition of success, your sole safeguard, is the moral and intellectual clearness of the individual citizen."

In all branches of business, in all lines of human endeavor, we have been taught to strive for efficiency. Indeed, so greatly has the idea been stressed that we have been in danger of regarding it as an end in itself. We have forgotten why we are thus striving. We have often had our attention directed to the efficiency of the German people as something well worth imitating. Dr. Nicholas Murray Butler, in a most scholarly address on "Education after the War,"[1] has given us a timely warning: "The war has taught the lesson that the proper place of efficiency is as the servant of a moral ideal, and that efficiency apart from a moral ideal is an evil and a wicked instrument which in the end can accomplish only disaster. In other words, we should encourage efficiency not for its material results, not simply for the greater amount of wealth in dollars and cents, in bushels of wheat, tons of ore or yards of cloth thereby produced, but for its value in the development of character, and for its aid in the achievement of our ideals and the guidance of the individual and the race in their progress toward fuller self-expression and more complete self-realization."

Now, the only road to efficiency is education. In all departments of life, in business, in government, in commerce and trade, education must precede efficiency, and the broader and more widely disseminated the real education of a people, of a class, of a community, the higher its efficiency. And this brings me down to the suggestions which I wish to make this evening.

Scarcely a day passes that the newspapers do not contain notices of strikes, of lockouts, of labor dissatisfaction and disturbances in this or that industry. The strife between the employer and the laboring man is incessant and irritating to both. It is world-wide and apparently everlasting and its cost is beyond calculation. It has received the careful consideration of the world's greatest economist, and is a problem whose solution requires far greater intellect than mine. Nevertheless, it vitally affects the interests of everyone of us here tonight, for it contains possibilities which one shudders to contemplate, with the tide of Bolshevism rolling westward, and other forces at work to mold our civilization into new and different and untried forms. Here and now are demanded more than ever before that "moral and intellectual clearness of the individual citizen" referred to by Huxley; now, more than at any period in our history do we appreciate the worth of that poise and stability which are provided and acquired by education; and now do we

---

[1] *Ed. Rev.* (Jan., 1919.)

feel more fully than ever the importance of extending education to those of all classes who are in any way responsible for the industrial turmoil existing and impending.

It has always seemed to me that labor troubles must, in large measure, result from a failure on the part of the protagonists to understand each other's respective situations and motives; from a lack of comprehension of the simple elements of economics on the part of the masses, and the failure on the part of the employer to explain how and why conditions have arisen which have made necessary or inadvisable a readjustment of any accepted or demanded situation. In other words, through lack of education. And here, it seems to me, is an opportunity and a duty for the engineer. He occupies a peculiar relation to the capitalist and the laborer. He is customarily one of the employed; he is, on the other hand, the adviser and trusted representative of capital. He comes into frequent and close contact with the laborer, and is in a position to understand his difficulties, to win his friendship and confidence, and to impart advice and information which would go far to explain the difficulties of any given situation. He is a sort of middle man, who might easily acquire such influence with those above and below him as to be of very great aid in time of industrial crises. If this be indeed true, then it must follow that in not exercising this privilege, in not performing this service, he is not meeting fully his responsibilities as an engineer and a citizen. As the boys would say, he is not strictly "on his job."

Let us consider for a moment the situation in the average mining camp. A few hundred or few thousand miners are employed. Here they come, muckers, mule skinners, trammers, nippers, timbermen, machine men, track layers, powder monkeys, station tenders, pipe men, carpenters, electricians, shift bosses, blacksmiths, and helpers, of different nationalities and varying degrees of intelligence and education. They are checked in and checked off by the time keeper; except in case of accident they spend their alloted time on the job and disperse without receiving as much personal attention as the mules underground. When not at work, they spend their time idling around saloons or other shady resorts where they not only learn nothing to their advantage, but spend their substance and sap the foundations of their health and strength, mentally, morally, and physically. Under such conditions they afford fertile and receptive soil for the seeds sown by the demagogic agitator. They attend open or secret meetings of the union and are constantly taught the program of violence and disrespect for law and order. In those camps where club houses are provided and reading matter and forms of amusement furnished, they are seldom visited and cultivated by anyone connected with the mine management. The men are still left to their own devices, and no advantage is taken of the opportunity to gain the friendship and confidence of those who are approachable, to aid those who are worthy and in need of some sort of encouragement or assistance, or to educate those eager for knowledge.

In recent years, it is true, many mining companies have arranged for moving picture shows two or three times each week; in some states like Nevada, the State educational institutions, such as the College of Mines, have of late years conducted night schools for instruction in scientific and technical subjects; and the Federal Bureau of Mines has sent its car around and taught the men the principles and methods of safety first. In all of these matters, the mining engineers have taken a part and shown sympathetic interest; but it seems to me we have fallen far short of the full measure of our duty and our opportunity. We have failed on the social and humanitarian side; we have done little to counteract the deliberate spread of socialistic and bolshevistic doctrines; we have permitted the raising of a crop of noxious weeds on soil which might well have yielded the fruits of thrift, industry, loyalty and patriotism.

With the coming of prohibition and the closing of the saloons, those old-time haunts of the miner, we find the men increasingly in need of comfortable recreation quarters, and in better condition to be interested in opportunities for entertainment which shall be at the same time instructive and uplifting. There are many forms of such entertainment, and many methods by which such instruction may be given without the appearance of officiousness or pedantry. I have never yet visited a mining camp where there were not some men of ideas, of travel, of wide experience and observation, of talent in some form of entertaining, where there are not frequent visitors who could be pressed into service for the benefit of the general cause of education and good fellowship. And there are ways and means by which such service could be organized and carried forward. There are national organizations and societies; there are State institutions; there are our own Local Sections; and there are the mining companies themselves which would quickly appreciate the value of such work among their men. It is probable that our Committee on Industrial Organization, with its sub-committee on Education, may be already at work along these lines. But we should not leave it entirely to small committees. It is work for every one of us, to be performed daily and perpetually, wherever we may be. It is not only a moral duty, but is one of the best expedients for making sure of permanent tenure of our positions and the salaries thereto attached. How many of our engineers are now enjoying princely honorariums in Russia? None. How many are at work in Mexico? Very few. What is the fundamental reason in both countries? Illiteracy. How many engineers will be thrown out of employment if bolshevism carries out its declared program in the United States? Untold thousands. Is it not then, in the most literal sense, our job to protect our jobs by dissolving and dispersing the clouds of prejudice and ignorance which threaten the existence of all kinds of employment and the destruction of all industry? Yea, verily.

Nor is the laboring man the only one who needs educating. We need it ourselves; and so do Congress and capitalists, managers, and the general public. The engineer's own education is too often defective in economics and politics, and he suffers thereby. He is wounded in his most sensitive parts by bad mining laws, by wrong principles of taxation, by ill-advised governmental regulation. In the conduct of all these matters he often has a keen sense of defective functioning, but is not sure of the remedy, nor skillful and earnest in urging its adoption; and in his perplexity often decides that he is not the doctor nor the plumber to stop up that particular leak! And when the trouble has grown irritating and chronic and no one provides a cure he is apt to become cynical and to blame society and the government for a situation which he himself should have helped to alleviate.

Now, Brother Engineers, let us have more confidence in the essential fair-mindedness of our fellows; let us believe in the future of our country; let us feel some responsibility for the condition of society; let us individu-

ally and through our Institute and kindred organizations cultivate
simplicity of analysis, clearness of thought, expression in words of one
syllable, and let us seek means by which we may increase our moral efficiency, improve the relations between the citizen and the government,
between the employer and the employed; and so add to the sum of human
happiness.    The task may appear great but there are many workers and
there will be an abundant reward in the resultant and well justified
consciousness of service rendered combined with a swelling sense of pride
of profession, pride of race, and pride of country.

### EXCURSION TO FEDERAL SHIPBUILDING PLANT

On Thursday, Feb. 20, about 240 attended the excursion to the
Federal Shipbuilding Plant, at Newark, N. J., where through the courtesy
of the manager, Mr. Robert McGregor, and his assistant, Mr. W. A. Bush,
the entire plant and the ships under construction were open to the visitors.
An unusual feature of the trip, worthy of note at the beginning, was
the manner in which the party was guided about the plant.    Thirty or
more guides were provided, each of whom wore a large badge bearing
a number and was provided with ten pasteboard tags, bearing the same
number as that on his badge, which were given to the guests.    These
constituted his party, for which he was responsible during the trip.    These
tags were given out when the party went through the gate and the guides
kept their parties intact and the specified number of minutes behind the
party ahead.    This not only prevented crowding, but prevented the
bunching of visitors around some prominent member of the Institute or
some fair feminine attraction, and gave a party of suitable size so that
the guide could give descriptions and explanations so that all could hear.
A specified route for the visitors had been laid out and it had been arranged that in all of the departments through which they were taken,
there would be few or no obstructions.    Great care was taken to prevent
accidents and, thanks to the foresight of the management, none occurred,
although it was an extremely busy place.

There are twelve ways at the plant with a ship being built on each,
which were in various stages of construction, from one with an uncompleted keel to another that will be launched in about a week.    Three
recently launched ships were floating in the basin, almost completed,
and one completed ship left the plant and started out to sea as the
visitors' train arrived.

The Federal Shipbuilding Co. was formed in July, 1917, by the United
States Steel Corporation, in order to assist in the great shipbuilding
program being carried out by the United States Shipping Board for war
purposes.    The plant is situated on the west bank of the Hackensack
River, and comprises about 185 acres of land, originally marsh, practically covered by water at high tide.    About 750,000 cu. yd. of fill was
obtained by the excavation of the wet basin and dredging of the Hackensack River and a final top fill of cinders was applied.    The first pile was
driven on Aug. 6, 1917, and the construction work was prosecuted in a
most energetic manner, despite the severe winter and other difficulties.

All of the facilities and organization of the United States Steel Corporation were generously placed at the service of the Federal Shipbuilding
Co., which accounts for the remarkable and substantial progress made.
The plant is entirely self-contained, having twelve ways, and a fitting-out
basin capable of accommodating eight steamers at one time, beside the
necessary workshops for carrying out every detail of manufacture required

in ship construction. Special attention has been paid to welfare work and safety appliances as carried out in the other subsidiary companies of the Corporation.

The first keel was laid in November, 1917, and up to the present time nine boats have been launched and five delivered. These steamers are 410 ft. over all, 55 ft. beam, with a deep draft of 27 ft., carrying 9600 tons of cargo, and will make a speed of 11½ knots loaded. Satisfactory results have been given by all the vessels delivered, and the boats are a fine type of the modern economical cargo carrier.

In order to make the greatest possible progress under war conditions. about 60 per cent. of the first thirty steamers was fabricated by the American Bridge Co. and sent into the plant all ready for erection, the remaining portions, consisting of the two ends, were fabricated in the plant.

During the early part of 1918, extensive preparations were made for carrying out experimental electric welding with the view of building steamers without rivets. This work was under the charge of the Emergency Fleet Corporation's Experimental Section, but on the signing of the armistice, operations were suddenly stopped, and the equipment is now dismantled.

After the plant had been inspected, the party was conducted to the old office building, now used as a restaurant, where a very elaborate luncheon, comprising soup, celery, olives, roast chicken, candied sweet potatoes, peas, salad, sandwiches, dessert, and coffee, was served. After the luncheon, two very brief and interesting speeches were made by Mr. Bush of the plant and Mr. Henry of the excursion committee, and an enthusiastically unanimous vote of thanks was given to the management of the plant for their most generous courtesies and a most interesting trip.

## LADIES ENTERTAINMENT AT THE MEETING

Immediately after lunch Monday noon, the Ladies' Committee entertained the visiting ladies in a short but very interesting sightseeing trip—first to Columbia University; then to the Cathedral of St. John the Divine; Grant's Tomb; the American Museum of Natural History, where the building being closed after hours to the public, was given over entirely to the ladies. After inspecting the exhibits, particularly those of precious stones and the mine model of the Copper Queen mine, the ladies were entertained at a very elaborate tea in one of the halls of the building. About one hundred ladies were present.

A novelty of this sightseeing trip was that private automobiles were supplemented by double-decked Fifth Avenue 'buses for conveying the party over the route. The weatherman had been subsidized by the Ladies Committee and the upper deck of the bus was much in demand by the visitors.

On Tuesday afternoon, Feb. 18, about 100 ladies visited the art galleries of Senator W. A. Clark, who has so frequently opened his collections to the Institute on former occasions. After the galleries had been inspected, tea was served.

On Feb. 19, the ladies were entertained at a theater party at the Gaiety Theatre, where "Lightnin'" was very much enjoyed.

A large number of ladies attended the evening entertainment Tuesday, Feb. 18, and the trip to the Federal Shipbuilding Plant on Thursday, which are mentioned in more detail elsewhere.

## WOMAN'S AUXILIARY

### SECOND ANNUAL MEETING

In the absence of the President, Mrs. R. Gemmel, Mrs. Louis D. Huntoon, First Vice-president, presided at the second annual meeting of the Woman's Auxiliary. The session on Tuesday morning was devoted to the receiving of the reports of the secretary, treasurer and auditors; the Central Americanization, Emergency and Foreign Relief Committees, and from the Section Directors of Arizona, Columbia, Missouri, New York and Utah.

On Wednesday morning, the following officers were elected: President, Mrs. James F. Kemp; first Vice-president, Mrs. Louis D. Huntoon; second Vice-president, Mrs. H. P. Henderson; third Vice-president, Miss M. P. Stone; Secretary, Mrs. Sidney J. Jennings; Treasurer, Mrs. H. K. Masters. The New York Section at this time also elected Mrs. W. Y. Westervelt as Section Director instead of Mrs. H. P. Henderson, who was elected second Vice-president.

The new President, Mrs. James F. Kemp, then took the chair, and in her inaugural address said:

I want to say just a word to you this morning. This organization has a large field before it and we want to make it worth while; we want it to be as good as the men's organization, we want to become so valuable that they will feel they can't get along without us.

Just a word on some of the problems: As long as we have people here who do not speak the English language, who read newspapers printed in their own language and not in our own, we can never make good citizens of them. I think we must work among these people, we must not only make them see things as we see them, but we must see how they see things. We must make them feel their interests and ours are alike.

Of course, in different sections we have different ways of meeting these things. In the Western camps they must be met differently from the Eastern camps. I think we can get at the men through the women and children. In many camps we must have mothers' meetings. I know from experience that some of our foods and vegetables are entirely different from those which they use—they are very strange to them and they do not know how to make the best of them. We must work, not as different divisions, but as one big organization, and to do that we must keep in closer touch. We must get the different sections to write monthly letters to New York, to make this a clearing house. We should get our men to bring home the Bulletins in which those would be published, and as different problems arose, those who read them and have had experience along those lines could write in and give their yiews on them. We have an emergency committee which stands ready at any time to help if help is called for in the way of money, clothing or anything else. It seems to me that we must increase our membership very largely. We must form not only local sections, but we must try to get all women whose husbands or brothers are connected with the mining industry to join this organization. I think we should be a very large flourishing organization. Of course we all have to work—I don't want anybody to think they can belong to this organization and do nothing—because everybody must work and work hard. I think it would be very encouraging if the ladies from the different parts of the country, the ladies who are here, would give us a little idea of what they have been doing or what they think shall be done, because we want to be united.

Mrs. JENNINGS.—I think all the wives ought to insist on their husbands bringing home the Bulletins.

Mrs. KEMP.—I think we have a tremendous work ahead of us if we are going to live up to the work of our men. I think Americanization should be our work, of course, and we have all got to do it—we must go to these people and do the very best we can to make them feel that we. are all one and the same.

Mrs. L. D. HUNTOON.—Madam President, members of the Woman's Auxiliary of the American Institute of Mining Engineers, and Guests: We have with us this morning Mrs. William D. Sporburg, President of the Jewish Women's Council, Third Vice-President of the New York City Federation of Clubs, and a member of the Women's Committee of the National Council of Defense. Knowing how vitally interested Mrs. Sporburg is in the question of Americanization and how inspiringly she always puts over her message, I have asked her to come this morning and give us some practical suggestions as to how we may carry on our work. It is with great pleasure that I present her to you.

### ADDRESS OF MRS. WM. D. SPORBURG

Madam President and Members: After this elaborate and very gracious introduction I feel like Governor Whitman felt when he was called on to speak at a big convention of women. The presiding officer introduced him with a big flattering introduction, and when Governor Whitman got up he said, "I endorse everything you said."

Mrs. Huntoon asked me if I would come and bring a message and I told her I would be happy to do it; I would be happy to meet a group of earnest women who were thinking along the lines of Americanization. That is a subject that should interest us as Americans above and beyond everything else. However, I was completely intimidated when she told me the type of organization before which I was to appear. It reminded me of a story I had heard some time ago, of a woman, oh, it happened a long time ago, who was asked to writer a paper (you remember those meetings where they used to read papers, it usually did us who wrote them a good deal more good than it did those who had to listen [to them);' but in this story it is recorded that the presiding officer asked this particular woman to write on "The Cliff Dwellers of Tasmania."

She hadn't ever heard of Tasmania, but she quite made up her mind that she would simply go into research work and find out about the Cliff Dwellers of Tasmania. She started to write her paper and she wrote at the top, "The Cliff Dwellers of Tasmania," and then started her first line, "I will now tell you about the cliff dwellers," but that is as far as she got. Her intentions of going to the library were very good, but she was the mother of a family, and the first week little contagious children's diseases kept her home, and the second week something else detained her, and lo! the day of the meeting was upon her head and she hadn't anything but one sentence, "I will now tell you about the cliff dwellers."

She went to the meeting and to her great astonishment the presiding officer said, "The subject of the afternoon is 'The Cliff Dwellers of Tasmania,' which will be read by Mrs. Samuel Johnson. Mrs. Johnson will be pleased to know that the Rev. John White, who has been a missionary in Tasmania, is with us and will be glad to corroborate any statement she may make."

And Mrs. Johnson, with the usual presence of mind that women can call upon, said, "I am sure, Madam Chairman, that I would love to give the time allotted to me to the Rev. John White, because he will be able to tell you first hand all about the Cliff Dwellers of Tasmania."

I know there isn't a woman in this room that could not tell more than I could about what Americanization should be so far as your particular institution is concerned. I do feel that perhaps by asking questions, if you will ask me questions and let me ask you questions we can exchange some helpful ideas.

However, as to the subject itself, there is a certain general aspect to be considered which is very important, whether we are organized and working in a great big cosmopolitan city or whether we are working in a small city, and that is the policy with which we shall approach the foreigner.

Years and years ago, just as a matter of human kindness and because we were interested in helping the strangers within our gates, all sorts of organizations were formed to help them, and of course since we have been plunged into war this subject has become a very paramount subject, something which we have had to take account of for very obvious reasons. My contention is that the great trouble with the women's organizations doing Americanization work in this country was that they did it from the standpoint of welfare work. You know we never looked upon them as quite up to our level, that is, while we did not look down upon them, yet we never considered the foreigners within our own group or our own status. We did do that as welfare work. Now, we have come to realize that the only way that we can really do effective, telling work with the foreigners is by stretching out a hand of real sisterly fellowship and making it clear to them that their contribution to America is as great as America's contribution to them.

We have some very serious things to consider. We have been told that there is not one piece of industry in the United States of America that could go on for fifteen minutes without the aid of foreigners, that factories and industries would have to be closed down if we barred foreign labor. We are absolutely dependent upon them. What we want to do is to really and truly assimilate them. There has been a great deal of talk about Americanization. We felt that the very standards of our country were in themselves a melting pot, that we could turn them out of this melting pot believing in the Constitution and waving an American flag. We have learned recently that Americanizing a foreigner, and making him live up to the standards of our country, the country of his adoption, is a very solemn thing. It is not a mechanical process at all. The mere fact that he signs Americanization papers does not mean that he is a real American in spirit. It is only when he accepts the standards for which America stands and believes in them that he is really, truly Americanized.

I have been perfectly horrified at the process of Americanization. Within the last two years or so we have given it a great deal of consideration. We know that in the courts where the naturalization process is going on that the actual final step itself has been treated with a great deal more solemnity and thought than has been given to it heretofore. I visited one of the courts with Mrs. Pennybacker. There were nine Italians standing in line waiting for their final test. The Judge turned to one of them (they are supposed to study up certain standards and be able to answer certain questions) and said, "Do you believe in polygamy?" The Italian looked a little nonplussed, poor soul, because I presume he felt he must not be anti anything, so he said, "Yes." The man sitting next to the Judge leaned over and said, "I don't think he quite understood that question. Do you believe in plural marriages?" The Italian still looked a little nonplussed, but he thought a moment and then he said, "Yes."

We felt the pathos, the real pathos of these men taking a step as solemn as a naturalization step without the full knowledge of what the obligations meant and what the standards of American life meant. I think that is a tremendous problem which we can help to solve. In the interim between the time they take out the papers and the final test we can teach them our language and help them understand these things, help them understand what real citizenship means. And the way to help them to understand is to make them as familiar with the English language as possible. It also gives them an opportunity to understand us and it gives us an opportunity to understand them. We are going to make them realize what American standards are by our own actions, by the example we set—for that is the thing that is going to be followed most closely, the example and precept of our own lives. If we American women will conduct ourselves in a thoroughly American spirit, the foreigners are going to imitate us very quickly. We must point the way to them, showing them loyalty to Government institutions, loyalty to the people who are elected to represent them in those institutions, which does not mean, women, when the people themselves have put men in power, whether it is the Presidency or a Governorship of state, Mayor of municipality or even the lesser offices, it does not mean that loyalty, respect and support of them bars us from honest criticism, because we need honest constructive criticism, but we must not fall into pitfalls and object and criticise men in power on partisan issues, whether we are Republican in tendency or whether we are Democratic in tendency, we should never believe that an issue, which is an important issue, and in which we can rightfully believe, is not a good issue because a Republican or a Democrat has stood for that particular measure. We must think and vote on measures on their face value. And when foreign-born men and women see us doing that, it is going to be one of the best forward steps in real Americanization that can happen.

We ourselves, we Americans, must be educated to the proper spirit of Americanization. We must be thoroughly American and in dealing with the foreigners in our ranks we must do it with toleration for each other and with mutual understanding.

I come from a very small community; there are about 14,000 inhabitants in Port Chester of whom 71 per cent. are foreigners. Twenty-two nationalities are represented. Through our district nurse, who gets the confidence of these people, we get together, once every two weeks, a group of these foreign-born mothers. At the first meeting we had eleven different nationalities and I certainly thought the Irish woman would scratch out the eyes of the Italian woman. The second and third time they became just a little more tolerant of each other and now there has grown that splendid understanding, that though they are different and apart, they have traditions and things and ideals that are worth while. They are working in harmony. I think that is the solution—we must tolerate each other and we must tolerate different conditions.

We have no right to go to an Italian, Greek, Russian, or anybody else and say, "Here, America offers you everything. You must forget everything else that has ever happened in your life, you must forget your folk-songs and everything about your country." If we make them realize their traditions are beautiful that we respect them just as we want them to respect us, we will accomplish a great deal. We must make them appreciate that we are going to be helpful to each other and that they mean as much to us as we mean to them. We want them to become Americans not only in name but in spirit.

I want to impress upon you that you must not depend on your Chairman of the Committee to do your work, no matter how efficient she may be; each individual must feel that you have a very important part to play in your organization and in your own community. You know very well that the war has taught us that lesson that it is the spirit of cöoperation that counts. We think with dread upon what might have happened to England if she were warring with Germany alone, we know what would have happened to France if she had been alone, we know what happened to Belgium and we know what might have happened to America and we know what happened by coöperation and organization. If every woman would get in back of just one immigrant family, think of the tremendous amount of good. Your individual effort counts.

> " It ain't the individual, or the
> Army as a whole
> But the everlastin' teamwork,
> Of every bloomin' soul."

QUESTION.—You spoke about the District Nurse. Isn't that a practical way of starting?

MRS. SPORBURG.—It is the most practical way, Madam. The foreigner is a little bit suspicious, she wonders if you are coming to buy the wage earner's vote or to disturb the routine of her housing methods. But a district nurse, a woman who comes to these people in times of trouble, gets their confidence. We have found that the most effective work we can do in smaller communities is through a nurse. She knows their problems and they trust her, and she leads them to the other people in the community who are ready to work for them.

A MEMBER.—I have had considerable experience with that and I have found that the women do feel that you are trying to influence them or trying to get the better of them—they cannot understand why American women are trying to get into their houses. What would you suggest in New York City outside of working through settlement organizations or through large organizations? How would you suggest that the individual should go about it?

MRS. SPORBURG.—I firmly believe in organized effort. Frank Lane is now planning a nation-wide unified program for Americanization. Congress has a bill to appropriate $500,000,000 for education. I do believe, aside from following a program of organized effort, that individual effort does count. I lived in New York City before I ever thought of myself in connection with this work and I was intensely interested in the Italian who had a fruit stand on the corner. I was interested in others, I went right into their homes and worked with them. I made them feel that at any time there was a problem in their lives that they could come to me and I would help them to the best of my ability; whether it was the man drinking or the child who was becoming so Americanized that it poked fun at its foreign mother—which I think is indeed a tragedy. I think it is an indictment against us that we have neglected the foreign mother so long. The foreign father gets his Americanization through industry, through his contact with the business world and the child gets his through the school.

I do believe that those of us who live in New York City or other large places and who come in daily contact with the foreigners—we meet them wherever we go—if we would show and direct a little personal interest in them and their families I believe it would bring returns. As I said before, just suppose every American woman made it her object to get behind one family, think of the tremendous amount of good that could be accomplished.

A MEMBER.—In doing some work in New York City during the food conservation period I came across several cases where the children could not speak with their own mothers. Another thing, do not you think there is a danger in becoming over-organized—that over-organization tends toward destroying individual effort and initiative and that it has a tendency to destroy the soul of Americans just as it destroyed the soul of Germany?

MRS. SPORBURG.—There is always a danger, but I think if the individual women will work from a personal standpoint they will avoid that. But there must be some organization, otherwise there would be such a lack of efficiency that a good deal of time would be wasted. As I said before, I think it is tragic that the foreign mother cannot speak the English language. Very often the foreign men especially the Italians, do not allow the women to leave their homes. They think her place is in the home and, before the franchise was granted, they were afraid she would have too much right. To overcome that and bring about confidence, in New York City, groups of women meet in one another's homes, and in that way they come in touch with the problems of the other woman, whether it is the food question, the question of citizenship, or the question of an unmanageable child, or whatever it may be.

A MEMBER.—Your feeling then is that at the present time the greatest need lies with the mothers. You feel that the children and fathers have had, so to speak, their share in a way through schools and agencies which are already established?

MRS. SPORBURG.—Yes, through unconscious contact with business and the people around them in business they learn certain things about the American standards.

A MEMBER.—If you were supporting any new work your feeling would be to largely get in touch with the mothers and leaving the children and men to what is already started?

Mrs. Sporburg.—My first step would be to bring up the status of the mother to that of the child and father.

A Member.—What is needed now is to bring the mother up to the level of the children and the father.

Mrs. Sporburg.—My experience has been that. As I said before it is appalling, it is an indictment against us that something has not been done before to educate the foreign mothers. Do you know there were 700,000 soldiers in the American army who could not read nor write English? Do you know that some of that 700,000 were American born, southern born, who never had the opportunity of obtaining an education? There are 400,000 women in New York state who cannot read, write, nor speak the English language and there are 300,000 who cannot read nor write any language. I say it is an indictment against us women to allow such conditions to exist.

A Member.—You were telling us about the gatherings where the Irish woman came to understand the Italian woman. Would you tell us how you conduct those gatherings?

Mrs. Sporburg.—Through the district nurse an invitation is extended to the various women she visited. It could be extended through other agencies if you have no district nurse. Our district nurse, when visiting these women would say, "At my rooms"—which is in the heart of the village—"on Thursday afternoon we usually have a little party. We have some clothing there which has been given to us by people who no longer need it. You are welcome to come and see if there is anything there you want. You can bring your child, or if there is anything in your own home you want to make over and need some assistance on, bring it along and we will help you." That is the way to get the group together and ostensibly they come to sew. In that way they get in touch with one another and become more tolerant of one another. You know there are always feuds between these various little colonies. It is also well to make the foreigners understand that we have something to learn from them; in music, folklore, cooking and preparation of foods they can teach us a great deal.

I would like to cite an instance: Last year in our desire to help the Food Administration, we had in Port Chester, as they had in many communities in West Chester County, the problem of dehydration. We all learned about using substitutes and swapping recipes that had proved successful (we did not say much about those that were not successful) but when it came to the subject of dehydration we were very much puzzled, they told us that even the commercial institutes were not particularly successful. I remembered one day the large rows of red peppers and onions, etc., that I used to see while driving through the Italian colonies. I drove down there and told these women to come to our demonstration, I said you can probably help us because your boys are in the service with our boys, they are all fighting together, and we are all going to stand together with hands outstretched. We can tell you a great deal about certain flours and things and I am sure you could tell us something about drying foods. We got an Italian woman to demonstrate their process and she was more successful than any woman who had made a study of it for eight years in Cornell. If you make them appreciate their contribution as well as ours they are willing to help. If you want any woman to think your way you do not approach her with a hammer, make as gracious and as kind an approach as you possibly can;

it is the understanding heart of the woman who is doing that work that really truly counts.

A great many of the foreign-born women signed the food cards, not knowing what they were signing—they never refused a food pledge. I made a test of this one time and I went from house to house and out of all the places only one woman stopped to ask me a question. I did this simply as a test. They are frightened stiff when you say you are from the Government and want them to sign a pledge. It is absolutely the wrong way to approach them.

A MEMBER.—Do you not think that as Charity begins at home Americanization also begins at home with our children? I feel that in these days of Kindergartens and Montessori Methods where the children are taught to take the line of least resistance, we are losing sight of the old Plymouth Rock ideals of duty and hard work and the necessity of doing the right thing no matter how disagreeable or irksome it may be.

MRS. SPORBURG.—That is what I meant when I said we Americans must be educated to the proper spirit of Americanization.

MRS. KEMP.—I feel that we owe Mrs. Sporburg a very great debt. There is a great deal which these people can give us which we should incorporate in our own lives, but if we are to be an example, which we must be if we want them to follow our ideals and ideas, we must be thoroughly American ourselves.

A rising vote of thanks was tendered to Mrs. Sporburg.

## REPORT OF TREASURER OF WOMAN'S AUXILIARY

### RECEIPTS

| | |
|---|---|
| Balance on hand year beginning Feb., 1918.................... | $226.20 |
| Entrance fees and dues received during the year ending Feb., 1919 | 65.10 |
| Total............. : ..................................... | $291.30 |

### DISBURSEMENTS

| | |
|---|---|
| Stationary, printing, etc.............................................. | $27.75 |
| Secretary's expenses—typing, postage.......................... | 48.85 |
| Money refunded to Sections on dues........................... | 17.00 |
| Total expenditures...................................... | $ 93.60 |
| Balance on hand........................................ | 197.70 |
| | $291.30 |

MRS. HARRIS K. MASTERS, *Treasurer.*

## REPORT OF EMERGENCY COMMITTEE

This Committee was formed with eighteen members in addition to the Chairman, the Treasurer, Mrs. Karl Eilers, and the Secretary, Miss Olga Ihlseng, in March, 1918, and owing to the cessation of hostilities its activities ceased last December.

Two knitting machines were purchased and installed in the Ladies' Reception Room in the Engineering Building, which room was placed at the joint disposal of this Committee and the Foreign Relief Committee through the courtesy of the United Engineering Societies.

During the period of its activities your Emergency Committee obtained 550 pounds of wool and material for 72 wool cloth sweaters purchased from funds supplied by Mr. R. W. Ingalls, Treasurer of the Comfort Fund of the Association of the 27th Engineers.

Thanks to the energies of the members of the New York Committee and their friends, and also the Salt Lake City, Colorado, and Columbia Sections, 170 sweaters, 773 pairs of socks, 89 wristlets, 85 helmets, 24 comfort kits, and 3 mufflers, were provided for the Comfort Fund.

To raise funds for general expenses a Concert was held in April, the return from which will be noted on the attached Treasurer's Annual Report.

It was with the deepest regret that your Committee learned of the death of Mr. A. L. Gresham, who gave his untiring assistance to your Committee since its inception.

LAURA M. SPICER, *Chairman.*

### REPORT OF TREASURER OF EMERGENCY COMMITTEE

Our income was $204.01 and our expenditures were $41.05, leaving a net cash on hand as of December 31, 1918 of $162.96.

RECEIPTS

| | |
|---|---:|
| April 30 Mrs. Spicer concert................................... | $167.21 |
| April 30 Mrs. Thurston (donation)............................ | 5.00 |
| April 30 Donation concert.................................... | 2.00 |
| May 10 Mrs. Prosser concert................................. | 2.20 |
| May 20 Mrs. Porrier concert................................. | 2.20 |
| May 20 Miss Stone concert.................................. | 11.00 |
| May 20 Stationary refund................................... | 10.00 |
| July 20 Geo. A. Schroder concert (Mrs. Spicer)............... | 4.40 |
| Total receipts....................................... | $204.01 |

EXPENDITURES

| | |
|---|---:|
| April 30 Olga K. Ihlseng (postage, etc.)...................... | $3.37 |
| May 13 Mrs. H. Hardinge (postage, etc.)..................... | 2.17 |
| May 13 Mrs. Mann (addressing and mailing).................. | 5.30 |
| June 28 Olga K. Ihlseng (postage, etc.)...................... | 1.94 |
| Sept. 27 Collector Internal Revenue (concert).................. | 27.20 |
| Nov. 8 Marion M. Shields (postal cards).................... | 1.07 |
| Total expenditures..................................... | $41.05 |

MRS. KARL EILERS, *Chairman.*

### REPORT OF THE CENTRAL FOREIGN WAR RELIEF COMMITTEE

The Foreign War Relief Committee has been very active during the past year, raising funds to the amount of $6207.11 for the relief of devastated France. Reference to the Treasurer's report will show that $5686.51 was disbursed through the American Fund for French Wounded, at a total operating cost to the Foreign War Relief Committee of $331.60. Four war lectures were given to provide funds for the operating expenses of the committee, and every dollar raised by subscription was sent to France for relief work. $966.00 was contributed for the purchase of sheep for stocking reclaimed districts in France, this sum being disbursed through the civilian committee of the A. F. F. W.

Early in May, the Foreign War Relief Committee, Mrs. Henry H. Knox, chairman, carefully considered the various channels for relief in France, and launched the project of establishing a dispensary there for the care of women and children. How this work prospered has been told in reports from time to time, but it may be repeated that $3000 was sent to France and a dispensary bearing the name of the New York

Section of the Woman's Auxiliary to the A. I. M. E. has been established at Briey. The committee closed its dispensary fund in January, with a contribution to the American Fund for French Wounded of $1720.51, to meet their urgent need for emergency dispensary work among the refugees returning to certain distressing centers.

Tuesday was established as "Engineer's day" at headquarters of the A. F. F. W. The committee has had six full working days and two half days there, and Mrs. Percy E. Barbour, director of work room, reports an enrollment of twenty workers. The average attendance was ten; largest attendance for any one day, thirteen; smallest, eight.

Finished garments turned in to the A. F. F. W. include thirty-nine hospital robes and eight children's garments. Contributed to this work— one infant's layette, three pairs shoes, ten refugee garments.

Mrs. J. P. Hutchins reports that letters received from the mothers of the four fatherless children, war orphans of France who were adopted by the Woman's Auxiliary for two years, express sincere appreciation of what has been done for them. Photographs of the little ones indicate that they more than merit the help which they are receiving.

MRS. JESSE SCOBEY, Chairman.

## FINANCIAL REPORT FOREIGN WAR RELIEF COMMITTEE OF NEW YORK SECTION

### RECEIPTS

Funds raised to meet operating expenses of the Committee:

| | | |
|---|---:|---:|
| By four war lectures | $515.60 | |
| Contribution | 5.00 | $520.60 |

Funds raised for purchase of sheep for stocking reclaimed districts of France. Disbursed by Civilian Committee, American Fund for French Wounded:

| | |
|---|---:|
| By subscription | 966.00 |

Funds raised for French dispensary work. Disbursed by American Fund for French Wounded:

| | |
|---|---:|
| By subscription | 4,720.51 |
| Total receipts | $6,207.11 |

### DISBURSEMENTS

Operating expenses of Committee:

| | | |
|---|---:|---:|
| Four war lectures | $129.36 | |
| Printing, postage, etc. | 53.04 | |
| Raising sheep fund | 22.67 | |
| Raising dispensary fund | 112.74 | |
| Work room charges | 13.79 | $331.60 |

| | |
|---|---:|
| Donated to sheep fund of Civilian Committee, A. F. F. W. | 966.00 |
| Donated for dispensary at Briey, France, under name of Woman's Auxiliary | 3,000.00 |
| For maintaining dispensary work | 1,720.51 |
| Total disbursements | $6,018.11 |
| Balance in bank | $189.00 |

MRS. JESSE SCOBEY, Chairman.

### REPORT OF THE CENTRAL AMERICANIZATION COMMITTEE

During the Spring of 1918, the Chairman of the Committee held various meetings among the foreign-born women, living on the East Side. These were discontinued during the Summer, when considerable work was organized and supported on Long Island among the various factories, several addresses being arranged for during the noon hour.

In the Fall, Americanization work was continued at the various Settlements in the Greenwich Village Section, much helpful constructive work being done during the recent influenza epidemic in advice to the foreign-born women, as to better methods of caring for their families.

The Committee has been able to get in close touch with the foreign-born families during the past month in connection with the City's "Back to School Drive." Many of the foreign families not realizing the importance of their children remaining in school until the age of sixteen, nor the advantages thereby gained, until presented to them by members of the Americanization Committee. Such Committees would prove of great assistance to teachers in all parts of the Country, as they could visit the families of delinquent children, and ascertain the cause of the delinquency, thereby getting in personal touch with the families.

With the signing of the armistice began a new era for America. We stand as never before in the eyes of the world and the duty of every American should be to consider themselves a member of an Americanization Committee.

LAURA G. BURGER, Chairman.

---

## MEMORIAL SERVICE TO DR. ROSSITER W. RAYMOND

All technical sessions Monday afternoon were brought to an end in time for the members to gather in the Auditorium as the Institute paid its tribute to Dr. Rossiter W. Raymond. In opening this meeting, President Jennings said:

We have gathered here to render our tribute of honor and affection to the memory of one who was for 47 years the guiding genius of this Institute. One of its founders, and at that early day one of the foremost in his profession, he saw it grow from infancy to the great body it is to-day. At the beginning, as now, its membership comprised the leaders in geology, mining, metallurgy, and technical education. Because so many were qualified to lead, and because ambition is an essential qualification for leadership, the most momentous of the problems coming before them for solution was the selection of the one to whom they could confide the care and direction of the institution which was to record their proceedings and to stand as an enduring monument of their accomplishments. Their decision would determine whether the members of this group of leaders were to be coöperators or competitors—associates with a common purpose or rivals for individual advancement.

The selection of Rossiter Worthington Raymond for Vice-president, President, and finally Secretary; his retention in that office for 27 successive years; his elevation to the office of Secretary Emeritus and to Honorary Membership, constitute a testimonial greater than any honor that we can offer to his memory. In holding these exercises to-day, we simply voice our confirmation of the wisdom displayed by his colleagues in placing in his hands the guidance of their enterprise.

The Resolutions passed by the Directors, and printed on page liv of this Volume, were then read. Afterward, Dr. Henry S. Drinker, president of Lehigh University and one of the two survivors of the twenty-two who attended the first session of the Institute, was then introduced.

## ADDRESS OF H. S. DRINKER

A friend, whom we loved, has gone from among us. He was a man who by his genius dominated any assembly in which he stood. He ·was a teacher of teachers, a leader in all the many lines in which his energetic able personality led him.

Of his eminence as an engineer, and of his ability, learning, and surpassing power in argument and presentation as an expert and as a lawyer, I will not speak—the tributes paid him by Mr. Rickard and Mr. Ingalls are so well studied that they should stand as the record of our friend's professional reputation. He was a wonderful man in the absolute absence of pretense in all that he said and did. If Raymond said it you could rely it was so—and his mind was so encyclopedic, his learning so vast, that association with him was an education, intensive and broad.

It was my privilege to know him for a life-time. We were associated with the founding of our Institute at Wilkes-Barre in May, 1871. I was then a young fellow just stepping out into practice from college training under Rothwell in the Lehigh School of Mines. and Raymond and Rothwell, Coxe and Coryell, the men who organized the first coming together of the Institute, were men in the leadership of the profession, earnest, enthusiastic—early exponents of the profession they dignified and in fact introduced into this country.

From the beginning, Dr. Raymond's trained mind, inexhaustible energy, and wonderful aptitude of expression enhanced by his personal charm of manner, meant everything in the early setting and development of our Institute, which has grown into such a power in the engineering progress of our land.

We all pay tribute to Dr. Raymond's recognized ability and power of leadership—but there are today but few of us left who can personally turn and look back over a half century of actual association with him, a precious privilege filled with memories of a man of whom it may well be said he was typical of " Whatsoever things are true, whatsoever things are honest, whatsoever things are just, whatsoever things are pure, whatsoever things are lovely, whatsoever things are of good report," for he was of virtue—and we may well, in thinking of him, think of these things. Dr. Raymond was generous in his encouragement and aid to younger men. I can personally, with all my heart, echo the words of Ingalls in his recent splendid tribute to Raymond where he speaks of having in his early association with the *Engineering and Mining Journal* looked on Raymond as " a guide, philosopher, and friend," trite words—but never more aptly or better or more truthfully applied.

Dr. Raymond's history has been recorded, and his engineering record has been and is being given by men far better fitted than I to do technical justice to so large a subject. It is for me as one of Raymond's many friends and admirers, one of his old friends, yet speaking from the standpoint of one younger than he and ever looking up to him as a leader and teacher, to pay tribute to his personal qualities that so endeared him to all who were privileged to know him.   I owe a great personal debt to him for encouragement and aid to me as a young man, and I am moved to speak of it only as an instance of what was common to so many, for he was ever ready with counsel and cheering words of uplift and practical suggestion to the younger men who came under his observation, and in this he typified in person, what our Institute has done as an association.   Founded as it was by men of large heart and human sympathy, such as Raymond and Eckley B. Coxe, the Institute, particularly in its younger days when our membership was small, and the friendships engendered among members were intimate and common to all, did, and indeed has ever continued to do, a great work in giving to young engineers who came into its fold opportunity for betterment by association with older and eminent men, with an opening for the publication and discussion of their engineering experiences, and theories.   In the development of this practice, and as the able Editor for many years of our Transactions, Dr. Raymond ever showed his kindly sympathetic helpful nature, and the men, and their number is legion, whom he so aided, pay tribute today to his memory with loving gratitude and appreciation.

He was a wonderful man in his faculty of doing so well so many different things.

Did his record rest only on his professional work as mining engineer, metallurgist and mining lawyer, his friends might be content, but he was not content with this.   Dr. Hillis has told us in his beautiful tribute to our friend, of Dr. Raymond's leadership in religious work in Plymouth Church, and how after Mr. Beecher's death Dr. Raymond was asked to retire from his engineering and editorial work and take up the pastorate of Plymouth Church (and how beautifully his reply reflects Dr. Raymond in his sincerity, good judgment, and never-failing humor)—Dr. Raymond said that the Providence of God, through his fathers, had lent him certain gifts, and by His providence guided him into an appointed path, and now that his life journey had been two-thirds fulfilled, he did not believe that the Lord was going to return to the beginning of that path, and reverse Himself, and he would, therefore, follow the way appointed to the end of the road.

And in Plymouth Church and the friendships he made and cherished there, we can see how, while laboring for the good of his fellow-men, and for their souls' good, he yet rested from his professional work, and took pleasure and solace in his touch with the Church and Sunday School in which his heart delighted.

His addresses in the Church, of which many have been published, show a vivid and ever fresh and inspiring flood of wise helpful admonition and teaching—and his annual Christmas stories to the Sunday School children—fifty in all, ending with the one given on Sunday December 29th, only two days before his death on December 31st, are a unique and beautiful illustration of the faculty he possessed of using his great gifts for the young.   The fiftieth and last of his Sunday School addresses is as vivid in interest as its predecessors, among which those who read them can never forget the delicious talks chronicling the woodchuck who inhabited the Doctor's garden at Washington, Connecticut, and who is introduced with the words "At our place in the country, where we spend five or six months of the year, we have among other fascinating attractions, a woodchuck of our own.   That is nothing very remarkable.

The whole region is full of woodchucks, and the difficulty is *not* to have one.* * * Our Garden is not far from his hole on the lawn, yet he never comes into the garden—for which reason we call him Maud, after the lady in Tennyson's poem. That lady did come into the garden; but then she was invited. If the gentleman had sung to her "Don't come into the garden, Maud," or even if he had never mentioned the garden, I am sure she would have stayed away politely, just as our Maud does,"— and then the address goes on with Raymond's never ending sense of humor, deliciously emphasizing the wise words on current events and international politics that are voiced by the woodchuck in his conference with his host.

As Ingalls has well said, Dr. Raymond was one of the most remarkable cases of versatility that our country has ever seen—sailor, soldier, engineer, lawyer, orator, editor, novelist, story-teller, poet, Biblical critic, theologian, teacher, chess-player—he was superior in each capacity. What he did he always did well.

In his writings and poems his ever-present sense of humor shone out—and yet always there was an adumbration of wise reflection or suggestion—often a direct emphasis of advice on current questions of the day. In his wonderful story of "The Man in the Moon," published over forty years ago, and doubtless reflecting some of his own personal experiences as an officer in the Civil War, Dr. Raymond recorded in his inimitable way what today may well be read as a prophetic utterance on the folly and the wickedness of the world war, in his account of the way that the opposing soldiers in the ranks came together on Christmas Day—and how a sentiment in favor of peace spread from the ranks to the peoples concerned until the Generals in charge of the war, and the governing authorities of the countries concerned, awakened to the folly of the contention in which they had been striving and came together in a peaceful solution.

The story is an immortal one, and those of you who have not read it, have a great treat in store when you find it. "The Man in the Moon —A War Story."

Dr. Raymond's home-life was ideally beautiful and loving. On Christmas Day just passed this little poem—so characteristic of him, and so expressive of the love he bore Mrs. Raymond, accompanied his gift to her of a bond:

> "Tis strange, Oh Lady! fair and fond
>  Of me (as likewise I of you)
> That there should be another bond
>  Between us two!
>
> "You do not need this thing to make
>  Your life more full of hope and zest;
> And yet sometimes you well might take
>  More interest!
>
> "And there is nothing better serves
>  For weary hearts and hands to droop on
> And stimulate exhausted nerves
>  Than a good coupon."

Dr. Raymond suffered a great sorrow in the loss of the son of whom he was so justly proud, a loss that he bore with a man's fortitude, and in which he was upheld by the faith and hope that his life so strikingly exemplified.   That he should have been first taken, leaving here the wife to whom he devoted so many years of loving care is a part of that great mystery into which we cannot look, but she at least has the comfort of the memory of her knight as one "Without fear and without reproach" a Bayard among warriors—a Sir Percival among knights.

Dr. Raymond belonged to many societies and his abilities received due recognition in many honorary titles from societies, universities and colleges.   Among them it was the pleasure and honor of Lehigh University to confer on Dr. Raymond in June, 1906, the first Doctorate of Laws ever granted by the Institution.   When, in 1905, I was asked by my fellow-alumni of Lehigh to lay aside my professional work and take on the responsibility of the Presidency of Lehigh University, it was to Dr. Raymond I went for advice on my course.   He urged me to take it up and during the years since then I have reason to be grateful for his steady counsel and support, and his visits to speak to our student body have ever been welcome and uplifting.

He and our honored Dr. Drown and I had a close and common bond in the association we all three had with Lehigh, and I know of no words more fittingly applicable to Dr. Raymond than those he spoke of Dr. Drown at the time we laid the foundation of Drown Memorial Hall on our Lehigh campus.   Dr. Raymond said: "How well I remember that sunny afternoon at Philadelphia, when, in the sacred stillness of 'God's Acre,' ringed with the noisy life of the metropolis, we buried in flowers and evergreens the body of our beloved friend, while overhead branches, like these, waved their solemn murmurous benediction, and all around us white fingers pointed upward, mutely saying, 'He is not here; he is risen!'—and in our ears sounded that deep, dear message of the Spirit, chanting how the blessed dead rest from their labors, while their works do follow them!

"Methinks we do not always perceive the full meaning of that message.   Too often we interpret it as saying, 'They depart; they cease from their labors; and the work they have done takes their place, as their only representation on earth, as all that is now left of their fruitful power.'   Surely, this is not all.   To rest is not to cease; to follow is not to remain behind forever separated from the leader, but rather to abide with the leader, though he be on the march.

"Our human experience is not without interpreting analogies.   We know what it is to rest from our labors for a few happy summer weeks, laying upon other shoulders the daily burden and upon other hearts the daily anxiety, yet still in forest solitudes or up shining summits or by the boundless sea, carrying with us in a higher mood our work—weighing it more accurately, because we are not too tired; seeing it more clearly, because we are out of the dust of it; realizing its proportions and purpose, because distance gives us a perspective view; tasting its full sweetness, because its bitter cloudy precipitate has had time to settle; and renewing our high ambitions for it as we renew our strength for it.   We rest from our labors, but our work goes with us, inseparably—only now we bear it, not as weight, but as wings.

"So, it seems to me, we are to think of our absent dead; they rest, but do not cease; they go on, and their work goes on with them.   Indeed, the interpretation is yet deeper. To my ears, the Spirit says, 'Blessed are they who have labored so earnestly as to deserve the rest of a higher sphere of labor, and who have left behind them works which deserve to follow them, and to receive, even in that higher sphere, their continued remembrance and interest.'"

How more fittingly can I close this tribute to the memory of our beloved friend than by these his own words, spoken of a friend dear to him, and honored by us all—words that today we may cite as a requiem and fitting thought of Rossiter W. Raymond himself, loved by us, whose name will go down in the annals of our Institute as that of a super-man of many parts to whom we owe much.

### Address of T. A. Rickard

"Brethren"—it was thus that he addressed us on an occasion that many of you will remember: in 1893, at Chicago, at the closing session of the International Engineering Congress. Other men, representing other nations, had spoken—some of them in poor English—before he was called upon to reply for the arts of mining and metallurgy in America. When he said "Brethren," the audience was startled into lively attention, which was maintained throughout his speech; for then, as always, he knew how to reach the minds of men, and their hearts too. I remember his saying that those present had taken part in numerous scientific discussions; that they had evolved new ideas and had discovered new principles, but that they had done something much better: they had "discovered one another." So saying he put his finger on the distinctive feature of all such conventions. His mode of salutation also reminded those of us who were his personal friends that he was an evangelist as well as an engineer, and that he could instruct a bible-class in Job or St. Paul with the same power of exposition as he could deliver a lay sermon on mining or metallurgy. Indeed Rossiter Raymond was a deeply religious man, and no sympathetic understanding of his extraordinarily versatile character is possible without appreciating this fact. He was not only a prominent member of Plymouth Church, Brooklyn; he was superintendent of the Sunday-school for 25 years, he led in prayer-meeting and in bible-class, he interpreted the Old Testament during the period when the so-called higher criticism was undermining the faith of the churches, and he aided Henry Ward Beecher in steering his congregation through the storm of biblical exegesis that crossed the Atlantic forty years ago. The eminence that he attained as a religious teacher is measurable by the fact that when Beecher died the trustees asked him "to give up his work as editor, lawyer, and mining engineer, and take the pastorate of Plymouth Church," as recorded by the Rev. Dr. Dwight Hillis. He declined the honor, thinking it better "to give his life and strength to the vocation of an interpreter, chronicler, guide, and assistant to engineers, rather than to that of a creative and constructive leader." I quote the words he himself used on the occasion of the dinner celebrating his 70th birthday.

Not many in the mining profession knew this phase of his character, although during his journeys through the West he would occasionally take the pulpit in some mining community and surprise a congregation that knew him only as the most distinguished of the experts engaged during the previous week in an important apex litigation. I have spoken of the part he played in the history of Plymouth Church, but his deeply religious nature was never so brought home to me as when his son Alfred died in 1901. He was a son of whom any father might feel proud; gifted and amiable, and on the threshold of a brilliant career. When he died Dr. Raymond proved, if it were necessary, the sincerity of his religious convictions, for his glad way of speaking of his departed son, showed his confidence in a future reunion. I never saw a more convincing expression of the belief in immortality than in the attitude of Alfred Raymond's father and mother. It were improper for me therefore on this occasion to speak of the passing of our honored friend in a lugubrious strain. I shall speak of his life and career as an inspiring memory to be treasured as a heritage of our profession; and in doing so, I shall abstain from flattery. To extol the honored dead with honeyed words is an impertinence. Rossiter Raymond's career was so rich in performance as to require none of the insincerities of conventional biography.

To the profession, Dr. Raymond's work as Secretary of the American Institute of Mining Engineers was the outstanding feature of his supremely useful life. When the Institute was founded, in 1871, he was elected vice-president, with the understanding that he would perform the duties of president, which David Thomas, by reason of his age, could not discharge. Thus from the beginning Raymond was the real president, and, on the resignation of Mr. Thomas, a few months later, he became president, in name as well as in fact; thereafter to be elected again and again, until an amendment to the rules, proposed by himself, provided that no president could serve more than two years. Soon afterward, in 1884, he became Secretary, a post that he held for 28 years—until his retirement from active service in 1912. He was Secretary Emeritus until the end.

The duties of the Secretary included the editing of the Transactions. For this he was well prepared. He had been the writer of successive volumes of the "Mining Statistics West of the Rocky Mountains;" he had been editor of the *American Journal of Mining* for one year, in 1867, and for the seven following years the editor of its successor, the *Engineering and Mining Journal*, of which he continued to be associate editor with Richard P. Rothwell until they had a friendly disagreement over the "silver question" in 1893, after which he withdrew from editorial responsibility, becoming a "special contributor," in which capacity he assisted the editors that succeeded Rothwell. Thus he took a notable part in the development of technical journalism in this country; but I regard his share in the early editing of the *Journal* as important chiefly because it was a training for his life-work, that of Secretary of the Institute. It is noteworthy that as the owner of the *Journal* in its early days he found the work of writing and editing far more to his taste than the management, for in financial affairs he was too kindly to be a shrewd business-man.

As Secretary of the Institute he performed divers duties; he invited written contributions and revised them before publication; he organized the meetings; he was the administrator. In course of time his ebullient personality so dominated the Institute that he was allowed a free hand to do as he thought fit. Presidents came and went; although nominally Secretary, he exercised complete control. The personnel of the board of management, or "council," of the Institute changed from year to year, but Dr. Raymond managed its affairs, practically without let or hindrance.

The Institute became identified with him.  For a period longer than a generation he was the mainspring of the activities of the Institute, its presiding genius, its chief spokesman.  Those who participated in the meetings of ten or twenty years ago will retain a vivid impression of the way in which Dr. Raymond stamped his individuality on the organization. Courteous and friendly to all, resourceful and tactful in steering the discussions, witty and eloquent whenever he rose to his feet, he was the managing director of the proceedings; he gave point and distinction to them; he infused them with his keen enthusiasm; he lighted them with the brilliance of his mind.  His versatility was unlimited.  All knowledge was his patrimony and nothing human was alien to his understanding. Whatever the subject of a paper, he could add something to it; nay more, on many occasions when some new phase of geology or engineering was presented for discussion, he would rise to supplement the speaker's remarks and show himself so well informed on the subject as to eclipse the specialist.  He did this not unkindly, but out of super-abundance of knowledge and sheer exuberance of spirit.  On the other hand, no member engaged in preparing a paper for the Transactions failed to obtain his whole-hearted assistance in collecting the necessary data or in hunting for the needed references.  When the member's manuscript arrived, the Doctor went through it with painstaking care.  Before the use of the typewriting machine came into vogue, and even after, he would send letters in long-hand of as much as ten pages, explaining or suggesting improvements in the text.  As a beneficiary of his conscientious industry, I can testify to the instruction in the art of writing that he gave to those who contributed to the Transactions.  He was a delightful helper and a stimulating teacher.  If any criticism is to be made, I venture to suggest that he over-edited; that is to say, the writings of the inexperienced were so much revised as to be practically re-written by him.  He would take the half-baked production of a semi-literate engineer and subject it to the warmth of his intellectual combustion until it emerged a wholesome biscuit.  I recall a valuable metallurgical paper, written by a professor now recognized as an authority, that was so full of German idioms that Dr. Raymond had to re-write it.  Shortly before the Colorado meeting of 1896 I persuaded a Cornish mining engineer to contribute a paper on the lode-structure of Cripple Creek.  He was a keen observer, but a poor writer; when the paper arrived it was quite unsuitable for publication.  Dr. Raymond showed it to me and said, "What am I to do with this?"  I replied, "Don't accept it."  "No," said he, "that would not be fair; we asked him to write it."  "Yes," I said, "but I am responsible for asking him; let me lick it into shape."  "No," he insisted, "that is my job, I'll see what I can do with it."  He did, and he did it so thoroughly that my Cousin Jack friend obtained credit for an informing and well-written contribution to the Transactions.  The result of such revision was to lessen the value of the paper as scientific evidence.  The authenticity of the testimony, it seems to me, suffered by being given through the mouth of a skilled advocate.  On the other hand, this overplus of editorial labor gave the Transactions a level of style that no other technical society could claim either then or since.  All technical writing in the English language has felt, and long will continue to feel, the inspiration to excellence that he gave while editor of the reference library that we call the Transactions of the American Institute of Mining Engineers.

He left an enduring mark on the jurisprudence of mining. A keen observer and a clear expositor, he achieved distinction as an expert witness in the litigation arising from attempts to apply the law of the apex, a subject on which he wrote a series of essays that exercised a strong influence on the interpretation given by the highest courts to that Congressional statute. In the first big case in which he took part, the famous Eureka-Richmond lawsuit, he gave the term 'lode' a definition that not only swayed the decision in that controversy, but influenced all later mining litigation. On one occasion he was invited to address the United States Supreme Court on a point of mining law, and his exposition is said to have been accepted by the Court in its subsequent opinion. At that time he had not qualified as a lawyer, but in 1898 he was admitted to practise in both the State and the Federal courts. Five years later he was appointed lecturer on mining law at Columbia University.

As an expert witness, he was, as he said of Clarence King, approvingly, "an honest partisan." He used the gift of exposition with great effect when addressing the jury, under cover of giving evidence. I recall the explanation of the formation of mineral veins with which he began his testimony in the Montana-St. Louis case. Fortunate was the jury that had the opportunity of listening to such a fascinating lecturer. He was not only an able witness in chief and extremely dexterous in circumventing cross-examination, but he was a great general. He was quick to recognize the important features of a case and skilful in marshalling his forces to the discomfiture of the enemy. In forensic duels he displayed characteristic wit and versatility. This legal practice was a source of honor and profit to him, but I venture to say that he helped geology more in other ways.

In 1868, when only 28 years of age, he was appointed U. S. Commissioner of Mining Statistics, and in that capacity he visited the mining districts of the West, which was then at the beginning of an era of widespread exploration. He was quick to appreciate the economic value of geology and to utilize the opportunities for study afforded by his official travels. In 1870, he was appointed lecturer on economic geology at Lafayette College, which appointment he held for twelve years.

When he became Secretary of the Institute he transferred his keen interest in economic geology to the Transactions. As Secretary, he persuaded the engineers to record observations made underground and at the same time he induced the officers of the Geological Survey to present their scientific inductions to the Transactions in a form that rendered them attractive to the mining profession. Thus he brought the official geologist into touch with the mine manager and consulting engineer, greatly to the advantage of all. He also did much to diminish the self-sufficiency of the Survey and to lessen the shyness of the so-called practical man. By his understanding of geology, his knowledge of Western mining conditions, and the zest with which he pursued the application of geology to mining, he aided greatly in exciting intelligent interest in the genesis of ore deposits. The Posepny volume proves that; so does the volume dedicated to the memory of his friend Emmons. In 1893, he translated Posepny's treatise from the German into his own vigorous English, and organized a discussion that enhanced the value of the original paper. By means of another treatise, by Van Hise, presented to the Institute seven years later, in 1900, he gave a fresh impetus to the study of ore deposits, the general result being to make the mining geologists of this country the leaders in a branch of study in which European scientists had theretofore held pre-eminence.

On his skill as a writer it is pleasant to dwell. He wrote out of the fulness of a rich mind, an alert imagination, and an abundant vocabulary, aided by the knowledge of several modern languages. He knew not only how to select *le mot juste*, but also how to weave words into ingenious phrases and to construct balanced sentences, following each other in logical order within well-proportioned paragraphs. He liked to number his paragraphs, in order to emphasize successive points at issue. He wrote with pen or pencil, usually the former, because it is less rigid and therefore less fatiguing to the fingers. He did not like to dictate anything except ordinary correspondence, but he could dictate a long article or legal testimony, punctuation included, with remarkable clearness and continuity. He wrote easily, with all the joy of the practised hand and the disciplined brain. He twitted one of his contributors with having "an inveterate fluent profuseness of speech" and the happy victim protested that the phrase exactly fitted *him*, not the lesser writer. He was fluent and profuse, but not to redundance or verbosity; on the contrary, his style was marked by force and consecutiveness, and, not infrequently, by those "saber thrusts of Saxon speech" that are the delight of the critical.

His literary ability was partly inherited from his father, Robert Raikes Raymond, who was editor successively of the *Free Democrat* and the *Evening Chronicle*, at Syracuse, New York, from 1852 to 1854, and later Professor of English in the Brooklyn Polytechnic Institute and Principal of the Boston School of Oratory. It is also a safe surmise that Rossiter Raymond owed much of his fine feeling for the language of Shakespeare to his daily draughts from that well of English undefiled, the King James version of the Bible. There is no better schooling in our language than familiarity with The Book. A third aid to the cultivation of a good prose style was his frequent exercise in versification. The expression of simple ideas in verse by means of short words is excellent training for the effective construction of logical sentences in prose; moreover, the sense of rhythm incites assonance. On his return from life at the German universities, he brought with him many old folk-songs and student-songs, some of which he adapted to Sunday-school use. Thousands of children sang his hymns with delight because he knew how to present pretty thoughts in simple guise. That he could write serious poetry we know; for example, the lines to the Grand Canyon engraved on the silver tray that formed part of the gift presented to him on his 70th birthday. He wrote merry rhymes for our Institute meetings and for other occasions of a similar kind, making good-natured fun for himself and his friends. This playing with words in rhyme and rhythm gave him facility of expression in the more serious business of prose, and also in public speaking.

He was a delightful speaker. Our profession has never had a more eloquent spokesman. He seemed as little at a loss for ideas as for words; his enunciation was clear, he had a resonant voice, and his gestures were natural. Owing to his retentive memory and easy delivery, it was difficult to distinguish a speech that he had written from one that was extempore.

At any gathering he was individual—a distinguished figure. The wearing of a black silk cap and an old-fashioned way of trimming his beard gave him a striking appearance. Clear eyes wide apart, an aquiline nose, and a square chin indicated imagination, perception, and determination. His military training had taught him to stand upright. His pose was that of a captain of men. When he made a humorous hit he would tilt his head and smile, as if eager to share the fun with his audience. He never touched anything without giving it human interest. He found

"Tongues in trees, books in the running brooks,
Sermons in stones, and good in everything."

Rossiter Raymond exercised an immense influence in his day and generation—nay more, two generations felt the force of his personality. How he stimulated his religious co-workers has been recorded by the successors of Henry Ward Beecher. Both Lyman Abbott and Dwight Hillis have testified to the courage that he imparted to them during the troublous times of Plymouth Church. To the geologists who broke the trail for the scientific investigations of a later day he was a guide, philosopher, and friend. Such men as Clarence King, James D. Hague, and S. F. Emmons have recorded their gratitude for his support and advice. Among his engineering contemporaries were scores to whom he was an ever-ready source of information, a wise counselor, a cheery friend—for them he did many unselfish and kindly things. To those of us who were young when he was at his prime he was the very embodiment of scientific attainments. We looked up to him as the exemplar of effective writing and polished speaking, the pattern of engineering culture, the leader in everything that concerned the welfare of our profession. As Secretary of the Institute we found him a lovable man, full of natural kindness and that helpfulness, without condescension, which the young appreciate so keenly when shown by a senior whom they admire. We—for I was one of them—found him an inspiring leader and a loyal friend. Loyalty—yes, that was one of his qualities. It got him into trouble more than once, for in friendship, as in apex litigation, he was unmistakably partisan. He stuck to his friends through thick and thin; he gave them the benefit of the doubt if they did wrong; he championed them when they were set upon. Lucky was the man on whose side he fought.

He was pre-eminently a publicist and an educator; he declined the pastorate of Plymouth Church to become the pastor of a bigger congregation; he resigned his professorship at Lafayette to be a teacher in a bigger school; he was the dean of the mining profession in the United States. For fifty years the force of his personality was felt among the men that were organizing and directing the mining industry of a continent; for fifty years he did not fail to write a Christmas story for the children of his Sunday-school; he was a friend to the old and to the young. Age could not wither him nor custom stale his infinite variety. He influenced those that today are influencing others; his spirit still moves among men. Blessed be his memory.

## BRIEF BIOGRAPHY OF DR. RAYMOND

Rossiter Worthington Raymond, Ph.D., LL.D., mining engineer, metallurgist, lawyer and author, was born in Cincinnati, Ohio, April 27, 1840, son of Prof. Robert Raikes and Mary Ann (Pratt) Raymond;

grandson of Eliakim and Mary (Carrington) Raymond, of New York City, and of Caleb and Sally (Walker) Pratt, of Providence, R. I.

He was of English descent, his earliest American ancestor on the paternal side, Richard Raymond, having emigrated from England to this country and settled at Salem, Mass., about 1632; while on his mother's side he was descended from well-known New England families. His great-grandfather, Nathaniel Raymond, was an officer in the Revolutionary army; and his grandfather, Caleb Pratt, served in the war of 1812.

His father (born 1817, died 1888), a native of New York City, was a graduate of Union College in 1839, editor of the Syracuse *Free Democrat* in 1852, and the *Evening Chronicle* in 1853–4, and afterward Professor of English in the Brooklyn Polytechnic Institute and Principal of the Boston School of Oratory. His mother (born 1818, died 1891) was a native of Providence, R. I. They were married at Columbus, Ohio, in 1839, and Rossiter W. was the eldest of a family of seven children, of whom four were sons.

He received his early education in the common schools of Syracuse, N. Y., and in 1857 entered the Brooklyn Polytechnic Institute, of which his uncle, Dr. John H. Raymond (afterward President of Vassar College), was then President, graduating from that institution, at the head of his class, in 1858. He spent the ensuing three years in professional study at the Royal Mining Academy, Freiberg, Saxony, and at the Heidelberg and Munich Universities.

Returning to the United States in August, 1861, he entered the Federal army and served as aide-de-camp, with the rank of captain, on the staff of Major-General J. C. Frémont, by whom, during his campaign in the Valley of Virginia, he was officially commended for gallant and meritorious conduct.

From 1864 to 1868, he engaged in practice as a consulting mining engineer and metallurgist in New York City; and in the latter year was appointed United States Commissioner of Mining Statistics, which position he held until 1876, issuing each year "Reports on the Mineral Resources of the United States West of the Rocky Mountains" (8 vols., Washington, 1869–76), several of which were republished in New York, with the titles of "American Mines and Mining," "The United States Mining Industry," "Mines, Mills and Furnaces," and "Silver and Gold." These reports contained descriptions of the geology, ore deposits and mining enterprises of the United States public domain, discussions of metallurgical processes adapted to American conditions, and observations and criticisms concerning the practical operation of the United States mineral land laws of 1866 and subsequent years. In 1870, he was appointed lecturer on economic geology at Lafayette College, which chair he occupied until 1882, and for one year during that period gave the entire course on mining engineering.

In 1873, Dr. Raymond was appointed United States Commissioner to the Vienna International Exposition and as such delivered at Vienna addresses in the German language at the International Convention on Patent Law, and the International Meeting of Geologists; and an address in English at the meeting of the Iron and Steel Institute, in Liége, Belgium. From 1875 to 1895, he was associated, as consulting engineer, with the firm of Cooper, Hewitt & Co., owners of the New Jersey Steel & Iron Company, the Trenton Iron Company, the Durham and the Ringwood Iron Works, as well as numerous mines of iron ore and coal. As President of the Alliance Coal Company, and director of the Lehigh & Wilkesbarre Coal Company, as well as a personal friend of Franklin B. Gowen, he became acquainted with the inner history of the memorable campaign against the "Molly Maguires," and has since been known as a fearless opponent of all tyranny practised in the name of labor. His articles on "Labor and Law," "Labor and Liberty," etc., published in the *Engineering and Mining Journal* at the time of the Homestead riots, attracted wide attention and for these, as well as similarly frank discussions of the operations of the Western Federation of Miners in Montana, Idaho and Colorado, he received special denunciations and threats from the labor unions thus criticised. While connected with Cooper, Hewitt & Co., he also assisted Abram S. Hewitt in the management of Cooper Union and for many years directed the Saturday Evening Free Popular Lectures on Science, etc., which constituted the beginning of what has since become a vast lecture system in the city of New York.

From 1885 to 1889, he was one of the three New York State Commissioners of Electric Subways for the city of Brooklyn, and served as member and secretary of the Board, preparing its final report, which was generally regarded as the best statement of the problem of municipal engineering and policy involved in the distribution of electric conductors. At the close of his official term as Commissioner, he became consulting engineer to the New York and New Jersey Telephone Company, which position he retained for many years.

In 1898, Dr. Raymond was admitted to the bar of the Supreme Court of New York State and of the Federal District and Circuit Courts, his practice being confined to cases involving either mining or patent law, in the former of which he was a leading authority. In 1903, he was lecturer on Mining Law at Columbia University, New York. He has also delivered numerous addresses at other colleges and universities, including Yale, Cornell, Pittsburgh, Lehigh, Lafayette, Union, California, the Worcester Polytechnic and the New York College of Physicians and Surgeons.

An original member of the American Institute of Mining Engineers, he served as its Vice-president in 1871, 1876 and 1877, President from 1872 to 1875, and Secretary from 1884 to 1911. In the latter capacity he edited 40 of the annual volumes of *Transactions*, to which he liberally contributed essays, especially pertaining to the United States mining laws, as well as other articles of importance. In 1911, Dr. Raymond resigned his position as Secretary of the American Institute of Mining Engineers, of which he has been since that time Secretary Emeritus.

Dr. Raymond was the editor of the *American Journal of Mining* from 1867 to 1868, of the same periodical under the title *Engineering and Mining Journal* from 1868 to 1890, and thereafter was a special contributor to that journal. In 1884, he prepared for the United States Geological Survey an historical sketch of mining law which was subsequently translated into German and published in full by the *Journal Des Bergrechts*, the only periodical in the world devoted exclusively to the subject of mining jurisprudence, and for which he received high praise.

In addition to the official works previously mentioned he was the author of "Die Leibgarde" (1863), a German translation of "The Story of the Guard" by Mrs. Jessie Benton Fremont (1863) "The Children's Week" (1871); "Brave Hearts" (1873); "The Man in the Moon and Other People" (1874); "The Book of Job" (1878); "The Merry-go-

Round" (1880); "Camp and Cabin" (1880); "A Glossary of Mining and Metallurgical Terms" (1881); "Memorial of Alexander L. Holley" (1883); "The Law of the Apex, and Other Essays on Mining Law" (1883–95); "Two Ghosts and Other Christmas Stories" (1887); "The Life of Peter Cooper" (1897); various technical works and papers on mining law, as well as numerous addresses and magazine articles, and contributions to several American dictionaries and encyclopedias. In 1916, Dr. Raymond published a volume of poems, entitled "Christus Consolator and Other Poems." At the time of his death he was at work upon a history of the American Institute of Mining Engineers, which he hoped to finish this year.

In 1909, in collaboration with W. R. Ingalls, he contributed to the First Pan-American Scientific Congress, held at Santiago, Chile, a paper on "The Mineral Wealth of America," and at the Second Congress, which assembled at Washington, D. C., in 1915, he was represented by a paper entitled "The Value of Technical Societies to Mining Engineers." "The Conservation of Natural Resources by Legislation" was delivered in 1909 before a joint meeting of the four national engineering societies.

In 1910 the 70th birthday of Dr. Raymond was celebrated by a dinner at which all branches of the engineering profession, the scientific and learned societies, and the prominent institutions of learning were represented. On this occasion the gold medal of the Institution of Mining and Metallurgy was awarded to Dr. Raymond "in recognition of eminent services and lifelong devotion to the science and practice of mining and metallurgy and of his numerous and valuable contributions to technical literature."

In 1911, during the visit of the American Institute of Mining Engineers to Japan, Dr. Raymond received from the Mikado the distinction of Chevalier of the Order of the Rising Sun, fourth class—the highest ever given to foreigners not of royal blood—"for eminent services to the mining industry of Japan." These services consisted in advice and assistance rendered in America to Japanese engineers, students and officials throughout a period of more than 25 years.

Dr. Raymond was an honorary member of the Society of Civil Engineers of France, the Iron and Steel Institute and the Institution of Mining and Metallurgy of Great Britain, the Mining Society of Nova Scotia and the Australasian Institute of Mining Engineers. He was a Fellow of the American Association for the Advancement of Science and the American Geographical Society, a member of the American Philosophical Society, the Brooklyn Institute of Arts and Sciences, the American Forestry Association and various other technical and scientific organizations both at home and abroad. He received the degree of Ph.D. from Lafayette College in 1868, and that of LL.D from Lehigh University in 1906. On the latter occasion, speaking as an adopted alumnus of the University, he delivered to the graduating classes an address on "Professional Ethics" which has been widely quoted and approved.

In February, 1915, Dr. Raymond delivered the commemorative address on the 150th anniversary of the foundation of the University of Pittsburgh, and received from that Institution the honorary degree of LL.D.

He married in Brooklyn, N. Y., March 3, 1863, Sarah Mellen, daughter of William R. and Mary (Fiske) Dwight of that city. Of their five children two survived to adult years; Alfred (b. 1865, d. 1901), an architect and engineer of thorough training and great promise; and Elizabeth Dwight (b. 1868), since 1892 the wife of H. P. Bellinger of Syracuse, N. Y.

He died suddenly, of heart failure, at his home in Brooklyn, N. Y., on the evening of December 31, 1918, and was buried in Greenwood Cemetery.

## RESOLUTIONS

The Board of Directors of the American Institute of Mining Engineers would place upon its minutes its profound sense of loss and sorrow in the death of Rossiter Worthington Raymond, Ph. D., LL. D., Secretary Emeritus of the Institute. Both as one of its founders and as its Secretary for 27 years, his was the guiding spirit of the Institute for more than a generation. During the greater part of this long period, it might almost have been said that the Institute was Dr. Raymond—and Dr. Raymond the Institute. When with the progress of growth and development, great changes were introduced, Dr. Raymond acquiesced in these in spite of some misgivings, such as those with which a father might contemplate the emergence of his child from the careful supervision of the home; but as Secretary Emeritus for the past eight years, he was always ready with valuable advice and helpful suggestion.

His presence at the annual meetings was an inspiration, which his rare ability as a speaker further enhanced. Among the most versatile of men of genius, among the most distinguished as a mining engineer; a scholar, editor and authority on mining law, yet to his personal friends he revealed a simplicity, a loyalty, and a steadfastness which held his intimates and bound them to him in spite of time and change.

With his death there closes an epoch in the history of American mining and metallurgy. The Institute thereby loses one of its great leaders, but his example will live as an inspiration to those who survive, within its councils, and his name will be long an inspiration for many who knew him only through our transactions and by his other writings.

## MEMORIAL MEETING FOR THE MEN WHO DIED IN SERVICE

On Feb. 19, a meeting was held in memory of the members of the Institute who died in the service of the United States and the Allies. At this meeting Captain Percy E. Barbour, Acting Assistant Secretary of the Institute, said: We have received word of the death of twenty-three of our members in service. The fifteen named in the following list were mentioned (and biographical notices were read) at the memorial meeting held in Colorado last September: Lewis Newton Bailey, Lieut. Louis Baird, William Morley Cobeldick, Ralph Dougall, Lieut.-Col. Alfred Winter Evans, Lieut. Thomas C. Gorman, Lieut. William Hague, Capt. William T. Hall, Lieut. Bernhardt Heine, Capt. John Duer Irving, Lieut. Edward H. Perry, Lieut. Frank Remington Pretyman, Capt. Fred B. Reese, Soren Ringlund, Lieut. George Roper, Jr.

Since September, we have heard of the deaths of Capt. John H. Ballamy, Lieut. Martin F. Bowles, Andrew Burt, Corp. Sheppard B. Gordy, Serg. Herbert M. Harbach, Sidney A. Lang, Lieut. Norman L. Ohnsorg, Raymond W. Smyth. Captain Barbour read brief biographies of these members as their portraits were shown upon the screen. Then two buglers from the army post on Governor's Island sounded taps for their comrades. (Biographies will be found at the end of this volume.)

## REPORTS FOR THE YEAR 1918

### THE SECRETARY'S REPORT

*Rossiter Worthington Raymond, Ph. D., LL. D.*—1840 *to* 1918.—Dr. Rossiter W. Raymond, Past President, Honorary Member and Secretary Emeritus, died suddenly of heart failure at his home, 123 Henry St., Brooklyn, N. Y., on the evening of Tuesday, Dec. 31, 1918. He was one of the founders of the Institute and its second President. A Memorial Service will be held on Monday afternoon, Feb. 17.

*James Douglas.*—The Institute suffered a second very severe loss by the death of Dr. James Douglas, Honorary Member, Past President (having served as President for two years), and in three separate respects the greatest benefactor of the Institute: first, in raising funds by voluntary subscription and his own personal gifts to pay the Institute's share of the money owed on the land on which the Engineering Societies Building now stands; second, as original donor to Engineering Societies Library; and third, by his bequest of $100,000 for the maintenance of the library of the American Institute of Mining Engineers.

A resolution regarding the death of Dr. Douglas was passed by the Board of Directors and at the meeting of the Institute on September 4, 1918, in Colorado, at which time a memorial service was held in the theater of the Hotel Broadmoor, attended by about 400 members and guests. A copy of the resolution was prepared and sent to the family. A bronze tablet is now being prepared and will be placed in the Members' Room and unveiled at the February meeting. A biography and portrait of Dr. Douglas was published in the September, 1918, *Bulletin*. There is also placed in the Engineering Societies Library an oil portrait of Dr. Douglas which had his own approval and was presented by him.

*American Institute of Metals.*—During the Spring of the year, plans were completed with the American Institute of Metals whereby this society became the Institute of Metals Division of the American Institute of Mining Engineers. This body was a dignified aggregation of metallurgists which had been in existence for about 11 years. A bronze tablet commemorating the American Institute of Metals is now being placed in the Members' Room of the Institute, and during the year 1919 the *Bulletin* will bear the following title: "Bulletin of the American Institute of Mining Engineers, with which is consolidated the American Institute of Metals."

*Visits to Local Sections.*—During the year President Jennings made visits to six of the Local Sections, on three of which occasions he was accompanied by the First Vice-president and the Secretary. First Vice-president Goodale made five visitations and the Secretary made eight.

At Washington, D. C., a Local Section of the Institute was formed at a meeting held on June 21, 1918, and is described in the August *Bulletin*. There has also been a request received for the formation of a section in the Lake Superior region, to be known as The Upper Peninsula Section. The President of the Institute, accompanied by Mr. Horace V. Winchell and the Secretary, visited the members of Duluth, Minn., and Houghton, Mich., on their return from the Colorado Meeting.

*Meetings.*—Three meetings were held during the year; namely, the Annual Meeting in February, the Colorado Meeting in September, and the Milwaukee Meeting in October.

*Expulsion of Enemy Aliens.*—All enemy aliens were expelled from the Institute. This included Honorary Members and those in enemy countries as well as all persons residing in other countries and known to be enemy aliens. The Committee on Membership was requested to report regarding the latter class of persons.

*Remission of Dues of Men in the Service.*—By action of the Board of Directors the dues of all Members, Associates, or Junior Associates in the service of the United States or its Allies were remitted upon receipt of request.

*Remission of Dues of Older Members.*—By action of the Board of Directors the dues were remitted of certain members who had paid dues for the past 30 years and had attained the age of 70 years. In individual cases, the dues were remitted of members who had not yet attained the age of 70 years but who had paid dues for 35 years or more and were no longer active.

*Membership, Finance, Publications, Library.*—The details as to the activities during the year of the Committees on Membership, Finance, Papers and Publications, and Library of the Institute are set forth in the reports of the committees on pages lvii to lxi of this volume.

*Canadian Mining Institute.*—To strengthen the bonds of friendship and coöperation that have always existed between the Canadian Mining Institute and the American Institute of Mining Engineers, members of the Canadian Mining Institute not residing in the United States were privileged to subscribe to the *Bulletin* of this Institute at one-half the regular subscription price. This is the same price at which it is sold to members of the American Institute of Mining Engineers. Furthermore, the members of the Canadian Mining Institute were all invited to attend the 119th meeting of the Institute, and Feb. 18, 1919, will be known as "Canadian Mining Institute Day" at this meeting. Returning the courtesy, all members of the American Institute of Mining Engineers are invited to attend the meeting of the Canadian Mining Institute on Mar. 6, 1919, which day will be known as "American Institute of Mining Engineers Day."

*Employment Department.*—Employment activities of the Institute during 1918 have been on an important scale. Owing to the shortage of engineers, it has been difficult to fill all positions that were open. This situation was reversed on the signing of the armistice, when a great many engineers were discharged from the service. Coöperation with the Government Department was maintained through a committee of Engineering Council. In November, this Committee was superseded by a bureau of employment headed by the Secretaries of the Four Founder Societies, which bureau is now known as the Engineering Societies Employment Bureau. This Bureau is also coöperating with the United States Employment Service, maintained by the United States Department of Labor. At the time of this writing (January, 1919), the supply of engineers returning from the war is very much larger than the number of situations offered.

*National Research Council.*—With the approval of the President, in May, 1918, the Secretary assumed the duties of Chairman of the Section on Metallurgy of the Engineering Division of the National Research Council. In this way the Institute was enabled to be of service to the National Research body in connection with many problems in metallurgy for the benefit of the Army, Navy, Aircraft Board, Emergency Fleet Corporation, and other war activities. At the 119th meeting of the Institute, in February, 1919, nine papers on metallurgy will be presented, all of which are reports of researches under the auspices of National Research Council.

The National Research Council was originated in 1916 at the request of the National Academy of Science. On May 11, 1918, National Research Council was made a permanent body by executive order of the President of the United States. The President and Secretary of the Institute joined with the Presidents and Secretaries of the other Founder Societies in giving a dinner to the Chairman and officers of the National Research Council, at the Engineers' Club. In this way the Institute has been performing one of the important objects of its incorporation, namely, the promotion and encouragement of engineering research.

---

### COMMITTEE ON MEMBERSHIP

The total number of applications brought before the Committee during the year 1918 was 675; the total number of persons who were elected and became members of the Institute during the same period was 666.

The total membership of the Institute on Dec. 31, 1917, was 6528, consisting of 20 Honorary Members, 5832 Members, 237 Associates, and 439 Junior Associates. The changes in membership during the year are shown in the accompanying schedule:

| | | |
|---|---:|---:|
| Total Membership, Dec. 31, 1917. . . . . . . . . . | | 6528 |
| Loss by resignation. . . . . . . . . . . . . . . . . . . . . | 60 | |
| Loss by suspension. . . . . . . . . . . . . . . . . . . . | 162 | |
| Loss by death. . . . . . . . . . . . . . . . . . . . . . . | 90 | 312 |
| | | 6216 |
| Elected. . . . . . . . . . . . . . . . . . . . . . . . . . . . . | 666 | |
| Reinstated. . . . . . . . . . . . . . . . . . . . . . . . . . | 56 | |
| By affiliation with American Institute of Metals. . . . . . . . . . . . . . . . . . . . . . . . . . . | 220 | 942 |
| Membership, Dec. 31, 1918. . . . . . . . . . . . . . | | 7158 |

Change of Status:

| | |
|---|---:|
| Associates to Members. . . . . . . . . . . . . . . . | 1 |
| Junior Associates to Associates. . . . . . . . . | 8 |
| Junior Associates to Members. . . . . . . . . . | 21 |

## REPORT OF TREASURER AND FINANCE COMMITTEE

### ASSETS

*Cash on Hand and in Banks:*
General funds...................................... $6,752.27
Special funds...................................... 1,506.10    $8,258.37

*Investment of Life Membership Fund:*
$2,000 Interborough Rapid Transit 5 per cent. bonds 1966  $1,974.31
$1,000 Illinois Central, Chicago, St. Louis & New Orleans
   5 per cent. bonds due 1963....................... 1,010.56
$2,000 Chicago, Milwaukee & St. Paul R. R. 4 per cent.
   bonds due 1934................................. 1,877.63
$1,000 Chicago, Milwaukee & St. Paul 4½ per cent. bond
   due 2014...................................... 824.25
$650 U. S. Government Third Liberty Loan 4¼ per cent.
   due 1928...................................... 650.00    6,336.75

*Liberty Bond:*
$100 U. S. Government Third Liberty Loan 4¼ per cent.
   due 1928...................................... · 100.00
*Interest in United Engineering Building:*
Land and building 29 West 39th Street—¼ of $1,947,171.16........ 486,792.79
*Library:*
Books and periodicals in Library belonging to American Institute
   of Mining Engineers.................................... 40,000.00

$541,487.91

### RECEIPTS

*Initiation Fees*....................................... $6,000.00
*Annual Dues:*
Current dues.......... .................. $66,135.43
Arrears................................. 2,321.14
Advance................................ 1,706.86    70,163.43    $76,163.43

*Receipts from Other Sources:*
Sale of Transactions............................... $4,673.75
Sale of binding................................. 11,997.15
Sale of advertising............................... 10,448.96
Sale of special editions........................... 794.02
Sale of Bulletins and pamphlets..................... 3,060.90
Sale of pins and fobs.............................. 237.55
Interest on investments and bank deposits.............. 957.23
Sundry refunds from Societies........................ 269.57
Sundry refunds from members.......... .. ........... 950.24
Sundry receipts.................................. 213.99    33,603.36

*Special Funds:*
Life memberships................................ 450.00
Liberty Bonds purchased by employees................ 1,250.00
Proceeds from sale of Dr. Williams' book "The Diamond
   Mines of South Africa"........................... 776.82
From the Dinner Committee ...................... 19.00
Hadfield prize—interest............................ 33.87
Thayer prize—interest............................. 1.71    2,531.40

*Total receipts*....................................... $112,298.19
*Cash on Hand January 1, 1918*...................... 10,465.13

$122,763.32

## LIABILITIES

*Special Funds:*

| | | |
|---|---:|---:|
| Hadfield prize and interest | $1,154.51 | |
| Thayer prize and interest | 58.14 | |
| Dinner committee | 115.20 | |
| Dr. Williams's book | 15.00 | $1,342.85 |
| *Reserve for Life Membership Fund* | | 40,000.00 |
| *Life Membership Fund:* | | |
| Balance January 1, 1918 | $6,150.00 | |
| Additions during year 1918 | 450.00 | 6,600.00 |
| *United Engineering Society:* | | |
| Balance due additions to building $12,500, less $7,500 paid | | 5,000.00 |
| *Surplus:* | | |
| As at hand January 1, 1918 | $546,924.13 | |
| *Deduct:* | | |
| Adjustment of equity in land and building as authorized by chairman of Finance Committee due to one additional founder society being taken in and sharing in the equity | 59,053.83 | |
| | $487,870.30 | |
| *Add:* | | |
| Net income year 1918 | 674.76 | 488,545.06 |
| | | $541,487.91 |

## DISBURSEMENTS

*General Funds:*

| | | | |
|---|---:|---:|---:|
| Bulletin | | $30,750.08 | |
| Year Book | | 2,185.02 | |
| Transactions of 1917 | $1,490.42 | | |
| Transactions of 1918 | 7,697.07 | 9,187.49 | |
| Binding Transactions 1917 | $ 277.57 | | |
| Binding Transactions 1918 | 8,220.14 | 8,497.71 | |
| Special editions | | 2,036.43 | |
| Editorial and office | | 27,399.27 | |
| Treasurer | | 1,154.27 | |
| Library | | 3,999.97 | |
| Advertising | | 5,454.01 | |
| Meetings | | 3,705.88 | |
| Local sections | | 2,898.54 | |
| Technical committees | | 58.22 | |
| Committee on increase of membership | | 2,118.11 | |
| Back volumes 1–53—binding, etc | | 1,813.59 | |
| Circulars | | 452.76 | |
| Engineering Council | | 4,000.00 | |
| Delegate to Council of Engineers, Paris | | 1,500.00 | |
| Sundry charges to members (to be refunded) | | 747.54 | |
| Pins and fobs | | 164.95 | |
| Sundry disbursements | | 968.19 | |
| | | $109,092.03 | |
| Addition to Engineering Building: Payment on account | | 2,500.00 | $111,592.03 |

*Special Funds:*

| | | |
|---|---:|---:|
| Liberty Bonds purchased for investment and for employees | $2,000.00 | |
| Proceeds sale of Dr. Williams's book donated to American Red Cross Society | 761.82 | |
| Dinner Committee | 151.10 | 2,912.92 |
| *Total Disbursements* | | $114,504.95 |
| *Cash on Hand December 31, 1918* | | 8,258.37 |
| | | $122,763.32 |

## REPORT OF TREASURER AND FINANCE COMMITTEE

To the Chairman and Members of the Finance Committee:

We have audited the books and accounts of the American Institute of Mining Engineers and have prepared therefrom a statement of cash receipts and disbursements for the year ended December 31, 1918, and a balance sheet at the latter date. A summary of cash receipts and disbursements follows:

| | |
|---|---:|
| January 1, 1918—Balance in banks and on hand.... | $ 10,465.13 |
| December 31, 1918—Receipts for the year.......... | 112,298.19 |
| | $122,763.32 |
| *Deduct:* Disbursements for year.................. | 114,504.95 |
| December 31, 1918—Balance on hand............. | $8,258.37 |

Distributed as follows:

| | |
|---|---:|
| National Bank of Commerce..................... | $1,093.09 |
| Brooklyn Trust Company....................... | 4,261.51 |
| Fifth Avenue Bank of New York................. | 2,249.66 |
| Fifth Avenue Bank of New York (special account)... | 454.11 |
| Petty cash in office........................... | 200.00 |
| | $8,258.37 |

During the year investment was made in $2000 United States Government 4¼ per cent. Third Liberty Loan Bonds due 1928. Of this amount $650 was used as a further investment on account of the Life Membership Fund and $1250 was purchased by employees and $100 is held in the safe.

We examined the securities as set forth in the balance sheet and found them as there stated. The market values as at December 31, 1918, as quoted by Messrs. Lee, Higginson & Company, are shown below:

| | |
|---|---:|
| $2,000 Interborough Rapid Transit 5 per cent. bonds due 1966 @ 72........................................ | $1,440.00 |
| $1,000 Illinois Central, Chicago, St. Louis & New Orleans 5 per cent. bond due 1963 @ 94.................... | 940.00 |
| $2,000 Chicago, Milwaukee & St. Paul R. R. 4 per cent. bonds due 1934 @ 76............................. | 1,520 00 |
| $1,000 Chicago, Milwaukee & St. Paul 4½ per cent. bond due 2014 @ 72........................................ | 720.00 |
| $650 United States Government Third Liberty Loan 4¼ per cent. due 1928 @ 96............................... | 624.00 |

A further payment of $2500 was made during the year on account of the Institute's proportion of the cost of the addition to Engineering Building leaving a balance of $5000 still unpaid.

The change in valuation of the equity of the American Institute of Mining Engineers in the United Engineering Society land and building is made by the Finance Committee, on our recommendation, the total valuation having been obtained from the United Engineering Society and shown on the books of the American Institute of Mining Engineers at one-fourth of this total figure. This change was brought about by the admission of one additional Founder Society. The same method, we understand, has been adopted by at least one of the other Founder Societies.

Vouchers and cancelled checks were produced for all cash disbursements. The cash on hand was counted and found correct and certificates obtained verifying the bank balances.

All cash received as shown by the books was deposited to banks. The footings and postings to general ledger were checked and found correct.

We are pleased to state that we find the books well kept and the records of the Institute in good order.          Yours very truly,

BARROW, WADE, GUTHRIE & Co.

## COMMITTEE ON PAPERS AND PUBLICATIONS

During the past year 130 papers were submitted to the Committee and 107 were accepted and printed in the *Bulletin*. The aim of the Committee has been, not only in the acceptance of the technical papers but also in the selection of the other matter, to make the *Bulletin* of great interest and value to the members. To carry out this work it has been necessary for the Committee to ask many members to give considerable time to the work of the Institute, which all have freely done. This coöperation has been a great factor in the success attained by our meetings and publications.

For the 116th meeting, held in New York last February, 64 papers were received of which 50 were accepted and printed. Of the 57 submitted for the 117th meeting, held in Colorado last September, 42 were accepted and printed. All of the papers submitted for the 118th meeting, held in Milwaukee in October, 15 in number, were accepted and printed. For the 119th meeting, to be held in New York, February 17 to 20, 58 papers have been submitted, 50 of which have been accepted and printed.

In addition Volumes LVII, LVIII, and LIX of the *Transactions* were printed and sent to the members. Also the collective index of Volumes XXXVI to LV inclusive, which was prepared during 1917, was printed.

---

## LIBRARY COMMITTEE

In accordance with the requirements of By-Law LX, the report of the Library Committee is herewith appended.

The total accessions of the Library for the year were as follows: Gifts, 16,258; purchase, 663; total 16,921. All of this material has been examined and catalogued and is ready for the use of readers.

The largest single gift received during the year, comprising about 800 pieces, was a library of electrical literature presented jointly by the Westinghouse Electric & Manufacturing Co. and the General Electric Co., which the United Engineering Society has agreed to keep intact for five years.

It is proposed to recatalog all the books in the library as soon as practicable.

By the will of Dr. James Douglas, the Institute was the legatee of $100,000 for the maintenance of the Library. Arrangements have been made to turn this money into the treasury of the United Engineering Society for the use of the General Library.

The attendance in the Library showed a total for the year of 15,063, which is more than 1000 over any previous record.

*Engineering Index.*—Among the many plans for increasing the usefulness of the Library, it has been proposed to publish an index of mining and metallurgical literature, which was started in the September, 1918, *Bulletin* of the Institute.

*Sale of Publications.*—The return from the sale of volumes during the year amounted to $4673.75; special editions, $794.02; and *Bulletins* and pamphlets, $3060.90. Volumes XXXI, LI, and LII have all been sold and very few copies are in stock of Volumes IX, XIV, XVII, and LII. The special edition of the Posepny volumes has also been exhausted.

# PAPERS

# Mechanics of Vein Formation

BY STEPHEN TABER,* PH. D., COLUMBIA, S. C.

(Colorado Meeting, September, 1918)

A VEIN may be defined as an aggregation of mineral matter, more or less tabular or lenticular in form, which was deposited from solution and is of later origin than the inclosing rock. This definition differs from the one found in many text-books in that no assumption is made as to the origin of the space occupied by the vein or the manner in which the mineral matter was deposited. In the present paper the writer purposes to discuss the mechanics of vein formation, and does not wish to begin by begging the question.

The origin of ore deposits has been diligently investigated in recent years, but, while much has been written concerning the chemistry of the process, the physical side of the question has received less attention. The source of the ore minerals; the chemical composition of ore-bearing solutions; the relative importance of magmatic and meteoric waters as agents of concentration; and the causes of mineral precipitation have all been discussed at length. On the other hand, little has been written concerning the mechanics of vein formation, and the theories given in text-books today are essentially the same as those developed by the early Cornish and Saxon miners.

Early investigators often regarded both veins and dikes as igneous in origin, but today the theory that vein minerals have been deposited from solutions, either liquid or gaseous, is firmly established. The latter view is supported by the fact that many vein-forming minerals are not found in unaltered igneous rocks; by evidence obtained in the laboratory regarding the solubility of minerals; by the presence of many ore minerals in hot-spring waters; by the occurrence of some veins in regions consisting exclusively of unaltered sedimentary rocks; by evidence as to the temperature at which some minerals are formed; and by the absence of mineral segregation due to difference in specific gravity. There is still, however, much difference of opinion as to whether certain tabular-shaped masses should be classed as veins or as dikes. Some of the earlier investigators have also advocated the theory that veins are contemporaneous in origin with the inclosing rock, while others have believed the vein minerals to represent a recrystallization or transforma-

---

* Professor of Geology, University of South Carolina.

tion of the inclosing rock; but these views have now been generally discarded by geologists, and, since they are unsupported by evidence, it is not necessary to discuss them here.

## METHODS OF VEIN FORMATION

There are six different ways by which it is conceivable that mineral matter transported in solution may reach its resting place in a vein:

It may enter along an open fissure, and be deposited on the walls until the fissure is more or less filled.

It may enter along a fracture, bedding plane, or similar passage, and, as it is deposited, force the wall rock apart, thus making room for the growing vein.

It may be introduced into an open fissure through small openings in the walls, and thus more or less fill the fissure.

It may [be introduced through the walls of a narrow passage or incipient fracture, and, as it is deposited, force the walls apart.

It may enter along a fracture or other passage, and replace part or all of the minerals of the wall rock.

It may be derived from the surrounding material and be concentrated by locally replacing part or all of the minerals in the country rock.

Before estimating the relative importance of these methods of vein formation, it is necessary to establish criteria by which the origin of a given vein can be recognized. In studying the larger metalliferous veins having commercial value, it is sometimes difficult to discover evidence bearing on the mechanics of their formation, for only the final results are available for examination, and evidence that probably existed during the early stages of vein formation may have been largely obliterated by replacement or later alterations. The problem is rendered more difficult by the fact that many veins are the result of two or more methods of ore deposition.

In searching for evidence bearing on the mechanics of vein formation, the writer has not confined his attention to the large complex metalliferous veins, but has also made a detailed study of the small and relatively simple veinlets of a region of unaltered sedimentary rocks.[1] Natural deposits of mineral matter found elsewhere than in veins were also examined, and several different types of veins were successfully produced in the laboratory, where the origin and development of the more important vein structures could be observed in detail. A discussion of the six ways in which it is possible for veins to be formed, together with such

---

[1] Stephen Taber: The Origin of Veinlets in the Silurian and Devonian Strata of Central New York. *Journal of Geology* (Jan.–Feb., 1918) **26**, 56–73.

evidence as has been thus far collected, is given below. The present paper is preliminary in nature, and some vein phenomena are not discussed. The investigation is being continued, but it is only through the collaboration of many observers that such problems can be solved.

*The hypothesis that veins have been deposited from solutions circulating along open fissures until filling was more or less complete* is extremely old, and is to be found, in one form or another, in writings on ore deposits from the time of Agricola to the present day. Drusy cavities or vugs, and comb-structure, banding or crustification, are the evidences commonly cited in recent text-books as proof that veins were deposited in open fissures.

Drusy cavities are most common in veins recently formed at shallow depths, and, if we eliminate the openings due to solution in the belt of weathering, they are very rare, or entirely absent in veins formed at great depth. The gold-quartz veins of the Southern Appalachian region are believed to have been formed at depths of three miles or more; and, so far as the writer has been able to ascertain, they contain no vugs. If veins were formed through the filling of open fissures, there is no reason why vugs should not be as common in depth as in higher levels.

The close similarity in structure between drusy cavities and geodes, as emphasized by both Posepny[2] and Shaler,[3] is proof that open fissures are not essential for the development of the former. Shaler observed that in the process of enlargement, which geodes undergo, they sometimes condense and deform the inclosing rock strata. The texture of the vein matter surrounding vugs is often such as to indicate that the crystals lining the cavities have been deposited, as in geodes, from infiltrating solutions rather than from solutions circulating along an open fissure. In many instances the vugs are lined in part with minerals that are secondary in origin.

It has been demonstrated experimentally that pre-existing fissures are not essential for the development of drusy cavities; such openings are commonly present in veins, produced in the laboratory, which have made room for themselves by pushing their walls apart. A careful consideration of all the facts leads to the conclusion that the crystal lining of druses was deposited on the walls of open cavities, and that druses may be formed in veins that were deposited in open fissures, but the mere presence of these druses in a vein does not prove that the vein as a whole was thus deposited.

Werner seems to have been the first to call attention to banding and to explain it as being due to differences in the composition of the solutions from which the minerals were deposited; but Fournet observed

[2] Franz Posepny: The Genesis of Ore Deposits. *Trans.* (1893) **23**, 207, 218, 255.
[3] N. S. Shaler: Formation of Dikes and Veins. *Bulletin*, Geological Society of America (1899) **10**, 260–262.

that banding was not always parallel to the walls of a vein, since fragments of the adjacent rock were sometimes surrounded by layers of vein
minerals, as, for example, in the "ring-ore" of the Hartz.[4]  De la Beche
discussed the formation of comb-structure and of vugs or druses, and cited
these phenomena as tending to prove that some veins owe their width
to repeated reopenings of the fissures.[5]

Posepny, in his treatise on "The Genesis of Ore Deposits," devoted
many pages to "the filling of open spaces," and referred to banding or
"crustification" as the characteristic feature of cavity-filling; but he
acknowledged that the genesis of the non-crustified deposits was more
difficult of determination, and concluded that "they were not deposited
in pre-existing spaces."[6]

Veins of the latter class are, however, very much more common
than those showing crustification.  Sometimes banding is very imperfect,
and again one part of a vein may show distinct crustification, while in
other parts this structure is entirely absent.  The crustified portion
of some veins "is surrounded by solid quartz possessing no banded or
comb-structure."[7]  Banding, as well as vugs, may result from the filling
of open fissures; it would then necessarily be symmetrical with respect to
the center, but in many veins it is asymmetrical.  Banding can be formed
in other ways, however, and therefore it can not be regarded as proof of
the manner of vein deposition.  In veins produced in the laboratory,
which were not deposited in open spaces, both symmetrical and asymmetrical banding were obtained by changing the composition of the vein-
forming solutions.  A banded structure may also result from replacement
or from metamorphic processes.

When crystals are slowly enlarged by additions of material from
supersaturated solutions, they develop crystal faces on their exposed surfaces; prismatic crystals, such as quartz, are commonly normal to some
supporting surface from which they project, and therefore often have a
parallel orientation.  This is well illustrated by the crystals lining geodes,
drusy cavities in veins, and similar openings in rocks; but in vein quartz
crystal faces are relatively rare or entirely absent except in the immediate vicinity of drusy cavities, and microscopic examination proves that
parallel orientation is likewise uncommon in massive quartz.  There is
unquestionably some recrystallization of vein minerals contemporaneous
with their deposition, as a result of which certain crystals may be enlarged at the expense of others, but this would not change the crystal-

---

[4] Henry T. De la Beche: "Report on Geology of Cornwall, Devon, and West
Somerset," London (1839) 370–371.

[5] Op. cit., 339–342.

[6] Franz Posepny: Op. cit., 262.

[7] J. E. Spurr: Geology of the Tonopah Mining District, Nevada.  U. S. Geological
Survey, Professional Paper No. 42 (1905) 171.

lographic orientation. Crystal faces and parallel orientation are most common in veins formed at very shallow depths, and seem to be practically absent from the deeper veins. Veins formed at shallow depths occasionally show a spheroidal texture with prismatic crystals of quartz radiating outward from numerous central nuclei. Vein quartz usually has a granular texture similar to that shown by many igneous rocks, being due to mutual interference of the growing crystals. The tendency to assume a regular polyhedral form is stronger in pyrite than in any of the other minerals commonly present in veins, and yet pyrite crystals seldom show idiomorphic outlines except where they are embedded in older minerals which they have partly replaced.

If veins were formed through deposition of minerals on the walls of open fissures, there should be a suture line near the center of the veins, marked by occasional vugs where filling was incomplete; but most veins show no trace of such a structure; in many of the smaller veins individual crystals extend from wall to wall, and vugs, when present, are not always confined to the central portions of veins. Rickard has described and figured such a cavity which extended for several feet along the foot-wall of the Jumbo vein in the Enterprise mine of Rico, Colorado.[8]

A soft gouge or selvage is often found along one or both walls of a vein, and, if this were present before the deposition of the vein minerals, it could never have remained in place on the walls of an open fissure even while supporting no additional load of vein minerals. Occasionally the gouge matter is found in the center of a vein.

The change in mineral composition when a vein passes from one rock to another has been attributed in several instances to substances present in the wall rock that act as precipitants;[9] but this action could hardly occur if the vein-forming solutions were separated from the country rock by layers of impermeable mineral matter deposited on the walls of the fissure. Most veins are accompanied by some replacement and mineralization of the wall rock, but there is no evidence that the process of replacement is completed before fissure-filling begins. The mineral composition of some veins varies greatly along their strike; this is difficult of explanation under the assumption that the minerals were deposited from solutions filling continuous open fissures.

It is inconceivable that large fissures could have remained open at great depth below the surface during the long period of time required for their filling by mineral deposition. As a result of his ingenious experiments, Adams concludes that small cavities may exist in granite to a depth of at least 11 miles, but he distinctly points out that the walls of

[8] T. A. Rickard: Vein-walls. *Trans.* (1896) **26**, 224–226.

[9] Richard Beck: *The Nature of Ore Deposits;* Translation by W. H. Weed. New York (1909) 384–395.

W. P. Jenney: The Chemistry of Ore Deposition. *Trans.* (1903) **33**, 456–457.

larger openings would collapse under much less pressure.[10] The latter fact has apparently been overlooked by some geologists. Data obtained from tests on small specimens of stone free from fractures and flaws can not be used in computing the depth at which fissures would be closed by collapse of their walls. In the vicinity of many veins the country rock shows evidence of having been greatly shattered prior to the entrance of vein-forming solutions, for branching veinlets sometimes extend out from the main vein and ramify in all directions. It is especially difficult to understand how open fissures could be maintained in rock that has been shattered to the extent shown by the "horse-tail" structure of veins at Butte[11] and in the Basin District,[12] Montana.

The hydrostatic pressure of solutions occupying an open fissure would of course help to support the walls, but, on the other hand, there are factors that would aid in closing them. Deep-seated veins are formed under high-temperature conditions, which, together with the presence of mineralizers, would reduce the differential stress necessary for rock flowage. Veins. are often far from vertical, and some are nearly horizontal.

It is only with great difficulty that openings are maintained in some of the deeper mines during the extraction of ores, and yet veins have been formed at very much greater depth than any reached in mining. Most of the deeper mines are practically dry in their lower levels, which indicates that even the smaller openings in rocks are largely confined to relatively shallow depths.

Most veins are found in mountain regions or in other areas that have been subjected to orogenic stresses, where, because of lateral pressure, gaping fissures are least likely to form; veins are relatively rare in regions of epirogenic movements where normal faults and monoclinal folds prevail. Moreover, the ore deposits that do occur in regions of important normal faulting, as, for example, at Clifton, Arizona,[13] are not, as a rule, found occupying fault fissures. The presence of slickensides and stria on the walls of veins formed along faults are evidence indicative of closed rather than of open fissures.

The walls of many veins contain angular irregularities of such a nature that the opposite walls would fit perfectly into one another. Such veins evidently do not occupy fault fissures, nor can they be explained by any process of replacement. Many veins also split up and

[10] Frank D. Adams: An Experimental Contribution to the Question of the Depth of the Zone of Flow in the Earth's Crust. *Journal of Geology* (1912) **20**, 115–117.

[11] Reno Sales: Ore Deposits of Butte, Montana. *Trans.* (1913) **46**, Fig. 2.

[12] Paul Billingsley and J. A. Grimes: Ore Deposits of the Boulder Batholith of Montana. *Trans.* (1918) **58**, 309.

[13] Waldemar Lindgren: The Copper Deposits of the Clifton-Morenci District, Arizona. U. S. Geological Survey, *Professional Paper* No. **43** (1905).

finger out into the country rock in such a way as to prove that the spaces were not formed by faulting.

Yawning chasms in any way comparable in magnitude with the many barren and metalliferous veins are not known today at any place on the surface of the earth, and they are less likely to occur at the depths where most veins have formed; there is no reason, therefore, for assuming their existence in the past, for veins are formed in rock strata of all ages. Neither are partly filled fissures to be found, if the veins containing drusy cavities are excepted. Some veins are enormous in size; according to Suess, the Great Pfahl in the Bavarian Forest is somewhat over 150 km. in length and averages 70 to 115 m. in width. He states that this great quartz vein represents the infilling of a long cleft produced by dislocation.[14] When we consider the multitude of veins, barren as well as metalliferous, and the extremely rare occurrence of open fissures of any kind, it is difficult to believe that many veins have been formed through the filling of such openings.

In the Bendigo gold fields of Victoria, Australia, and also in Nova Scotia, large masses of auriferous quartz, known as saddle-reefs, occur in the crests of anticlinal folds, but open spaces, with the exception of solution caverns in limestone, have never been observed in anticlinal folds, although folded rocks in all stages of dissection are found in many parts of the earth. Simple filling of open fissures of tectonic origin does not explain a change in the width of veins on passing from one wall rock to another, such as occurs in the veins of the San Juan Mountains, Colorado.[15]

The lenticular form of veins is also evidence opposing the hypothesis of deposition in open fissures. Most veins are more or less lens shaped; this form is often accentuated, and in extreme cases the lenses are bulbous or almost spherical. Some veins alternately swell and pinch; some consist of small separate lenses strung out along the same line of strike and connected by an almost imperceptible seam; others are made up of small and apparently disconnected parallel lenses irregularly distributed along a narrow belt or zone. Sometimes the lenses are remarkably symmetrical (Fig. 5); often they are irregular. The lenticular form is usually best developed in veins formed at depth, or where the wall rock is a highly laminated schist, slate, or shale. The walls are usually sharply defined and there is often other evidence that replacement has not been an important factor in the formation of the lenses.

De la Beche[16] showed that the relative displacement of the walls of an

---

[14] Eduard Suess: "The Face of the Earth;" Translated by H. B. C. and W. J. Solas, 1, 208–209. Oxford (1904).

[15] C. W. Purington: Ore-horizons in the Veins of the San Juan Mountains, Colorado. *Economic Geology* (1906) 1, 129–133.

[16] Henry T. De la Beche: "Researches in Theoretical Geology," 207–208. London (1837).

irregular fracture could result in the formation of open spaces which thin
out where the opposite walls are in contact, and this has frequently been
advanced as an explanation of lenticular veins; but no method of faulting
can account for the space occupied by apparently disconnected lenses,
and, in some instances, there is positive evidence that the walls have been
separated without other displacement.   Billingsley and Grimes have
recently described short lenticular veins in granite which show no evidence
of faulting, "such as shearing in the intervening country rock or displacement of intersected dikes of aplite." They attribute the drawing apart
of the walls to tension resulting from the cooling and contraction of the
granite;[17] but the distribution of the veins is not such as to relieve tensile
stresses in all directions in a shrinking mass, many granites show no evidence of lenticular openings either filled or unfilled, and lens-shaped veins
are  more  common  in  sedimentary  and  metamorphic  rocks  than  in
igneous rocks.

The bedding planes of sedimentary rocks and the folia of schists and
slates, where in contact with lenticular veins, are commonly curved to
conform with the curvature of the lenses.   The later origin of the veins
is indicated by the fact that they occasionally transect the rock structure
and contain included fragments of the wall rock.   Such displacement
of structural features antedating the formation of the vein can be explained only on the theory that the walls have been pushed outward by
some force acting from within, for the effects of tectonic forces acting
parallel to the rock structure could not be limited so as to confine the
formation of lenticular openings to a single plane.

Lenticular veins were believed to be igneous intrusives by some of the
earlier geologists, but true dikes do not show such extreme forms, and
the theory is also open to many other objections previously cited.   Graton[18] concluded that the lenticular gold-quartz veins of North and South
Carolina were deposited from solutions which were under such great pressure as they ascended toward the surface that they were able to force
apart the walls of small fractures and thus form the lenticular openings.
This explanation was advocated by Lindgren[19] who opposed the theory
that the veins were formed through the "force of crystallization."
If the enlargement of vein spaces were due to the pressure of fluids, very
flat lens- or tabular-shaped openings would be formed, and not a series of
thick lenses more or less isolated and disconnected; moreover, the pressure of the fluid would have to be uniformly maintained until the openings were filled, or the walls would collapse.   It is obvious that this

[17] *Op. cit.*, 309–310.
[18] L. C. Graton: Reconnaissance of Some Gold and Tin Deposits of the Southern
Appalachians.   U. S. Geological Survey, *Bulletin* No. 293 (1906) 60.
[19] Waldemar Lindgren: The Relation of Ore-deposition to Physical Conditions.
*Economic Geology* (1907) **2**, 107–108.

theory is inapplicable to many lenticular veins, such as those of calcite or gypsum, formed near the surface in unaltered sedimentary strata. The pressure of fluids cannot account for the formation of lenses that are subdivided by thin partitions of included schist detached from the walls, as illustrated in Fig. 1. In view of all the facts, it must be concluded that the typical lenticular veins were not deposited in preëxisting openings.

The peculiar banded structure, sometimes referred to as "book

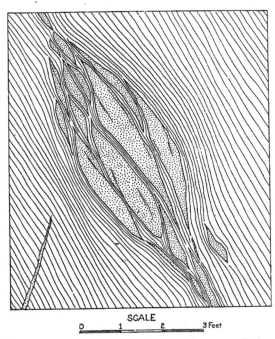

SCALE
0      1      2      3 Feet

FIG. 1.—QUARTZ VEIN IN SERICITE SCHIST NEAR COLUMBIA, S. C. THE QUARTZ LENS IS SUBDIVIDED BY THIN PARTITIONS OF SCHIST, AND IN PLACES CONTAINS DETACHED INCLUSIONS OF THE WALL ROCK.

structure" or "ribbon-ore," resulting from the alternation of narrow veinlets of ore with parallel sheets of slate or schist, is difficult or impossible of explanation under any theory of ore deposition in preëxisting openings. Becker and Day[20] have called attention to the fact that in some cases "it can be shown conclusively that the distribution of the slate is not due to faulting." This structure is well developed at many places along the Mother Lode of California, and also in some of the

[20] G. F. Becker and A. L. Day: The Linear Force of Growing Crystals. *Proceedings,* Washington Academy of Sciences (1905) 7, 283–284.

gold and copper veins of the Southern Appalachians. Usually the veinlets, consisting of quartz or calcite, are lenticular in form, and pinch out
to be replaced by others a little to one side or farther along the strike;
sometimes layers of slate are wholly included within the veins.

The commonest evidence indicating· that veins were not entirely
deposited in open fissures is furnished by the presence in them of angular
inclusions of the country rock. These fragments often show by their
shape that they have been detached from an adjacent wall and displaced
laterally without falling. Where the inclusions are unaltered by replacement, they could refit perfectly into the positions which they formerly
occupied. Inclusions that have suffered altération are usually more or
less rounded or subangular. Posepny states that upon search he always found points of contact between such fragments; but, in many instances, the writer, by making parallel sections, proved that the inclusions were not in contact with the walls or with other fragments.

Inclusions are rare in some veins and plentiful in others, in some cases
making up the larger portion of the ore mined. The inclusions also
vary in size up to masses so large that it is a question whether they
should be classed as "horses" or as country rock inclosed between two
branches of a vein. There is no essential difference between some breccia
veins and certain ore deposits that consist of a network of anastomosing
veinlets which ramify in all directions through the country rock. In
veins that are horizontal or dip at flat angles, the inclusions are no more
plentiful along the foot-wall than near the hanging wall.

Deposition of mineral matter in opening filters with gas is practically limited to the belt of weathering. Such deposits are, as a rule,
easily recognized; they are often finely banded, and usually exhibit stalactitic, mammillary or analogous forms. As they are almost exclusively
secondary in origin and obviously of little importance in unaltered
veins of primary ores, further discussion is unnecessary here.

The evidence outlined above leads to the conclusion that deposition
of mineral matter in widely open fissures is of minor importance in
the formation of veins. Minerals deposited in open spaces are present in
many veins, especially those formed at shallow depths, though in most
cases they make up a relatively small proportion of the total vein mass.
Some veins have possibly been formed entirely through the simple filling
of open fissures, but they are believed to be extremely rare.

*The hypothesis that veins may be deposited from solutions entering
along extremely small passages and that the growing veins have made room
for themselves by forcing apart the inclosing walls* has been suggested by
investigators from time to time; but it has never received general acceptance, and is practically ignored in text-books treating of the origin
of ore deposits. Comparatively little evidence in support of the hypothesis has been published, and the mechanics of the process by which

growing crystals exert pressure has not been clearly understood. According to Andrée,[21] who has compiled an excellent bibliography of the subject, von Weissenbach[22] was the first to assume a "force of crystallization" in explaining geologic phenomena. Lavalle,[23] however, appears to have been the first to record experimental proof of pressure exerted during crystal growth.

It has been demonstrated beyond question that growing crystals may exert force and actually grow in directions in which their growth is opposed by adjacent solid bodies, but some geologists[24] have doubted that this force is sufficient to separate the walls of veins at great depths. In replying to this objection, attention is called to the fact that *all* crystals have been formed under pressure, and many mineral crystals must have developed under enormous pressure. The effect of pressure or of strain is to increase slightly the solubility of most substances, but the change is very small as compared with that due to difference in temperature. So long as a crystal surface is in contact with a solution supersaturated with respect to that surface, growth will continue no matter how great the pressure or strain may be. The supersaturation of a solution in contact with a growing crystal is maintained by circulation and diffusion through the solution; and, when a crystal grows in a direction in which growth is opposed by an adjacent solid, it is due to the fact that the material necessary for growth is able to diffuse between the crystal and the other solid.

The film of solution separating the crystal from the other solid is not expelled by the pressure developed; otherwise, growth would cease. If two perfectly smooth parallel surfaces were separated by a thin layer of liquid which wets them both, capillarity would tend to bring the two surfaces close together, and reduce the thickness of the separating film to a minimum; but it is improbable that this film could be completely expelled by capillarity or even by external force that does not rupture the solids. When a crystal, in growing, approaches another solid, the surfaces that are closest together tend to become parallel, because deposition is most rapid where diffusion is least restricted.

There is abundant evidence that solutions, especially when under great

[21] K. Andrée: Die geologische Bedeutung des Wachstumsdrucks kristallisierender Substanzen. *Geologische Rundschau* (1912) **3**, 7–15.

[22] C. G. A. von Weissenbach: Abbildungen merkwürdiger Ganverhältnisse aus dem sächsischen Erzebirge, 22, 23, 26. Leipzig (1836).

[23] Jean Lavalle: Recherches sur la Formation Lente des Cristaux à la Température Ordinaire. *Comptes Rendus* (1853) **36**, 493–495.

[24] Waldemar Lindgren: *Op. cit.*, 107.

F. L. Ransome: Discussion sur l'Influence de la Profondeur sur la Nature des Gisements Metallifères. *Congrès Géologique International, Canada* (1913), *Compte Rendu de la XIIe Session*, 305.

pressure and high temperature, can penetrate even the densest rocks. Penetration takes place between the mineral crystals and along cleavages and microscopic fractures in the minerals themselves. It has been suggested that diffusion may take place directly through individual·crystals, but this is doubtful, and certainly has not been proved. Evidence of the enormous quantities of mineral matter which may diffuse for considerable distances through small capillary and sub-capillary spaces is furnished by the size of replacement deposits. The presence of ore minerals having idiomorphic surfaces, where in contact with older minerals which they have partly replaced, proves that in such instances the solution and removal of the replaced minerals has resulted from the growth of the ore minerals, and therefore did not precede the precipitation of the latter so as to form open spaces for their reception.

The pressure effects accompanying the growth of some crystals are evidently, in part at least, due to the tendency to develop crystal faces, but this "linear force of growing crystals" is believed by the writer to be of minor importance. In previous papers[25] he has expressed his belief that these pressure effects are to be attributed chiefly to the molecular forces associated with the separation of solids from solution. Laboratory experiments prove that the pressure effects are independent of the cause of precipitation; for they have been obtained where precipitation was induced by cooling, by evaporation, by reactions between liquids, by reactions between liquids and solids, and by reactions between gases and solids. When a single molecule of the solid passes by diffusion into the thin film of solution separating a growing crystal from a foreign body, and attaches itself to the surface of the crystal, a slight displacement of the entire crystal relative to the foreign body results. According to this conception, the mechanism of the process is somewhat analogous to that of a hydraulic jack. It is inconceivable that the addition of a single molecule could displace the crystal in any other way, and there is no evidence of periodicity in the growth of most crystals. The diffusion of a solid through a solution is ascribed to osmotic pressure, and its separation therefrom to the relation between osmotic pressure and solution pressure. The writer thinks it preferable not to use the term "force of crystallization" until it has been definitely established that the pressure effects accompanying the separation of solids from solution are limited exclusively to crystalline solids. There is some evidence that the deposition of non-crystalline substances, such as limonite, may also result in the development of pressure under favorable conditions. This question is now being investigated experimentally, but as yet with only negative results.

---

[25] Stephen Taber: Pressure Phenomena Accompanying the Growth of Crystals. *Proceedings*, National Academy of Sciences (1917) **3**, 297–302.

Lindgren[26] thinks that a platy or schistose structure due to development under one-sided pressure would necessarily characterize veins which have made room for themselves by forcing their walls apart, and that the absence of this structure is evidence indicating that the veins were not so formed. There is, however, much evidence opposing this view. A crystal developing at depth below the surface, completely surrounded by interlocking minerals, is probably under nearly uniform pressure in all directions. Pseudophenocrysts of biotite which have developed in highly schistose rocks under mass static conditions do not show parallel orientation, although in growing they have forced apart the folia of the schist.[27] Calcareous concretions which develop in shales and push apart the bedding planes, as shown in Fig. 4, may be even-granular in texture. The writer has examined, microscopically, thin sections cut normal to the surface of these concretions and others cut parallel to the surface without observing any difference in the appearance of the calcite grains. In some concretions that are relatively free from impurities the calcite crystals are elongated normal to the surface.[28]

Even in cases where growing crystals are subjected to much greater pressure in one direction than in others, they are not necessarily flattened normal to the direction of greatest pressure, for the shape of crystals is often determined by the relative accessibility, in different directions, of the material for growth. This has been proved experimentally by growing slender columnar, and even fibrous, crystals which were elongated in the direction of greatest pressure.[29]

In laboratory experiments conducted during the last four years, the writer has been successful in growing veins which have made room for themselves by pushing apart their walls; and in these veins such structures as drusy cavities and banding have been obtained. The drusy cavities were due to a local deficiency of the material for vein growth, resulting either from insufficient concentration or from relative inaccessibility; and evidence has been elsewhere cited in support of the hypothesis that at least some of the drusy cavities occurring in mineral veins are similar in origin.[30]

[26] Waldemar Lindgren: Sur l'Influence de la Profondeur sur la Nature des Gisements Metallifères, *Congrès Géologique International, Canada* (1913), *Compte Rendu de la XIIe Session*, 305; and " Mineral Deposits," 155: New York (1913).

[27] Stephen Taber: Geology of the Gold Belt in the James River Basin, Virginia. Virginia Geological Survey, *Bulletin No. 7* (1913) 29–33.

[28] R. A. Daly: The Calcareous Concretions of Kettle Point, Lamberton County, Ontario. *Journal of Geology* (1900) **8**, 138.

[29] Stephen Taber: The Growth of Crystals Under External Pressure. *American Journal of Science*, Ser. 4 (1916) **41**, 546, 552.

[30] Stephen Taber: The Origin of Veinlets in the Silurian and Devonian Strata of Central New York. *Journal of Geology* (1918) **26**, 63, 65.

Banding was produced in laboratory veins by merely changing the composition of the solutions during the growth of the veins. Mineral veins showing asymmetrical banding could have been formed in open fissures only through successive reopenings of the fissure, and usually there is no evidence supporting such an assumption. Asymmetrical banding, as demonstrated by laboratory experiments, may develop in growing veins that make room for themselves whenever the circulation of the vein-forming solutions is limited to a single wall. There is much evidence indicating that the banding found in metalliferous veins is often due to deposition from solutions circulating along the walls of the growing veins, and that the minerals deposited last are sometimes concentrated near the walls rather than in the center of the veins, as would be necessary on the assumption of ore deposition in open fissures.

Many large veins show a regular arrangement of quartz in the center and chalcedony near the walls. On the assumption that such veins are referable to the activity of hot-spring waters, Beyschlag, Vogt and Krusch[31] attribute this regular arrangement to the cooling effect of the walls on the ore-bearing solutions, for, as is well known, quartz usually separates from solutions at a higher temperature than chalcedony. It is just as logical, however, to assume that the vein-forming solutions gradually became cooler because of the slow cooling of the igneous rocks from which the heat was derived.

The expiring stages of igneous activity are often marked by the formation of veins through deposition from hot ascending solutions, and in such veins the commonest gangue minerals are quartz and calcite. Similar deposits are formed about hot springs where the waters reach the surface. Calcite is usually deposited later than quartz, but in some instances this order is reversed, and then there is more or less replacement of calcite by quartz. According to Spurr,[32] the calcite veins of Ararat Mountain in the Tonopah District are in places beautifully banded, with quartz in the center and carbonates near the walls. These veins are locally as much as 20 ft. thick, but are exceedingly irregular and non-persistent. The rhyolite forming the walls is in places silicified by the vein-forming solutions, but there is no mention of replacement of calcite, and in many cases the carbonates were observed occupying cavities in the silicified rhyolite, thus indicating a later deposition for the carbonates. Spurr states that "these are fine examples of veins which have filled open fissures," but the facts cited above would indicate that the veins had made room for themselves by forcing their walls apart, and this hypothe-

[31] F. Beyschlag, J. H. L. Vogt and P. Krusch: "The Deposits of the Useful Minerals and Rocks"; Translated by S. J. Truscott. **1**, 104–105. London (1914).

[32] J. E. Spurr: Geology of the Tonopah Mining District, Nevada. U. S. Geological Survey, *Professional Paper* No. 42 (1905) 101–104.

sis is also supported by the presence of numerous angular rhyolite fragments within the veins.

Small roughly banded veins with calcite in the center and later pyrite along the walls have been described by the present writer. The pyrite is partly in the form of idiomorphic crystals which replace both the earlier vein calcite and the wall rock.[33]

An irregular and more or less interrupted banding, as in specimens of ore from Silverton, Colorado, is much more common in veins than an evenly banded structure such as is usually found in travertine, agate, and other deposits known to have been formed in open spaces. Very imperfect banding or absence of banding may result when mineral deposition, instead of being confined to the walls, takes place partly or entirely within the growing vein, the later minerals making room for themselves by replacing or mechanically displacing the earlier minerals. The later minerals may be localized, through deposition along well-defined channels of circulation, or they may be deposited throughout the vein mass. Where the latter occurs, the material for vein growth must be supplied, at least partly, by diffusion between the earlier minerals. There is much evidence that this method of mineral deposition is of the utmost importance in the formation of ore deposits.

Many veins are intersected by numerous small but well-defined veinlets, consisting chiefly of the later minerals, though often the growth of these veinlets has been practically contemporaneous with the formation of the vein as a whole. Minute veinlets are often revealed by microscopic examination of ore when their presence would not be suspected from a megascopic examination. These microscopic veinlets frequently cut directly through the older minerals, and separate the fragments without other displacement. Occasionally the veinlets may be traced considerable distances, but usually they are non-persistent, and often they are confined to the limits of a single crystalline individual. When thus limited, the veinlets sometimes follow crystallographic directions in the older minerals.

The deposition of mineral matter within a growing vein is not, however, limited to intersecting veinlets, and there is also evidence of more or less contemporaneous recrystallization of some minerals during the enlargement of veins. When very small veinlets are examined microscopically, it may be observed that the individual mineral crystals, usually extending from wall to wall, tend to vary in size with any variation in the thickness of the veinlets. This can be explained only on the assumption that the enlargement of the veinlets has been accompanied by an enlargement of some of the vein minerals and the elimination of others. In other words, the crystals which are less stable because of smaller size, or

---

[33] Stephen Taber: *Op. cit.*, 65–66.

unfavorable orientation or location, tend to redissolve, and thus furnish additional material for the growth of their more fortunate neighbors.

Evidence of mineral deposition within growing veins is furnished by the separation of included fragments of the country rock from one another as well as from the walls. In such cases the fragments often show by their shape that they would fit perfectly together. Inclusions are occasionally found in most veins formed by simple deposition unaccompanied by appreciable replacement, and in some veins the inclusions make up a large percentage of the vein mass. Where the

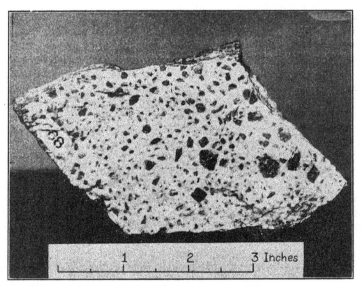

Fig. 2.—Vein specimen from the Swauk district, Washington, showing inclusions of wall rock in a matrix of quartz.

inclusions are abundant, they are sometimes rather uniformly distributed (Fig. 2), as in the so-called "bird's-eye" quartz found in gold veins of the Swauk district of central Washington. The angular inclusions are of black shale, much silicified, while the matrix is of quartz and calcite. The quartz is partly in the form of small prisms which radiate outward from the separated fragments. At their ends some of these prisms mutually interfere with those radiating from neighboring inclusions, while others project into drusy cavities. According to Smith,[34] the wall rock is in places much shattered, with many small

[34] George Otis Smith: Gold Mining in Central Washington. U. S. Geological Survey, *Bulletin* No. 213 (1903) 79.

quartz veins traversing it in all directions, so that it is difficult to draw any limits to the vein itself, and there is a transition from this shattered wall rock into the typical "bird's-eye" quartz. Russell described these veins, and suggested that the vein minerals "in crystallizing have exerted a force analogous to the expansion of water in freezing, which crowded the rock fragments asunder."[35]

Von Cotta discussed breccia structure in veins, and stated that in some cases it seems to be due to a special force of crystallization of the principal vein mass, while in others it is to be explained by the opening of vein fissures more than once.[36] The latter explanation, however, is obviously inapplicable to such veins as those described above in which the spherulitic texture is well developed with quartz prisms radiating in all directions from the nucleal fragments. In some veins having a spherulitic texture the nuclei of the spheroids consist of earlier vein minerals instead of fragments of the country rock. This is well illustrated by some of the veins in the Telluride[37] and Silverton[38] districts. Less frequently the nuclei are surrounded by several layers or concentric shells differing in mineral composition. This texture, variously known as "cockade ore," "ring ore" and "concentric ore," was described by von Weissenbach as early as 1836. He noted that there was no contact between the original fragments, and attributed their separation to the crystallizing force of the later minerals.[39] The spherulitic textures seem to be limited to veins formed at comparatively shallow depths.

The laboratory veins, which made room for themselves by forcing their walls apart, were sometimes branching, and frequently contained inclusions of the wall material—a structure which could not be duplicated experimentally by any method of vein deposition in open spaces.

In a manner similar to that described above, certain vein minerals of early deposition are commonly ruptured, and the fragments separated by the growth of later minerals. This is well illustrated by many tourmaline crystals found in quartz veins, and also by pyrite and other sulfide minerals present in ore deposits. Pyrite is usually one of the first sulfide minerals deposited, and microscopic examination shows that the

[35] I. C. Russell: A Preliminary Paper on the Geology of the Cascade Mountains in Northern Washington. U. S. Geological Survey, 20th Annual Report, Pt. II (1900) 207.

[36] B. von Cotta: "Die Lehre von den Erzlagerstäten," 25. Freiberg (1855).

[37] C. W. Purington: Preliminary Report on the Mining Industries of the Telluride Quadrangle, Colorado. U. S. Geological Survey, 18th Annual Report, Pt. III (1897) 798–799.

[38] F. L. Ransome: A Report on the Economic Geology of the Silverton Quadrangle, Colorado. U. S. Geological Survey, Bulletin No. 182 (1901) 91–92.

[39] C. G. A. von Weissenbach: Abbildungen merkwürdiger Gangverhältnisse aus dem sächsischen Erzgebirge, 22. Leipzig (1836).

pyrite crystals are commonly ruptured and the fragments separated by later minerals, such as chalcopyrite, bornite, etc. Graton and Murdoch, after an extensive investigation, state that

"this is the characteristic structure of the average primary copper ore. . . . In such ores there can be no doubt that the pyrite has undergone actual mechanical deformation, for in some instances it is found that somewhat separated fragments would fit perfectly together, even to the minor irregularities, and would collectively form a compact grain if the intervening cement were removed. Such an occurrence gives the appearance of an exploding bomb. The cause of this crushing and distortion is not yet clear, but it is plain that it took place in an early stage of crystallization of the ore, because no trace of such a thing is commonly to be found in the later sulfides or in the gangue, in the latter of which it would be expected that crushing or at least strain-shadows would be seen in thin sections if these materials had been subjected to stress."[40]

Emmons[41] discussed this structure of sulfide ore deposits, and attributed it to regional metamorphism, but it is also a peculiarity of deposits found in non-metamorphic areas, and this peculiar structure is not characteristic of any type of metamorphic rock. The ore deposits described by Emmons are, it is true, situated in a belt of intensely dynamo-metamorphosed rocks, but the orebodies, while roughly lenticular in form and elongated parallel to the rock structure, are very wide in comparison with their length. Emmons, in commenting on this fact, states that "as a class they seem to be wider than deposits not metamorphosed. This is just the opposite of what should be expected, for, from internal evidence, one gets the impression that they have been squeezed thin."[42]

The separation of fragments of pyrite and other early minerals is evidently due to the growth of the later minerals, but the manner in which the fractures were first formed is not always clear. In some cases the fracturing of crystals appears to have resulted from external pressure, often due to the growth of later minerals, and in other cases it is probably caused by the development of later crystals in the small cavities present in most crystals or along planes of incipient cleavage or parting. The latter view is supported by the presence in some crystals of dots and stringers of the later minerals arranged chiefly in parallel lines along crystallographic directions.

While some minerals found in veins are commonly fractured and the fragments separated, this structure appears to be rare in other minerals, such as galena, having equal or greater porosity and cleavability. This difference cannot always be attributed to later crystallization. In

---

[40] L. C. Graton and Joseph Murdoch: The Sulphide Ores of Copper. Some Results of Microscopic Study. *Trans.* (1913) **45**, 37.

[41] W. H. Emmons: Some Regionally Metamorphosed Ore Deposits and the So-called Segregated Veins. *Economic Geology* (1909) **4**, 755–781.

[42] *Op. cit.*, 777.

some cases it is possibly due to a difference in the surface tension existing between different minerals and the ore-bearing solutions, but no definite conclusion can be drawn until more data are at hand.

Minerals of early crystallization are occasionally bent as well as fractured, and then the evidence of deformation by external forces is conclusive. Emerson[43] has described large crystals of spodumene apparently bent, and in some cases broken, by the growth of later minerals, for the vein in which they occur is not crushed or sheared. The forces that bent the large spodumene crystals through angles of 45° and 90° without fracturing must have been applied gradually through a considerable period of time while the crystals were imbedded in a matrix of other minerals. Bent and broken crystals of covellite, the intervening spaces being filled with "later, but nevertheless primary, bornite," have been described by Graton and Murdoch.[44] The flakes of mica and chlorite found in some veins are often bent and shredded by the growth of sulfides.

The hypothesis that veins have made room for themselves by forcing their walls apart furnishes a ready explanation of lenticular form and of "book structure" or "ribbon ore," features characteristic of veins formed in highly laminated rocks, and these features are difficult or impossible of explanation under any other theory of vein formation. According to this view, "book structure" has resulted where the ore-bearing solutions have entered along closely spaced parallel planes, and deposited mineral matter in the form of veinlets, which, in growing, have gradually separated the laminæ of the country rock. Becker and Day[45] were the first to suggest this origin for the so-called "ribbon-ore" of the Mother Lode district in California. The separation of the silicified shale fragments in the ore breccia forming the Enterprise blanket of the Rico district, Colorado, should probably be attributed to a similar process. Ransome has reproduced a photograph of a beautiful specimen of this breccia from the Newman Hill mine.[46]

Dunn states that the saddle-reefs and leg-reefs of the Bendigo district, Australia, are often laminated by the inclusion of thin films of slate and occasionally of "thick substantial flakes." The laminations are farther apart where the reefs are widest, and closest together where the reefs are narrowest. The laminations are also more closely spaced near the walls. Dunn cites these facts as evidence that the quartz has forcibly made room

---

[43] B. K. Emerson: Geology of Massachusetts and Rhode Island. U. S. Geological Survey, *Bulletin* No. 597 (1917) 257.

[44] L. C. Graton and Joseph Murdoch: *Op. cit.*, 51.

[45] G. F. Becker and A. L. Day: *Op. cit.*, 284.

[46] F. L. Ransome: The Ore Deposits of the Rico Mountains, Colorado. U. S. Geological Survey, 22nd *Annual Report*, Pt. 2, 279, and plate XXXV.

for its accumulation, and concludes that the growth of the quartz layers was continuous during the period of reef formation.[47]

Lenticular form is characteristic of most aggregations of mineral matter that develop in highly laminated rocks, such as shales or schists, and displace rather than replace the older minerals. Eye-like lenses may result from the growth of a single crystal, in which case the outer circumference of the lenses is filled in with other minerals, as in Fig. 3. The single crystal may be of such mineral as garnet, biotite, pyrite, galena, or sphalerite, while the remainder of the lens is filled with quartz or calcite. Larger lenses are formed by the growth of crystalline aggregates. The

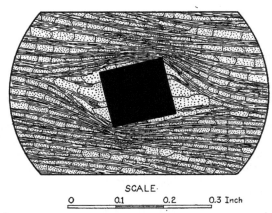

SCALE

0          0.1          0.2          0.3 Inch

FIG. 3.—SKETCH SHOWING EYE-LIKE LENS IN QUARTZ MICA SCHIST. THE LAYERS OF MICA HAVE BEEN DISPLACED BY THE GROWTH OF THE PYRITE CRYSTAL WHILE QUARTZ HAS BEEN LARGELY REMOVED FROM POINTS OF GREATEST PRESSURE AND REDEPOSITED WHERE THE PRESSURE WAS LEAST.

calcareous concretions shown in Fig. 4 are essentially similar to the gold-quartz lenses shown in Fig. 5, and in both cases the form has obviously resulted from similar processes. In a previous publication,[48] the writer has cited the lenticular form of certain auriferous-quartz veins as evidence that the vein minerals had made room for themselves by forcing the walls apart. One of these veins, exposed in the Tellurium mine, Virginia, consisted of lenses which were in some cases wonderfully symmetrical in form. In laboratory experiments, lenticular veins were obtained through the slow growth of crystalline masses of salts between layers of heavy cardboard coated with paraffine.

[47] E. J. Dunn: Reports on the Bendigo Gold-field Nos. I and II.  *Victoria Department of Mines* (1896) 24, 25, 27.

[48] Stephen Taber: Geology of the Gold Belt in the James River Basin, Virginia. Virginia Geological Survey, *Bulletin* No. 7 (1913) 222–231.

According to this hypothesis, the veins have been deposited from solutions which have penetrated slowly along lines of least resistance, such as faults, fractured zones, planes of bedding or schistosity, or through porous strata; where appreciable openings existed, these have been filled, but the veins have chiefly made room for themselves by forcing apart the inclos-

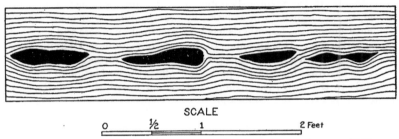

SCALE

0    ½    1              2 Feet

FIG. 4.—CALCAREOUS CONCRETIONS IN SHALE, NEAR UNION SPRINGS, N. Y.

ing country rock. The form assumed by the growing deposits is determined by the nature of the opening along which the solutions enter, by the accessibility of the ore-bearing solutions during the growth of the ore-body, and by the forces resisting enlargement. When the material for growth is everywhere equally available, the shape produced is that which

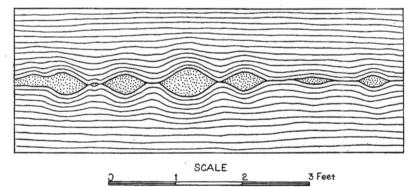

SCALE

0    1    2              3 Feet

FIG. 5.—SYMMETRICAL LENSES OF A GOLD-QUARTZ VEIN EXPOSED IN THE TELLURIUM MINE, VIRGINIA.

requires the least expenditure of energy. This furnishes an explanation of the swelling of some veins where they pass into less rigid rocks, and of the growth of "saddle-reefs" in the crests of anticlinal folds where the rigidity of arching strata relieves the underlying beds of part of their load.

After carefully considering all the facts, there seems to be no escape from the conclusion that it is possible for growing veins to make room for themselves by forcing their walls apart; and the evidence indicates that many veins have been formed in this manner.

*That veins have been formed of material leached from the inclosing rock, transported in solution and introduced into open fissures through their walls,* is the old lateral-secretion hypothesis advocated by Bischof, Sandberger, and others. It was vigorously opposed by Posepny, and has been abandoned, in its original form at least, by practically all geologists. Much evidence against the hypothesis has been recorded by others and need not be repeated here. It is also open to the same objections as all hypotheses postulating gapping fissures as a prerequisite for the formation of veins.

In its narrower form this hypothesis confines the derivation of the vein minerals to the wall rock in immediate contact with the deposits, and, therefore, is especially applicable to such veins as the calcite veins in limestone and calcareous shales, gypsum veins in gypsiferous strata, quartz veins in siliceous rocks, and chrysotile veins in serpentine. The veins belonging to this class are commonly horizontal or nearly so, and frequently ramify in all directions through the country rock so as to make up a large proportion of the total mass; hence the difficulty of explaining the formation and maintenance of open fissures is even greater for such veins than for others.

Laboratory investigations indicate that two different types of deposits may be formed in open spaces when mineral matter is introduced through small openings in the walls. One type is formed when mineral matter is deposited from solutions subsequent to their entrance into cavities, in which case the deposits are similar to those formed in open spaces by circulating solutions. The minerals deposited on the inside of geodes, and also in many of the drusy cavities of veins, have evidently been deposited in this way, though there is no evidence that large veins have had such an origin. In the belt of weathering, the deposits usually show stalactitic or analogous forms.

The other type is due to the deposition of mineral matter from supersaturated solutions which do not enter the cavities. In this case additions are made to growing crystals only at their bases, where they are attached to the walls, thus making them columnar, acicular, or fibrous. Because of inequalities in the supply of material, the resulting fibers are unequal in length and in places entirely absent. Where the fibers are in close contact with one another, they are usually parallel and normal to the surface from which they grow; where the fibers are isolated or in small groups, they are commonly curved, and sometimes form tangled coils. Thin crusts are often formed on the walls and later pushed outward by the growing crystals, which also occasionally detach fragments of the wall-material and displace them in a similar manner. Mineral deposits of this type may sometimes be seen forming on the walls of caverns and in other favorable places. Vugs and smaller openings found in veins often contain minerals, usually secondary in origin, which exhibit this structure; but there is no evidence of vein fissures having been filled with such deposits. The filiform varieties of the native metals have probably developed in this way, and some intermediate forms may be due to a modification of wire-like varieties by solutions which have occupied the cavities.

*That the material for vein formation has been supplied through small, closely spaced openings in the walls, as the growing veins have made room for themselves by forcing their walls apart,* is the hypothesis advocated by the present writer, in previous papers,[49] as offering an explanation of cross-fiber veins. Most of the evidence supporting this view has been published, and therefore it need not be discussed in detail here.

A somewhat similar explanation of the origin of certain fibrous veins was suggested by von Weissenbach in 1850, but this was not known to the writer until after the publication of the papers referred to above. Von Weissenbach[50] described the growth of fibrous gypsum crystals on the walls of old buildings and of the needle ice that forms on clayey soils, comparing it with the formation of the fibrous veins. He thought that vein walls of soft material might be pushed back, or that the growth of the needle-like crystals was simultaneous with the widening of the vein fissures which resulted from shrinkage due to drying. As evidence in support of this hypothesis, he states that fibrous veins are never persistent and never show banding or drusy cavities; but this statement is not altogether true, for color banding parallel to the walls is sometimes present in veins of fibrous celestite, and drusy cavities have been observed in similar veins of calcite and of gypsum.

Cross-fiber veins are composed essentially of a single mineral, which is also usually an important constituent and often the dominant mineral in the wall rock. For example, there are fibrous veins of quartz in sandstone, of calcite in limestone, and of chrysotile in serpentine. In rare instances the vein mineral is not found in the wall rock, but in such cases it is a secondary mineral which has evidently been derived from minerals that are present in the wall rock. Veins of fibrous chalcanthite an inch in thickness have been found in the Yavapai mine at Morenci, Arizona.

All cross-fiber veins, irrespective of mineral composition and kind of wall rock, are characterized by the same structural peculiarities. The mineral fibers are parallel to one another, and in many instances there is direct evidence that they extend in the direction in which the walls

[49] Stephen Taber: The Genesis of Asbestos and Asbestiform Minerals. *Trans.* (1917) **57**, 71–84.

The Origin of Veinlets in the Silurian and Devonian Strata of Central New York. *Journal of Geology* (1918) **26**, 61–65.

[50] C. G. A. von Weissenbach: Über Gangformationen vorzugsweise Sachsens; Cotta's Gangstudien, **1**, 66. Freiberg (1850).

have been displaced during the process of separation. In most veins the fibers are approximately normal to the walls, but frequently they are oblique, and, if the course of the vein is not straight, the fibers may be normal to the walls at one place and oblique at another. This is well illustrated in specimens from Hoboken, New Jersey, showing nemalite veins in serpentine. The fibers are usually straight, especially when the veins are small, parallel, and not very numerous; often they are curved or abruptly bent, and this may be observed most frequently where the veins branch, intersect, and ramify in all directions. Sometimes there are several bends in the fibers, thus giving the veins a banded appearance due to unequal reflection of light where the fibers run in different directions, and ordinarily this pseudo-banding is symmetrical with respect to the central portion of the veins. A central parting is present in some veins, several partings are present in others, and in many veins the fibers extend from wall to wall without a break. The partings are often very irregular, appearing as though displaced parallel to the fibers, and similar irregularities may occasionally be noted in the bands that run parallel to the vein walls. Angular inclusions are common, especially along lines of parting, and in many instances it is evident that these fragments have been merely detached from the adjacent walls and displaced laterally. Fibrous veins as a class are non-persistent, and usually have a marked lenticular form.

The structural features here given are common to all kinds of cross-fiber veins; and, with the exception of lenticular form, all of these features have been duplicated in veins grown in the laboratory. All of the features may be satisfactorily explained under the hypothesis outlined above, while many of them are inexplicable under any other hypothesis so far advanced.

The formation of veins through lateral secretion is not necessarily confined to the zone of vadose circulation, as argued by Posepny[51] and Raymond,[52] for it has been proved that the vein minerals may be transported to the walls of the growing vein by diffusion through solutions occupying minute spaces, instead of by circulation.

The fibrous structure has been attributed by the writer to a mechanical limitation of crystal growth through the addition of new material in only one direction. In veins of the asbestiform minerals, the fibrous structure is accentuated by a normal prismatic habit and cleavage. In studying cross-fiber veins of calcite and of gypsum, it was found that the diameter of the fibers varied with the spacing of the openings through which the material for growth was supplied, and that the veins were non-

[51] Franz Posepny: The Genesis of Ore-deposits—Discussion. *Trans.* (1894) **24**, 967.
[52] R. W. Raymond: *Idem*, 985–986.

fibrous where they passed through relatively impervious chert nodules which forced the vein-forming material to diffuse between the walls. Therefore, if material reached a growing vein through openings in the walls, spaced sufficiently far apart, it would be impossible to determine from the structure whether the vein had been formed through a process of lateral secretion or not. A gradation is, therefore, possible from fibrous veins to non-fibrous veins of class II. The fibrous structure of some veins may be made coarser or even completely obliterated as a result of recrystallization, since, for many minerals, a fibrous form is not the most stable.

It has been suggested elsewhere that minimum resistance to growth may often be the factor which has determined the location of chrysotile veins in serpentine, and there is considerable evidence that cross-fiber veins in general tend to develop in those places where the least expenditure of energy is required in making room for them. The cross-fiber veins in stratified rocks are commonly parallel to the bedding, especially where the beds have not been greatly disturbed; and in folded strata the veins in places may be largely confined to the crests of anticlines. These facts are well illustrated by the crocidolite veins of West Griqualand, South Africa, which are briefly described below. For this description[53] the writer is indebted to Dr. A. W. Rogers, Acting Director of the Geological Survey of the Union of South Africa.

The cross-fiber crocidolite is interbedded with cherty and magnetite layers, as a rule, but it breaks across these beds where the latter are bent and broken, though the amount of cross-fiber mineral obtained from these cross-fractures or veins is quite insignificant. . . . . . Where the beds are sharply folded the cross-fiber mineral thickens very much in the arches and troughs, and it may disappear completely in the connecting limbs. . . . . . The folding and fracturing of the beds was completed before the crocidolite assumed the cross-fiber form; I have not seen the fiber disturbed by these processes, and it appears that its formation took place very soon after, or during, the time when the beds were disturbed. . . . . . Minute fibers of crocidolite are distributed, more or less abundantly, through the sedimentary beds lying between the layers of cross-fiber mineral.

The formation of veins through replacement of country rock by ore minerals, introduced in solution through small cracks, was suggested by Charpentier as early as 1778.[54] Little evidence in favor of the hypothesis could be cited at that time, and therefore it received almost no attention for nearly a century. Facts learned from the study of pseudomorphs were gradually applied to the formation of ore deposits, and, as evidence accumulated, it became evident that metasomatism or replacement is an important factor in the formation of ore deposits. In 1910

[53] Written communication, dated Mar. 5, 1917.
[54] S. F. Emmons: Theories of Ore Deposition Historically Considered. *Bulletin*, Geological Society of America (1904) **15**, 8-9.

Lindgren published his essay, "Metasomatic Processes in Fissure Veins,"[55] in which he described in detail the various changes that take place in wall rock as a result of vein formation, and suggested a classification of veins based on characteristic metasomatic processes. It is now generally recognized that while the wall rock of many veins shows absolutely no evidence of metasomatism, most of the larger metalliferous veins, on the other hand, are bordered by altered zones, varying in width and intensity of alteration, and some veins have been formed almost entirely through processes of replacement.

Criteria for the recognition of replacement have been given by Lindgren,[56] Irving,[57] and others, and therefore a discussion of this phase of the subject may be omitted here. In regard to the mechanics of replacement, the views of the present writer are slightly at variance with those of the authorities just cited, and it is this question that is discussed in succeeding pages.

The changes which take place in wall rocks as a result of vein-forming processes are often complex, for most rocks are composed of several minerals differing in chemical composition and physical properties, while the ore-bearing solutions may also contain many constituents. The problem can be simplified somewhat, however, by considering separately the various ways by which changes may take place in each individual mineral of which the rock is composed.

One mineral may take the place of another in at least three different ways: (1) by solution of the original mineral and subsequent deposition in the space vacated; (2) by solution of the original mineral and simultaneous deposition of the replacing mineral without chemical reaction between them; (3) by a chemical reaction involving the addition, subtraction, or interchange of one or more elements.

(1) While minerals are sometimes deposited in spaces of dissolution, this process is believed to be of little importance in the formation of veins, and, strictly speaking, it is not a process of metasomatism. With such a method of replacement it would be impossible for the replacing material to inherit and preserve details of internal structure, regular arrangement of inclusions or the structure of organic remains. It would also be impossible for the replacing mineral to show idiomorphic boundaries, where in close contact with the partly replaced mineral. Microscopic examination of minerals that are partly altered or replaced always shows the replacing mineral to be in close contact with the earlier mineral, and, even under the highest magnification, no trace of intervening space is observable.

---

[55] Waldemar Lindgren: *Trans* (1900) **30**, 578–692.

[56] *Op. cit.*, 595–597.

[57] J. D. Irving: Replacement Orebodies and the Criteria for their Recognition. *Economic Geology* (1911) **6**, 619–662.]

It has been suggested that mechanical deposition may follow so closely after dissolution as to make the two processes appear as one, but it is highly improbable that two processes could be independent and at the same time so closely synchronized. Such results might be obtained, however, if solution and deposition were interdependent, and therefore simultaneous; *i.e.*, if solution of the replaced mineral induced the deposition of the replacing mineral, or if deposition of the latter forced the former into solution.

(2) Mineral crystals in contact with supersaturated solutions must grow, and, if surrounded by other minerals, the latter will be mechanically displaced, or, if they are rendered more soluble by pressure, they may be removed in solution and deposited elsewhere. The mechanical displacement of individual minerals in a compact rock would necessitate enormous forces; the pressure per unit area would have to be very much greater than that required in separating the walls of a growing vein. The solubility of nearly all substances, so far as known, is increased by pressure and strain, and this appears to be especially true of quartz and calcite, minerals that are common in rocks subject to replacement. Therefore, when crystals develop in compact rocks, the necessary space is usually obtained through solution rather than mechanical displacement.

In his paper on "The Nature of Replacement," Lindgren draws a distinction between replacement in rigid rocks and crystallization in yielding substances. He acknowledges that in soft material "the stresses produced by crystallization may become of direct importance," but he does not believe that these forces are sufficient to make room for new crystals "in a rigid medium like quartz."[58] In support of this view he states that "the general absence of fracturing or breaking of the host mineral is evidence enough; even optical anomalies and undulous extinction are uncommon around crystals developing in comparatively rigid minerals."[58] Under his summary of previous views, Lindgren refers to the occurrence of clear gypsum crystals in clay as proof of a "force of crystallization," and then on the other hand cites "instances where the growing crystal has failed to exert this mechanical force, as in the well known calcite crystals in the sand of Fontainebleau, which include so much foreign material as to appear like crystals of sand."[59]

Now these apparently contradictory phenomena are susceptible of direct experimental investigation; and the laboratory experiments of the writer demonstrate that mechanical resistance to crystal growth is only one of several factors determining whether foreign material be included or excluded by a growing crystal. The size of the pore spaces in the inclosing material is usually the controlling factor; and, in some cases per-

[58] Waldemar Lindgren: The Nature of Replacement. *Economic Geology* (1912) 7, 533.

[59] Waldemar Lindgren: *Op. cit.*, 521–522.

haps, the surface tension between the solids and the solution is also of importance.

When a growing crystal is surrounded by foreign material in which the pore spaces are large, as for example in sand, the pressure is unevenly distributed over the crystal surface, growth is relatively rapid on the exposed areas while there is insufficient time for diffusion to supply additional material to the surfaces that are in contact with the foreign bodies. Therefore the latter are gradually surrounded and included in the growing crystal. On the other hand, when the pore spaces are small, as in clay, pressure is more uniformly distributed over the crystal surface, growth is slow, and new material may reach the entire surface of the crystal by diffusion. Therefore, the foreign bodies are mechanically displaced by the growing crystal. In the so-called rigid rocks, the pore spaces are small and the growth of new crystals extremely slow; therefore, new crystals usually exclude most foreign material.

The absence of fracturing and of optical anomalies in partly replaced minerals is not, in the writer's opinion, to be regarded as proof that the growth of the new minerals was unopposed by mechanical force. Distortion and fracturing of crystals can be produced only by a differential stress exceeding the elastic limit of the substance. Pressure resulting from the growth of new minerals is developed very slowly, and in general it is transmitted through a thin layer of solution which supplies the material for mineral growth. Each mineral grain in compact rocks is surrounded by others, which increases its resistance to distortion. Moreover, strained crystals are less stable than those that are not in a state of strain; and, therefore, the replacing solutions tend to dissolve such crystals in preference to others.

New minerals that make room for themselves by exerting pressure on the surrounding material must have begun their growth in small openings filled with gas or liquid. The replacing minerals often appear as microscopic grains, completely surrounded by the older mineral. Sometimes minute fractures, through which the solutions probably entered, may be recognized, but often they are invisible. In many such cases, however, the presence of sub-microscopic passageways may be inferred from the fact that dots and stringers of the new mineral develop along cleavage directions in the older crystals or along the planes of fluid inclusions so common in quartz. The enlargement of new crystals is sometimes accompanied by a breaking up of the older inclosing crystals, and the substitution therefor of interlocking aggregates of smaller grains. This fact has been recognized by Lindgren,[60] though he does not attribute the fracturing to pressure from the growing crystal.

In some rocks the development of new minerals is accompanied by

[60] Waldemar Lindgren: Metasomatic Processes in Fissure-veins. *Trans.* (1900) **30**, 615.

solution and removal of the less stable minerals and mechanical displacement of those that are more stable. Rocks in which replacement has barely begun commonly contain disseminated crystals of metasomatic pyrite, and, when these crystals are examined microscopically, it may sometimes be observed that in their immediate vicinity the more soluble minerals of the country rock, such as quartz, are relatively scarce, while the less soluble minerals, such as muscovite and chlorite, are crowded together, and in some cases bent or broken (Fig. 3).

It is difficult, if not impossible, to explain well developed crystal form in metasomatic minerals under any hypothesis except that the crystals in growing have exerted pressure on the surrounding material. The isolated pyrite crystals that develop in wall rocks usually show idiomorphic outlines; but, where pyrite has completely replaced the older minerals, crystal form is much less common, because the growing crystals have mutually interfered with the growth of one another.

The mechanical replacement of one mineral by another, in the manner outlined above, would be, necessarily, volume for volume, and not molecule for molecule. Pseudomorphs might be formed through the replacement of crystals imbedded in a matrix of more stable minerals, but it is improbable that the more minute details of internal structure could be preserved.

It is conceivable that mechanical replacement may also result when solutions dissolve the mineral that is being replaced, and thus lower the pressure on the replacing mineral sufficiently to permit additions of new material. This method of replacement may be applicable to some deposits formed in the belt of weathering, but it is improbable that primary ores in veins have been formed in this way. Primary vein minerals are deposited from solutions which, as a whole, are moving toward regions of lower temperature and less pressure, and for this reason the solutions are dominantly agents of deposition rather than of solution. This assumption is borne out by the general absence of solution cavities in primary replacement deposits below the belt of weathering. Even under the microscope, it is impossible to detect any intervening space between a partly replaced mineral and the replacing mineral, or other evidence that solution is dominant over deposition.

(3) When one mineral replaces another as a result of a chemical reaction, a definite number of atoms is added, subtracted, or interchanged, the number depending on the quantity of mineral replaced and the nature of the reaction. But the number of molecules that are deposited and left behind in solid form upon the completion of the reaction depends also on the solubility of the new mineral under the conditions existing at the time of its formation. In some cases the original substance is entirely removed, as when copper is replaced by silver from a solution of silver sulfate.

The new mineral formed by the reaction may be practically equal to the old one in volume, but usually the change is accompanied by either a decrease or an increase in volume. In case of a decrease, the replacing mineral is commonly porous or cellular in texture, and sometimes shows drusy cavities lined with small crystals. This is well illustrated in pseudomorphs of native silver after argentite and of smithsonite after calcite. The voids may be filled, however, through deposition of other minerals, especially in primary ores deposited by ascending solutions. When the volume of the new mineral is greater than that of the old, the surrounding material may be mechanically displaced in order to make room for the additional volume, or, more commonly, the pressure resulting from expansion may force into solution either a part of the inclosing rock or a part of the new mineral, and thus remove enough material to compensate for the increase in volume. In many cases adjustment probably results from two or more processes, and the problem may be further complicated by contemporaneous reactions involving neighboring minerals.

Irving[61] and Lindgren[62] have referred to the absence of optical anomalies in adjacent mineral grains and the lack of distorted structural and textural lines as evidence that pressure, due to increase in volume, is of little importance in the process of replacement, but with this view the writer does not agree. Pressure resulting from chemical change in the minerals of a compact rock is developed slowly, the resistance to stress is great and the solubility of most minerals is increased by pressure and strain. Therefore, it is only when relatively insoluble minerals are present that evidence of mechanical displacement or distortion of mineral crystals may be expected. Most of the arguments given above in support of the view that the growth of new minerals in rocks may result in the development of pressure are equally applicable here.

It has been demonstrated experimentally that chemical reactions, similar to those producing changes in the minerals of rocks, may be accompanied by the development of pressures due to expansion in volume; and that these pressures may be many times the crushing strength of the material as ordinarily determined. An experiment has been described in which a porous porcelain cup containing anhydrous cupric chloride was ruptured through the formation of the hydrous salt.[63] Similarly, a glass bottle of 150 c.c. capacity, having walls with a minimum thickness of 1.5 mm., was fractured by the expansive force resulting from the hydration of zinc nitrate. Although such reactions are accompanied by an increase in the volume of the solid, there is usually a decrease in the volume of the system (the anhydrous salt plus water); and, where this is true, pressure

[61] J. D. Irving: *Op. cit.*, 292–293.

[62] Waldemar Lindgren: The Nature of Replacement, *Economic Geology* (1912) 7, 531.

[63] Stephen Taber: *Op. cit.*, 76.

can have no direct effect in preventing the reaction.  The pressure developed is therefore limited solely by the resistance offered to expansion.

In many metamorphic rocks, quartz has disappeared from places subjected to maximum stress, and recrystallized where the stresses were least, while at the same time minerals, such as mica, which are less soluble, have been bent and displaced.  Evidence has been cited elsewhere in support of the view that chrysotile veins may be due to the solution of serpentine under pressure resulting from its formation and the deposition of the material as chrysotile at places where the forces opposing expansion were sufficiently feeble.  The general absence of vugs in chrysotile veins may be due to the pressure under which they are formed, since drusy cavities are not uncommon in fibrous veins of other minerals, such as calcite and gypsum.

The metasomatic replacement of a rock is often the result of extremely complicated processes, partly mechanical and partly chemical. In the case of individual minerals it may be impossible to determine whether replacement was the result of mechanical processes, or of chemical interchange, or, perhaps, of both combined.  It is realized that such processes can never be classed as wholly chemical or wholly physical, for replacement of any kind involves the removal or addition of atoms; but, nevertheless, it is believed that the distinction outlined above is important.  Silica is pseudomorphic after calcite, and commonly replaces fossil shells in such a way as to preserve the minutest details of internal structure; but, on the other hand, perfectly formed quartz crystals sometimes develop in crystalline limestone.  In both cases calcite is replaced by silica, though the two processes are evidently different.  It is suggested that the first is primarily a chemical process and due to chemical interchange, while the second is chiefly mechanical.

*The segregation of mineral matter originally dispersed through the surrounding material and its deposition in relatively concentrated masses through some process of replacement* is of common occurrence.  In this way nodules and layers of chert or flint are formed in calcareous rocks, and pyrite is segregated as single crystals, or as larger crystalline aggregates.  It is conceivable that some bedded veins and veins paralleling contacts might be formed in this manner, but it is highly improbable that well defined tabular deposits could develop transverse to the rock structure.  Such veins have been recognized by LeConte,[64] and classed under the heading of "veins of segregation."  They can be distinguished from the replacement veins described above only by determining that the vein minerals were derived from the inclosing rock and were not introduced along fractures or similar channels.  For most veins the evidence is overwhelmingly against the hypothesis that the ore minerals were

---

[64] Joseph Le Conte: "Elements of Geology," 234.  New York (1891).

derived from the immediate wall rock.   Many of the arguments used by Posepny and others in combating the old lateral secretion theory of Sandberger are applicable here, and need not be repeated.   Most veins are also formed along fractures or other structural planes which have facilitated the circulation of ore-bearing solutions, and thus controlled the form of the deposit.   The irregular and nodular form of some of the magnesite veins occurring in serpentine suggests that they may have been formed partly in this way.

Some tabular and lenticular ore deposits, inclosed in igneous rock, and commonly regarded as magmatic in origin, should possibly be classed as veins, and included in the present discussion; for there is no essential difference between (1) the transfer of ore minerals through the agency of mineralizers set free during the final stages in the solidification of an igneous magma, and (2) the solution and transportation of ore minerals originally disseminated through a rock mass.   In both cases the ore minerals may be deposited through the partial or complete replacement of earlier minerals.   Through their comprehensive investigations, Tolman and Rogers[65] have established the general law that sulfide minerals are among the final products of magmatic differentiation rather than the first.

### SUMMARY AND CONCLUSIONS

It is obvious that all veins have not been formed in the same way, and that many veins, particularly the larger metalliferous ones, are the result of more than one method of deposition.   There are gradations between the different types of veins and also between veins and other types of epigenetic ore deposits.   It is the opinion of the writer that the methods of ore deposition herein discussed are applicable to all such deposits.

All minerals deposited in veins or other orebodies occupy space previously filled by other material, either gaseous, liquid, or solid. Minerals are deposited in openings filled with gas or liquid, where such spaces are available, but only very small open spaces are believed to exist at the depths where most ore deposits have been formed.   It has been demonstrated that yawning fissures are not necessary for the formation of veins, and the evidence indicates that very few veins have been formed wholly by deposition in preexisting openings.   All minerals are formed under pressure, and no pressure could prevent the growth of a mineral crystal the surfaces of which were in contact with a supersaturated solution.   Mineral-bearing solutions, especially under conditions of high temperature and great pressure, are able to enter the densest

[65] C. F. Tolman and A. F. Rogers: A Study of the Magmatic Sulfid Ores.   *Leland Stanford Junior University Publications* (1916).

rocks by penetrating between adjacent crystals and even along mineral cleavages and sub-microscopic fractures. The introduction of mineral matter through very small openings is probably due chiefly to diffusion through thin films of solution. The circulation of solutions is greatly facilitated by fault fractures, planes of bedding or schistosity and by porous strata; therefore, most veins are formed along such passages.

Growing veins may make room for themselves either by mechanically displacing or by replacing the wall rock with which they are in immediate contact, and there is every gradation between these two extremes. When the minerals of the country rock in contact with a vein are forced aside, other adjacent minerals are likewise displaced, and final adjustment is attained either wholly or in part through condensation of porous material, rock flowage to places of less pressure, solution and removal of unstable minerals, or recrystallization to minerals of higher density.

All veins having cross-fiber structure are lateral secretion veins which, in growing, have forced apart their own walls, but all lateral secretion veins are not necessarily fibrous. The fibrous structure is a direct result of the manner of deposition, which has mechanically limited crystal growth to a single direction.

The replacement of one mineral by another may be primarily a mechanical or a chemical process. When the process is mechanical, the solution and removal of the old mineral is induced by pressure resulting from the growth of the new; and, therefore, the volume of the replacing mineral equals the volume of the mineral replaced. When replacement results from a chemical reaction, the ratio between the number of atoms added and the number removed is constant for a given set of conditions, and the change may be accompanied by either an increase or a decrease in volume.

The importance of the fact that crystals may grow in directions in which growth is opposed by adjacent solid bodies has not been generally recognized by geologists, but it is a fact that must be taken into consideration in framing an adequate theory of ore genesis. The hypothesis that ore minerals have made room for themselves by mechanically displacing older minerals and rocks elucidates many phenomena which otherwise are contradictory. It apparently solves some of the questions which have resulted in such widely divergent views as to the origin of certain sulfide orebodies.

Inclusions of wall rock, often angular in shape, are common in most sulfide orebodies, including those which have been classed as magmatic in origin. The fragments vary in size from microscopic specks up to blocks weighing thousands of tons. In some instances their position suggests that they have been pried off the adjacent wall. Occasionally the inclusions are bent, twisted or split into smaller fragments. The high fluidity of the ore-bearing solutions (or magmas ?) is indicated by the manner in which the minerals have penetrated along minute fractures in the inclusions and in the wall rock, and yet there appears to be no tendency for the fragments to segregate under the influence of gravity. Usually there is evidence of considerable replacement, but often replacement is of minor importance. These facts are all in accord with the hypothesis that the concentration and deposition of the ore minerals has been brought about through the agency of mineralizers, and that the ore minerals have made room for themselves, in part at least, by mechanical displacement of older minerals. The facts are apparently irreconcilable under a hypothesis that attributes the introduction of the ores to magmatic injection,[66] magmatic stoping,[67] or any similar process.

As evidence accumulates, it is becoming more and more certain that the formation of ore deposits is essentially a process of orderly and progressive concentration through the agency of mineralizers, and that the different types of deposits are due largely to difference in the original magmas. The writer believes that the ores make room for themselves, in most cases, by mechanically displacing or by replacing the older minerals and rocks.

## DISCUSSION

BLAMEY STEVENS, Nogales, Ariz. (written discussion*).—This subject should be approached boldly from the purely physical standpoint. There are usually many known ways of making chemical deposits of any particular mineral and probably many more ways that are still unknown. With physical effects, however, there are, in general, only one or two, or possibly three, ways that will pass muster before the physicist, when he realizes the geological conditions to be harmonized with the physical ones.

Let us consider the second way that Mr. Taber describes: "It (the mineral matter transported in solution) may enter along a fracture, bedding plane, or similar passage, and, as it is deposited, force the wall rock apart, thus making room for the growing vein," because it is the one Mr. Taber chooses for the great bulk of metalliferous veins. By "forcing the wall rock apart" Mr. Taber indicates that he means that the process of crystallization or solidification of the mineral in the vein forces the walls apart. To prove this, he offers a great variety of geological evidence, which, from the physical standpoint, is entirely superficial. From the geological standpoint the evidence is the best to be had, but it is entirely

---

[66] H. Ries and R. E. Somers: The Pyritic Deposits Near Röros, Norway. *Trans.* (1918) **58**, 244.

[67] H. M. Roberts and R. D. Longyear: Genesis of the Sudbury Nickel-copper Ores. *Trans.* (1918) **59**, 27.

* Received Oct. 30, 1918.

negative and one single grain of really positive evidence will entirely offset it.

Suppose it be granted that there is a considerable force of crystallization. It is well known among physicists that, other circumstances being equal, the crystal would rather form where it does not have to exert that pressure; or, to be more exact, where it is not necessary for the crystal to do work on external objects. It follows, therefore, that the available channels which Mr. Taber calls "a fracture, bedding plane, or similar passage" would be first filled as far as the available solution would last. But as these passages filled up, they would be exterminated as solution carriers, for certainly Mr. Taber does not consider that merely capillary passages can carry, laterally, enough solution to make a vein. This, then, is positive physical evidence that metalliferous veins are not formed in that way.

My conviction as to how they are formed is contained in my paper, "The Laws of Igneous Emanation Pressure,"[1] and, at a later date, in a less technical article published in the *Mining Magazine*.[2] These were preceded by Mr. Graton's paper mentioned by Mr. Taber. The veins are considered to be opened by the pressure of the emanations. These do not necessarily open up a large space all at once; indeed this would be opposed to the physical explanation, for the pressure of the emanations would be sufficiently relieved by small openings being made to the surface. The fractures are forced open a little at a time as they clog up with the material deposited. The illustrations were chosen in an effort to include as great a variety of deposits as possible and no doubt one or two of the cases are outside of the limits allowable to this class of deposit. The only case the physical aspect of which has been since mentioned is that of the great salt domes of Louisiana and Texas. The cause of deposition in this case was the partial evaporation of upward-flowing brine solutions by their own heat and a continually decreasing pressure. The less soluble salts must have remained in the brine, because the solution could hardly be imagined to carry enough temperature in the beginning to make a complete evaporation to saturated steam at atmospheric pressure. This discussion does not mean that Mr. Taber's well known work on asbestos veins is considered incorrect. But it is plain to see that he tries to draw all other veins as nearly as possible into his own particular category.

STEPHEN TABER (author's reply to discussion*).—In my paper, the geological evidence was summarized in some detail; the physical evidence

---

[1] *Trans.* (1912) **43**, 167.

[2] Intrusive Pressure of Mineralizing Solutions. *Mining Magazine* (November, 1914) **11**, 313.

* Received Nov. 6, 1918.

was largely omitted as this phase of the problem had been investigated experimentally and discussed in earlier papers.[3] The hypothesis that large quantities of mineral matter may diffuse through small capillary or subcapillary passages is supported by experimental proof of the permeability of rocks and also by the enormous size of some replacement deposits. In many such deposits, the ore minerals have idiomorphic surfaces where in contact with the partly replaced minerals, thus proving that the ore minerals may be introduced and deposited without the formation of perceptible openings. In laboratory experiments, I have grown small veins, non-fibrous as well as fibrous; and, if these veins could make room for themselves in a few weeks, I see no physical objection to the hypothesis that the larger metalliferous veins may be formed by similar processes.

The hypothesis of igneous emanation pressure advocated by Mr. Stevens is obviously inapplicable to veinlets of calcite and gypsum in regions of unaltered sedimentary rocks, and yet such veins frequently furnish similar evidence that they could not have been deposited in preëxisting open fissures. Certainly Mr. Stevens does not consider that the calcareous concretions found in shale have been formed through the agency of igneous emanations, and yet in such instances the bedding planes of the shale have been disturbed by the growth of the concretions in exactly the same manner as in the case of some vein-walls. The larger concretions may even contain numerous veinlets that do not extend into the surrounding rock.

ADRIEN GUÉBHARD,* St. Vallier de Thicy, A. M. (written discussion†).—Whatever may have been the mode of formation and the constituent substance of the first crust of the earth,[4] one thing is certain; namely, that, as soon as complete, this solid shell formed a division between the substances that were contained in its interior in a liquid state, and the vapors that remained incompletely purified, in suspension, in the atmosphere. The former tended to extrava-

---

[3] The Growth of Crystals under External Pressure. *American Journal of Science*, Ser. 4 (1916) **41**, 532–556; and Pressure Phenomena Accompanying the Growth of Crystals. *Proceedings*, National Academy of Science (1917) **3**, 297–302.

* Prof. of Physics of the Faculty of Medicine in Paris.

† Received Jan. 3, 1919. Translated from the French.

[4] See series of my notes at the Academy of Sciences in my quarterly review, *Notes Provençales*. But I insist upon the entire independence of my present thesis from any other hypothesis (if indeed it can be called one), as to the phase of fluidity through which our globe passed before taking on a solid form. Even if one does not accept the remaining part of the hypothesis of Laplace, it is this detail, almost universally considered valid, that is my starting point and from it I deduce an argument by incontestable physical facts.

sate,[5] owing to the increase in volume; the latter tended to precipitate in liquid layers, due to cooling. Therefore, a distinction which seems not to have been sufficiently considered has to be made between the two classes of deposits of which the lithosphere is formed, one being of endogenous magmatic origin and the other simply of exogenetic atmospheric origin. Of the first class, as yet the only one considered with regard to the genesis of metals, everything has been said, even the improbable; but the second is the one on which to lay stress.

Unfortunately, it is difficult to define the distinction, as it would be necessary to know exactly not only the conditions of temperature and of pressure at which the crust was formed, but also the critical constants of all the substances in question at that critical moment of astral genesis. Then only would it be possible to make out accurately a table dealing with the formation of bodies, of which we have knowledge only through the extravasation of magma, and of the great number which can never have had any connection with the magma.

Even of the heavy metals, the great density and very low fusibility of which in a free state ought seemingly to have precipitated them very early at the center to form there a solid barysphere, one cannot know whether some light alloys, formed by the enormous pressures in spite of the temperature were not maintained for a long time in a volatile state in the external atmosphere rather than in a dissolved state in the interior melting pot. Under ordinary pressure, the pentachloride of tungsten and of tantalum remain in a vaporized state up to 275.6° and 242° respectively; the tetrachloride of titanium, up to 136°; the peroxide of osmium, up to 100°; and tetrachloride of nickel, at 43°. But what can one conclude from that, except that there are many chances that, in the high temperature in which the pyrosphere gathered into crust, there may have remained in the atmosphere, in one form or another, some samples of nearly all the metals, even of the heaviest, which later fell upon the earth, in a rain of fire, as they condensed, and to attempt to determine the order of these condensations from the simple list of the normal points of ebullition, without knowing the critical melting point, would be sheer temerity.

When, for argument, we say that, according to what takes place in

---

[5] It is this idea of the necessary increase of the magma that has been the foundation of my entire geophysical synthesis. The experiences of Fleischer (Untersuchungen zum Beweise der Ausdehmung des Basalts, bez. der Silicaten, beim langsmen Erstarren. *Zeit. Deut. Geol. Gesell.* (1907) **59**, 122–131, and (1908) **60**, 244–78) demonstrate that in the ordinary consolidation of volcanic rocks there is an increase in volume. These observations agree with those of Stuebel (1897) and render it easy to understand certain anomalies in the results obtained by Barus (1893) and Doelter (1901) on the same subject. Michel Longchambon: Considérations sur la Formation des Colonnes Prismatiques dans les Coulées de Roches Éruptives. *Bull.* Soc. Geol. de France, Ser. 4 (1913) **13**, 33–81.

our blast furnaces, the first pellicle[6] must have detached itself from the siliceous slag at about 1850°, below which temperature or about 1500°, according to my ideas, the ferruginous elements of the magma would have become solidified, we are certainly far away from the facts. Also, it is well below 1000° and preceded, among other things, by certain sulfides such as the sulfides of lead, tin, etc., or perhaps accompanied by drops of zinc or cadmium, if they existed in the metallic state, that the anhydrous rains of haloid salts must follow each other, and these rains, under ordinary pressure,[7] would gradually range themselves between 1000° and 600°, followed by still others (salts of bismuth, antimony, etc.), before the appearance of water. Only the last one is strictly provided for in the thermo-chronometry of geogenesis, as head of the list (after iodine), from the too small number of critical points known among inorganic substances.[8]

| Substance | Degrees C. | Substance | Degrees C. | Substance | Degrees C. | Substance | Degrees C. |
|---|---|---|---|---|---|---|---|
| $H_2O$ | 365.0 | $CS_2$ | 273.05 | $H_2Se$ | 137.91 | $PH_2$ | 52.8° |
| $SnCl_4$ | 318.7 | $N_2O_4$ | 171.2 | $NH_3$ | 132.3 | $HCl$ | 52.0 |
| $Br$ | 302.2 | $So_2$ | 157.2 | $COS$ | 105.0 | $H_2N_2O_2$ | 38.8–36.5 |
| $PCl_3$ | 285.5 | $HI$ | 150.7 | $H_2S$ | 100.0 | $Xe$ | 14.7 |
| $GeCl_4$ | 276.9 | $Cl$ | 146.0 | $HBr$ | 91.3 | $CO_2$ | 31.35 |

However incomplete this little table[9] may be, it suffices to show that water had to perform a very long clearing process in the atmosphere.

[6] H. Douvillé: Les Premières Époques Géologiques. *Comptes Rendus* (1914) **159**, 221. Another author, M. E. Belot (L'origine des Formes de la Terre et des Planètes, 1915) gives 1100°; but all probability points to the augmentation rather than to the elimination of the aforesaid number.

[7] If we take the usual pressures, we would see that after the liquefaction of vapors of iron, at 2450° (the highest temperature of ebullition ever measured), those of Cu, Ca, Ag, Mn, appear below 1850°; in other words, before the assumed completion of the magmatic melt. On the other hand, Al, Tl, Pb, Sb, Bi, Te, Mg (supposing that none of these substances existed in the shape of vapor in the free state) have become condensed subsequently between 1800° and 1120°, then Zn, Cd, Na, from 918° to 742°, that is to say, shortly before the alkaline rains, they in turn being followed shortly after by Se, K, As, S, from 689° to 444°, while Hg, P, and even I, are supposed not to have been precipitated until below 357°, which means after the water (365°). This surely demonstrates clearly how fallacious it would be to base seriation on only these facts, since the critical point of iodine is known to be far higher than that of water (512° vs. 365°), and consequently its liquefaction must necessarily take place well in advance of it.

[8] I have borrowed the elements of this table from the valuable "Recuire de Constantes Physiques," of the Société Française de Physique, Paris, 1913. I have left out only iodine, before water, and the gases of negative critical points destined to remain floating in the atmosphere in so far as they escape capture by affinity, below free hydrogen, capable of braving the highest interplanetary cold.

[9] It is a pity that organic substances have been almost the sole objects of research from a point of view that ought to be of supreme interest in mineralogy.

But there can scarcely have been left any metal, properly speaking, which would not have left the clouds for the earth in a simple or compound state, in the course of the long period (by far the most important from the viewpoint that concerns us here) when fiery streams never ceased to sweep the surface of the earth and mix at the bottom of the synclines their detrital products with those from the volcanic eruption.

Much better than water, these heavy showers, more or less intermittent and localized, were made to enter under pressure, caused by gravitation and capillarity, into the smallest fissures of the earth's crust and there to develop, with a mechanical and chemical energy, amplified by the heat, actions similar to those that induced Prof. Taber, after experimenting and observing, to attribute to them the formation of the veins. They, like water, were apt to produce these stalagmitic forms that bear witness to liquid rather than gaseous actions and to the preponderance of concretion over sublimation. Even if this appears to be true, does it not explain itself quite naturally by the small amount of accretion of the geothermic kind, sufficient to have redistilled the liquid, which has dropped to a temperature only a trifle lower than at the point of its ebullition? All metalliferous deposits that one finds enclosed, .more or less modified, in the rocks, can have, even must have, originated without water; not only without water, but also without any intervention of interior fire, to which we owe the rare metals directly brought forth by the lavas.

The only part water played in it was to alter physically and chemically these deposits in the same manner in which it altered the surface of the protosphere and the pyrospheric ejections, to form, as a secondary matter, our thin lithosphere.

This is the viewpoint, neither plutonic nor neptunic, but trying to reconcile one with the other in the domain of pure physics, which I would like to bring to the attention of professional mineralogists, who, like Prof. Taber, know how to bring experimentation to the assistance of observation and who, either on account of their great knowledge of practical mining or by directing their research work in the laboratory toward determinations of the critical points destined to add to the distinctive marks of geogenesis, will be able to concede the possibility that many of these minerals have come from above instead of from below, as has been so persistently claimed and finally will be able to do real justice to the multitude of theories that attribute exclusively to either water or magma a part which often has nothing to do with the one or with the other, but only with that great era of primitive, anhydrous and metallogenetic sedimentation, from above 1850° to 365°, the importance of which I first pointed out.[10]

---

[10] See my *Note VIII.*

## Certain Ore Deposits of the Southwest

BY W. TOVOTE, E. M., TUCSON, ARIZ.

(New York Meeting, February, 1919)

THIS paper is based upon 12 years' experience in the Southwest, including three years that were spent in constant traveling as examining engineer for the Phelps-Dodge Corporation. The material was gathered during rather extensive work in Bisbee, Clifton-Morenci, Globe, and Tombstone, and also on short and rapid trips into most of the mining camps and prospect fields of New Mexico, Arizona, the desert country of California, and as much of Sonora as was safe at the time. My methods of observation and my deductions are not put forward as new or scientific; they are simply rough field methods, based upon short notes, rough sketches, and largely upon memory, which I have found useful for coördinating geological data for the purpose of passing quick judgment on mining possibilities in certain districts exhibiting uniform characteristics. The opportunities for observation have been better these last few years than ever before, due to the unprecedented activity in mining, which has caused the reopening of hundreds of old mines and the starting of even more numerous prospects all over the mining states.

### POSSIBLE SYSTEMS OF CLASSIFICATION

#### By Geological Provinces

There are a number of well defined geological provinces in the territory under discussion, and a study of their individual features is of interest. The following are a few of those that are most striking in their uniformity.

1. The "Desert province," extending roughly from Wickenburg, Ariz., to and beyond Barstow, Cal. In its north-south extent it reaches up to the Colorado River and down into Sonora, but overlaps other provinces. It is characterized by a worn-down basement of schistose and granitic rocks, on which rest remnants of Paleozoic and pre-Paleozoic sediments, including very old limestones. The sediments are disrupted and contorted. Alteration in the mineralized areas is very prominent, especially in the schists and limestones. Cenozoic volcanic rocks, principally of semi-basic composition, have been intruded into this complex and sometimes cover it as lava flows.

2. The "Yavapai schist" province, similar to the one first mentioned, consists of metamorphosed basic and semi-basic rocks. It lies in northern

Yavapai and Gila counties, having Jerome and Mayer as principal mining centers.

3. The territory of the great lava fields in north and central New Mexico.

4. The "Red Beds" in the same district.

The outlines and the characteristics of these provinces help somewhat toward understanding the mines and prospects in them, but I have found it to be of only slight advantage to classify ore deposits according to their geological provinces, because too many widely differing deposits are likely to be found in the same geological district..

## By Form of Ore Deposits

Classification according to the form that an ore deposit has assumed may serve for purposes of subdivision, but it is of little practical use because the form is merely accidental, depending upon the structure and chemical propèrties of the invaded country-rock, and the factors of stress and depth of burial. In consequence, a number of different forms may occur in a single camp. Some of the most prominent forms are: (1) Veins. (2) Lodes, intermediate between veins and disseminated deposits. (3) Disseminated deposits. (4) Infiltration deposits. (5) Contact-metamorphic deposits. (6) Replacement deposits. (7) Detrital deposits.

The ore-forming agencies react differently in different rocks. The same fracturing force would produce a trunk-channel in a granitic massive, but a stockwork of interlocking fissures, or a lode, in porphyry; it would cause lamination in schist, or devious and quickly cemented breaks in limestone; it might almost disappear in plastic shale but break through brittle quartzite in a series of compound cracks. Thus the same mineralizing activity might produce a vein in granite, a disseminated or lode deposit in porphyry, an infiltration deposit in schist, or contact metamorphic or replacement deposits in limestone; it might form rich ore zones at the contact with shale, but remain barren in them, and cause irregular stringer veins and replacements in quartzite.

## By Composition of Originating Magma

Disregarding detrital deposits, all ore deposits in the Southwest show an intimate connection with igneous rocks, and the different types of intrusives have produced certain unmistakable characteristics in the resulting ore deposits, no matter in which geological province they may occur. A classification according to the igneous mineralizing rocks has more elements of universal adaptability for this limited area than any other system that has been tried. It would not be applicable, of course, to a wider territory, because well established types are known elsewhere which, according to the best authorities, do not reveal any connection with igneous sources.

## Igneous Rocks

In field work, and for the purpose of this paper, I dispense with the standard appellations of the igneous rocks, except in rare unmistakable cases, and use instead the following classification:

· 1. Ultra-acid rocks: quartz dikes and quartz-pegmatites.

2. Acid rocks: having prominent quartz as a constituent, like granites, quartz-monzonites and quartz-diorites.

3. Semi-acid rocks: light-colored, with little or no quartz.

4. Semi-basic rocks: dark-colored, corresponding roughly to the monzonite and diorite families.

5. Basic rocks: dark colored, embracing diabases, gabbros, basalts, and related rocks.

6. Ultra-basic rocks: amphibolites, pyroxenites, peridotites, and rocks abnormally rich in iron.

All these groups of rocks have been found in connection with ore deposits, but No. 2, 3, and 4 are by far the most important.

Structural variations are designated as: Holo-crystalline, or granitic; semi-crystalline, or porphyritic; dense, or rhyolitic; aphanitic, or glassy.

### Periods of Volcanism, Periods of Ore Deposition and Systems of Fracturing

Igneous activity in the Southwest was concentrated in several distinct periods, accompanying and following great continental changes. The oldest of these, in pre-Paleozoic times, is of little interest in this discussion, because no ore deposits can unmistakably be connected with it; even those the origin of which antedates the Mesozoic era are extremely rare, and proof of their age is not above doubt. Two periods of volcanism stand out preëminently:

1. The Mesozoic Era, ranging from late Cretaceous into early Tertiary.

2. The Cenozoic Era, or Tertiary period.

### *Mesozoic Igneous Rocks*

The Mesozoic period was inaugurated by the "great Cretaceous transgression" which swept over an old continent and buried it beneath the sea. The igneous rocks of this period are mostly granitic, less frequently porphyritic, in texture, and the horizons now exposed were at the time overlain by strata, long since removed. All types of rocks are complemented, but acid, semi-acid and basic magmas prevail. The acid 2. have the widest distribution as well as the greatest importance as consists mineralizers.  Ultra-acid rocks appear with some of the less important deposits.  Basic and semi-basic intrusives seem to have produced ore only in a few places, and even there their importance has been questioned. For these reasons, I shall not attempt a further division of the Mesozoic ore deposits according to their mineralizers.

### Tertiary Igneous Rocks

The Tertiary period witnessed a gradual recession of the ocean and the emerging of a new continent.  The change was not as abrupt and universal as that of Cretaceous time, but was the result of a series of oscillations. The volcanic activity accompanying it was very widespread and intense. Its products are the mighty lava fields, still impressive after ages of disruption and erosion, besides volcanic necks, bosses, and intrusive dikes.

Magmas of all compositions are represented, but acid, semi-basic and basic magmas prevail.  The texture is principally rhyolitic, less frequently porphyritic or glassy.  The sequence is generally: 1. Semi-basic; 2. Acid; 3. Basic.  In some localities, the cycle has been duplicated, in others only a part of it was developed or has survived.  This produces some apparent or actual inconsistencies in the relative age of the different rocks.  The varying acidity of the mineralizing magmas has afforded strikingly different characteristics of their respective ore deposits; I therefore subdivide the Tertiary ore deposits as follows:

1. Andesitic deposits, caused by semi-basic mineralizers.  2. Rhyolitic deposits, caused by acid mineralizers.  3. Basaltic deposits, caused by basic mineralizers.

There is a marked difference between Mesozoic and Tertiary deposits, even when produced by mineralizers of similar composition.  This is due partly to different conditions in temperature and depth, and partly to the varying length of time during which modifying agencies have been able to work on the deposits.

### Systems of Fracturing

The forces which produced both the volcanism and the great continental changes radiated, in all probability, from certain well defined centers, and acted in the same general direction over wide sections of the country. It is therefore not surprising that the different periods show a marked preference for certain directions of fracturing, as now indicated by the respective porphyry dikes and veins.  Of course this is not an invariable rule, because great uniform rock masses, like batholiths, caused deviation of the fissuring.  In other places the fracturing force may have had a local origin, for instance, a detached volcano.  In other places, the exception may be only apparent, a dike having been deflected from its normal course by a preëxisting fault, or ore-forming agencies deviated by such channels or by favorable rocks.   In spite of many exceptions, actual and apparent, I have time and again noticed a coincidence between direction of fracturing and age of the occurrence, and, in otherwise doubtful cases, have even considered the direction a factor in determining the age of a deposit.

Mesozoic fracturing is principally in northeast-southwest direction; Cenozoic-andesitic fracturing favors northwest-southeast; and Cenozoic-rhyolitic favors north-south and east-west fracturing.

## Mesozoic Ore Deposits

The Mesozoic ore deposits of the Southwest include most of the largest and well known mines, especially those of copper.   Bisbee, Globe, Clifton-Morenci, Miami-Inspiration, and Ray, in Arizona, Chino in New Mexico, La Cananea in Sonora, belong in this group; also, probably, the Burro Mts., Organ Mts., and San Pedro districts in New Mexico and the Twin Buttes district in Arizona, but in these latter cases a younger mineralization has possibly supplemented the original Mesozoic mineralization.

Acid intrusives of the quartz-monzonite type are most prominent. Basic mineralizers are indicated in Globe and in parts of the Clifton-Morenci district.   The fracturing is normally in N.E.-S.W. direction. The volcanic activity was of long duration; in Clifton-Morenci and in the Twin Buttes district, Cretaceous strata have been invaded by the monzonite; in Bisbee they have not.   Many of the districts do not show the intrusives and the Cretaceous strata in juxtaposition, and the exact age of the intrusion cannot be determined.

The ore deposits are principally those of moderate temperature, but a few high-temperature deposits are known.   Many of the districts began by working contact deposits, probably of high-temperature origin, but their present-day importance depends on ores which do not indicate deposition under conditions of either great stress or high temperature. The mines are scattered over a series of islands of older rocks, emerging from the surrounding lava fields or gravel-covered valleys and mesas. The prospective area is much smaller than that of the more recent periods of mineralization.   The ratio of successful mines to the total number of prospects is high.   Remineralization by descending waters has proceeded for a long time, and oxidation, kaolinization, and kindred processes, are far more complete than in the younger deposits.   The mineralogy is generally very simple, often almost uninteresting, except for the changes wrought by secondary processes.

### Pegmatitic Deposits

Pegmatitic deposits connected with ultra-acid dikes are known, but are not of great economic importance; for instance, the tungsten veins near Johnson Camp, copper-molybdenite and tungsten veins in the Hualpai Mts. and Aquarius Range, Ariz., in whose vicinity stream-tin has been found. Bismuth is encountered in the veins occasionally; tourmaline is frequent in the veins, but not in the ore shoots. Closely related are copper-zinc veins in the Hualpai Mts., in granite-schist country-rock with actinolite as prominent gangue mineral. The universal gangue is naturally quartz, sometimes with orthoclase and muscovite. Rare minerals reported from these deposits are: Arizonite, possibly only a gadolinite, well-crystallized molybdite, and huebnerite intergrown with bismutite. Deposits of this type are considered as of high-temperature origin. Bismuth deposits in the San Andres Mts., New Mexico, probably also belong in this group, but are not decidedly pegmatitic in origin. They are found in laminated schist, and a number of rare bismuth minerals are reported. Ore high in bismuth has been stoped, but the economic importance of the occurrence remains to be proved.

### Contact-metamorphic Deposits

Deposits of this nature gave the start to a number of the greatest mining centers in the Southwest. Some of these, after exhausting the contact deposits, now work on different types of ore; others continue in the original type. The mineralogy of the contact-metamorphic deposits is the most variegated among the Mesozoic deposits so far as gangue minerals are concerned; the metallic mineralization is uniform and simple. Both have been exhaustively described. Exceptional instances are the occurrence of bismutite and bismutinite in Organ, New Mex., in connection with contact-metamorphic copper ores (Memphis mine), fluorite gangue (Bennett-Stephenson mine), and intimate intergrowth of galena and epidote (Modoc mine); but the mineralization in the Organ Mts. district has been decidedly influenced by porphyritic intrusions younger (possibly Tertiary-rhyolitic) than the main quartz-monzonite which produced the contact metamorphism. How closely contact deposits are interlinked with other types, and therefore how unsatisfactory are former systems of classification, is shown in most of the principal districts. The following incomplete list of districts shows the prominent forms in which the ore deposits occur:

*Bisbee.*—Contact deposits, metasomatic, or replacement, and disseminated deposits.

*Clifton-Morenci.*—Contact, replacement, disseminated, and vein deposits.

*Pinal Mts. district,* comprising Globe, Miami-Inspiration, Ray, and Christmas.—Contact, replacement, vein, and disseminated deposits.

*Chino.*—Contact, replacement, disseminated, and vein deposits.

*Twin Buttes.*—Contact, replacement, and lode deposits.

*San Pedro.*—Contact and replacement deposits.

*Organ Mts.*—Contact, replacement, and vein deposits.

All these districts are principally copper producers. The only straight lead contact deposit is in the Modoc mine at Organ. Lead-zinc mineralization is found in most of these copper districts, sometimes important, sometimes so overshadowed by the copper production that it remains practically unknown. Where lead and zinc occur, their ores usually follow the outer edge of the mineralized area. Extensive deposits of lead or zinc, or lead and zinc, principally as replacement deposits, are known in Bisbee (Copper Queen and Shattuck); Globe (Irene mine), here also in veins (Powers Gulch); Chino (Hanover and Vanadium); Cananea; San Pedro; Twin Buttes; and Organ Mts.

The distinctive metallic minerals of all the deposits of this group are pyrite and chalcopyrite, frequently bornite in copper deposits, and galena and sphalerite in the lead-zinc deposits. Tetrahedrite and enargite are found in the veins, but not often. Gangue minerals are primarily quartz and sericite; secondarily, kaolin, sericite and sometimes calcite. Hematite is found in the rare deposits which might be ascribed to basic mineralizers. Gold and silver are low, as a rule, and seldom have an important bearing on the value of these deposits, even in the lead mines, where silver is normally most prominent. Secondary enrichment of copper ores is strongly developed. Chalcocite is the principal product; covellite is mentioned from several places and I have seen it in Bisbee, Morenci, and Twin Buttes. Secondary concentration of the precious metals is practically unknown.

### Disseminated Copper Deposits

These, which are probably the best understood examples of Mesozoic ore deposition, deserve separate mention, not because they are genetically an independent type, but simply to emphasize their characteristics in contrast to those of the less understood disseminated copper deposits of more recent (Tertiary-rhyolitic) age.

The basis of these deposits is pyrite, and pyrite with a slight admixture of chalcopyrite; the mineral chalcopyrite itself is rare in these ores, although it occurs at Ray, Ariz. Primary mineralizing agencies caused silicification and deposition of white vein quartz and sericite. Primary ore is not of economic grade except in narrow veins or feed-channels. The economic value of the deposits is confined to a horizon of limited vertical extent, bounded above by the oxidized zone and below by the primary ore, and is the product of long-continued secondary enrichment. This horizon is defined by the partial or complete replacement of the pyrite by secondary chalcocite. The outcrops of these deposits are very prominent; the surface shows rusty discoloration, quartz seams, and copper-stained quartz cliffs. Skeleton quartz weathers out frequently over large areas.

Below this is a leached and thoroughly kaolinized horizon, usually bare of copper; this grades into lower and less impoverished zones, which have retained part of their copper in oxidized form, highly concentrated locally. The upper part of the leached kaolinized zone is usually white. Iron oxide remains in the cap-rock and is found in veins and seams in the kaolin. Pyrite, probably of secondary origin, but usually considered residual, is found quite frequently immediately above the chalcocite horizon, in conjunction with hydrated quartz. The values are in the copper only; gold and silver are practically negligible. The ground is very soft, and dilution of the ore by mass-stoping becomes a serious factor. The ore worked ranges from less than 1 per cent. up to slightly over 2 per cent. copper.

Considering the Mesozoic deposits as a whole, what impresses me most is that their genesis, the pegmatitic deposits excepted, must have been an exceedingly slow and long-continued process. While the first ore may have been formed under high temperature, deposition continued under steadily diminishing heat, until a very moderate temperature prevailed during the last stages of primary deposition. Almost immediately thereafter, secondary processes began, and continued active through long geological ages. The present-day aspect is, therefore, very complex.

## Cenozoic or Tertiary Ore Deposits

The ore deposits of this period originated close to the surface and their genesis is intimately connected with the volcanic activity which produced the great lava flows. They are rarely an integral part of such flows (basaltic deposits) and usually originated after the lava had consolidated and undergone fracturing. Still, even in these cases the ore deposition must have been compressed into a very short time and while the mineralizing magma, at least in its deeper parts, whence the metallic constituents rose, was far from cooled. Many deposits were completely formed before the next lava masses were poured out above them. Base metals are sometimes important, but are always accompanied by considerable values in precious metals; by far the greatest number of deposits owe their importance almost exclusively to their gold and silver contents. Secondary enrichment of base metals is almost unknown, but a secondary concentration of the precious metals is found frequently. The deposition must have proceeded rapidly and under active volcanic conditions. This is one reason why I cannot fully accept Lindgren's argument for the moderate-temperature origin of these ores; most of the accompanying gangue minerals are stabile over a very wide range of temperature.

Three types of deposit are clearly distinguishable, differing in age, origin, and characteristics: the basaltic, the andesitic and the rhyolitic.

Representatives of these types cover a much larger territory than the Mesozoic deposits and, while their present importance does not equal that of the older group, their prospective possibilities are very great. In the early mining days in the Southwest, their importance was preëminent, but the old bonanzas have been worked out, and the abandoned mines are now mostly inaccessible. Most of the prospects seen in the ordinary run of field work belong in these groups.

### Basaltic Ore Deposits

These are the youngest and the least important of the Tertiary deposits. They really are of geological importance only, not of economic value. I have seen only a few representatives of this group, mostly in New Mexico, near Steeple Rock, Grant County, and near Bland, Sandoval county. Large areas of basaltic copper deposits are reported from the Mogollon Plateau, from which I have seen specimens.

The primary mineralization consists of nuggets and flakes of native copper, imbedded in usually vesicular, basaltic lavas. The vesicles are either filled with chalcedony and quartz, or are empty and lined with a greenish mineral, an iron compound (probably kraurite) which is frequently mistaken for impure chrysocolla. This type of deposit can assume economic importance only in exceptional circumstances, such as secondary concentration in gash veins; here, chrysocolla and malachite, subordinately azurite, cuprite, native copper, or even chalcocite are formed. Calcite is the principal gangue. The veins give out at shallow depth.

### Andesitic Deposits

Andesitic deposits are found, in conjunction with rhyolitic, over the entire area under discussion, from the Colorado line into southern Sonora, and from Texas into California. Very frequently, both types are represented in the same general district, but many districts belong distinctly to one type to the exclusion of the other.

Andesitic deposits are usually connected with fracturing in N. 60° W. and N. 20° W. directions. The fracturing seems practically contemporaneous and the veins frequently drag one another, producing intermediate directions. Intersection points and dragged sections are loci for ore shoots. The andesite is often metamorphosed to semi-schistose structure in the ore-bearing areas, but usually can be traced back to fresh and unaltered dikes.

Prominent districts representing this type are: The "Parker Cut-off" country in Yuma county, Ariz., and San Bernardino and Riverside counties, Cal.; the lead-zinc belt near Kingman and Chloride, Mojave county, Ariz.; the Casa Grande section in Pima and Pinal counties,

Ariz.; the Tularosa country in Otero county, New Mex., and others.
I believe that most of the copper deposits in the Red Beds in New Mexico
derive their metallic minerals from andesitic sources.

Important mines of this type are: the Tennessee, near Chloride;
the Golconda, in Union Basin, near Kingman; the Clara Consolidated,
near Swansea, Yuma county; and the Planet mine, a little farther west
on the Bill Williams River; the famous Monte Christo mine, near Wickenburg; and the Lakeshore mine near Casa Grande.

Gangue minerals are: hematite, chlorite, baryte, frequently calcite
and dolomite. Quartz is subordinate; where found, it is often very recent. Fluorite is rare. Kaolinization and sericitization occur, but are
not nearly so important or conspicuous as in the Mesozoic deposits.
Serpentinization and propylitization are important. Kaolinization, if
found, is usually accompanied by deposition of chalcedony in roughly
banded structure. Copper, while often important, is never so predominant as in the Mesozoic deposits. Many districts are prevailingly lead-
zinc producers. Gold and silver are always present in economically
important quantities, and may become the main object of mining,
always accompanied by base-metal values.

The mineralogy of these deposits is variegated and interesting. It
comprises, besides the sulfides of the base metals and their oxidation
products, the sulf-arsenides, sulf-antimonides and sulfides of silver,
with their derivatives. Beautiful specimens of flaky, or even crystallized,
free gold have been found, sometimes intergrown with hematite or
calcite.

The andesitic deposits frequently have very conspicuous outcrops.
The neighborhood of the Planet mine, in Yuma county, for example,
shows some of the most wonderful outcrops to be found anywhere, being
actual mountains of hematite and limonite, stained green by the salts of
copper. In other places, outcrops are marked by the bleaching of normally purplish andesite. Here, high-grade oxidized copper ore is not
an exception close to the surface, underlain by barren, partly kaolinized
andesite. As a general rule, copper ore and copper stain at and near the
surface have led to the prospecting and discovery of many mines, which
ultimately were found to contain copper only as a negligible accessory.
Some outcrops are misleading; there are wide areas of copper-stained
andesite, which have even been considered worth testing by churn-
drilling, but have been found to contain only tight and narrow veins
below. Many an outcrop, standing out as an imposing reef, covers
insignificant and unremunerative veins; while, on the other hand, very
inconspicuous croppings hide some of the biggest orebodies. This is
especially true where chlorite is the main gangue mineral. Chloritization
and propylitization are the safest indicators of deep-seated mineralization; hematite, while attractive and conspicuous, is not always reliable.

## Rhyolitic Deposits

The rhyolitic deposits comprise a wide variety of both precious and base-metal orebodies. Some of the most important representatives are:

*Tombstone, Ariz.*—Silver and gold derived from tellurides and sulfo-salts; also combined in the sulfides of lead, zinc, and copper. Alabandite is found; quartz and fluorite are prominent gangue minerals.

*Prescott and Vicinity, Ariz.*—Chalcopyrite and pyrite, seldom bornite, with quartz and orthoclase, and such carbonates as dolomite and ankerite.

*Crown King, Ariz.*—Sphalerite, galena, and pyrite, with very little chalcopyrite, in quartz-carbonate gangue; high silver values are probably due to admixture of sulf-antimonides.

*Oatman, Ariz.*—Free gold, with practically no base metals, in calcite and chalcedony gangue. Adularia is a frequent gangue mineral.

*Mogollon, New Mex.*—Argentite, finely impregnating quartz, accompanied by gold in some form, and sometimes associated with bornite and chalcopyrite. Fluorite and carbonate minerals are the gangue.

*Black Range, New Mex.*—Bornite and tetrahedrite, with high-grade silver minerals, like argentite and stephanite, in quartz-fluorite-carbonate gangue. Cuproplombite is reported from here, and I found crystallized bornite.

*Nacozari, Sonora*, is a prominent example of the rhyolitic ore deposits; and the *Patagonia-Nogales* country furnishes more representatives.

Most of these rhyolitic deposits owe their origin to volcanic disturbances, far more violent and abrupt than those connected with the previously described deposits. Brecciation is usually very pronounced and in many cases has been repeated at least once. Ore deposition began apparently under high temperature, but probably not high pressure, and continued through periods of diminishing temperature. The fracturing is north-south and east-west, usually a little east of north and north of east. Ore shoots are almost universally due to intersections, and very seldom to a single factor.

Gangue minerals are: Quartz and chalcedony, often replacing older carbonate gangue minerals. Carbonates, like dolomite, ankerite, rhodochrosite; seldom siderite. Calcite is among the first (pre-mineral) as well as among the last (post-mineral) depositions. Fluorite is very frequent; baryte rather rare. Zeolites are typical, especially stilbite; orthoclase or adularia, often in well defined crystals, and rhodonite are common. Molybdenite, usually in large flakes or even sharp crystals, is a common metallic associate; nephelite has been found, and zinnwaldite and other rare micaceous minerals, as well as amphiboles, sometimes occur.

The metallic minerals comprise most of the sulfo-salts of both precious and base metals and their alteration products. High-grade silver minerals frequently seem associated with amethyst, the latter being a good indicator.

In Santa Cruz county, Ariz., I found a type of deposit which I believe has not been described. It is characterized by the somewhat rare micaceous mineral zinnwaldite (silvery white, lithia-bearing and crystallizing in rosette-shaped crystals). The zinnwaldite occurs in masses of crystalline aggregates as well as in well defined columnar crystals. Sometimes it forms narrow ribbons in veins of banded structure. It is always closely associated with metallic sulfides (chalcopyrite principally) and sometimes with rutile, both massive and in splendid crystals. The country-rock is the younger quartz-monzonite.

In another place in this same vicinity, I found a vein in the monzonite which had, besides zinnwaldite and coarsely crystalline molybdenite, an intimate mixture of magnetite, apatite, pyrite, and chalcopyrite.

Many deposits of this period are pockety, but some mines have developed large and persistent ore shoots. Sudden changes are the rule, unexpected swelling or pinching, and fabulously rich bonanzas surrounded on all sides by poor and weak vein sections.

The disseminated copper deposits of rhyolitic age are sure to become important. Nacozari, Sonora, is the principal representative, followed by Red Mountain, near Nogales, and Copper Basin, near Prescott, Ariz. Possibly Ajo, Ariz., belongs in this group, but I am not familiar with the latter district. The ore is found in fractured and brecciated rock, rich ore occurring in the breccia zones. Metallic sulfides have pervaded the rock generally along narrow cracks and fissures, and as impregnations. The country-rock is generally acid (quartz-porphyry to quartz-monzonite). Mineralization begins with silicification, closely followed by pyritization, these processes often being duplicated. Toward the end of the first period of pyritization and during the second period, chalcopyrite is deposited. Molybdenite is a frequent accessory, and carbonate minerals, zeolites, orthoclase or adularia, form the principal gangue, besides the more conspicuous quartz. Amphibole, rare micas, and nephelite are found occasionally.

The ore is strictly primary; secondary enrichment (chalcocitization) is practically unknown. The copper is mostly deposited as a very pure chalcopyrite, sometimes with bornite, but is also occluded microscopically in the pyrite. Gold and silver are always present in economically important quantities. The sulfide remains unaffected by oxidation close to, and even at, the surface. Superficial indications commensurate with those of the Mesozoic "porphyries" are the exception. Secondary rock alteration is confined to a shallow belt near the surface, and along prominent fissures; elsewhere the rock remains hard and fresh.

The grade is rather low, but I feel confident that these deposits will prove economically valuable in the near future. At this time, disseminated ore of 1 per cent. Cu, and less, without accessory gold and silver, is treated in some of the older "porphyries," a grade far below that of some of these rhyolitic deposits.

## JEROME DISTRICT

My previous omission of the Jerome district has been intentional, because I have not enough data from my own observation to define the age of these deposits. Generally, the ore at Jerome is considered very old. F. L. Ransome apparently considers it pre-Cambrian, because he mentions eroded pyritic deposits in the basal schist as the probable source of the ferruginous coloring in the pre-Cambrian Tapeats sandstone. This view is supported by the appearance of the 600-ft. level of the United Verde Extension, where the Paleozoic sediments seem to rest upon an eroded schist surface, containing remnants of an oxidized orebody. Still, this evidence does not seem conclusive to me, because the oxidized ore here might possibly be a later deposition, derived from mineralized surface waters, which found a ready precipitant along this contact.

The fact stands out preëminently that the pyritic deposits in the schist have escaped the metamorphism undergone by the country-rock. They do not show any stratification or lamination, and are therefore later than the metamorphism which formed schist out of the original basic and semi-basic volcanic rocks.

While no ore has been found in the Paleozoic sediments overlying the schist, a strong contact metamorphism and some indication of mineralization are noticeable on the south side of the Black Hills, on whose north slope Jerome is situated, and this contact metamorphism is closely connected with minor ore deposits, identical in type with the Jerome ore.

In the United Verde Extension, the ore is found in veins striking N. 10° to 20° E., but the bonanza orebody is at the intersection of such veins with a northwest-southeast cross-vein, or fault. This fault seems to have had a decided influence upon the mineralization, and the fault material suggests derivation from semi-basic intrusive rocks. From the information available at present, I would hesitate to state the age of the Jerome deposit, but am inclined to believe that it is much younger than usually assumed.

## FUTURE OF THE SOUTHWEST MINING INDUSTRY

As to the outlook for the mining industry in its relation to the three great groups of ore deposits, I venture the prediction that very few new mines of Mesozoic origin will be found in the Southwest. Those that will come into prominence eventually will be mines known today, but enhanced in value by the progress of mining and metallurgy, or extensions of known and operating districts, or new deposits, discovered by accident.

The Tertiary deposits, on the other hand, are by no means developed to their full capacity. I look for many mines of andesitic origin and even more of rhyolitic derivation to become prominent in the near future. Still, very few of these will compare in size and richness with the great Mesozoic mines, so far as base metals are concerned. This seems to be in keeping with the general trend of recent development in the mining industry here, where great mines are more and more concentrated in the hands of a few big companies, while beside them has sprung up a healthy undergrowth of small operators. The small mine and the combination of small mines seem to be destined for increasing importance.

Prospecting of the old days is dead. How closely the whole country has been searched is brought home, time and again, when in places where hardly a rattlesnake could live, and where there is no indication of human habitation for miles around, remnants of old workings and prospect holes are found. Good discoveries are still made quite frequently, but as the result, usually, of highly intelligent development work. It is regrettable that, on the one hand, the big companies do not more frequently develop small but promising prospects while, on the other, no safer and more business-like method of development appeals to the public at large than the floating of 1,000,000-share development companies.

## DISCUSSION

PHILIP D. WILSON,* Warren, Ariz. (written discussion†).—Mr. Tovote's idea of attempting to classify according to their broad geologic relations the ore deposits and prospects of the Southwest is an admirable one. Attempts have been made to evolve some comprehensive scheme of classification that would be of practical assistance in determining the possible value of prospects, but the ore deposits are too varied in type, form, genesis, and distribution to lend themselves readily to any definite, scientific pigeon-holing. The classification Mr. Tovote recommends, combining the age of the mineralization with the type of the rock supposed to be its source, is open to criticism even on the evidence he presents. The age of many deposits is not nor can be definitely determined because of the absence of sedimentary rocks in their vicinity. Furthermore, the character of the mineralizing solutions has, in a large majority of cases, no relation to the composition of the associated igneous rock. While the solutions may have originated in the same magma, they

---

* Geologist, Calumet & Arizona Mining Co. † Received Dec. 12, 1918.

were usually a much later manifestation of it, often appearing after the intrusive rock had completely solidified and been subsequently crushed and fractured, and in many cases the chemical difference is very marked.

It is true that igneous activity in the Southwest is definitely related to certain geologic periods.    The statement, however, that no òre deposits can unmistakably be referred to periods earlier than late Cretaceous is not supported by fact.    The bonanza copper orebodies of the Jerome District are definitely pre-Cambrian.    As suggested by J. R. Finlay,[1] they are probably post-Beltian or late Algonkian in age, perhaps related to the same great mountain-making period as the copper deposits in the Keweenawan series of Lake Superior and the Sudbury, Porcupine, and Cobalt deposits of Ontario.    The proof that they are pre-Paleozoic may be summarized as follows:

1. The pre-Cambrian complex in which they occur is overlaid by a great thickness of Paleozoic sediments with the Tonto or Tapeats sandstone at their base.    The only primary mineralization in these sediments is an economically unimportant one connected with very recent basalt.

2. In the United Verde Extension mine, the gossan over the great orebody extends to the old pre-Cambrian surface where it is directly overlaid by unmineralized Tapeats sandstone.    This would further indicate that the oxidation and enrichment of this orebody was completed before the Tapeats sandstone was deposited.

3. In the United Verde mine, included fragments of sulfide ore are found in later andesite dikes that cut the orebodies but do not invade the Paleozoic sediments.

4. The characteristic dark, rusty, red color of the basal Tapeats sandstone is attributed, by F. L. Ransome,[2] to ferruginous gossan material from the pre-Cambrian pyrite deposits in the schist.

5. It is true, as Mr. Tovote states, that the pyritic deposits in the schist do not show the result of the regional metamorphism that the neighboring rock has undergone for the reason that they are later than the probable source of the ore, the United Verde diorite, which in turn is massive and shows none of the evidence of the stresses that have produced the schistose structure in the older complex.    The diorite is definitely pre-Cambrian but very evidently much younger than the greenstone and quartz porphyry schists that it intrudes.

Besides the Jerome district, there are several of less importance in Arizona in which the ore deposits may be pretty certainly referred to the pre-Cambrian, notably the copper deposits in the Yavapai schist

---

[1] Jerome District of Arizona.  *Engineering & Mining Journal* (Sept. 28, 1918) **106,** 557.

[2] Some Paleozoic Sections in Arizona and their Correlation.   U. S. Geological Survey *Prof. Paper* 98-K (1916).

belt in the Bradshaw Mountains from Stoddard south to Canyon, and the many copper specularite deposits in the pre-Cambrian schist in the vicinity of the Bill Williams and Colorado rivers. It is not logical to include Bisbee, where the mineralization is definitely pre-Cretaceous in age, in the same group with younger Globe-Miami, Clifton-Morenci, Ray, and Chino deposits when the entire period of Cretaceous sedimentation intervenes. It obviously belongs to a distinct period of igneous and mountain-making activity. There is some reason to believe that the Cananea deposits are of the same age as those in Bisbee, although there is no positive means of determining it. There is reason to question the classification of the "Parker Cut-off" deposits in pre-Cambrian schist, such as the Clara Consolidated and the Planet, as Tertiary-andesitic. The original intrusions with which the ore is in places associated have been very thoroughly altered and sheared to amphibolite and Howland Bancroft,[3] in common with most other observers, has considered them and the associated mineralization as definitely pre-Cambrian.

It seems dangerous and of little practical advantage to correlate such dissimilar and widely separated deposits as the disseminated copper deposits of Pilares, Ajo, and Red Mountain with the silver-lead orebodies of Tombstone, the silver-zinc lead veins of Crown King, and the gold veins of Oatman. In some of these cases, the igneous rock that can be regarded as the source of the mineralization is uncertain or unknown while in the case of the Ajo deposit the rock is a monzonite porphyry with decidedly basic phases. In the latter example, the unimportance of chalcocite enrichment, one of Mr. Tovote's criteria, can readily be explained by the geologic history, climatic conditions, and mineralogy of the ore.

In connection with Mr. Tovote's classification according to geologic provinces, in which he assigns Jerome to the "Yavapai schist" province, it may be interesting to note that no Yavapai schist, as it is defined by Jaggar and Palache,[4] has been found in the Jerome district. The schist in Jerome is composed entirely of ancient intrusive and extrusive volcanic rocks, flows, and bedded tuffs, while the Yavapai schist is typically a fine-grained quartz muscovite of sedimentary variety, intruded by granitic rocks. Both are probably older than the belt rocks but evidently do not belong to the same series and cannot be correlated.

If the final test of any scheme of classification proposed is its practical value in simplifying and facilitating the estimation of the probable

---

[3] Ore Deposits in Northern Yuma Co., Arizona. U. S. Geological Survey *Bull.* 451 (1911).

[4] T. A. Jaggar, Jr. and Charles Palache: U. S. Geological Survey *Bradshaw Mountains Folio* (1905).

value of a mine or prospect offered for sale or development, it should be based upon those qualities that actually determine the value of the ore deposit. These factors are neither the age of the mineralization nor the character of associated igneous rock, but rather the form of the deposit, which Mr. Tovote dismisses with scant consideràtion. It is with the fact that a deposit is a large disseminated body or a narrow vein or a garnet contact type that a prospective purchaser is interested rather than that it is Tertiary rhyolitic or Mesozoic semi-acid.

A rough physiographic separation of Arizona into the three divisions of Mr. Ransome[5] and Mr. Finlay[5]—the plateau region of prevalent subsidence in the northeast, the desert region of relative uplift in the southwest, and the zone of stress, fracturing, and igneous activity between the two, characterized by the great parallel mountain ranges and containing most of the important ore deposits—is valuable in indicating the most likely place to look for valuable mines. But even this broad scheme is no infallible guide, for one of the really great mines of the state, the New Cornelia, and many less important ones are found in the isolated desert ranges.

In short, while Mr. Tovote's paper shows much thought and a wealth of close observation and the idea that prompted it is excellent, there is ground for criticism both of the fundamental scheme proposed, which is impracticable, and of the attempt to include in too small a compass so many deposits differing widely in age and character.

W. G. MITCHELL,* New York, N. Y.—I quite agree with Mr. Wilson in the statement that the Bonanza copper orebodies of the Jerome District are definitely pre-Cambrian. You might go much further than that. It is an interesting fact and, I think, well substantiated, that the chalcocite enrichment is pre-Cambrian, and in that particular the deposits are unique, so far as I know, among copper-ore deposits. It is certainly true, under my scheme of classification, at least, and I think it has been supported by the work of Mr. Reber of the United Verde Extension, that chalcocitization was entirely accomplished prior to the laying down of the Cambrian sediments. The statement of Mr. Wilson that no Yavapai schist, as defined by Jaggar and Palache, has been found in the Jerome District and that the schist in Jerome is composed entirely of ancient intrusive and extrusive volcanic rocks and bedded tuffs is one with which I am not entirely in accord. I believe that it is true that the schists at Jerome are largely composed of bedded volcanic material, I am also certain that there is a considerable proportion of plastic material in those schists. I have seen a good many slides, and it is quite common to find quartzites and limestones in the schists. At a short distance from Jerome,

⁵ *Op. cit.*                    * Mining Engineer, R. Martens & Co.

at the Arizona Binghamton, there are limestone strata in the schists several feet in thickness, and unmistakably a part of the original sediments as laid down.

Another statement of Mr. Wilson is supported by my own observations. In the United Verde mine included fragments of sulfide ore are found in later anthracite dikes, which cut the orebodies but do not invade the chalcocite sediments. This is not only true, but those fragments of sulfide ore that are of small size, from the size of a walnut to fragments 2 and 3 ft. in diameter, are of the original unenriched pyritic material. I have never found high-grade ore, either chalcopyrite or chalcocite, in any of those dikes.

## Petrographic Notes on the Ore Deposits of Jerome, Ariz.

BY MARION RICE, NEW YORK, N. Y.

(New York Meeting, February, 1919)

THE copper-mining district of Jerome, Ariz., is of such economic importance that the following brief notes may be of interest.

The ore deposits are said by Ransome[1] to be pre-Cambrian, and are contained in the pre-Cambrian schists of the region.

In the vicinity of the mine (the United Verde) the schist stands nearly vertical and strikes a little west of north. At least three varieties are distinguishable—(1) a green rock, schistose on its margins but grading into massive material, which is evidently an altered dioritic intrusive; (2) a rough gray schist with abundant phenocrysts of quartz, apparently an altered rhyolite; and (3) a satiny, greenish gray, very fissile sericitic schist that may be a metamorphosed sediment. The ore occurs in varieties (2) and (3), the main belt of dioritic rock (1) lying just west of the orebodies. The ore is said to follow as a rule the layers of fine sericitic schist.

T. A. Rickard[2] says that the ore at the United Verde Extension mine is found at the contact of diorite and schist, that both diorite and ore are earlier than the regional metamorphism, and that the quartz porphyry ("rhyolite" of Ransome) is of post-Cambrian age.

The material considered here, a part of which was gathered by the writer, came from four of the mines of the district, and its study was undertaken at the suggestion of Dr. Berkey of Columbia University. The field relations and exact locations of the different specimens are not known, but a microscopic study of the thin sections gave some evidence as to the origin of the ore, which will be discussed below. The indications are that both ore and porphyry were introduced at the closing stages of the metamorphism, this being in accord with Ransome's view of the genetic relation of porphyry and ore rather than with the opinion that the ore is related to the diorite. Certain field occurrences mentioned in the literature also support this theory. Rickard, in the same article to which reference has been made, speaks of a vein of chalcocite several inches wide at the contact of one of the porphyry dikes, and Provot,[3] although he regards the porphyry as subsequent to the ore, says: "Acid dikes encountered underground should be followed as they have been found in practice to lead to orebodies."

---

[1] F. L. Ransome: U. S. Geological Survey, *Bull.* 529 (1913) 192.
[2] T. A. Rickard: *Mining and Scientific Press* (Jan. 12, 1918) 116, 47.
[3] F. A. Provot: "Geological Reconnaissance of Jerome District."

## Rock Types

### *Andesite Porphyry*

This name as used here includes the variety sometimes referred to as diorite porphyry. In the hand specimen, this rock is typically fine-grained, of a dark green color, veined with carbonate and chlorite, and showing general chloritization. In thin section, it is seen to be an andesite of diabasic structure, with the feldspars very much crushed and sheared (Fig. 1). Hydrothermal alteration has partly converted the feldspars to epidote, and changed the ferro-magnesian minerals to chlorite and serpentine. Fractures have been healed with carbonate and chlorite (Fig. 2), and some contain small pyrite cubes. Some speci-

FIG. 1.                                FIG. 2.

FIG. 1.—ANDESITE PORPHYRY, SHOWING CRUSHED CHARACTER OF ROCK. LIGHT GRAY MINERALS ARE FELDSPARS, DARK GRAY IS CHLORITE. ORDINARY LIGHT. × 30.

FIG. 2.—ANDESITE PORPHYRY, SHEARED AND VEINED. LIGHT GRAY, LATH-LIKE MINERALS ARE ALTERED FELDSPARS; DARK GRAY IS CHLORITE AND SERPENTINE. LARGE, LIGHT GRAY VEIN IS OF CARBONATE AND QUARTZ. ORDINARY LIGHT. × 30.

mens are coarse-grained, but in general the type is that of a small intrusive. The history indicated is one of not very deep-seated dynamic disturbance, followed by hydrothermal alteration with slight mineralization.

### *Schist*

Two types of schist were observed. In the hand specimen both were very fine-grained, dark green to grayish colored, with small, whitish patches somewhat sheared out.

The first type would be more properly classed with the phyllites. It is an extremely fine-grained rock, now composed of an aggregate of quartz, chlorite, carbonate, and sericite, which has not been subjected to any very profound metamorphism. The shearing has developed abundant flakes of sericite but there has been no thorough reorganization of the material, with development of the high-density minerals.   Later cracks are healed with carbonate, and some specimens show tiny disseminated grains of pyrite.   The history of this rock is essentially the same as that of the andesites.   It was originally a fine-grained fragmental, possibly a shale, but more likely, in view of its associations, an ash.

The second type is similar in composition and mineral make-up,

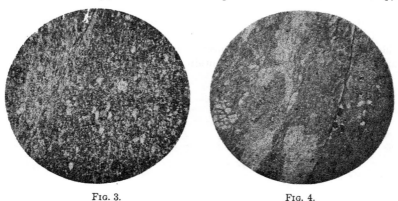

FIG. 3.                              FIG. 4.

FIG. 3.—SCHIST.   AGGREGATE OF SECONDARY MINERALS.   LIGHT GRAY STREAKED PATCH IS SERICITE;  LIGHT GRAY GRAINS ARE QUARTZ.   CROSSED NICOLS.   × 30.
FIG. 4.—SCHIST. APPARENTLY FORMED FROM IGNEOUS ROCK.   LIGHT GRAY PATCHES SUGGEST FELDSPAR PHENOCRYST;  WHITE, CRUSHED FRAGMENTS ARE QUARTZ; BLACK CUBES IN VEINLET ARE PYRITE.   ORDINARY LIGHT.   × 30.

but was formed by the more intense shearing of coarser material (Figs. 3 and 4).   Grains of original quartz and scattered dense sericitic and chloritic patches, such as would result from the alteration of feldspar and hornblende crystals, suggest that the original rock was an igneous type, but it has been so much sheared that nearly all the primary structure has been obliterated and the determination of the origin is uncertain. The rock may have been a tuff formed from small rock fragments.

### Quartz Porphyry

In the hand specimen, this rock is of a gray-green color, fine-grained, with large quartz phenocrysts, and occasional red and green patches of feldspar and chlorite.   Under the microscope, it is seen to be a very siliceous quartz porphyry with abundant quartz phenocrysts and an occasional striated feldspar (Fig. 5).   The groundmass is typically a very fine intergrowth of quartz and feldspar, so that the rock is of the type known as graphophyr, characteristic of small intrusions from a siliceous magma. The feldspar is partly altered to sericite, epidote, and carbonate (Fig. 6), and the original small ferro-magnesian content is now in the form of chlorite. Chlorite is also developed in the sheared zones, and seems here to be a vein mineral. These porphyries, as might

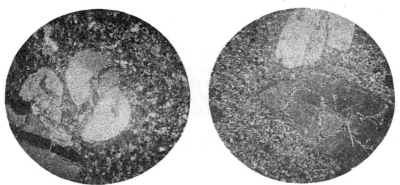

FIG. 5.                                              FIG. 6.

FIG. 5.—QUARTZ PORPHYRY SHOWING CRUSHED QUARTZ AND FELDSPAR PHENO-CRYSTS AND FINE-GRAINED GROUNDMASS. CROSSED NICOLS. × 30.

FIG. 6.—QUARTZ PORPHYRY SHOWING FRACTURED FELDSPAR PHENOCRYSTS VEINED WITH CARBONATE. CROSSED NICOLS. × 30.

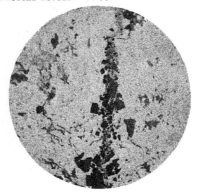

FIG. 7.—QUARTZ PORPHYRY, SHOWING PYRITE CUBES (BLACK) IN VEINLET CUTTING ROCK. ORDINARY LIGHT. × 30.

be expected of a series of small intrusions, are somewhat variable in composition and habit. They also show different amounts of dynamic disturbance Some show no trace of fracturing, while others have quartz, chlorite, and carbonate veins and crush effects in the phenocrysts. Most of the specimens contain pyrite in the veins (Fig. 7); some as perfect cubes and some as crushed fragments. Black rims on the pyrite result from secondary processes.

The quartz porphyries, then, were intruded during the closing stages of the regional disturbance, and also after movement had stopped.   The mineralization follows fractures in the porphyry and therefore is later in each case than the rock in which it is found, but is not necessarily later than the entire porphyry series, so that ore may be found cutting the dikes, and likewise dikes cutting the ore.

## ORE SPECIMENS

In addition to the wall rocks, some ore specimens, (1) and (2) from the United Verde, and (3) and (4) from the United Verde Extension mines, were examined microscopically.   Unfortunately, all the material at hand was of the high-grade ore, so that the process of mineralization could not be studied.

### *United Verde Ore*

1. The ore consists of bands of crushed pyrite and of sphalerite, the whole specimen veined and partly replaced by covellite (Fig. 8). A very small amount of chalcopyrite appears in the sphalerite, and chalcocite is entirely absent.   The banding may be due to replacement preserving the structure of a schist, or to deposition along the weaknesses

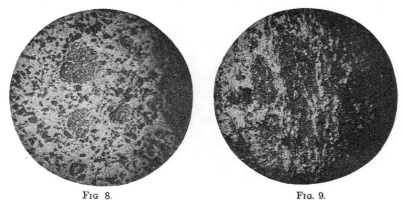

FIG 8.                    FIG. 9.

FIG. 8.—MASSIVE UNITED VERDE ORE.   DARK GRAY, ROUGH, CRUSHED MINERAL IS PYRITE; LIGHT, SMOOTH MINERAL IS SPHALERITE.   COVELLITE CANNOT BE DISTINGUISHED FROM SPHALERITE IN PHOTOGRAPH.   REFLECTED LIGHT.   × 34.

FIG. 9.—UNITED VERDE EXTENSION HIGH-GRADE ORE.   GRAY, ROUGH MINERAL IS PYRITE; GRAY, SMOOTH MINERAL IS QUARTZ; SMOOTH, WHITE MATRIX IS CHALCOCITE. REFLECTED LIGHT.   × 34.

of a sheared rock; probably the former, as it is quite fine and regular. The crushing of the pyrite appears to be a result of continued movement parallel to the schistosity.   This is regarded as additional evidence that the mineralization began before the conclusion of the dynamic modification of these rocks. The covellite belongs to a different and later period, as it occurs in veinlets cutting the original structure and independent of it.

2. This ore is more massive pyrite, considerably crushed but not banded, containing chalcopyrite interstitially and in veinlets. An occasional large grain of chalcopyrite is in primary relation to the pyrite, so that it would appear that the chalcopyrite is, in general, later than the pyrite, but that some overlapping of these two constituents occurred.

### United Verde Extension Ore

3. This is a massive pyrite with secondary chalcocite in veinlets, interstitially and in occasional large grains. The chalcocite is thus in the same relation to the pyrite as the chalcopyrite is in specimen (2), and is most likely the result of secondary enrichment (replacement) of the chalcopyrite. It has not at all the distribution to be expected if it were an enrichment of the original pyrite.

4. This ore is made up of small crushed pyrite remnants in a matrix of secondary chalcocite, and represents almost the final stage of the enrichment process (Fig. 9).

### SUMMARY

A general summary would be somewhat as follows: A complex of tuffs, dikes, flows, etc., not very deeply buried, was intruded by diorite and the whole subjected to regional metamorphism, developing schistose and crush rocks of great variety. At the closing stages of the metamorphic period, intrusions from a deeper-seated, very siliceous magma cut the older rocks. The same paths of weakness were followed by the mineralizing solutions, which were high in silica and replaced the wall rocks extensively. The abundant chlorite and sericite, and absence of the pneumatolytic minerals, such as tourmaline, would class this deposit with the intermediate-temperature type described by Lindgren.

The primary ore minerals observed were pyrite, chalcopyrite, and sphalerite. Secondary enrichment has converted the chalcopyrite and part of the pyrite to chalcocite. The covellite is probably secondary.

## Pyrite Deposits of Leadville, Colo.

BY HOWARD S. LEE,* LEADVILLE, COLO.

(Colorado Meeting, September, 1918)

IN central Colorado is a great belt of intrusive porphyry nearly 100 miles long (160 km.), extending from the Clear Creek district on the north to Aspen on the south, which includes many of the well known mining camps of the state. Near the southern end of this belt, at Leadville, the occurrence of these intrusives is most marked, resulting in intensive mineralization and the formation of a great variety of large orebodies. During the past 40 years, Leadville has been a steady producer of gold, silver, lead, zinc, manganese, iron, and bismuth, in the forms of sulfide, carbonate and oxide ores. It is the purpose of this paper to discuss pyrite alone, which, under present economic conditions, is valuable chiefly for its sulfur contents.

The ore-bearing formation of the Leadville district consists of two beds of limestone separated by a layer of quartzite. The upper limestone is carbonaceous, and is known locally as the blue limestone. It varies in thickness from 50 to 300 ft. (15 to 91 m.). The lower limestone is Silurian, known locally as white limestone, and is about the same thickness as the upper bed. Underneath the Silurian limestone is a layer of Cambrian quartzite from 100 to 250 ft. (30 to 76 m.) thick resting on Archean granite, which forms the limit of ore deposition. As a rule, there is a sheet of white porphyry above the upper, or blue limestone; a sheet of gray porphyry is associated with the blue limestone, either in or below it, but the porphyries are not so regular as the quartzites and limestones and are not always found in their expected positions.

Throughout the district there is an elaborate system of faults and intrusive porphyry dikes, which have an important bearing on ore deposits. The writer has seen two or three orebodies which were not closely associated with either a fault or a porphyry dike, but if the minor fractures and water courses are considered as part of the fault system it is safe to say that all orebodies are directly traceable to one or both of these causes.

The distribution of pyrite in the Leadville district is quite general; but, eliminating the smaller bodies and those containing prohibitive quantities of zinc or lead, and considering only the larger deposits commercially valuable for their sulfur, we find the chief pyritic orebodies extending in a zone approximately 6000 ft. (1828 m.) long and several hundred feet wide, from the Yak and Moyer, on the south; through the Louisville, Greenback, Mahala, Adams, and Wolftone, on Carbonate Hill; across Yankee Hill, and into the Quadrilateral, Denver City, and Tip Top, on Fryer Hill.

* Manager, Leadville Unit, U. S. Smelting, Refining and Mining Exploration Co.

The orebodies are replacement deposits in the limestone along some minor fracture, or contact metamorphic deposits along some porphyry dike or fault. Replacement deposits closely related to intrusive gray porphyry dikes are the most important and most prevalent. This is due to chemical conditions, such as the presence of carbon in the blue limestone, acting as a precipitant; and to structural conditions, such as the action of the porphyry dikes as dams, interrupting the flow of ore-bearing solutions, the general formation of depressions near faults, forming basins favorable to orebodies; and other minor conditions, all of which are discussed in detail by Emmons.[1]

One feature of interest to a geologist is the universal presence of a shell of manganiferous siderite enclosing iron orebodies, which is particularly noticeable in the carbonaceous limestone. These siderite deposits are fully discussed by Philip Argall,[2] and the fact is noted that they are always closely associated with iron sulfide orebodies. The practical importance of these siderite casings is that they very often indicate the presence of ore, when approached from the outside, and mark the limits of the ore, when approached from the inside.

The pyrite, and the iron and manganese forming the siderite, undoubtedly came from the porphyries in the form of acid sulfates. As these solutions entered the limestone they became carbonated, and the well known self-destructive action of the limestone continued. During the disintegration of the limestone, the self-liberation of large volumes of carbonic acid also liberated a constantly renewed supply of carbonaceous matter which, in turn, removed free oxygen and reduced the sulfates to sulfides. This action would continue as long as any sulfur was present, after which any remaining iron and manganese would unite with the carbonic acid, forming siderite and causing the destructive action of the limestone to cease. The fact that the pyrite is surrounded by siderite would indicate that the pyrite had been deposited first, while the destructive action in the limestone was continuing away from the center of the deposit; and that the formation of siderite, and disintegration of the limestone, ceased only when all sulfur was consumed and the carbonic acid was neutralized by the remaining iron and manganese.

After the pyrite was deposited, there were at least two periods of further action in which other minerals were introduced. Two periods are apparent at the Yak mine, where the pyrite was evidently deposited along a fracture system having a general northeast direction. At a later period, there was movement resulting in fracturing along a N. 15° W. course, and it is along these fractures that secondary zinc enrichment is en-

---

[1] S. F. Emmons: Geology and Mining Industry of Leadville, Colorado. U. S. Geological Survey, *Monograph* No. 12 (1886).

[2] Siderite and Sulfides in Leadville Ore Deposits. *Mining and Scientific Press* (July 11 and 25, 1914) **109**, 50, 128, 148.

countered.   Still later there was movement along the original northeast system where secondary silver enrichment took place.   At the Denver City mine, only secondary enrichment of silver has occurred, and the ore contains no zinc or lead.   The enrichment here is quite similar to conditions at the Louisville mine, in that lines of weakness developed along small dikes of porphyry.   The mechanical action in both cases would be the same, the only difference being that the precipitating action of the porphyry and pyrite together would be greater, and higher values would be found along the small dikes.

The ore is a compact, fine-grained, fairly hard pyrite.   Along enrichment zones, the ore is fractured and granular, while the barren pyrite is much harder.   This would naturally form ribs of harder material between ore shoots, which in many cases have been left in lieu of timbers.

## Method of Mining

In the past, the method of mining has been to drive exploratory drifts to locate the zones of enrichment, and the walls.   As soon as this preliminary information was obtained, stoping was started, using the ordinary square-set method with selective mining.   If it was found too expensive to hold the more or less barren pyrite in place, it was taken out and sold to the smelters for what little gold or silver it might contain as well as for the iron.   In two or three cases, notably at the Yak and Leadville Unit, it was found more profitable to mine the entire orebody, since the zones of enrichment were small but frequent, and the mass was quite heavy.   In this way the unit costs were reduced, the tonnage increased, and the total profit returned was greater than if the richer parts had been gouged out and the larger part allowed to remain in place.

## Value in the Manufacture of Sulfuric Acid

Because of the small amount of sulfuric acid manufactured and consumed in Colorado, and the high freight rates to the eastern part of the country, which precludes competition with eastern and Spanish pyrites, no great amount of Leadville pyrite has been used in the manufacture of acid, although small shipments have been made for a number of years to the Western Chemical Manufacturing Co. of Denver, and to E. I. Du Pont de Nemours & Co., at Louviers, Colo.   Occasional trial lots have been shipped east, which have roasted satisfactorily.   Unfortunately, it is not the policy of the Western Chemical Manufacturing Co. to give out information concerning its practice, so that no comparison of methods and results with Leadville pyrite and with Spanish pyrite can be made through this company, but the fact remains that it has used and is still using a good deal of Leadville pyrite satisfactorily.

The following information was given by D. S. Robinson, Superintendent of the Du Pont plant at Louviers, Colo.:

Our experience with eastern pyrite has been somewhat limited, but, as a general statement, we would not hesitate to say that the Leadville pyrite roasts fully as well as eastern pyrite, both as to being free running and in ease of obtaining almost a dead roast. Our ore is crushed to a maximum size of ¼ in. diameter. The small amount of sulfur left in the cinder is contained in those lumps which do not disintegrate to any extent during roasting. The burners are the five-hearth Wedge type with central shaft and stirring arms or rakes. The pyrite maintains its own combustion.

We have burned pyrite from a number of mines in the Leadville district, and in one case only was the arsenic high enough to interfere with our process for the manufacture of sulfuric acid. Any arsenic in excess of 0.2 per cent. makes the ore unfit for the manufacture of sulfuric acid as carried on at our Louviers plant.

All pyrite shipments from Leadville in the past have been run-of-mine and no attempt has been made to deliver a sized product. The ore was crushed to suitable size for sampling and subsequent roasting in "fines" burners. For this reason, no reliable information is available concerning the roasting qualities of lump pyrite. The writer made certain tests in an assay furnace with lumps as large as 6 in. in diameter, with which more or less violent decrepitation occurred, particularly in the larger lumps. This may have been due to excessive heat and to the fact that the heat was not applied gradually. It is also probable that an excessive amount of fines would be made during the handling of lump ore. At any rate, larger experiments under proper conditions are necessary before anything definite can be said regarding the value of Leadville pyrite as lump ore.

During 1917, about 2000 short tons of run-of-mine pyrite were shipped to the Western Chemical Mfg. Co. from the Tip Top mine, the average sulfur content of which was 48.31 per cent. Since the ore was sold for sulfur only, no further analyses of the control samples were made, but a number of mine samples were analyzed, showing less than 1.0 per cent. zinc, trace of lead, less than 0.1 per cent. arsenic, and 1.5 per cent. insoluble. During 1918, 5000 short tons of pyrite were shipped to Denver and to smelters from the Denver City and Quadrilateral claims. The average contents of this ore was:

| | Per Cent. | | Per Cent. |
|---|---|---|---|
| Sulfur | 48.0 | Arsenic | Trace |
| Iron | 45.0 | Lead | Trace |
| Manganese | 2.0 | Lime | Trace |
| Insoluble | 2.2 | Moisture | 1.24 |
| Zinc | 0.6 | | |

Two lots, of approximately 100 short tons each, contained 50 per cent. sulfur, and one lot contained 51.2 per cent. sulfur.

At the Greenback, Louisville, and Yak, several carloads of pyrite have been shipped to various acid plants, containing from 42 to 49 per cent. sulfur. The ore in the southern part of the district contains more zinc and insoluble than that in the northern part, but not enough to interfere with its use in the manufacture of sulfuric acid, provided care is exercised in mining to eliminate portions containing zinc.

Following are freight rates from Leadville to some of the eastern and Middle West cities where pyrite is used:

*Valuation of $5 and under*

                                                      PER TON
                                                      (2000 LB.)

Leadville to Colorado common points........................... $1.50

(Colorado common points are Denver, Colorado Springs, and
Pueblo.)

*Valuation of $20 and under*

Colorado common points to Chicago........................... $4.45
To Mississippi River common points........................... 3.65
To Memphis............................................... 4.45
To Nashville.............................................. 6.10
To Pensacola.............................................. 4.00

It has been a general practice with local acid manufacturers to sell
calcined residues to Colorado smelters. This could be done profitably
because the smelters have always had an excess of siliceous ores and iron
has commanded a premium.[3] Unfortunately, there is no place in the
Middle West where residues can be sold, and if there is any metal value
in the residue it must be returned to Colorado smelters for treatment.
This is now being done with certain residues from Kansas and Oklahoma zinc smelters. An average pyrite containing 40 per cent. iron in
excess of silica, and 45 per cent. sulfur, will roast with a loss of 30 per
cent. in weight and an increase in the iron to 56 per cent. The proportion of silver and gold in the residue will also increase in the same ratio.
The iron credit will pay the treatment charges, so that the net cost of
treatment is the return freight to Denver, Salida, or Pueblo. Since all
Leadville sulfides contain some gold and silver, the sale of residues is to
be considered as entirely practical, where the original ore contains 4 oz.
or more silver, or its equivalent in gold.

## SUMMARY

There are large deposits of pyrite in Leadville, containing a minimum
of 43 per cent. sulfur, suitable for the manufacture of sulfuric acid.
In these deposits, there are enrichment zones containing gold, silver, and
zinc, but the system of mining in use will permit selective methods
whereby the portions more valuable for other metals may be stoped
separately and shipped to the smelters while the pyrite is kept clean
for acid use.

The only material available now is run-of-mine ore suitable for crushing and roasting in "fines" burners. The problem of lump ore, both as
to suitability and preparation, will be worked out only when there is
sufficient demand to warrant experiments.

The returning of residues from a pyrite originally containing precious
metals to the value of, say, $2 per ton, after calcination in the Middle
West, for smelting in Colorado, is financially practicable.

---

[3] See Report of Smelter and Ore Sales Investigation Committee—State of Colorado, 1917.

## Molybdenite Operations at Climax, Colorado

BY D. F. HALEY,* B. S., DENVER, COLO.

(Colorado Meeting, September, 1918)

THE molybdenite deposits at Climax, Colo., have recently attracted considerable notice, because of their great size, as compared with other known deposits of the same mineral.

Climax station, on the Colorado Southern narrow-gage railway, was at one time the Fremont Pass station on the Denver & Rio Grande. It is 14½ miles northeast of Leadville, and exactly on the divide between the headwaters of the Arkansas and Blue Rivers. Climax station is 11,300 ft. (3444 m.) above sea level. The working tunnel of the molybdenum mine is about 1 mile from the railroad, and 900 ft. above it, at an elevation of 12,200 feet.

The climatic conditions are severe, and snow can be expected during any month of the year. Winter sets in not later than Nov. 1, the snowfall is very heavy, and the ground remains covered with snow until June 1, and sometimes even later. From June 1 to Oct. 1, there is much rain and a little snow falls in the late afternoon nearly every day. In October, 1917, there was one of the biggest snowstorms and blizzards of the year. Being just on the divide, constant strong winds blow the fine, dry snow with such force that, during most of the winter, outside work of any kind is accomplished under most trying conditions.

Within a few miles of Climax, some of the richest gold placers of Colorado occur. One of them, McNulty Gulch, was just across on the northeastern flank of Bartlett mountain. This mountain, therefore, was well prospected, and the presence of a bluish mineral and a yellowish one resembling sulfur was known to the old prospectors. They thought for many years that the blue mineral was graphite, and it is difficult to say when it first became definitely known that it was molybdenite. Claims had been staked as early as 1896, but it is evident that the prospectors at that time were looking for gold. In 1902, several claims were acquired by H. Leal, who drove a tunnel about 900 ft. (274 m.) long, probably to intersect a fault, which shows on the surface, and was thought to be a possible gold carrier. Gold was not found in paying quantities, but throughout its length the tunnel passed through a fractured silicified granite, showing a fairly uniform distribution of a fine-grained bluish mineral, which was finally determined as molybdenite. For several

---

* Assistant Manager, American Metal Co., Ltd.

years no work was done on the property, although it had been examined by several engineers.

In 1916, the property was brought to the attention of the Denver office of The American Metal Co., Ltd., whose engineers made a thorough examination. They recognized the extent of the deposit and its tonnage possibilities, but had great difficulty in getting concordant results from the different assayers to whom their samples were submitted, the values varying between 0.4 and 1.2 per cent. Pop shots were then put into the sides of the tunnel for its entire length, and the ore was broken down on canvas. In addition, three short crosscuts were run east and west from the tunnel, and car samples were taken from the broken muck. The ore thus obtained was cut down to a ton sample and sent to the General Engineering Co. at Salt Lake. After the usual preliminary tests, the whole sample was put through flotation machines and a concentrate was obtained, assays of which showed concordant results, and indicated that a recovery of 12 lb. of $MoS_2$ per ton of ore could reasonably be expected.

### Geology of the Deposit

The orebodies thus far exposed are found on the southwestern slope of Bartlett mountain, which forms the northern spur of a horseshoe-shaped glacial cirque, the southern spur of the horseshoe being Mt. Ceresco. Both of these mountains consist of alaskite, or highly siliceous granite. The high ridge forming the cirque is a coarsely crystalline biotite-gneiss. The alaskite is unquestionably an intrusion into the gneiss, and the latter at one time probably covered the whole area, but has been eroded from Bartlett and Ceresco mountains. In the alaskite, many inclusions of the gneiss are found; one inclusion several feet in diameter has been found in the tunnel workings 350 ft. (106 m.) below the surface. On the western flank of Bartlett mountain, thick limestone beds are found, tilted at such an angle as to indicate that they at one time covered the area but have been removed by erosion. The gneiss may be an altered sedimentary rock.

Glacial action has been severe, and the divide is covered by a heavy moraine. Great boulders of alaskite containing molybdenite are found in the moraine, for at least 2 miles from Bartlett mountain. At the base of the mountain, several hundred thousand tons of talus have collected, which carries commercial quantities of $MoS_2$, and will be milled.

The ore occurs as an intricate system of fine stringers quite evenly distributed throughout the alaskite. The original granite has been much altered and silicified, and appears now to consist only of a quartz groundmass in which some phenocrysts of feldspar still remain. Throughout the main ore-bearing rock are found masses of a typical quartz-porphyry; it is not yet determined whether this is a segregation or a later intrusive. It is significant that very little ore mineral is found in the quartz-porphyry. The greater portion of the ore is found in tiny veinlets, crisscrossing the alaskite, and showing out clearly from the light groundmass. At the center of each veinlet very fine grains of molybdenite are visible to the eye, but toward the edges the grains become so fine as to appear simply as a dark coloration.

The orebody has not yet been thoroughly studied with reference to its genesis. It is believed, however, that when the igneous granite mass was intruded into the overlying gneiss and sediments, an outer shell was formed, due to rapid cooling of the intruding magma. Shrinkage cracks and fractures were formed in the shell by this quick cooling, and the cracks were then filled with silica and molybdenite expelled as a final phase of the cooling magma. The whole granitic mass carries considerable iron pyrite as crystals and films, part of which was probably an original constituent of the granite. Many large fractures occur, running in all directions, with no apparent relation to one another. Along them a considerable amount of oxidation has taken place, the iron and molybdenite being wholly changed to oxides; the oxidation is still prominent down to 350 ft. (106 m.) below surface. A foot or so away from a fracture, however, oxidation has ceased. The molybdenite is fine grained, very little flaky material being found, which is unusual for this mineral.

The only associated minerals are iron pyrite and a very small amount of copper. Occasionally a small pocket of hübnerite is found, but not in appreciable amounts. The ore going to the mill, to date, has been taken from development in all parts of the workings, and has averaged 0.92 per cent. $MoS_2$, and about 2.7 per cent. iron, with a trace of copper.

## Mining System.

Two systems of mining are being used. The thin outer edge of the orebody, bounded by the horizontal plane of the tunnel and the steep sloping surface, will be removed by open-pit glory-hole methods during the summer months. For this purpose, drifts have been run about 40 ft. (12 m.) below the surface and parallel to the contours of the hill. From the drifts, raises will be driven to the surface, and the ground will be broken down through them, then drawn into cars, and trammed out to the bins.

Further in, the ground has been laid out in a series of parallel drifts running diagonally across five parallel patented claims—claims are only 150 ft. (45 m.) wide; the drifts are 50 ft. (15 m.) apart and 860 ft. (262 m.) long. From them, stope raises are put up every 30 ft. (9 m.). At a height of 20 ft. (6 m.) above the back of the level, they are widened out to a cone-shaped pocket, at the bottom of which a chute is placed. This widening connects all of the pockets, and, when completed, a stope 400 ft. (121 m.) long and 25 ft. (7.6 m.) wide is ready for shrinkage operations.

While the stope drift is about 860 ft. long, the stope itself is carried up as two separate stopes, each 400 ft. long and 25 ft. wide. In the center, between the stopes, a pillar 40 ft. long is left standing. A two-compartment raise is carried up ahead of the stope through the center of this pillar, and about every 16 to 20 ft., small drift holes are extended from the raise to the stopes on each side. All steel, air lines, etc., are carried up this raise. At the end of each stope, a similar pillar is left, and a one-compartment raise is carried up through it, this also being connected with the stopes by small drifts. By these three raises, the stope becomes readily accessible and good ventilation is obtained. The stopes are to be carried up to the surface as shrinkage stopes, and when the surface is reached, the pillars left between the stopes, both end and side pillars, will be broken down by glory-hole methods into the stopes. In this connection, it must be borne in mind that all of the surface rock within the five claims mentioned is commercial ore.

All of the tunnels have been made large, and 30-lb. rails have been laid in anticipation of long-time operations. At present, a jig-back tramway takes the ore from the tunnel to the crusher bins. A lower tunnel of large size is now being driven, from which a rock raise will be brought up to the working level, and thereafter all haulage will be done through the lower level. A two-compartment raise for men and supplies will also be driven a short distance from the rock raise, which will obviate the present steep climb from quarters to workings.

### Milling Plant and Practice

The ore from the mine is dumped into a 300-ton bin, from which it is drawn into the 1-ton buckets of a 700-ft. Leschen jig-back tram. This delivers it to a pair of bins of 400-ton capacity, so designed that both bins feed into the 10 by 20-in. Blake crusher. From the crusher, the ore drops onto a 16-in. (40.6-cm.) conveyor belt, and is elevated into a 400-ton tramway bin. A 4700-ft. (1432.5-m.) Leschen aerial tramway carries the ore from this bin to the mill bin, of 500-ton capacity. From any part of the mill bin the ore can be fed by plunger feeders onto a 16-in. flat conveyor belt, which delivers to a 6 by 6-ft. Allis-Chalmers ball-mill. The ore going to the ball-mill ranges up to 3 in. (76 mm.) in size. From the ball mill, the ore (now crushed to about 20-mesh) goes to a standard duplex Dorr classifier. The overflow goes direct to the flotation machines, while the sands drop to a 6 by 10-ft. Allis-Chalmers ball-mill and return to the same classifier until fine enough for overflow. Crushing is to 120-mesh.

The flotation plant is equipped with five Janney air-mechanical machines, three Callow standard roughing cells, and two Callow half-size cleaner cells. In operation, the feed is proportioned between the Janney and the Callow rougher cells, one machine appearing about as efficient as the other for roughing purposes. The first four Janney cells deliver rough concentrates to the fifth cell, for partial cleaning; cell No. 4 discharges tails, while the discharge from No. 5 consists of middlings, which are returned to the circuit. The concentrate from the Janney cleaner (No. 5) goes to the first Callow cleaner; the concentrate from this goes to the second Callow cleaner; and recently, in order further to improve the grade, we have converted one of the Callow roughers to a third cleaner. The concentrate from the Callow roughers is finished in the same cleaners as the Janney concentrate. The mill is putting through 300 tons daily, and the four Janneys can treat about that tonnage; the Callows easily treat 100 tons per rougher per day.

The flotation operation is very delicate, as there is nearly three times as much pyrite in the ore as molybdenite, and with a concentration ratio of 115 or 120 to 1, a very large portion of the iron must be dropped, in order to get a concentrate carrying from 65 to 70 per cent. $MoS_2$. Another difficulty is that when treating, say, 300 tons of ore per day, carrying 0.92 per cent. $MoS_2$, the resulting concentrate, on present recovery, amounts to only 5100 lb. of 65-per cent. product, or only about 4 lb. per minute from the cleaner. We have greatly reduced the size of the cleaner, but it is still difficult properly to control the grade of concentrate. We have, however, averaged 65 per cent. $MoS_2$ in our product.

A vanner is being installed, to treat the concentrate for the removal of pyrite. Each unit of iron removed will increase the grade of concentrate 2.2 per cent., the concentrate now carrying about 10 per cent. iron. All assays are on a basis of $MoS_2$, but the feed to the mill carries from 0.25 to 0.3 per cent. of $MoO_3$. In reporting assays, this is figured as $MoS_2$, but the oxide is not amenable to flotation. The mill recovery is now 60 per cent. of the total Mo in the feed, which is equivalent to a sulfide recovery of 82 per cent. A great many tests are being made, with encouraging results, for the recovery of the oxide, molybdite. It is interesting to note that the feed to the mill averages almost exactly the figure originally estimated, the recovery being about $11\frac{1}{2}$ lb. of $MoS_2$ per ton, as against 12-lb. estimated recovery. This recovery will undoubtedly be greatly improved when the process for the recovery of molybdite is installed. ·

### Living Accommodations

On Aug. 1, 1917, Climax was a barren waste, with not even a tent. By Feb. 1, 1918, the mill, boiler house, bins and crusher, a 1-mile tramway and an 800-ft. jig-back, had been completed and were in operation.

Although there was a great amount of work to be done in a short time, as much as possible was done for the health and comfort of the men. There is a well equipped hospital of four beds, in charge of a high-class safety-first man, who devotes all of his time to looking after the health of the men and the sanitary conditions at the camp.  Excellent bunk houses have been constructed, steam-heated, with shower baths and modern plumbing.   There is also a library.   Every effort is made to keep the camp clean and the men healthy and contented, and as a result, there is a good supply of labor.

## MARKETING

Until it was demonstrated that an assured supply of molybdenite concentrate could be depended on from Climax, the supply on this continent was variable and, as a result, it was in demand only for chemical manufacture and other limited uses.   Much private work had been done in steel laboratories, and some of the advantages of adding molybdenum were known, but the scarcity and uncertainty of supply did not encourage any great expenditure by steel makers in thorough testing of ferromolybdenum.   The present output at Climax is about 60 per cent. of the world's output prior to the operations at Climax.   The small amount of work done by the steel makers indicated that molybdenum could be used to advantage in combination with vanadium and tungsten, or as a substitute for these metals, giving the steel the same properties at . a much reduced cost.   At present the market for molybdenum is very limited, and the quoted prices are nominal and do not at all represent the prices at which the material is marketed.

## Air Blasts in Kolar Gold Field, India

BY E. S. MOORE,* M. A., PH. D., STATE COLLEGE, PA.

(Colorado Meeting, September, 1918)

THE Kolar gold field has been for a long time the most important gold-producing area of India. It is situated in the State of Mysore, southern India, and not far from the City of Bangalore. The productive field is about 3 miles long and in it a gold-bearing quartz vein varying in width from 1 in. to over 10 ft. is worked. The vein carries high values to great depths and many of the mines have paid large dividends.

### GEOLOGICAL FORMATIONS IN KOLAR REGION

The rocks of the Kolar region are pre-Cambrian in age and bear a close resemblance to some of the pre-Cambrian formations of America. The oldest rocks seen are basic lavas, now mostly altered to hornblende-schists, with which is associated some banded iron-formation; these rocks are similar to the Keewatin schists and iron-formation in the Lake Superior region. The schist is cut by the gold-bearing quartz vein.

In certain parts of the area there is also a conglomerate containing pebbles of granite, jasper, and schist. The matrix resembles a hornblende-schist and is intruded by small granite dikes, which in some cases have been pinched off by squeezing of the rock. This has led Dr. W. F. Smeeth, Chief Inspector of Mines of Mysore, to whom the writer is greatly indebted for information regarding this area and for his hospitality while visiting the field, to regard this conglomerate as a breccia due to crushing of granite dikes in the schist. Dr. J. W. Evans has regarded the conglomerate as probably a squeezed glacial boulder clay. From analogy with pre-Cambrian conglomerates in America, the writer regards it as a metamorphosed rock similar to our basal Huronian conglomerate. However, the main bearing which it has upon the problem under discussion is the apparent fact that it is younger than the schist and that it indicates the presence of a syncline in the rocks cut by the quartz vein.

Surrounding the area of schist and conglomerate is a large mass of granite-gneiss resembling the Laurentian gneiss of the Lake Superior region. Still later than the rocks mentioned above are basic intrusions which cut the quartz and schists; in the mines these are known as trap. One dike of this type is 50 to 60 yd. wide.

---

* Professor of Geology and Mineralogy.

The quartz vein consists of a dark, translucent, rather opalescent to chalcedonic type of quartz, almost everywhere showing, by its refraction of light and by its fractures, that it has been subject to molecular strain. It generally turns whitish on exposure, owing to the admission of air into the numerous small fractures. Streaks of schist·are common in the vein, and their arrangement suggests that the solutions depositing the quartz were thrust into the cleavage planes of the schists under great pressure, and probably during some shearing in the schists. In some places the vein is highly folded and the schist in the vicinity of the quartz is, as a rule, fine-grained. It likewise shows evidence of a strained condition.

### DESCRIPTION OF AIR BLASTS

The term "air blasts" has been used not only in India but also in America and elsewhere to describe certain disturbances which occur in mines and are accompanied by strong rushes of air through the workings.

Such occurrences have been described from the Lake Superior copper and iron districts; they are caused by the falling of large masses of roof in stopes or by the sudden crumbling of pillars under the superincumbent weight of the rock above the mine workings, thus producing a rapid movement of air in the partly enclosed spaces. There is thus a reasonable justification for the use of the term to designate such phenomena. As the term is now used in India, by some writers, it must be regarded as a misnomer since the "air blasts" are distinguished from the larger disturbances in the mine, which are called "quakes."[1] The name was apparently applied to them originally by the miners, owing to the similarity between the explosions in the rock and those which might be caused by occluded gas. Air blasts and quakes are doubtless very closely related, and are similar to the phenomena described in Australia,[2] Bohemia,[3] and England,[4] as "explosive rock" and "air blasts."

In the Kolar field the air blasts occur in the quartz vein, trap dikes or the hornblende-schist, being most frequent in the quartz. They bear a strong resemblance to the explosion of a small charge of powder placed in the wall of the workings, because in some cases the rock blows out from the solid face of the drift or stope as a puff of rock powder, while in others small fragments are shot out with sufficient force to scratch and severely cut the miners. Often a continuous crackling and snapping

---

[1] W. F. Smeeth: Air Blasts and Quakes on the Kolar Gold Field. Mysore Geological Department, *Bulletin* No. 2 (1904). ·

[2] J. B. Jaquet: Explosive Rock. *Engineering & Mining Journal* (Mar. 30, 1905) **79**, 605–6.

[3] Hugo Stefan: The Cause of Air Blasts in the Mines at Pribram, Bohemia. *Mining & Scientific Press* (Dec. 29, 1906) **93**, 789–90.

[4] Aubrey Strachan: *Geological Magazine* (1887) Decade 3, **4**, 400–408.

is kept up at the fresh working face, interspersed with explosions like those of small detonators. The explosions do not bear any definite relation to the depth of the mine, after a few hundred feet in depth has been reached, and they are therefore independent of the superincumbent weight of the rock in the mine workings. They have occurred during the sinking of the large circular shaft on the Champion Reef property, in very compact hornblende rock, and less than 700 ft. below the surface. In some of these explosions large masses of rock were blown out with loud reports, and one was mentioned in which it was estimated that 80 tons of rock were precipitated from the wall, making it necessary to employ a shield to protect the workmen while sinking the shaft.

The term "quakes" has been employed by Dr. Smeeth in his excellent work on these phenomena to describe the heavier shocks which occur in the Kolar mines and produce effects at the surface in all respects like those of local earthquakes. In some cases these shocks have been sufficiently severe to be felt at a distance of nearly 4 miles from their point of origin. They have been quite destructive to mine structures and in some cases to the lives and limbs of the miners. One case is described by Smeeth in which a block of rock, the weight of which was estimated at $\frac{1}{2}$ ton, was hurled with a low trajectory from end to end of a stope 30 ft. in length. In some cases the footwall of the vein, which usually lies on a slope of 50° to 60°, buckles up and large masses peel off with considerable violence.

As a rule these larger shocks, or "quakes," occur in the deeper levels and in areas where considerable stoping has been done. They are particularly prevalent where quartz pillars have been left to support the roof of the workings and they seem to owe their origin partly to the superincumbent load of rock in the workings, but their violence can only be ascribed, like that of the air blasts, to some latent energy or strain in the rocks of this area.

A marked similarity is seen between these air blasts in India and the explosive rocks of Australia and Bohemia. Jaquet describes explosions in a block of slate called the "kicking-ground" in the Hillgrove mines, New South Wales.[5] This explosive rock is a more or less silicified and altered slate traversed by numerous joints, which are coated with thin films of calcite. The rock is liable to split off at any time, particularly just after blasting, and the more serious explosions occur when a chain pillar of rock is being removed between a stope and the level above. In the same article Jaquet quotes J. R. Godfrey (Inspector of Mines), who describes an explosion in which a fragment flew from the face of a stope where two men were drilling and blinded one of them; while in another explosion the whole floor over a section of a stope split up into

---

[5] J. B. Jaquet: *Op. cit.*

thousands of fragments with the sound of breaking crockery. In still another case a fragment of rock flew from the face of a stope and cut a man in two. It is stated, further, that the explosions do not occur in the Hillgrove mines in shafts or crosscuts off the lines of the reef.

Another case of explosive rock has been reported from the Genowlan and Hartley mines, in the kerosene-shale area of New South Wales, by J. E. Carne, who states that the fragments split and fly when the shales are being worked, so that the miners wear wire-gauze guards for protection of their eyes.[6]

An example of explosive coal is described by Fry[7] who mentions a colliery in the south coast district of New South Wales in which small accidents have been caused to the miners by fragments of coal flying from the face with great violence. He ascribes the phenomenon to the presence of large quantities of gas under high pressure, since methane is given off in these mines in copious amounts; after a fresh face stands for a time it becomes "winded" and much more difficult to break down.

Air blasts have also been described by Stefan[8] from the silver-lead mines of Pribram, Bohemia. In these mines, according to this writer, the rocks consist of sandstones and hard siliceous graywackes alternating with coarse- and fine-grained clays in dense schistose layers. The veins consist of calcite and siliceous vein material and the outbursts occur mostly in them and in the silicified graywackes, but at times in the sandstone also. The veins are said to occur in a syncline of Cambrian sandstone which is cut by dikes of diabase and greenstone; the veins are frequently closely associated with these dikes, being found both in them and in the sandstone, in some cases. The syncline is very steep on one side and flat on the other and it is cut by a fault fissure. A number of barren veins of calcite run out from this fault fissure and many barren calcite veins show slickensides. In these mines fragments of rock are often thrown off from the face of the drift, causing painful injuries to the miners. At one time a mass of rock weighing about 200 lb. burst from the sandstone footwall in the stope and broke into a number of pieces. No air blasts occur in these mines above 1000 m., and they all occur in the flat limb of the syncline; these are rather significant facts in considering their origin.

Still another case of explosive rock has been described by Strachan from the Derbyshire lead field of England.[9] In this area it is said that slickensided rocks sometimes burst when struck with the hammer.

In the quarries of the United States some cases are known where the

[6] Quoted by Jaquet: *Op. cit.*, 605.

[7] Sidney Fry: Letter regarding explosive rock. *Engineering & Mining Journal* (Aug. 19, 1905) **80**, 310.

[8] *Op. cit.*

[9] Aubrey Strachan: *Op. cit.*

rocks burst during quarrying operations but probably none have been encountered which equal the Kolar rocks in their explosive violence.

## ORIGIN OF BLASTS

Various explanations have been offered to account for these explosions. For the Hillgrove mines Jaquet has mentioned the following hypotheses:

1. Molecular strain.
2. Occluded gases.
3. Compression due to intrusion of granite, or other causes.

He concludes that the rocks must be under great strain, and, being unable to bend, they must break suddenly under certain conditions. He apparently does not favor the granite intrusion hypothesis and dismisses entirely the idea that occluded gases might cause the explosions. Strachan considers the rocks in the Derbyshire lead mines as under molecular strain due to the pressure exerted when · the slickensides were produced.

Regarding the air blasts in the Pribram mines in Bohemia, Stefan states that "the conclusion seems justified that the cause of these explosions is not only the pressure of the superincumbent rock-mass, which is at right angles to the strata, but is due also to a stress parallel to the bedding planes and to the axis of the syncline."

For the Kolar field, Bosworth Smith[10] considers that the air blasts are due to a molecular strain in all three of the rocks, trap, schist, and quartz, and he likens the explosion which occurs during mining operations to the breaking of a Prince Rupert's glass drop, which is a mass of glass under severe strain caused by sudden cooling. As to the cause of the strain, he considers that the quartz is compressed by the walls of the vein through regional pressure, and not simply by the weight of the overlying rocks while the trap is under strain because of sudden cooling and the hornblende-schist because of metamorphic changes which have given rise to the calcite, forming stringers through the rock.

Smeeth agrees with Smith regarding the origin of the strain in the trap dikes since, as he states, there is no evidence that the dikes have suffered extensive compression since their solidification. As to the possibility that metamorphic changes in the schist produced the strain, Smeeth points out that the chief change has been the alteration of augite to hornblende, and since this change is facilitated by pressure there is reason to suppose that the compressional strain would be relieved rather than increased. Further, the compressional strains in the schists produced by the intrusion of the surrounding granite and the injection of the

---

[10] Quoted by W. F. Smeeth: *Op. cit.*, 16.

dolerite dikes would be relieved by the later cooling and shrinkage of these rocks, and the strain in the schists would be tensional. Regarding the strain in the quartz, Smeeth is of the opinion that it is also tensional, partly because of the fact that there are secondary stringers of calcite, quartz, and metallic minerals in the reef suggesting that the rocks have suffered sufficient tensional strain to open joints and permit the filling of these with mineral matter.

## CONCLUSIONS REGARDING BLASTS IN KOLAR FIELD

After observing the rocks in the Kolar field and considering the descriptions of the air-blast phenomena in the other regions which have been mentioned, the writer has come to the following conclusions regarding their origin.

They are due to a stress on the rocks, which has produced a strain, and under mining operations this strain results in a violent rupture. In the case of the larger shocks, which Smeeth calls "quakes," the violence with which pillars and other supporting masses give way in the workings under the superincumbent load is due to a large extent to this internal strain.

As to the cause of the stress, the only satisfactory explanation is the application of pressure resulting from crustal movements. At Pribram the blasts occur in the flat-lying strata of the syncline and not in the upturned beds in which the strain has been relieved by faulting and tilting. In the Derbyshire region the rocks are slickensided by movement and in the Hillgrove mines, while the geological description by Jaquet is not very detailed, it is evident that the slates have been intruded by masses of granite. In the Kolar region, the older rocks, in which the quartz vein lies, are surrounded by granite and nipped in by it in a closely compressed syncline. There thus seem to be in all these fields certain similar conditions pointing to the fact that the rocks in which the blasts occur have been subject to great compressional forces. As to the ability of the cooling dolerite dikes to produce sufficient tensional strain to cause the explosions, this seems very doubtful. The spheroidal weathering and fracturing described by Smeeth as probably supporting this view is a very characteristic feature of such basic rocks in many regions and does not seem to warrant much consideration. The greater facility with which the fine-grained edges of the dikes will explode is what would be expected under normal conditions, since the rock would be more brittle in those portions of the dikes, just as the silicified slates and schists in the other regions described would be more brittle and would crack up more readily than the other portions of the same rocks. Nor does the tensional hypothesis adequately explain how the heating of the schists and quartz by the intrusion of the later rocks could produce tension in these rocks unless it can be shown that a complete change in the molecular character of the rocks was produced by this heating; such evidence appears to be lacking.  If the heating converted augite into hornblende, the resultant increase in volume of over 4 per cent.[11] would cause expansion rather than tension; whereas, if rocks simply expanded by heating. they would contract again to their normal condition on cooling.

It also seems probable that the crystallizing of the quartz vein may have exerted some expansive pressure on its walls.  As to the occurrence of the small secondary quartz and calcite veins, they may be accounted for by torsional movements in the rocks which are still, on the whole, under great compressive stress.

The rocks in the mines are not uniformly explosive and such an explanation would account for this condition.  In some places the rocks have opened through torsional movements and in others they have yielded to compressional forces, which have produced the movements indicated by the slickensides, while in still others they have not been compressed sufficiently to cause them either to shear or to rupture, and in these spots the potential energy gives rise to the blasts.  Such a condition is particularly likely to arise in a region where heavy dikes intrude rocks of varying compressive strength and brittleness.

The force which compresses these rocks may be due to two or more causes:

1. Epirogenic movements in the earth's crust, due to adjustments of stresses over considerable areas of the earth's crust with accompanying igneous activity.

2. General settling back of considerable areas of the crust during adjustment after the eruption of large amounts of igneous rock from certain portions of the crust.

Such forces might easily obliterate the results of all contraction and relief of pressure in the rocks of the area by cooling, and to such forces as these the strain producing the air blasts is attributed.

In this field a well-equipped seismological laboratory has been established and complete records of all shocks are kept.  The effects of the heavy shocks are similar to those of local earthquakes.  Some of the officials at the mines have been very anxious to have a number of tests made on the rocks of the region to determine their elasticity, crushing strength, and other properties, and the writer has endeavored to have such tests made, but so far, partly owing to war conditions, he has been unable to accomplish this work.

---

[11] C. R. Van Hise: A Treatise on Metamorphism.  U. S. Geological Survey, *Monograph* 47 (1904) 378.

## SUMMARY

The Kolar gold field, which is situated in the State of Mysore, southern India, is noted for peculiar and violent explosions in the mine workings. These may occur in the quartz vein, or in the hornblende schist, which forms the bulk of the country rock, or in the dolerite dikes, which cut the vein and the schists. They are most frequent and violent in the quartz and they are due to a strain in the rocks which causes violent fracturing when the potential energy is permitted to act, as occurs when mining operations have removed the supporting rock on one or more sides of a mass.

In addition to the minor explosions, which occur at nearly all depths, and in shafts, stopes, and drifts, there are heavy shocks caused by the giving way of pillars or other supporting masses of rock, with great violence. This unusual violence is doubtless due to the fact that the rocks are already subject to internal strain. Several hypotheses have been suggested to account for these explosions, but the only one which, in the opinion of the writer, is adequate is that the rocks of this field have been subjected to a great compressive force by crustal movements associated with igneous activity and that the rocks have never been relieved from the resulting strain except near the surface and in local areas in various parts of the mines.

## DISCUSSION

W. F. SMEETH,* Bangalore, Mysore, India (written discussion†).— Before dealing with the air blasts, I would like to clear up a geological point. The author states that the matrix of the conglomerate resembles a hornblende schist. This is a misapprehension. The conglomerate band, or series, is composed essentially of granitic gneiss with sporadic patches containing pebbles—or lumps—of granitic material and fragments, or included patches, of the overlying hornblende schists and banded iron-ore formation. The conglomeratic gneiss is oblique to the banding of the schists and penetrates them in tongues. Toward the south of the field it cuts across the schist bands for about 2 miles, and similar granitic material is found in the mines, at a depth of 3000 ft. or more, in bands or tongues which give evidence of intrusive action in relation to the hornblende schists. It is not understood in what sense the author regards this material as similar to the basal Huronian conglomerates of America, but I gather that he considers it to be sedimentary in origin and that he differs from my view that it is an intrusive gneiss with autoclastic pebbles and included fragments of the schists. Also, it is not clear in what way this conglomerate, which is admitted

---

*Chief Inspector of Mines, Mysore, India.     †Received Aug. 16, 1918.

to be younger than the schists, can be regarded as basal or how it can indicate the presence of a syncline in those schists. Perhaps the author would kindly further elucidate his views.

The author criticizes the use of the term "air blast" and in this everyone will, I think, agree with him. Many years ago the term was in use in the Cornish mines where rock was encountered which showed a tendency to fly, or split off violently in small fragments; this was referred to by the miners as "airing." Some 15 or 20 years ago similar phenomena became prevalent in the Kolar Field and were referred to in the same terms, but about the same time more than usually violent effects, which involved the displacement of considerable masses of rock, the smashing of large timbers and production of violent shocks, which were distinctly and alarmingly felt at surface, began to attract attention.

A study[1] of a number of cases led me to divide these phenomena into two classes, one of which I called "quakes" and for the other I retained the term "air blast," then in common use.

The distinction which I endeavored to make was that quakes were larger and more serious phenomena, involving considerable destructive effects underground, accompanied by perceptible shocks at surface, and due, essentially, to the sudden fracture of pillars or blocks of quartz left standing in the workings during the course of mining operations. The failure of such pillars was considered to be the result of compressive or shearing stresses due to superincumbent weight, which was a function partly of the depth from surface and partly of the extent to which the surrounding quartz had been stoped out, thereby increasing the weight to be supported by the pillar. The suddenness and violence of shock depends on the physical characters of the quartz and of the overlying and underlying rock (hard crystalline hornblende schist and epidiorite) which do not permit of quiet deformation, or adjustment, with the result that disruption is correspondingly violent when the breaking stress is reached.

Air blasts, on the other hand, were restricted to bursts of rock, whether of quartz, hornblende schist or trap dyke, which occurred in circumstances which seemed to preclude superincumbent pressure as an essential factor, and were considered to be due to intrinsic strain in the particular mass of rock involved. I may add that the term "air blast" has been practically discarded for some years, and these occurrences are now commonly referred to as "rock bursts."

At the time the Bulletin quoted above was written, these rock-bursts were much less serious phenomena than the quakes. Small, and sometimes moderately large, pieces of rock were thrown from the

---

[1] Air Blasts and Quakes in the Kolar Field. By W. F. Smeeth, M. A., D. Sc. Mysore Geological Department, *Bulletin* No. 2 (1904).

working faces with disruptive violence accompanied by considerable detonation, but the effects were very local, timbers were not affected and it was not considered that any perceptible shocks were felt at surface. Since that time these rock bursts have been recognized as being of greater magnitude and accompanied by damage to timbers and' perceptible shocks, so that the visible and sensible distinction between quakes and rock bursts has almost ceased to exist, although I am of opinion that the genetic distinction is still valid. Such distinction can be recognized, however, only when the scene of a particular occurrence is known and can be carefully inspected; and in many cases this is not practicable.

In his paper, Prof. Moore agrees that the rock bursts are due to a condition of strain in the portions of rock in which they occur, and considers that the strain is the result of regional compression which operated in former times, and that the rocks are still, on the whole, under great compressive stress. In the next paragraph (p. 83) he states: "in some places the rocks have yielded to compressional forces, which have produced the movements indicated by the slickensides, while in others they have not been compressed sufficiently to cause them either to shear or to rupture, and in these spots the potential energy gives rise to the blasts."

The latter statement appears to contradict the former, but I presume that what the author means is that in some places the strain has been relieved by fracture and movement while in others the rock is still under great compressional stress. This is quite a reasonable view and I shall refer to it later. With regard to the larger shocks, or quakes, due to the failure of pillars of quartz in stoped areas, the author admits that they seem to owe their origin partly to the superincumbent load but adds that "their violence can only be ascribed, like that of the air blasts, to some latent energy or strain in the rocks of this area." With this view I do not agree. If we have a pillar of quartz under great stress due to superincumbent weight, and if, in addition, some of the adjacent rock is in a condition of intrinsic strain due to other causes, we may obviously get any combination of a quake and a rock burst; the one may set the other off and the resulting effects may be greater than would have been the case with either acting alone. But, in my opinion, the quake, or shock due to failure of a pillar under weight, may, and often does, take place without any of the rock concerned being in a condition of what we may call intrinsic strain, and therefore, without the production of what I have defined as an air blast or rock burst. In so complex and difficult a problem it would be a pity to confuse or obscure any legitimate and reasonable distinction which can be detected among the very varied phenomena which are known to occur, and it appears to me that Prof. Moore's contention, that the violence of the quakes is really due to some condition of strain resulting from original compression, tends to such

confusion and merely obscures the issues.  In my opinion, as already
stated, the quakes are due simply to superincumbent weight and their
violence is a function of the natural physical constants of the rock
involved.

As an illustration of this we may take the case of the Rand mines of
the Transvaal where, in recent years and with increasing depth, shocks
due to failure of pillars have become troublesome and have been reported
upon by a special Commission.  In those mines, rock bursts in shafts,
drives or crosscuts, far removed from stoped areas, are practically unknown, and the strained rocks, with which we have to deal at Kolar,
do not appear to exist.  This is useful evidence that violent quakes may
occur as the result of pressure due to weight alone, quite independently
of strain due to other causes, and the evidence is clearer than that obtainable at Kolar where both sets of conditions are frequently associated.
Such association is, however, no justification for allowing our ideas to
become confused.

My object in again drawing attention to the genetic difference between
rock bursts and quakes is not for the purpose of offering criticism, or for
the sake of establishing a mere academic distinction, but rather with a
view to a practical working policy in dealing with these dangerous phenomena.  When I started my investigation, some 17 or 18 years ago,
rock bursts were not very serious features and little was done to guard
against them.  Quakes were comparatively rare and were regarded as
bigger air blasts or as falls of ground.  Comparatively little timber was
put in, except in shafts and to support heavy ground, and systematic
filling of stopes was practically unknown.  The shafts were nearly all
inclined and carried down along the course of the quartz veins or shoots,
and pillars of quartz or poor rock were numerous throughout the stopes.
The recognition that pillars were a source of danger, which was bound
to increase with the extension of stoping and with increase of depth,
led to a general change of policy.  Inclined shafts have, in recent years,
been carried in the solid hanging or footwall rock.  In the case of several
of the old shafts, which had been smashed up, time after time, owing
to the failure of some of the supporting pillars, the bold step was taken
of entirely removing these pillars and replacing them by timber and packs
of waste rock, with the result that most of the trouble has ceased.

Stoping is planned, as far as possible, to avoid the leaving of small pillars by taking out the ore in large blocks as rapidly as possible and leaving
large blocks of untouched ground between them.  In the leaving of
these large blocks advantage is taken of poor zones between the payable
shoots, but where these are not available large blocks of payable ore may
be left for future exploitation.

Systematic filling with waste rock, some of which is sent down from
surface, has been extensively developed and, in the steeper mines,

methods have been devised for securing considerable consolidation of the packing. At first, it was customary to carry a stope down for a depth of 30 ft. (9 m.) and then put in a stull and fill the excavation with waste rock. Another 30 ft. was then stoped below the first stull and a second stull, with packing on top, was put in. The final 30 ft. was then taken out and similarly packed. The result of this method was that the three separate sets of packing were poorly consolidated and afforded little support for the hanging until the latter had closed down considerably and caused dangerous strain in adjacent pillars or in the more massive bands of rock in the hanging itself. These strains were relieved by sudden fractures accompanied by severe shocks, the smashing of heavy timbers and the filling up of levels and stopes with broken timber and rock. In other words, although this filling was doubtless of some service, and was preferable to the practical absence of filling previously in vogue, it failed to prevent a marked increase in the number and intensity of the quakes with increase of depth and extension of stoped areas.

In this connection it may be as well to note that failure of pillars under excessive weight is not the only source of quakes, and it is remarkable that quakes have been much less frequent in the Mysore mine, where the stopes dip at 45° to 50°, than in the Champion Reef and Ooregum mines where the dip is from 60° to 80°. In the latter mines the stress on pillars, due to weight, will be very oblique to the axis of the pillars, and failure will be due, largely, to shearing forces. Where the dip is very steep, much of the weight will go to produce a longitudinal thrust, in the direction of the dip, on the beds or bands of rock in the hanging, and as these vary in texture and schistosity we may expect that some of the more massive bands will take upon themselves the greater part of the burden; and if allowed to bend, to any great extent, they are liable to snap suddenly with the result that a disastrous quake is produced. The importance of supporting the hanging as early as possible is obvious, and the following method has been adopted, for some years, with a view to getting in the filling before any wide extent of hanging is exposed, and of obtaining as much consolidation as possible.

A heavy packed stull is put in at the back of the lower level. From the level above, a stope, 30 ft. (9 m.) in length, is carried down to the bottom stull as rapidly as possible, and stulls, covered with 4 by 4-in lagging, are put in at depths of 30 and 60 ft. to protect the workers from loose fragments or rock bursts during this operation. The ends are also boxed in. When the stope is finished, the laggings are removed, and sometimes also the intermediate stull pieces, and waste rock is shot down the whole depth of the stope from the level above. The next stope is similarly treated, and so on until the entire block has been removed. The method has proved satisfactory in operation and there is little doubt that it has saved many quakes though it has by no means entirely obviated them. It is impossible to get any filling which will take up its work at once without material shrinkage, whether the filling be waste rock or the water-borne sand which is now so largely used in the Transvaal. The hanging will therefore settle down to an appreciable extent, and in the case of hard rocks with low elastic limits the pressure will be unequally and dangerously distributed on certain pillars, or the more solid bands in the hanging will reach their limits of tensile or shearing stress and yield with sudden violence.

Other methods of stoping and filling are at present under trial, which I will not refer to here.

I may now turn for a moment to rock bursts and endeavor to point out how they differ from quakes, and to supplement the illustrations given by Prof. Moore. There is no doubt that they occur in stopes, but in such places the evidence as to their nature and origin is complicated and obscured by the occurrence of quakes, falls of ground, and other effects of movement and superincumbent weight. For more distinctive evidence we must go to drifts, crosscuts, and shafts, which are being excavated in solid rock far removed from stopes, in which the superincumbent weight is merely that due to depth. As the rock bursts have been known to occur at all depths from about 600 ft. (183 m.) below surface downward, it is obvious that the mere weight of the overlying rock can contribute little, if anything, to the causes that produce these violent disruptions.

They occur most frequently in the quartz of the veins—probably due to the fact that a greater footage is driven in quartz than in other rock— but, contrary to what Prof. Moore states, they are not of any considerable magnitude or violence in this material. Occasionally they have been known to occur in the trap (dolerite) dykes which cut across the formation from east to west or north to south, and they occur fairly frequently in the more massive hornblende schists and epidiorites which form the mass of the country.

Sometimes they occur at or close to the working end while excavation is in progress; more frequently, some little time after the excavation has been made, as though the internal strain took some time to reach a critical point, and occasionally the effect is delayed until several years after the excavation (such as a shaft) has been completed and used.

The bursts are quite sporadic. A working may be driven for many feet, and even for thousands of feet, with no sign of trouble when, without warning, a burst or a series of bursts may occur, or a zone or patch of rock may start firing itself off for some considerable time, and the working may have to be suspended until the display is finished. A few specific cases may be quoted in illustration.

In driving the 1940 level, Champion Reef, annoyance was caused by small pieces of quartz bursting from the roof. In one section, where the quartz was 10 ft. (3 m.) wide, the trouble became so bad that work had to be stopped and the quartz fired itself out in a wedge extending upward for 8 or 9 ft. above the roof of the level, and then stopped.   Except for the levels at every 100 ft. (30.4 m.) there was no excavation below this section nor for 1300 ft. (396 m.) above it.   Superincumbent weight appears therefore to be excluded as a cause of these bursts and, moreover, no such bursts were experienced in driving the 1840 level above nor the 2040 level below.   The circumstances appear to exclude existing regional compression, also, as a cause and suggest that the bursts were due to local internal strain in the section of quartz affected.

Instances of bursts in the trap dykes are not numerous.   A couple of years ago I was passing along a level in the Mysore mine which was driven through a zone of poor, unstoped, ground.   The level passed through a transverse dyke of a few yards in width, the surface of which was much cracked, owing, doubtless, to long exposure.   It looked safe enough and I dug out a piece as a specimen.   I had not proceeded more than 10 yd. or so along the level when there was a loud bang and about 100 lb. of the dyke was thrown out across the level.   I was told that similar bursts had occurred in this dyke on previous occasions.

A severe burst occurred in Gifford's shaft, Champion Reef, at a depth of less than 1000 ft. (305 m.) from surface.   It is a vertical, circular shaft, lined with brick, and was being sunk in solid rock some thousands of feet from any mine workings.   Its depth, at the time, was 1090 ft. (332 m.), the brick lining was down to 34 ft. (10.3 m.) from the bottom, and no bursts or other trouble had been experienced.   On the day in question, the exposed surface below the bricking was carefully cleaned down and all loose removed.   The stage was lowered and drilling started. About 20 min. afterward there was a loud report, accompanied by shock, and some 80 tons of rock burst from the west side and fell into the sink. In this case the depth from surface was small and the noise and shock, which are stated to have been heard and felt, are against the explanation that the occurrence was merely the breaking way of a large mass of loose material.

We may next take a case in which there could be no suggestion that the burst was due to a fall of loose.   Carmichael's shaft, Champion Reef, was being sunk, on the underlie, at a depth of 4700 ft. from surface (nearly 4000 ft.—1219 m.—vertical).   The hanging-wall of the shaft was timbered and lagged to within 10 or 12 ft. of the sink.   On the footwall the skip road had been completed to within 25 ft. of the bottom and below that bearers had been put in.   A severe burst then occurred, accompanied by a shock which was felt and registered at surface.   The footwall was broken up for 50 ft. (15 m.) above the bottom, many tons of rock were thrown up and fell to the bottom of the shaft and the skip road and timbers were much damaged. Neither the hanging, nor its timbers, nor the sides of the shaft showed the least evidence of disturbance or injury. The shaft was well within the solid hanging, some 70 ft. west of the lode, and there was no stoping or excavation (except the shaft itself and its crosscuts) within 1000 ft. (305 m.) of the point where the burst occurred. Moreover, no trouble of this kind had been experienced, previously, during the sinking of the shaft. All the circumstances point to intense local strain in some particular mass of rock at the point where the burst occurred.

One more case may be given, in which the burst occurred several years after the shaft had been completed. Garland's shaft, Champion Reef, is an underlie shaft the upper portion of which followed the lode and was therefore surrounded by pillars and stopes and had suffered, from time to time, from quakes. In the lower levels it passed into the solid hanging and at the point where the burst occurred it was 70 ft. (21 m.) west of the lode in solid rock. Little trouble had been experienced in the portion carried with the hanging wall. About noon of the day in question I had occasion to pass up the shaft, from the 35th to the 30th level, in company with Mr. Henry Gifford, the superintendent of the mine, and we were discussing the satisfactory results of keeping these shafts well within solid rock and away from the stopes. The shaft was fully timbered with square sets and complete lagging and the timbers showed no sign of pressure or movement. A few hours later a severe burst occurred between the 32d and 34th levels; I myself felt the shock at a distance of nearly 2 miles from the shaft and the shock was duly recorded on the seismographs at Ooregum. I went down next morning and found that the footwall of the shaft had burst for some 30 or 40 ft. above the 33d level; the 9-in. timbers were smashed and thrown into the shaft along with tons of broken rock. Further, a considerable portion of the hanging below the 33d level was also blown out and the timbers and lagging smashed. Probing with a stick showed that the roof of the shaft had been blown out to a height of some 8 ft. (2.4 m.) above the original timbers. There was a crosscut from the shaft and a level and some stopes 70 ft. (21 m.) east of it, and in these, beyond the shaking down of some pieces of loose, there were no signs of movement or fracture. In this case, also, the bursts appear to have been due to the release of intense local strain within the solid rock without any evidence of pressure or movement in the immediate environment.

The foregoing cases will serve to show the character of these rock bursts, which appear to be more frequent and violent in the Kolar mines than anywhere else. It does not seem possible to suggest any means of preventing their occurrence, but a great deal has been done to protect workmen against the more numerous bursts which are not of exceptional violence. In vertical shafts a movable protective shield is provided below the point to which the brick lining has been put in. In underlie shafts and in many winzes, drives and crosscuts, either permanent or temporary timbering is carried as close to the working faces as possible. When a patch of ground which shows a tendency to burst is being driven through, work is frequently suspended to give the strains a chance to relieve themselves. Some bursts are so unexpected or so violent that no amount of care or protection is of any use and regretably large loss of life, or injury, occasionally occurs. Efficiency of protection must, however, be judged by average results and, considering the large number of quakes and rock bursts which occur, the average results must be regarded as indicating a considerable measure of success which will, I hope, still be increased. For some years past a well-equipped seismological observatory has been installed on the field by the mining companies under the management of Messrs. John Taylor & Sons. This is in charge of H. M. A. Cooke, Superintendent of the Ooregum Company, who has devoted much time and ability to securing records and subjecting them to critical scrutiny. The number of shocks recorded depends on the adjustments of the instruments, and for some years past they have varied from about 13,000 to over 17,000 per annum —the tendency being toward an increase. In the majority of shocks the displacement of the stylus is less than 1 mm., representing a rock movement of about 0.01 mm. In a few cases the movement of the stylus is greater, up to, if I remember correctly, about 7 mm. The period of vibration is too short to permit the recording of any wave motion, and the stylus makes a single tick, the length of which is taken as an indication of the violence of the shock.

Of the shocks recorded, about 2 per cent. are reported as having been felt and of these less than 10 per cent. are usually classed as heavy. In other words some 250 to 300 shocks are felt during a year, of which 15 to 25 are classed as heavy. The exact location of the majority of the shocks felt, and of many of the heavy ones, is never discovered; probably they occur in stoped out and abandoned areas.

As to the causes of the shocks, I believe that they are many and varied and that each requires to be examined and discussed on its merits before any sound conclusion can be drawn.

I have defined quakes as fractures or ruptures of pillars or other masses which are called upon to support a superincumbent weight in excess of their limits of endurance. These weights are not due to great depth, but depend upon the extent to which the adjacent support has been removed by stoping operations, while the sudden violence of the shock is a function of the physical constants of the mass of rock involved. I am satisfied that this is a genuine and sufficient cause and that it is unnecessary to conclude that the rock is under any additional stress due to other causes. Such additional (intrinsic) strain may or may not be present in any given case and may add to, or perhaps even detract from, the net result, but it is not an essential feature.

In the case of rock bursts, exactly the opposite would appear to hold good. The cases quoted seem to show that, in places, the rock is in a condition of intrinsic strain which is practically independent of the pressure due to depth or to the removal of support by extensive excavation. The origin of such strains and their sporadic character are undoubtedly difficult problems. Prof. Moore regards them as due to great compressive stresses in former times and considers that the rocks are still under great compression. He suggests that the strains may have been partially relieved, or distributed, by movement, but portions of highly squeezed and strained rock remain which are liable to burst disruptively when encountered in the mine workings. This is practically the view put forward some 16 years ago by Mr. Bosworth-Smith, and I am by no means prepared to deny that such highly squeezed patches may exist.

I am not, however, prepared to admit that these rocks are, on the whole, under great compression at the present time, as asserted by Prof. Moore, and I know of no evidence to support such a view. No doubt they were highly compressed when the schists were folded, as well as after the introduction of the quartz veins which are crushed, faulted and intersected by great faults with slickensided walls. Much of the pressure and movement probably accompanied the intrusion of the Peninsular gneiss in Archaean times, and since Archaean times there is no evidence of pressure or differential movement in the rocks of the state. In pre-Cambrian times the rocks, both schist and gneiss, were penetrated by a great series of dolerite dykes which traverse them in east and west and north and south directions. The walls of the dykes show no signs of faulting, brecciation or slickensides, and the dykes show no signs of pressure, faulting or folding. It seems probable that the dykes welled quietly into a great rectangular series of contraction rifts due partly to contraction, resulting from the cooling of the great mass of the Peninsular gneiss, and partly to buoyancy which accompanied extensive denudation. Since that time, denudation has continued uninterruptedly and many thousands of feet of schist and gneiss must have been removed with consequent relief of compression, which would be further relieved by secular cooling and shrinkage as the depth from surface was reduced.

In the absence of any evidence of existing regional compression, it seems reasonable to conclude that the original compression which produced folding, faulting, slickensides, etc. has been largely, though unevenly, relieved. If some portions of rock still remain in a state of squeeze it is open to doubt if they would burst out in large pieces, though some shaling might occur, as in the case of pillars under great weight. If there are residual strains remaining from the period of great compression and folding it is probable that the more important of these are connected with the bending of the more massive layers and local inequalities of adjustment.    Parts of the rock will be in tension and parts in compression and the former will be those to give way first and most frequently.    There may also be bending moments due to the regional elevation of the Mysore Plateau, though we have no evidence of this at the present time, or there may be a certain amount of stretching due to buoyancy resulting from the removal of a great superincumbent layer. If we add the shrinkage due first to the cooling of large igneous masses of gneiss and later to reduction of temperature from degradation of the surface, there appears to be ample opportunity for the production of tensional strain, which will be locally and variably distributed according to the varying character of the rock layers and the local adjustments previously effected.    Such earlier adjustments are suggested by the cross joints filled with calcite and quartz, and present-day adjustments, apart from the rock bursts themselves, are suggested by the frequent tendency of the footwall to loosen itself in large rectangular blocks and the occasional occurrence of fissures in both the foot and hanging walls which show no tendency to close.    It may be that the excavation of the mine workings, followed by lowering of temperature due to ventilation or compressed air, is often the last straw which carries the tensile strain past the critical moment.

I have pointed out that these rock bursts occur in the trap dykes and that there is no suggestion or evidence that the dykes have been subject to compression since their solidification.    On the other hand, we cannot doubt that they have cooled and contracted and as they are firmly frozen to the enclosing rock it is probable that they are under tension, with a tendency to shrink and rupture when the enclosing envelope is pierced.    I do not suggest that the tendency of the dykes to contract produces a condition of tension in the schists, but the fact that rock bursts occur in them suggests that the similar phenomena in the crystalline schists may likewise be due to tensional stresses in spite of the fact that schists were, obviously, at one time under great compression.    I have endeavored to show that the compression of the schists as a whole does not preclude a state of tension in some members, and that there are grounds also for believing that active compression has long ceased and that the rocks have since been subject to contraction and stretching.

In conclusion, I may state that this note has been written while traveling and that figures given are quoted from memory and may be subject to some revision; they are, however, substantially correct.    I fear I may have exceeded the customary limits of discussion but I trust that the more recent details which I have been able to furnish may add to the interest and value of Prof. Moore's paper.

E. S. MOORE (author's reply to discussion*).—I have read with much interest Dr. W. F. Smeeth's criticism of my article on the air blasts in the Kolar Gold Field, India. However, before answering the questions he asked, I should like to suggest that possibly the phrase "difference of opinion" might be substituted for the word "misapprehension" when applied to my statement that the matrix of the conglomerate resembles a hornblende schist. This conglomerate is certainly regarded as a metamorphosed sediment. Dr. Evans, who has had wide experience with metamorphic rocks, has regarded this as a clastic rather than an autoclastic rock and mentions it as a probable glacial deposit[2] and similar to the Lower Huronian conglomerate of North America. I was greatly impressed by the similarity between this rock and the bands of Huronian conglomerate in the United States and Canada where metamorphism has rendered the matrix of the rock schistose. It was scarcely intended that the word basal, as used in my text, should be extended to the Kolar field, although possibly it may be equally applicable there. The word is so frequently used on this continent for the Lower Huronian conglomerate, which in many places is the basal formation of the predominantly sedimentary Proterozoic group, overlying the more largely igneous Archeozoic rocks, that it was inadvertently used without special stratigraphic significance for the Kolar area. In reply to Dr. Smeeth's question as to how the conglomerate indicates a syncline, it would seem that if there is present in a closely folded area a band of sediment that is younger than the surrounding rocks, it must be in a syncline in those rocks.

As to an apparent contradiction in the statement on page 83 regarding the compressive stress, it does not seem that such contradiction exists; I think every one will take the same meaning as Dr. Smeeth has taken, which is exactly what was intended.

As to the difference between the minor explosions, or rock bursts, and the larger disturbances, or so-called quakes, shocks, or earth tremors, I feel like reiterating the opinion that there is no real sharp distinction between them in many cases, because the same agent that produces the former may aid in producing the latter and in causing them to be of greater intensity. The results of the larger disturbances are very much greater than those of the smaller, but the same strain, inherent in the rocks, causes the larger explosions to occur more frequently and at shallower depths than those at which they would occur if this unusual strain were not present. However, the fact is not lost sight of that

---

* Received Nov. 4, 1918.

[2] J. W. Evans: *Proceedings*, Geologists Association (1910) **21**, 448.

heavy shocks have occurred in deep mines in South Africa[3] and in other regions where the superincumbent weight of the rocks alone causes violent ruptures in pillars and other supporting bodies of rock left in mine workings. The physical character of the rock will, of course, have great influence on the intensity of the blast.

As to the cause of the strain in the rocks of the Kolar region, it must still be regarded as due to movements which cause compression in the earth's crust. This seems to be well illustrated by the bursting of rocks, observed by some engineers in the quarries of this country, as well as by the other cases cited in my article.

In concluding, I wish to thank Dr. Smeeth for his discussion of the paper and the very valuable data which he has submitted in describing in detail the effects of notable shocks and the methods employed to prevent them.

---

[3] Since the writing of this article, R. N. Kotzé, Government Mining Engineer of the Union of South Africa, has very kindly sent the writer a copy of the Report of the Witwatersrand Earth Tremors Committee, which concluded that the shocks in the mines of that region were entirely the result of mining operations and that they were liable to occur after such a depth was reached that the superincumbent weight of the overlying rocks was sufficient to crush the pillars in the mine workings.

## Limonite Deposits of Mayaguez Mesa, Porto Rico*

BY CHAS. R. FETTKE,† PH. D., PITTSBURGH, PA., AND BELA HUBBARD, NEW YORK, N. Y.

(Colorado Meeting, September, 1918)

DURING the summer of 1916, while on a visit to the United States Agricultural Experiment Station at Mayaguez, Porto Rico, the writers were told by D. W. May, the director, that an occurrence of manganese ore had been reported from the mesa southeast of Mayaguez. A visit to the locality showed that the mesa is covered by a mantle of bright red soil resting upon serpentine. Here and there over its surface concretionary masses of limonite have been washed out of this soil. These were

FIG. 1.—MAYAGUEZ FROM NORTHWEST END OF MESA.

apparently mistaken for manganese ore. On account of the similarity of this occurrence to the extensive iron deposits of northeastern Cuba, it was thought that a brief description based upon a two days' reconnaissance survey of the area would be of scientific interest.

Mayaguez is on the west coast of Porto Rico, at Mayaguez Bay, which is the third in importance of the harbors of the island. The city itself is about ¾ mile inland on the low-lying plain at the mouth of the

---

\* Published with the permission of the New York Academy of Sciences, under whose auspices, in its undertaking of the natural history survey of Porto Rico, the data for this paper were gathered.

† Geologist, Carnegie Institute of Technology.

Mayaguez River.   With its population of about 16,500, it ranks third in
size among the cities of the island.   The narrow-gage railroad of the
American Railroad Co. of Porto Rico, running along the south, west,
and north coasts from Guayama to San Juan, the capitol, passes through
Mayaguez.   Fig. 1 is a view of the city and harbor as seen from the
northwest end of Mayaguez Mesa looking toward the bay.

### SUMMARY OF PHYSIOGRAPHY AND GEOLOGY

The mesa, lying to the southeast of the city, is an erosion remnant
of a former peneplain developed during the period just preceding the
early Eocene.   Dr. Chas. P. Berkey[1] has traced this old erosion surface

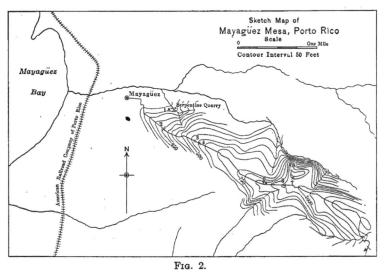

FIG. 2.

northward to a point where it dips gently underneath the Tertiary limestone series of the island.

In the vicinity of Mayaguez this surface now has a maximum elevation of from 1100 to 1200 ft. (335 to 365 m.) above sea level.   The uplift
of the early Tertiary peneplain probably occurred toward the close of
that period.   Since that time numerous streams have incised their valleys
into it, so that at present very little of the original surface remains.
Fig. 2 is a sketch map from a pacing and compass traverse, with elevations determined by aneroid barometer.

[1] Geological Reconnaissance of Porto Rico.   *Annals*, New York Academy of
Sciences (1915) **36**, 51.

Outcrops of the underlying serpentine rock are abundant along the steep northern and southern sides of the mesa. The appearance of this. rock, as exposed in a quarry for road material at the northwest end of the mesa, is shown in Fig. 3; the numerous cracks, slips, and checks

FIG. 3.—QUARRY IN SERPENTINE ROCK.

FIG. 4.—SERPENTINE ROCK, CROSSED NICOLS.

characteristic of this rock are well shown. In thin section, it is seen to be composed mainly of serpentine, having a light yellowish-green tinge and, under crossed nicols, showing the characteristic mesh structure of serpentine derived from olivine (Fig. 4). No remnants of unaltered olivine were detected. Bastite occurs in numerous grains and crystals, pseudomorphous after pyroxene; extinction is parallel to the cleavage, double refraction is weak, the interference colors being pale grays, and the optic axial plane is at right angles to the cleavage. Fig. 5 (crossed nicols) shows a portion of a crystal of bastite and its relation to the serpentine. Occasionally fibrous serpentine has developed along some of the cracks. Here and there throughout the section are minute grain, of magnetite and chromite. Secondary iron oxides, principally limonites indicated by their rich brown color, are also present along minute cracks between and in the former olivine grains (Fig. 6), having separated out

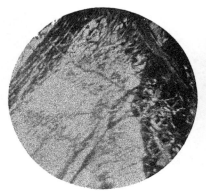

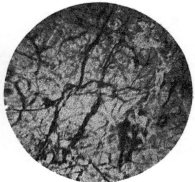

FIG. 5.—PART OF A BASTITE CRYSTAL, CROSSED NICOLS.

FIG. 6.—IRON OXIDES IN SERPENTINE ROCK.

during the serpentinization process. The original rock was a peridotite composed largely of olivine and a pyroxene, probably an orthorhombic one.

The following analysis by the senior author (Table 1) shows the chemical composition of the serpentine rock:

TABLE 1.—*Analysis of Mayaguez Serpentine Rock*

| | PER CENT. | | PER CENT. |
|---|---|---|---|
| $SiO_2$............... | 38.41 | $K_2O$ ................ | 0.10 |
| $Al_2O_3$............ | 4.96 | $H_2O$ at 110° C....... | 0.83 |
| $Fe_2O_3$............ | 6.32 | $H_2O +$ ignition...... | 13.40 |
| FeO............... | 1.27 | $TiO_2$................ | 0.08 |
| MgO.............. | 33.32 | NiO................ | 0.72 |
| CaO.............. | 0.04 | $Cr_2O_3$.............. | 0.42 |
| $Na_2O$............. | 0.27 | | |

On recasting this analysis, the following mineralogic composition is indicated (Table 2):

TABLE 2.—*Mineralogic Composition of Mayaguez Serpentine Rock*

PER CENT.

| | | |
|---|---|---|
| Magnesium serpentine, $3MgO.2SiO_2.2H_2O$ | 76.69 | 80 per cent. serpentine |
| Iron serpentine $3FeO.2SiO_2.2H_2O$ | 2.01 | |
| Nickel serpentine, $3NiO.2SiO_2.2H_2O$ | 1.30 | |
| Kaolinite, $Al_2O_3.2SiO_2.2H_2O$ | 4.38 | |
| Gibbsite, $Al_2O_3.3H_2O$ | 4.21 | |
| Limonite, $2Fe_2O_3.3H_2O$ | 7.30 | |
| Rutile, $TiO_2$ | 0.08 | |
| Chromite, $FeO.Cr_2O_3$ | 0.45 | |
| Albite, $Na_2O.Al_2O_3.6SiO_2$ | 2.10 | 2.38 per cent. plagioclase |
| Anorthite, $CaO.Al_2O_3.2SiO_2$ | 0.28 | |
| Orthoclase, $K_2O.Al_2O_3.6SiO_2$ | 0.56 | |
| Absorbed water, $H_2O$ | 0.83 | |

In making this calculation, the magnesia has all been considered to be present as the mineral serpentine. Some of it is actually present in bastite, but as this is believed to have the same composition as serpentine it is included under the latter mineral. No allowance has been made for spinels, as none were detected in the thin sections. After deducting enough ferrous iron to satisfy the chromic oxide in chromite, the remainder has been accredited to iron serpentine. No allowance has been made for magnetite, although small percentages are present in the rock. Following Kemp,[2] the nickel oxide has all been calculated as though it were present as a nickel serpentine. The alumina has been distributed among feldspar, kaolinite, and gibbsite. Whether any of the latter mineral is actually present could not be determined from the thin sections, but there is not enough silica in the rock to combine with all of the alumina as a silicate. The ferric oxide has all been calculated as limonite, although some of it is present as magnetite. Hematite and other hydrated iron oxides besides limonite may also be present.

At one point along the road, southeast of locality 6 on the map, Fig. 2, a residual boulder of augite-andesite porphyry was observed; this rock probably occurs as a dike cutting the serpentine.

The serpentine was derived from a peridotite which was intruded as a stock into the older series of formations that underlie Porto Rico at some time prior to the close of the Cretaceous. This rock was exposed during that interval of erosion which led to the development of the early Tertiary peneplain which extended over a large portion of Porto Rico at the beginning of Tertiary time. Since then, weathering has given rise to the limonite deposits that now cover it as a mantle.

---

[2] *Trans.* (1915) **51,** 21.

#### Distribution and Mode of Occurrence of Limonite

The limonite occurs as a mantle of brown to reddish-brown soil overlying the top of the mesa and extending part way down its sides. Over most of the steep northern and southern flanks, however, it is absent, the underlying serpentine rock coming to the surface here. The material does not furnish a very fertile soil, so that vegetation on the mesa is scanty as compared with the luxuriant growth of tropical plants covering the adjoining hills. Over most of the mesa, the total thickness of the limonite is not exposed, and as this cannot be determined without drilling, a few measurements only were obtained during the hasty reconnaissance made by the writers.

Fig. 7.—Contact of limonite and serpentine.

At locality 2 on the map (Fig. 2) 6 ft. (1.8 m.) of limonite is exposed in a cut along the road and no serpentine shows. At locality 3, there are 10½ ft. (3.2 m.) of limonite resting on the serpentine, while at 4 there are only 7 ft. (2.2 m.). At locality 6, there are 21 ft. (6.4 m.); although the serpentine does not show at the bottom of this exposure, the appearance of the material indicates that it is not far beneath. At 8, a total thickness of 32 ft. (9.7 m.) of iron ore was observed, resting upon the serpentine.

The reddish-brown soil continues along the road following the crest of the mesa nearly to the point where the small stream crosses it near the southeast corner of the map, but no further measurements of its thickness were obtained in this locality.

As a rule, the contact between limonite and the underlying serpentine is very irregular. This can be seen to best advantage in cuts along the mesa road between localities 3 and 4; it was here that the photograph shown in Fig. 7 was taken; the light-colored rock is partly weathered serpentine, while the darker-colored material is the iron ore.

The gradation from weathered serpentine into limonite, wherever the contact is exposed, is a sharp one. As a rule, the serpentine just beneath the iron ore has been altered from the dark green, dense rock, already described, to a light yellowish-green, soft, porous mass, abundantly stained by brown hydrated oxide of iron. The iron ore itself usually has a reddish-brown color, but intervening between it and the altered serpentine there are, in most cases, at least a few inches of brown material.

## Character of Iron Ore

Most of the limonite deposit consists of a loose, porous, earthy mass, varying in color from light yellowish to a dark reddish-brown. Usually the red portions occur nearest the surface, while the brown rest upon the serpentine. The relative percentage of iron in the ore cannot, however, be determined from the color, as frequently the brown varieties contain a higher percentage of ferric oxide than the darker red. The shade of color is undoubtedly due to the degree of dehydration of the ferric oxide; toward the surface, the limonite tends to lose its combined water and gradually passes into less hydrated forms of ferric oxide, and finally to hematite.

Incomplete analyses of the limonite from a number of localities on the mesa (see Fig. 2) gave the following results (Table 3):

TABLE 3.—*Iron and Water Contents of Mayaguez Iron Ores*

| Locality | 1 | 2 | 3 | 4 | 5 | 6 | 7 | 8 |
|---|---|---|---|---|---|---|---|---|
| $Fe_2O_3$ | 56.17 | 44.53 | 63.29 | 66.36 | 41.35 | 37.85 | 24.70 | 58.63 |
| $H_2O$ | 13.86 | 12.74 | 14.81 | 11.35 | 16.74 | 17.61 | 15.71 | 16.05 |
| Fe | 39.30 | 31.16 | 44.28 | 46.43 | 28.93 | 26.49 | 17.28 | 41.01 |
| Fe calculated on water-free basis | 45.62 | 35.71 | 51.98 | 52.37 | 34.75 | 32.03 | 20.50 | 48.85 |

The sample from locality No. 1 represents material that has been washed down the sides of the mesa and has accumulated at its foot. At this place, a gully 5 ft. (1.5 m.) deep has been cut into it without exposing the bed rock; the color is a dark reddish-brown. No. 2 comes from a cut along the mesa road where 6 ft. of the limonite is exposed without uncovering any serpentine at the bottom. This material is dark red. No. 3 represents 10.5 ft. of limonite resting on top of partly weathered serpentine to the southeast of No. 2; it has a dark reddish-brown color. No. 4 comes from nearby, where 7 ft. of dark reddish-brown iron ore overlies the serpentine. No. 5 represents a sample of the dark reddish-brown soil exposed in the hurricane cellar at D. W. May's summer home on top of the mesa; no bed rock was reached in this excavation, which is 5.5 ft. deep. No. 6 was taken from an exposure 21 ft. thick, which probably represents the total amount of this material present; although the bed-rock serpentine does not show, the lower part of the soil contains partly weathered fragments of it. The color here is reddish-brown, but somewhat lighter in shade than over most other portions of the mesa. No. 7 represents a sample from a slide which exposed the upper 10.5 ft. of limonite at locality 7. The soil at this place is red in color, but of considerably lighter shade than elsewhere; the total thickness could not be determined. No. 8 represents an average sample from 32 ft. of the iron ore overlying the serpentine along the steep northeastern flanks of the mesa at locality 8; the material has a brown color.

A complete chemical analysis of sample 8 is given in Table 4.

TABLE 4.—*Complete Analysis of Sample 8*

| | PER CENT. | | PER CENT. |
|---|---|---|---|
| $SiO_2$ | 2.44 | $H_2O$ at 110° C | 1.09 |
| $Al_2O_3$ | 20.21 | $H_2O$ + ignition | 14.96 |
| $Fe_2O_3$ | 57.69 | $TiO_2$ | 0.26 |
| FeO | 0.85 | NiO | 1.00 |
| MgO | 0.61 | $Cr_2O_3$ | 1.57 |
| CaO | trace | | |
| | | | 100.68 |

Recasting this analysis gives the mineralogic composition shown in Table 5

TABLE 5.—*Mineralogic Composition of Sample 8*

| | PER CENT. |
|---|---|
| Limonite, $2Fe_2O_3 \cdot 3H_2O$ | 22.07 |
| Goethite, $Fe_2O_3 \cdot H_2O$ | 42.90 |
| Magnetite, $FeO \cdot Fe_2O_3$ | 0.46 |
| Bauxite, $Al_2O_3 \cdot 2H_2O$ | 25.67 |
| Kaolinite, $Al_2O_3 \cdot 2SiO_2 \cdot 2H_2O$ | 3.02 |
| Absorbed water, $H_2O$ | 1.09 |
| Magnesium serpentine, $3MgO \cdot 2SiO_2 \cdot 2H_2O$ | 1.38 |
| Nickel serpentine, $3NiO \cdot 2SiO_2 \cdot 2H_2O$ | 1.60 |
| Rutile, $TiO_2$ | 0.26 |
| Chromite, $FeO \cdot Cr_2O_3$ | 2.25 |
| | 100.70 |

In recasting this analysis the magnesia and nickel oxides were assumed to be present in the ore as serpentine. This is probably true for the magnesia; as to the nickel oxide, while this is undoubtedly present also as a hydrated silicate, this may not have the formula of a serpentine.

Only a small percentage of the alumina in the ore is present as hydrated aluminum silicate, as shown by the low silica content; the remainder of the alumina has been calculated as bauxite; possibly other forms of hydrated aluminum oxide are also present. After allowing for the chromite, the remainder of the ferrous oxide has been calculated as magnetite. That the chromium is present as chromite was determined by treating the material with hydrochloric acid and then separating the black metallic particles in the residue. These gave the typical emerald-green bead of chromium with salt of phosphorus. After allotting a small percentage of ferric oxide to magnetite, the remainder was distributed between limonite and goethite in such proportions as to account for the remainder of the combined water. It is probable that at least a portion of the ferric oxide and water are present in other combinations, but it is not supposed that hematite occurs to any extent in the sample on account of the absence of red color.

FIG. 8.—SECTION OF LIMONITE BOULDER, DIRECT LIGHT.

In addition to the loose earthy material, numerous boulders and masses of hard ore occur scattered over the limonite area. These boulders retain the porous structure of the soft ore, but contain little veinlets running in all directions and filled with the botryoidal form of limonite with varnish-like luster. These boulders range in diameter from a few inches up to several feet; apparently they have been washed out of the upper layers of loose material.

A thin section from such a boulder, under the microscope, showed that the hydrated iron oxide is present in two forms, cryptocrystalline and amorphous. The former variety retains the structure of the serpentine, while the amorphous has been deposited afterward in little veinlets running in every direction throughout the former; these show a banded structure.    Small grains of magnetite and chromite appear here and there, just as in the serpentine rock.    Fig. 8 is a photomicrograph of a thin section of the hard ore taken by transmitted light.    The banded veinlet represents the amorphous botryoidal hydrated oxide while on each side can be seen the cryptocrystalline variety, retaining the structure of the serpentine.    The white spots are holes in the section.    This section should be compared with Fig. 6, illustrating the serpentine rock under similar conditions; the light mineral is the serpentine, while the dark one along the cracks is limonite.

An incomplete analysis of the hard ore gave the following results: $Fe_2O_3$, 74.92 per cent.; $H_2O$, 15.03 per cent.; Fe, 52.42 per cent.

True limonite has the formula $2Fe_2O_3 \cdot 3H_2O$.    To satisfy 74.92 per cent. of $Fe_2O_3$, therefore, would require 12.67 per cent. of $H_2O$ if all of the ferric oxide were present in this form.    Since the analysis shows 15.03 per cent. $H_2O$, it is possible that both the amorphous and the cryptocrystalline variety have this composition.    In general, where the term limonite is used in this paper, unless specifically stated otherwise, it includes all of the hydrated forms of ferric oxide that are usually classed under limonite in the iron-ore trade.

The retention of the original structure in the limonite derived from the serpentine affords an interesting parallel to a similar phenomenon observed by W. J. Mead[3] in the pisolitic bauxite of Arkansas which, in thin section, shows the granitoid texture of the nepheline syenite from which it was derived by a process of weathering similar to that which caused the formation of the limonite from serpentine at Mayaguez.

## ORIGIN OF LIMONITE DEPOSITS

From the preceding description, it is seen that the limonite deposits of the Mayaguez mesa are almost exactly similar to those in northeastern Cuba, which have been studied by a number of geologists, all of whom have agreed that they are residual in origin, being derived from the underlying serpentine by the weathering of this rock.[4]

---

[3] Occurrence and Origin of the Bauxite Deposits of Arkansas, *Economic Geology* (1915) **10**, 42.

[4] A. C. Spencer: Three Deposits of Iron Ore in Cuba.    U. S. Geological Survey, *Bulletin* No. 340 (1908) 318–329.

C. M. Weld: The Residual Brown Iron Ores of Cuba.    *Trans.* (1909) **40**, 299–312.

J. S. Cox, Jr.: Iron Ores of the Moa District, Oriente Province, Island of Cuba. *Trans.* (1911) **42**, 73–90.

C. K. Leith and W. J. Mead: Origin of the Iron Ores of Central and Northeastern Cuba.    *Trans.* (1911) **42**, 90–102.

A. C. Spencer: Occurrence, Origin, and Character of the Surficial Iron Ores of Camaguey and Oriente Province, Cuba.    *Trans.* (1911) **42**, 103–109.

It will, therefore, be sufficient to summarize briefly here the main lines of evidence that point to the residual origin of the limonite. These have already been ably discussed by the geologists who have described the Cuban deposits.

The close association of the limonite and the serpentine points strongly to such mode of origin; where other rocks lower in iron underlie the surface, the soils are relatively poorer in iron oxides. That the original structure and texture of the serpentine can still be detected in some of the hard ore furnishes almost indisputable evidence that the limonite has been derived from it in this manner. The irregular contact of the iron ore with the underlying serpentine is also characteristic of residual deposits derived from underlying rocks by the process of weathering. Likewise, the extremely porous structure of the limonite, where slumping has not occurred, shows that it has been formed by a leaching process in which the more soluble constituents have been dissolved by circulating waters, leaving the less soluble constituents behind. The relatively high percentage of chromite in the ore also indicates the same origin. Chromite is a relatively insoluble mineral, and therefore remains behind in the residual soils.

Finally, a comparison of the analyses of the serpentine and the limonite furnishes further evidence for supposing that the latter was derived from the former (Table 6).

TABLE 6.—*Comparative Composition of Serpentine and Limonite*

|  | SERPENTINE, PER CENT. | LIMONITE, PER CENT. |
|---|---|---|
| $SiO_2$ | 38.41 | 2.44 |
| $Al_2O_3$ | 4.96 | 20.21 |
| $Fe_2O_3$ | 6.32 | 57.69 |
| $FeO$ | 1.27 | 0.85 |
| $MgO$ | 33.32 | 0.61 |
| $CaO$ | 0.04 | trace |
| $Na_2O$ | 0.27 | not det. |
| $K_2O$ | 0.10 | not det. |
| $H_2O$ at 110° C. | 0.83 | 1.09 |
| $H_2O$ + ignition | 13.40 | 14.96 |
| $TiO_2$ | 0.08 | 0.26 |
| $NiO$ | 0.72 | 1.00 |
| $Cr_2O_3$ | 0.42 | 1.57 |
|  | 100.14 | 100.68 |

C. W. Hayes: The Mayari and Moa Iron-ore Deposits of Cuba. *Trans.* (1911) **42**, 109–115.

W. L. Cummings and B. L. Miller: Characteristics and Origin of the Brown Iron Ores of Camaguey and Moa, Cuba. *Trans.* (1911) **42**, 116–137.

J. F. Kemp: The Mayari Iron-ore Deposits, Cuba. *Trans.* (1915) **51**, 3–31.

C. K. Leith and W. J. Mead: Additional Data on Origin of Lateritic Iron Ores of Eastern Cuba. *Trans.* (1916) **53**, 75–78.

The recasting of the analysis of serpentine rock brought out the fact that 80 per cent. of its weight consists of the minerals serpentine and bastite.   Both minerals succumb fairly readily to the decomposing action of circulating waters carrying oxygen, carbon dioxide, and organic acids, particularly in a moist tropical climate.   The magnesia is probably first converted into a carbonate, in which form it is sufficiently soluble to be leached out by downward-moving waters.   The silica, being liberated in a finely divided and colloidal form, likewise appears to be susceptible to solution in the circulating groundwaters.   The iron, on the other hand, becoming oxidized to the ferric condition, is precipitated as a hydrated ferric oxide.   The silicates containing alumina are also attacked; a hydrated aluminum silicate is probably first formed, which then undergoes further change, the silica being liberated and dissolved, leaving behind only a relatively insoluble hydrated oxide of aluminum.   This last step in the complete decomposition of aluminum silicates does not usually occur in temperate climates but is widely characteristic of the tropics.   Soils rich in hydrated aluminum and iron oxides are called laterites.

Various theories have been advanced to account for the more complete decomposition of aluminum silicates under tropical weathering.   On account of the warmer climate, the chemical decomposition of the minerals is naturally accelerated during the process of weathering.   This is accompanied by a leaching out of the soluble compounds and naturally the rock becomes more and more porous.   As Leith and Mead[5] have pointed out, in a tropical or subtropical region where no frost occurs, slumping due to freezing will not take place as in a temperate climate; the open porous structure will therefore tend to be retained in the resulting soil and thus increase the active downward circulation of the water and allow the process to go on until decomposition is complete.

At Mayaguez, the conditions first became favorable for the weathering of the serpentine rock probably during the early Tertiary.   While there is a bare possibility that this area was submerged during a portion of the Tertiary, it does not seem likely that this actually occurred.   With the uplift of the area at the close of the Tertiary, erosion became very active. At the present time, on account of the topography of the mesa, more material is probably being removed by erosion than is being added by further weathering.

A comparison of the analyses of the iron ore from different portions of the mesa (Table 3) shows a considerable variation in iron contents. This is probably due, in part at least, to variations in the amounts of alumina and iron oxide in different portions of the serpentine.

[5] *Economic Geology* (1915) **10**, 35.

### OTHER SIMILAR DEPOSITS IN PORTO RICO

The writers were informed, by residents of Mayaguez, that deposits of red soil similar to those overlying the Mayaguez mesa exist in other parts of western Porto Rico, but none of these places was visited.

Later in the summer, the junior writer came across three small areas of serpentine south of Aquada, in the northwestern part of the island. These have apparently been derived from a peridotite intruded into an amygdaloidal andesite showing distinct flow structure. An incomplete analysis of a sample of the reddish-brown soil overlying one of them gave the following results: $Fe_2O_3$, 15.17 per cent.; $H_2O$, 10.62 per cent.; Fe, 13.76 per cent. Its iron content is far too low to class it as a possible ore of iron.

Professors Berkey and Crampton[6] report the occurrence of massive serpentine near Yauco in southwestern Porto Rico; near Comerio in the central part of the Island; and near Maricao, east of Mayaguez. No statement, however, is made in regard to the nature of the soils overlying them.

### DEPOSITS OF LIMONITE ASSOCIATED WITH SERPENTINE IN OTHER PARTS OF WORLD

Deposits of limonite associated with serpentine and peridotite similar to the Mayaguez mesa occurrence have been described from a number of widely separated localities. A number of analyses of this type of ore and of the rocks from which they are derived are given in Tables 7, 8, and 9.

Of these deposits, those of northeastern Cuba are by far the most important from an economic standpoint. The development work by the Spanish American Iron Co. has shown that there are 600,000,000 tons[7] of commerical iron ore in the Mayari district alone. In the Moa district, Jennings S. Cox, Jr., has estimated 1,251,000,000 tons[8] of ore with 50 per cent. metallic units of iron and nickel, while Willard L. Cummings and Benjamin L. Miller place the available reserves of the Camaguey district at 400,000,000 tons of 40-per cent. ore and 50,000,000 tons of 45-per cent.[9] In addition to these three main areas of iron ore, several smaller deposits are also known.

The Cuban deposits are probably contemporaneous in origin with those of the Mayaguez mesa. The serpentine from which they are

[6] C. P. Berkey: Geological Reconnaissance of Porto Rico. *Annals*, New York Academy of Sciences (1915) **26**, 34.

[7] *Trans.* (1911) **42**, 139.

[8] Iron Ores of the Moa District, *Trans.* (1911) **42**, 86.

[9] Characteristics and Origin of the Brown Iron Ores of Camaguey and Moa, Cuba. *Trans.* (1911) **42**, 125.

TABLE 7.—*Analyses of Serpentines and Derived Limonites*

|  | Mayaguez Mesa,[10] Porto Rico | | Mayari, Cuba[11] | | | |
|  | | | | Iron Ore | | |
|  | Serpentine | Limonite | Serpentine | Bottom | Middle | Surface |
|---|---|---|---|---|---|---|
| SiO₂........... | 38.41 | 2.44 | 37.28 | 7.54 | 2.70 | 2.26 |
| Al₂O₃.......... | 4.96 | 20.21 | 2.45 | 4.97 | 7.13 | 14.90 |
| Fe₂O₃......... | 6.32 | 57.69 | 5.14 | 64.81 | 71.89 | 68.75 |
| FeO........... | 1.27 | 0.85 | 5.14 | 1.49 | 1.29 | 0.77 |
| MgO.......... | 33.32 | 0.61 | 34.59 | 1.50 |  |  |
| CaO.......... | 0.04 | Trace | None |  |  |  |
| Na₂O......... | 0.27 | Not det. |  |  |  |  |
| K₂O.......... | 0.10 | Not det. |  |  |  |  |
| H₂O at 110° C.. | 0.83 | 1.09 | 0.91 | (a) | (a) | (a) |
| H₂O+ignition . | 13.40 | 14.96 | 12.80 | 12.75 | 12.90 | 11.15 |
| TiO₂......... | 0.08 | 0.26 |  |  |  |  |
| P₂O₅.......... | ..... | ...... | 0.013 |  |  |  |
| NiO.......... | 0.72 | 1.00 | 0.30 | 2.75 | 1.60 | 0.74 |
| Cr₂O₃........ | 0.42 | 1.57 | 1.68 | 3.66 | 3.17 | 1.89 |
|  | 100.14 | 100.68 | 100.303 | 99.47 | 100.68 | 100.46 |

(a)  Analyses of ore made on samples dried at 110° C.

derived also represents an altered peridotite intruded into an older series of formations which are overlain unconformably by early Tertiary limestones of the upper Eocene or lower Oligocene periods.  The weathering of the serpentine that resulted in the formation of the iron ores occurred under tropical conditions during the Tertiary.  A tropical climate still prevails in this area.

The Staten Island, New York, deposits were likewise formed during Tertiary times, probably largely during the earlier portion of that period when a humid semi-tropical climate prevailed, as shown by the fossil flora found in nearby formations of this age.  They are now covered in part by Pleistocene glacial deposits.  These ores represented only a very small tonnage which had escaped erosion during the later Tertiary and Pleistocene periods.[12]  They are practically exhausted today.

The Clealum iron-ore deposits occur on the Clealum River, a tributary of the Yakima River in the eastern spurs of the Cascade Range of Washington.  The deposits lie on the former surface of an extensive area of serpentine, at the contact and in the base of a sandstone, known as the Swauk, which is of early Tertiary age, forming the base of the

[10] Analyses by Chas. R. Fettke.

[11] J. F. Kemp: The Mayari Iron-ore Deposits, Cuba.  *Trans.* (1916) 51, 4–30.

[12] C. R. Fettke: Limonite Deposits of Staten Island, New York.  *School of Mines Quarterly* (1912) 33, 382–391.

TABLE 8.—*Analyses of Serpentines and Derived Limonites*

| | Staten Island, N. Y. | | Clealum, Wash.[15] | | Surigao[16] Province, Mindanao, P.I. |
|---|---|---|---|---|---|
| | Serpentine[13] | Iron Ore[14] | Serpentine | Iron Ore | Lateritic Iron Ore |
| $SiO_2$ | 36.72 | 14.19 | 39.00 | 7.50 | 1.04 |
| $Al_2O_3$ | 1.06 | 8.59 | 1.75 | 21.90 | 10.56 |
| $Fe_2O_3$ | 6.59 | 55.95 | 5.16 | 37.10 | 66.80 |
| $FeO$ | 1.53 | 0.73 | 1.71 | 21.30 | 0.36 |
| $MgO$ | 29.09 | 3.60 | 38.00 | 2.30 | |
| $CaO$ | 9.95 | 0.18 | } 0.10 | | |
| $Na_2O$ | .... | 0.40 | | | |
| $K_2O$ | .... | 0.19 | | | |
| $H_2O$ at 110° C | 0.52 | 2.21 | } 13.74 | } 6.80 | 13.50 |
| Ignition | 14.02 | 10.20 | | | 6.60 |
| $TiO_2$ | .... | .... | Trace | 0.70 | |
| $CO_2$ | .... | 0.22 | None | 0.15 | |
| $P_2O_5$ | .... | 0.135 | Trace | 0.09 | Trace |
| S | .... | 0.40(b) | Trace | 0.03 | Trace |
| $MnO$ | .... | 0.52 | 0.15 | Trace | |
| $NiO$ | .... | 0.59(b) | 0.10 | 0.20 | None |
| $Cr_2O_3$ | 0.49 | 1.91(b) | 0.47 | 2.20 | 1.15 |
| C | .... | 0.13 | | | |
| | 99.97 | 100 145 | 100.21 | 100.27 | 100.01 |

(b) $FeS_2$, $NiS$, and $CoS$ of original statement recalculated as $FeO$, $NiO$, and $CoO$; latter included under $NiO$. Sulfur given as S.

Eocene in central Washington. The serpentine had been extensively eroded before the sandstone was laid down on it. The ore occurs in the form of lenses which represent erosion remnants of a mantle of limonite which covered the serpentine area prior to the deposition of the sandstone. These deposits were formed during the Cretaceous period when the climate was humid semi-tropical to tropical, as is indicated by the fossil flora of this age in the western United States. Since their burial beneath the Eocene sandstones, the iron oxides have undergone dehydration and partial reduction so that the ore now consists largely of hematite and magnetite.

[13] D. H. Newland: The Serpentines of Manhattan Island and Vicinity, and Their Accompanying Minerals. *School of Mines Quarterly* (Apr., July, 1901) **22**, 307–317, 399–410.

[14] B. T. Putnam: Mining Industry of the United States, Excluding Precious Metals. *Tenth Census of the United States* (1880) **15**.

[15] G. O. Smith and B. Willis: The Clealum Iron Ores, Washington. *Trans.* (1900) **30**, 956–366.

[16] W. E. Pratt: The Iron Ores of the Philippine Islands. *Trans.* (1916) **53**. 30–105.

TABLE 9.—*Analyses of Peridotites and Weathered Products,*
*French Guinea*[17]

|  | Peridotite | Peridotite | Laterite | Iron Capping | Iron Ore |
|---|---|---|---|---|---|
| $SiO_2$ | 40.01 | 38.32 | 12.67 |  |  |
| $Al_2O_3$ | 2.54 | 2.66 | 12.59 | 4.80 | 8.70 |
| $Fe_2O_3$ | 1.00 | 4.35 | 46.84 | 83.50 | 77.20 |
| FeO | 11.70 | 11.78 |  |  |  |
| MgO | 39.90 | 36.22 | 1.26 |  |  |
| CaO | 1.68 | 2.74 | 0.04 |  |  |
| $Na_2O$ | 1.07 | 0.16 |  |  |  |
| $K_2O$ | 0.52 | 0.06 |  |  |  |
| Ignition | 1.10 | 3.38 | 15.32 | 10.18 | 11.40 |
| $TiO_2$ | ..... | 0.28 | 0.55 |  |  |
| $Cr_2O_3$ | 0.16 | 0.16 | (c) | 0.20(c) | Trace |
| Insoluble | ..... | ..... | 10.73(c) | 1.70(c) |  |
|  | 99.68 | 100.11 | 100.00 | 100.48 | 100.10 |

(c) Insoluble matter contains chromic oxide in picotite.

The lateritic iron ores of the Surigao Province of northern Mindanao, Philippine Islands, also overlie a serpentine rock, in the form of a residual mantle. The available reserves of iron ore here have been estimated at 430,000,000 metric tons.[18] This area has a humid tropical climate at the present time. The French Guinea deposits are also located in the tropics.

From the above descriptions, it is seen that wherever peridotite or serpentine rocks are exposed at the surface in a moist tropical or semi-tropical climate the tendency is for lateritic soils to develop from them, which are sufficiently rich in iron content to be classed as possible ores of iron. The lower the alumina content of the original rock, the higher will be the grade of the iron ore, since, during the process of weathering, the alumina tends to remain behind in the residual soil in the form of bauxite or other hydrated oxides of alumina to an even greater extent than the iron oxides.

[17] Alfred Lacroix: Les Latérites de la Guinée. *Nouvelles Archives du Muséum National d'Histoire Naturelle* (1913) **5**, 255–356.

[18] W. E. Pratt: Iron Ores of the Philippine Islands. *Trans.* (1916) **53**, 105.

## Recent Geologic Development on the Mesabi Iron Range, Minnesota

The following correspondence relating to a paper bearing the above title, presented by J. F. Wolff, at the New York meeting in February, 1917, and published in the *Transactions*, Volume LVI, page 142, has passed between the author of the paper, and Mr. Anson G. Betts, and is interesting, although it was received much too late for incorporation in its proper place in Volume LVI.

ANSON G. BETTS, Asheville, N. C.—1. I should like to ask Mr. Wolff if he finds from analysis of the iron formation and of the iron ore, and from the relative amounts of each in each unit area of the formation, unleached and after leaching, that the iron remains entirely insoluble and the total quantity still remains present; or whether he finds that some of the iron is removed, or possibly even added by deposition from the water passing through the formation.

2. The water that passed into the iron formation had to go somewhere; was it absorbed by underlying formations through hydration, or, after passing through the ore formation, did it come to the surface in springs at a lower elevation, or how did it escape from the ground?

3. I should like to ask Mr. Wolff whether he finds that the iron formation in the sedimentary series does not come next to the quartzites and generally between the quartzites and slates.

J. F. WOLFF, Duluth, Minn.—1. In common with others, you appear to have some hesitancy in accepting the fact of the removal, by solution, of the enormous quantity of silica which has been leached out of portions of the Biwabik iron formation, thus forming orebodies of merchantable iron ore. I believe that even Dr. Waldemar Lindgren once hesitated to accept this fact. However, the evidence is so simple and absolutely conclusive that the fact is established beyond all question. At the contact of orebodies with their rock walls, the bands of iron ore in the orebody can be seen in contact with the corresponding bands of original iron oxide in the rock wall, and the chert layers in the rock wall can be seen to give place to fine granular material (fine silica and iron oxide) and pore space. This is especially noticeable where slumping at the rock walls has not been so great as to displace the layers in the orebody from their contact with their corresponding layers in the rock wall.

Although complete analyses of the entire original rocks in the iron series have not been made in any place, to my knowledge, from the best data at hand as to the average analysis of the Biwabik series and the corresponding average analysis of orebodies derived from them, I have computed that the total iron per unit of area has remained the same in the orebodies as it was in the original rocks from which the orebodies have been derived. In other words, there has been practically no removal of iron, although of course secondary iron, principally in the form of the hydrated oxides, is found in nearly all orebodies.

2. Regarding the circulation of water in the ore formation, no theory need be advanced, for there are facts at hand which indicate quite certainly what that circulation was. In a series of articles published in the *Engineering and Mining Journal*, July 17 to August 7, 1915, you will find a great deal of information regarding the detailed geology and engineering practice of the Mesabi Range. In the issue of Aug. 7, 1915, you will find a diagram showing the probable circulation of water in the Biwabik iron formation, which circulation resulted in the development of orebodies. I have pointed out therein that drilling through the Virginia slate, which caps the Biwabik formation on the south, in several places has developed artesian wells, showing that water is impounded under the Virginia slate.

I have also pointed out that toward the south side of the outcrop of the iron formation many fissure orebodies have been found and individual drill holes have penetrated a large thickness of ore material of rather low-grade, and that these fissure orebodies and isolated drill holes penetrating decomposed iron formation are the places where the underground water, which has effected the concentration of the orebodies, has risen again to the surface in an artesian circulation. Moreover, recent explorations have shown that the large trough orebodies become narrow toward the south, or Virginia slate, side of the outcrop, and these channels represent the lower ends of the artesian circulation. Furthermore, these channels contain ore or ore formation of a very inferior grade, as a rule, as compared with the ore in the larger or main orebodies farther up the dip of the iron formation. Of course these facts as to the original circulation of ground waters are not susceptible of absolute demonstration but rather are inferred from knowledge of character and structure of orebodies, and they are recognized as the most probable explanation of underground circulation in the iron formation by those best qualified to judge of them on the Mesabi Range.

I do not believe that the area was cut up extensively by erosion, because at the present time the local relief of the top of the iron formation is practically negligible. In other words, there are no marked hills and valleys such as would account for underground drainage. There has been a considerable amount of hydration, the average loss on ignition being perhaps 6 per cent. However, such hydration probably would result in throwing the silica out of solution and therefore we should have large amounts of secondary silica present in the orebodies. There is no great amount of secondary silica, so I conclude that the amount of water taken up in hydration of the ores is relatively very small as compared with the amount necessary to effect the solution and transportation of the silica. Where the silica has gone to, of course, no evidence at hand tells, but we can only assume that it flowed out onto the surface at the point of discharge of the artesian water.

3. From what we know of the pre-Cambrian, Algonkian metallographic series of rocks in different parts of the world, I believe that most geologists who are familiar with these rocks will look for iron formation in any sedimentary rocks, whether they be slates, quartzites, or limestones, or any combination of these three which can be determined to be of Algonkian age.   In other words, to answer your last question specifically, if I found quartzite and slate in an area of Algonkian rocks, I would examine the area closely for evidence of iron formation, because we know that, especially in North and South America and Africa, and probably in Eastern Asia also, iron formations occur in sedimentary series in pre-Cambrian, Algonkian rocks.

## Certain Iron-ore Resources of the World [*]

(Milwaukee Meeting, October, 1918)

AT a meeting of the New York Section, on May 23, 1918, the sole subject of discussion was the nature and occurrence of iron ores in certain parts of the world.[*]  Owing to the importance of this subject, it has been deemed advisable to publish the remarks made at that meeting in the form of a paper, for presentation at the Milwaukee Meeting.  The remarks related to the following districts:

### Brazil

BY E. C. HARDER[†]

During the years immediately preceding the war, the iron ores of Brazil were attracting the attention of iron operators both in Europe and in the United States.  This was due to the large quantities of practically undeveloped ore available, and to its great purity.  It is reasonably safe to state that had capital not been diverted from the exploitation of these deposits to war purposes, Brazilian iron ores would probably now be offered in European and American markets.  It is certain that after the war these ores will play a prominent part in the economic reconstruction of Europe and perhaps in the development of America.

The Brazilian iron-ore field ranks among the six great iron-ore districts of the world, the others being: (1) The Lake Superior district of the United States; (2) the Lorraine ore field of northern France and southern Germany; (3) the magnetic deposits of northern Sweden; (4) the ore fields of Oriente, Cuba; and (5) the Wabana ores of Newfoundland.  It is doubtless the greatest known undeveloped iron-ore district in the world.

The Brazilian iron ores are situated in the State of Minas Geraes, the center of the district being located about 250 miles in a direct line west of

---

[*] For additional information on this subject, readers are reminded of the exhaustive work, "Iron Ore Resources of the World," published by the XI International Geological Congress, Stockholm, 1910.  Two volumes and Atlas.

[†] U. S. Geological Survey.

Victoria and about 225 miles directly north of Rio de Janeiro. The district is roughly 100 miles square and within this area are in the neighborhood of 30 known deposits of high-grade iron ore, having an aggregate tonnage variously estimated up to 3½ billion. The largest of the deposits contains, at a conservative estimate, at least 350,000,000 tons. There are numerous orebodies containing from 10 to 50 million tons. These estimates of ore tonnages have been checked by many engineers. They are based to some extent on exploration work but more largely on natural exposures, for the covering over the orebodies is scant.

The iron-ore district is reached by way of the Central of Brazil Railway, which runs northward from Rio de Janeiro through the iron-ore region to Pirapóra, on Rio São Francisco. It cuts the district into two approximately equal parts. The Central of Brazil has sharp curves and steep grades. It crosses several divides and was not planned as an ore-carrying road. A new road, the Victoria a'Minas Railway, is now under construction westward from Victoria up Rio Doce into the iron-ore district. When completed it will be used primarily for iron-ore transportation.

The principal orebodies occur as huge lenses interbedded with extensive layers of siliceous sedimentary iron-bearing formation, known as *itabirite*, which covers many square miles. There is, however, another type of ore occurring in surface blanket deposits formed by the weathering of the iron-bearing formation. This ore is known as *canga*.

The interbedded ore lenses consist of both hard ore and soft ore. The former is of uniformly high grade, averaging about 69 per cent. metallic iron, being low in silica, and generally containing less than 0.02 per cent. phosphorus. The soft ore contains from 60 to 68 per cent. metallic iron, depending on the amount of silica present, and from 0.01 to 0.07 per cent. phosphorus. The canga ores are of lower grade, ranging up to 65 per cent. in metallic iron and carrying from 0.1 to 0.3 per cent. phosphorus. More than 500 analyses of various classes of ores and covering many deposits have been made by one company alone.

That iron ores existed in Brazil has been known for many years. The early gold discoveries in Minas Geraes, made more than 200 years ago, were in the iron-ore district. In fact, many of the gold-mining operations were in the iron-bearing formation itself, in which gold occurs disseminated through soft itabirite. Later the manganese ores of Brazil also were discovered in the same region, and locally were directly associated with the iron-bearing formation.

That the iron ores themselves might be of value, however, was not suspected until recent years, due to a number of factors, such as: (1) The small demand for iron and steel products in Brazil, due to its still largely undeveloped industrial condition; (2) lack of coal suitable for iron manufacture in Brazil and neighboring countries; (3) the distance of the deposits from the coast, making the exportation of the ore a difficult problem.

About 8 years ago, Dr. Derby, then director of the geological bureau of the Brazilian Government, called attention to these great reserves of iron ore at the Stockholm Geological Congress. Shortly before this, English capitalists had made some initial purchases of iron-ore lands, and immediately thereafter, doubtless in part due to the publicity referred to, engineers of various nationalities began actively exploiting the deposits. In less than 5 years all but a few of the deposits were in the hands of English, American, French, and German interests, whereas previous to 1910 nearly all had been owned by native Brazilians. When the war began the development of the deposits was going on actively, but it soon began to decrease in vigor, and gradually stopped.

This being the general situation with regard to the Brazilian iron ores, the question arises as to what use will eventually be made of these ores and how they will affect the iron and steel industry of the world. In general, the Brazilian ore region may be compared to the Lake Superior district as it was 50 or 60 years ago. There is one great difference, however, that the Lake Superior ores were then in the midst of a rapidly developing country within easy reach of some of the richest coal fields in the world, while the Brazilian ores are in a region where there is no coal suitable for iron manufacture, and in a country which has not yet reached the stage of development when iron and steel in large quantities are being consumed. These factors necessitate the exportation of the iron ores to countries where coal is abundant and where manufactured products of iron and steel are used in large amounts.

The absence of rich, extensive coal fields in South America has been a great handicap to the development of Latin American countries. In only a few districts are coal mines being operated to any extent. The principal coal regions are the sub-bituminous coal fields of southern Chile, the sub-bituminous and bituminous coal basins of northern Peru, Colombia and Venezuela, and the bituminous coal district of southern Brazil. The latter is the only coal region within reasonable distance of the iron-ore district and, unfortunately, the coal is rich in pyrite. Frequent rumors of extensive coal beds in the Amazon valley have so far not been shown to be well founded.

The past history of the iron industry has shown that wherever iron-ore fields have been developed the movement of the ore has been in the direction of the fuel-producing centers. It is less wasteful to carry iron ore to fuel than to carry fuel to iron ore, and thus practically all of the great iron and steel manufacturing centers of the world have been established in or near coal fields. At times, after years of progress, there is a movement in the other direction, such as we notice today in the establishment of manufacturing plants in the upper Great Lakes region, but such a movement is

usually more or less artificial and does not form part of the natural course
of development in any large degree.

The Brazilian iron ores, by being exported, can therefore never
benefit Brazil as they would if fuel could be brought to the iron-ore
field. The Brazilian Government is much concerned about the future
exportation of this great natural resource, but it is now generally recognized that this is inevitable. Some years ago concessions were granted
giving large bonuses on manufactured products to persons who would
establish iron and steel producing plants in Brazil. In this manner it was
hoped that a domestic iron industry might be started. But no capitalists have availed themselves of these offers, as it is generally recognized
that an industry based on such bonuses is purely artificial. Perhaps,
however, it may be possible to operate, without loss, plants of sufficient
capacity to supply the domestic needs if such operations are linked with
the exportation of ores.

As the absence of suitable coal makes the utilization of the ores in
Brazil impracticable, so the inaccessibility and the transportation difficulties have delayed their development for export purposes. Even now
prominent engineers do not believe that Brazilian iron ores can be landed
in Europe or on the Atlantic seaboard of the United States at a profit.
Most of the engineers who have carefully studied the situation, however,
are confident that by efficient management it can be done.

As has been stated, English, American, French, German, and Brazilian capital is largely concerned in the ownership and exploitation of the
deposits. The English and Americans are most heavily interested and,
with one or two exceptions, control all the deposits in the upper Rio Doce
basin which will be tributary to the new Victoria a'Minas Railroad. The
Germans own two or three deposits in the Rio São Francisco Valley
tributary to the Central of Brazil Railway. French and Brazilians
own scattered deposits.

The Central of Brazil Railway is hardly capable of handling the man-
ganese-ore traffic at the present time, amounting to approximately 500,000
tons annually. That it will be able to haul iron ore in any considerable
quantity, besides the manganese ore, does not seem possible. The only
ore which will therefore come out of Brazil for many years to come is the
English, American, and perhaps some of the French ore, and this ore will
probably go largely to English markets to replace the gradually decreasing imports of Spanish and other high-grade iron ores. The English
process of steel manufacture demands an ore low in phosphorus.

In the United States, in recent years, the basic open-hearth process of
steel manufacture has rapidly increased in importance over the acid
Bessemer process, making it possible for American furnaces to utilize
increasingly larger quantities of ores moderately high in phosphorus.
German steel works, by using the basic Bessemer process, have been able

also to utilize high-phosphorus ores; in fact, a premium is paid in Germany for ores containing 1 per cent. and over of phosphorus.    In both United States and Germany, however, there is a strong demand for the cheaply smelted low-phosphorus ores.    In England, acid Bessemer furnaces have long been used more largely than basic furnaces and as the supply of Bessemer ores in Europe is gradually decreasing English furnace practice will have to change unless another source of ore low in phosphorus is found.

England has imported annually, during the last few years, about 4,500,000 tons of iron ore from Spain, about 1,000,000 tons from northern Africa, and about 800,000 tons from Scandinavia.    It is hoped that the Brazilian output may eventually reach 10,000,000 tons annually, which will be sufficient to supply England's demands for foreign ores and leave a surplus for the United States and for other countries.

In the United States, the Brazilian ores will probably be found to be very desirable for mixing with low-grade ores and more refractory ores in the furnace.    They will doubtless also be used in the Bessemer process to replace the gradually decreasing supply of domestic Bessemer ores.    At the eastern seaboard furnaces they will compete with imported Cuban, Chilean and European iron ores.

The great fleet of vessels that will be necessary to carry the Brazilian iron ore to the United States and to Europe will be used to carry return cargoes to Brazil.    This will offer cheap transportation for such products as coal, iron and steel, manufactured articles, and cement.    It will aid greatly in the industrial development of that country and perhaps eventually it may be possible to operate small iron and steel plants in Brazil to supply the domestic needs of iron and steel products.

### Scandinavia

#### BY WALDEMAR LINDGREN[*]

Scandinavian countries have long been known for their supply of iron ores, and Sweden more particularly has been renowned for a long time for the purity and excellence of its iron.    I will discuss the Swedish iron ores first, referring first to the southern part of the country, and then to the northern part.

The southern central part of Sweden is the land of the old iron mines, which for hundreds of years have furnished a most excellent ore, being extremely low in sulfur and phosphorus.    There are a great number of mines there, the old celebrated mines of Dannemora, Persberg, and a number of others.    They all lie in one district extending from east to west across the country at about the latitude of Stockholm.    The high-grade

---

[*] U. S. Geological Survey.

ores, generally magnetites, occur in a great number of deposits, but the resources are relatively small; the ore is utilized for local smelting into very high-grade iron.

Geologically considered, these deposits are of a very interesting nature. They occur mainly as beds in lenses of limestone or dolomite, which, again, are imbedded in dense schistose siliceous rock. They are pre-Cambrian in age; most of the deposits were formed by replacement in limestone and dolomite, at very high temperatures. The nearest analogy in this country are the contact metamorphic deposits which we find in southern parts of the West. Along with these ores, and imbedded in the same gneiss, are also hematites, in alternating layers. These have never furnished any great production, and probably never will; they are of importance, but not for export. A great deal of discussion has been carried on regarding the origin of these ores; at the present, they are generally considered as of sedimentary origin, though highly metamorphosed.[1]

In central Sweden there is a third group, represented by only two or three deposits, comprising hematite ores. The leading producer is the great Grängesberg mine from which a considerable export is now carried on. These deposits contain considerable reserves, perhaps 100 million tons. These ores could not be utilized formerly, but lately a railroad has been built from the deposits down to the Baltic coast, and, until a few years ago, they were very largely exported to Germany. They occur in the same kind of rocks but in a different manner, so that, although subsequent metamorphism has pretty thoroughly veiled the original sources, it is believed that these ores are of igneous origin. They contain a considerable amount of apatite.

Summing up for the middle part of Sweden, we may say that the annual production amounts to perhaps 1,800,000 tons, but much of this comes from Grängesberg. This is not a big production and it is very widely scattered. The total reserves are not very large, either; they have been calculated at 122,000,000 tons, of which a large proportion belongs to the Grängesberg deposit.

Still another place should be mentioned: the old locality of Taberg, which has great geological and possibly economical interest. It is an enormous mass of diabase in which magnetite is contained. It forms a knob sticking up to a height of about 300 ft. and is visible for a long distance from the surrounding lower country. Unfortunately, the ore contains a few per cent. of titanium, which has interfered with its utilization, and at present it is not used. Little furnaces were there, however, as much as a hundred years ago. The celebrated chemist Sefström discovered vanadium in those iron ores. The percentage of titanium is not excessive and it is

---

[1] H. Sjögren: The Geological Relations of the Scandinavian Iron Ores. *Trans.* (1907) **38,** 766.

believed that the material can be concentrated and worked. That reserve, which amounts to a great many million tons, is included in the total given above. For the world as a whole, this deposit presents no particular interest.

It is altogether different with respect to the iron ores in the northern part of Lapland, north of the Arctic Circle. Here, too, the iron ores are contained in old pre-Cambrian rocks, but their relations are not so obscured by metamorphism as in the iron ores of southern Sweden, and their mode of origin can be more easily interpreted. These iron ores are of the greatest importance and contain reserves or developed ore to enormous extent. Altogether these deposits comprise some 10 or 15 different localities.

The first is the celebrated Kiirunavaara deposits, forming a mountain of iron ore which rises to a height of about 300 ft. above the lake, of the same name. The second is the deposit at Gellivare, which lies about 50 miles further south, about on the Arctic Circle; this is also of great size. Then there are about ten other places, many of them containing developed iron ores to the amount of 10 or 20 or 50 million tons. Altogether the iron-ore resources of Lapland are calculated at 1,150,000,000 metric tons; of this amount, something like 750,000,000 tons is credited to Kiirunavaara, about 250,000,000 tons to Gellivare, and the rest of it is divided among the other minor deposits. A characteristic of all of these iron ores is that they are rich in phosphorus. They all contain apatite, sometimes in very beautiful crystals. At Kiirunavaara the apatite is widely scattered through the ore, but is quite inconspicuous in form.

At Kiirunavaara[2] is a mass of practically pure magnetite, mixed with some apatite, which is exposed continuously from north to south for 4 km., rising about 30 ft. (91 m.) above the lake. The orebody has been bored to a depth of at least 300 m. (984 ft.) below the lake, and contains, above the lake, something like 215,000,000 tons and below the lake about 500,000,000 tons. Apparently the deposit does not contract or pinch out in that depth. Extensive magnetic observations have been made, which seem to show that the center of the body lies at a depth of about 1000 m. below the surface so there seems to be considerably more magnetite than is contained in the prisms calculated to a depth of 300 m. This plate of magnetite lies between two sheets of igneous rocks, both referred to as syenite.

The mode of origin of the iron ore has been the subject of a great deal of discussion. At present, most geologists consider it to be a dike, developed from a magma which rested far below the surface, and there had time to differentiate into an iron-rich and an iron-poor part; the iron-

[2] See "Iron Ore Resources of the World," XI International Geological Congress Stockholm (1910) 2, 558.

rich part was then forced up between these sheets of syenite porphyry and formed this great ridge now standing up above the surrounding country. The ores differ in their phosphorus contents; a little of it is of Bessemer grade (0.05 per cent. phosphorus), but the bulk of the ore contains about 58 per cent. of iron and from 1.5 to 3 per cent. of phosphorus, as apatite. There are no other minerals present. The richest ores run about 68 per cent. in metallic iron.

The development of this great deposit has been made possible by a railway which runs down to Lulea on the Baltic in the east, and down to Narvik, one of the Norwegian fiords, on the west side. Immense docks have been built at both places, and export was proceeding at an intense rate, both to Germany and to England. The last accounts indicate that the annual production of iron ores in Sweden amounts to 6 or 7 million tons, of which the great majority, say 5,000,000 tons, comes from northern Sweden, of which, in normal times, perhaps one-third went to England and two-thirds went to Germany.

The second of these great deposits, Gellivare, is in many respects similar to Kiirunavaara. It is contained in rocks which, though of the same general kind, are yet different, inasmuch as they have been metamorphosed and made schistose. Gellivare consists of a series of lenses of magnetite in schist, which has been subjected to great pressure. The reserves are about 250,000,000 tons and the deposits are now being worked. The other deposits are of less interest; some of them contain as much as 50,000,000 tons, but they are small compared to these two giants.

Mining was formerly conducted at both Kiirunavaara and Gellivare by open cuts, but more recently they have substituted underground work for open cuts at Kiirunavaara, which were quite a hardship in the wintry climate, and at present most of the working is underground; this same plan of operation was in contemplation for Gellivare.

Passing over to Norway, until recently the production of iron has been small; the latest statistics show that only about 600,000 tons were produced in a year, but this does not prove that the resources are small. In southern Norway there are some deposits which correspond to those of middle Sweden, but they are of no international importance. In the northern part of Norway are three districts which are of special interest, and each of them differs a great deal from any of those which have yet been mentioned.

The first one is located at Sydvaranger, a part of Norway which adjoins Finland, on one of the fiords which extends to the Arctic Ocean. Here the rocks also are pre-Cambrian. The prevailing formation is granite, but in this are imbedded, for a distance of about 20 km., a great number of lenses of magnetite. These are peculiar; they consist of an alternation of thin layers of quartzose gangue and magnetite, which have been folded and compressed to a most extraordinary degree. Here, also, the geologists have differed considerably; some consider these magnetites to be of magmatic origin, but I think they were probably originally in a sedimentary formation which has been surrounded by granite and gradually metamorphosed  The ores are low-grade, but there are about 100,000,-000 tons available, and it is probable that they will be able to concentrate this material by magnetic processes; in fact, I have reason to believe that the last production from Norway, about 600,000 tons, was largely obtained by that method.  The mining practice is to take the 35-per cent. ores by open cuts, and to mine underground those parts which contain 50-per cent. iron, which will amount to about 15,000,000 tons.

To the south or southwest of Sydvaranger, on the Lofoten Islands, are a number of hills that jut out from the coast of Norway, and a great many deposits are found very similar to those that I have just described, but none of them is of very great importance.

The last of the important deposits of Scandinavia are those of Dunderlandstal; this is one of the fiords that cut into the coast of Norway. Here we have about 150,000,000 tons of iron ore available by open-cut workings.  They probably will not average more than 35 per cent. iron, and as the ore contains hematite, a great deal of capital has been sunk on experiments to discover a method of concentration.  Of the quantity available there can be no doubt.

The Dunderlandstal hematite ores are of sedimentary origin, of Cambrian age, and are younger than any of the previously described deposits. They consist of well defined beds imbedded in limestone and somewhat metamorphosed sandstone.

In conclusion, I repeat that the total prospective iron-ore supply of Norway is about 218,000,000 tons, and that of Lapland 1,200,000,000 tons.  The most important Scandinavian deposits, therefore, are Gellivare and Kiirunavaara, in Sweden.

## Cuba

### BY C. M. WELD*

Cuba, with its 3,000,000,000 tons of iron ore, holds a very important place in the world's iron-ore resources.  Nearly 90 per cent of this enormous tonnage is comprised by the soft ores of the north coast of Oriente Province.  There are less important areas of soft ores in Camaguey Province and in Pinar del Rio, while the reserves of hard ore found on the south coast of Oriente are estimated to be only about 5,000,000 tons.

### Occurrence and Extent of Ore

The broad mountainous belt of country which follows the north coast of Oriente from Nipe Bay to Baracoa is underlain by serpentine, with an

* U. S. Bureau of Mines.

occasional coastal fringe of limestone.   Lateritic weathering of the serpentine has produced very extensive surface blankets of residual iron ore. The direct derivation of this ore out of the serpentine has been repeatedly described[3] and need not be further discussed here.

Tremendous as these blankets appear to be in horizontal extent, they do not in fact cover more than a small proportion, much less than 10 per cent. of the entire serpentine area.   They have been allowed to accumulate here and there on more or less flat or gently sloping tablelands; elsewhere erosion has removed them as fast as they have been formed.

There are three localities where the topographic conditions have permitted very extensive accumulations of ore.   These have been recognized as so-called iron-ore districts and have been named, from west to east, the Mayari, the Levisa, and the Moa fields.   In my paper of some years ago[4] I distinguished also the Taco and Navas fields, but I now include these two in the Moa field.   Between these fields there are immense stretches of barren serpentine, with occasional *caitos* or islands of ore having little importance.

The ore blankets sometimes extend horizontally for several miles without a break.   In thickness they vary from a foot or less up to 60 or 70 ft. (18 or 21 m.); one drill hole was over 100 ft. deep in ore before reaching the underlying serpentine.   The average depth is from 15 to 25 ft.; there is absolutely no cover.   While the ores naturally vary greatly from place to place, owing not only to their degree of laterization but also to the nature of the immediately underlying parent rock from which they have been derived, they all show essentially the same characteristics.   These are, more particularly, high contents of both hygroscopic and combined water, high alumina, and persistent contents of nickel and chromium. The nickel is always at least from 0.5 to 1 per cent. and may rise as high as 1.5 per cent.   The chromium hovers with equal persistence around 1.5 per cent.   Iron, silica, and alumina vary between wide limits.   In what is generally ranked as ore, the iron (sample dried at 212° F.) will run from 40 to 50 per cent., averaging about 46 per cent.; alumina, from 6 to 14 per cent.; and silica from 2 to 6 per cent.   Phosphorus is always well below the Bessemer limit, and sulfur is negligible.

The Mayari field lies 15 miles south of Nipe Bay.   Nearly 40,000 acres (16187 ha.) have been taken up, all by the Spanish-American Iron Co., a subsidiary of the Pennsylvania Steel Co., and therefore now owned by the Bethlehem Steel Co.   These 40,000 acres are estimated to contain 600,000,000 tons of ore, of which about 530,000,000 carry over 40 per

---

[3] See p. 106, this volume, for discussion of this subject, and a bibliography of other papers relating to Cuban iron ores.

[4] *Trans.* (1909) **40**, 299–312.

cent. metallic iron.    This field is the only one that has been operated.
. I will return to it shortly.

The Levisa field lies some 15 miles (24 km.) east of the Mayari field
and from 5 to 10 miles south of an excellent deep-water harbor known as
Levisa Bay.    The Guantanamo Exploration Co. holds all the ore-bearing
areas of importance on this field, its estimated tonnage being about
75,000,000 averaging over 45 per cent. iron.

The Moa field is by far the largest of the three.    It lies immediately
south of the splendid harbor of Moa Bay, the ore at many places extending to the water's edge.    The field stretches 12 to 15 miles inland, and
possibly 30 miles from west to east.    It lies roughly 50 miles east from
Nipe Bay.

It is not to be understood that this entire area of some 400 square
miles is covered by iron ore, although well over half of it is actually under
denouncement.    The mining claims, however, include a great deal of
barren territory.    They also include large areas of very lean ore, locally
called salmon ore on account of its pink color, which runs only about 30
per. cent. in iron.

There are five large interests on the Moa field and a number of small
isolated holdings.    The following summary of their estimated respective
reserves is compiled from several sources:

1. The Bethlehem Steel Co., which includes the Spanish-American
Iron Co. and the Bethlehem Mines Co., holds 78,500 acres, containing
1,170,000,000 tons of ore.

2. The Midvale Steel Co., by virtue of its ownership of the Buena
Vista Iron Co., holds 19,840 acres, containing 300,000,000 tons of ore.

3. The Guantanamo Exploration Co. holds 7152 acres, containing
155,000,000 tons of ore.

4. The United States Steel Corporation holds 15,000 acres, estimated
to contain 210,000,000 tons.

5. The Eastern Steel Co. owns 10,188 acres, estimated to contain 50,-
000,000 tons; and various individuals hold some 50,000 acres, containing
probably about 100,000,000 tons of merchantable ore.

This makes a grand total of 1,985,000,000 tons of ore for the Moa
Bay field.    I have mentioned that these mines in some cases include a
great deal of lean ore which would not be classed as over 40 per cent.    Of
this nearly 2,000,000,000 tons I estimate that there are about 1,572,000,-
000 tons which would run over 40 per cent.    If we add the three places
together, Moa, Levisa and Mayari, we find a total, in Oriente Province, of
approximately 2½ billion tons of all ores and about 2,000,000,000 tons of
ore exceeding 40 per cent. iron.

I have referred to other areas of these ores in Camaguey and Pinar
del Rio Provinces.    The ores of these two provinces resemble, in all their
essential characteristics, the Oriente soft ores.    Of the Camaguey (San

Felipe) ores there are estimated to be 400,000,000 tons with 40 per cent. metallic and 50,000,000 tons with 45 per cent. metallic iron. The Pinar del Rio field contains about 40,000,000 tons of ore which will probably average slightly over 40 per cent. metallic iron.

The grand total for the Island of Cuba thereby becomes 3,000,000,000 tons of all ores, of which approximately 2¼ billion tons exceed 40 per cent. of metallic iron (dry basis). The total tonnage in its undried condition carries about 33 per cent. metallic iron. We may therefore say that Cuba's iron-ore resources represent 1,000,000,000 tons of metallic iron.

These are enormous figures and very naturally doubt at once enters one's mind regarding their reliability. The answer must be that the estimates are based on abundant prospecting, including thousands of drill holes and many more thousands of samples. I have explained that the ores lie as a surface blanket with absolutely no cover. The horizontal measurements are therefore easily determined; and the depth and quality are determined by a very simple method of drilling. Wherever areas, previously estimated, have been re-checked by close drilling or by actual mining, the reliability of the prospecting has been amply confirmed. The drilling is done with hand augers, at a cost of from 2 to 2.5 c. per foot. A party of three men will progress at the rate of about 30 ft. per hour with average conditions.

### Mining and Marketing

Since mining operations have been conducted only at Mayari, that field must furnish us with our information.

It will readily be appreciated that these soft clay-like ores offer an ideal subject for machine mining. The Spanish-American Iron Co. is now using Bucyrus drag-line excavators with 3-yd. buckets and 70-ft. booms. These machines, working continuously, can load fourteen 50-ton cars per hour. Needless to say, this capacity is never actually reached in practice, but the actual digging costs are very low. It seems not impossible that still better might be done, and at the same time some inherent difficulties of transportation eliminated, by hydraulicking these ores. I have myself made some preliminary small-scale tests in this direction, which have given promise.

I have spoken of the large content of water which is typical of these ores. The hygroscopic water amounts to from 20 to 30 per cent., generally about 25 per cent., and the combined water, at 212° F., is about 13 per cent. The total water is therefore about 35 per cent., or over one-third the weight of the ore in the ground. It is, of course, highly desirable to eliminate this water before shipping. For this purpose, and also to improve the physical condition of the ore so as to minimize dust losses in the blast-furnace, the Spanish-American Iron Co. adopted nodulizing kilns of large dimensions, resembling cement kilns.. Since that time, very successful nodulizing or sintering tests have been carried out, both with Greenawalt and Dwight-Lloyd machines, and it seems not improbable that considerable economies might result by using a sintering machine rather than a nodulizing kiln.

The present practice at Mayari is to produce nodules with 90 per cent. on 10 mesh. The composition of these nodules averages about as follows: iron, 55 per cent.; silica, about 4.5; alumina, about 13; chromium, about 2; nickel, about 1 per cent. Phosphorus is way below the Bessemer limit and sulfur is practically negligible.

Up to 1917, about 2½ million tons of nodules had been produced and shipped to Sparrows Point; this represented 7 years' operation. The maximum output was in 1913, when 491,713 tons were shipped. The present maximum capacity of the plant is about 500,000 tons yearly. Since 1913, however, bad times and more recently lack of ships have restricted the output. Before the war, sea-freight cost 85 c. per ton, including stevedore charges.

As to costs, I may quote Mr. Rand's testimony given in the Steel Corporation suit. His statement was that, with Mayari working to capacity, the nodules could be delivered on Atlantic seaboard for less than 5 c. per unit; which may be assumed to be about $2.50 per ton. This, of course, was several years ago, and labor costs have risen in Cuba as well as elsewhere. The rise in the cost of fuel would also have to be considered in any estimate of today's probable costs. Sea-freights constitute a problem in themselves.

### Metallurgical Practice

The metallurgist who is not familiar with what has actually been done with these ores might look askance at the high alumina and the chromium. On the other hand, he would no doubt readily admit the peculiar values which the nickel and chromium would give to the steel made from them.

The 2,500,000 tons of nodules which have been shipped to this country have all been smelted and converted into iron and steel, chiefly at Sparrows Point, but also to some extent at Steelton. That entire success had been attained with them was shown by Richard V. McKay[5] some 3½ years ago. At that time he recorded that a mixture containing a little over 60 per cent. of Mayari nodules was being smelted on a fuel ratio of practically 1 to 1, and he had nothing but praise for the slag. Continuous runs have been made for months with 100 per cent. Mayari ore.

The resulting pig iron is very similar to spiegel in appearance, showing a white crystalline fracture with no grain. A peculiar feature is the

[5] *Year Book*, American Iron and Steel Institute (1914) 85.

high amount of combined and the low amount of graphitic carbon. It contains practically all the nickel and most of the chromium in the ore, low silicon, very low phosphorus, and small amounts of titanium and vanadium. These latter increase with the silicon.

In its natural condition, this pig is useful only for steel making. It has been shown, however, that as a mix with other irons it is extremely useful for castings, improving the chill and giving increased hardness and strength, while increasing only slightly the power required to machine it. With all sorts of chilled castings, particularly rolls, guides, crusher plates, car wheels, and other products subject to hard wear, mixtures including from 15 to 40 per cent. Mayari pig have given tremendously increased service.

### Mayari Steel

In making steel with Mayari pig, the duplex process is used. This process as carried out at Sparrows Point has been fully described by F. F. Lines.[6] It will be enough to say here that the results have been entirely satisfactory. A large part of the chrome is slagged off and a steel of superior quality is produced, containing from 1 to 1.5 per cent. of nickel, and from 0.2 to 0.7 per cent. chromium, as desired. Sulfur and phosphorus are below 0.04 per cent. This steel may be forged, rolled, and machined as easily as carbon steel, and has from 8000 to 10,000 lb. per sq. in. higher tensile strength and elastic limit. Mayari rails have been tested in positions where the service was particularly hard and have given excellent results.

In Paper No. 108 of the International Engineering Congress of 1915, there is an interesting discussion of the use of Mayari steel in the Memphis bridge. A series of tests is given showing an elastic limit for large-size angles of 59,000 lb. and a tensile strength of 94,500 lb. per sq. in. Elongation and reduction of area also generously exceeded the specifications, which were 1,600,000 ÷ T.S. for the former, and 30 per cent. for the latter.

But Mayari steel shows its excellent qualities more particularly when heat-treated. By simple quenching in oil, track bolts are made with a tensile strength of 100,000 lb. For eye-bars, springs, crank-shafts, and all sorts of automobile parts, this steel is particularly useful. Being a natural alloy, there is not the same degree of segregation as when nickel or other constituents are artificially added. S. W. Parker has published[7] an interesting comparison between heat-treated Mayari and 3.5-per cent. nickel steels, which is most favorable to Mayari steel. He finds that the one may generally be substituted for the other.

---

[6] *Trans.* (1915) **53**, 357.     [7] *Iron Age* (June 7, 1917) **99**, 1380.

In closing, I may emphasize the peculiarly excellent qualities of the materials which may be produced out of these Cuban soft ores, of which there are such enormous reserves. Mayari has but served as an example. The other fields will give exactly the same results. The ores themselves can be cheaply produced and offer no metallurgical difficulties which have not long been overcome in actual practice. When the world's shipping problem has once more been settled, these Cuban ores will be among the most available iron resources of the several countries bordering on the Atlantic Ocean, including the United States.

### Southern Europe

#### BY A. C. SPENCER*

In presenting a rude picture of the iron-ore reserves of the region surrounding the Mediterranean, I will not touch on the Lorraine problem, but will include the remainder of France. The International Geological Congress, in 1910, gives as an estimate of the iron ores in France 3,300,000,-000 tons, of which 3,000,000,000 are assigned to French Lorraine, leaving 300,000,000 for the remainder of France.. Since those estimates were made, geologists and engineers have been searching France with more or less care, and other estimates have been made—how reliable I am not able to judge. Recently, in one of our technical papers, I noticed that a German iron master, in addressing a meeting of colleagues, made a statement which, to use my own words, was to the effect that France was a dog in the manger, as she had 10,000,000,000 tons of iron ore at her command, including those of the home country and of the colonies, whereas Germany had but 3,000,000,000; hence France was standing in the way of Germany in refusing willingly to give up the 3,000,000,000 tons of ore in French Lorraine, which would give Germany only 6,000,-000,000 and still leave France 7,000,000,000. The conclusion was that it would be only fair for France to divide up. Presumably, in those circumstances, a country with which we are at war would be inclined to exaggerate, so I suspect he has combined estimates of French engineers and German engineers with exaggeration enough to indicate that France is playing a mean part.

Let us scale down that estimate and allow that, outside of French Lorraine, France and her colonies may have 2,500,000,000 tons of iron ore, and it is perhaps within the realm of possibilities. The new ore reserves, now partly developed, are in Normandy and Brittany. They consist of bedded iron ores of moderate grade, and rather siliceous. The most extensive development took place just prior to the war, largely under the influence of the German iron masters of Westphalia. Aside from that,

* U. S. Geological Survey.

iron ore on a small scale has been mined in nearly every department of France. They are scattered about the plateau of France and in the foothills of the Pyrenees and the Jura Mountains.

As we approach the Pyrenees we find a different type of ore, namely those occurring in veins and in many places as replacements of limestone. At the surface the ores are limonite, but as depth is gained they grade into carbonate or a mixture of limonite and hematite. It appears that ores of this class may have more industrial importance than they have been assigned in the inventory of the iron resources of the world. They range along the slope of the Pyrenees and extend into Northern Piedmont. If we go to the west face of the Jura Mountains, we find deposits of rather low-grade surface ores, and moderate deposits of metamorphosed sedimentary ores, the latter occurring also in Switzerland.

In Spain, in the Pyrenees, there are iron ores of two types; metamorphic ores in limestone near igneous intrusions, and the carbonate type carrying pyrite and other sulfides in minor amounts. In Spain, the ores that have been inventoried comprise the deposits adjacent to the Bay of Biscay, which have produced the largest amount of iron ore up to the present time, if I remember correctly, an aggregate of some 25,000,000 tons. The maximum annual production of Spain in 1913 was nearly 10,-000,000 tons. Elsewhere in the Pyrenees there has been an aggregate production, including the total of Spain, of somewhere around 33,000,000 tons of iron ore—not a very large amount. The reserves were estimated in 1910, together with 18,000,000 assigned to Portugal, at above 700-000,000 tons, so that to Spain we can hardly look for a long supply of the high-grade iron ores which has made Spain a principal contributor, on the one hand to Great Britain, whose Bessemer steel industry she has largely supported, and on the other hand to Germany, to whom she has furnished a very large part of such low-phosphorus iron ore as she has required.

Considered with reference to the world's supply, the iron resources of Italy are almost negligible. In the center of Italy are widely separated deposits. Then in Northwestern Italy, in the Aosta district, there are magnetite deposits of a considerable size, of about equal importance with those of Elba, which thus far have been the only deposits to be worked on a large scale. These are ores of direct igneous origin. Then in Lombardy there are some deposits of a sedimentary nature, but from the world's standpoint we can ignore them. Italy has neither fuel to meet the requirements of the blast furnace nor iron ores to support a long-lived industry.

The iron resources of Austria-Hungary are also moderate, being rated at some two or three hundred million tons in the aggregate. In Bosnia and Herzegovina there are small deposits, and presumably in Servia some moderate deposits, whereas Greece has a goodly supply for a small country, stated at 100,000,000 tons of nickeliferous chromiferous ores of a class very similar to the Cuban ores. In passing, it may be said that some of the essential points in the metallurgy of these chromiferous ores were worked out in connection with the Grecian ores. These ores do not occur as surface blankets, but I believe that it is not amiss to state that they are of precisely the same type as those of Cuba. They have been covered by the limestone and sandy shale, but they were formed as a result of the weathering of serpentine rock precisely as in Cuba and in various other parts of the world.

This short outline can now end with a consideration of the iron ores of Algeria, and Tunis. All I can say of them is that iron ores occur at various points along the Atlas Mountains, and are of various origins, largely siderite replacing limestone and, like the carbonate ores of France and Spain, carry varying proportions of sulfide minerals. They have been rated at between 100 and 150 million tons of ores, averaging about 50 per cent. iron, and in large part of Bessemer grade.

On the whole, it is fair to say that, aside from French Lorraine and parts of Brittany and Normandy, the iron reserves of the countries surrounding the Mediterranean do not measure up to those of the Lake Superior region or those of Newfoundland, Cuba, or Brazil.

## China

### BY H. FOSTER BAIN[*]

It has been customary for many years to look upon China as containing one of the great iron-ore reserves of the world. This notion probably came about through the writings of Von Richthoven, who, in his journeys through China, passed through many regions in which iron was made; particularly in Chih-li he saw large numbers of furnaces, and excellent coal, and came to the conclusion that that would be one of the great iron-ore districts of the world. At the time he went through this country, the world was still using small furnaces, such as those of which you will find ruins in the Mississippi Valley. Places we have forgotten about were running then, and also small local furnaces were running all over Europe, so the conditions were such as to give some justification for the opinions he gave.

Recently the iron ores of China have been studied with considerable care by the Japanese, by the Chinese themselves, and by various foreign engineers, and we can say with some assurance that the old idea of iron-ore reserves is wrong, that the iron-ore reserves of China are very moderate in comparison with the previous notions or opinions. In a general way, I may say that, making full allowances for the iron ores suitable for

---

[*] U. S. Bureau of Mines.

treatment by the native methods, China still has to be classed with Spain rather than with Brazil, United States, or the Lorraine district.    The best estimate that I am able to give you for the total of the iron-ore bodies which are of such type, character, and situation as to be suitable for modern blast-furnace work is in the neighborhood of 400,000,000 tons. V. K. Ting, the head of the Chinese Geological Service and a very capable man, estimates the additional amount of iron ore which may be worked by Chinese local furnaces as 300,000,000 tons, but for practical purposes we are clearly safe in assuming that the iron ores of China are in the neighborhood of 400,000,000 tons.

Of this, approximately one-third was still in the possession of the Chinese Government last year; that is, the title belonged to the Chinese Government, although titles are in quite a dubious condition.    The Chinese Government has been attempting for some years to reëstablish its right to the mineral under the land, and it has succeeded fairly well so far, at least, as the big iron-ore deposits are concerned.    Chinese-owned companies held something less than one-third, and the Japanese and Sino-Japanese companies had more than one-third.    In the whole of China there was not one single deposit that belonged to the nationals of any other country.

As to the character of the deposits, five types can be distinguished.

1. Ancient banded deposits.    These contain both hematite and magnetite; they are of uncertain origin and the grade is also open to dispute. They occur mainly in Manchuria, and they are the ones of which we have heard some very large estimates.    Those estimates seem to include the lean ore which will have to be concentrated to make it workable, and in nearly every case the actual amount of ore which can be used directly in a furnace is relatively small.    Whether the lean ore can be worked on a large scale and concentrated is a technical question and such work runs into cost rapidly.    Some preliminary tests of magnetic concentration in one case did not work out favorably, but others may be more successful.

2. Sedimentary carbonates and hematites, such as are in Shan-si. This is the type which Von Richthoven saw.    When the Pekin Syndicate studied the question and W. H. Shockley investigated the field, he pointed out that the individual masses of ore are so small that they are unfitted to feed any modern furnaces.    The individual lenses are 3 to 6 in. thick and up to 16 ft. in diameter.    This does not look very attractive to a man who owns a modern furnace.    In Pao-king and Hu-nan they have the same type of ore and some pig is made; this is generally marketed in the form of castings.

3. Sedimentary Oolites.—One such deposit occurs in Chih-li.    These ores are extensive and probably valuable.    A second deposit is in Kiang-si and it is probably too lean; that is, while there are perhaps 30,000,000 tons of ore in a particular place, the bulk of it is probably too lean to work.

The story of genesis there seems to be exactly the same as in the Mesabi districts but the leaching and enriching has not gone so far, so that we can hardly look for any considerable deposits of ore there.

4. The most important is the contact metamorphic type in the Yang-tze Valley. In the opinion of Dr. J. G. Andersson, these ores are found only around the edge of lacolites. The deposits vary considerably in character and grade, and a number of them are workable. It is difficult to make exact estimates of contact deposits without very accurate prospecting, which has not taken place.

5. Residual Deposits.—There are brown ores in the Yang-tze Valley and probably there is a larger amount of brown iron ore in the southern part of China than has been appreciated, but it is very likely in small scattered deposits and in conditions not very favorable for development.

I can cite one or two examples of the way these estimates scale down on examination in the field. You have all seen estimates of an iron deposit of Fu-kien; it is usually given as 10,000,000 tons of iron ores. Examination of that deposit showed a maximum of 2,000,000 or 2,500,000 tons of ore, and that a large amount of black porphyry had been confounded with iron ore. There are three recent estimates of the Tayeh deposits (at one time supposed to contain 300,000,000 tons) which seem to check fairly well and indicate perhaps 30,000,000 to 40,000,000 tons of workable iron ore. The best information seems to be that there is just about enough iron ore there for the company to fulfil its existing contracts. There is, of course, the possibility of finding more, but in contact beds that is an uncertain resource.

We have had an example in this last year of how important iron ore is to a nation. At the time that the United States went into the war and had temporarily to shut off exports of steel, the Japanese shipyards were in very bad shape, because the steel consumption in Japan had gone far beyond its capacity to produce, and it was found that if the shipyards were to be kept going it would be necessary to cut off absolutely all other use of steel, for the maximum of their local consumption was barely enough for their shipyards alone.

Originally the Japanese and the Chinese had about the same amount of iron ore *per capita*, but the Japanese woke up before the Chinese did and they proceeded to acquire additional quantities of iron ore. Japan had been depending largely on agricultural resources. She has an expanding population and has not had the same opportunity to spread out in new territory that we had in our great West, or as the British had in Australia, Canada, and South Africa; hence the Japanese had found that it was necessary, in order to maintain themselves, to change the character of their civilization and become a manufacturing and shipping nation instead of an agricultural nation. In order to do that, it was necessary to secure large deposits of iron ore. It is from this fundamental point of view of necessity that Japan has been so insistent on acquiring additional iron ore. The Chinese have not felt that necessity yet, [because they have not yet, in any large number, come to the conclusion that they must change the character of their fundamental industries. '

## Alsace-Lorraine

BY SIDNEY PAIGE*

About 6 months ago I became interested in the Alsace-Lorraine matter from the standpoint of the policy of the United States in its relation to these areas, and the possible peace between the Allies and Germany. I feel that ultimately the problem of solving this difficulty must rest, at least I hope it will rest, on geologists, engineers, metallurgists, and technologists.

Never before, perhaps, in the history of the world has the policy of a great nation been fraught with more momentous consequences than is the policy of the United States at present. To any thoughtful man impressed with the "velocity" of modern civilization, it must be evident that "direction" of movement is of vast importance, if the best interests of humanity and of that which is best in modern civilization are to be preserved and fostered. It is a significant fact that today there remains no great unpopulated territory. The westward movement of peoples, a movement impelled by economic pressure—a search for resources and later for markets—has slowed down. The great resources of the world are known. No longer may nations cramped by expanded industrial systems seek relief, without meeting unparalleled opposition, or without upsetting the industrial balance of the entire world. Trade, the cause of many wars, now links together the nations of the entire globe, and there is every reason to believe that science, which in a brief past has so marvelously obliterated time and space, will in the future accomplish even more. It follows that the political task of modern life must relate to equitable economic adjustments of nations. The waves of peoples inundating the world have met around its circumference, and man is confronted with the responsibility of seeing to it that some measure of quiet is established.

Problems of great difficulty confront modern civilization, even though the Utopian assumption be made that the last great war is being waged. These difficulties arise from the patent inequalities in resources, nature of peoples, ethics, education, and all that this implies. The spirit of self-preservation, persisting as selfishness, the desire for surplus and power, are apparently innate in human nature. But there are other traits in human nature which, it is the hope of many men, may dominate the world.

---

* U. S. Geological Survey.

No one who analyzes the "regulation," adopted by the great industrial nations involved in this war, can fail to realize that in its very essence it is unselfish.   It has to do with the "greatest good for the greatest number" and has been promulgated by free peoples.   In, fact, by *the people*.   He who imagines that these changes have been born but to die an early death, lacks vision.   They will be nurtured, cultivated, and will thrive—to grow into a body of procedure of vast importance to the wel- · fare of man.

War is modifying our views of labor, of distribution, of public finance, and production.   In fact, it is shaking the whole traditional structure of our economic life.[8]

The principle is being established or will be established "that the essential commodities are subject to control in the public interest precisely as are the utilities."[9]

Nations now at war are applying these ideas.   In the coming period of peace, it is vital that something of this wisdom be applied *among* nations.

There are two great aims in this war, and their importance justifies any sacrifice.   First, the arrogant, autocratic, military despotism of Germany must be crushed.   Nothing is clearer than this—that control in Germany has been in the hands of a ruthless group—so strong indeed that they have been able to carry on a biological experiment on a vast scale.   A people have been imbued, from the cradle to the grave, with a philosophy which appears to us no less false than cruel.   The people exist for the State, and the particular State in question is a relic of barbarism and the philosophy built upon a decadent idea that the State can do no wrong.   Our second aim of the war has to do with such redistributions of resources and territory as will tend to preserve peace and lead to the normal industrial progress of nations.   And it is in this connection that the iron ores of Lorraine are involved.   The position of the United States is unique.   Her resources are the greatest in the world; her peoples are imbued with the virility of all great peoples, and her position geographically with respect to trade is, to say the least, remarkable.  She lies between Europe and Asia.   Great trade routes will meet and pass through her territory.   And she has entered the war at a stage either when her weight will compel decision (and thus determine the nature of the coming peace) or else she will go down in defeat, which is unthinkable.

Practical men realize that, in wars, agreements usually follow a decision by arms; but never before in war has a decision by arms so completely involved the entire genius of peoples, so thoroughly tried their powers of coöperation or been so fraught with possible unexpected developments.

[8] William S. Culbertson: *North American Review* (Jan., 1918) **207**, 61.
[9] Charles R. Van Hise: *Science*, N. S. (Jan. 11, 1918) **47**, 36.

Therefore, never before has it been so important to have well defined policies. It is well recognized that without the iron ores of Lorraine, Germany could not have waged the present war; and therefore that if these ores be taken by force of arms peace will follow almost at once. The general staffs of all the great nations are no doubt aware of this fàct; likewise the Germans.

It must be recognized at once that in Lorraine (leaving Alsace out of the discussion for the moment) and contiguous France, there is today the greatest iron-ore reserve of Europe. No single factor, perhaps, no group of facts involved in this war, deserves more thoughtful consideration. Upon the proper disposition of these reserves, upon the nature of the barriers that may be set up or torn down in their utilization, upon the spirit of coöperation or competition which enters into their disposal, depends in large measure the future peace of Europe.

It is of interest to analyze briefly the present situation with respect to Alsace-Lorraine.

Alsace-Lorraine has been the battle ground of Europe ever since the days of Caesar when, in 72 B. C., a German tribe invaded this territory and settled down. It is useless except for moral discipline when feelings of self-righteousness become acute to dwell upon the number of times this fair territory has changed hands. Enough, that in 1871, France lost half of it to Germany. And it is significant to note that before this date it belonged wholly to France, left to her by the preliminary agreement of Versailles at the end of the Franco-German war, but that the German geologist, Hauchcorne, pointed out to Bismarck the potential value of this area, and France was persuaded to cede a strip of it by the subsequent treaty of Frankfort in exchange for land of military value near the fortress of Belfort. Geologists, therefore, were already, in 1871, of use to their governments at war. Since that time any lack of foresight on the part of French geologists has been corrected. They realize today the value of Alsace-Lorraine.

Coal and iron are so important in modern war, and the Germans, so well supplied with certain other necessities, are so ingenious in devising substitutes, that a consideration of such other necessities must take second place. Germany has vast resources in coal but poor resources in iron; but if the resources of Lorraine be included the resources of Germany in iron become vast. And if there be included French territory now held, she possesses by far the greatest resources on the eastern hemisphere. Of a production of 28,600,000 tons of iron ore in 1913 by Germany, 21,000,000 came from Lorraine; of coal, Germany, without French territory, possesses more than half the resources of all Europe. England follows her in resources, Russia next, then Austria-Hungary, and fifth comes France. The Saar Valley alone contains more coal than is known in France today.

One does not need a vivid imagination to picture the strength of this combination which at present she commands, the greatest coal and iron resources of Europe.

France, on the other hand, is deficient in coal. In iron, before the war, her resources were only slightly less than those of Germany, and of a total production of 21,700,000 tons by France, 19,500,000 came from the Longwy-Briey field, the identical field from which Germany draws her main supply.

Briefly then, what was the situation before the war? Simply this: Germany, by unparalleled activity in the development of great coal resources, was using not only all the iron she could command, but importing an increasing amount year by year from France. France, on the other hand, short of coal, was importing some 23,000,000 tons and exporting iron ore to the amount of one-tenth of the production of the Longwy-Briey field, to Germany.

Then came the war, and with it changes of the utmost significance, for not only does Germany now hold the entire Briey-Longwy iron field, but also, in the north of France, the coal field extending from Valenciennes to beyond Lens.

The battle line on the western front offers opportunity for speculation. After the great advance which threatened Paris (German troops were within 20 miles of the city) the invading armies withdrew to the line of July, 1916. Recall the position at that time of the St. Mihiel salient and recall the line around Verdun. Note also that the great manufacturing center of Lille and Lens are in German hands and that the tip of the Flanders coal field alone remains in French possession. From that date the line moved to that of September, 1917, and again recall that the St. Mihiel salient has not changed; recall that the ferocity of the attacks upon Verdun were unparalleled and note that Lille and the Flanders coal field have not been relieved.

Much has been written to account for the German onslaughts at Verdun; political reasons have been assigned. Only recently, it would seem, has the true significance of this maneuver been understood. The St. Mihiel salient and Verdun are in the path to the Lorraine iron fields. That the Germans should desire to straighten this line is natural and that it remains nearly stationary means that it is well defended. Indeed, a great German offensive may be expected here. To regain the Flanders coal field would relieve a great need in France. To regain French iron fields and drive the Germans from Lorraine would change the aspect of the war.

Let us turn now to the war aims of French geologists and metallurgists. DeLaunay foresees France at the end of the war "triumphant, happy. * * * We impose our wishes, we impose them completely. I do not consider any other hypothesis as possible, as worthy of discussion." And the conditions that he expects to be imposed include not only the restitution of Alsace and Lorraine, but the annexation of sufficient German coal and coal fields to redress the mineral deficiencies of France.

In 1913, France produced 40,000,000 tons of fuel, while her consumption reached 60,000,000 tons. A deficit of 23,000,000 tons, therefore, growing year by year, required imports from England, Germany, and Belgium to supply her needs. Of coke, which is the principal element entering into the cost of pig iron, the situation before the war was even more unfavorable. The 3,000,000 tons which were imported approached the figure of domestic production.

As has just been said, France before the war faced a deficit each year of 23,000,000 tons of fuel. This figure represents 19,000,000 tons of coal and 3,000,000 tons of coke (4 tons of coal to produce 3 tons of coke). Lorraine alone produces 4,000,000 tons of coal but no coke, and consumes 6,000,000 tons of coal and 4,500,000 tons of coke (made from 6,000,000 tons of coal). A total of 12,000,000 tons of coal, therefore, is consumed in Lorraine, while this territory produces only 4,000,000 tons. There is a deficit, therefore, of 8,000,000 tons of coal in Lorraine. Therefore, if Lorraine is returned to France, the deficit of that country in coal will rise from 23,000,000 tons to 31,000,000 tons. The French, therefore, propose that they also be given the coal of the Saar Valley in Germany. Should this wish be realized, the consequences are as follows: The Saar Valley produces 10,000,000 tons of coal, which is consumed crude, and 3,000,000 tons of coal, which is made into coke, 13,000,000 tons in all. This region consumes 5,000,000 tons of coal (2,000,000 as crude coal and 3,000,000 transformed into coke). There is a surplus, therefore, of 8,000,000 tons of coal, that is to say, precisely the amount deficient in Lorraine.

But this apparent balance of resources takes on a different aspect when analyzed with respect to coke.

Lorraine and the Saar Valley together produce, as set forth above, 17,000,000 tons of coal, divided as follows:

|  | Tons |
|---|---|
| Coal consumed crude | 14,000,000 |
| Coal made into coke | 3,000,000 |

This territory likewise consumed 17,000,000 tons of coal, but this amount is divided thus:

|  | Tons |
|---|---|
| Coal consumed crude | 8,000,000 |
| Coal made into coke | 9,000,000 |

Now, while the deficit of 23,000,000 tons which France faced before the war will not be augmented if Lorraine and the Saar Valley are taken, her situation with respect to coke will be worse. Before the war the

23,000,000 tons deficit of France was divided as follows: 19,000,000 tons of crude coal; 4,000,000 tons of coal for coke. With the new arrangement these figures will become: 13,000,000 tons of crude coal; 10,000,000 tons of coal for coke. The importations of metallurgic coke into France in 1913 were divided as follows:

|  | Tons |
|---|---|
| From Germany | 2,393,000 |
| From Belgium | 547,000 |
| From other countries | 130,000 |
| Total | 3,070,000 |

There is therefore a deficit of nearly 7,000,000 tons of coke which France must make up by import. The Saar Valley does not produce good coking coal and France wants coking coal, not coke, for she wishes to obtain the byproducts. Eng`and, before the war, consistently refused to send coking coal and would sell only coke. France after the war will be forced to buy fuel. Where can she buy coking coal? And just here the basic principle is illuminated, that nations must cöoperate.

If the situation in France with respect to ore is examined, assuming that she possess, after the war, Lorraine and the Saar Valley, it will be found that she faces an equal predicament—that is, a market for her ores. This is clearly set forth in the analysis by Robert Pinot.[10] But the crux of the matter is reached when the situation with respect to the manufacture of steel is analyzed.

In 1913, France produced 5,311,000 tons of pig iron; 957,000 tons were consumed by French foundries or exported crude. The remainder, 4,354,000 tons, was made into iron and steel. Lorraine and the Saar Valley in 1913 produced 5,241,000 tons of pig iron, of which 4,502,000 tons were transformed into steel. France after then would have a capacity of 11,000,000 tons of pig iron, of which 9,000,000 tons would be made into steel. Furnaces erected during the war would supply another million tons of steel. France thus would be compelled to dispose of 10,000,000 tons of steel. Where would this steel go? It may be shown that, even if a roseate view is taken of the French industrial situation after the war, France will be in a position (always assuming that she possesses Lorraine and the Saar Valley and ample coal from Germany) to produce 4,000,000 more tons of steel than she can dispose of. Pinot estimates as follows:

|  | Tons |
|---|---|
| French consumption before the war | 5,000,000 |
| Lorraine-Saar consumption | 400,000 |
| Exportation and reduction of temporary importation. | 600,000 |
| Total | 6,000,000 |

[10] Robert Pinot: La métallurgie et l'après guerre. *Bulletin et Comptes Rendus Mensuels*, Société de L'Industrie Minérale, 1re livraison de 1917, 36.

Four million tons approximately remain unaccounted for.

|  | TONS OF STEEL |
|---|---|
| In 1913 Germany exported | 6,500,000 |
| In 1913 England exported | 5,000,000 |
| In 1913 United States exported | 2,500,000 |
| In 1913 Belgium exported | 1,700,000 |

France cannot hope to develop exportation at the expense of the United States, a country which has enormously increased her capacity during the war. Nor can she hope to compete with Great Britain, who will, after the war, be in a strong position metallurgically. And as for Belgium, she will have some 1,184,000 tons for export. Again, therefore, the French are driven into competition with the Germans.

Pinot outlines a plan and presents figures, to the end that Germany lose her markets. There is no need to review these figures here. Their importance lies in the fact that this entire question deserves the most painstaking research on the part of the United States, England, and France, that some equitable basis for peace may be reached.

The analysis above brings out clearly the need of an economic policy with regard to the disposal of the Lorraine and contiguous French ores. It must be constantly borne in mind that this iron-ore reserve is of supreme importance to all the great industrial nations of Europe. If the war is concluded with the military autocracy of Germany victorious, there is no need of further thought to this matter. Such a Germany will see to it that an enslaved Europe will pay tribute.

I will close in repeating one sentence. Upon the proper control of these resources, upon the nature of the barriers that may be set up or torn down in their utilization, upon the spirit of competition or cöoperation which enters into their disposal, depends in large measure the future peace of Europe.

## DISCUSSION

THE CHAIRMAN (J. W. RICHARDS, South Bethlehem, Pa.)—The first point on which I wish to speak is that sufficient attention has not been given to the quality of the ores, in estimating the reserves. I made this same criticism of the work on the iron-ore resources of the world issued by the XI International Geological Congress. They listed only the ores having present metallurgical value and left out of consideration all others.

However, since metallurgical practice is steadily improving, the worthless ores of today gradually become the commercial ores of tomorrow; hence an estimate to the effect that we shall run out of good iron ore inside of 50 years, if the present rate of consumption is maintained, is of small interest to the metallurgical industry. The world will never run short of ores for producing iron, although it may reach the end of the rich ores such as we now think we cannot get along without.

The second point is that, while the world will never be without iron ore, we cannot get iron out of its ore without fuel; hence the supply of fuel becomes the controlling factor.    For this reason, at the XI International· Geological Congress, as a companion to the magnificent work on iron-ore resources, I proposed that a similar volume on the coal resources of the world should be compiled.    I did not realize at the time .that I was unwittingly promoting the designs of the military Junkers. Those countries which have the supply of fuel will control the production of iron.    The supply of iron ore is practically unlimited, compared with the supply of fuel by which it must be smelted.

## Relation of Sulfides to Water Level in Mexico

BY P. K. LUCKE,* SAN ANTONIO, TEX.

(Colorado Meeting, September, 1918)

ONE of the interesting features connected with the great continental uplift, which formed the table land of Mexico, is the great depth to which oxidation and secondary enrichment of orebodies occurred. This table land lies at altitudes ranging from 4000 to 7000 ft. (1219 to 2133 m.) above sea level. Several cases could be cited of oxidized zones in copper and lead mines which reached depths of upward of 1500 ft. (457 m.) below the surface, and it is universally admitted that the oxidized zone in a mine is limited by the permanent water level.

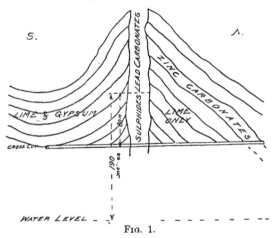

FIG. 1.

Mexico, however, like many other countries in which deserts occur, is visited at rare intervals by tremendous floods, which sometimes cause variable water levels over long periods. The following cases bear out the idea, and lead to interesting conclusions of commercial value. My paper demonstrates that carbonates of ores may occur below water level and sulfides above, and I will endeavor to explain the causes of the phenomena in certain specific instances.

*Case I.*—Fig. 1 in this case shows that sulfides of lead remained

---

* Consulting Mining Engineer.

unoxidized 190 m. (623 ft.) above water level, while carbonates of zinc will probably continue down to water level.   The cross-cut bisecting the anticline affords the following data: The south side of the anticline shows gypsum and limestone interbedded, while the north side contains only limestone.   Immediately contiguous to the lead orebody, massive gypsum occurs, and the body of ore is in sulfide form at the level of the cross-cut and for a distance of 80 m. (262 ft.) above it.

The zinc orebody, on the other hand, has extended to the cross-cut level in carbonate form, and will probably remain in oxidized condition to the water level.   The explanation of this condition is probably that the gypsum resulting from the oxidation of the lead orebody has cemented

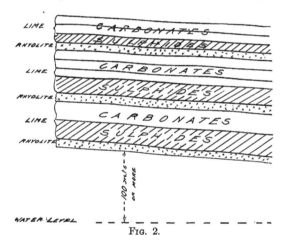

Fig. 2.

the bedding planes of the limestone underneath and to the south side of the anticline, thus damming back the surface waters, which under normal conditions would have oxidized the ore down to the permanent water level.

On the north side of the anticline, as stated, no cementing of the bedding planes by gypsum occurs, and it was therefore possible to predict that oxidation of the zinc orebody would probably continue down to the water level, and development has proved that the prediction was justified.

Other cases of local occurrences of sulfides above water level could be mentioned, but one more will suffice for the purposes of this paper.

*Case II.*—The lead-silver orebodies occur in this instance in a series of more or less vertical fractures cutting interbedded limestone and rhyolite. Fig. 2 illustrates the example.   While the water level has not been reached, it probably lies at a depth of from 100 to 200 m. (328 to 656 ft.) below the deepest workings in this mine, sulfides remain unoxidized immediately above the interbedded igneous rock, but owing to the surface waters having found their way on the line of ore shrinkage represented by the footwall of that rock at the various horizons, oxides of lead and zinc are found overlying the sulfides. This curious alternation of oxidized ore and sulfides will probably continue in depth until water level is reached, and an increase in the thickness of sulfides in comparison to that of the carbonates may be looked for at each successive horizon. When water level is encountered, no more oxides will be found unless Case III is duplicated in this camp, which is possible.

*Case III.*—Lead and copper ores occur in this example in irregular chimneys which originated in fractures in Cretaceous limestone, and the present water level lies at a depth of 350 to 400 m. (1148 to 1312 ft.) below

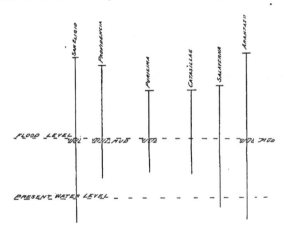

Fig. 3.

the surface. In the Aranzazu, San Eligio and Salaverna mines, carbonates have been found at depths as great as 100 m. (328 ft.) below the present water level. In three mines in the district under discussion, and 100 m. (328 ft.) above water level, good orebodies have bottomed in valueless mud. The mud in one case was 20 m. (65 ft.) thick, and, in each case below the mud, valuable ore has been found again and continues downward.

It is fair to conclude, therefore, that the mud represents the highest level reached by the floods, and that the present water level is receding to greater depths, as otherwise carbonates of the ores would not persist below the present water level. It is well to note, also, that no marked zone of enrichment has yet been encountered, which would certainly be the case if the permanent water level had been reached.

The above conclusions are of commercial importance to the district, and probably to others also, as the original sulfides, when encountered will probably be of low grade, although certain lead mines may become paying copper mines when the sulfide zone is reached.

As my statement with regard to the probability of the original sulfides being of low grade may be questioned, I will qualify it by stating that in my opinion the majority of the large oxidized orebodies in the table land of Mexico owe their high values to secondary enrichment, which takes place under ideal conditions in respect to water level and climate. The value of carbonate ores in Mexico is therefore no indication of the value of original sulfides.

It is also reasonable to conclude from the foregoing that:

1. The discovery of original sulfides, showing no secondary enrichment, above permanent water level is possible but not usual.

2. The discovery of a local water level high above the known water level of the district is a fairly common phenomenon, and when found indicates the presence of local sulfides.

3. Even if a high local water level be found, it is still possible, under favorable conditions, to find that oxidation occurs beneath the local sulfide zone indicated by it.

4. Many present water levels, especially those in desert countries, may rise and fall over considerable distances and periods of time, and do not necessarily decide the limits of the carbonate zones in mines, although they may warn us of the proximity of sulfides.

## A Study of Shoveling as Applied to Mining

BY G. TOWNSEND HARLEY,* TYRONE, N. M.

(New York Meeting, February, 1919)

### CONTENTS

### INTRODUCTION

STOPING methods in which shoveling plays an important part are gradually being replaced by other and cheaper methods. But there will always be considerable shoveling done underground in stopes as well as in drifts, tunnels, winzes, and shafts. At the mines of the Phelps-Dodge Corporation at Tyrone, N. M., the cost of shoveling in all stopes in 1917 amounted to 24 c. per T. In the top-slice stopes for the same period, it cost 27 c. per T. or 16 per cent. of the total cost of these stopes. The tonnage for shovelers from all stoping was 9.3 T. per man, and for top-slicing 8.2 T. per man. These stopes were not unduly hot, and there was not more than the usual amount of timber to interfere with the work of the men.

The tonnages obtained per shoveler were considered low; first, because of a poor grade of Mexican labor, many of the men having come in from railroad grading camps; and second, because of a poor spacing of raises, especially in the top-slice stopes, where, in general, they were spaced 25 ft. by 66 ft. (7.6 by 20.1 m.) centers. The average wage per laborer shift was $2.67 during the year. It was thought, however, that even under

---

* Efficiency Engineer, Phelps-Dodge Corp., Burro Mountain Branch.

these conditions the men were not producing the tonnage that they should, so, with the consent of the management, the writer undertook to determine how the general efficiency of the underground shoveling could be improved. No predetermined plan for conducting these experiments was arranged because we had no definite ideas as to the scope of the work or the number of elements into which the investigation would resolve itself, before all of its phases could be determined. We were sure, however, that any work that would thoroughly cover the ground would have to be in the nature of a systematic time study, combined with a course of instruction in correct shoveling methods and adequate and intelligent supervision of the work.

Two or three companies in the Southwest have done some work to determine the proper shovel to be used underground, but so far as is known the work has been limited to equipping certain parts of their mines with a particular type and size of shovel and thereafter watching the cost and efficiency records. In each case it seems to have been the shovel that held the 21-lb. (9.5-kg.) load that gave the best results. Excepting personal communications from these companies, the only data available on scientific shoveling are contained in F. W. Taylor's book, "The Principles of Scientific Management," and D. J. Hauer's article, in *The Contractor* for March 29, 1918, "A Hundred Hints for Shovelers."

This paper discusses and draws conclusions from several thousand time-study readings, taken both on the surface and in the mines for nearly a year. A sufficient number of readings were taken on each factor in the problem for the plotting of curves of the performances and to obtain accurate indices of the work to be expected from this class of labor. The work is not as complete in all of its details as we would like to have it, because we were forced to stop the work temporarily, owing to the number of men going into the National service and our inability to get others who could make time studies. The results obtained so far, however, have been of such a startling nature that we have decided to submit them at this time, subject to future modification. It is hoped, also, that a free discussion of this paper will lead to a disclosure of any errors that may have been embodied and offer some valuable suggestions for the conduct of future work.

## Preliminary Work

As a first step, several weeks were spent underground making a general survey of the field and making time studies on various men, in order to see what points would need to be determined for a full consideration of the subject. The following factors were soon recognized:

1. The type, weight, size, and design of shovel giving the greatest shift tonnage without too much wear and tear on the man would have to

be determined. This work would also determine the point at which a shovel should be discarded as worn out.

2. A standard of comparison would be necessary if the ill effects of mine air, powder gas and smoke, temperature, humidity, and poor light were to be estimated.

3. The layout and spacing of chutes would have to be studied with regard to their effect on shoveling directly into the chutes, or loading into wheelbarrows or cars and tramming to them. This work would determine the proper distance at which shoveling into a chute should leave off and loading into a wheelbarrow or car be taken up. The information would also be of very great use in planning the development of a stope.

TABLE 1.—*Weights and Volumes of Broken Ore*

| Cubic Feet of Broken Ore per Ton | Pounds of Ore in 1 Cu. Ft. | Pound of Ore in 1 Cu. In. | Cubic Inches of Ore in a 21-lb. Load | Cubic Feet of Broken Ore per Ton | Pounds of Ore in 1 Cu. Ft. | Pound of Ore in 1 Cu. In. | Cubic Inches of Ore in a 21-lb. Load |
|---|---|---|---|---|---|---|---|
| 10 | 200 | 0.1157 | 182 | 26 | 77 | 0.0446 | 471 |
| 11 | 182 | 0.1053 | 199 | 27 | 74 | 0.0428 | 491 |
| 12 | 167 | 0.0966 | 217 | 28 | 71 | 0.0411 | 511 |
| 13 | 154 | 0.0891 | 236 | 29 | 69 | 0.0399 | 526 |
| 14 | 143 | 0.0828 | 254 | 30 | 67 | 0.0388 | 541 |
| 15 | 133 | 0.0769 | 273 | 31 | 65 | 0.0376 | 558 |
| 16 | 125 | 0.0723 | 290 | 32 | 63 | 0.0365 | 575 |
| 17 | 118 | 0.0683 | 307 | 33 | 61 | 0.0353 | 595 |
| 18 | 111 | 0.0642 | 327 | 34 | 59 | 0.0341 | 616 |
| 19 | 105 | 0.0608 | 345 | 35 | 57 | 0.0329 | 638 |
| * 20 | 100 | 0.0579 | 363 | 36 | 56 | 0.0324 | 648 |
| 21 | 95 | 0.0549 | 383 | 37 | 54 | 0.0313 | 671 |
| 22 | 91 | 0.0527 | 398 | 38 | 53 | 0.0307 | 684 |
| 23 | 87 | 0.0504 | 417 | 39 | 51 | 0.0295 | 712 |
| 24 | 83 | 0.0480 | 438 | 40 | 50 | 0.0289 | 727 |
| 25 | 80 | 0.0463 | 454 | | | | |

* This was the ore handled in these tests.

4. Hindrances to work, such as timber standing in line of throw or very closely spaced, men and supplies passing back and forth through working space, etc.

5. Manner of placing the shovelers to obtain maximum results from them, number of men in one working place, and size of working place required per man.

6. The hours of actual work and the cause and amount of delays, such as shoveler interrupted to help in other work, etc.

7. The capacity of a man for work as the day progresses.

8. Proper rest periods for men to maintain maximum efficiency.

9. Best means for instructing men and supervising work.

10. Payment received and manner of paying.

At the time this work was started, three types of shovels were in general use at the mines; a No. 2 scoop, used principally by contractors in development work, but favored by some of the shift bosses for use in the stopes; a No. 2 or a No. 3 square-point D-handle shovel for shoveling off of a mat in the stopes; and a No. 2 round-point long-handle shovel, for

TABLE 2.—*Capacity in Cubic Inches of Various Types and Sizes of Shovels*

| Capacity in Cubic Inches | Size of Shovel Blade, Inches | | Square Point, Plain or Back Strap Number | Round Point, Plain or Back Strap Number | Scoop, Plain or Back Strap Number | Scoop, Hollow Back Number |
|---|---|---|---|---|---|---|
| | Width | Length | | | | |
| 249 | 9⅞ | 11⅞ | ...... | 2 | | |
| 278 | 10¼ | 12½ | ...... | 3 | | |
| 302 | 9⅞ | 11⅞ | 2 | | | |
| 308 | 10⅞ | 12¾ | ...... | 4 | | |
| 340 | 10¼ | 12½ | 3 | | | |
| 340 | 11⅜ | 13¼ | ...... | 5 | | |
| 363 | 10 | 13 | * | | | |
| 373 | 10⅞ | 12¾ | 4 | | | |
| 384 | 11¾ | 13¾ | ...... | 6 | | |
| 414 | 11⅜ | 13¼ | 5 | | | |
| 446 | 11¾ | 13¾ | 6 | | | |
| 457 | 11¼ | 15 | ...... | ...... | 2 E.P. | 2 E.P. |
| 521 | 11½ | 15½ | ...... | ...... | 3 E.P. | 3 E.P. |
| 564 | 11¾ | 16 | ...... | ...... | 4 E.P. | 4 E.P. |
| 579 | 12¾ | 15½ | ...... | ...... | 1 W.P. | 4 W.P. |
| 622 | 12⅜ | 16½ | ...... | ...... | 5 E.P. | 5 E.P. |
| 633 | 13 | 16¼ | ...... | ...... | 2 W.P. | 5 W.P. |
| 665 | 12⅞ | 16¾ | ...... | ...... | 6 E.P. | 6 E.P. |
| 706 | 13⅜ | 17 | ...... | ...... | 7 E.P. | 7 E.P. |
| 721 | 13¾ | 17 | ...... | ...... | 3 W.P. | 6 W.P. |
| 761 | 13¾ | 17½ | ...... | ...... | 8 E.P. | 8 E.P. |
| 802 | 14¼ | 17¾ | ...... | ...... | 4 W.P. | 7 W.P. |
| 832 | 14½ | 18 | ...... | ...... | 9 E.P. | 9 E.P. |
| 881 | 14⅝ | 18½ | ...... | ...... | 5 W.P. | 8 W.P. |
| 910 | 15⅛ | 18½ | ...... | ...... | 10 E.P. | 10 E.P. |
| 934 | 14⅞ | 19 | ...... | ...... | 6 W.P. | 9 W.P. |
| 1042 | 15¾ | 19¾ | ...... | ...... | 12 E.P. | 12 E.P. |

* Specially Made Test Shovel.

E.P. = Eastern Pattern Scoop.          W.P. = Western Pattern Scoop.

scraping down a muck pile, shoveling off of a rough bottom, cleaning up, etc.   The first task was to determine the average load that the various types and sizes of shovels would handle, in order to be able to determine whether the 21-lb. load, as advocated by Dr. Taylor, applied to underground work as well as to the surface work, and whether it was the best load for the average Mexican laborer of the Southwest.   These average capacities were obtained by repeatedly shoveling a weighed pile of ore with each of the shovels and counting the number of shovel loads required to move it. Table 1 gives the number of cubic inches of ore in a 21-lb. load, for ore breaking to different volumes per ton; and Table 2 gives the sizes and types of shovels that will average up to any given content. Owing to the variety of conditions in underground shoveling, such as the material of which the shoveling platform is made, whether of wood, iron, or natural bottom; the unsized material shoveled; and the amount of moisture in the ore, causing it to be sticky at times; these

| Test No. | | | | Shoveling Test | | | Shoveling into | Chute | | |
|---|---|---|---|---|---|---|---|---|---|---|
| Date | | | | | | | | Wheelbarrow Car | | |
| Type of Shovel | | | | Size of Car | | | Distance Trammed | | Kind of Muck | |
| Serial No. | | | | Condition of Car | | | Total No. of Cars | | Started Work | |
| Weight of Shovel | | | | Capacity of Car | | | Total No. of Tons | | Lunch Period | |
| Capacity of Shovel | | | | Height - Car above Rail | | | No. of Cars per Hour | | Quit Work | |
| Distance - Chute to Shoveler | | | | Distance - Car to Shoveler | | | No. of Tons per Hour | | Total Hours Worked | |

| No. of Shovels per Min. | No. of Shovels per Car | Penetrating Mass. Seconds | Lifting Mass. Seconds | Throwing Mass. Seconds | Return Time Seconds | Picking Down Min. | Resting Time Min. | Tram Time Min. | Dump Time Min. | Return Time Min. | Remarks | Time |
|---|---|---|---|---|---|---|---|---|---|---|---|---|
| | | | | | | | | | | | | |
| | | | | | | | | | | | | 7:30 - 9:30 A.M. |
| | | | | | | | | | | | | 9:30 - 11:30 A.M. |
| | | | | | | | | | | | Lunch 11:30-12:03 | |
| | | | | | | | | | | | | Tot. Avg. |

FIG. 1.

average shovel capacities were found not to accord with actual practice, except over test periods of long duration; for short periods they would vary as much as ¾ lb. (0.34 kg.) from the average, while single shovel loads would vary as much as 3 lb. (1.36 kg.). For Burro Mountain ore, the tables show that it requires a specially made shovel with a 10- by 13-in. (25.4- by 33-cm.) blade to hold the 21-lb. (9.5-kg.) load, or 363 cu. in. (5948 cu. cm.). In practice, however, we are using at the present time a No. 4 square-point shovel holding 373 cu. in. (6111 cu. cm.), and a No. 5 round-point shovel holding 340 cu. in. (5571 cu. cm.).

At this time also, a time-study sheet, shown in Fig. 1, was developed, which was used for all tests. In addition to the data placed on this sheet, an extensive log of the work was carried on, which undertook to explain, in detail, all delays, changes of work, rest periods, changes in conditions that would affect speed, high and low efficiency periods during the day, etc.   Each night, the study sheets and the log for the day were turned in at the office and the writer totaled, averaged, and checked

FIG. 2.—POSITION WHEN STARTING TO "PENETRATE THE MASS."          FIG. 3.—MASS HAS BEEN PENETRATED.

all readings, and entered them on special record sheets and charts in a convenient manner for comparison and study.

During the period of preliminary work, it was discovered that the

FIG. 4.—POSITION JUST PRIOR TO LIFTING MASS.          FIG. 5.—MASS HAS BEEN LIFTED.

work of a shoveler can be classified into the following divisions, each susceptible to comprehensive study and analysis, and to each of which can be given a definite relative time value:

Time spent actually shoveling, which may be divided into: Penetrating mass, lifting mass, throwing mass, and return to start of first motion.

Time spent picking down, considered as a rest.

Tramming and dumping time, with wheelbarrow or car.

Time spent resting, divided into: Definite rest periods and delays due to interferences, blasting, men and supplies passing, etc.

FIG. 6.—PREPARING TO THROW MASS. A SHORT STEP HAS BEEN TAKEN IN DIRECTION OF THROW.

FIG. 7.—MASS HAS BEEN THROWN. NOTE STRAINED POSITION IN EFFORT TO THROW A LONG DISTANCE WITH SHORT-HANDLE SHOVEL.

FIG. 8.—POSITION WHEN STARTING TO PENETRATE MASS WITH LONG-HANDLE SHOVEL.

FIG. 9.—MASS HAS BEEN PENETRATED. NOTE EASY POSITION OF MAN.

Time spent other than in shoveling, not counted in shoveling time, but included delays before starting to work, lunch period, quitting early at end of shift, and time spent on other work, helping machine man, timbermen, etc.

The motions listed are illustrated in Figs. 2 to 17. By studying
each motion separately, it was possible to establish a standard time for
each and, consequently, a standard of performance for the whole.   It

FIG. 10.—WRONG WAY TO START LIFTING:
RIGHT FOOT AND HAND MOVED FORWARD.

FIG. 11.—MASS HAS BEEN LIFTED.

FIG. 12.—PREPARING TO THROW MASS.
A SHORT STEP HAS BEEN TAKEN IN DIREC-
TION OF THROW.

FIG. 13.—MASS HAS BEEN THROWN·
NOTE EASE OF POSITION AS COMPARED
WITH FIG. 7.

was possible, also, to discover which were the most tiring motions and
how each was affected by length of time worked, length and distribution
of rest periods, size of shovel, design of shovel, and length of throw.

It was, of course, impossible to time all the motions made with any one shovelful; consequently these figures had to be obtained in rotation, each figure set down on the sheet being an average of 10 consecutive readings.  For example, the day would be started by taking 10 readings of the number of shovels per minute and the average of these set down on the first line of the column of the time-study sheet.  Ten readings would next be taken of the time for penetrating the mass, the average of which would be set down on the first line in the third column, and so on through-out the day.  All delays and rest periods were timed and all wheel-barrow and car loads counted.  As a check on the tonnage handled, a record was made of the number of shovelfuls making up a load, the average capacity of the shovel and of the wheelbarrow or car, an estimate was

Fig. 14.—Correct manner of throw-ing ore into a car.  Car is only 4 ft. from toe of ore pile.

Fig. 15.—Penetrating mass with a high-lift shovel.  Man not crouching as low as in Fig. 3.

made of the tonnage in the original pile and, in many cases, the tonnage drawn out of the chute into which the man was shoveling or dumping the ore.  At intervals during the day, a line was drawn across the time-study sheet and all subsequent readings were recorded below this line. This was done to watch the relative efficiency of the men at different times of day.

### Investigations Made on Surface

In order to obtain some standard of comparison for the underground work, some of the mine shovelers were brought to the surface and a record made of their work under ideal conditions; that is, with good air, good light, no timber to interfere, steady shoveling for various lengths of time, and standard lengths of throw for the muck.  A platform was built on the side of the mine-waste dump, about 12 ft. (3.6 m.) below the yard level, with a slide from the track above so arranged that no matter what quantity of muck was in the slide the toe of the pile was always in the same place on the platform and the shoveler did not have to move up as shoveling progressed.    At several places on the platform trap doors

Fig. 16—Correct manner of throwing ore into car with long-handle shovel. Car is 4 ft. from toe of pile.

Fig. 17.—Increased effort necessary to throw ore a long distance into car.

were installed so as to obtain any desired length of throw into what corresponded to a chute in the mine.    The muck thrown through these doors rolled down the side of the waste dump, out of the way, so that the opening was always clear.    A track was laid along the side of the dump at the platform level, so that tests could be conducted in which the shoveler loaded the muck into a car, which he then had to tram a distance of about 100 ft. (30 m.), dump, and return to the muck pile again.

In addition to obtaining the comparison standard, it was possible to form some definite conclusions, which were later checked very satisfactorily under actual conditions in the mine, as to the most advantageous size, type, weight, and design of shovels for general mine use, under the various conditions to be met with there.

Tests were carried on for 2 mo., three different shovelers, taken from the mines, being observed. Each of these men was warned that he had to work at his best speed, all during the job, but that he was not to overtax himself. He was told that when he became tired he was to take a few moments rest, as it was better for him to rest at intervals than to try to work all the time, at the expense of speed and capacity. Later, we undertook to regulate the rest periods, to obtain the proper intervals at which they should occur, and their length. The results of these tests will be discussed in conjunction with the underground work

### SHOVELING DIRECTLY INTO A CHUTE

All of the underground shoveling tests may be classified under one of three headings, shoveling directly into chutes, shoveling into wheelbarrows and tramming to chutes, and shoveling into cars and tramming to chutes. Each of these series was conducted independently of the others, and was complete in itself. The men under observation worked for periods varying from 1 to 8 hr., and for each length of job they threw or trammed the muck over a wide range of distances, with various types and sizes of shovels. In all the underground tests, the work was done under the actual mining conditions, with the one exception that the men were always under observation and consequently were working at a good speed for the full period of the test. In no case did the men overtax themselves and we feel confident that all tonnages obtained and indicated on the charts are easily obtainable by a good, but not exceptional, Mexican laborer after he has been properly instructed, and under close and intelligent supervision, together with a wage paid in such a manner that he has a constant incentive to do good work.

It soon became evident that the great majority of shovels being tested were not suitable for efficient work, and although we continued to work with them to some extent, we have charted only the work of the No. 4 shovel, which handles the 21-lb. load, together with the No. 2 scoop, which was held in high esteem by many of the men in the operating department. In each of the charts of this series, Figs. 18 to 22, the results obtained during the surface tests are plotted alongside of corresponding results from underground, in order to accentuate the adverse effects of underground conditions on shoveling capacity.

In Fig. 18 will be found the number of shovels per minute thrown
into a chute at a distance of 8 ft. (2.4 m.) from the ore pile for jobs varying
in length from 1 to 8 hr.   In all of these charts, the length of job should
be understood to mean the total working time, and when it is said that
the length is 4 hr., the man was actually occupied at shoveling ore for
4 hr., and then his work was finished.   All points on the curves are
corrected averages for the time periods to which they correspond.

In connection with Fig. 18, the following facts will be noted:   For
all lengths of job, the number of shovels per minute is greater with the
No. 4 shovel than with the No. 2 scoop.   Both on the surface and under-

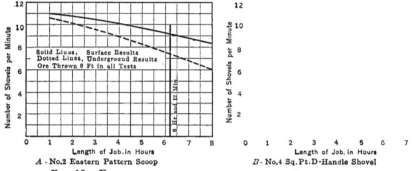

FIG. 18.—EFFECT OF LENGTH OF JOB ON NUMBER OF SHOVELS PER MINUTE.

ground, the speed of shoveling decreases more rapidly with the scoop
than with the shovel, as the length of the job increases.   A man working
with a scoop underground can perform at only 72 per cent. of his speed on
surface for 8 hr., while with a No. 4 shovel he can work at 82 per cent. of
his surface speed.   The percentage reduction in speed between surface
and underground work is the measure, in part, of the effect of mine air,
powder gas and smoke, temperature, humidity, and poor light.   Under
the same condition of work, the difference in speed between the No. 4
shovel and the No. 2 scoop is due to the difference in the load handled.
For short lengths of time, the difference in working speed between a
scoop and a shovel is so small that, disregarding rest periods, the scoop
is a slightly greater tonnage mover than the shovel; but for longer periods,
the difference in speed is such that the shovel with its smaller capacity
actually moves more muck than the scoop.

Fig. 19 indicates the manner in which the length of throw will affect
the speed of the shoveler.   These charts were worked out for a uniform
length of job of 6 hr. and 12 min., and for varying distances, as shown.
The same relationship exists between the lines of these charts as on the
charts of Fig. 18.   The decrease in shoveling speed on the surface

amounts to an average of 2.5 per cent. for every foot increase in distance thrown in the case of the scoop, and 1.8 per cent. for the No. 4 shovel. Underground, the working speed is decreased more rapidly, being

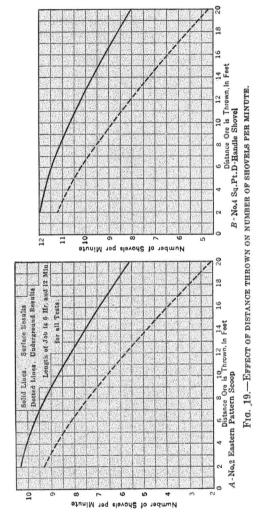

FIG. 19.—EFFECT OF DISTANCE THROWN ON NUMBER OF SHOVELS PER MINUTE.

respectively 4.4 per cent. and 3.2 per cent. per foot increase in throw. The rate of decrease in shoveling speed, both on the surface and underground, is greater for the heavily loaded scoop than for the shovel.

To find the average shoveling speed for any length of job and for any distance that the ore has to be thrown, the number of shovels per minute for a throw of 8 ft. (2.4 m.), for the proper period, can be obtained from Fig. 18; this can be increased or diminished by the proper percentage obtained from Fig. 19, depending on whether the distance is less or greater than 8 ft. As an example, suppose that it is desired to determine how many shovels per minute a man will average, throwing a distance of

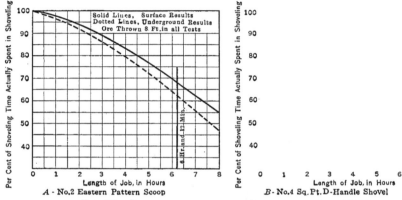

FIG. 20.—REST PERIOD REQUIRED FOR VARIOUS LENGTHS OF JOB

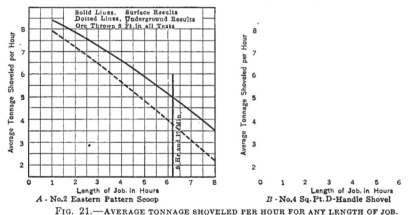

FIG. 21.—AVERAGE TONNAGE SHOVELED PER HOUR FOR ANY LENGTH OF JOB.

12 ft. (3.6 m.) for a length of a job of 6 hr., using a No. 4 shovel underground. Chart B, Fig. 18, shows that for 6 hr., throwing a distance of 8 ft., a man will average 9.6 shovels per minute. Chart B, Fig. 19, shows that to increase the length of throw from 8 ft. to 12 ft. diminishes the shoveling speed 15.7 per cent.; therefore for a 6-hr. job, at 12 ft., the man will be expected to handle about 8.1 shovelfuls per minute.

Fig. 20 shows the amount of rest required for shoveling jobs of various lengths. Here again will be noted the same relations between lines that exist on the other charts in which times are given. The scoop again has a negative effect both on surface and underground, causing a man to use up more time in resting than with a No. 4 shovel. The rest period, as considered here, is made up of the time consumed in delays, the time actually

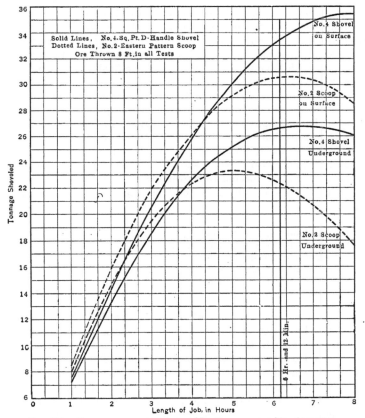

FIG. 22.—COMPARISON OF WORK OF No. 2 SCOOP AND No. 4 SHOVEL.

spent in resting, during which the man may smoke a cigarette and sit down for a few minutes, and the time used in loosening the muck pile, scraping up the dirt on the shoveling plat, or doing other light work, not actually shoveling, but closely related to it.

Over a long period it was possible to demonstrate the feasibility of accurately determining the percentage of the working day that a man

will actually devote to shoveling. The working day at the Burro
Mountain mines is 8 hr., ½ hr. of which is given up to the lunch period,
leaving 7½ hr. as the total possible working time. It was found that
of this 7½ hr., the man actually worked at shoveling for 82.5 per cent.
of the time, or for 6 hr. and 12 min., and on all the charts involving
time, this particular length of job has been designated by a special
line. The remainder of the possible working time, or 17.5 per cent.,
is spent on other work, quitting early for lunch or to leave the mine
or commencing to work late at beginning of the shift or after lunch.
Observations of this character were gathered by some of the shift bosses,
on several hundred-man shifts, and it is surprising how little the figures
obtained by each have varied from the average finally obtained.

Fig. 21 gives the average tonnage per hour to be expected of a man
throwing the muck a distance of 8 ft. over any period of time; and
Fig. 22 gives the total tonnage shoveled for any period, over the same
distance. These graphs are corrected to agree with Figs. 18, 19 and 20,
but as an illustration of how closely the actual results obtained agree
with the charts, the following figures are presented: For a job lasting
5 hr. and 30 min., with a No. 4 shovel underground, according to Fig.
22, a man would be expected to shovel 26.0 T. a distance of 8 ft. Five
representative tests actually gave the following tonnages under average
conditions:

|                                           | TONNAGE |
|-------------------------------------------|---------|
| Length of job 5 hr. and 40 min.           | 28.0    |
| Length of job 5 hr. and 10 min.           | 23.0    |
| Length of job 5 hr. and 25 min.           | 26.0    |
| Length of job 5 hr. and 50 min.           | 34.0    |
| Length of job 5 hr. and 30 min.           | 25.0    |
| Average, 5 hr. and 30 min.                | 27.4    |

Fig. 23 shows how the time of shoveling one shovelful of ore is influenced by the length of the job, with the length of throw remaining
constant. In chart B, the total time of handling one shovelful of muck
has been divided into its component movements. The lines representing
the work of penetrating mass, lifting mass, and return, show a constant
increase as the length of job increases. The actual increase in the time
of each movement is not due so much, we think, to a decrease in the speed
of making the move, which probably is fairly constant, as it is to an
ever-increasing period of rest taken at the beginning and end of each
movement, which, however, was too short to be accurately timed.
Throwing mass is not influenced as much as the other moves, as the muck
must be thrown along a definite path, which is limited to distance and
height, and hence a constant speed must be maintained to carry it over.

The reader will be well repaid by a careful study of Fig. 22 and the
following points should be noted:

1. The difference in tonnage handled by the same shovel, on the surface and underground, for any length of job, is the measure of the bad effects of underground conditions. For a job of 6 hr. and 12 min., with a No. 4 shovel, the underground work is 20.5 per cent. less than on surface.

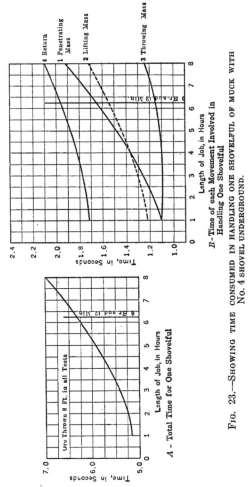

Fig. 23.—Showing time consumed in handling one shovelful of muck with No. 4 shovel underground.

2. The difference between the amounts shoveled with the No. 2 scoop and the No. 4 shovel, under same conditions, is the measure of the effect of the difference in load handled by the man.

3. Each line on this chart shows a peak at some particular length of job, and the total tonnage shoveled for any greater period than this

is actually less. The point at which this peak occurs should be termed the "economic shoveling day," and a company should not require its men to work at shoveling any longer than this, except in emergency cases.

4. The presence of this peak accords with the experience of many superintendents and managers, who state that their men do more work in an 8-hr. day than they did on an old 10-hr. basis.

5. The "economic shoveling day" is about $6\frac{1}{3}$ hr., with a No. 2 scoop on the surface, and $5\frac{1}{3}$ hr. underground. With a No. 4 shovel, on the surface 8 hr. is about the proper length of day, while underground $6\frac{2}{3}$ hr. seems to be about correct. As the men actually shovel only $6\frac{1}{3}$ hr. per day on an average and as their other work is generally of a very light nature, the 8-hr. day with the correctly proportioned shovel is probably the best; but with a scoop it is certainly too long.

6. For work on the surface, on jobs lasting longer than $4\frac{2}{3}$ hr., the No. 4 shovel is superior to the scoop. Underground the No. 4 shovel demonstrates its superiority for jobs longer than $3\frac{2}{3}$ hr. The scoop then may be considered as a task shovel for short-time jobs, but even here, its value is only slightly greater than the No. 4 shovel, beside tiring the man so that he is unfit for other work when the shoveling task is finished. There is also the additional danger of having some men continually trying to use the scoop for the full shift, thinking that the amount of work (and hence the amount of pay in the case of contract and bonus systems) is greater as the size of the shovel increases.

The following formulas show the manner in which use is made of the figures presented in the preceding diagrams:

Let $W$ = weight of load on shovel, in pounds;
$N$ = number of shovels per minute, Figs. 18 and 19;
$P$ = per cent. of time actually shoveling, Fig. 20;
$L$ = length of job, in minutes;
$T$ = total tonnage shoveled;
$n$ = number of shovels per minute for an 8-ft. throw, Fig. 18;
$p$ = per cent. increase or decrease due to various lengths of throw, Fig. 19.

$$\frac{W \times N \times P \times L}{2000} = T \qquad\qquad N = n(1.00 \pm p)$$

*Example* 1.—What will be the total tonnage handled, using a 21-lb. load shovel, throwing the ore 8 ft., underground, for a job of 5 hr. duration? Chart B, Fig. 18, shows that for 5 hr. a man will average 10.1 shovels per minute, and chart B, Fig. 20, shows that he will actually shovel 79 per cent. of the 5 hr. period, therefore:

$$\frac{21 \times 10.1 \times 0.79 \times 300}{2000} = 25.1 \; T$$

*Example 2.*—What will be the total tonnage handled, using a 21-lb. load shovel, throwing the ore 15 ft. underground, for a job of 6 hr. duration? Chart B, Fig. 18 shows that for 6 hr. a man will throw a distance of 8 ft. at the rate of 9.6 shovels per minute, Chart B, Fig. 19, shows that increasing the distance to 15 ft. reduces the capacity by 27.4 per cent., therefore:

$$N = n(1.00 \pm p) = 9.6(1.00 - 0.274) = 7;$$

hence,

$$\frac{21 \times 7.0 \times 0.73 \times 360}{2000} = 19.3 \text{ T.}$$

### SHOVELING INTO A WHEELBARROW

The charts presented in this series, Figs. 24 to 27, follow as closely as possible the series just discussed, but offer only the results obtained with the No. 4 D-handle shovel. It was soon discovered that a throw of 3 ft. to the wheelbarrow gave the best results as far as number of shovels per

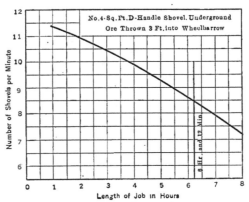

FIG. 24.—EFFECT OF LENGTH OF JOB ON NUMBER OF SHOVELS PER MINUTE.

minute and rest periods required were concerned, and in all subsequent work the ore was thrown into the wheelbarrow from this distance. It will be noted that, in Fig. 24, for any length of job, the number of shovels per minute are less than when throwing 8 ft. into a chute; this is due to the fact that the shoveler must place each shovelful carefully to keep wheelbarrow from spilling its contents and to make it ride easily.

Chart A, Fig. 25, shows the length of time consumed in tramming and dumping a wheelbarrow over any distance and chart B shows the average tramming speed developed for any distance. For this chart, careful determinations were made of the distance in which it takes a man to acquire full speed and the distance in which, after having attained full speed, he can make his stop. The full-speed rate of travel in stopes will average 165 ft. (50 m.) per minute. The wheelbarrow in use is the No. 7, which holds 3 cu. ft. (0.085 cu. m.) and stands 21 in. (53.34 cm.)

above the floor at point of maximum height. The maximum load in a wheelbarrow should be about 300 lb. (136 kg.) as larger loads are too exhausting, and lighter loads consume too much time in tramming and dumping.

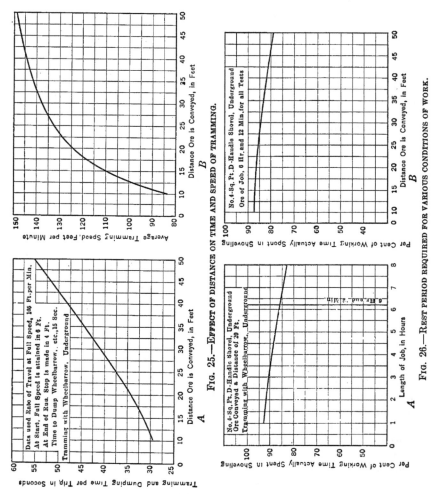

FIG. 25.—EFFECT OF DISTANCE ON TIME AND SPEED OF TRAMMING.

FIG. 26.—REST PERIOD REQUIRED FOR VARIOUS CONDITIONS OF WORK.

Chart A, Fig. 26, shows the per cent. of time a man will work during any given working period, the length of tram in each case being 20 ft. (6 m.). The rest period is practically a constant proportion of any length of job, as that part of the tramming time in which the man brings the empty wheelbarrow back to the ore pile is virtually a rest period. For long trams, the work of tramming the loaded barrow is so heavy that a greater rest is required than is obtained on the return, and chart B shows how the rest period increases for a constant length of job, as the length of tram increases.

Fig. 27 shows the tonnage to be expected of a man, based on Figs. 24 to 26, for any length of job, the length of tram being constant at 20 ft. This chart shows that the shoveler has not quite reached his maximum

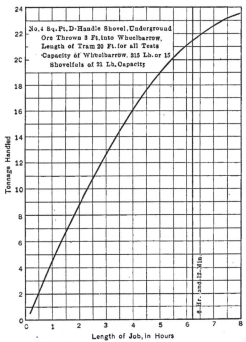

FIG. 27.—SHOWING CAPACITY OF A SHOVELER USING A WHEELBARROW FOR VARIOUS JOBS.

capacity at the end of 8 hr. Two reasons are advanced for this: (1) As long as a man can throw the ore into a chute, he has a fairly direct throw from the ore pile to the chute, and with a car he has a definite path to traverse each trip. With a wheelbarrow, however, the direction and length of tram is constantly varying, as is also the amount of interference from other trammers, timbermen, machine men, etc. The retarding influence of these factors increases as the length of the tram increases. (2) The sequence of operations, shoveling, tramming, dumping, etc. is of such short duration and changes so often from one to the other that it is very hard to keep up any pace that may be set and probably an unnecessary amount of rest is indulged in for all periods.

### SHOVELING INTO A CAR

Fig. 28 shows the number of shovels per minute thrown into a car for any length of job.   In this series of tests the ore was thrown a horizontal distance of 4 ft. (1.2 m.) into a mine car 42 in. (106.68 cm.) high; 4 ft.

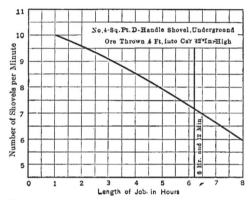

FIG 28.—EFFECT OF LENGTH OF JOB ON NUMBER OF SHOVELS PER MINUTE.

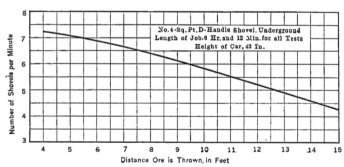

FIG. 29.—EFFECT OF DISTANCE THROWN ON NUMBER OF SHOVELS PER MINUTE.

seems to be the best distance to maintain between car and ore pile, for a man to work to the best advantage.   Due to the height of the car, the capacity of a shoveler is decreased, as compared to his capacity in shovels per minute, when loading into a wheelbarrow.   This decrease in shoveling speed amounts to about 8 per cent. per foot of height.   The best type of car for a shoveler to use holds about a ton of ore, is as low as is consistent with good design, certainly not over 45 in. in height, and is equipped with roller bearings, which should be kept in the best of condition.   Cars much larger than this are too hard to tram and cars much smaller use up too much of the shovelers' time tramming back and forth.

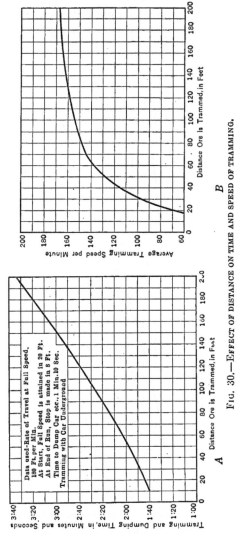

Fig. 30.—Effect of distance on time and speed of tramming.

Fig. 29 shows the effect of having to throw the ore a greater distance than 4 ft. into the car, for a given length of job, using a No. 4 D-handle shovel. For every additional foot between the car and the ore pile, the height of the car remaining constant, the decrease in shoveling speed amounts to about 3.6 per cent.

In Fig. 30, chart A shows the time consumed in tramming, dumping, and returning with the car, over various distances.   Chart B shows the

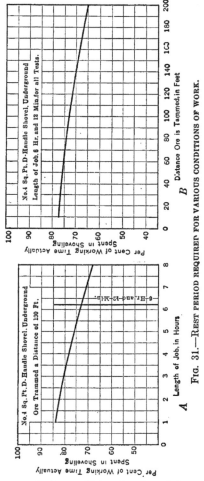

Fig. 31.—Rest period required for various conditions of work.

average tramming speed that will be developed for any distance over which the ore has to be conveyed.

Fig. 31A shows the amount of rest required for various lengths of jobs under constant conditions of length of tram, and distance and height through which the ore is thrown by the shoveler.   For any length of job, as the length of tram increases, the amount of rest needed is increased, as shown in chart B. Both of these lines, however, are quite flat, for a man can get very nearly enough rest as he returns each trip with the empty car.

Fig. 32 shows the tonnage to be expected of a man mucking into a car and tramming a constant distance, for various lengths of jobs. It will be noticed that the economic shoveling day is between 7 and 8 hr. and that the maximum average results to be expected of a mine shoveler under the given conditions have probably been reached.

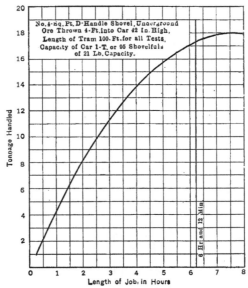

No.4-Sq. Pt. D-Handle Shovel, Underground
Ore Thrown 4-Ft. into Car 42 In. High,
Length of Tram 100-Ft. for all Tests,
Capacity of Car 1-T. or 95 Shovelfuls
of 21 Lb. Capacity.

FIG. 32.—SHOWING CAPACITY OF A SHOVELER USING A CAR FOR VARIOUS JOBS.

Fig. 33 is made up for a uniform shoveling day of 6 hr. and 12 min. and shows the tonnage to be expected under average shoveling conditions for any distance that the ore must be thrown or trammed. The line representing the tonnage to be expected of a man with a wheelbarrow may not be entirely correct, especially as the length of tram increases. It is thought that up to 15 ft. (4.5 m.), the line is about correct but that it may slope off a little too rapidly beyond this point. On the other hand, the wheelbarrow is generally used where neither direct shoveling nor the use of a car, with its attendant track expense, etc., is feasible, consequently, the wheelbarrow is always at work under adverse conditions in a stope and no improvements over the results here tabulated are to be expected. The writer thinks that the work with a wheelbarrow in a stope has been closely approximated but that a greater efficiency could be obtained in a clear and unobstructed way, such as a drift. The

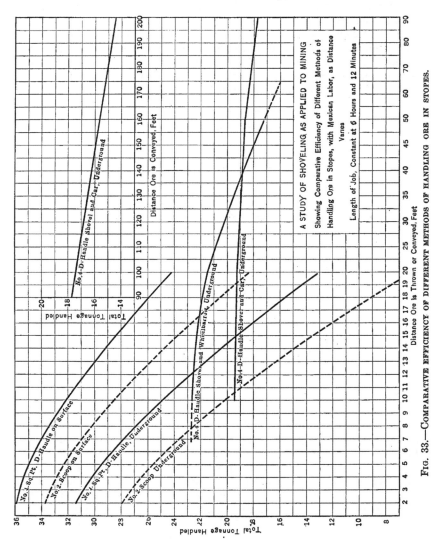

FIG. 33.—COMPARATIVE EFFICIENCY OF DIFFERENT METHODS OF HANDLING ORE IN STOPES.

calculation of the tonnage expected when tramming either with a car or wheelbarrow, for any length of job and distance trammed, is expressed in the following formulas:

Let $W$ = weight of load on shovel, in pounds;

$\quad\quad N$ = number of shovels per minute, Figs. 24 and 28;

$\quad\quad P$ = per cent. of time actually shoveling, Figs. 26 and 31;

$\quad\quad L$ = length of job, in minutes;

$\quad\quad T$ = total tonnage shoveled;

$\quad\quad a$ = time to load one car or wheelbarrow;

$\quad\quad b$ = time to tram and dump one car or wheelbarrow, in minutes, Figs. 25 and 30;

$\quad\quad c$ = load on one car or wheelbarrow, in pounds;

$$\frac{c}{W \times N} = a \qquad\qquad \frac{\left(\dfrac{L}{a+b}\right) \times c \times P}{2000} = T$$

*Example* 1.—What will be the total tonnage handled, using a 21-lb. shovel and a wheelbarrow, and tramming the ore 20 ft. for a job of 5 hr. duration?

It is necessary first to find the time required to load one wheelbarrow,

or $a = \dfrac{315}{21 \times 9.3} = 1.61$; then the total tonnage handled is

$$T = \frac{\left(\dfrac{300}{1.61 + 0.56}\right) \times 315 \times 0.88}{2000} = 19.16$$

*Example* 2.—What tonnage will be handled in 6 hr. and 12 min., using a car of 1 T. capacity and a shovel of 21-lb. capacity, tramming a distance of 100 ft.? As the time required to load one car is $\dfrac{2000}{21 \times 7.2} = 13.3$; the total tonnage handled is

$$T = \frac{\left(\dfrac{372}{13.3 + 2.43}\right) \times 2000 \times 0.73}{2000} = 17.26$$

TABLE 3.—*Schedule of Wear of Shovels*

| Type of Shovel | Used on | Tonnage Handled by Blades Made of | | | |
|---|---|---|---|---|---|
| | | Chrome-nickel Steel | Secret Composition Steel | Common Carbon Steel | Extra Light Carbon Steel |
| No. 2 scoop | Iron sheet | 1220 | 950 | 750 | |
| No. 2 scoop | Rough bottom | | | | |
| No. 2 scoop | Wooden mat | 1500 | 1100 | 900 | |
| No. 4 shovel | Iron sheet | 990 | 770 | 620 | 250 |
| No. 4 shovel | Rough bottom | 1075 | 870 | | |
| No. 4 shovel | Wooden mat | 1168 | 1000 | 730 | 340 |
| Gage of steel in blade ............... | | 15 | 13 | 14 | 16 |
| Cost of shovel per ton handled, cent. .... | | 0.0015 | 0.0018 | 0.0019 | 0.0026 |

## WEAR OF SHOVELS

To determine the relative wearing qualities and the cost per ton for supplying the men underground with new shovels, different places in the mines were equipped with different makes and styles of shovels and the results carefully noted. At frequent intervals, these shovels were measured to detect the wear of the blade, and checked up to see that all were being used in the proper places underground; the tonnage coming from each place and the number of shovelers employed were also noted. Table 3 gives the results obtained with the different shovels.

The shovels made of chrome-nickel and special steel were excellent implements but the special steel shovel was considerably heavier than the other. Cracks developed along the form line of the chrome-nickel steel blade, on each side, but these did not impair the shovel's usefulness. The blades of the three other shovels bent easily with rough usage; while the blade made of extra light carbon steel wore very rapidly and the edges curled up almost immediately. The No. 2 scoop was used until its capacity had been reduced 25 per cent. and the No. 4 shovel until its capacity had been reduced 9 per cent. The cost of shovel per ton handled includes the cost of the shovel, freight, supply-house handling, handling new shovels into mine, and disposal of worn shovels. We had hoped to be able to detect a difference in the man efficiency on account of the different styles and weights of shovels in use at this time, but owing to the constantly changing conditions in the working places selected for the trials, no conclusive evidence was available.

The wearing quality of any shovel used on an iron sheet varies from 74 per cent. to 86 per cent. of the wearing quality of the same shovel on a wooden mat, the average being 82 per cent. The wearing quality of a shovel on a rough bottom is about 90 per cent. of that on a wood mat. These figures are based on about 50 observed shovels underground.

## TYPE OF SHOVEL ADOPTED

Tests were conducted with square- and round-point shovels varying in size from No. 2 to No. 6 and with standard No. 2 scoops, to determine what size of shovel was best adapted to the work. For short jobs of less than 4 hr. duration, the No. 2 scoop and the No. 5 and 6 shovel were slightly the best from the standpoint of tonnage handled; but for jobs requiring more than 4 hr. for their completion, the No. 4 shovel was greatly superior, see Fig. 22. From the standpoint of "number of shovels per minute," work with a scoop is at all times slower than with a No. 4 shovel, see Fig. 18, and as the day progresses the percentage of time required for resting becomes greater with the scoop than with the shovel, see Fig. 20. The result is that although for short work periods, the larger capacity of the scoop brings the total tonnage handled above that of a

No. 4 shovel, for long periods the increased amount of rest required when handling the heavier load serves to put the No. 4 shovel considerably in the lead as a tonnage mover. With shovels smaller than the No. 4, the number of shovels per minute was not increased and the amount of rest required was not decreased enough to make the smaller shovel superior for any working period. It may be stated as a generalization, that for shovels smaller than the 21-lb. load shovel, the tonnage handled per shift is approximately directly proportional to the shovel capacity; that is, if a man using a No. 4 shovel will handle 26 T. in an 8-hr. shift, with a No. 3 shovel which holds 91 per cent. of the load of a No. 4 shovel, he would be expected to shovel about 24 T. a shift. If the increased cost of shoveling with a smaller shovel, or one that has been worn, is balanced against the cost per ton of putting a new shovel underground and discarding the old one, it will indicate the economic limit of wear of the shovels in use. We have, for the present, rather arbitrarily selected as the limit, a shovel of size No. 4, which has been worn to about 12 in. (30.48 cm.) in length, or roughly a 9 per cent. to 10 per cent. reduction of capacity.

The lift of a shovel is very important. By "lift" is meant the amount of rise in the handle just behind the blade. A handle that is not very much bent at this point, but which goes off straight, is said to have a low lift, while one that arches steeply is said to possess a high lift. To work with a shovel having a low lift the man must stoop down more each time to take a grip on his shovel after it has penetrated the mass, and the added movement takes longer, and requires a greater effort to lift the weight of the body and the load through a greater space. As a result, more rest is required in the course of a day. With a very low lift, the shovel is not well balanced and there is a tendency for it to turn over in the hand, especially if it is not loaded evenly. With a high lift, the man does not have to stoop so far to grasp his shovel, the amount of rest period is decreased, and the loaded shovel is better balanced, because its center of gravity is well below the line of the handle. A lift of 8 in. (20.32 cm.) is the best, as with greater lifts the awkwardness of the throwing movement is considerably enhanced. The height of the end of the handle above the floor when the shovel blade is flat on the floor is of considerable importance, too; this is the measurement that the manufacturers call the "lift." In a short-handle shovel, the end of the handle should strike just above a man's knee, a height of 23 in., to give the most effective effort in penetrating the mass. With a long-handle shovel the height should be the same at a distance back of the blade, corresponding to the length of the short handle.

It is an advantage to have the weight of the shovel as low as is consistent with good material and length of life. Increasing the weight of the shovel slows up every motion involved in shoveling and increases the amount of resting required.   However, it is not wise to go to extremes in the matter, as a very light shovel does not possess the strength and wearing qualities, and the cost of constant replacement is greater than the advantage gained in shoveling speed.   In a personal communication, Mr. Frank B. Gilbreth says:   "The 21-lb. load refers to shoveling any kind of material anywhere, above ground or below ground, and this is the live load upon the shovel and does not include the weight of the shovel. I make this statement after having asked this question of Mr. Taylor. Obviously it would have been better if the data had been obtained on the basis of having the load, live and dead, combined in one figure."

As we lacked information, we assumed that Mr. Taylor experimented, at least in part, with stock shovels, which weigh about 6 lb. (2.7 kg.) in the No. 4 size.   Our experiments were conducted with shovels of both regular and special design, varying in weight from 4 lb. 10 oz. (2.09 kg.) to 6 lb. 5 oz. (2.86 kg.) and we found that shovels weighing between 5 lb. 8 oz. (2.49 kg.) and 5 lb. 10 oz. (2.55 kg.) give the greatest per man capacity.   This is a total combined live and dead weight of 26 lb. 8 oz. to 26 lb. 10 oz.   A shovel of this weight can be made very sturdily, the gage of blade being No. 15 of some composition steel and the handles of best selected XX second-growth northern white ash.   With heavier shovels, there is a distinct falling off in capacity; while for lighter shovels, although we could detect no difference in capacity, the wearing quality was poorer, due to lack of strength..

The use of the scoop is not advocated except where the material to be moved is so light that the scoop holds only 21 lb.   Even for short jobs its use offers only a doubtful advantage, see Fig. 22.   For shoveling on any sort of a mat or platform, the square-point shovel is the better; while for scraping down and working on a rough bottom, the round-point shovel should be used.   Where there is plenty of room for men to work, the long-handle shovel of both square- and round-point pattern is superior to the short-handle.   This is true for all distances and heights to which the ore has to be thrown, and the farther or the higher the ore has to be thrown, the more pronounced is the superiority of the long-handle type. D. J. Hauer says that where men are working in a free space they can do, on an average, 10 per cent. more work with a long-handle shovel than with the short-handle; that the limit of throw, taking only one step, is 12 ft. (3.6 m.) with a long-handle shovel and 9 to 10 ft. with a short-handle shovel.   We have checked Hauer on these points and find his statements on the relative efficiency of the two types of shovels to be substantially correct; but we found that it is economy to throw as far as 12 ft. with a short-handle shovel and 14 ft. with a long-handle.   Unfortunately, however, most working places underground are very restricted in area and it is necessary to use the short-handle shovel.   Where one man is shoveling in a drift, or one to each set of timber in a stope, the long-handle shovel can be used to advantage. Where men are working $2\frac{1}{2}$ to 3 ft. apart, a short-handle shovel should be used. According to D. W. Brunton and J. A. Davis, in U. S. Bureau of Mines' *Bull.* No. 57, the minimum spacing of men working side by side in a drift should be 2.5 ft. (0.76 m.). Calling the performance of a square-point shovel on a wooden mat 100 per cent., the efficiency of a square-point shovel working on a rough bottom is only 60 per cent. while with a round-point shovel an efficiency of about 70 per cent. may be maintained.

Fig. 34 shows the design considered best adapted to mining work. The blade should hold 21 lb. of broken ore as an average load. The

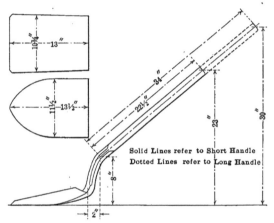

Fig. 34.—Design of shovel best adapted to mining work.

approximate dimensions of blades for various ores are given in Table 2; the dimensions on the illustrations are for Burro Mountain ore. Both the square- and the round-point blades should be of standard shape, of No. 15 gage at the point, and of such composition that the shovel will handle not less than 1100 T. of medium hard ore when shoveled off a wooden mat. All blades should be of the plain back type without rivets, the back strap being welded to the blade. Only best-grade, second-growth, northern white ash should be used for the handle, which should be bent to the shape and dimensions shown. On short-handle shovels, the Dirigo, or split D, handle is preferred, as it is much stronger than the ordinary D handle.

## Correct Shoveling Methods

A right-hand shoveler throws the ore from him from his right side. When using a short-handle shovel, he grasps the D handle with his left

hand, the cross of the D being in the palm of the hand to obtain a good
hold, and with the right hand takes a grip on the handle just back of the
straps.   Standing close to the material to be shoveled, with feet placed
as in Fig. 35, he bends his back, shoulders, and knees, and assumes a
squatting position so as to remain well balanced on his feet.   The left
hand grasping the D handle rests against the left leg just above the knee,
and the right arm below the elbow rests on the right leg.   Without moving the feet, the whole body is lunged forwards from this position, thrust-
ing the shovel blade forcibly under the muck pile, and heaping it full.
To elevate the full shovel, the knees, back, and shoulders are simultaneously straightened, the feet remaining motionless.   To throw the ore

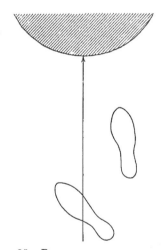 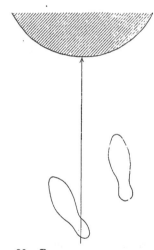

FIG. 35.—POSITION OF FEET OF RIGHT          FIG. 36.—POSITION OF FEET OF RIGHT
HAND MAN USING SHORT-HANDLE SHOVEL.          HAND MAN USING LONG-HANDLE SHOVEL.

into a car, after the shoveler has reached a nearly erect position, the
shovel is raised farther by drawing up the arms, the left hand acting as a
moving fulcrum, and the load is cast directly over the right shoulder
without turning the body or moving the feet.   To cast in a horizontal
direction for any distance, the body must be turned part way around to
the right and a short step made in the direction of the throw; the load is
cast by a swing of the arms, first slightly backward to obtain momentum
and then forcibly forward to deliver the load.   ·

     When using a long-handle shovel, a right-hand man grasps the shovel
close to the end of the handle with the left hand and places the right
hand just back of the straps.   The feet are placed as in Fig. 36 and the
body assumes a crouching position with knees bent and the right elbow

resting on the right thigh just above the knee. The handle of the shovel lies across the left thigh close to the groin and the left hand falls into position against the body near the waist. With a lunge of the body the shovel is then thrust under the mass of ore without moving the feet. To lift the mass on the loaded shovel, the back and shoulders are straightened and the load is brought up by using the left thigh as a fulcrum, over which the shovel handle works as a lever; the knees are then straightened to bring the shoveler into an erect position where the ore is cast directly over the right shoulder into a car as with the short-handle shovel. To cast the load horizontally, a turn to the right is made and a short step in the direction of the throw, exactly as with the short-handle shovel.

It is surprising what a small proportion of the men underground know how to use a shovel to the best advantage, and all sorts of tricks are resorted to in an effort to lighten the work. Among these are: Taking less than a shovelful each time, using the foot in an effort to force the shovel into the muck pile in the manner of using a spade, skimming a thin layer of loose dirt off the sides of the muck pile instead of energetically penetrating the mass to obtain a full shovel load, not holding the shovel properly, and taking two or three steps with each load.

To obtain the highest shoveling efficiency underground, every man hired as a shoveler should be placed in a particular stope or working place that is directly in charge of a shoveling boss. This boss should have had a large experience in shoveling, have learned correct shoveling methods, and should be able to instruct men and gain their confidence. Each man should be taught: (1) The necessity of using the correct type of shovel for any given purpose; (2) the proper way to handle a shovel; (3) the range of usefulness of wheelbarrow and car; (4) the advantage of using a platform to shovel from; when shoveling has progressed beyond the platform time should be taken to advance the boards or iron sheet and to scrape the broken ore forward on to the platform; (5) the mass of broken ore should be thoroughly loosened with a pick; it is waste of effort to try to shovel material that has become packed; (6) shoveling should be done at a good steady pace, the speed depending on the length of job; it is waste of time and energy to try to rush through the work; (7) in addition to the amount of rest inherent in the work itself, that is, the rest gained while picking down, tramming, etc., definite rest periods should be maintained during the day. When each man has been thoroughly instructed in the methods of shoveling, he should be placed in general run-of-mine work among the more experienced shovelers, so that another new man may take his place for instruction.

When possible to avoid it, a shoveler should never be made to work alone. Shovelers working in pairs produce the best results, as they set the pace for one another and compete to a large extent, besides, any laxness can be detected almost at once. Shovelers should be placed in groups of two, four, six, etc., so that the men can work in pairs. It is best not to have more than four men in any group, as with larger groups it is hard to watch the work of each man separately and they try to put the work off on one another. Best results are always obtained when the tonnage shoveled by each man, or small group of men, can be accurately measured at the end of every shift. When men are shoveling in a stope where there is room, they should be so placed that each can use a long-handle shovel; where the area is restricted, a short-handle shovel should be given to them and they should be placed so that a right-hand and a left-hand man can work together.

The ideal shoveling day is the period during which a man can rest at stated intervals and can produce the maximum tonnage by working at a steady pace for the full period, and yet not wear himself out, so that his health is impaired. It is obvious that a steady working pace for the full period is physically impossible unless the rest periods are excessive, in which case the total tonnage handled falls off; in other words, a man cannot do a good day's work and leave the job feeling as fresh as when he arrived. In all of the tests, the shovelers showed a decreasing efficiency, as the day advanced; rest periods brought up the efficiency, but after each successive rest period the efficiency did not advance to quite the same point as after the preceding period, and at the end of the shift it had reached its lowest ebb. Much work can still be done on this point, but after considering the amount of rest that is inherent in the work itself and the amount of added supervision necessary to maintain shoveling on a scientific basis, the tentative statement is made that, in addition to the lunch period, there should be two 10-min. periods of complete relaxation, one midway between the beginning of the shift and lunch time and the other midway between lunch and quitting time.

## Manner of Paying

In all discussions of efficiency methods and wage payments, one fact must be recognized: unless a man feels that he is to be well paid for his work, he is going to use every subterfuge to keep those in charge from knowing just how much he is capable of doing. Almost without exception the workman will be most interested in his wage; if he can earn the same wage with less effort he will do so, and if an increase depends directly on his extra effort he will work harder. It is not the purpose to discuss wage payment and benefit work except in so far as the relation of wages to an efficiency campaign must be clearly recognized; but it is necessary that some thought be devoted to the method applying these standards so as to obtain the best results from day to day without antagonizing the men.

Three fundamental laws of management are: (1) the man must be contented with his surroundings and conditions of work; (2) he must have confidence in the desire of the men over him to be fair in their dealings; (3) the payment for the work done must be proportionate to the performance.

It must be obvious to any one that such things as accident insurance and compensation, good hospital service, old-age pensions, steady work for old and faithful employees even in times of financial stress, good housing and living facilities, improved sanitary conditions, good educational features, such as schools and libraries for both old and young, plenty of entertainment, such as a club, moving pictures, and lectures, and a sincere interest in the men by the management, tend to make the men happy and contented with their surroundings and to put them in a receptive mood for any bonus and efficiency systems that may be contemplated. Unless the men are contented, it is not possible to attempt to increase the efficiency by scientific study, as the men will at once become antagonistic to the time-study man and the efficiency engineer and will do all they can to mislead them.

No campaign for scientific management will be successful as long as the men feel that the company is trying to determine what each man can do merely to cut his wages and to make him work harder for a wage only slightly in advance of his old rate. The company should make each man understand that it is out to determine what a fair day's work constitutes and that for a fair day's work it is ready and willing to pay a fair day's wage and to reward any ambitious and energetic man who does more than the average, by paying him accordingly. The company must supply the best tools for the work and have the conditions such that the man can work to the best of his ability. When first installed, a schedule of payment should always be a little too low rather than too high; then as the men fail to make a good bonus the schedule can be increased from time to time as occasion requires. A schedule should never be lowered when once it has been put into effect, as the men will immediately become suspicious of the intentions of the management. The new system should be introduced by applying it to one man at a time, picking out only those who are willing; after several of the best men are working under it, with demonstrated advantage to themselves, most of the others will volunteer to come under the new system. In general, it takes 2 or 3 yr. to institute a new system and to put it on a successful operating basis without making the men suspicious, and the utmost patience and tact is necessary at all times.

The Phelps-Dodge Corporation has not as yet introduced a system of payment other than the regular daily wage and, in the case of development work, the contract system, in which the shovelers share by receiving a higher wage than the men in the stopes.

## Daily Wage

In the daily-wage system, the supervision is generally held at a minimum and the men have an ideal opportunity, and use it to the full extent, to do as little for their pay as possible. It is a system of payment that is by no means adapted to efficiency methods, as it puts the men into large classes and takes no account of individual effort.

## Task Work

The task-work system merely designates a piece of work to be done after which the laborer may go home. In some cases the laborer gets a full day's wage for the period he works, while in others he divides with the company the amount of wage for the time he saves. In either case his only reward is shorter hours; he does not make any more by working harder, in which respect the system falls short.

## Profit Sharing

The men receive a set wage for a day's work and at intervals (quarterly, half yearly, or yearly) a certain proportion of the company's profits are divided among them, either according to wage rates or to length of time in the service of the company. This system has the disadvantage of not giving immediate reward to the men for the work they are supposed to do in return, and as a result their interest is not stimulated.

## Piece Work

In the piece-work system, the men receive a set price per ton handled or per foot advanced, etc. The system is an ideal one when the conditions of work vary only within narrow limits, but underground it has very grave drawbacks. When a piece-work price or a contract is given, very often the workman will make too much money or not enough; the price set is seldom one that enables the man to make what the company thinks is fair. In the first case, the company is tempted to cut the price; in the second, it is forced to help the man out by giving him enough to live on. Many unions forbid the piece-work or contract system. The harm in cutting rates is very largely caused by poor judgment on the part of the management in first setting the rates. The standards are not based on scientific data, but on the previous best records, an ordinary try out, or on the foreman's off-hand estimate of what should be done, all of which are generally from 50 to 150 per cent. too low. Conditions underground vary to such an extent from day to day that no true contract or piece-rate system can be entirely fair and either the man or the company is bound to suffer. Just as soon as the men in charge of such a system try to anticipate the conditions in making contracts, the men raise the cry of favoritism and become dissatisfied. Such a system presents great opportunities for evil in that the effort to equalize results actually does produce favorites and offers a field for a dishonest man to profit from the laborer.

## Bonus System

The bonus system guarantees a fair wage to the man for everything that he does up to a certain standard performance and beyond that a bonus is paid for each unit of work done. In the writer's opinion, it is an ideal system for getting around all the difficulties and uncertainties that exist in mining. To install such a system requires much preliminary study and unremitting care during its existence, but if carefully carried out it creates a bond of sympathy and good feeling between the men and the company that is worth far more than the expense. It simply means that for any class of work the company must determine through the medium of past performance and time study what an average day's work shall consist of and must guarantee the men a fair day's wage for that amount of work. In order to keep the men interested and up to this standard, the company must say that it is ready and willing to pay for any work done in excess of the required standards.

The wage in force at present at the Burro Mountain mines for Mexican shovelers is $3.40 a day. Assuming that this is a fair wage for this class of labor, to put the men in the proper frame of mind to take kindly to the bonus system, the wage should be raised to $3.75 a day, an increase of 10.3 per cent.; this wage will be paid to them whether they make the required tonnage or not. If any man makes the tonnage required of him, his wage may be raised to $4 for that shift, an increase of 17.6 per cent. above the $3.40 rate, although this has not been designated in Table 4. If he produces anything over the required amount, he should be paid a bonus per ton, depending on the original task allotted to him. In order to make an increase of 50 per cent. over his old $3.40 rate, a shoveler will have to produce between 25 and 30 per cent. more than his allotment. Experience has shown that when a laborer receives an advance in wages of more than about 50 to 80 per cent. above a fair wage, he tends to become shiftless and the increase does him more harm than good.

Fig. 33 shows that under good working conditions a laborer moving ore 20 ft. (6 m.) should use a wheelbarrow, and should move 21.5 T. According to Table 4, this man will receive $3.75 a day for any work up to 21.5 T., or at the rate of $0.175 a T. If he reaches the required 21.5 T., his wage will become $4, or at the rate of $0.185 a T., and for every ton over the required amount, he will receive $0.185 a T. In each individual case, the man setting the standard of work should make

sure whether there are any interfering elements that will prevent the
man making the standard tonnage, in which case he should make a
fair reduction.

TABLE 4.—*Bonus Payments to be Applied to Shovelers.* Base Rate of
$3.75 per Day is Guaranteed to Man

| Distance Ore is Thrown or Conveyed Feet | Shoveling into Chute | | | Shoveling into Wheelbarrow | | | Shoveling into Car | | |
|---|---|---|---|---|---|---|---|---|---|
| | Average Tonnage Expected | Payment Rate for Average | Bonus per Ton Above Average | Average Tonnage Expected | Payment Rate for Average | Bonus per Ton Above Average | Average Tonnage Expected | Payment Rate for Average | Bonus per Ton Above Average |
| 2 | 31.5 | 0.120 | 0.130 | | | | | | |
| 3 | 31.0 | 0.121 | 0.131 | | | | | | |
| 4 | 30.0 | 0.125 | 0.135 | | | | | | |
| 5 | 29.5 | 0.128 | 0.138 | | | | | | |
| 6 | 29.0 | 0.130 | 0.140 | | | | | | |
| 7 | 28.0 | 0.134 | 0.144 | 22.7 | 0.165 | 0.175 | | | |
| 8 | 27.0 | 0.139 | 0.149 | 22.7 | 0.165 | 0.175 | | | |
| 9 | 26.0 | 0.144 | 0.154 | 22.7 | 0.165 | 0.175 | | | |
| 10 | 25.0 | 0.150 | 0.160 | 22.6 | 0.166 | 0.176 | 19.5 | 0.192 | 0.202 |
| 11 | 24.0 | 0.156 | 0.166 | 22.6 | 0.166 | 0.176 | 19.5 | 0.192 | 0.202 |
| 12 | 23.0 | 0.163 | 0.173 | 22.6 | 0.166 | 0.176 | 19.5 | 0.192 | 0.202 |
| 13 | 22.0 | 0.170 | 0.180 | 22.5 | 0.167 | 0.177 | 19.4 | 0.193 | 0.203 |
| 14 | 20.0 | 0.188 | 0.198 | 22.4 | 0.167 | 0.177 | 19.4 | 0.193 | 0.203 |
| 15 | 19.0 | 0.197 | 0.207 | 22.3 | 0.168 | 0.178 | 19.4 | 0.193 | 0.203 |
| 16 | 18.0 | 0.208 | 0.218 | 22.1 | 0.169 | 0.179 | 19.4 | 0.193 | 0.203 |
| 17 | 17.0 | 0.220 | 0.230 | 22.0 | 0.170 | 0.180 | 19.3 | 0.194 | 0.204 |
| 18 | 16.0 | 0.235 | 0.245 | 21.8 | 0.172 | 0.182 | 19.3 | 0.194 | 0.204 |
| 19 | 14.0 | 0.268 | 0.278 | 21.6 | 0.173 | 0.183 | 19.3 | 0.194 | 0.204 |
| 20 | 13.0 | 0.290 | 0.300 | 21.5 | 0.175 | 0.185 | 19.3 | 0.194 | 0.204 |
| 25 | | | | 21.0 | 0.178 | 0.188 | 19.2 | 0.195 | 0.205 |
| 30 | | | | 20.3 | 0.184 | 0.194 | 19.1 | 0.196 | 0.206 |
| 35 | | | | 19.6 | 0.191 | 0.201 | 19.0 | 0.197 | 0.207 |
| 40 | | | | 18.8 | 0.199 | 0.209 | 18.9 | 0.198 | 0.208 |
| 45 | | | | 18.0 | 0.208 | 0.218 | 18.8 | 0.199 | 0.209 |
| 50 | | | | 17.3 | 0.217 | 0.227 | 18.7 | 0.200 | 0.210 |
| 60 | | | | 16.4 | 0.228 | 0.238 | 18.4 | 0.204 | 0.214 |
| 70 | | | | | | | 18.2 | 0.206 | 0.216 |
| 80 | | | | | | | 18.0 | 0.208 | 0.208 |
| 90 | | | | | | | 17.8 | 0.210 | 0.220 |
| 100 | | | | | | | 17.5 | 0.214 | 0.224 |

In 1917, in a stope with raises spaced 25 by 65 ft. (7.6 by 19.8 m.)
from which was mined 145,000 T., the shovelers average 8.5 T. a
man.    With wages at $3.40 a day, this shoveling cost $0.33 a T., assuming
that the men were on other work for 17.5 per cent. of the day.    Charts
show that under the new system these men should have averaged 22.9 T.

a man, for which they would have received $4 a day. This would be an average shoveling cost of $0.175 a T., or a gross saving of $22,475 for the year. Out of this gross saving would have to come the cost of, say, five special men, at an average salary of $160 a month, or a total of $9600 a year, to take care of this branch of the work. This sum deducted from $22,475 leaves a saving of $12,875, or a total shoveling cost of $0.239 a ton.

If, under the new system, the men consistently fail to make bonuses, it is either the company's fault or the men's fault. If it is the fault of the company, the standard may have been set too high on account of a faulty analysis of the time-study figures. The men may not be receiving the proper shovels to work with; or the men in charge of the work may not be making fair reductions from this standard, when working conditions are poor. If the men are to blame, either they are not physically fit to be shovelers, in which case they should be tried out at some other work, or they are suspicious of the company's motives, which would lead one to suspect that the system had been introduced too hastily, or that the men have organized to force the company to raise the rate of payment or to lower its standards.

### Summary and Conclusions

Shoveling ore cost so much at Tyrone during the year 1917 that it was deemed advisable to conduct tests to see where the fault lay. These tests were in the nature of time studies and extended over a period of very nearly a year. No data were available on this work to supply a starting point, so it was necessary to begin in a very elementary way, to find out the capacity of the average Mexican laborer.

To obtain a basis for comparison, it was necessary to determine the capacity of various types and sizes of shovels, so as to know whether the 21-lb. (9.5-kg.) load was the best load underground as well as on surface. A list of essential factors influencing shoveling was also made out during the course of a short series of preliminary time studies, and the motions involved in shoveling were analyzed and subdivided for convenience in studying. The necessary forms and record sheets were also decided upon at this time. A second short series of tests was then conducted on the surface, so as to be able to estimate the negative effects of underground work on the shovelers, and to minimize, as far as possible, the need for studying obviously poor types and sizes of shovels under the difficult conditions to be encountered in the mine.

The test work underground consumed by far the greater proportion of the time and was divided into three series: Shoveling directly into a chute; shoveling into a wheelbarrow and tramming to a chute; and shoveling into a car and tramming to a chute. In each series the following points were determined for various lengths of time worked: Number of shovelfuls handled per minute; effect of distance thrown on shoveling speed, amount of rest required, amount of time consumed in tramming and dumping, proportion of working day occupied in shoveling, total tonnage handled during various working periods, effect of distance on tonnage handled during the day.

These tests showed that the design of the shovel has a very marked effect on the shoveling efficiency and that with the proper weight and size of tool the man's efficiency will be increased in spite of himself. It was demonstrated that a man handling a total load of 26 lb. 8 oz. (12 kg.) did the greatest day's work, other things being equal, and that this load is divided into a live load of 21 lb. (9.5 kg.) and a shovel weight of 5 lb. 8 oz. (2.5 kg.). A shovel weighing 5 lb. 8 oz. made out of the best composition steel gives very excellent service, and a lower cost per ton than a shovel having a lower first cost.

Further marked increases in shoveling efficiency are to be gained by instructing the shovelers in the proper methods of using a shovel, a thing that very few laborers know in spite of the fact that they may have been shoveling for a living for years. ·

Probably the most important part of all efficiency work lies in the wage that is paid to the men, the manner of paying the wage, and the feeling of confidence and good will that exist between the men and the company. Various types of wage payment are briefly discussed and the conclusion is reached that the bonus system in some form, if properly handled and carefully watched with a sufficient number of specially trained men to instruct the laborers and keep them working to the best advantage, is the preferable system of payment. A bonus schedule is proposed as a starting point for the work and the benefits that will probably be derived from it are outlined, using some actual stoping experience as an example.

Scientific management systems applied to certain eastern factories have demonstrated conclusively that the efficiency of the average workman can be increased by from 50 to 250 per cent. and that at the same time the men will receive better wages, will remain in better health, and will be happy and contented, while the company will actually produce its products at a lower cost.

F. W. Taylor, working at the Bethlehem Steel Co. plant several years ago, increased the capacity of the iron-ore shovelers from 16 T. to 59 T. per day per man, raised the men's wages from $1.15 to $1.88 a day, and at the same time decreased the operating cost from $0.072 to $0.033 per ton shoveled. The increase in shoveling capacity amounted to 269 per cent. Figures obtained at Tyrone indicate that the tonnage in at least one of the large stopes can be raised from 8.5 T. per man to 22.9 T. per man, an increase of 169 per cent., the wages can be raised from $3.40 a day to $4 a day, and the cost to the company reduced from $0.33 a ton to $0.24 a ton, after taking care of all extra supervision needed. In addition to this actual saving in shoveling costs, the overhead and general mining costs will be reduced per ton by mining a greater tonnage per man; and the other classes of labor such as machine miners and timber men will be able to do more work, as they will not be hampered and delayed by the shovelers to such an extent as at present.

Labor should not be given too much of an increase in wages over what is recognized as a fair living wage, as the men will at once tend to become shiftless and prosperity does them more harm than good.

The modern manager is beginning to see that the relation of standard times to the other features of organization is very close and vital. The determination of the standard, or the shortest time in which a job can be done, is the starting point for all work that has as its aim the establishing of an equitable wage system.

### PARTIAL BIBLIOGRAPHY

Principles of Scientific Management, F. W. Taylor.
Shop Management, F. W. Taylor.
Cost Keeping and Management Engineering, Gillette and Dana.
Modern Business, Vol. 2, Alexander Hamilton Institute.
Safety and Efficiency in Mine Tunneling, U. S. Bureau of Mines, *Bull.* 57.
Increasing Human Efficiency in Business, W. D. Scott.
Scientific Management and Labor Welfare, Robert F. Hoxie, *Jnl. Pol. Econ.* (Nov., 1916) **24**, 833–854.
A Hundred Hints for Shovelers, D. J. Hauer, *The Contractor* (Mar. 29, 1918) **25**, 135.
*Industrial Management*, various issues.
Pamphlets of Alexander Hamilton Institute.

The writer wishes at this time to acknowledge the help derived from the above sources, and from personal communications, from many of which material has been freely drawn, and particularly to express his gratitude to those manufacturers of shovels who gratuitously sent him both standard and specially made shovels.

### DISCUSSION

B. F. TILLSON,* Franklin, N. J.—A preliminary study of this paper did not, in my case, succeed in checking with the shovel dimensions of one of the large shovel manufacturers. I therefore raise a question as to there being some difference in standards used by different manufacturers in regard to the dimensions of the shovels for similar sizes and styles, so it may be rather confusing to some of us when we attempt to use the table in this paper and classify the capacity of the shovels as given here with some that are on the market.

*Discussion continued on p.* 706

---

* Mining Engineer, New Jersey Zinc Co.

## Mining Methods of United Verde Extension Mining Co.

BY CHARLES A. MITKE,* A. B., PH. B., BISBEE, ARIZ.

(New York Meeting, February, 1919)

THE United Verde Extension mine is located in the Jerome mining district, on the eastern slope of the Black Hills, approximately northeast of the town of Jerome, Yavapai County, Arizona. The ore deposit may be termed a replacement in Yavapai schist. This schist is one of the oldest formations in the district, being probably of volcanic origin, and has been folded and faulted and contains intrusions of diorite and quartz porphyry. It is believed that the mineralization followed the porphyry intrusions and replaced the country rock; this was followed by erosion and a period of secondary enrichment, all completed in pre-Cambrian time. Later sediments of sandstones and limestones, to the depth of approximately 600 ft. (182 m.), were laid down, followed by volcanic flows. Faulting and erosion again took place, resulting in the present complicated arrangement of the strata. Prospecting, therefore, is very difficult, because the sediments practically cover the mineralized areas. This necessitates the sinking of deep shafts, cross-cutting, and, in some instances, diamond drilling to learn something of the various relations of the different formations and the occurrences of intrusions associated with ore deposition. The locating of new orebodies, therefore, is beset with innumerable difficulties.

### EARLY MINING METHODS

The two orebodies constituting the principal resources of the company are known as the small orebody and the Bonanza, or large orebody. These are about 150 ft. (45 m.) apart, and while they appear to be distinct orebodies may later prove to be connected. The character of the ore, the general associations of the porphyries adjoining the orebodies, and other characteristics, both in the ore and general environment, are very similar. The small orebody was the first to be discovered and was mined for some time before the Bonanza was found.

At first, a large amount of ore was extracted by means of development work, but as the smelters treating this class of ore were located from 60 to 400 mi. (96 to 643 km.) away, only the highest grade yielded a fair profit and it became necessary to do selective mining. The main workings

---

* Consulting Mining Engineer.

of the mine were on the 1300-ft. (396-m.) and 1400-ft. (426-m.) levels. Square-set stopes were located on these levels in the high-grade ore in approximately the center of the orebodies; and at this time only the very richest ore (averaging 20 to 50 per cent. carload lots) was taken. This ore was hoisted up the Edith shaft, transported over the aerial tramway to loading bins on the Verde Tunnel & Smelter Railway, where it was loaded into cars and shipped to the smelters. During the 2½ years it was necessary to continue selective mining, the company's smelter in the Verde Valley, some 5 mi. away, was being designed and constructed; a new main hoisting shaft was put down and concreted; and a haulage tunnel driven, connecting directly with the company's railroad to the smelter. The average monthly production during this period was about 4,000,000 lb. (1,814,360 kg.) copper.

On July 18, 1918, the blast furnace was put in operation and the reverberatory was completed on Oct. 2. Owing to the difficulty experienced in obtaining cars, all ore shipments to distant smelters were discontinued when the first unit was put in operation, so that the monthly production dropped somewhat, due to the temporary limited capacity of the smelter. Now that both units are completed, it is expected, unless labor difficulties interfere, that normal production will be continued.

Prior to the blowing in of the smelter, a definite plan was carefully worked out for mining according to a regular system, and for stoping the ore as it exists in place, without regard to grade, but with emphasis on complete recovery as far as possible, on account of the general high copper content, and with special reference to fire protection. No plan for stoping an orebody is complete which does not include provision for adequate ventilation in all the working places nor protection from fire from either natural or artificial causes. Therefore, in planning a stoping method for the average grade of ore and to open up sufficient stopes to furnish a daily tonnage that would ultimately amount to the capacity of the smelter, selective mining had to be abandoned and a number of considerations taken into account.

## Considerations Influencing Choice of Mining Methods

It was of prime importance that the method selected rank high among the most approved safety-first methods. The character of the ore made it necessary that the stoping method permit of maximum ventilation in order to provide satisfactory working conditions. It was important that the efficiency be unimpaired when changing from selective mining to a system for mining the average grade of ore. The ore contains approximately 25 per cent. sulfur; therefore there is extreme danger from fire. One fire has occurred in the small orebody, due to friction caused by caving ore and pyrite. The mining method should not leave pillars, as these are likely to be crushed and generate heat.

The ore is very rich, so that a high extraction was necessary; consequently, a low mining cost per ton was of secondary importance.  Small bodies of high-grade silver ore occur in the capping, which it is most desirable to recover.  Intermediate drifts from the stopes into the walls have encountered scattered bodies of rich copper ore, containing a high percentage of sulfur; provision had to be made for the recovery of this ore.  The ore is very heavy; the accepted figure for ore in place is approximately 9 cu. ft. (0.25 cu. m.) per ton.  It has the same cleavage as the schist it replaced; it has therefore many seams and caves readily when not supported, especially at the junction of the ore and capping, 60 ft. (18 m.) above the 1300-ft. level, where the ore simply falls, leaving the capping suspended.

Owing to the unusual size and shape of the Bonanza orebody, methods containing stopes extending vertically rather than horizontally could be used to advantage.  In some places the capping is quite soft and would cave if not supported.  Practically the whole capping above the Bonanza is a mass of silica.  As siliceous material is necessary for smelting purposes, certain silica stopes must be operated from time to time.  The walls surrounding the orebodies are, in many places, composed of altered quartz porphyry, which is very soft, and, as a consequence, the stopes near them were very heavy.

### Mining Methods Suggested
#### Caving System

The ordinary block caving systems were at once ruled out, for while the cost of mining is low, the great friction that results from caving an orebody containing a high percentage of sulfur invariably results in a mine fire; the stopes would also be so hot as to be unworkable.  A large proportion of the ore would become diluted with waste, while some of it would actually be lost.  The small bodies of silver ore in the capping would suffer considerable dilution and probably be lost.  The scattered bodies of copper ore occurring in the walls surrounding the orebody would cave with the capping, as the walls in places are decomposed quartz porphyry, which is very soft and would readily follow a general cave

#### Shrinkage Method

In drawing off a stope of this character, the ore would become very hot; this is shown by the heat generated when a chute full of ore is emptied.  The greatest difficulty would be experienced in drawing the pillars, as these would generate sufficient heat to cause fire.  The ore drops off in large blocks, which would make it extremely dangerous for miners to attempt drilling in shrinkage stopes.  No stope, from which a permanent supply of silica might be obtained, could be located in the capping, as this would all be caved within a short time.

## Top-slice Methods

Neither horizontal nor incline top-slicing were considered applicable to either the small or large orebody, for the following reasons:

1. Fire from natural causes. Small bodies of pyrite have been located above the Bonanza, and there is a possibility of other bodies of pyrite existing around the edges of this ore. Bodies of high-grade copper with high sulfur content have also been found around its sides. If, therefore, the capping and the walls around the orebody, which are very soft, cave, some of the pyrite and high-sulfur ore will come in contact with the timbered mat, and ultimately result in a mine fire.

Bodies of pyrite in the soft hanging wall of the small orebody would cave upon the timbered mat if top-slicing were used. In the early part of 1917, a cave occurred in this orebody, causing considerable friction and heating the ore until the sulfur took fire. The entire stope had to be put under pressure before mining operations could be resumed.

2. Fire from incendiarism. Top-slicing lends itself more readily to this danger than any other method of mining. Numerous fires in top-slice stopes in this state during the past 2 yr. have originated from this cause. These mine fires are very dangerous, expensive, and difficult to get under control. They cannot be sealed off from the rest of the mine, as the ground above is caved and gases may pass through these caves into the live workings. These fires are hard to handle and generally require a temporary shutdown of that particular orebody. During the past year, a large orebody mined by this method was closed for 5 mo. because of an incendiary fire.

3. In every top-slice stope, a small percentage of ore is lost in the mat. As the mat becomes thicker and the weight from above increases, this ore, which is held in the timbered mat under great pressure, will generate heat because of its high percentage of sulfur and will develop into a mine fire.

4. In case of fire in the timbered mat, it would be necessary to drive drifts and raises above the mat in order to turn water on the fire. This would be extremely dangerous, as it would require spiling through caved ground, with the probability of having the working drifts closed by movements in the extensive cave.

5. In the majority of cases, in top-slicing high-grade copper ore in deep mines in Arizona, where there is at least 1000 ft. (304 m.) of overburden, an excessive amount of timber must be used in order to make the recovery of ore as great as possible. This adds considerably to the mining cost.

6. Where top-slicing is used in deep mines in the Southwest, the overburden, as a rule, does not follow readily. Consequently, after the top-slice has been carried down several floors and a large area opened up,

huge boulders, perhaps hundreds of tons in weight, are likely to drop from the back and close up the top-slice. In one mine containing a stope of this character, 1200 ft. (365 m.) underground, the crash came unexpectedly and closed up the slice. Providentially, this occurred on change day when the mine was idle.

7. Top-slices in deep mines are more difficult to ventilate than other stopes, because there is no connection from the stoping level to the level above, whereas when they are located near the surface the ground is so thoroughly broken that the warm air from the slice can escape through the capping.

8. Top-slicing is not as flexible a method as square-setting, as it does not permit the mining of disconnected stopes in different parts of the orebody in order to obtain the necessary daily quantities of proper fluxing material for the smelter, such as iron, silica, etc. In top-slicing, mining would necessarily be commenced on parallel horizons at the top of the orebody and the secondary, or high-grade, ore would be exhausted before the primary, or low-grade, ore could be stoped. Under these circumstances, a constant production could not be maintained throughout the life of the mine.

### Cut-and-fill Method

Examples of caves in large square-set stopes in this mine have shown that the ore does not support itself as it would have to do in a cut-and-fill stope. Furthermore, even if cut-and-fill stopes could be carried up on the 1400-ft. (426 m.) level, they would have to be carried up under old square-set stopes that are filled with waste and would later have to be changed over to square-setting. In every instance, it has been found difficult to make the connection at the 1300-ft. level on account of the heavy overburden. Therefore, if cut-and-fill stopes should be put in they would have to be taken up extremely narrow, lagging put in very close, and preferably double-lagged on the sides to prevent loss of ore. Lagging would also have to be put in the back, which is not practical in a cut-and-fill stope and, eventually, these would have to be changed over to square sets, in order to make connection with the filled stopes on the 1300-ft. level. Under these circumstances, the amount of timber required in the extremely small and narrow cut-and fill stopes, with cribs to support the back, would be equal to that used in square-setting.

### Square-setting

The first method employed was the square set. In the beginning, before much was known regarding the nature of the ore, these stopes were taken up large and some caves resulted. Experimentation showed that a better way was by narrow sections of square-set stopes, about two sets in width and taken any convenient length. This proved entirely satisfactory as regards mining and costs, the only objection that might be raised being that a direct timber connection would exist between all the old and new square-set stopes. Should fire occur in any part of the large orebody, it might spread to the old and new square-set sections.

### Mitchell Stoping Method

In order to eliminate the foregoing objection and to introduce a more fire-proof method of mining, in the early part of 1917, the writer recommended the use of the Mitchell stoping method. One of these stopes was immediately started on the 1300-ft. level, which proved very successful from a mining point of view and also from the standpoint of costs, as the expense was even less than that of square-setting. It was also very satisfactory as regards fire protection, inasmuch as in mining by this system large barriers of waste, which act as fire breaks, were left between the square-set sections.

In some camps this mining method is known as the "Mitchell slice." Under this name, however, it has frequently been confused with top-slicing, which is a caving method, whereas the Mitchell system is really a combination of square-setting, mining pillars, and filling with waste, thus keeping the overburden intact. Therefore, in order to avoid any misunderstanding, it is referred to in this paper as the "Mitchell stoping method."

### Square-set and Pillar Method

It was suggested to mine the orebody by this method, which consists of alternate rows of square sets and pillars through the entire orebody, the rows of square sets to be from three to ten sets wide, according to the weight of the ground, and the pillars between the rows 12½ ft. (3.8 m.) in width.

After the square-set stopes were carried up, the pillars were to be mined by the overhand stoping method, beginning on the level and mining upward. When the pillar stope had advanced two sets, the stringers beneath were to be reclaimed as waste filling was turned in, this procedure to be followed until the stope was carried up to the floor above. The supposition was that a stope of this character would hold heavier ground than the Mitchell (although not quite so heavy as that supported by a square set), thus acting as an intermediate stoping method between the Mitchell and square-set systems.

The objection to this method was that work would be started on the 1400-ft. (426-m.) level, beneath a pillar of ore two sets in width by ten sets long by 100 ft. (30 m.) high, which, aside from the weight of the capping, had 60 ft. (18 m.) of gob above it on the 1300-ft. (396-m.) level. It would be necessary, therefore, for the timbers in this stope (which would be started on the 1400-ft. level) to support this total weight.

A pillar of ore two sets by ten sets by 100 ft. high weighs approximately 6722 tons. The weight of 60 ft. of gob on top of this ore is in the neighborhood of 2016 tons, making a total weight of 8738 tons (aside from the capping), which would have to be supported when work was commenced on the 1400-ft. level.

In using the Mitchell system, the timbering immediately below the 1300-ft. level had to support only a weight of 2016 tons, which is a little less than one-quarter of the weight to be sustained by the timbers when using the other method. This 2016 tons (in addition to the capping) was the entire weight to be supported until the pillar was taken down 50 ft. (15 m.), when the timbering was again thoroughly braced before waste filling was turned in; and while this filling naturally added to the weight, it was gob, rather than ore in place. Aside from this objection, ore could be mined by the Mitchell stoping method a great deal faster and at a lower cost than by this overhand modification.

### Underhand Stoping Method

This system was used in the small orebody to mine out caved ore, also in places in the large orebody, where the ground showed signs of caving. It is used only in exceptional cases, but has proved very successful in all places where the ore is badly broken. ·

### GENERAL STOPING PLAN

After a consideration of the various systems, the following stoping methods were adopted: Overhand square-set method, for very heavy ground, approximately 50 per cent.; Mitchell stoping method, wherever weight of ground will permit, approximately 50 per cent. In exceptional cases, caved and badly broken ground will be mined out by the underhand square-set method. The principal stopes and workings of both orebodies are located on the 1300-ft. and 1400-ft. levels. Mining operations have not as yet been extended to the lower levels. The small orebody will be mined by the overhand square-set method, carrying up sections only two to three sets in width.

### Bonanza, or Large Orebody

The principle followed on both levels is that the old filled square-set stopes are taken as the center of a circle and the newer stopes proceed from the mined area toward the circumference. On the 1300-ft. level, the Mitchell stopes will be used immediately surrounding the old stoped area while square-set stopes will be used in very heavy ground, or where the walls of the orebody are not self-sustaining. On the 1400-ft. level, narrow square-set sections, two sets in width, will principally be used, the weight of the ground being too heavy to permit of the Mitchell system

being used extensively.   However, in exceptional places this system will
be used.   In broken ground, where a great weight must be sustained,
the underhand square-set system will be used.

The stoping on the 1300-ft. level is always to be kept ahead of that on
the 1400-ft. level so that the stopes on the 1300-ft. are finished before
those on the 1400-ft. are brought up to this level.   The sills are to be left

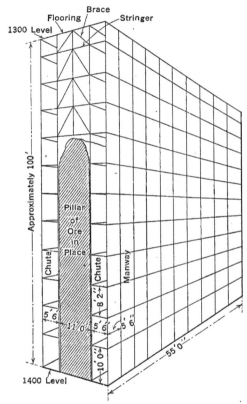

FIG. 1.—TWO ROWS OF SQUARE SETS WITH PILLAR OF ORE BETWEEN.

in on both the 1300-ft. and 1400-ft. levels in order to keep down the repair
cost and so that drifts may be kept open for waste filling in the stopes
beneath.   A waste raise has already been put through to the surface and
a large pit started to supplement the waste filling obtained from ordinary
development work in the mine.

The Mitchell stopes consist of parallel leads of square sets, one set in
width, separated by solid pillars of ore 11 ft. to 16 ft. (3.4 m. to 4.8 m )

(two to three sets) thick. These narrow sections of square sets are carried up to the capping and contain chutes every second or third set, through which the ore from the square sets, and later from the pillars between, will be taken out. The pillars of ore are left intact until the leads of square sets on either side have been mined out, and, with the

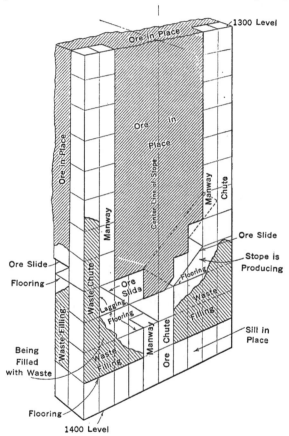

FIG. 2.—METHOD OF USING SQUARE SETS.

exception of the chutes, filled with waste. The pillars are then mined out. Timbers with angle braces are put in at the top of the pillar to hold the capping and mining is commenced at the top, using Jackhamers. After the ore in the pillar is removed, and as the waste filling is being turned in from above, tugger hoists are used to reclaim the long stringers, which can be used again. When the stope is completed, a barrier of waste, containing no timber (other than the floor), will exist between the leads of square sets originally put in and serve as a fire break.

Mitchell stopes will also be started on the 1400-ft. level, directly under stopes on the 1300-ft. level which have previously been mined by this system, in order to carry continuous barriers of waste from one level

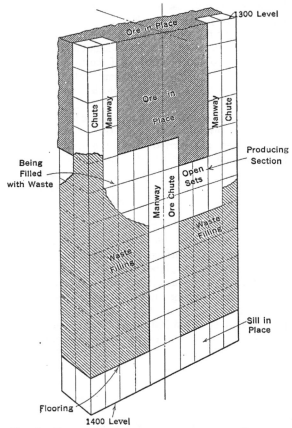

FIG. 3.—SQUARE-SET STOPE IN EXTREMELY HEAVY GROUND.

to the next. These stopes will necessarily be heavier than those above the 1300-ft. level as the overburden consists of 60 ft. of waste filling, in addition to the weight of the capping. In order, therefore, to hold the weight, they will be carried up as shown in Fig. 1, which illustrates a section consisting of two rows of square sets with an 11-ft. (3.4 m.) (two sets wide) pillar between. After the pillar has been taken down approximately 50 ft. (15 m.), a floor will be laid, the space filled with waste, and the timbers reclaimed; then the remainder of the pillar will be taken down and the rest of the timbers reclaimed as the waste is poured in.

· Where square-setting is used, the stopes will be carried two to three sets in width and about five to ten sets in length. Sufficient chutes will be put in, and in certain cases, slides built to these chutes, to eliminate mucking as far as possible (see Fig. 2). All the square-set stopes in extremely heavy ground will be two sets in width and taken any convenient length, as shown in Fig. 3. These small square-set stopes, even though they are only two sets in width and from five to ten sets in length, are taken up at least 50 ft. before the adjacent section is started, in order to prevent the ore from breaking in large blocks and causing side weight against the newer stopes.

The main raise is first driven through from one level to another. This contains standard manway, chute, and timber compartment. After that, chutes, or raises, are carried up as the stope progresses. With slides, waste filling may be turned into one-half of the stope while the other half is producing. Later, the work is reversed, and this process is followed continuously until the level above is reached, each half of the stope alternately producing and being filled with waste.

## Ventilation

There are three shafts on the property, Little Daisy, Edith, and Audrey, and two raises to the surface, one for ventilation and the other for waste. · The Audrey, when completed, will be the main hoisting shaft; the Edith, the supply shaft; while the Little Daisy is used entirely for ventilation. One suction fan, having an air capacity of 40,000 cu. ft. (1120 cu. m.) per min. is located at the top of the Little Daisy shaft, and a second, having a capacity of 135,000 cu. ft. (3780 cu. m.), will be installed at the top of the air raise, which will become the main air-shaft. A connection will be made from the long tunnel directly to the large orebody on the 1300-ft. level. A raise will be driven from the 1400-ft. level, connecting at this point, so as to force all the fresh air entering the large tunnel down to the lower workings. Doors have been put in at the different shaft stations on the 800-ft., 1100-ft., and 1200-ft. levels in order to force all the air to pass through the different stopes.

Measures will also be taken for the proper coursing of the air through the Bonanza orebody, where it is most needed. Gangways, immediately beneath the capping and connecting old raises, have been preserved by building walls of waste rock. These are maintained solely for ventilation purposes and will remain open for some time in order to afford a means of escape for the heated air from the old square-set sections

While at present the temperature, humidity, and velocity of the air in the different working places fall somewhat below the requirements for a good working atmosphere, when the necessary development work is completed, standard raises built, and the large mine fán installed on the surface, the volume of air passing through the workings will amount to at least 400 cu. ft. (11.3 cu. m.) of air per man per minute and the United Verde Extension will be one of the best ventilated mines in the Southwest.

## Fire Protection

A spraying system has been installed in the Edith shaft. In case of fire in this shaft, the fire-doors on the 1300-ft. and 1400-ft. stations will be closed immediately by means of compressed air and the sprays, which are controlled from the surface, put in operation. The doors on the 800-ft., 1100-ft., and 1200-ft. levels are ventilating doors and always kept closed. The mine fans would also be closed down at once, but not under any other circumstances.

The Little Daisy shaft, which is used entirely for ventilation, has a concrete connection from the fan to a depth of one set below the collar. From there down it is naturally so wet as to make the installation of a fire protection system unnecessary. The new Audrey shaft is fire-proofed by means of concreting. The barriers of waste left by the Mitchell stoping method will serve as fire-breaks in that part of the Bonanza orebody in which this system is used.

A number of electric blowers, and a sufficient supply of bulkheading material and fire-fighting equipment, are kept on hand so that the fire area can immediately be put under pressure and steps taken to extinguish the fire. Should these measures fail, the general procedure will then be similar to that followed in mining out the fire area in the small orebody, from which, for 1½ yr., ore that was actually on fire was mined successfully under pressure without interfering with the production or the daily mining operations of the rest of the mine. During this time not a single serious injury was reported.

## Haulage

A series of parallel drifts 50 ft. apart traverse the entire orebodyon the 1400-ft. level, which is the main haulage level, and connect into a large motor haulage drift, which extends to the Audrey, or main hoisting shaft. The stopes are located at right angles to these parallel drifts in order to facilitate the handling of ore.

The raises between the 1300-ft. and 1400-ft. levels will be used as storage chutes for all ore above the 1300-ft. level. All ore above the 1400-ft. level will be drawn out in motor cars on the main haulage level, taken to the Audrey shaft, and dumped into ore pockets. In the course of time, when the 1500-ft. and 1600-ft. levels are developed, the 1600-ft. will be the next main haulage level.

At present, the ore is being hoisted in skips to surface and transported over the aerial tramway, but when the large haulage tunnel on the 1300-ft. level (now being driven) is completed, the ore will be hoisted to the 1100-ft. and dumped into pockets, whence it will be drawn out on the 1300-ft. level into standard-gage railroad cars and taken direct to the smelter.

## CONCLUSIONS

The general stoping plan just outlined presents the following advantages:

*Safety.*—Stopes will be carried small and the process of filling with waste will keep pace with the stoping in order to avoid a possible cave.

*Ventilation.*—Gangways, immediately under the capping, will permit the escape of heated air from the old filled stopes. Sufficient raises, containing standard manways and timber compartments, will be carried up in all working stopes to furnish an adequate supply of fresh air in each working place. All the air entering the mine will be coursed through the stopes.

*Fire Protection.*—Adequate means of fire protection has been provided for all shafts. In the Mitchell stopes, barriers of waste separate the square-set sections; these are absolutely necessary in order to localize a fire area in the event of an outbreak. Equipment for an auxiliary ventilating system will always be kept on hand in order that any section of the orebody may be operated under pressure in case of necessity.

*Efficiency and Economy.*—As all stopes must be filled with waste, a large waste pit on the surface is available, from which a supply may be obtained at any time. Mucking will be largely eliminated by the use of extra raises, slides, etc.

In the Mitchell stopes, considerable timber will be saved by reclaiming the stringers when the stopes are filled. The ore will also be extracted from the pillars more rapidly and at a lower cost than with ordinary square-setting.

A large number of stopes will always be available from which the smelter may be furnished with a continuous supply of ore containing the necessary chemical contents. By leaving the sills intact, drifts can be maintained without an excessive expenditure for repairs. Where these have been removed, the drifts are very heavy and are kept open with great difficulty.

As all the ore above the 1400-ft. level will be dumped direct into the regular ore chutes between the 1300-ft. and 1400-ft levels, and from there be drawn out into motor cars, it will not be necessary to have large separate ore pockets between these levels. This general plan of stoping, arrangement of chutes, and motor drifts, may also be continued in the deeper levels as the work progresses downward.

## Fireproofing Mine Shafts of Anaconda Copper Mining Co.*

BY E. M. NORRIS, BUTTE, MONT.

(Colorado Meeting, September, 1918)

IN the summer of 1917 it was decided to fireproof the main Tramway hoisting shaft of the Anaconda Copper Mining Co. at Butte, Mont. The shaft has three hoisting compartments and one pump compartment; it is timbered with 12 by 12-in. (30.5 by 30.5-cm.) fir timber, and is 2475 ft. (754 m.) deep.

Subsidence and displacement of the surrounding country rock had produced exceedingly heavy ground, and had carried the shaft out of line, in several places, by as much as 2 ft. (0.6 m.) displacement. Constant repairing and realigning of the timbers had been necessary in order to maintain clearance for the cages. The most feasible method of fireproofing, therefore, seemed to be to cover the timbers with a coat of concrete applied with the cement gun.

The cement gun, which is operated by compressed air at ordinary mine pressures, feeds a mixture of sand and cement through a hose to a nozzle having a water connection. The mortar in fluid form is thus sprayed upon the prepared surface in thin layers, which can be built up to any desired thickness. The cement gun can be set up at any convenient point; satisfactory results have been obtained with the nozzle a distance of 500 ft. (152 m.) from the gun.

The shaft was thoroughly overhauled and the timbers put in the best possible state of repair. Between the 1000 and the 1400 levels, much loose ground lay against the shaft timbers on the north side; this was breast-boarded back and a 30-in. (76-cm.) reinforced-concrete retaining wall was erected, leaving a space of 12 in. (30.5 cm.) outside the shaft timbers to allow for future movements of the ground.

For convenience in handling men and materials, it was arranged to concrete the auxiliary hoisting-cage compartment and the pump compartment first, using the cages of the main hoist to serve the cementing crews. The auxiliary cage was then available while the remainder of the shaft was being concreted.

A tight partition of 2-in. (5.08-cm.) plank was erected between the auxiliary and the adjoining hoisting compartments, for better protection of men riding on the cages. The application of concrete made this

---

* Originally presented at a meeting of the Montana Section.

partition air-tight, which should prove a valuable feature of the fireproofing measures.

It was considered necessary to guard against the spreading of fire in the timbers behind the concrete covering, as once happened in a Michigan shaft. A set of shaft lagging was therefore removed just above and below

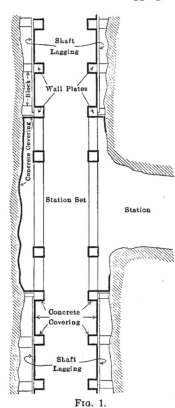

FIG. 1.

each station, and the concrete casing was built back to the walls so as to form an air-tight seal at these points. Another seal was made at the rear end of each station in a similar manner (see Fig. 1).

Experiments were made to determine whether reinforcing material was necessary to hold the concrete to the timber and, if so, what kind of material was best. It was found that if the timbers were wet down thoroughly the concrete would stick without reinforcement, but subsequent distortion of the timbers caused the concrete coating to shell off in large slabs. Tests of chicken wire, herringbone metal lath, and diamond-mesh metal lath showed that all these materials made equally satisfactory reinforcement for the concrete coating, the only difference being the matter of cost. Chicken wire was the cheapest to buy but the labor of nailing it to the timbers was much greater than with metal lath. After using several thousand square yards of each of these materials it was found that the 27-gage diamond-mesh metal lath, 24 by 96 in. (61 by 154 cm.), was the most economical and satisfactory reinforcing material; it was also determined that 6-d wire nails driven two-thirds of their length into the timber, and bent over, made the best fasteners.

The compartments to be concreted were covered with timber bulkheads at each level and lathing was begun. The lathing crews consisted of six men to a lift, two or more lifts being lathed at one time according to the number of men available. With the diamond-mesh material the rate of lathing was 225 sq. ft. per man per 8-hr. shift.

Two cement guns, type N-1, were set up on mine trucks so that they could be pushed on the cages and moved from level to level as required

With an air pressure of 75 lb. (34 kg.) per square inch it was found that 200 ft. (61 m.) above, and 75 ft. (23 m.) below, were the greatest vertical distances from the gun at which satisfactory work could be done with the nozzle. Where lifts greater than 275 ft. (83 m.) occurred, it was necessary to set the guns on bulkheads in the shaft.

The sand was dried, when necessary, and screened through a 1/4-in. (6-mm.) screen. It was then mixed with the cement, on the surface, and put into old cement sacks for convenience in handling. It was found necessary to screen the cement also, as lumps blocked the discharge and caused frequent delays. Mixtures of 3, 3 1/2, and 4 parts of sand to 1 of cement were tried. Where thin coatings (3/4 in. or less) were applied, the 3 to 1 mixture was the most satisfactory as it went on more evenly and formed a tougher coating.

Each crew consisted of four men, two feeding the gun, and two on the nozzle. The nozzlemen worked from stage planks in the shaft and were provided with rubber gloves, safety goggles, and respirators. The surface to be coated was washed off thoroughly with water sprays. Concrete was applied in two successive layers 1/4 in. (6 mm.) thick. After the coating had become firm it was sprinkled often enough to keep it damp for several days. It was found that a gun crew could cover from 800 to 1200 sq. ft. (74 to 111 sq. m.) of surface in 8 hr. Repairs on the gun were very slight, wear being taken by the rubber liners. Nozzle liners lasted 48 hr., discharge liners about 6 weeks, and the material hose several months.

The shaft and station timbers were repaired and concreted from surface to the 2000 level in 94 days, 175,465 sq. ft. of surface being covered. The average number of men employed was 54, including superintendence and all surface labor connected with the job. Material used was 6102 sacks cement, 1500 tons sand, 165,495 sq. ft. of lathing, 2600 lb. nails and staples.

Four samples were prepared for a fire test. Pine timbers, 6 by 10 in. by 5 ft. were covered on all sides with lathing; three pieces were covered with chicken wire and one with metal lath. These timbers were then coated with 1/2 in. of concrete put on with the cement gun. Three of the samples were sprinkled daily for 3 days; the fourth was not sprinkled. After 6 days of hardening, the samples were placed in a large bonfire, until the concrete coating was heated to a dull red. The sample that had not been sprinkled shelled off with loud explosions, but the others were not affected. After 30 min. the samples were pulled out and allowed to cool. Inspection showed that the only apparent effect of the baking was a slight charring of the wood on the edges, where the concrete had cracked while drying.

## Engineering Problems Encountered During Recent Mine Fire at Utah-Apex Mine, Bingham Canyon, Utah*

BY V. S. ROOD,† B. S., AND J. A. NORDEN,‡ BINGHAM CANYON, UTAH

(Colorado Meeting, September, 1918)

### UNDERGROUND CONDITIONS BEFORE FIRE

THE general system of workings at the Utah-Apex is similar to that found in many of the western metalliferous mines. There is a vertical three-compartment shaft extending to the surface, from which the various levels have been driven to the orebodies. The 1000-ft. (305-m.) level, about ¾ mile long, was driven as a haulage and drain adit.

### *Geology*

The formation is a flat north-dipping series of alternating limestone and quartzite beds. Extensive north-south faulting and fissuring have taken place, resulting in the formation of large replacement orebodies in the crushed zones in the limestone. The orebodies are irregular in shape and size, but follow closely the trend of the mineralizing fissures. The ore consists of galena, pyrite, sphalerite and small values in silver and gold. Large masses of non-commercial pyrite are frequently found in or adjacent to the orebodies, and more or less pyrite is found for some distance in the walls of the orebodies. Lenticular bodies of a black pyritic shale also occur in the zones where extensive faulting has taken place.

### *Mining Methods and Condition of Workings*

Geological conditions, mentioned before, have made the ground extremely heavy. In mining above the 1000-ft. level, square-set and filling methods were used, but not altogether successfully. The crushing of timbers extended even into the filled portions of the stopes. The filling material itself, consisting of limestone, shale, and pyrite, possessed swelling qualities which added to the difficulty of maintaining

---

* Presented before the Utah Section, Apr. 4, 1918.
† Manager, Utah-Apex Mining Co.
‡ Asst. Supt., Utah-Apex Mining Co.

chutes and manways. The oxidation of the filling absorbed much oxygen and produced considerable heat, making artificial ventilation necessary in the stopes. Plans had been made to adopt the slicing system on these orebodies above the 900-ft. level, but they did not extend far enough in this direction to make the proposed change advisable, and, thereafter, work was devoted solely to following the downward extension of the bodies.

The slicing system was adopted when work was started on the 1150-ft. (350-m.) level. In this system, raises are driven at short intervals through the ore to the level above and then beginning at the top, horizontal slices, 10 ft. (3 m.) thick, are taken across the body, the roof being allowed to settle as soon as the slice has been completed. The

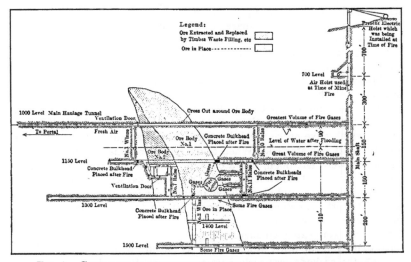

FIG. 1.—CROSS-SECTION THROUGH WORKINGS IN VICINITY OF FIRE ZONE.

system was successfully applied. The mat of timber and caved material absorbed some oxygen, producing heat, but as it became firmly packed soon after each successive drop, the amount of circulating air through the mat was materially reduced, and ventilation easily maintained by the exhaust air from the drills with compressed air and slight blowing between shifts. The raises through the ore gave little trouble, and it was found to be an easy matter to keep one slice open until the ore had been mined. The saving in timber over former methods was marked; little or no ore was lost through caving, and accidents from falling rock were practically eliminated.

At the point where the orebody was first encountered on the 1300-ft.

level, its width was much less than on the level above.   Ore was urgently needed to keep up normal production, and, as the walls looked firm, it was decided to try again the square-set stoping and filling methods. Stoping operations had not proceeded far before it became evident that work had been started on a faulted portion of the orebody.   Filling drifts, driven in the walls of the stope, encountered other portions. Stoping was carried on, however, with results similar to those obtained in previous work above the 1000-ft. level.   Raises were driven through the ore to the 1150-ft. level for ventilation.   The free circulation of air caused the filling to become hotter than ever before.   Temperatures of 98° F. were obtained in the working places where the current of air was good.   Temperatures in the closely packed fillings were not obtained, but they were probably higher.   While this stoping was in progress, it was decided that the other portions of the orebody should be mined by the slicing system.   Raises marked No. 7 and No. 11 were then driven between the main levels, with sub-levels extending to the orebodies. While this work was under way, the original stope was finished, and the timbers soon crushed so that further inspection was impossible.

Mining went ahead in the new stopes off the raises, under the slicing system, and the block of ore adjacent to No. 11 raise was quickly finished, but not without considerable difficulty.   The slices in these stopes continually broke through to the old stope, exposing the hot filling, which gave off so much heat that the slice had to be abandoned and the next one started.   When the stopes were finished, the raise and sub-levels were left open, and a current of hot steamy air continued to pass out onto the 1150-ft. level.   This condition caused no inconvenience or alarm, as all the old stopes gave off more or less steam and foul air. The last work was done at this point late in the fall of 1916, and no further attention was paid to the place.   The stopes in the vicinity of No. 7 raise were extended successfully and with little heat, as they were separated from the other stopes by 50 ft. (15 m.) of waste rock.   Early in 1917, the orebody was picked up on the 1500-ft. level, and in March raises had been started to block it out for stoping.   Ventilation of the lower levels at this time was effected by natural air currents controlled by doors, which were kept closed on the 1000 and 1150-ft. levels, causing the current of air to pass to the lower levels and thence up the main shaft to the surface.

### Discovery of Fire

During the latter part of March, 1917, a strange odor was noticed in the air coming up No. 11 raise.   For several days no attention was paid to the matter, and then it was noted that the odor was becoming stronger. This tainted air came up to the 1000-ft. level through No. 10 raise, then flowed along this level and up the shaft. At this time the main hoisting engine was on the 700-ft. level, and the auxiliary or sinking hoist was on the 1000-ft. level. The engineers on the hoists, and the station tenders, began to complain that the bad odor was making them sick.

The master mechanic had spent many years in Butte, and from the first he maintained that the mine was on fire. On the afternoon of Mar. 26, the men who had worked in the foul air all day came off shift unmistakably sick, and it was evident that something was seriously wrong. The Yampa, a neighboring mine, had had considerable trouble from a mine fire. The foreman of this mine was taken underground, and he at once expressed the opinion that the mine was on fire.

## CONDITIONS DURING THE FIRE

The position of the fire zone was known only approximately. It was near the sub-levels, extending from No. 11 raise, in a caved and tightly filled gob, and therefore inaccessible. To reach the fire by attempting to drive through the caved stope might have taken a long time, and the fire would have spread rapidly. It was possible that the fire zone would be so large by that time that one heading would open only a small part of the zone. It seemed unlikely that direct fighting with a hose would be feasible, even if the fire were reached, and there would be a constant risk of life on the part of the men engaged in the work.

Confining the fire zone by placing concrete bulkheads at all openings to the old workings was considered. Such a plan meant turning over to the ravages of fire a space from which 500,000 tons of ore had been mined and in which several million feet of timber had been used. It would have been necessary to abandon about 100,000 tons of reserves. Much exploration work was in progress at that time on the 1150 and 1300-ft. levels, and was giving encouraging indications that new orebodies would be found on these levels along the same general ore-bearing channel which contained the old orebody. It seemed probable that the fire would spread throughout this channel in a short time, and make it necessary to confine future operations solely to shaft sinking and exploration below the 1500-ft. level, a long and expensive undertaking, with assured loss of profit while the work was being done. In this connection, the probable course of the fire was considered. In a fire in a building, the trend is usually upward or following the path of the gases generated; in fires underground, where the exit of gases is confined, the trend is invariably downward or toward the supply of fresh air. Under these conditions, it was by no means improbable that the downward rate of burning would keep pace with stoping operations, or perhaps exceed them, in which case large amounts of ore would have to be abandoned from time to time. Furthermore, a confined fire is a constant menace to the lives of the miners working in other parts of the mine, and there is constant expense for watching, repairing and building new bulkheads. All operations are placed on an uncertain basis, as there is no knowledge of the time or place at which the fire will get·beyond the confining bulkheads and cause a complete shut-down.

Confining a fire may be practicable in a mine having many and widely separated orebodies or in places where flooding is impossible, but the ultimate cost of such a method of control or the amount of ore in danger cannot be estimated with any degree of certainty until the system has been tried for several years. In the case of a new fire, it appears that every plan for the extinguishment of the fire should be considered before resorting to confining methods.

Had the fire started above the 1000-ft. level, or in some other place having no outlet at the top, flooding would probably have been out of the question, because an air pocket would have been formed and water prevented from reaching the fire. Had the fire been burning for a considerable time, there would have been the possibility of local air pockets in the edges of the stope. The smoke seemed to be derived entirely from the combustion of wood, which indicated with a fair degree of certainty that the fire was somewhere in the caved timbering and that by prompt flooding it could be extinguished, as the caved stope and filling would permit almost complete saturation.

It seemed certain that flooding would do little damage. The shaft and levels were in hard rock, and the timbered portions of the levels near the orebodies were in good condition. No stoping had been started below the 1300-ft. level. Above the 1300-ft. level there were several top-slicing stopes near No. 7 raise. One slice was open in these stopes, and therefore, at the worst, only this one floor could close up, and all ore remaining upon it could be obtained from the next slice below upon resumption of operations. Main drifts and raises in the orebody were closely lagged, and there appeared little likelihood of serious caving occurring at these points. The workings were lower than those of the neighboring mines and were separated from them by such distances that it appeared certain the mine would be water-tight and that flooding could be accomplished rapidly with the 600 or 700 gal. (2271 or 2649 l.) per minute available. It was estimated that 15,000,000 gal. would be required to fill the mine to the 1100-ft. level, and that the time consumed would be from two to three weeks. It was believed that if the new hoist which had been installed during the winter were equipped with bailing tanks, it could unwater the mine faster than it had been filled.

Since mining operations had stopped and the workings had been abandoned above the 1000-ft. level, it had been noticed that the air in these places had rapidly lost its oxygen through oxidation of the pyrite and timber.  The heat had also diminished, and samples of air taken in these old stopes by the Bureau of Mines had shown the following composition:

| SAMPLE NO. 1, PER CENT. | SAMPLE NO. 2, PER CENT. |
|---|---|
| $CO_2$............. 4.11 | $CO_2$............. 3.39 |
| O.............. 1.68 | O.............. 5.19 |
| N.............. 94.21 | N.............. 91.42 |

This was apparently an inert atmosphere and one in which spontaneous combustion could not occur.  If the flooding should be successful in extinguishing the fire, it appeared possible that conditions above the 1000-ft. level could be duplicated below by sealing off all openings to old workings by concrete bulkheads and driving new ventilating raises at places away from all contact with the old stopes.

Numerous other details were considered, but the above-mentioned reasons were the main ones discussed in conferences with Mr. Shilling, Superintendent of the Utah Copper Co., and Mr. Sheehan, Superintendent of the Tintic Mining and Development Co.  These men recommended flooding the mine, as did also Mr. Pope Yeatman, who had recently examined the mine and was familiar with conditions.  On the morning of Mar. 27, 1917, orders were received to flood the mine, and by 5 o'clock in the afternoon all tools, cars, electric locomotives and pumps had been removed from the lower levels.

### Flooding

The main air line was disconnected near the shaft, and the outside end of the line connected with the surface fire pump.  As soon as possible, the station pumps which had been removed were also put in operation on the surface, and wherever possible all natural drainage was diverted to the lower levels.  On the morning of Mar. 28, the volume of smoke coming up No. 10 raise was so great that it was impossible to go to the shaft.  This condition grew rapidly worse, and on Apr. 1 the top of this raise resembled a smokestack, emitting great quantities of dense wood smoke.  Some time during the night the water must have reached the fire, for on the following morning the atmosphere in the mine cleared rapidly, and about noon it was possible to go to the shaft and measure the water level for the first time.  Pumping was continued steadily, and daily measurement taken until Apr. 10, when the mine had been filled to approximately the 1100-ft. level.

As soon as the flooding started, a bailing tank was designed and two were ordered.  Unwatering was begun on the evening of Apr. 20.  The new hoist was used for the first time, and the tanks were discharged at the tunnel level   Two shaft men on each shift, using the sinking hoist and bucket in the end compartment, looked after the tanks, attending to the removal of floating timber and other details.

Good progress was made until the water was lowered to about the top of the 1300-ft. level station. At this point the tanks rested on the landing chairs and would go no deeper. Bailing was continued as carefully as possible, in the hope that, even with partly filled tanks, it might be possible to lower the water sufficiently to enable the shaft men to get chains around the chairs and pull them up, but this could not be done. One of the tanks stuck at the bottom and could not be pulled up. When the shaft men went down to see what had happened, they were suddenly plunged into a strange gas about 50 ft. (15 m.) above the water. Their lamps were put out, and it was only through the best of fortune that they were able to give the signals to be pulled up. Later, this gas was found to be mostly nitrogen and carbon dioxide, the latter having been formed at the time of the fire and confined in blind headings and stopes beyond the fire zone. Bailing was continued with the other tank; in a few hours this also stuck, but at the 1150-ft. level, safely out of the water. Air was turned down the regular lines, and the column pipe, but on the following day no change in the level of the gas nor in its strength was noticed. The sinking bucket would bring the gas up just like water, but blowing had no perceptible effect.

On the day the fire was discovered, six sets of Draeger 2-hr. breathing apparatus, and a month's supplies for these sets, had been ordered. This material had arrived on Apr. 20, and a crew was slightly trained in its use. Storage-battery lamps had also been secured. The helmet men did splendid work, considering the difficulties. Guides were torn out continually, and finally one compartment of the shaft was fitted with long pieces of 4 by 6, which extended a short way under the water. These pieces were placed at short intervals all around the compartment, and the tanks was fitted in so tightly that it was impossible for it to miss sliding onto the guides when pulled out of the water. It was then possible to drop the tank with sufficient force to give it a pile-driving effect, and the chairs were soon knocked out and fell out of the way. Bailing was continued with this tank, the lining being extended from time to time until the water was several feet below the station. Permanent repairs were then made to the other compartment, and bailing was resumed in it until the water had been lowered far enough to make repairs in the compartment first used.

As soon as the tanks were working smoothly again and the level was cleared of water, the helmet crew was able to walk into the level and open air valves near the stopes. This rapidly drove the gas out to the shaft, where the tanks carried it up into the fresh-air current. In a few days, it was possible to begin cleaning up the level without the use of breathing apparatus. Two helmet men, however, were held in readi-

ness each shift, to render aid in case further bad air or gas was encountered by the clean-up gang.

Bailing was rapid between the 1300-ft. and the 1500-ft. levels, and
no gas was encountered in opening the latter level. A bulkhead had
been left in the shaft, 4 ft. (1.2 m.) below the rails of the 1500-ft. level,

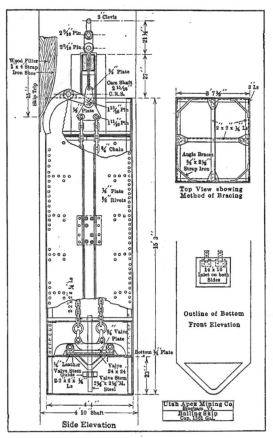

FIG. 2.—BAILING SKIP.

as a safeguard against lowering men into the sump during former operations. Several tons of rock and mud had settled on this bulkhead, and
the tanks could not lower the water below the 9-ft. mark. Three No. 5
sinkers were started, but made no headway, because much water was
still draining from the workings. These were replaced by a No. 9 sinker,
and the water was lowered below the level by May 27, exactly 61 days

from the time flooding was begun. Had it not been for the delays on the two stations, the flooding and unwatering would have taken only a few

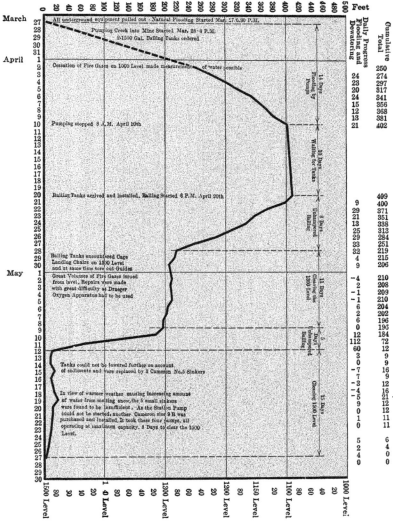

FIG. 3.—MINE FLOODING AND DEWATERING.

days over one month. The accompanying chart shows graphically the progress of this work.

## Conditions After Fire

As soon as the station pumps were again in place, the tanks were removed and the regular cages placed in service. Cars were sent to all the levels, and cleaning up started. Most of the mud was found on the 1500-ft. level, averaging 1 ft. thick. The greater part of this was a thin slime and was washed out to the pumps. Little mud was found on the other levels. Most of the mud had come from the muddy creek water which was pumped into the mine at the time of flooding. In the neighborhood of the orebodies, the drifts and raises were found practically blocked with fine ore which had run through the cracks in the lagging. The slices in the stopes had practically run full of ore, but the timbering everywhere was in fine condition, and the working places had only to be shoveled out. Four thousand tons of ore, mostly clean-up material, the cost of which was charged to the fire, was shipped in June. Nine thousand tons were shipped in July and 16,600 tons in August, the latter being the largest tonnage month of the year. The accompanying statement shows in detail the costs of fighting the fire and cleaning up the levels and stopes. It covers a period from Mar. 27 to June 30. Undoubtedly some of the post-fire conditions hampered July operations, but as the tonnage was nearly normal and the work difficult to distribute, no charge was made against the fire in this month.

### Preventive Measures

As fast as the levels were cleaned up, concrete bulkheads were placed at all openings leading to the fire zone. In August, samples of the air behind the bulkheads were taken, and showed the following analysis:

| 1150 Level, Per Cent. | | 1300 Level, Per Cent. | |
|---|---|---|---|
| $CO_2$ | 10.00 | $CO_2$ | 8.53 |
| O | 1.70 | O | 4.59 |
| N | 88.00 | N | 87.00 |

Flooding cooled everything to the natural temperature of the water, and there has apparently been no heating behind the bulkheads since they were put in place. New stopes, started below the 1300-ft. level since the fire, have become warm, but as soon as a stope has been finished all openings will be sealed, and it is hoped that in this way future trouble will be avoided.

In conclusion, the thanks of the company are extended to Messrs. Shilling and Sheehan for their good counsel and suggestions relative to fire-fighting methods. All the neighboring mines were prompt to offer any aid in their power, and this action was greatly appreciated. To the helmet crew, the company is particularly indebted, and has endeavored to express its appreciation in the form of bonuses and better positions. Its thanks are also given to all its employees who through their loyalty and energetic efforts helped greatly in the time of need.

### EXPENSES INCURRED DURING PERIOD OF MINE FIRE AT UTAH-APEX MINE

Mar. 27 to June 30, 1917

| | LABOR | SUPPLIES | TOTAL |
|---|---|---|---|
| Preparing to flood: | | | |
| Taking out machinery.................... | $482.38 | | |
| Construction of dam and flume............. | 62.63 | $4.85 | |
| | $545.01 | $4.85 | $549.86 |
| Flooding: | | | |
| Direct operating....  ..................... | $462.31 | $238.84 | |
| Watching dam........................... | 117.00 | | |
| Steam power........................... | 497.30 | 344.00 | |
| Electric power........................... | ...... | 950.54 | |
| Miscellaneous charges.................... | 238.50 | 81.89 | |
| | $1,315.11 | $1,615.27 | $2,930.38 |
| Preparing to unwater: | | | |
| Construction of flume and tunnel under portal of Parvenu tunnel to Bingham Creek...... | $307.57 | $52.15 | |
| Raising 1000-level station and installation of tank trips............................ | 392.67 | 84.47 | |
| | $700.24 | $136.62 | $836.86 |
| Unwatering: | | | |
| Auxiliary hoists........................... | $725.22 | $4.05 | |
| Surface hoist (actual tanking).............. | 832.71 | 331.39 | |
| Electric power........... .............. | ...... | 3,791.79 | |
| Steam power.... ...................... | 545.85 | 635.26 | |
| Bailing tanks........................... | 364.37 | 3,308.77 | |
| Direct operating......................... | 1,112.94 | 681.14 | |
| Hazardous work requiring the use of breathing apparatus........................... | 1,105.48 | 647.57 | |
| Repairs to shaft and equipment............. | 827.78 | 887.30 | |
| Miscellaneous charges.................... | 367.50 | 274.33 | |
| | $5,881.85 | $10,561.60 | $16,443.45 |
| Preparing to resume mining: | | | |
| Cleaning up levels......................... | $8,183.71 | $847.97 | |
| Construction of concrete bulkheads.......... | 738.12 | 200.19 | |
| Surface hoist............................ | 91.56 | 20.56 | |
| Steam power........................... | 205.62 | 239.12 | |
| Electric power........................... | ...... | 1,000.26 | |
| Replacing and connecting pumps............. | 301.21 | 95.81 | |
| Miscellaneous charges.................... | 186.29 | 115.28 | |
| | $9,706.51 | $2,519.19 | $12,225.70 |

Overhead expenses:

| Supervision | $1,361.47 | | |
| Timekeeping | 356.00 | | |
| Barn expense | 183.25 | $140.16 | |
| | $1,900.72 | $140.16 | $2,040.88 |

Total fire expense............................................. $35,027.13

*Equipment Installed, Value of Which was not Consumed During the Fire*

| 3000 ft. of 6-in. pipe line | $959.68 | $4,160.69 | |
| Cement gun | 21.94 | 1,345.51 | |
| 6 sets Draeger breathing apparatus | 9.00 | 1,360.07 | |
| | $990.62 | $6,866.27 | $7,856.89 |

Total fire and equipment expense............................. $42,884.02

## Tailing Excavator at Plant of New Cornelia Copper Co., Ajo, Ariz.

BY FRANKLIN MOELLER,* M. E., CLEVELAND, OHIO

(Colorado Meeting, September, 1918)

CONSIDERING the really short time that has elapsed since hydro-metallurgical processes of extracting copper from ores have been extensively developed, and the large scale on which this method is practised at Ajo, the successful operation of the plant must excite the admiration of every visitor.

The application of the mechanical unloader to the purpose for which it is employed at Ajo is but an extension of the use for which it was originally designed, some 20 years ago, for unloading iron-ore vessels at the lower lake ports. At a time when the maximum output of a single machine, with men shoveling the ore into buckets, did not exceed 40 tons per hour, G. H. Hulett invented and constructed a mechanical unloader with a bucket of 10-ton capacity. This machine was of the steam hydraulic type, and it is still in operation after nearly 20 years of service.

Several more machines of the same type were built, but the difficulties of power transmission, and the rapid development of electrical machinery, soon led to the adoption of electrically operated machines, and practically all of the later installations have been thus equipped. This electrically operated unloader has proved so well fitted for the work that the design of the lake boats has been practically revolutionized, with special view to the ease and rapidity of unloading. It is now possible to unload a 12,000-ton vessel with four machines in a little over 2 hr. This has necessitated an increase in the capacity of the machine, and the largest yet built handles an average load of 17 tons, with a possible maximum of 20 to 21 tons.

The Ajo excavator is somewhat simpler in design than the iron-ore unloader, in that it is not provided with a larry. In the unloader, the iron ore, picked up by the bucket, is delivered to a larry or transfer car which runs in the lower chord of the bridge; the larry then delivers the ore either into railroad cars, standing on several parallel tracks, or into the stock pile, thereby saving the time that would be required for the bucket to make the trip. At Ajo, the bucket delivers the tailing

---

* Engineer, Power and Mining Dept., Wellman-Seaver-Morgan Co.

directly into cars, and as the distance is short, the capacity of the machine is limited only by the regularity with which the cars can be supplied.

The leaching tanks from which the tailing is to be removed are 12 in number, arranged in two parallel rows. Each tank is 88 ft. (26.8 m.) square, and, when filled to a depth of 14 ft. 6 in. (4.42 m.), holds 5000 tons of ore; one tank is emptied every day. The excavator has a span of 106 ft., just sufficient to clear the buttresses on both sides of the tanks. Beyond the bridge runway, on one side of each set of tanks, are the railroad tracks for the cars which take the tailing to the dump. To move the excavator from one row of tanks to the other, a transfer table is provided, having a runway at right angles to the bridge runway of the excavator.

The excavator consists essentially of four parts: the bridge, the trolley, the walking beam, and the bucket.

The bucket is of the two-part type, and each shell is given a considerable sliding action during the closing movement, thus acting as a hoe and insuring a more nearly complete filling of the bucket when approaching the bottom of the pile. When open, the distance between the cutting edges of the shells is over 20 ft. (6 m.). The shells are U-shaped, without back-closing piece, so that the bucket can

FIG. 1.—LEACHING PLANT, NEW CORNELIA COPPER CO., AJO, ARIZ.

always be closed without choking.  The opening and closing chains are attached to cam-shaped drums, rotated by gearing and rope drum located in the vertical leg to which the bucket is attached.  The bucket

FIG 2.—TAILING EXCAVATOR AT WORK ON A TANK.

FIG. 3.—TAILING EXCAVATOR DISCHARGING.

can be rotated about the axis of the bucket leg through an arc of about 325° in either direction, and as the center of the bucket is offset about 2 ft. from the center of the bucket leg, it can clean up, in any given position, a space 24 ft. (7 m.) in diameter. The driving mechanisms for all the bucket movements are located on the end of the walking beam opposite to that where the bucket leg and bucket are attached. The mechanism for raising and lowering the bucket is also located in the walking beam and consists of a motor-operated drum from which four ropes lead around sheaves, to the framework of the trolley, back again to the walking beam, and are anchored there.

The use of the walking beam has several advantages: it provides a ready means for balancing the weight of the bucket; it eliminates the hinged cantilever of the bridge; and its length can be adjusted to suit

FIG. 4.—TAILING EXCAVATOR WORKING AT THE BOTTOM OF THE TANK.

FIG. 5.—TAILING EXCAVATOR WORKING AT THE TOP OF THE TANK.

special requirements. In order to insure vertical movement of the bucket, the bucket leg is fitted with parallel rods. The forward or bucket end of the walking beam is not entirely balanced, and the downward movement of the bucket is due to gravity, but is always under the control of the operator. A solenoid brake on the motor holds the bucket in any position.

The main trolley to which the walking beam is pivoted is mounted on two six-wheel spring-bearing trucks located immediately under the pivot posts. The rear end of the extended trolley is fitted with upper- and under-running wheels to prevent this end of the trolley from lifting.

The trolley is moved by a rack and pinion operated by a motor-driven mechanism, and is supported on a moving bridge, both legs of which are mounted on two four-wheel equalizing spring-bearing trucks, one of each set being provided with driving gears. The wheels are double flanged, and the trucks have ball-and-socket connections to allow for full equalization. One of the trucks is also provided with an expansion bearing to permit the bridge structure to adjust itself to inequalities of the track

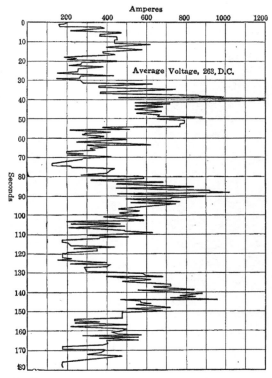

FIG. 6.—POWER DEMANDS OF EXCAVATOR DURING THREE MINUTES.

gage. Power is applied to the trucks in the customary manner, the motor for driving the bridge being located in its lower chord.

All of the movements of the bucket are controlled from one point by a single operator, who is situated in the bucket leg, immediately above the bucket. He can thus manœuvre the bucket with perfect accuracy and assure the maximum output of the machine. One movement, that of the bridge, is not controlled from the bucket leg, but the operating levers for this are so located that they can be reached by the operator when at about the middle of the bridge. It is customary for an assistant to look after the oiling, and have general oversight of the various mechanisms. All told, about 500 hp. of direct-current motors are required for operating the machine, and the several switchboards are grouped in the main trolley. Safety appliances are provided wherever possible, both electrical and mechanical, and every precaution has been taken to avoid interruptions to service. The machine will handle 500 tons of tailing per hour, which is equivalent to a bucketful every 72 seconds.

To avoid injuring the bottom of the tanks, the tailing is removed only to within 6 in. (15 cm.) of the bottom. Men are then sent down to shovel the remainder into windrows so that at the next excavation the greater portion of this tailing will be removed. The tailing is delivered into 20-ton cars, running on a track parallel to the line of the tanks, and the operators become so adept in manipulating the discharge of the bucket that practically nothing is spilled outside of the cars or back into the tanks.

The actual amount of current required per ton of material is extremely low, and the accompanying chart gives a clear picture of the power demands. It frequently happens that the operator combines the closing, lifting, and traveling movements and by doing so creates a heavy peak demand.

It is interesting to note how the cost of discharging ore has been cut by improved methods. When steam hoists were first used, the cost was 40 c. to 50 c. per ton. With the introduction of the bridge system, this was reduced to 18 c. to 20 c. per ton, and with the present type of machine, the cost varies from 4 c. to 5½ c. per ton. This cost includes superintendence, repairs, maintenance of apparatus and dock, and sundry contingent funds.

The transfer table is of simple construction, consisting of two eight-wheel trucks connected by a steel structure, and operated by an 80-hp. motor, giving the table a speed of 75 ft. per minute. The transfer table is fitted with latches to insure proper registering of the table tracks with the excavator tracks, and its movements are controlled from a cab located on the table.

## DISCUSSION

E. P. MATHEWSON, New York, N. Y.—I would like to call attention to the excellent plan for protecting the lining of the tanks from rough handling by the excavator. Many engineers, when considering mechanical excavators for lead-lined tanks, have been afraid that the machinery might get out of order and rip out the lining. This difficulty seems to have been entirely overcome in this installation, and those who have been responsible for its success deserve a great deal of credit.

A. F. CASE,* Cleveland, Ohio (written discussion†).—The application of the automatic unloader to the copper leaching process at Ajo was a very radical departure from the common use of this machine, inasmuch as the only previous use of this device has been in connection with the unloading of coal and ore from vessels on the Great Lakes.

There are many phases in connection with the unloading of ore from lake vessels that are not presented in the extraction of ore tailings from leaching vats. For instance, in unloading from boats, it is necessary to provide means to reach under the hatches to parts of the cargo hold which do not lie directly under the hatches, and to work in inaccessible places that do not exist in a leaching vat. These necessities require complications that could be very easily dispensed with in this new application. On this account the unloader, as applied to the New Cornelia plant, represents what is probably the simplest development of the machine that has ever been produced, and, manifestly, in its simplest form it could be expected to show results corresponding to its performances when installed with all of its complications.

One of the requirements of vessel unloading is extreme vertical travel, when the bucket is required to reach the bottom of the cargo hold, often 12 or 13 ft. below water-level, and discharge into a receptacle 30 ft. above the water. This was materially simplified at Ajo for the reason that the vertical travel was never more than 20 ft. This condition immediately resulted in the possibility of shortening the walking beam which carries the bucket leg, and, in fact, in entirely reducing the size of all the working parts of the trolley, walking beam, and bucket leg, as far as reaches were concerned.

The bucket is interesting on account of its mechanical operation. The shells of this bucket are carried on heavy cast-steel supporting arms which are attached to the bucket pins to which the shells are in turn attached; the arms, in turn, being carried on roller-mounted pins which are guided by horizontal and vertical guides in the lower end of the bucket leg. The purpose of this arrangement is to control the path of the cutting edge of the bucket shell so that a sharp edge is at all times presented to the material in which the bucket is working, thus minimizing the effort required to close the shells.

In this machine, three chain drums are required for opening and closing the bucket; two closing drums located on the outside of the bucket leg, and one opening drum, located in the center of the leg, all on the same shaft. As the chains lead to these drums from opposite sides, the opening and closing of the bucket is accomplished by simply reversing the direction of rotation. In order to compensate for the

* Engineer, Coal & Ore Handling Dept., Wellman-Seaver-Morgan Co.
† Received Oct. 7, 1918.

changes of the relative lengths of the chains in opening and closing, the drums are made cam-shaped.

The immense closing pressure required at the cutting lips is obtained by the introduction of gearing connecting the chain drums to a power wheel located in the bucket leg. The upper shaft, carrying the power wheel, is rotated by ropes which lead to the bucket-operating mechanism in the rear end of the walking-beam.

In addition to the bucket-operating mechanism, all of the machinery for operating the rotating and beam-hoist motions are also located in the back of the walking-beam. In this position they act as a counter-weight to counterbalance the weight of the bucket parts. This feature of the machine is one that is important, as it is possible in this way to secure accurate control because the balance of the walking beam can be delicately adjusted.

We have mentioned the motion of rotation, which is also applied to the bucket. This permits the operator to rotate the bucket at right angles to the normal position, thus enabling the bucket to reach into places that would otherwise be inaccessible.

It might be interesting to note the power requirements, in terms of motor capacities, of this equipment. The motors have the following ratings:

|  | Hp. |
|---|---|
| Bucket hoist | 150 |
| Bucket closing | 100 |
| Trolley travel | 100 |
| Bridge travel | 100 |
| Rotating | 15 |

These motors are all provided with magnetic control, with the exception of the rotating motor.

The conditions existing at Ajo required some provision for moving the machine transversely, in order that it might be used over a duplicate set of vats. For this purpose a transfer table was introduced. The table consists simply of a structural frame carrying a section of the excavator runway, and provided with trucks which travel at right angles to the main runways. The transfer runways are about 5 ft. below the main tracks. Electric conductors are provided for supplying current to the motor for moving the table, as well as short sections of main conductor for supplying current to the excavator when running on and off the table. An 80-hp. motor is provided for moving the transfer table.

## Hand-sorting of Mill Feed*

BY R. S. HANDY,† KELLOGG, IDAHO

(Colorado Meeting, September, 1918)

DOES hand-sorting of mill feed pay? The fact that the practice is so general would seem to indicate that there must be good reasons for following it; yet, to my mind, the advantage in many cases is doubtful enough to invite a thorough discussion of the subject, in an effort to determine under what conditions hand-sorting does pay.

A typical sorting plant in the Cœur d'Alene district handles about 800 tons of mine-run ore per day. The material is run over a grizzly with about 4-in. opening; the oversize is crushed to about 4-in. size and joins the undersize, and the whole product is washed and screened on trommels with about 1½-in. openings. The oversize from these trommels is spread on a belt and the waste and first-class ore are sorted out by hand, the residue being further crushed, and passes, together with the trommel undersize, into the mill feed bins.

About 50 tons of shipping ore and 150 tons of waste are sorted out per day, leaving about 600 tons of mill feed. To sort out this material requires 20 ore sorters and five bosses and repairmen. The normal cost of hand-sorting is about 16 c. per ton of run-of-mine ore, or 65 c. per ton of material sorted out. I believe that straight crushing and milling of the run-of-mine ore, in this case, is not only cheaper, but is metallurgically more economical and efficient than hand-sorting followed by milling of the residue.

At one of the Bunker Hill rock-houses, 1250 tons of run-of-mine ore are reduced to pass 30-mm., or about 1¼-in., opening in six hours, requiring five men. The equipment and the power consumed are not greatly different from that required in a sorting plant of the same capacity, and the normal cost of thus preparing the feed for milling is under 3 c. per ton of mine product. The relative cheapness of this method as against hand-sorting is thus clearly established, the saving being about 13 c. per ton. If there is any virtue in hand-sorting, then, it must lie in the metallurgical results achieved.

Possibly the favor that hand-sorting has found is due to a misconception as to what happens to the clean shipping ore if it is sent to the mill. If a mill makes, say, an 80 per cent. recovery on its feed, the hand-sorting advocate may assume that if the hand-sorted product be sent to the mill, 20 per cent. of the metals contained in it will be lost. This, I believe,

---

* Originally presented at a meeting of the Columbia Section in November, 1917.

† Mill Superintendent, Bunker Hill & Sullivan M. & C. Co.

is a false assumption. There is no ore of shipping grade getting past the first two cups of the jigs in the Bunker Hill mill, down to at least 7 mm. in size, and very little gets into the middlings in the sizes below 7 mm. Of course, none gets into the tailings. The only loss, then, in milling the hand-sorted product must be the losses due to sliming in the crushing operation and the slime lost on account of attrition of the soft galena particles in the jig beds. Crushing through 1½-in., in our experience, produces less than 2 per cent. of material below 40-mesh and the saving on this product nowadays is practically the same as on the coarser material.

I have tested our jigs time and again for slime losses, and have found that the slimes in the tailings of all the jigs down to 7 mm. size, including those from the re-concentrating jigs, does not exceed 0.1 per cent., by weight, of the tailings. This slime assays approximately the same as the mill feed. Attributing the whole of this slime loss to the 50 tons of hand-sorted ore in the sorting plant mentioned would account for the loss of about 50 lb. of lead per day, against which is the cost of $30 or $40 per day for sorting.

Our bull jigs, handling material between 1¼ and ½ in. in size, make no shipping product—only middlings, which are crushed through rolls and re-concentrated, and tailings. Therefore, there are no smooth, worn "marbles" of galena in the beds of our bull jigs, such as occur when jigs are forced to produce a higher-grade product than the conditions warrant. For example, if we try to make a shipping product in our bull jigs, about the best grade we can get is 55 per cent. lead, and under these conditions there is considerable wear of the soft material on the jig beds. By taking off a middling product assaying about 25 per cent. lead, and crushing and re-jigging this, we can easily produce a concentrate in the re-concentrating jigs assaying 65 per cent. lead with very little wear on the bed.

Considering the slight mechanical losses which occur in milling the hand-sorted ore, unless I am much in error, there is little to be said in favor of hand-sorting from this standpoint.

If there is no advantage in sorting out shipping product, there must be less advantage in sorting out waste, when the cost is 50 or 60 c. per ton of waste rejected. Each of our bull jigs will treat 150 tons of feed per 24 hr. and throw off 90 tons of tailings assaying less than 0.4 per cent. lead. I am satisfied that a jig handling larger sizes and designed primarily to discard tailings could throw off waste of lower grade and do it more cheaply than any method of hand-sorting in use.

There is an economical advantage in the milling of ore as against hand-sorting that seems to be little considered. This is in the grading of the products, and is illustrated in the accompanying chart, Fig. 1. Beginning at the left, a curve is drawn showing the net smelter value of mill product derived from conditions shown in the other curves. The

top scale indicates the tons of product shipped, and the scales on the left
and right indicate the units on the curves. For example, if 1400 tons
of product (or the mill feed, assaying 10 per cent. lead and 4 oz. silver
per ton) are shipped to the smelter under the conditions stated in the
note, the recovery of lead and silver is 100 per cent. and the net smelter
value is $2800. If nothing but the clean galena is shipped, about 50 tons
can be produced, assaying 86 per cent. lead and 26 oz. of silver per ton.
The recovery of lead would be about 30 per cent. and the net smelter
value of the product would be $4500. If a concentrate assaying 33 per

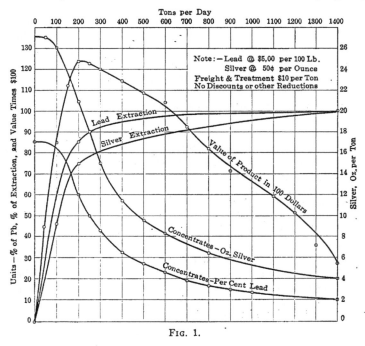

FIG. 1.

cent. lead were produced, a recovery of 95 per cent. of the lead and 84
per cent. of the silver would have to be achieved to get approximately
the same net smelter value in the product as that recovered when the
grade of the product is about 60 per cent. lead and the recovery 85 per
cent. of the lead and 75 per cent. of the silver. The latter condition
seems to be the most economical one possible under the conditions stated.
That is, there seems to be a definite economic peak in the milling of ores
where the various factors of recovery, grade, and cost combine to give a
maximum net value, or money return, to the operator.

When hand-sorting was practised at the Bunker Hill plant, it was not practicable to sort out a product containing over 45 per cent. lead, and I understand the average grade was less than that.. In order to estimate the possibilities of hand-sorting at the Bunker Hill plant under the conditions existing at the sorting plant first referred to, I have used 45 per cent. lead and 16 oz. of silver per ton for the hand-sorted product and the economic relationships in the various treatments possible for this product have been indicated graphically in Fig. 2. This is based on a

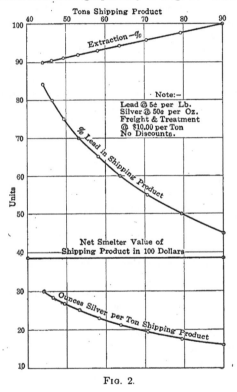

Fig. 2.

production of 90 tons per day of hand-sorted shipping product, which would be the Bunker Hill production under the conditions stated. The net smelter value of these 90 tons, under the conditions stated on the note in the chart, would be $3870 and the recovery of the metals, from a milling standpoint, would be 100 per cent. If this material were milled and a concentrate produced containing 84 per cent. lead and 29.8 oz. of silver per ton, 43.5 tons of such concentrate would have to be produced to equal the net smelter value of the original ore. This would mean a recovery of the metals of 90 per cent. and the discarding of 46.5 tons of

tailings assaying 8.5 per cent. lead and 3.1 oz. of silver per ton. If the
grade of the concentrates produced is 60 per cent. lead and 21 oz.
of silver per ton, the recovery is about 94 per cent., and 26.5 tons of
tailings are produced assaying the same as before. The production of a
concentrate, from the treatment of this material, assaying 60 or even 70
per cent. lead, would be an easy thing to do, but I think it would not be
necessary to discard a tailing assaying over 8 per cent. in lead in producing this concentrate. In fact, it could probably be done with a tailing
running less than 2 per cent. lead, in which event the economic advantage
over hand-sorting would be very large.

The crux of the whole matter is, "What can you do with the hand-
sorted product in the mill;" and not, "What does the mill do on the feed
it is getting?" The only sure test is to run the mill on a hand-sorted
product long enough to determine what grade of concentrate and of tailings can be achieved. I am sure a test of this kind would surprise many
of the advocates of hand-sorting by demonstrating the enormous economic advantage to be gained by the opposite method of treatment.

## DISCUSSION

A. STANLEY HILL, Gem, Idaho.—Mr. Handy's paper is most interesting and instructive, but I do not believe he has submitted enough data
to prove his case. As he has taken a particular plant, the Bunker Hill
& Sullivan mill, to prove his statement, in discussing the opposite side of
this question, I too will consider a particular plant, the Hecla company's
sorting plant at Burke, and its mill at Gem.

In the two cases, the ores are nearly the same and the same general
method of milling is employed. The grade of the concentrates, however,
is different; while the Bunker Hill produces a concentrate containing 60
to 65 per cent. lead, the Hecla concentrate contains about 53 per cent.
lead.

While Mr. Handy's sorting plant is not exactly like the one at Burke,
it is approximately the same. Taking the figures for 1915, I wish to
submit a few sorting and milling costs which I believe to be the only fair
basis upon which to determine the cost per ton of shipping product.

### Ore Sorting Plant, 1915

| | |
|---|---|
| Tons material hoisted to sorting plant | 146,675.00 |
| Tons waste sorted out | 22,689.00 |
| Tons first-class sorted out and shipped | 11,340.00 |
| Tons mill dirt (crushed) | 112,646.00 |
| Percentage material sorted out | 23.20 |
| Percentage material sent to mill | 76.80 |
| Ton costs of material sent through plant | $0.147 |
| Ton costs of material sorted out | 0.635 |
| Ton costs of first-class ore shipped | 1.907 |

## *Milling*, 1915

| | |
|---|---|
| Total tons milled...................................... | 112,646.00 |
| Total tons dry concentrates shipped................... | 15,345.00 |
| Ton costs of ore milled.............................. | $0.357 |
| Ton costs of concentrates shipped.................... | 2.62 |
| Extraction, per cent................................ | 82.87 |
| Ratio of concentration (approximate)................. | 7 to 1 |

At the above tonnages, the mill was doing its best work, and the addition of any excess tonnage naturally would decrease the extraction if the feed assayed about the same. In this case, the ore-sorting plant is doing 23.2 per cent. of the concentrating work at less cost than the mill. While the cost of labor for treating the whole tonnage in the mill will not increase 23.2 per cent., the wear and tear will do so. In order to put the total tonnage through the mill, it would mean enlarging it at a greater cost than the cost of the sorting plant, as the only machinery in the sorting plant not required in a crushing plant of the same size is the picking belt and feeder, two pieces requiring practically no repairs. Thus, when increasing the capacity of the mill we do not decrease the size of the sorting plant if we convert it into a crushing plant. This means a greater annual depreciation.

If in putting the total feed through the mill and making the same extraction, we make the same total quantity of shipping product, our ton cost has changed. No exact figures can be given on this cost; the additional expenses involved would be those for crushing, for extra freight charges on 34,029 tons, now sorted at the mine, and for mill supplies, fuel, and power.

As may be seen, the question of cost cannot be determined offhand, as Mr. Handy has done in comparing the sorting and crushing costs, without considering the effect of eliminating 20 to 25 per cent. of the run-of-mine ore from the mill. In the crushing plant, no material is obtainable for any purpose other than milling, whereas in the sorting plant we obtain waste suitable for mine filling (an acute problem, as tailings could not be shipped from the mill except at prohibitive rates) and crude ore from which we get returns from the smelter at the same rate as from the mill concentrates.

From the metallurgical standpoint there is also opportunity for disagreement. I agree with Mr. Handy in his statement that with a mill extraction of 80 per cent., 20 per cent. of the shipping ore sent to a mill would not be lost, but at the same time there is a loss, and a considerable one, if the sorted ore and the concentrates are of the same grade. If we can ship a sorted product carrying 52 per cent. lead and only a 52 per cent. lead product from the mill, why lose 1 per cent. of the shipping product?

This leads us to the discussion of the relative merits of a 50 to 55 per cent. lead product and a 60 to 65 per cent. lead product from the mill. In order to obtain the higher-grade product, Mr. Handy re-treats his first jig product, a middlings, by re-crushing and re-jigging. He states that less loss is due to sliming in the re-crushing than occurs on the bed of the jig when bringing up the grade of the concentrates. I do not agree with him, and believe that he should submit sizing tests and assays in support of his declaration. It is in the bed that wear occurs on the ore particles in a jig and the small particles are at once drawn through the hutch as concentrates. I believe that this re-treatment sends a greater proportion of the initial feed to the slime section of the mill, and in order to compare methods it would be necessary to know the proportion of slimes and the grade of slime concentrates as well as the extraction. Mr. Handy does not state that all the products shipped, namely jig, table, and flotation concentrates, assay 60 per cent. lead. A middlings is shipped to the smelter and it would be interesting to know how the average is kept up to 60 per cent. lead.

To raise the grade of lead from 50 to 60 per cent., it would be necessary to enlarge the mill and increase labor, supplies, power, etc. The depreciation would also be greater. To make this increase, more water is required; an important item in some localities. In the case of the Hecla, to do as Mr. Handy advises would mean a 25 per cent. larger mill on the basis of a 55 per cent. product and no increase in the total tonnage of concentrates; or a 35 to 45 per cent. larger mill on the basis of a 60 per cent. product.

As to the value of the shipping product, all the various costs of production must be considered, and also the cost of producing a ton of concentrates of various grades. A curve on Mr. Handy's Fig. 1, showing these costs at the Bunker Hill, would be a great aid in making comparisons.

When considering the advisability of using a sorting plant, it would be necessary to get more detailed data than either Mr. Handy or I have given. The character of the ore, local conditions, and finances necessarily have the greatest influence. I doubt that a new property would primarily build a mill large enough to treat and re-treat the ore as the Bunker Hill does, if they could put up a sorting plant and a small mill at much less cost which would handle the same tonnage as the large mill.

Considering the absence of important details in Mr. Handy's comparisons of cost, I think his summing up of the matter does not cover the subject fully nor prove his point conclusively.

W. L. ZEIGLER, Sunset, Idaho.—The possible advantages of hand-sorting depend mainly upon the physical characteristics of the ore in question. At the Interstate Callahan, it is possible to hand-sort both crude lead and zinc, also waste, the metallic contents of the mill feed remaining about the same as before sorting. Approximately half the total shipping output is obtained from the sorting belt, and these products assay higher than the concentrates made in the mill, as a complete separation of lead and zinc on jigs or tables is impossible.

While it is not possible to sort crude lead or zinc at the Success mine, a large tonnage of waste is taken from the sorting belt, assaying high in iron and low in lead and zinc.    The rock is extremely hard and the crushing cost is high.    If all this ore were jigged, the iron, which is disseminated through the waste, would form a middling more readily than the zinc, resulting in a high-zinc tailing and a high percentage of iron in the zinc concentrate.

L. O. HOWARD, Miami, Ariz.—If I understand the paper and the preceding discussions rightly, Mr. Handy has excluded just those cases where hand-sorting of mill feed is applicable, and confines his argument to the narrow field in which hand-sorting is not desirable.    He has made an excellent case for procedure in the Bunker Hill & Sullivan mill, and his array of evidence leaves little doubt that he has found the best practice for his particular ore.    Lacking long acquaintance with Cœur d'Alene ores, I am unable to form a definite opinion as to whether other mills in the district would be warranted in following the same procedure.    One of the axioms of mining is that no two ores give equally good results when concentrated in the same way.    I can conceive of some occurrences in the district—for instance, an ore carrying sulfide of lead and zinc in distinct aggregates—that would certainly be treated only with excessive losses did not hand-sorting precede mechanical concentration.    It is conceivable that finely disseminated lead and zinc products could readily be sorted out by hand, which could be separated mechanically only at great expense and loss of mineral if sent to the mill.    Where galena occurs in large and very friable crystals, as at Park City, Utah, it is folly not to hand-sort the mill feed.

No one who has shipped fine concentrates need be told of the greater desirability of shipping coarse ore having an equal or higher percentage of lead.    I have observed that in the case of some Cœur d'Alene mines, the galena is remarkably hard and massive, and scarcely more brittle than the gangue.    Under certain conditions it might be profitable not to hand-sort feed of this character.    The whole question, it appears to me, can be summed up by saying that each ore must be considered on its own merits.

CLARENCE A. WRIGHT, Moscow, Idaho.—I have not spent sufficient time in the mills of the Cœur d'Alene district to ascertain whether or not hand-sorting of the mill feed should be practised at all mills.    I do believe, however, that Mr. Handy, through his careful study of the treatment of the Bunker Hill ore, is correct in deciding that the character of that ore is such that it is best to treat the mill feed direct, without sorting.

It may be of interest to describe the method of hand-sorting as practised in the Wisconsin and Joplin lead and zinc districts.    At the mines in these districts, the main shaft, as a rule, is connected directly with the mill hopper.   The ore hoisted from the mine is delivered onto a chute leading to a grizzly, consisting usually of heavy rails or pipes spaced 4 to 5 in. apart, laid horizontally over the hopper.   The oversize mineralized pieces are broken with sledges so as to pass through, while the boulders, which are practically barren, are sorted out by hand and trammed to the waste-rock pile.   These boulders vary in size from 5 to 15 in. in diameter. The proportion of waste rock thus sorted out varies from a fraction of 1 per cent. to 15 or 20 per cent.

Those operators in the Joplin and Wisconsin districts who practise this method of hand-sorting claim that the elimination of waste rock before the ore passes into the mill makes a higher recovery possible, and that the productive capacity of the mill is increased.   They also claim that by eliminating the relatively large boulders of hard flint, which is the principal gangue material in the ore, there is a saving in power and in the wear and tear on rolls, elevator belts, and other equipment.   It should also be mentioned that the average lead and zinc contents of the ores mined in the Joplin and Wisconsin districts are considerably less than the average for the Cœur d'Alene district.

D. C. BARD, Seattle, Wash.—There are several conditions under which sorting of ore before milling would probably be considered advisable by Mr. Handy.   For instance, if there is a considerable transportation expense between mine and mill, these would be reduced by a sorting belt at the mine.   Then again, where ore-sorters are cheap and efficient, as in southern Mexico, and power and crushing costs are high, sorting might be cheaper than crushing.   Also, of course, there are certain types of ore which will not concentrate to advantage in a mill but can be satisfactorily sorted before reaching the mill.

S. A. EASTON, Kellogg, Idaho.—Hand-sorting of mill feed has been practised at the Bunker Hill since the very earliest operations.   Underground hand-sorting of waste from the mill feed is still performed in order to raise the grade of mill feed, and furnish waste for stope filling.   This sorting is done by shoveling the broken ore into a wheelbarrow; after each shovelful the broken waste, not smaller than an egg, is thrown out into the waste pile; occasionally the sorting is done by throwing out the coarse waste as the ore runs from the chute lip into the car.

For years prior to 1903, it was customary to sort clean, high-grade ore underground for direct shipment, without concentrating.   The movement of this sorted high-grade, both in the stopes and in special tramming and binning, was always an expense and a care, and with deepened and greatly extended workings and largely increased output, these items became more burdensome; and to provide facilities for such sorting, a picking belt conveyor was put into commission early in 1903.   This appliance closely resembled those now in general use throughout the Cœur d'Alene, and was the first of its kind in that district. Sorting out of first-class ore for direct shipment, underground, was then abandoned and all the mill feed that passed over a 2-in. grizzly was hand-picked for first-class ore on this picking belt conveyor. In special cases, as in the cutting out of the easily accessible sill floors of stopes, before the passing of the ore through chutes became necessary, a considerable tonnage of straight shipping ore was won directly, without going through the sorting plant. One other desirable result was accomplished by such picking, namely, the relieving of our concentrator, which was badly crowded by increasing mine production.

As soon as our concentrator capacity was increased to take proper care of the mine output, the advantages of hand-sorting of high-grade ore became less clear. The objections were its cost of operation and the lower grade of the product recovered, as compared with concentrator product. Also the desirability, to the smelters, of such high-grade crude ores became less pronounced with their increased facilities for handling fine mill product. Inasmuch as smelter and freight charges were at a flat rate per ton, without regard to grade of material, it was our aim to produce as clean a shipping product as possible, consistent with high metallurgical recovery.

The first detailed study of this work was done by Gelasio Caetani, then metallurgist for this company, and afterward mill superintendent; he confirmed in detail, and by conclusive physical tests, the belief of the management that no profit was obtained by the practice of sorting out high-grade ore for direct shipment. Subsequently, when Mr. Caetani had given up his work here to take up general professional work in San Francisco, R. S. Handy, his successor as mill superintendent, entirely confirmed and extended the first results. Hence, for the last 10 years hand-sorting at the Bunker Hill has been confined to the throwing out of coarse waste underground; all mineral is recovered in the form of mill product, yielding higher-grade material than can be sorted out by hand and giving a metallurgical recovery which is quite satisfactory and compares favorably with the best results anywhere.

F. A. THOMSON, University of Idaho, Moscow, Idaho.—Mr. Handy is to be complimented upon the excellent presentation which he has made of hand-sorting at the Bunker Hill & Sullivan plant, and he has satisfied me that, so far as that particular mine is concerned, the problem has been solved correctly. Even in the Cœur d'Alene district, however, I am sure that mines could be mentioned at which hand-sorting is not only necessary but indispensable. There are others, such as the National, in which the mineral is so finely disseminated as to make hand-sorting impossible. Between these two extremes there are doubtless other mines concerning which there is room for legitimate difference of opinion.

I trust it will not be considered presumptuous if I venture to suggest to the other mill operators who have so courteously taken issue with Mr. Handy, that they should devote the same careful scrutiny to their problem, and make the same thorough analysis of it, as Mr. Handy has done for the Bunker Hill & Sullivan problem.

W. H. LINNEY, Spokane, Wash.—The Bunker Hill ore is a very simple one to concentrate, and the figures presented by Mr. Handy are, I believe, entirely correct when applied to this or similar ores. There are, however, some ores to which I believe his conclusions would not apply; for example, an ore containing a mixture of galena, iron pyrites, and some very high-grade antimonial silver minerals, in a hard siliceous gangue. These antimonial silver minerals are great slimers and even a very small loss of slimes would be a serious loss because of its high silver contents. With ores of this character, I believe the removal of the high-grade silver minerals by hand-sorting is desirable and economical in many cases.

R. S. HANDY.—My paper should have been prefaced with the statement that the following conditions are outside the issue:

1. Where the mine product is as clean as that resulting from any treatment of the mine product in the mill.

2. Where the mine product is not amenable to mechanical concentration.

In either of these cases there is no excuse for a milling plant.

It should be remembered that the hand-sorted product is the material claiming our attention and not any material which is allowed to go to the mill. With this in mind, and considering the economic results, I believe I am safe in saying that if any process in the mill produces a higher-grade concentrate than it is practicable to sort out by hand, or if any process in the mill produces a tailing equal in grade to that sorted out by hand, then it is advisable to mill the hand-sorted material.

Slimes seem to be a bugaboo in the consideration of this question, but I think we use this term too loosely. I have heard the term slimes applied to material as coarse as 40-mesh, but I think it should be applied only to colloidal material. At the Bunker Hill, the most satisfactory material for table work, for economical results, is the sand below 200-mesh; it makes no difference how fine the ore is so long as it is crystalline.

Replying especially to Mr. Hill, I wish to thank him for his thorough discussion of my paper. It is fortunate that his discussion comes from the Hecla mill for, as Mr. Hill says, the ores of the Bunker Hill and of Hecla are similar in character and the bases for costs and other economic factors are comparable.

Elaborating Mr. Hill's figures, for comparative purposes, we have:

Total cost of crushing and sorting 146,675 tons @ 14.7 c...... $21,561
Total cost of milling 112,646 tons @ 35.7 c.................. 40,214
Total cost of producing 26,685 tons shipping product........ $61,775
or $2.31 per ton of shipping product.

I do not know whether or not Mr. Hill includes administrative and other overhead expenses in his costs, but I shall use the costs of labor, material, heat, light, power, supervision and depreciation of equipment—the only costs over which a mill superintendent has control. On this basis the costs at the Bunker Hill west mill in 1915 were approximately as follows, per day:

Crushing and screening through 30-mm. 1250 tons @ 3 c..... $ 37.50
Milling 1250 tons @ 32 c................................. 400.00
    Total............................................. $437.50

For a shipping product containing 50 per cent. lead, this is a cost of $2.18 per ton; but on a basis of 65 per cent. lead product it is $2.65 per ton. Therefore, I think the cost per ton of shipping product is a rather slippery basis for comparison.

Allowing for the difference in tonnage and the freight to the mill at the Hecla, all of their mine product could surely be prepared for milling for 7 c. per ton if it were not hand-sorted. I believe the only additional equipment necessary in the mill to handle the 90 tons per day of clean coarse material sorted out by hand would be a jig. There would be no middlings or slime to consider—just clean shipping ore and clean waste. If this is so, the Hecla could produce the same tonnage of shipping product by milling for $2 per ton. I think Mr. Hill has failed to consider that probably 98 per cent. of his milling cost is in the handling of middlings and fine material.

For example, at the Bunker Hill west mill there are 25 jigs, five of which are bull jigs handling 450 tons of feed and producing (when discharging shipping product) 70 tons of 53 per cent. lead concentrate and 250 tons of tailings per day. The cost of operating these 25 jigs is $50 per day or $2 per jig. Applying this to the five bull jigs we have a cost of $10 per day, or $0.143 per ton of shipping product. That is, these jigs are capable of disposing of 25.6 per cent. of the total final products at 2.3 per cent. of the total cost, while hand-sorting at the Hecla produces 23.2 per cent. of the total final products at, I judge, 18.3 per cent. of the total cost.

I cannot help believing that hand-sorting finds favor on account of the tendency to confuse the cost of the very simple mechanical separation of clean, coarse concentrate from clean, coarse waste with the very complicated process of freeing included galena from gangue and collecting it, together with the galena entangled with colloidal matter, into a shipping product. If Mr. Hill's mill feed consisted of nothing but the material which is sorted out in the sorting plant, his lead recovery and costs would put the milling world to shame.

Coming to the question of grading, the conviction seems to be fixed in the minds of most millmen that raising the grade of concentrates causes higher-grade tailings. I maintain that this is not necessarily so, and I believe we have proved it by raising the average grade of the concentrates at the Bunker Hill west mill about 12 per cent. in lead assay and by reducing, at the same time, the average grade of the tailings about 1.2 per cent. in lead assay.   A little over half this reduction in the tailings is due to the introduction of flotation, but the injection of a considerable tonnage of low-grade flotation concentrates necessitated raising the grade of the other shipping products in order to maintain the standard grade. Raising the grade of concentrates did not require the addition of much equipment, and, in fact, 13 tables, two elevators, one trommel and two Huntington mills were eliminated from the original equipment in the readjustment for grading, in 1912.   Grading is simply a process of selecting the products as they are produced of the grade desired.   If galena in a certain material, such as the normal coarse jig concentrate, is included in gangue, then this material must be crushed to free the galena. If the galena is free in a mixture with other materials then the product must be re-treated until the free galena is separated.   The tailing can be selected at the grade desired in the same manner.   These operations require a pretty thorough knowledge of the products from each machine in the mill, both as to their physical nature and mineral contents, and that they are profitable is indicated in the following example:

Taking lead at 5 c. per lb., and smelter, freight and treatment charge of $15 per ton, and applying it to the Hecla shipping product, we have:

| | |
|---|---|
| 53 units lead @ $10.................................... | $53.00 |
| Freight and treatment................................. | 15.00 |
| Net ton value....................................... | $38.00 |

For the same amount of lead in a 65 per cent. lead product we have:

| | |
|---|---|
| 65 units lead @ $10.................................... | $65.00 |
| Freight and treatment................................. | 15.00 |
| Net ton value....................................... | $50.00 |

or, $40.50 for the 0.81 ton corresponding to 1 ton of 53-per cent. lead product.

Against this difference of $2.50 per ton is the cost of grading and whatever losses there may be in the process, which I consider nil.

The grades of the various concentrates from the Bunker Hill west mill in 1915 were approximately as follows: Jigs, 72 per cent. lead; tables, 68 per cent. lead; flotation concentrates, 55 per cent. lead.   The middlings were not a concentrate and are shipped only periodically. The grade of the concentrate is maintained regardless of the production of middlings.   Ideal milling of lead ore would be the separation of the mill feed into perfectly clean galena and perfectly barren waste.   While many economic factors affect this work, to my mind our success is measured by the lengths to which our efforts carry us in the accomplishment of this object.

## Grinding Resistance of Various Ores

BY LUTHER W. LENNOX, E. M., VICTOR, COLO.

(Colorado Meeting, September, 1918)

DURING the last few years, one of the great problems in the milling of all ores has been that of grinding. This subject involves not merely the cost of the operation, but also the selection of the proper degree and character of grinding to yield the best metallurgical results on the given ore. Considering the diversity of machinery, one is led to wonder at the variety of grinding methods employed, even when realizing that metallurgy demands varying products.

In an attempt to investigate the relative grinding resistances of a number of ores now being milled in this country and Alaska, the management of the Portland mills requested samples of average mill-run of ore

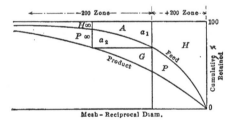

from various companies. The request met with a ready response, indicating the desire of all operators to gain information on the subject.

The difficulties involved in this investigation were numerous, lack of time for outside problems during this strenuous period being a serious one. Also, we were unable, in the time, to obtain a correct 150-mesh screen; in the screen analyses tabulated and plotted herewith, the 150-mesh screen was used merely to protect the 200-mesh screen, not to ascertain points on the curves.

The experimental grinder was a small iron tube-mill, 8 by 12½ in., in which was placed a definite number (74) and weight (3450 gm.) of ¾-in. steel balls, 1 lb. water and 1 lb. ore. To obtain uniform conditions, each ore was first crushed and screened, and from the sized products a "standard feed" was artificially prepared, the same standard being adopted for all tests in that particular series. The prepared feed was placed in the little mill, and this was allowed to revolve at 84 r.p.m. for a definite time; the sample was then dried and screened. The whole operation was then repeated for other periods of time.

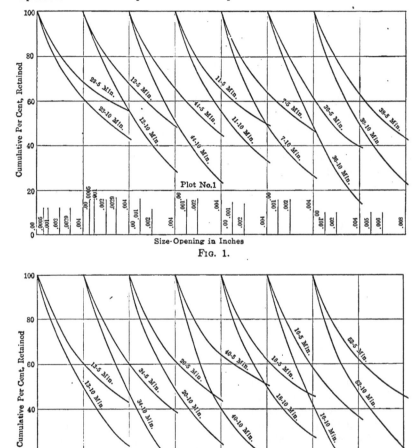

FIG. 1.

FIG. 2.

The mass of data accumulated by these tests is only partially given below, as time at present will not permit a complete analysis. I realize that the method here adopted has its drawbacks, and yet, with all conditions identical, the screen tests of the products should at least give an indication of the relative grinding resistances of the ores. One objectionable feature is the small size of the grinder, although the labor in-

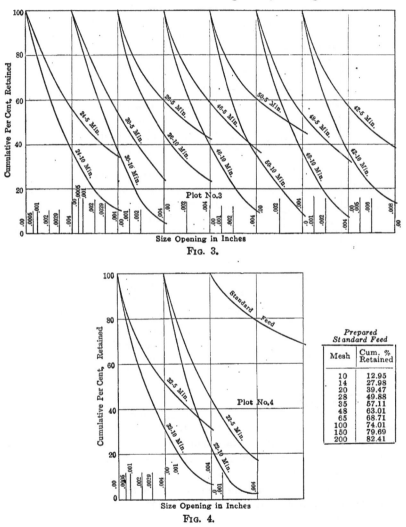

FIG. 3.

FIG. 4.

| Prepared Standard Feed | |
|---|---|
| Mesh | Cum. % Retained |
| 10 | 12.95 |
| 14 | 27.98 |
| 20 | 39.47 |
| 28 | 49.88 |
| 35 | 57.11 |
| 48 | 63.01 |
| 65 | 68.71 |
| 100 | 74.01 |
| 150 | 79.69 |
| 200 | 82.41 |

volved in drying, cutting down, and screening the product of a larger machine would be burdensome. The small size also necessitated a comparatively fine feed.

The method adopted for computing the results is that of Rittinger, based on the law that work done in crushing is proportional to the surface exposed by the operation, or to the reciprocal of the diameter. For plotting the screen tests, the mesh (reciprocal of diameter) and the cumulative per cent. of oversize are used as coördinates. The area

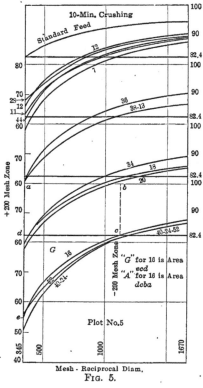

between the feed and product curves will be proportional to the "apparent work" done. The apparent work is in turn inversely proportional to the resistance of the ore to grinding; hence the area between the curves is inversely proportional to grinding resistance.

Unfortunately, we are unable to deal accurately with the material below 200 mesh, which is where most of the work done is represented. Microscopic measurements seem to offer the only accurate method. Curves plotted in the area below 200 mesh seem to follow no definite law, as that of a hyperbola. Many operators may feel but little interest in what happens after a certain fineness (say 48 mesh) is reached, but this does not alter the necessity of measuring the total work done on the whole of the ore.

As a method of measuring the work done in the region below 200 mesh, I offer the following suggestion, not asserting that it is accurate as to *absolute* results, but only that it yields additional evidence as to the *relative* grinding resistances of ores.

Product work (Rittinger's law) $= H + P + G + a_1 + a_2 + H\infty + P\infty - (H + a_1 + H\infty) = P + G + a_2 + P\infty$. (See diagram, p. 237.) For a given ore, assume that the material in the feed (say 20 per cent.) below 200 mesh has the same value in mesh-tons as the finest 20 per cent. in the product, that is,

$$a_1 + H\infty = P\infty + H\infty, \text{ or } a_1 = P\infty$$

Substituting this value for $P\infty$, and calling $(a_1 + a_2) = A$,

$$\text{Product work} = P + G + A$$

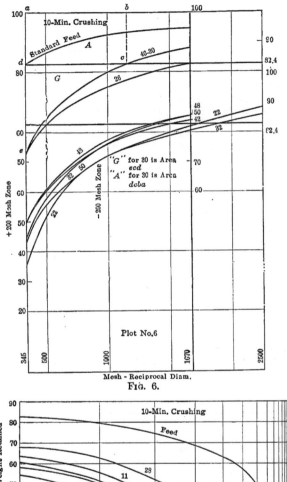

FIG. 6.

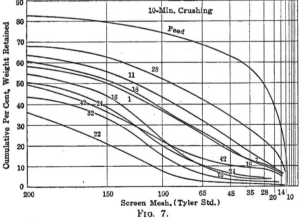

FIG. 7.

TABLE 1.—*Screen Analysis of Product from 5-min. Grinding—Cumulative Per Cent. Retained on Screen*

| Mesh | Opening, Inch | 28 | 12 | 44 | 11 | 7 | 36 | 38 |
|---|---|---|---|---|---|---|---|---|
| 10 | 0.065 | 4.07 | 4.07 | 3.88 | 6.57 | 5.85 | 1.79 | 3.70 |
| 14 | 0.046 | 19.08 | 9.35 | 9.51 | 14.72 | 12.78 | 4.55 | 8.39 |
| 20 | 0.0328 | 26.29 | 13.98 | 13.50 | 20.45 | 18.67 | 6.89 | 11.80 |
| 28 | 0.0232 | 34.06 | 20.81 | 19.15 | 27.10 | 24.12 | 11.39 | 16.91 |
| 35 | 0.0164 | 41.24 | 29.63 | 26.33 | 34.34 | 30.32 | 18.60 | 23.78 |
| 48 | 0.0116 | 48.40 | 38.83 | 35.79 | 41.58 | 36.87 | 29.44 | 33.56 |
| 65 | 0.0082 | 55.27 | 48.41 | 44.66 | 49.33 | 46.06 | 39.67 | 42.83 |
| 100 | 0.0058 | 62.29 | 58.11 | 55.03 | 57.78 | 55.19 | 51.12 | 52.90 |
| 150 | 0.0041 | 69.77 | 68.37 | 66.24 | 66.75 | 64.59 | 63.93 | 59.57 |
| 200 | 0.0029 | 74.32 | 74.14 | 72.95 | 71.77 | 70.09 | 70.93 | 70.35 |

| Mesh | 40 | 34 | 52 | 13 | 16 | 18 | 24 | 20 |
|---|---|---|---|---|---|---|---|---|
| 10 | 3.47 | 2.91 | 3.96 | 3.24 | 2.85 | 5.41 | 3.59 | 4.41 |
| 14 | 9.76 | 6.77 | 9.07 | 7.91 | 6.91 | 12.86 | 8.30 | 11.20 |
| 20 | 13.74 | 9.68 | 13.30 | 11.65 | 9.59 | 18.32 | 11.12 | 15.98 |
| 28 | 23.32 | 14.24 | 19.57 | 17.38 | 13.15 | 24.51 | 14.40 | 21.82 |
| 35 | 28.11 | 20.36 | 27.17 | 25.11 | 17.69 | 30.81 | 18.96 | 27.95 |
| 48 | 34.97 | 29.13 | 36.47 | 33.49 | 25.91 | 38.37 | 26.03 | 35.49 |
| 65 | 43.26 | 38.46 | 44.91 | 42.13 | 35.75 | 45.49 | 34.18 | 43.29 |
| 100 | 52.86 | 48.89 | 53.81 | 51.35 | 47.82 | 53.67 | 44.48 | 52.64 |
| 150 | 59.26 | 62.27 | 63.87 | 61.35 | 60.94 | 63.28 | 56.81 | 62.01 |
| 200 | 69.40 | 68.22 | 68.83 | 67.79 | 67.68 | 68.51 | 65.36 | 68.29 |

| Mesh | 26 | 46 | 42 | 30 | 50 | 48 | 22 | 32 |
|---|---|---|---|---|---|---|---|---|
| 10 | 5.18 | 1.77 | 4.99 | 0.44 | 1.17 | 0.71 | 0.22 | 2.15 |
| 14 | 12.10 | 4.87 | 11.76 | 1.41 | 2.63 | 1.81 | 0.60 | 5.14 |
| 20 | 16.82 | 7.35 | 16.26 | 2.05 | 3.96 | 3.63 | 0.97 | 6.96 |
| 28 | 22.38 | 11.71 | 20.57 | 2.86 | 6.83 | 5.44 | 1.67 | 9.67 |
| 35 | 28.00 | 18.66 | 25.14 | 4.79 | 11.92 | 9.50 | 3.43 | 13.71 |
| 48 | 35.25 | 28.79 | 31.42 | 11.86 | 21.59 | 19.35 | 8.71 | 21.84 |
| 65 | 42.23 | 37.39 | 38.02 | 23.78 | 33.19 | 30.96 | 16.64 | 30.28 |
| 100 | 50.87 | 48.74 | 47.50 | 40.59 | 45.82 | 43.77 | 29.27 | 41.72 |
| 150 | 60.97 | 61.64 | 53.55 | 56.84 | 53.92 | 50.89 | 46.53 | 53.77 |
| 200 | 66.24 | 66.38 | 65.68 | 64.86 | 65.97 | 61.26 | 55.80 | 59.88 |

TABLE 2.—*Screen Analysis of Product from 10-min. Grinding—Cumulative Per Cent. Retained on Screen*

| Mesh | Opening, Inch | 28 | 12 | 44 | 11 | 7 | 36 | 38 |
|---|---|---|---|---|---|---|---|---|
| 10 | 0.065 | 6.11 | 1.98 | 1.94 | 3.68 | 3.67 | 0.48 | 2.02 |
| 14 | 0.046 | 13.77 | 3.91 | 4.19 | 7.65 | 7.17 | 1.19 | 4.31 |
| 20 | 0.0328 | 18.12 | 5.05 | 5.34 | 10.01 | 8.97 | 1.50 | 5.34 |
| 28 | 0.0232 | 22.54 | 6.63 | 6.71 | 12.88 | 10.81 | 1.85 | 6.44 |
| 35 | 0.0164 | 27.53 | 9.86 | 9.35 | 16.90 | 13.27 | 2.53 | 8.22 |
| 48 | 0.0116 | 34.71 | 17.12 | 15.08 | 23.08 | 17.29 | 5.31 | 13.76 |
| 65 | 0.0082 | 42.57 | 28.04 | 23.13 | 32.14 | 25.65 | 14.15 | 22.49 |
| 100 | 0.0058 | 51.92 | 41.78 | 36.20 | 43.27 | 36.76 | 30.08 | 35.31 |
| 150 | 0.0041 | 62.91 | 56.83 | 52.46 | 55.90 | 50.22 | 51.07 | 44.86 |
| 200 | 0.0029 | 68.25 | 65.48 | 61.73 | 63.55 | 58.07 | 60.84 | 60.74 |

| Mesh | 40 | 34 | 52 | 13 | 16 | 18 | 24 | 20 |
|---|---|---|---|---|---|---|---|---|
| 10 | 2.20 | 0.93 | 1.66 | 1.22 | 1.32 | 2.65 | 1.50 | 2.35 |
| 14 | 5.65 | 2.11 | 3.47 | 2.79 | 2.71 | 6.19 | 2.96 | 5.70 |
| 20 | 7.23 | 2.64 | 4.40 | 4.25 | 3.24 | 8.07 | 3.62 | 7.44 |
| 28 | 8.42 | 3.19 | 5.39 | 5.56 | 3.68 | 10.17 | 4.06 | 9.16 |
| 35 | 9.76 | 4.08 | 7.05 | 7.90 | 4.01 | 12.93 | 4.55 | 11.40 |
| 48 | 12.77 | 6.79 | 12.13 | 13.13 | 5.44 | 18.45 | 6.16 | 15.87 |
| 65 | 17.92 | 14.04 | 20.90 | 22.18 | 10.34 | 27.22 | 10.54 | 24.12 |
| 100 | 28.05 | 28.45 | 34.20 | 35.08 | 23.22 | 37.79 | 20.36 | 35.39 |
| 150 | 35.81 | 50.06 | 48.22 | 50.60 | 44.24 | 51.59 | 38.04 | 49.52 |
| 200 | 50.19 | 58.91 | 55.89 | 60.60 | 54.82 | 59.58 | 50.44 | 58.21 |

| Mesh | 26 | 46 | 42 | 30 | 50 | 48 | 22 | 32 |
|---|---|---|---|---|---|---|---|---|
| 10 | 3.47 | 0.70 | 2.76 | 0.33 | 0.33 | 0.24 | 0.22 | 0.77 |
| 14 | 7.43 | 1.58 | 6.29 | 0.66 | 0.66 | 0.55 | 0.51 | 1.65 |
| 20 | 9.51 | 2.02 | 7.64 | 0.79 | 0.79 | 0.70 | 0.75 | 2.00 |
| 28 | 11.33 | 2.46 | 8.48 | 1.19 | 0.92 | 0.88 | 0.99 | 2.29 |
| 35 | 13.35 | 2.99 | 9.19 | 1.39 | 1.12 | 1.12 | 1.23 | 2.55 |
| 48 | 17.63 | 4.80 | 10.73 | 1.83 | 2.20 | 2.24 | 1.78 | 3.48 |
| 65 | 23.01 | 9.87 | 13.84 | 4.81 | 7.03 | 6.39 | 2.88 | 6.29 |
| 100 | 32.79 | 23.02 | 22.94 | 16.90 | 22.13 | 19.89 | 7.06 | 15.39 |
| 150 | 45.09 | 41.82 | 31.06 | 40.85 | 33.33 | 30.91 | 21.38 | 32.74 |
| 200 | 53.51 | 52.03 | 49.33 | 52.27 | 44.99 | 49.85 | 36.36 | 43.48 |

TABLE 3.—General Data

| Ore No. | Mesh-tons Above 200 Mesh P. ×100 | | Mesh-tons Area G. ×100 | | Mesh-tons Area A. ×100 | | Total Mesh-tons P+G+A. ×100 | | | Comparative Grinding Resistance | Source of the Ore |
|---|---|---|---|---|---|---|---|---|---|---|---|
| | 5 Min. | 10 Min. | 5 Min. | 10 Min. | 5 Min. | 10 Min. | 5 Min. | 10 Min. | Average | | |
| 28 | 3,840 | 6,830 | 800 | 2,900 | 4,400 | 8,980 | 9,040 | 18,710 | 13,870 | 1.33 | Calumet & Hecla jig tails. |
| 12 | 5,360 | 10,070 | 1,000 | 2,870 | 4,400 | 8,280 | 10,760 | 21,220 | 15,990 | 1.16 | British Columbia, Tonopah-Belmont Development Co. |
| 11 | 5,180 | 9,590 | 1,530 | 4,300 | 5,800 | 10,550 | 12,510 | 24,440 | 18,470 | 1.00 | Portland mill feed. |
| 44 | 6,200 | 11,420 | 1,200 | 4,350 | 5,000 | 11,000 | 12,400 | 26,770 | 19,580 | 0.94 | Smuggler Union. |
| 36 | 7,620 | 13,160 | 90 | 4,50 | 5,000 | 9,500 | 13,570 | 26,810 | 20,190 | 0.92 | Nr Cornelia Co. |
| 38 | 7,90 | 11,670 | 1,670 | 5,270 | 5,800 | 12,700 | 14,660 | 29,640 | 22,150 | 0.83 | Liberty Bell. |
| 7 | 6,020 | 11,300 | 2,050 | 6,150 | 6,950 | 13,000 | 15,020 | 30,450 | 22,730 | 0.81 | Butte & Superior. |
| 34 | 7,910 | 13,530 | 1,950 | 5,470 | 5,900 | 11,300 | 15,760 | 30,450 | 23,030 | 0.80 | Humboldt mine (Smuggler Union). |
| 13 | 7,300 | 11,850 | 2,280 | 5,760 | 6,300 | 12,700 | 15,880 | 30,310 | 23,090 | 0.80 | Tonopah-Belmont. |
| 18 | 6,380 | 11,050 | 2,050 | 6,400 | 6,200 | 14,400 | 14,630 | 31,850 | 23,240 | 0.79 | Goldfield Con, cyanide feed. |
| 20 | 6,960 | 11,420 | 2,200 | 7,880 | 7,220 | 16,200 | 16,380 | 35,500 | 25,940 | 0.71 | Goldfield Con., flotation feed. |
| 52 | 6,750 | 12,550 | 2,100 | 8,520 | 7,050 | 14,950 | 15,900 | 36,020 | 25,960 | 0.71 | Ia Treadwell, quartz & bit. |
| 16 | 8,450 | 14,360 | 2,100 | 8,050 | 5,990 | 13,550 | 16,540 | 35,960 | 26,250 | 0.70 | Mi mi Co. |
| 40 | 6,900 | 13,700 | 2,070 | 8,850 | 7,050 | 14,950 | 16,020 | 37,500 | 26,760 | 0.69 | Ala Gold Mines Co. |
| 24 | 9,070 | 15,000 | 2,890 | 8,850 | 7,560 | 14,950 | 19,520 | 38,800 | 29,160 | 0.63 | Homestake, unaltered ore. |
| 46 | 8,350 | 14,800 | 3,200 | 8,960 | 8,600 | 14,450 | 20,150 | 38,210 | 29,180 | 0.63 | Co. |
| 30 | 10,950 | 15,530 | 3,200 | 8,960 | 7,560 | 14,450 | 21,710 | 38,940 | 30,320 | 0.61 | Nevada Con., pit ore. |
| 48 | 10,240 | 15,600 | 4,380 | 10,800 | 10,550 | 17,800 | 25,170 | 44,200 | 34,680 | 0.53 | i, Phelps-Dodge. |
| 42 | 8,050 | 14,720 | 3,900 | 12,000 | 11,100 | 20,000 | 23,050 | 46,720 | 34,880 | 0.53 | Calumet & Arizona, roaster ore. |
| 26 | 7,470 | 12,900 | 3,520 | 12,190 | 10,750 | 23,500 | 21,740 | 48,590 | 35,160 | 0.52 | Alaska Treadwell, schist. |
| 50 | 9,350 | 15,830 | 3,300 | 12,400 | 9,330 | 29,400 | 21,980 | 57,630 | 39,800 | 0.46 | Copper Queen, mill feed. |
| 22 | 13,290 | 18,900 | 7,680 | 18,100 | 14,060 | 24,400 | 35,030 | 61,400 | 48,210 | 0.38 | r. |
| 32 | 10,390 | 16,650 | 7,550 | 17,800 | 18,160 | 28,000 | 36,100 | 62,450 | 49,270 | 0.37 | Ray d. |

I have plotted the diameters and cumulative per cents as coördi- nates. The last definite point in diameter is 0.0029 in. (200 mesh); the curves must also pass through the point corresponding to 100 per cent. retained and zero diameter. The last three known points—0.0082 or 65 mesh; 0.0058 or 100 mesh; and 0.0029 or 200 mesh—are shown

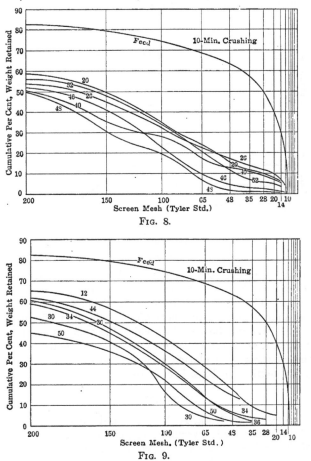

FIG. 8.

FIG. 9.

in plots No. 1, 2, 3, 4. The extension of the curves through the unknown area between 0.0029 in. and zero is made with reference to the known curve section. Subsequent investigation may require some of these curves to be slightly altered. The ratio between mesh-tons of the 5- min. and the 10-min. products does show a proportion fairly close to

**1:2.** From the curves thus drawn, I have taken points representing 0.0029 in. (345 reciprocal), 0.002 in. (500 reciprocal), 0.001 in. (1000 reciprocal), 0.0006 in. (1667 reciprocal), with their corresponding cumulative per cents, and have replotted these as No. 5 and 6' (only the curves for the 10-min. grinding being shown). From these curves, the so-called "product area" $G+A$, of material below 200 mesh, is

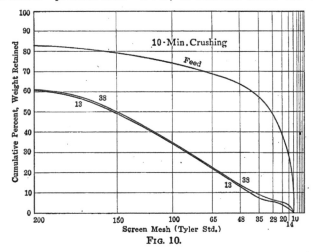

FIG. 10.

obtained; this, added to the product area $P$ of material above 200 mesh, gives the total mesh-tons to be used for comparison with the mesh-tons of other ores. These totals are recorded in Table 3.

Screen analyses of the products after 5-min. and 10-min. grinding, starting with the standard feed of which the analysis can be observed in the diagrams, are given in Tables 1 and 2. As a matter of further interest, these analyses are plotted according to the Tyler direct cumulative diagram method in Figs. 7, 8, 9, 10.

## DISCUSSION

C. Q. PAYNE, New York, N. Y. (written discussion\*).—The method adopted by Mr. Lennox is a very interesting test of the practical application of Mr. Gates' crushing-surface diagram to a great variety of gold and copper ores. His method of dealing with the material below 200-mesh is also a distinct contribution to the subject, although, as he suggests, it seems likely that some microscopic measurement of this material may be advisable before the mesh reciprocals of its group particles can be dealt with as accurately as can those coarser than 200-mesh.

\* Received Sept. 3, 1918.

In examining the general results shown in Table 3, it is somewhat puzzling to perceive why Calumet & Hecla jig tails should show a crushing resistance 3½ times greater than that of the ore from the Ray Consolidated copper mine. The latter is, I believe, an altered Pinal schist containing over 70 per cent. silica, and the composition of the former probably does not differ greatly in this respect. I also note that the crushing resistance of Calumet & Hecla jig tails compares with that of Miami ore in the ratio of about 2 to 1. Hardinge mills are now being employed to crush Calumet & Hecla conglomerate, and also Miami ores on a very large commercial scale, and it is interesting in this connection to note that M. K. Rodgers[1] gives the relative crushing duty of 1 hp. on C. & H. conglomerate and Miami ore as 1 : 3.75, the feed in both cases being ¼-in. size. The inverse of the above figures would represent the relative crushing resistances. At a later date Mr. Gates[2] pointed out that the crushing-surface diagrams of the two ores, on material coarser than 200 mesh, give for C. & H. conglomerate 12.5, and for Miami 34.0 mesh-tons per hp.-hr. This would indicate the relative crushing resistance of these ores to be 2.72 : 1. However, this difference in the crushing resistance of these two ores, as measured by Mr. Lennox's small tube-mill and by the Hardinge mill may be due to the fact that the crushing-surface diagrams in the two cases are not comparable.

A possible question in connection with these tests is whether the determination of the crushing resistance of an ore may not be difficult to measure owing to what may be called "screen resistance." Of two ores which may have the same crushing resistance when measured by a 4- or 8-mesh screen, for example, one may show a much greater crushing resistance than the other when measured by a 200-mesh screen, owing to the presence of a certain amount of mica or other flaky mineral unlocked at, say, 50 mesh, but only reduced to pass a 200-mesh screen by very prolonged grinding. The micaceous ore would thus show a different crushing-surface diagram from the granular ore, although the same amount of energy might have been expended in crushing it. The proximity on the scale of crushing resistance (Table 3) of such a hard and tough ore as the Homestake to a comparatively soft ore like that of the Nevada Consolidated for example, has suggested the idea that a difference in the "screen resistance" might perhaps here have counteracted a greater difference in the crushing resistance than the results actually show.

The interesting and valuable work done at McGill University by Professor Bell seems clearly to establish Rittinger's law as a better measure of the energy absorbed in crushing than Kick's law. The method of recording the work of crushing by the crushing-surface diagram, which has been developed and illustrated by Mr. Gates, also marks a very not-

---

[1] *Trans.* (1915) **52**, 944.    [2] *Mining and Scientific Press* (Mar. 11, 1916) **112**, 366.

able contribution to the subject. The crushing-surface diagram is fascinating from its very simplicity. But are we not expecting too much from it until we have corroborated it, and perhaps corrected it, by means of certain more exact physical measurements on a wide range of minerals and ores? Many ores consist of a more or less loose aggregate of different minerals, and it may help toward clear thinking if we subdivide the subject, and apply the laws of crushing only to the breaking up of aggregates, as performed by coarse crushing, in which surface areas can be accurately measured by screen analysis; and then apply the laws of grinding to the reduction of small-sized particles of homogeneous composition, the surface areas of which are difficult to measure by screen analysis alone.

For illustration, if we knew the number of heat units developed, and therefore the energy absorbed, in grinding 100 gm. of a given ore or mineral, so that it would all pass a 200-mesh screen, could we not then develop an energy unit which would be a physical constant for that particular ore or mineral? With a number of such physical constants determined for various ores, we would then have more accurate means of measuring not only the grinding resistance of the ores, but also the mechanical efficiencies of different machines employed in grinding them. Such an energy unit as I have suggested might require some other method than screen analysis to estimate the surface exposed, since the unit should be independent of the habit of crystallization of the component ore minerals, and should be directly related to their molecular structure, on which their coherence and mechanical resistance to grinding must ultimately depend.

The establishment of accurate methods for measuring the energy absorbed in crushing and grinding ores is a matter of great importance. Most mining engineers realize the backward state in which the art of crushing now lies, largely for lack of accurate units of measurement. Considering the subject broadly, and including the crushing of cement rock and clinker and pottery materials, it is probable that the average efficiency of crushing as a whole does not exceed 20 to 25 per cent. When we recall that the efficiencies reached in the concentrating, cyaniding, and flotation of ores frequently exceed 90 per cent., it is evident that the first step in the ore-dressing and allied industries is worthy of more serious attention than it has yet received.

LUTHER W. LENNOX (author's reply to discussion).—Mr. Payne calls attention to the relative grinding resistance of Calumet & Hecla and Miami ores, as pointed out by Mr. Rodgers and by Mr. Gates in comparison with the figures obtained in my Table 3. As pointed out by Mr. Gates, the relative grinding duty of one horsepower on Calumet & Hecla and Miami ores as 1 to 3.75 obtained by Mr. Rodgers takes into consideration only the tonnage and power, and ignores the mesh of the product and also the feed, except that it passes $\frac{1}{4}$-in. mesh.   After plotting, Mr. Gates arrives at figures that indicate a relative grinding resistance of 2.72 to 1 instead of 3.75 to 1.

In this connection I wish to call attention to work on "Hardinge Mill Data" by A. F. Taggart[3] in which are published tests on various ores, among them being Calumet & Hecla, Miami, and Arizona copper. Plotting feeds and product and taking into consideration only the plus 200-mesh zone, combined with the data on tonnage and power in these tests, we find the following relation: Card No. 34, Calumet & Hecla relative grinding resistance being assumed as 1.33.   From the three tests Nos. 107, 108, and 80 Miami become 0.41, 0.81, and 0.45, respectively, as the relative grinding resistance, while Arizona copper, card No. 142, gives 0.64.   The standing of these three ores in my Table 3 will be found to be 1.33 for Calumet & Hecla, 0.62 for Miami and 0.61 for Arizona copper, where only the plus·200-zone is. considered.

Two factors might easily enter into the tests as printed by Mr. Taggart and those referred to by Mr. Payne, either one of which could alter the apparent grinding resistance of these ores.   The ores might or might not be typical of the present ores.   Then again the tests referred to were not necessarily run under such conditions that it would be fair to compare the ores with reference to grinding qualities.   The results on Miami would indicate this, as they do not give consistent data.

---

[3] *Trans.* (1915) **52**, 932.

## Fine Crushing in Ball-mills

BY E. W. DAVIS,* MINNEAPOLIS, MINN.

(New York Meeting, February, 1919)

ON the eastern end of the Mesabi Range, in Northern Minnesota, is a large formation of siliceous rock which contains bands and fine grains of magnetite. The magnetite comprises about 35 per cent. of the rock, the remainder being chiefly quartzite and iron silicates. The rock has a specific gravity of 3.4, a hardness of 7, and is extremely tough.

This large deposit was located early in the history of the Minnesota iron-ore industry but has not been utilized because of its low percentage of iron as compared with the other Mesabi ores, and because of the difficulty and expense of any milling treatment that would concentrate the iron. An investigation, begun about 3 years ago, shows that the magnetite is finely disseminated throughout the entire formation and that there are bands or lenses of higher- and of lower-grade material, in which the magnetite and silicate are intimately mixed. As a result, the scheme of milling adopted must include a fine-crushing plant. As it is necessary to crush to 200 mesh in order to produce the desired grade of product, the fine crushing is one of the largest items of expense and for this reason has been given detailed study. It is the purpose of this paper to present some of the results of the work on fine crushing, as to both theory and practice.

Previous to fine crushing the part of the rock that contains little or no magnetite can be discarded by magnetic concentrators after each reduction in size in the dry-crushing plant. This makes it possible to fine-crush a minimum of rock and also establishes the feed to the ball-mill as below $\frac{1}{4}$ in. The fine-crushing problem, then, consists simply in crushing the rock from $\frac{1}{4}$ in. to 200 mesh at the minimum expense.

### EQUIPMENT OF THE TESTING PLANT

In order to determine, among other things, the operating conditions of the ball-mill when working on this rock, a test mill of about 300 T. daily capacity was erected at Duluth, Minn. The fine-crushing plant contains a Hardinge 8-ft. by 22-in. (2.4-m. by 55.8-cm.) conical mill, a 6- by 27-ft. (1.8 by 8.2-m.) Dorr duplex bowl-type classifier, a $4\frac{1}{2}$-ft. (1.37-m.) standard Akins classifier, and the auxiliary machinery necessary to handle the products. Each machine is driven by an individual motor, each of which is provided with meters for measuring the power required. Over 150 tests have been made in the ball-mill, varying in

* Superintendent, School of Mines Experiment Station, University of Minnesota.

duration from a few hours to several weeks, in every case being continued until operating conditions became steady. The plant is so constructed that good samples of all products can be secured, both by automatic samplers and by hand. An apron feeder governs the feed rate and the tonnage is checked in every test. Water is metered into the circuit and every precaution is taken to make the data accurate and reliable.

It is a little hard to secure a basis upon which to compare crushing results. Neither Kick's nor Rittinger's law of crushing is of much use in this case. This is evident when it is considered that the average size of a particle finer than 200 mesh is a matter of opinion, and that in this crushing problem practically all the ore must be crushed to pass a 200-mesh screen. The comparisons have therefore been made on the basis of kilowatt-hours per ton of material finer than 200 mesh actually produced. This, of course, does not give a scientifically exact basis for comparison, but since only the material below 200 mesh is considered finished product, in this case this is a suitable method for comparison.

## OPEN-CIRCUIT CRUSHING

The object of this test was to determine the crushing efficiency of the ball-mill when operating in open circuit. The conditions were as follows:

'Feed rate, variable from 3 to 18 T. per hr.
Ball load, 28,000 lb. of 5-, 4-, 3-, and 2½-in. balls.
Speed, 19.7 r.p.m.
Ball-mill power, 88 kw.
Feed, minus ¼ in., containing 6.52 per cent. minus 200-mesh material.
Amount of solids, about 50 per cent.

TABLE 1.—*Data Obtained in Open-circuit Crushing*

|  | Test No. | | | | |
|---|---|---|---|---|---|
|  | 1 | 2 | 3 | 4 | 5 |
| Feed, tons per hour................... | 3.66 | 7.40 | 11.00 | 15.00 | 18.00 |
| Tons of −200 mesh actually produced per hour................... | 2.24 | 3.77 | 4.85 | 6.03 | 7.67 |
| Kilowatt-hours per ton of −200 mesh produced........................... | 38.55 | 23.53 | 18.04 | 14.62 | 11.55 |

Fig. 1 shows graphically that the tonnage of minus 200-mesh material produced varies directly with the tonnage fed to the mill. There is undoubtedly some limit to this relation, but there seems to be no indication of it at 18 T. per hr. Some of the conclusions drawn from this test are that: (a) The ball-mill is naturally a machine of very large capacity; (*b*) if it is not possible to deliver a large tonnage of original feed to the mill, a closed circuit should be provided so that the mill may crush its own oversize.

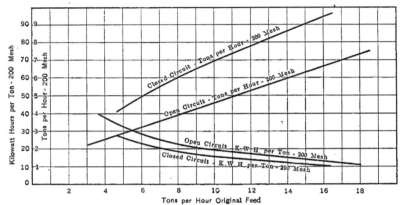

FIG. 1.—COMPARISON OF OPEN AND CLOSED CIRCUIT CRUSHING.

### CLOSED-CIRCUIT CRUSHING

The object of these tests was to determine the crushing efficiency of the ball-mill when crushing in closed circuit with a classifier. The conditions were as follows:

Feed rate, variable from 4 to 15 T. per hr.    Ball-mill power, 108 kw.
Ball load, 28,000 lb. of 3- and 2-in. balls.    Feed, minus ¼-in. material.
Speed, 23.8 r.p.m.    Amount of solids, about 60 per cent.

TABLE 2.—*Data Obtained in Closed-circuit Crushing*

|  | Test No. | | | | | | | |
|---|---|---|---|---|---|---|---|---|
|  | 108 | 110 | 111 | 112 | 113 | 149 | 150 | 154 |
| Feed, tons per hour.......... | 4.63 | 5.00 | 5.50 | 6.50 | 7.37 | 11.00 | 12.57 | 15.31 |
| Tons per hour of −200 mesh actually produced.......... | 4.01 | 4.34 | 4.62 | 5.34 | 5.63 | 6.82 | 8.07 | 9.11 |
| Kilowatt-hours per ton of −200 mesh produced........ | 26.95 | 24.90 | 23.40 | 20.20 | 19.20 | 16.00 | 13.10 | 11.92 |

The results of these tests also are shown in Fig. 1. It is interesting to note that the curve showing tons per hour of minus 200-mesh material does not tend to flatten out as the tonnage to the mill is increased. The power per ton is also continually decreasing. It is, of course, impossible to state how much further this condition will continue, but it seems evident that it will continue for tonnages considerably beyond 15 T. per hr. As the two curves are slowly converging, at some large tonnage the amount of minus 200-mesh material produced per kilowatt-hour will be the same for either open- or closed-circuit crushing.  The real advantage then gained by the closed-circuit system lies in the fact that the product consists of particles much more uniform in size.  Although the average reduction in both systems may be the same, the closed-circuit will deliver a product in which the maximum-size particle will be much nearer the average size than will the open-circuit system.

The following conclusions may be stated from these two series of tests: (a) For equal tonnages of original feed, the closed-circuit crushing system produces the greater tonnage of minus 200-mesh material per kilowatt-hour.  (b) For equal tonnages of original feed, the closed-circuit system of crushing shows the greater average reduction.  (c) There is no indication that the mill was operated at, or even near, a tonnage that would give the greatest number of tons of minus 200-mesh material per kilowatt-hour.  (d) Closed-circuit crushing will always have the advantage over open-circuit crushing, in that the maximum-size particle produced will be much nearer the average size.  This is a desirable condition since the size of the balls making up the charge must be computed on the maximum-size particles in the feed rather than the average size.

## 5. SINGLE-STAGE CRUSHING No. 113

The object of this test was to determine the capacity of the ball-mill crushing in closed circuit and producing 200-mesh material.  The conditions were as follows:

Feed rate, 7.37 T. per hr.                      Ball-mill power, 108 kw.
Classifier, Dorr duplex bowl-type.              Feed, minus ¼-in. ore.
Ball load, 28,000 lb. of 2- and 2⅜-in. balls.   Amount of solids, about 70 per cent.
Speed, 23.8 r.p.m.

TABLE 3.—*Data Obtained in Single-stage Crushing*

| Mesh* | Opening, Mm. | Ball-mill Feed | | Classifier Overflow | |
|---|---|---|---|---|---|
| | | Per Cent. | Cum. Per Cent. | Per Cent. | Cum. Per Cent. |
| On    4 | 4.70 | 29.18 | 29.18 | | |
| On    8 | 2.36 | 27.07 | 56.25 | | |
| On   14 | 1.17 | 15.41 | 71.66 | | |
| On   28 | 0.59 | 7.93 | 79.59 | | |
| On   48 | 0.295 | 4.51 | 84.10 | | |
| On  100 | 0.147 | 3.66 | 87.76 | 1.45 | 1.45 |
| On  200 | 0.074 | 2.85 | 90.61 | 11.75 | 13.20 |
| On  300 | | 1.61 | 92.22 | 14.95 | 28.15 |
| Through 300 | | 7.78 | 100.00 | 71.85 | 100.00 |
| | | 100.00 | | 100.00 | |

*Tyler standard testing screens were used throughout this investigation.

There were 5.63 T. of minus 200-mesh material actually produced per hour and 19.2 kw.-hr. per ton of minus 200-mesh material produced were required. The classifier delivered 33 T. of sand per hour. The total ball-mill feed was therefore 40.37 T. per hr. or 550 per cent. of the original feed.

## TWO-STAGE CRUSHING

The object of these tests was to determine the capacity of ball-mills when crushing in two stages. The conditions of the test were as follows:

FIRST STAGE

Feed rate, 15.31 T. per hr.
Classifier, Dorr duplex with baffled overflow.
Ball load, 28,000 lb. of 3- and 2-in. balls.
Speed, 23.8 r.p.m.
Ball-mill power, 108 kw.

SECOND STAGE

Feed rate, 6.54 T. per hr.
Classifier, Dorr duplex.
Ball load, 28,000 lb. of 2- and 1-in. balls.
Speed, 23.8 r.p.m.
Ball-mill power, 108 kw.

TABLE 4.—Data Obtained in Two-stage Crushing

| Mesh | Opening, Mm. | First Stage | | | | Second Stage | | | |
|---|---|---|---|---|---|---|---|---|---|
| | | Ball-mill Feed | | Classifier Overflow | | Ball-mill Feed | | Classifier Overflow | |
| | | Per Cent. | Cum. Per Cent. | Per Cent. | Cum. Per Cent. | Per Cent. | Cum. Per Cent. | Per Cent. | Cum. Per Cent. |
| On 4 | 4.70 | 29.17 | 29.17 | | | | | | |
| On 8 | 2.36 | 26.22 | 55.39 | | | | | | |
| On 14 | 1.17 | 14.13 | 69.52 | | | 0.30 | 0.30 | | |
| On 28 | 0.59 | 8.62 | 78.14 | | | 0.32 | 0.62 | | |
| On 48 | 0.295 | 4.67 | 82.81 | 4.30 | 4.30 | 5.40 | 6.02 | | |
| On 100 | 0.147 | 3.32 | 86.13 | 15.80 | 20.10 | 39.00 | 45.02 | 2.45 | 2.45 |
| On 200 | 0.074 | 3.28 | 89.41 | 18.80 | 38.90 | 37.98 | 83.00 | 13.25 | 15.70 |
| On 300 | | 1.68 | 91.09 | 11.35 | 50.25 | 8.24 | 91.24 | 17.70 | 33.40 |
| Through 300 | | 8.91 | 100.00 | 49.75 | 100.00 | 8.76 | 100.00 | 66.60 | 100.00 |
| | | 100.00 | | 100.00 | | 100.00 | | 100.00 | |

In the first stage, 9.11 T. of minus 200-mesh material were produced per hour and 11.98 kw.-hr. were required for each ton of minus 200-mesh material produced. The classifier delivered 32 T. of sand per hour. The total feed to the mill was therefore 47.31 T. per hr. or 308 per cent. of the original feed. The classifier overflow was reclassified, most of the material below 200 mesh being discharged from the crushing circuit while the sands were fed to the second stage. In this stage 3.73 T. of minus 200-mesh material were produced per hour and 28.9 kw.-hr. were required for each ton of minus 200-mesh material produced.

The classifier delivered 5 T. of sand per hour. The total feed to the ball-mill was therefore 11.54 T. per hr. or 179 per cent. of the original feed. It was evident that the ball-mill was greatly underloaded in this test, but so much trouble developed in the classifier, due to the tendency of the sands to slip down the slopes, that a more rapid feed was not attempted at this time. The classifier had been set at a slope of 1½ in. per ft. (125 mm. per m.) and conditions were such that the slope could not be decreased. It was impossible also to use the bowl overflow at this flat slope without rebuilding the classifier. At the present writing this work has not been completed. It seems certain, however, that the ball-mill will crush to 200 mesh a considerably greater tonnage when the proper classification is provided. Since in previous tests the mill has crushed 7½ T. per hr. from ¼ in. to 200 mesh, it seems possible that it will crush at least 8 T. per hr. from 48 to 200 mesh.

Comparing single- and double-stage crushing on the basis of these two tests, it appears that the single-stage crushing produces a ton of minus 200-mesh material for 19.2 kw.-hr. while double-stage crushing produces a ton of minus 200-mesh material for 16.8 kw.-hr. These figures, though, do not show the real relative efficiencies of the two systems, for the second stage of the two-stage system was so obviously underloaded. The conclusions drawn from these tests are that (a) two-stage crushing shows a greater efficiency than single-stage, (b) two-stage crushing is much more flexible and offers greater possibilities for improvement than does single stage. In addition, a considerable amount of tailing can be discarded between the stages.

## LARGE VS. SMALL BALLS

Two tests with closed-circuit crushing are reported to show the efficiency of the ball-mill when charged with large and with small balls.

Test No. 12 shows a production of 6.3 T. of minus 200-mesh material per hour, which is 17.15 kw.-hr. per ton of this material actually produced. Test No. 150 shows a production of 8.07 T. of minus 200-mesh material per hour, which is 13.10 kw.-hr. per ton of this material actually produced. These two tests clearly indicate the superiority of small balls. It is instructive to compare the classifier sands in these two tests (see Table 0).

It appears that the small balls produced a much more uniform sand than did the large balls. The evident crowding of material at certain sizes is almost entirely absent in the small-ball test. Since this classifier sand is composed of the particles of ore that have passed through the mill at least once without being crushed, it appears that the large balls

TABLE 5.—*Data Obtained in Closed-circuit Crushing with Large and Small Balls*

| Operating Factors | Test No. 12, Large Balls | Test No. 150, Small Balls |
|---|---|---|
| Feed rate, tons per hr. | 10.8 | 12.57 |
| Classifier. | Drag type | Dorr duplex |
| Ball load, lb. | 28,000 | 28,000 |
| Size of balls, in. | 5, 4, 3, 2½ | 2⅜, 2 |
| Max. size of feed, in. | ¼ | ¼ |
| Speed, r.p.m. | 23.8 | 23.8 |
| Ball-mill power, kw. | 109 | 108 |
| Per cent. solids in feed. | 60 | 60 |
| Circulating load, tons per hr. | 40 | 27 |

| Mesh | Opening, Mm. | Test 12, Large Balls | | | | Test 150, Small Balls | | | |
|---|---|---|---|---|---|---|---|---|---|
| | | Ball-mill Feed | | Classifier Overflow | | Ball-mill Feed | | Classifier Overflow | |
| | | Per Cent. | Cum. Per Cent. | Per Cent. | Cum. Per Cent. | Per Cent. | Cum. Per Cent. | Per Cent. | Cum. Per Cent. |
| On 4 | 4.70 | 22.70 | 22.70 | | | 47.80 | 47.80 | | |
| On 8 | 2.36 | 31.98 | 54.68 | | | 19.08 | 66.88 | | |
| On 14 | 1.17 | 14.64 | 69.32 | | | 10.19 | 77.07 | | |
| On 28 | 0.59 | 6.60 | 75.92 | | | 6.03 | 83.10 | | |
| On 48 | 0.295 | 6.55 | 82.47 | | | 3.32 | 86.42 | | |
| On 100 | 0.147 | 4.13 | 86.60 | 11.50 | 11.50 | 2.64 | 89.06 | 8.65 | 8.65 |
| On 200 | 0.074 | 3.90 | 90.50 | 20.80 | 32.30 | 2.41 | 91.47 | 18.70 | 27.35 |
| On 300 | | 1.07 | 91.57 | 10.85 | 43.15 | 1.50 | 92.97 | 12.80 | 40.15 |
| Through 300 | | 8.43 | 100.00 | 56.85 | 100.00 | 7.03 | 100.00 | 59.85 | 100.00 |
| | | 100.00 | | 100.00 | | 100.00 | | 100.00 | |

TABLE 6.—*Classifier Sands in Tests of Table 5*

| Mesh | Opening, Mm. | From Large Balls | | From Small Balls | |
|---|---|---|---|---|---|
| | | Per Cent. | Cum. Per Cent. | Per Cent. | Cum. Per Cent. |
| On 4 | 4.70 | | | 10.26 | 10.26 |
| On 8 | 2.36 | 0.74 | 0.74 | 9.64 | 19.90 |
| On 14 | 1.17 | 2.80 | 3.54 | 9.86 | 29.76 |
| On 28 | 0.59 | 12.50 | 16.04 | 14.38 | 44.14 |
| On 48 | 0.295 | 31.60 | 47.64 | 18.10 | 62.24 |
| On 100 | 0.147 | 36.60 | 84.24 | 19.78 | 82.02 |
| On 200 | 0.074 | 7.48 | 91.72 | 10.08 | 92.10 |
| On 300 | | 4.20 | 95.92 | 2.36 | 94.46 |
| Through 300 | | 4.08 | 100.00 | 5.54 | 100.00 |
| | | 100.00 | | 100.00 | |

reduce the coarse particles very readily but have trouble in crushing the finer particles. On the other hand, the small balls appear to crush all particles equally well. From this it would seem to be possible, by an analysis of the classifier sands, to determine whether or not the balls are too large or too small for the work they are doing. If the screen analysis of the sands is crowded on the upper end, the balls are too small; if it is crowded at the approximate size of the overflow, the balls are too large. The best results have been obtained when the screen analysis of the sands is about uniform between the size of the original feed and the size of the overflow.

## RELATION BETWEEN SPEED OF MILL AND SIZE OF BALLS

Improper mill speed seems to be indicated in the same way. Table 7 shows screen analyses of classifier sands from tests in which only the speed of the mill was changed.

TABLE 7.—*Classifier Sands from Balls of 5-, 4-, 3-, and 2½- in. Diameter*

| Mesh | Opening, Mm. | 16.6 r.p.m. | | 19.7 r.p.m. | | 23.8 r.p.m. | |
|---|---|---|---|---|---|---|---|
| | | Per Cent. | Cum. Per Cent. | Per Cent. | Cum. Per Cent. | Per Cent. | Cum. Per Cent. |
| On 4 | 4.70 | 3.78 | 3.78 | | | | |
| On 8 | 2.36 | 6.32 | 10.10 | 1.02 | 1.02 | | |
| On 14 | 1.17 | 5.56 | 15.66 | 4.30 | 5.32 | 3.54 | 3.54 |
| On 28 | 0.59 | 7.35 | 23.01 | 15.34 | 20.66 | 12.50 | 16.04 |
| On 48 | 0.295 | 14.71 | 37.72 | 30.43 | 51.09 | 31.60 | 47.64 |
| On 100 | 0.147 | 35.41 | 73.13 | 33.24 | 84.33 | 36.60 | 84.24 |
| On 200 | 0.074 | 20.16 | 93.29 | 10.08 | 94.41 | 7.48 | 91.72 |
| On 300 | | 2.51 | 95.80 | 1.38 | 95.79 | 4.20 | 95.92 |
| Through 300 | | 4.20 | 100.00 | 4.21 | 100.00 | 4.08 | 100.00 |
| | | 100.00 | | 100.00 | | 100.00 | |

The balls were so much too large that a reduction in speed to 16.6 r.p.m. could not compensate for them. In the next tests the balls were more nearly of the proper diameter for the work to be done.

In the tests shown in Table 9, in which balls of 2½-, 2-, and 1½-in. diameter (63.5, 50.8, and 38.1 mm.) were used, the tonnage also being increased, the effect of a change in speed is much more marked.

In the preceding test, at 19.8 r.p.m. the circulating load became so large, over 60 T. per hr., that operation had to be discontinued, for the

TABLE 8.—*Classifier Sands from Balls of 4-, 3-, 2½, and 2-in. Diameter*

| Mesh | Opening, Mm. | 19.8 r.p.m. | | 21.1 r.p.m. | | 23.8 r.p.m. | |
|---|---|---|---|---|---|---|---|
| | | Per Cent. | Cum. Per Cent. | Per Cent. | Cum. Per Cent. | Per Cent. | Cum. Per Cent. |
| On 4 | 4.70 | 1.84 | 1.84 | 1.46 | 1.46 | | |
| On 8 | 2.36 | 1.32 | 3.16 | 0.82 | 2.28 | 0.66 | 0.66 |
| On 14 | 1.17 | 1.86 | 5.02 | 1.14 | 3.42 | 1.34 | 2.00 |
| On 28 | 0.59 | 4.26 | 9.28 | 3.28 | 6.70 | 4.19 | 6.19 |
| On 48 | 0.295 | 9.20 | 18.48 | 8.82 | 15.52 | 9.65 | 15.84 |
| On 100 | 0.147 | 25.84 | 44.32 | 26.06 | 41.58 | 29.44 | 45.28 |
| On 200 | 0.074 | 34.86 | 79.18 | 35.64 | 77.22 | 36.22 | 81.50 |
| On 300 | | 11.88 | 91.06 | 11.46 | 88.68 | 10.98 | 92.48 |
| Through 300 | | 8.94 | 100.00 | 11.32 | 100.00 | 7.52 | 100.00 |
| | | 100.00 | | 100.00 | | 100.00 | |

TABLE 9.—*Classifier Sands from Balls of 2½-, 2-, and 1½-in. Diameter*

| Mesh | Opening, Mm. | 19.8 r.p.m. | | 23.8 r.p.m. | |
|---|---|---|---|---|---|
| | | Per Cent. | Cum. Per Cent. | Per Cent. | Cum. Per Cent. |
| On 4 | 4.70 | 34.06 | 34.06 | 10.26 | 10.26 |
| On 8 | 2.36 | 14.06 | 48.12 | 9.64 | 19.90 |
| On 14 | 1.17 | 9.54 | 57.66 | 9.86 | 29.76 |
| On 28 | 0.59 | 10.24 | 67.90 | 14.38 | 44.14 |
| On 44 | 0.295 | 10.94 | 78.84 | 18.10 | 62.24 |
| On 100 | 0.147 | 11.56 | 90.40 | 19.78 | 82.02 |
| On 200 | 0.074 | 6.28 | 96.68 | 10.08 | 92.10 |
| On 300 | | 1.52 | 98.20 | 2.36 | 94.46 |
| Through 300 | | 1.80 | 100.00 | 5.54 | 100.00 |
| | | 100.00 | | 100.00 | |

mill was unable to handle the coarse ore. The obvious conclusion is that either the speed of the mill should be increased slightly or balls of a little larger diameter should be used. At 23.8 r.p.m. the sands were nearly uniform and at this speed the mill showed the greatest efficiency.

In order to show the marked effect of a slight change in the average size of balls, the two tests shown in Table 10 are reported, in which all conditions were the same except the size of balls.

TABLE 10.—*Effect of Slight Change in Size of Balls*

| Mesh | Opening, Mm. | Avg. Size, 2¼ in. | | Avg. Size, 2⅜ in. | |
|---|---|---|---|---|---|
| | | Per Cent. | Cum. Per Cent. | Per Cent. | Cum. Per Cent. |
| On 4 | 4.70 | 10.26 | 10.26 | 0.80 | 0.80 |
| On 8 | 2.36 | 9.64 | 19.90 | 3.22 | 4.02 |
| On 14 | 1.17 | 9.86 | 29.76 | 7.44 | 11.46 |
| On 28 | 0.59 | 14.38 | 44.14 | 18.84 | 30.30 |
| On 48 | 0.295 | 18.10 | 62.24 | 31.04 | 61.34 |
| On 100 | 0.147 | 19.78 | 82.02 | 25.36 | 86.70 |
| On 200 | 0.074 | 10.08 | 92.10 | 7.92 | 94.62 |
| On 300 | | 2.36 | 94.46 | 1.94 | 96.56 |
| Through 300 | | 5.54 | 100.00 | 3.44 | 100.00 |
| | | 100.00 | | 100.00 | |

From this series of tests the following conclusions may be drawn:

(a) If the balls are large or the speed of the mill is high, crowding will appear at the finer sizes in the classifier sands.

(b) If the balls are small or the speed is low, crowding will appear at the coarser sizes in the classifier sands.

(c) The indications are that best efficiency is obtained when the screen analysis of the sands shows a minimum of crowding at any size. This statement has not been proved conclusively, however.

### DEDUCTIONS FROM OPERATING TESTS

The foregoing tests are only a few of the more important ones that were made. Over 150 tests have been made altogether, and it may be of value to state some of the general conclusions from them.

1. In every case an increase in the tonnage fed to the mill resulted in an increase in the efficiency.

2. In every test the limiting factor of the test was not the ball-mill but some auxiliary apparatus, usually the classifier.

3. All tests point to the fact that the ball-mill is a machine of very large capacity, especially if it is provided with proper auxiliary apparatus.

4. Classifying and pulp-handling machines that will handle a circulating load of at least 500 per cent. of the original feed should be provided.

5. Closed-circuit crushing is more desirable than open-circuit.

6. The real advantage in closed-circuit crushing lies in the fact that the maximum size of particle is nearer the average size of particle discharged from the circuit.

7. Two-stage crushing is more efficient than single-stage.

8. The real advantage in two-stage crushing lies in the fact that the ball charges can be adjusted more nearly to the required conditions.

9. Two-stage open-circuit crushing does not present this advantage, as the maximum-sized particle in both stages is more nearly the same.

10. The proper adjustment between size of balls and speed of mill can be secured by an examination of the classifier sands.

11. If the balls are large or the speed is high, the screen analysis of the classifier sands will be crowded at the finer sizes.

12. If the balls are small or the speed is low, the screen analysis of the classifier sands will be crowded at the coarser sizes.

13. From the data at hand, the indications are that the best efficiency is obtained when the screen analysis of the classifier sands shows a minimum of crowding at any size.

14. Balls no larger than necessary should be used, as this makes it possible to charge the mill with the greatest number of balls.

15. Balls smaller than can crush the larger particles of ore should not be kept in the mill as they take up space, absorb power, and do inefficient crushing.

### DESIGN AND REGULATION OF A FINE-CRUSHING PLANT

In view of these conclusions and the test data at hand, it is interesting to outline the manner in which a fine-crushing plant may be designed. In this discussion, the following limitations are imposed:

(a) The first cost of the plant must not be excessive.

(b) Since the experiments were made with a Hardinge mill and a Dorr classifier, these are given first consideration herein, although not necessarily the best adapted for the work to be done.

(c) The plant is to receive a feed and deliver a final product approximately as shown in Table 11.

TABLE 11.—*Feed and Product of Plant*

| Mesh | Opening, Mm. | Feed | | Product | |
|---|---|---|---|---|---|
| | | Per Cent. | Cum. Per Cent. | Per Cent. | Cum. Per Cent. |
| On 4 | 4.70 | 29.17 | 29.17 | | |
| On 8 | 2.36 | 26.22 | 55.39 | | |
| On 14 | 1.17 | 14.13 | 69.52 | | |
| On 28 | 0.59 | 8.62 | 78.14 | | |
| On 48 | 0.295 | 4.67 | 82.81 | | |
| On 100 | 0.147 | 3.32 | 86.13 | 0.50 | 0.50 |
| On 200 | 0.074 | 3.28 | 89.41 | 6.65 | 7.15 |
| On 300 | | 1.68 | 91.09 | 13.75 | 20.90 |
| Through 300 | | 8.91 | 100.00 | 79.10 | 100.00 |

The flow sheet shown in Fig. 2 has been designed to meet these requirements. Its most conspicuous feature is the large number of classifiers; possibly there are too many, but in all tests the limiting factor has been the capacity of these machines. It is estimated that the capacity of this plant will be 720 T. per day, receiving a feed and delivering a product as shown. The plant will require 344 kw. at the switchboard, which will be 11.5 kw.-hr. per ton of ore crushed, or 14 kw.-hr. per ton of minus 200-mesh material actually produced. This is not an extremely low figure as better results have been obtained many times in the tests.

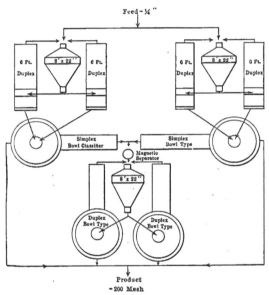

FIG. 2.—FLOW SHEET.

The 720 T. per day, or 30 T. per hr., of original feed will divide to the various units in the following manner: Each first-stage ball-mill will crush 15 T. per hr. to minus 48 mesh. This 15 T. will pass into the Simplex bowl-type classifier, where 10 T. will overflow as finished product; the 5 T. of sand, with the 5 T. of sand from the second Simplex bowl-type classifier, will constitute the feed to the second-stage ball-mill. The test data indicate that this can be accomplished with one classifier in closed circuit with each ball-mill. By adding the second classifier, as shown, it is expected that the tonnage can be increased at least 25 per cent. In order to maintain this feed rate, it will be necessary to maintain carefully the proper ball charge in each mill. A slight increase or decrease in the average size of the balls making up the charge will cause a large loss in efficiency and a corresponding reduction in tonnage.

If balls 3 in. (76.2 mm.) in diameter are fed to the first-stage mills and if all balls less than 2 in. (50.8 mm.) in diameter are removed from the mills regularly, the average size of the balls forming the working charge will be 2.499 in. (63.47 mm.), which is about the size indicated in the tests as giving the best results. The working ball charge in the mill will be as shown in Table 12.

TABLE 12.—*Working Ball Charge in Mill*

| Diameter of Balls, In. | Per Cent. of Weight | Actual Weight, Lb. |
|---|---|---|
| 3.00 to 2.75 | 32.63 | 9,136.4 |
| 2.75 to 2.50 | 27.21 | 7,618.8 |
| 2.50 to 2.25 | 22.32 | 6,249.6 |
| 2.25 to 2.00 | 17.84 | 4,995.2 |
| Total | 100.00 | 28,000.0 |

Suppose that once every week the mills are stopped and all balls less than 2 in. in diameter are removed. If the ball wear is 2 lb. per ton of ore crushed, this will amount to 720 lb. (326.5 kg.) per day for each mill. This is actual wear and takes no account of the small balls that are removed. Then, since the working charge is to comprise balls from 3 in. (76.2 mm.) to 2 in. (50.8 mm.) in diameter, the amount of wear secured from each ball will be 2.89 lb. (1.31 kg.) and the 720 lb. of wear per day will be taken care of by the reduction in diameter of 249 balls per day from 3 in. to 2 in. It will then be necessary actually to charge 249 three-inch balls, or 1020 lb. (462.6 kg.) per day.

It is evident that some of the minus 2-in. balls that will be removed after seven days will be smaller than others. It may be computed by methods hereinafter described that the smallest ball will be 1.92 in. (48.7 mm.) in diameter. Then if 249 three-inch balls are charged at the beginning of each day, at the end of each day 249 balls will have been worn to a diameter of less than 2 in. At the end of seven days there will be 1743 balls of diameter between 2 and 1.92 in., which will weigh 1995 lb. (904.9 kg.) and will be removed from the mill. As the operation of each of the first-stage mills will consist in charging 249 three-inch balls each day, the mill charge will gain in weight each day until at the end of seven days it will have gained 1995 lb. and will therefore weigh 29,995 lb. At the end of the seven days, however, the 1995 lb. of balls smaller than 2 in. will be removed, leaving the original 28,000 lb. working charge, as at the beginning of the week.

Since in this flow sheet there are two first-stage mills, there will be formed 3990 lb. (1809.8 kg.) of balls per week of average diameter 1.96 in. (49.7 mm.). These 3990 balls will be used up each week in the daily charges of balls to the second-stage mill. In order to have a balanced condition, it will be necessary to charge these balls at the same rate as that at which they are made, or 498 balls per day. These 498 balls, weighing 571 lb. (259 kg.), will constitute the daily charge to the second-stage mill. This mill is to handle 240 tons of ore per day and the steel consumption at 2 lb. (0.9 kg.) per ton will be 480 lb. (217 kg.) per day. Since 498 balls, weighing 571 lb., are to be added to the mill each day, 498 balls weighing 91 lb. (41 kg.) should be removed each day. These 498 balls will weigh 0.1827 lb. (0.08 kg.) each and will be 1.06 in. (26.9 mm.) in diameter. At regular intervals all balls less than 1 in. in diameter should be removed from the second-stage mill.

It is now possible to compute the screen analysis of the working charge of balls in the second-stage mill (see Table 13).

TABLE 13.—*Screen Analysis of Balls in Second-stage Mill*

| Size of Ball, In. | Per Cent. of Weight | Actual Weight, Lb. |
|---|---|---|
| 1.96 to 1.75............................. | 33.23 | 9,304.4 |
| 1.75 to 1.50............................. | 30.32 | 8,489.6 |
| 1.50 to 1.25............................. | 21.90 | 6,132.0 |
| 1.25 to 1.00............................. | 14.55 | 4,074.0 |
| Total............................. | 100.00 | 28,000.0 |

If the ball charge in the second-stage mill is screened once a month, there will be 14,940 balls less than 1 in. in diameter to remove. The smallest ball will be 0.85 in. (21.59 mm.) in diameter and the largest ball will be 1 in. The total weight of the balls removed at the end of the month will be 1825 lb. (824.8 kg.). The removal of this weight of small balls will again produce the original charge that is shown in the above screen analysis. Of course, if the ball wear is not 2 lb. per ton, as assumed, these figures will not hold. However, as soon as the correct ball wear is found, it will be possible to determine by this method the exact figures that will make it possible to maintain the proper ball charge and the balance between the different mills at all times. As a result in the design of this fine-crushing plant, provision should be made for sizing the ball charges of the first-stage mills each week and of the second-stage mill each month. It can then be done with a very small amount of lost time.

The chief advantages in this flow sheet are: Good efficiency as to the power expended; large tonnage for the capital invested; and flexibility.

By adjusting the overflow end of the classifier in the first stage, the load can be balanced perfectly between the two crushing stages.

## MECHANICS OF THE BALL-MILL

In the endeavor to determine the best working conditions for the ball-mill, a detailed mathematical study was made of the action of the ball charge. While the data taken from a properly conducted test are convincing, an engineer sometimes prefers a mathematical proof. Test data contain a large personal factor, not only of manipulation but also of the person reporting the results. In the case of ball-mill crushing the amount of available data is enormous and, by careful selection, nearly any statement can be "proved." For this reason, a consideration which is entirely theoretical and devoid of any personal element would seem to be desirable and instructive.

It is evident that the ball inside a revolving mill must act according to some exact regulating force which governs its every motion. There are three important variables to consider: the speed of the mill, its size, and the size of the ball charge, and it will be the aim of this discussion to show the relation existing between these variables.

## ACTION OF CHARGE AT SLOW SPEED

Any loose charge piled up in a cone will assume a certain definite critical angle, usually called the angle of repose. If more of the charge is added to the top of the cone, this will be increased in size but the same critical angle will be maintained. This is what happens inside a mill revolving at very low speeds. The charge is tilted until the critical angle is reached, after which the balls simply roll down the slope to the lower side of the mill. This critical angle is affected but slightly by a change in the speed of the mill, up to a certain point; the increase in speed simply increases the rapidity with which the charge is raised to the top of the incline. In this condition the balls are in contact with one another except as they may bounce in rolling down the slope of the charge; also, the balls must roll down the incline at the same rate, pounds per hour, at which they are raised to the top. Then, with a mill half full of balls, any particular ball will roll down the incline something less than twice per revolution of the mill.

As the speed of the mill is slowly increased, the time required to bring the ball back to the top of the pile is diminished, but the time required by it in rolling down remains practically the same. It would seem then that the whole problem of crushing would resolve itself into getting the balls to the top of the heap fast enough. This would be true if it were not for centrifugal force and inertia. As the speed of the mill is increased these two forces grow very rapidly in importance.

## ACTION OF CHARGE AT HIGHER SPEEDS

Consider the forces acting on a particle $p$, Fig. 3, in contact with the lining of the mill. The centrifugal force $c$ acts to press it against the lining while $w_1$, a component of the weight $w$, acts to pull it away from the lining. Then if $\alpha_1$ is the angle between the vertical axis and the radius $op$, $w_1 = w \cos \alpha_1$. It is possible for $c$ to be greater than, equal to, or less than $w_1$ or $w \cos \alpha_1$, for as $\alpha_1$ decreases $w_1$ increases. Then $c - w \cos \alpha_1 = f_1$, and $f_1$ may be positive, negative or zero. If $f_1$ is positive the particle will be held against the lining of the mill. As $\alpha_1$ decreases $f_1$ decreases and if, when $\alpha_1$ is zero, $f_1$ is still positive, it is evident that the particle will maintain its contact with the mill lining throughout a complete revolution.

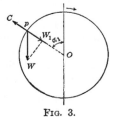

FIG. 3.

If $f_1$ becomes zero for some value of $\alpha_1$ the particles below $p_1$, having a greater angle $\alpha$, will be held against the lining of the mill by a positive force $f$. In other words, as the mill rotates, the force with which a particle $p$ is held in position decreases until it reaches zero. At this point the particle is being pushed on by the particles below it and is free to move in a path governed by this initial velocity and gravity. The path it takes will, of course, be parabolic. Then when the angle $\alpha$ is such that $f$ is zero, the particle $p$ will leave off contact with the mill and start on a parabolic path. In this position $c = w \cos \alpha$.

The centrifugal force $c = \dfrac{wv^2}{rg}$, where $w$ = weight of particle; $v$ = initial velocity; $r$ = radius; and $g$ = 32.2 ft. per sec. per sec. Also the initial velocity of the particle is its velocity in the circular path, or $v = 2\pi r n$ in which $n$ = speed of mill in revolutions per second. Then by substitution in the formula $c = w \cos \alpha$,

$$\frac{wv^2}{rg} = w \cos \alpha, \text{ or } \frac{w(2\pi r n)^2}{rg} = w \cos \alpha, \text{ or}$$

$$\cos \alpha = \frac{4\pi^2 r n^2}{g} \tag{1}$$

From this equation it is evident that an increase in either the speed of the mill or its radius will cause a decrease in the angle $\alpha$ and the parabolic path will not start until the particle is carried farther around in the direction of rotation. At any speed $n$,

$$\cos \alpha = kr \tag{2}$$

in which $k = 1.226n^2$. Then at constant speed, the particle $p$ nearer the center of the circle than $p_1$ will start on its parabolic path from a larger angle $\alpha$, or as $r$ decreases $\alpha$ increases, the relation between $r$

and $\alpha$ given in equation (2) always holding true when $\alpha$ is the angle at which the parabolic path starts. Equation (2), then, is really the equation of the curve above which all particles are following the parabolic path and below which all particles are following the circular path.

From equation (1) $n = \sqrt{\dfrac{g \cos \alpha}{4\pi^2 r}}$ and if the radius is considered as the constant, $\alpha$ must decrease as $n$ increases. Since the minimum value of $\alpha$ is zero, when $\cos \alpha = 1$, the speed has reached a point above which $c$ is always greater than $w \cos \alpha$ (Fig. 3), and the particle will cling to the lining of the mill throughout the complete cycle. Then the speed at which any particle of radius $r$ will cling is given by the equation $n = \sqrt{\dfrac{g}{4\pi^2 r}}$, in which $n$ is in revolutions per second and $r$ is in feet. If the speed is in revolutions per minute the equation will become

$$N^1 = \frac{54.19}{\sqrt{r}} \tag{3}$$

This equation shows the critical speed $N^1$ at which the particle of radius $r$ will cling to the lining of the mill or to the next outer layer of particles of radius greater than $r$. If $N^1$ is sufficiently large, $r$ will be sufficiently small to include all of the balls in the mill and the mill will rotate as a flywheel with no relative motion between the particles in the charge. Table 14 shows the speed at which the first particle will cling to the lining of the mill.

TABLE 14.—*Critical Speeds of Ball-mills*

| Diameter of Mill $(2r_1)$, Ft. | Critical Speed $(N^1)$, Rev. per Min. | Diameter of Mill $(2r_1)$, Ft. | Critical Speed $(N^1)$, Rev. per Min. |
|---|---|---|---|
| 0.125 | 216.76 | 5 | 34.27 |
| 0.250 | 153.30 | 6 | 31.29 |
| 1 | 76.63 | 7 | 28.97 |
| 2 | 54.19 | 8 | 27.10 |
| 3 | 44.25 | 9 | 25.55 |
| 4 | 38.32 | 10 | 24.23 |

PARABOLIC PATHS OF FALLING PARTICLES

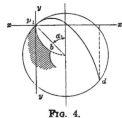

FIG. 4.

Below the critical speed given in equation (3), the particle of radius $r$ will reach the critical angle $\alpha$ and will then start on its parabolic path. The next consideration is to determine where the particle $p_1$ will strike the lining of the mill at the end of its parabolic path.

In Fig. 4, the equation of the parabola, origin at $p_1$, is $y = x \tan \alpha_1 - \dfrac{gx^2}{2V_1^2 \cos^2 \alpha_1}$, and the equation of the circle of the mill, origin $p_1$, is $x^2 + y^2 - (2r_1 \sin \alpha_1) x + (2r_1 \cos \alpha_1) y = 0$.

The simultaneous solution of these two equations will give the coördinates of the point $d$, where the two curves intersect. Substituting the value of $y$ from the equation of the parabola in the equation of the circle and then simplifying, the following equations are secured:

$$\frac{g^2}{4V_1^4 \cos^4 \alpha_1} x^4 - \frac{g \sin \alpha_1}{V_1^2 \cos^3 \alpha_1} x^3 + \left( \sec^2 \alpha - \frac{r_1 g}{V_1^2 \cos \alpha_1} \right) x^2 = 0.$$

Then $x^2 = 0$, and

$$\left( \frac{g^2}{4V_1^2 \cos^2 \alpha_1} \right) x^2 - g \tan \alpha_1 x + V_1^2 - r_1 g \cos \alpha_1 = 0.$$

The initial velocity $V_1 = 2\pi r_1 n$, and $V_1^2 = 4\pi^2 r_1^2 n^2$.

But from equation (1) $n^2 = \frac{g \cos \alpha_1}{4\pi^2 r_1}$; therefore $V_1^2 = r_1 g \cos \alpha_1$. Substituting this value for $V_1^2$ in the above equation and simplifying,

$$\frac{x^2}{4r_1 \cos^3 \alpha_1} - x \tan \alpha_1 = 0.$$

Then $x = 0$ and

$$x = 4r_1 \sin \alpha_1 \cos^2 \alpha_1 \tag{4}$$
$$y = - 4r_1 \sin^2 \alpha_1 \cos \alpha_1 \tag{5}$$

These are the coördinates of the point $d$, at which the particle $p_1$ will strike at the end of its parabolic path. From the above solution, it is seen that there are, in general, four points of intersection between the two curves. In this case, however, only one of these intersections requires consideration, the other three being zero. This is a very important fact, for if this condition did not exist the paths of travel of the various particles would cross and recross one another, resulting in a large friction loss above the pulp level in the mill and the performance of little or no crushing.

### APPLICATION TO A DEFINITE PROBLEM

It is now possible to draw the outline of the charge in the mill under operating conditions. Consider an 8-ft. mill running at 24 r.p.m. From equation (2), $\cos \alpha = \dfrac{1.226 \times 24^2 \times r}{3600} = 0.1962r$, which is the equation of a circle of radius $\dfrac{0.408}{n^2}$. The center is then on the vertical axis $\dfrac{0.408}{n^2}$ units above the center of the mill. From this the value of $\alpha$ in Table 15 can be computed:

TABLE 15.—*Data for Computing Values of* $\alpha$

| $r$ | Cos $\alpha$ | $\alpha$ | $x$ |
|---|---|---|---|
| 4 | 0.7848 | 38° 18' | 6,1072 |
| 3 | 0.5886 | 53° 56' | 3.3624 |
| 2 | 0.3924 | 66° 54' | 1.1328 |
| 1 | 0.1962 | 78° 41' | 0.1512 |
| 0 | 0.0000 | 90° | 0.0000 |

From the values of $r$ and $\alpha$, the curve $aO$, Fig. 5, can be drawn, which is the dividing line between the parabolic path and the circular path of the particles. Then by use of equations (4) and (5) it is possible to find any number of points on the curve $cd$. This may be done more simply by drawing the circle through the point $e$ and then measuring a distance $x$ to a vertical line which will intersect the circle at the desired point $f$ as in Fig. 5. For this purpose, the corresponding values of $x$ are added to the preceding table.

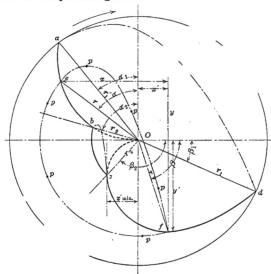

FIG. 5 —PATHS OF TRAVEL OF PARTICLES IN AN 8-FT. MILL MAKING 24 R.P.M.

It is then possible to draw the line $Od$, which is the dividing line between the parabolic path and the circular path for all particles of the charge. The complete cycle of any particle $p$ is then seen to be from $e$ to $f$ along the parabolic path, and then from $f$ to $e$ along the circular path. From Fig. 5 it is evident that the particle $p$ acts exactly as though it were in a mill of radius $r$, the lining of which is the layer of particles of radius next larger than $r$.

## END OF THE PARABOLIC PATH

Since, as has been shown,

$$x = 4r \sin \alpha \cos^2 \alpha \text{ and } y = - 4r \sin^2 \alpha \cos \alpha,$$

from Fig. 5,  $\quad x' = 4r \sin \alpha \cos^2 \alpha - r \sin \alpha,$

and  $\quad y' = 4r \sin^2 \alpha \cos \alpha - r \cos \alpha.$

As the value of $r$ at which a particle starts on its parabolic path and ends its parabolic path is the same, then

$$\sin \beta = \frac{y_1}{r} = \frac{4r \sin^2 \alpha \cos \alpha - r \cos \alpha}{r} = - (4 \cos^3 \alpha - 3 \cos \alpha).$$

Then $\sin \beta \doteq - \cos 3\alpha$; but $- \cos 3\alpha = \cos (180° - 3\alpha)$ and $\sin \beta = \cos (90° - \beta)$; then $\cos (90° - \beta) = \cos (180° - 3\alpha)$, or $90° - \beta = 180° - 3\alpha$; hence $- \beta = 90° - 3\alpha$,

or  $$\beta = 3\alpha - 90°. \tag{6}$$

From Fig. 5 it is evident that when $\alpha$ has increased beyond a certain large value, the parabolic paths of the balls near the center of the mill will overlap, thus causing interference. The balls will then be striking together at a point so near the maximum pulp level that little or no effective crushing can be done. It is thus obvious that the size of the charge should be so regulated that this interference shall not occur. It appears from Fig. 5 that this limiting condition occurs when $x'$ is a minimum, or is equal to its largest negative value.

Then since $x' = 4r \sin \alpha \cos^2 \alpha - r \sin \alpha$, or eliminating $r$ by use of the equation, $r = \dfrac{g}{4\pi^2 n^2} \cos \alpha$,

$$x' = \frac{g}{\pi^2 n^2} (\sin \alpha \cos^3 \alpha - \tfrac{1}{4} \sin \alpha \cos \alpha).$$

$$dx' = \frac{g}{\pi^2 n^2} (\cos^4 \alpha - 3 \sin^2 \alpha \cos^2 \alpha - \tfrac{1}{4} \cos^2 \alpha + \tfrac{1}{4} \sin^2 \alpha).$$

For maximum and minimum values of $x'$, $dx' = 0$.

Thus $4 \cos^4 \alpha - \tfrac{7}{2} \cos^2 \alpha + \tfrac{1}{4} = 0$;

whence  $\cos \alpha = 0.8925$ and $0.2801$,

and  $\alpha = 26°49'$ and $73°44' \tag{7}$

From equation (6), $\beta = - 9°33'$ and $131°12'$.

The larger result is obviously the one desired, the smaller one being the value that makes $x'$ a maximum. From this it appears that the largest charge that can be used in the mill without definite interference between the particles is when angle $\alpha_2 = 73°44'$ and $\beta_2 = 131°12'$. If the inner radius of the charge, corresponding to $\alpha_2 = 73°44'$, is $r_2$, then $\cos 73°44' = 1.226 n^2 r_2$ or $r_2 = \dfrac{0.2283}{n^2}$, which is the smallest value that $r_2$ should have at any speed.

### Blow Struck by Falling Particle

In order to determine at what angle $\alpha$ the maximum effective blow will be delivered by the particle when it strikes the surface of the mill, it is necessary to find the resultant velocity of the particle relative to the lining of the mill. Fig. 6 shows the resultant velocities and their components.

$V_p$ = velocity of point on parabola;
$V_c$ = velocity of point on mill;
$V_t$ = component of $V_p$ in direction of $V_c$, with mill stationary;
$V_m$ = component of $V_p$ in direction of $V_c$, with mill revolving;
$V_r$ = component of $V_p$ perpendicular to $V_c$;
$\theta$ = angle between $V_c$ and $V_p$;
$V_b$ = velocity that produces blow; or the velocity of particle relative to lining of mill.

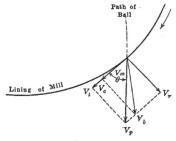

Fig. 6.

It has been shown that the initial velocity of any particle when it started on its parabolic path is given by the equation $V = \sqrt{rg \cos \alpha}$. This is the velocity in the circular path of a particle of radius $r$. The velocity of any point on the parabola is given by the equation[1]

$$V_p = \sqrt{V^2 \cos^2 \alpha + \left( V \sin \alpha - \frac{gx}{V \cos \alpha} \right)^2}$$

but $V = \sqrt{rg \cos \alpha}$ and $x = 4r \sin \alpha \cos^2\alpha$.

Then $V_p = \sqrt{9rg \cos \alpha - 8rg \cos^3 \alpha}$.

The velocity of any point on the circle being $V_c = \sqrt{rg} \cos \alpha$.
Then $V_p \cos \theta = V_t$; $V_p \sin \theta = V_r$; $V_t - V_c = V_m$; $\sqrt{V_m^2 + V_r^2} = V_b$.
Then $V_m = V_p \cos \theta - V_c$, and $V_b = \sqrt{(V_p \cos \theta - V_c)^2 + (V_p \sin \theta)^2}$

$$V_b = \sqrt{V_p^2 \cos^2 \theta - 2V_p V_c \cos \theta + V_c^2 + V_p^2 \sin^2 \theta}$$
$$= \sqrt{V_p^2 - 2V_p V_c \cos \theta + V_c^2}$$

$V_b^2 = 9rg \cos \alpha - 8rg \cos^3 \alpha + rg \cos \alpha - (2\sqrt{9r^2g^2 \cos^2 \alpha - 8r^2g^2 \cos^4 \alpha})$
$$\cos \theta.$$
$$= rg \cos \alpha [10 - 8 \cos^2 \alpha - (2\sqrt{9 - 8 \cos^2 \alpha}) \cos \theta].$$

In order to determine the angle $\theta$ it is necessary to find the angle at which the circle and parabola intersect. This is done by computing the slopes of the tangents to both curves at this point and then the angle between these tangents. As the slope of the tangent is the first derivative of the equation of the curve, the slope of the tangent to the circle is

$$\frac{d}{dx}[x^2 + y^2 - (2r \sin \alpha) x + (2r \cos \alpha)y] = 0,$$

whence $\dfrac{dy}{dx} = \dfrac{2r \sin \alpha - 2x}{2y + 2r \cos \alpha}$ at the point $(x, y)$.

Then since $x = 4r \sin \alpha \cos^2 \alpha$, and $y = -4r \sin^2 \alpha \cos \alpha$,

$$\frac{dy}{dx} = \frac{4 \sin^3 \alpha - 3 \sin \alpha}{4 \cos^3 \alpha - 3 \cos \alpha},$$

which is the slope of the tangent to the circle.

Likewise for the parabola, the slope of the tangent is

$$\frac{dy}{dx} = \tan \alpha - \frac{\text{$c$}\text{$v$}}{2r \cos^3 \alpha} \text{ at the point } (x, y);$$

then $\dfrac{dy}{dx} = -3 \tan \alpha$ is the slope of the tangent to the parabola.

The angle between two lines is expressed by the equation,

$$\tan \theta = \frac{m_1 - m_2}{1 + m_1 m_2},$$

in which $m_1$ and $m_2$ are the slopes.

Then $\qquad \tan \theta = \dfrac{-3 \tan \alpha - \dfrac{4 \sin^3 \alpha - 3 \sin \alpha}{4 \cos^3 \alpha - 3 \cos \alpha}}{1 - (3 \tan \alpha)\dfrac{4 \sin^3 \alpha - 3 \sin \alpha}{4 \cos^3 \alpha - 3 \cos \alpha}}$

$$= \frac{8 \sin^3 \alpha \cos \alpha}{-8 \sin^4 \alpha + 4 \sin^2 \alpha + 1};$$

but $\qquad \cos \theta = \dfrac{1}{\sqrt{1 + \tan^2 \theta}};$

then $\qquad \cos \theta = \sqrt{\dfrac{64 \sin^8 \alpha - 64 \sin^6 \alpha + 8 \sin^2 \alpha + 1}{8 \sin^2 \alpha + 1}}.$

It is now possible to write the complete formula for the effective velocity of the particle:

$$V_b^2 = rg \cos \alpha \left[ 10 - 8 \cos^2 \alpha - (2\sqrt{9 - 8 \cos^2 \alpha})\sqrt{\frac{64 \sin^8 \alpha - 64 \sin^6 \alpha + 8 \sin^2 \alpha + 1}{8 \sin^2 \alpha + 1}} \right]$$

This equation may be simplified to $V_b^2 = 16rg \cos \alpha \sin^4 \alpha.$ $\qquad$ (8)

But $\qquad \cos \alpha = kr$, and $\sin \alpha = \sqrt{1 - k^2 r^2}.$

Then $\quad \sin^2 \alpha = 1 - k^2 r^2$, and $\sin^4 = 1 - 2k^2 r^2 + k^4 r^4.$

and $\qquad V_b^2 = 16kgr^2 - 32k^3 gr^4 + 16k^5 gr^6.$

In order to find $r$ for the maximum velocity squared,

$$d(V_b{}^2) = 32kgr - 128k^3gr^3 + 96k^5gr^5 = 0,$$

or

$$3k^4r^4 - 4k^2r^2 + 1 = 0;$$

Whence $\quad r = \dfrac{1}{k}$ and $\dfrac{0.5775}{k}$

When $\quad r = \dfrac{1}{k}$ $\hspace{3cm}$ When $\quad r = \dfrac{0.5775}{k}$ $\hspace{1cm}$ (9)

$\cos \alpha = kr$ $\hspace{5cm}$ $\cos \alpha = kr$

$\cos \alpha = \dfrac{k}{k} = 1$ $\hspace{3.5cm}$ $\cos \alpha = k\dfrac{0.5775}{k} = 0.5775$

$\quad \alpha = 0$ for min. $V_t{}^2$ $\hspace{2.5cm}$ $\alpha = 54°44'$ for max. $V_b{}^2$ $\quad$ (10)

Thus when $\alpha = 54°44'$ the balls are striking with maximum velocity relative to the circular path in which they travel. Since the whole charge, near $ao$, may be considered as being concentrated at the radius of gyration, it would seem that the most effective conditions* would be obtained by placing $r$ (equation 9) equal to the radius of gyration of the charge. Since the radius of gyration is equal to $\sqrt{\dfrac{r_1{}^2 + r_2{}^2}{2}}$ when $r_1$ is the radius of the mill and $r_2$ is the inner radius of the charge, then

$$\sqrt{\frac{r_1{}^2 + r_2{}^2}{2}} = \frac{0.5775}{k} \tag{11a}$$

But $r_2 = Kr_1$, whence $\hspace{2cm}$ $\sqrt{\dfrac{(1 + K^2)r_1{}^2}{2}} = \dfrac{0.5775}{k}$

But $k = 1.226n^2$, whence $\hspace{2cm}$ $K = \sqrt{\dfrac{0.443}{r_1{}^2n^4} - 1}$ $\hspace{1cm}$ (11)

when $r$ is the radius of the mill and $n$ is its speed in rev. per second.

The value of $K$ given by this equation is the relation between $r_1$ and $r_2$ which should exist for best operating conditions and is really the measure of the quantity of the charge. This equation then gives the proper relation between the size of the charge, the size of the mill and its speed. It is the fundamental equation of the ball-mill and shows the relations mentioned on page 264.

Equation (11) may be restated in the following forms, which may be more convenient.

$$n^4 = \frac{0.443}{(1 + K^2)r_1{}^2}.$$

Then $\hspace{2cm}$ $n = \dfrac{0.8158}{\sqrt{r_1}\sqrt[4]{1 + K^2}}$ in rev. per sec., $\hspace{1cm}$ (12)

or $\hspace{2cm}$ $N = \dfrac{48.948}{\sqrt{r_1}\sqrt[4]{1 + K^2}}$ in rev. per min. $\hspace{1cm}$ (13)

*This is probably not exactly true. Detailed mathematical analysis shows it to be very near the truth, however, and it has been used here for the sake of simplicity

Since $\cos \alpha = 1.226rn^2$,

$$\cos \alpha = 1.226 \frac{r}{r_1} \sqrt{\frac{0.443}{1 + K^2}} = \frac{0.8165r}{r_1} \cdot \frac{1}{\sqrt{1 + K^2}} \tag{14}$$

in which $\alpha$ is the angle corresponding to any radius $r$.

Then
$$\cos \alpha_1 = \frac{0.8165}{\sqrt{1 + K^2}} \tag{15}$$

in which $\alpha_1$ is the angle corresponding to the radius $r_1$, which is the radius of the mill and is therefore the minimum angle $\alpha$.

$$\cos \alpha_2 = \frac{0.8165K}{\sqrt{1 + K^2}} = K \cos \alpha_1 \tag{16}$$

in which $\alpha_2$ is the angle corresponding to the radius $r_2$, which is the inner radius of the charge and is therefore the maximum angle $\alpha$.

Equations (12) to (16) all depend on the value of $K$, which is the real measure of the quantity of the charge. It is now necessary to get a better idea of the exact relation between $K$ and the volume of the charge. In order to do this the cycle of the charge must be known.

## The Cycle of Charge

Since $\beta = 3\alpha - 90°$, the angle passed through by a particle in the parabolic path is $\alpha + 90 + \beta$, or $\alpha + 90 + 3\alpha - 90 = 4\alpha$.

If the speed of the mill is $n$, the time per revolution is $T_r = \frac{1}{n}$. Then the time in the circular path is $T_r \left( \frac{360 - 4\alpha}{360} \right) = T_c$.

The time required by a particle passing through a parabolic curve is given by the equation, $T_p = \frac{x}{V \cos \alpha}$.

Then
$$T_p = \frac{2}{\pi n} \sin \alpha \cos \alpha. \tag{17}$$

Again consider the average of the whole charge as passing through the same cycle as the particle at the radius of gyration, which is $\left( \sqrt{\frac{1 + K^2}{2}} \right) r_1$ at the angle $\alpha = 54°44'$. (Equation 10.)

Then
$$T_r = \frac{1}{n}; \quad T_c = \frac{0.392}{n}; \quad T_p = \frac{0.3003}{n}.$$

But the total time of one cycle $T = T_c + T_p = \frac{0.392 + 0.3003}{n} = \frac{0.6923}{n}$

Then the portion of the total time in the circular path is

$$\frac{0.392}{n} \div \frac{0.6923}{n} = 0.5665; \text{ or } 56.65 \text{ per cent. of the time} \tag{18}$$

The number of cycles per revolution is given by the equation,

$$\frac{T_r}{T} = \frac{1}{n} \div \frac{0.6923}{n} = \frac{1}{0.6923} = 1.444 \qquad (19)$$

This means that when the mill is running at the proper speed for the ball charge, then the charge is passing through 1.444 cycles per revolution, and each ball in the charge strikes, on the average, 1.444 blows per revolution.

### VOLUME OF CHARGE

Equation (18) states that the balls spend 56.65 per cent. of the time in the circular path. Then it is apparent that 56.65 per cent. of the total charge is always in the circular path. In other words, the volume of the charge between $r_1$ and $r_2$ (Fig. 5) is only 56.65 per cent. of the total charge, the remaining 43.35 per cent. being spread out over the rest of the mill and following the parabolic path.

The exact analytical determination of the variation in the volume of the total charge as $K$ varies is very complicated and will not be gone into here. A very close approximation can be made, however, by use of the equation,

$$K = -0.024 + 0.39\sqrt{7 - 10P} \qquad (20)$$

in which $P$ is the fractional part of the entire volume of the mill that is occupied by the charge when the mill is stationary. It should be noted that the charge will contain a considerable proportion of voids. These are, of course, included in the space occupied by the charge.

### POWER

From equation (8) it appears that $V_b{}^2 = 16kgr^2 - 32k^3gr^4 + 16k^5gr^6$, but the kinetic energy $e = \dfrac{wv^2}{2g}$, or in this case, $e = w(8kr^2 - 16k^3r^4 + 8k^5r^6)$, which is the energy possessed by any particle of weight $w$, and radius $r$, at the end of its parabolic path, which is available for the purpose of crushing ore. Then the total energy possessed by the particles of radius between $r_1$ and $r_2$ is expressed by the equation,

$$\int_{r_2}^{r_1} e = w\left[ \left( \frac{8}{3}kr^3 - \frac{16}{5}k^3r^5 + \frac{8}{7}k^5r^7 \right) \right]_{r_2}^{r_1}$$

But $\qquad\qquad k = 1.226n^2$ and $r_2 = Kr_1$; then

$$E = w[3.269n^2r_1{}^3(1 - K^3) - 5.8968n^6r_1{}^5(1 - K^5) + 3.1656n^{10}r_1{}^7(1 - K^7)]$$

But $\qquad n = \dfrac{0.8158}{\sqrt{r_1}\sqrt[4]{1 + K^2}}$ for best theoretical efficiency; whence

$$E = wr_1{}^2\left[ 2.1756\frac{(1 - K^3)}{(1 + K^2)^{3/2}} - 1.7382\frac{(1 - K^5)}{(1 + K^2)^{5/2}} + 0.41333\frac{(1 - K^7)}{(1 + K^2)^{7/2}} \right]$$

In this equation, if $W$ is the entire weight of the charge, then $E$ represents the foot-pounds of energy delivered each time the charge passes through one cycle. Then since there are 1.444 cycles per revolution, and $n = \dfrac{0.8158}{\sqrt{r_1}\sqrt[4]{1+K^2}}$ rev. per sec., the number of foot-pounds per second is represented by the equation,

$$E = Wr_1{}^{3\!\!/\!\!2}\left[2.5472\,\frac{(1-K^3)}{(1+K^2)^{1\!\!/\!\!3}} - 2.0351\frac{(1-K^5)}{(1+K^2)^{3\!\!/\!\!6}} + 0.48394\,\frac{(1-K^7)}{(1+K^2)^{5\!\!/\!\!6}}\right] \quad (21)$$

$$\text{or Hp.} = Wr_1{}^{3\!\!/\!\!2}\left[0.004467\,\frac{(1-K^3)}{(1+K^2)^{1\!\!/\!\!3}} - 0.003700\,\frac{(1-K^5)}{(1+K^2)^{3\!\!/\!\!6}} + 0.000880\,\frac{(1-K^7)}{(1+K^2)^{5\!\!/\!\!6}}\right] \quad (22)$$

This formula gives the power output and therefore input to the mill when the weight of the charge is $W$, in pounds, the radius of the mill is $r_1$, in feet, and the value of $K$ and the speed $N$ are as given by formulas (20) and (13). In other words, formula (22) gives the power required to operate the mill at the most efficient speed for any ball charge. In Table 16 the horsepower required has been computed for certain operating conditions. Any table or set of curves covering all conditions would be too large and complicated to include in this paper; the formulas may, however, be applied to any particular condition.

TABLE 16.—*Horsepower Required per Foot of Mill Length*

| Internal Diam. of Mill, Ft. ($2r_1$) | Portion of Mill Volume Occupied by Charge | | | | | |
|---|---|---|---|---|---|---|
| | 0.1 | 0.2 | 0.3 | 0.4 | 0.5 | 0.6 |
| 1 | 0.0023 | 0.0099 | 0.022 | 0.042 | 0.065 | 0.093 |
| 2 | 0.0255 | 0.110 | 0.254 | 0.46 | 0.73 | 1.04 |
| 3 | 0.1087 | 0.463 | 1.07 | 1.94 | 3.07 | 4.38 |
| 4 | 0.288 | 1.24 | 2.87 | 5.22 | 8.25 | 11.78 |
| 5 | 0.64 | 2.72 | 6.29 | 11.43 | 18.07 | 25.74 |
| 6 | 1.20 | 5.16 | 11.88 | 21.47 | 34.14 | 48.69 |
| 7 | 2.06 | 8.85 | 20.43 | 37.06 | 58.61 | 83.53 |
| 8 | 3.26 | 14.12 | 32.54 | 59.08 | 93.46 | 133.32 |
| 9 | 4.96 | 21.32 | 49.09 | 89.31 | 141.24 | 201.33 |
| 10 | 7.13 | 30.81 | 71.06 | 129.00 | 204.07 | 291.10 |

NOTE.—In this table the mill is assumed to be operating at the most efficient speed, as given in Table 17, and the charge is assumed to weigh 325 lb. per cu. foot.

### CONSIDERATION OF A DEFINITE CASE

The actual application of the formulas to a definite problem may be instructive. Consider an 8 by 6-ft. cylindrical mill charged with 28,000 lb. (12,700 kg.) of steel balls. If the balls are made of steel weighing 500 lb. per cu. ft., and are all of one size, it may be shown that the charge will weigh 74.05 per cent. of 500 lb., or 370.25 lb. per cu. ft. The charge will then occupy 28,000 ÷ 370.25 = 75.6 cu. ft. (2.14 cu. m.) of space in the mill. The factor 75.05 per cent. is derived on the assumption that the spheres are equal; in the case of a mill, the space being limited and the balls not of one size, 65 per cent. is probably more nearly correct. Using this factor, the charge would weigh 325 lb. per cu. ft., and would therefore occupy 86 cu. ft. (2.23 cu. m.). The volume of an 8 by 6-ft. mill is 301 cu. ft. (8.5 cu. m.), hence the charge occupies 28.6 per cent. of the total volume. From formula (20), when the charge occupies 28.6 per cent. of the volume of the mill, the factor $K$ is 0.770. The speed of the mill, by formula (13), is 21.75 r.p.m., and the power, by formula (22), is 181.5 hp. Angle $\alpha$ will then be given by formula (15):

$$\cos \alpha_1 = \frac{0.8165}{\sqrt{1 + K^2}} = \frac{0.8165}{\sqrt{1 + 0.762^2}} = 0.6490; \; \alpha_1 = 49° \, 32'$$

$$\cos \alpha_2 = K \cos \alpha_1 = 0.4948; \; \alpha_2 = 60° \, 20'$$

$$r_2 = K r_1 = 0.762 \times 4 = 3.048 \text{ ft.}$$

$$\beta = 3\alpha_1 - 90° = 58° \, 36'; \; \beta_2 = 3\alpha_2 - 90° = 91°.$$

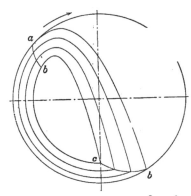

Fig. 7.—Paths of travel of particles in an 8 by 6 ft. ball-mill. 28,000-lb. ball load occupying 28.6 per cent. of mill volume. Speed of mill 21.75 r.p.m. 15,860 lb. of balls are traveling in the circular path. 12,140 lb. of balls are traveling in the parabolic path.

From these results and formula (2), which gives other values of $r$ for various values of $\alpha$, it is possible to plot the outline of the charge, as is shown in Fig. 7. The particular path of any ball can be plotted by use of equation (3a).

## Summary of Ball-mill Mechanics

### Notation

$r$ = radius to any particle $p$.
$r_1$ = radius of mill.
$r_2$ = inner radius of charge.
$R$ = radius of gyration of charge near $ao$.
$\alpha$ = angle from vertical to $r$.
$\alpha_1$ = angle from vertical to $r_1$.]
$\alpha_2$ = angle from vertical to $r_2$.
$\alpha_R$ = angle from vertical to $R$.
$\beta$ = angle from horizontal to $r$.
$\beta_1$ = angle from horizontal to $r_1$.
$\beta_2$ = angle from horizontal to $r_2$.
$n$ = speed of mill in rev. per sec.
$N$ = speed of mill in rev. per min.
$N^1$ = critical speed of mill in rev. per min.
$V_b$ = relative velocity of particle at $od$, ft. per sec.
$w$ = weight of portion of charge, lb.
$W$ = weight of entire charge, lb.
$P$ = fraction of mill volume occupied by charge.
$g$ = a constant = 32.2 ft. per sec. per sec.

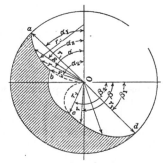

Fig. 8.

$k$ = a constant = $\dfrac{4\pi^2 n^2}{g}$ = $1.226n^2$.

$K$ = a constant = $\dfrac{r_2}{r_1}$.

$H$ = height of charge in mill at rest.
$E$ = kinetic energy, foot-pounds.
$C_n$ = cycles per revolution.
$rc$ = radius of circle $oa$.

### General Equations

(2) $\qquad \cos \alpha = kr = 1.226rn^2$

(2a) $\qquad rc = \dfrac{0.408}{n^2}$

(2b) $\qquad \cos \alpha_R = 0.8670 r_1 n^2 \sqrt{1 + K^2}$

(3) $\qquad N' = \dfrac{54.19}{\sqrt{r_1}}$

(3a) $\qquad y = x \tan \alpha - \dfrac{0.613 n^2 x^2}{\cos^4 \alpha}$ (equation of parabola)

(6) $\qquad \beta = 3\alpha - 90°$

(8) $\qquad V_b^2 = 16rg \cos \alpha \sin^4 \alpha$

(10) For $V_b$ max., $\alpha = 54° 44'$

(11a) $\qquad r_2 = Kr_1$

### Relations when Operating at Best Theoretical Efficiency]

(8a) $\qquad V_b$ max. = $\dfrac{7.88}{n}$

(8b) $\qquad V_b$ min. = $\sqrt{16Kr_1 g \cos \alpha_2 \sin^4 \alpha_2}$

(10) $\qquad \alpha_R = 54° 44'$

(11) $\qquad K = \sqrt{\dfrac{0.443}{r_1^2 n^4} - 1}$

(12) $\qquad n = \dfrac{0.8158}{\sqrt{r_1}\sqrt[4]{1 + K^2}}$

(13) $\qquad N = \dfrac{48.95}{\sqrt{r_1}\sqrt[4]{1 + K^2}}$

(14) $\qquad \cos \alpha = \dfrac{0.8165r}{r_1} \cdot \dfrac{1}{\sqrt{1 + K^2}}$

(15) $\qquad \cos \alpha_1 = \dfrac{0.8165}{\sqrt{1 + K^2}}$

(16)      $\cos \alpha_2 = \dfrac{0.8165K}{\sqrt{1 + K^2}} = K \cos \alpha_1$     (19) $C_n$ average $= 1.444$

(20)      $K = -0.024 + 0.39 \sqrt{7 - 10P}$  (very nearly)

(22)      $\text{Hp.} = Wr_1^{3/4} \left] 0.004467 \dfrac{(1 - K^3)}{(1 + K^2)^{3/2}} - 0.0037 \dfrac{(1 - K^5)}{(1 + K^2)^{3/4}} + \right.$

$$0.00088 \dfrac{(1 - K^7)}{(1 + K^2)^{5/4}} \right]$$

By use of these equations, with any given set of conditions, it is possible to determine the geometrical shape of the charge in the mill, the horsepower absorbed by it, the velocity of the blow struck by any ball, and the number of blows struck by any ball. The equations show also the relation that should exist between speed, diameter of mill and size of charge in order to secure the maximum theoretical efficiency.

## COMPARISON OF OBSERVED AND CALCULATED CURVES

In order to observe how closely theory and practice agree, a small model mill 3 in. (76.2 mm.) in diameter and 2 in. (50.8 mm.) long was made, having a bearing at only one end so that the other could be closed by a piece of glass through which the action of the charge could be observed. The method of comparison consisted in introducing a weighed charge of fine sand, computing the best operating speed, and drawing the outline of the charge according to the preceding theory. The mill was then operated at this speed, and photographed. The comparison between the photograph and the drawing shows how closely theory and practice agree. No accurate data as to the power required could be secured on this small model. A number of these photographs and the corresponding drawings are shown.

The similarity between the photograph of the mill in actual operation and the theoretical drawing is very striking, especially when the mill contains a large charge. When the charge is small, the difference appears greater; this is because all interference between the particles causes them to fall into the open space near the center of the mill. The photographs could not be made to show clearly the fact that these particles were accidental; it is apparent, however, when actually watching the mill run, that the particles in the central space are only occasional, as compared with the outer band. While the results of interference are more evident in the small charge than in the large one, the actual amount of interference is greater in the large charge than in the small. This is evident when the cause of this interference is considered.

The curve $a$–$b$ (Fig. 5) shows the boundary line between the circular and the parabolic paths. Each particle, as it passes the line $a$–$b$, starts on its parabolic path in a direction perpendicular to the radius of the

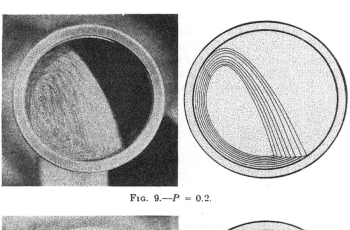

FIG. 9.—$P = 0.2$.

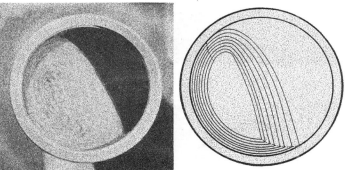

FIG. 10.—$P = 0.3$.

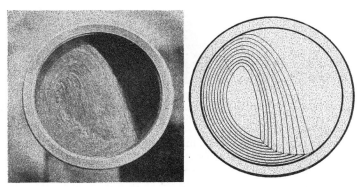

FIG. 11.—$P = 0.4$.

FIGS. 9–11.—SHOW RELATION BETWEEN ACTUAL AND THEORETICAL CURVES AT CORRECT MILL SPEEDS.

mill through that point. It is evident that the perpendicular to the
radius at $p$ would intersect the perpendicular to the radius at $a$ if it were
produced far enough. If the two particles considered are closer together
than $a$ and $p$ on the curve $a$–$b$, the intersection between the lines of initial
direction are closer to the points considered. If the two points are
adjacent on the curve $a$–$b$, it is apparent that the intersection will be

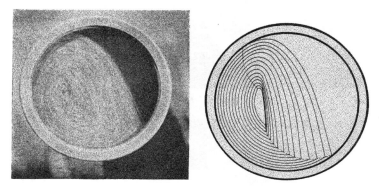

FIG. 12.—$P = 0.5$.

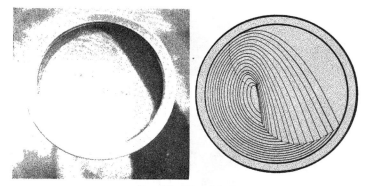

FIG. 13.—$P = 0.6$.

FIGS. 12–13.—SHOW RELATION BETWEEN ACTUAL AND THEORETICAL CURVES AT
CORRECT MILL SPEEDS

very close to the points and will, in fact, take the form of a slight crowding action between the particles. The result of this will be a slight
deformation of the curve through which the particle travels. This will
be just as prominent in a small charge as in a large one, although the
results will be more apparent in the small charge.

It may also be shown that when the size of the charge exceeds 0.4

the volume of the mill, there will be a tendency for the particles near the center of the mass to crowd one another. This is, however, not at all serious and could probably never be detected in operation, but as previously stated, when the mill is filled beyond 0.64 of its volume, the interference becomes quite important and the mill probably could not be made to operate efficiently when more than 0.6 full.

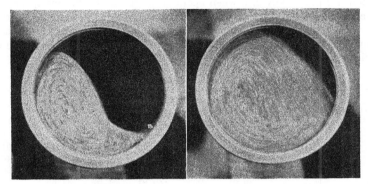

FIG. 14.—LOW SPEED.              FIG. 15.—HIGH SPEED.

FIG. 16.—HIGH SPEED.              FIG. 17.—HIGH SPEED
FIGS. 14–17.—MILL OPERATING AT INCORRECT SPEEDS.

It appears from the photographs that there is a certain amount of movement between the particles at the end of the parabolic path which causes the actual parabolic path to end sooner than the theoretical path. This motion between the particles is due to the fact that each particle must change its direction of travel and move at about right angles to its original path. This change in direction requires time and reacting forces; if the force is small the time will be long and the agitation great. If the force is large the time will be small and the agitation small. Therefore, when the mill speed is low, this zone of agitation is wide, as shown in Fig. 14. In this agitation zone grinding is done by attrition, but above it, the crushing is done by impact.

While these photographs and drawings were made to illustrate the action of the particles in the charge of a 3-in. mill, they also show the action in mills of any size when operated at the most efficient speed, as given by formula (13). In Table 17, the speeds have been computed for various sizes of mills and charges. In each case, the illustration that shows the position of the charge for the particular operating condition is indicated

TABLE 17.—*Speed of Mill for Best Theoretical Efficiency* (N)

| Inside Diam. of Mill, Ft. (2r₁) | Portion of Mill Volume Occupied by Charge at Rest | | | | | |
|---|---|---|---|---|---|---|
| | 0.1 | 0.2 | 0.3 | 0.4 | 0.5 | 0.6 |
| 1 | 59.19 | 60.45 | 61.82 | 63.36 | 65.10 | 67.01 |
| 2 | 41.88 | 42.75 | 43.72 | 44.81 | 46.04 | 47.40 |
| 3 | 34.19 | 34.90 | 35.69 | 36.58 | 37.58 | 38.69 |
| 4 | 29.61 | 30.22 | 30.91 | 31.67 | 32.55 | 33.51 |
| 5 | 26.48 | 27.04 | 27.64 | 28.34 | 29.11 | 29.96 |
| 6 | 24.18 | 24.69 | 25.24 | 25.88 | 26.58 | 27.36 |
| 7 | 22.38 | 22.85 | 23.36 | 23.94 | 24.60 | 25.32 |
| 8 | 20.94 | 21.38 | 21.86 | 22.40 | 23.02 | 23.70 |
| 9 | 19.74 | 20.15 | 20.60 | 21.13 | 21.70 | 22.34 |
| 10 | 18.72 | 19.12 | 19.56 | 20.04 | 20.59 | 21.20 |
| Illustrated by Figs....... | .. | 9 | 10 | 11 | 12 | 13 |

Table computed by use of Formulas (20) and (13).

TABLE 18.—*Speeds Corresponding to Given Charge Outline, R.p.m.*

| Inside Diam. of Mill, Ft. (2r₁) | Illustration Number | | | | |
|---|---|---|---|---|---|
| | 14 | 11 | 15 | 16 | 17 |
| 1 | 58.00 | 63.50 | 76.00 | 80.99 | 87.99 |
| 2 | 41.02 | 44.91 | 53.75 | 57.28 | 62.23 |
| 3 | 33.48 | 36.67 | 43.88 | 46.76 | 50.81 |
| 4 | 29.00 | 31.74 | 37.99 | 40.50 | 44.00 |
| 5 | 25.54 | 28.39 | 33.99 | 36.23 | 39.35 |
| 6 | 23.68 | 25.93 | 31.04 | 33.07 | 35.93 |
| 7 | 21.92 | 24.00 | 28.72 | 30.61 | 33.26 |
| 8 | 20.50 | 22.40 | 26.88 | 28.63 | 31.12 |
| 9 | 19.33 | 21.17 | 25.34 | 26.99 | 29.33 |
| 10 | 18.34 | 20.08 | 24.04 | 25.62 | 27.83 |

Portion of volume occupied by charge (P) is 0.4 in all caess.

In Figs. 14, 15, 16 and 17, the effects of high and low speeds are shown. In all of these photographs the mill was 0.4 full, the operation at the proper speed being shown in Fig. 11. Table 18 shows the speed at which the charge in mills of various diameters would appear as illustrated.

## CONCLUSIONS AS TO BALL-MILL OPERATION

In this whole discussion, only the force of gravity and centrifugal force have been considered. In a mill containing water and ore as well as balls, the force of adhesion is to be considered. This force not only tends to hold the balls and ore together but also tends to hold them against the lining of the mill. Just how important this force may be under ordinary operating conditions is hard to say. It was shown, however, in the little model mill, that adhesion tended to hold the particles together in their parabolic paths and almost entirely eliminated the accidental particles that fall near the center of the mass. It is impossible to apply the results secured in a small mill to a large mill in this respect, however, as adhesion varies inversely with the size of the particles considered.

Adhesion also varies with the moisture, so the mill operator has a convenient means of controlling this force so as to produce the best results. The amount of moisture in the pulp for best efficiency will vary with the nature of the ore, the nature of the ball charge, and the nature of the mill. With a mill charge of large steel balls, adhesion will not affect the operating conditions to any considerable extent unless the pulp is made very thick. The tendency would then be for the balls to stick to the lining and revolve with it.

It is very important to prevent slipping between the charge and the lining of the mill; the tendency to slip is much smaller with large charges than with small ones. If the friction between the charge and the lining of the mill is not great enough to carry the particles up to the curve $a$–$b$, the efficiency of the mill will be very greatly reduced and the lining of the mill will be rapidly worn away. Flat sides will also appear on the balls and the cycle of the charge will be slow and irregular. In an open-trunnion discharge mill, the pulp will not flow from the mill regularly but will come in pulsations. Lifters or roughened liners are therefore desirable, as they insure a greater coefficient of friction.

The following conclusions may be stated:

1. By use of formulas (20) and (13), the proper theoretical speed for the operation of the mill may be computed.

2. This speed is correct only when there is no slipping between the charge and the lining of the mill, and when the pulp is not so thick as to produce strong adhesion between the particles.

3. In actual practice it may be found that more effective crushing

can be done at some other speed, but it would appear that the variations above or below this theoretical speed should be small.

4. When operating at the speed shown by formula (13) the crushing is done largely by impact.

5. At speeds lower than $N$ (equation 13) the proportion of the crushing done by attrition is increased.

6. At speeds higher than $N$ (equation 13) the proportion of the crushing done by impact is increased.

7. The amount of crushing done depends upon the number of blows struck and the work done at each blow.

8. The number of blows struck can be increased by increasing the number and decreasing the size of the balls.

9. The work done at each blow can be increased by increasing the weight of the ball and by increasing the diameter of the mill.

10. From this it follows that mills of larger diameter should be charged with smaller balls and mills of smaller diameter with larger balls, when working on the same feed.

11. The proper operating speeds of ball-mills vary inversely as the square-root of the diameters.

12. The proper operating speed of a ball-mill increases as the size of the ball load increases.

13. Due to interference between the balls, the volume of the charge should not be over 60 per cent. of the volume of the mill.

14. Unless great care is exercised to prevent slippage, the volume of the ball charge should not be less than 20 per cent. of the volume of the mill.

15. It would then seem that a ball charge that occupies between 25 and 50 per cent. of the volume of the mill will give the most satisfactory results.

16. In an 8-ft. (2.4 m.) mill running at 22 r.p.m., with a 28,000-lb, (12,700 kg.) charge of 2-in. (50.8-mm.) balls, there will be an average of about 1,000,000 blows per minute; each of these blows will be equivalent to dropping a 2-in. ball 5 ft. (1.5 m.).

## BALL WEAR

In the previous discussion the fact was established that the work done by a ball when it strikes at the end of its parabolic path is proportional to its weight and velocity; then, since the velocity may be considered as constant for all the balls in the mill, the work done by a ball is proportional to its weight. Since the amount of ore crushed varies as the work done upon it, it seems reasonable that the amount of steel worn from the balls varies as the work done upon them; in other words, the ball wear is proportional to the work done. But it has been shown that the work done is proportional to the weight of the ball; hence, the wear is proportional to the weight of the ball, or

$$R = KW \qquad (23)$$

in which $R$ is the rate of wear of any ball of weight $W$; $K$ is a constant and depends upon the operating conditions of the mill and the resistance of the material from which the balls are made. In order to establish the accuracy of this equation, some transformations are necessary.

After a ball-mill has been in continuous and steady operation for, say, a year, during which time balls of only one diameter have been added at regular intervals, it may be assumed that the ball charge has reached a constant working condition. In this constant state, there is no change in the charge from day to day, either as to the weight of the charge or the average diameter of the balls composing it. Any particular ball enters the mill at a maximum diameter D, and gradually wears down until it is completely worn away; the residual charge of balls, however, always remains the same.

Suppose these balls composing the residual charge to be laid out in a line, beginning with the largest and ending with the smallest. Suppose intervals to be marked off along this line so that all of the balls of diameter from $D_m$ to $D_1$ are in the first interval, from $D_1$ to $D_2$ in the second interval, $D_2$ to $D_3$ in the third interval, and so on. Let $N_1$, $N_2$, $N_3$, etc., represent the number of balls in each interval, and let $D'_1$, $D'_2$, $D'_3$, etc., represent the mean diameters of the balls in their respective intervals.

In the constant state, the number of balls in each of these intervals does not vary, and when a new ball of diameter $D_m$ is added to the first interval, a ball of diameter $D_1$ passes into the second interval, and so on down the line, until in the last interval one ball is entirely worn away. Then, as time goes on, each ball of diameter $D_m$ which was added gradually passes along the line until its diameter becomes $D_1$ when it passes from the first interval into the second. If in time $T$, $N_w$ balls of diameter $D_m$ are added, then $N_w$ balls will pass from each interval into the next following interval. The time $T$ required for any ball to pass through any interval may then be expressed by the formula,*

$$T = \frac{N'_t}{N_w} \qquad (24)$$

or the time required for any ball to reduce its diameter from $D_a$ to $D_b$ is equal to the total number of balls in the interval divided by the number of balls added to the intervals, or passing from the interval during this time.

If, as previously suggested, the wear varies as the weight of the ball,

* The writer is indebted to Dr. F. H. MacDougall of the University of Minnesota for assistance in deriving these formulas.

then, $R = KW = K\frac{\pi}{6}SD^3$, in which $S$ is the weight of the material from which the balls are made and D is the diameter of the ball in question.

This formula shows the rate of wear at any instant. Since rate is always equal to the differential of space with respect to time, then $R = \frac{dw}{dt}$ in which $dw$ is the weight of material worn off in a very small interval of time, $dt$.

But $w = \frac{\pi S}{6} D^3$ and $dw = \frac{3\pi S}{6} D^2 dD$.

Then $\dfrac{\frac{3\pi S}{6} D^2 dD}{dt} = R = K\frac{\pi S}{6} D^3$,

or simplifying, $\frac{3dD}{dt} = KD$, or $\frac{3dD}{KD} = dt$.

Then $\int_{D_b}^{D_a} dt = \frac{3}{K} \int_{D_b}^{D_a} \frac{dD}{D}$, or $T = \frac{3}{K} \log_e \frac{D_a}{D_b} = \frac{6.9}{K} \log_{10} \frac{D_a}{D_b}$     (25)

In this formula, $T$ is the time required to reduce the diameter of the ball from $D_a$ to $D_b$.

Since     $T = \frac{N'_t}{N_w}$, $N'_t = TN_w = \frac{6.9N_w}{K} \log_{10} \frac{D_a}{D_b}$     (26)

This formula gives the total number of balls of diameter between $D_a$ and $D_b$.

The rate of wear is equal to the amount worn off divided by the time required; but the amount worn off in any interval is $\frac{1}{6}\pi S(D_a^3 - D_b^3)$.

Then $R = \dfrac{1/6\pi S(D_a^3 - D_b^3)}{T}$; or $K\frac{\pi}{6} S(D'_a)^3 = \dfrac{1/6\pi S(D_a^3 - D_b^3)}{\frac{N'_t}{N_w}}$;

or $K(D'_a)^3 = \dfrac{N_w (D_a^3 - D_b^3)}{N'_t}$; or $N'_t = \dfrac{N_w(D_a^3 - D_b^3)}{K(D'_a)^3}$

The total weight of the balls in any interval is equal to the number of balls in the interval times the mean weight of these balls,

or     $W'_t = N'_t \times \dfrac{(D'_a)^3 \pi S}{6}$;

then     $W'_t = \dfrac{N_w(D_a^3 - D_b^3)}{K(D'_a)^3} \times \dfrac{(D'_a)^3 \pi S}{6}$

or     $W'_t = \dfrac{\pi S N_w(D_a^3 - D_b^3)}{6K}$     (27)

But the sum of the weights of the balls in each section is the weight of the entire ball charge, or

$W_t = W_1 + W_2 + W_3 + W_4$, etc.

$= \dfrac{\pi S N_w}{6} \dfrac{(D_m^3 - D_1^3)}{K} + \dfrac{\pi S N_w(D_1^3 - D_2^3)}{6K}$, etc.

$= \dfrac{\pi S N_w}{6K} (D_m^3 - D_1^3 + D_1^3 - D_2^3, \text{etc.}) = \dfrac{\pi S N_w}{6K} (D_m^3)$     (28)

and
$$K = \frac{\pi S N_w D_m{}^3}{6W_t} = \frac{R_t}{W_t} \qquad (29)$$

Since
$$W'_t = \frac{\pi S N_w (D_a{}^3 - D_b{}^3)}{6K},$$

$$W'_t = \frac{W_t (D_a{}^3 - D_b{}^3)}{D_m{}^3} \qquad (30)$$

which is the weight of balls in one interval. The per cent. weight of balls in an interval is given by the formula,

$$\% \ W'_t = \frac{D_a{}^3 - D_b{}^3}{D_m{}^3} \times 100 \qquad (31)$$

The average weight of the balls of diameter between $D_a$ and $D_b$ is equal to the total weight of the balls divided by the total number of balls, or

$$W \text{ avg.} = \frac{W'_t}{N'_t} = \frac{\dfrac{W_t (D_a{}^3 - D_b{}^3)}{D_m{}^3}}{\dfrac{3N_w}{K} \log\epsilon \dfrac{D_a}{D_b}} = \frac{K W_t (D_a{}^3 - D_b{}^3)}{3 N_w D_m{}^3 \log\epsilon \dfrac{D_a}{D_b}} \qquad (32)$$

But since $W$ avg. $= (D'_a)^3 \dfrac{\pi S}{6}$,

then
$$D'_a = \sqrt[3]{\frac{2 K W_t (D_a{}^3 - D_b{}^3)}{D^3{}_M \ \pi S N_w \log\epsilon \dfrac{D_a}{D_b}}} = \sqrt[3]{\frac{1}{3} \frac{D_a{}^3 - D_b{}^3}{\log\epsilon \dfrac{D_a}{D_b}}} \qquad (33)$$

By use of formula (31) it is possible to compute the actual screen analysis of the balls in the mill. If the balls are removed from the mill when they reach a certain minimum diameter, $D_o$, formula (31) becomes,

$$\% \ W'_t = \frac{D_a{}^3 - D_b{}^3}{D_m{}^3 - D_o{}^3} \qquad (34)$$

in which $D_a$ and $D_b$ are the upper and lower limits of the diameter for any desired interval, $D_m$ is the diameter of the balls charged to the mill, and $D_o$ is the diameter below which the balls are removed from the mill.

It should be remembered, however, that these formulas (31) and (34) hold true only provided that the balls wear down at a rate proportional to their weight. It would seem that if the percentage weight computed by formulas (31) and (34) agreed reasonably closely with the actual results secured by carefully screening a ball charge after the mill had been in operation a sufficient length of time for the charge to become steady, and if this agreement could be secured in a number of cases under different conditions, it would be convincing evidence that the ball wear varies with the weight of the ball.

## DATA ON BALL CHARGES

In an endeavor to follow this plan, the attempt has been made to secure reliable data showing the screen analysis of ball charges which

have been in steady use for a long period. It seems to be difficult to
secure reliable information on this subject but the following results
seem to indicate the truth of this law of ball wear.

The screen analysis of the ball charge of an 8-ft. by 22-in. (2.4-m.
by 55.8-cm.) Hardinge conical mill shown in Table 19 was made at the

TABLE 19.—*Screen Analysis of Ball Load at Miami*

| Diameter, In. | Weight, Lb. | Actual Per Cent. Wt. | Computed Per Cent. Wt.* |
|---|---|---|---|
| 2.0 to 1.8.............................. | 4,080 | 27.46 | 27.10 |
| 1.8 to 1.6.............................. | 3,120 | 21.00 | 21.70 |
| 1.6 to 1.4.............................. | 2,630 | 17.70 | 16.90 |
| 1.4 to 1.2.............................. | 1,908 | 12.84 | 12.70 |
| 1.2 to 1.0.............................. | 1,344 | 9.04 | 9.10 |
| 1.0 to 0.8.............................. | 854 | 5.75 | 6.10 |
| 0.8 to 0.6.............................. | 500 | 3.36 | 3.70 |
| 0.6 to 0.4.............................. | 284 | 1.91 | 1.90 |
| Below 0.4............................. | 140 | 0.94 | 0.80 |
| Total........................... | 14,860 | 100.00 | 100.00 |

* Computed from equation, $\% \text{ Wt.} = \dfrac{D_a{}^3 - D_b{}^3}{D_m{}^3} \times 100$.

Miami Copper Co.'s plant after the mill had been operating for a year
with a ball load of 14,800 lb. (6713 kg.) which was maintained by the
addition of 400 lb. (181 kg.) of 2-in. (50.8-mm.) steel balls daily.

In this table, the actual per cent. weight obtained by weighing the
balls is compared with the theoretical per cent. weight computed by
use of formula (30). The two columns of figures are almost identical,
thereby showing the accuracy of the formula and the truth of the law of
ball wear.

At the Golden Cycle Mining and Reduction Co.'s plant a dry crushing
test was made on a 6 ft. 2 in. by 6-ft. (1.85 by 1.8-m.) Kominuter ball-
mill. The mill was operated for 694 hr., during which time 4825 lb.
(2188 kg.) of 5½-in. (139.7-mm.) balls were added. The original ball load
in the mill was 6614 lb. (3000 kg.) and the load at the end of the 694 hr.
was 6338 lb. (2874.8 kg.). During this time, 590 lb. (267.6 kg.) of balls
less than 3 in. (76.2 mm.) in diameter were discarded from the mill.
The screen analysis of the ball charge at the end of the operation is
shown in Table 20.

The slight irregularity in these results may be explained by the fact
that at one time during the test, after about 400 hr., twenty-two 5½-in.
balls were added at one time. This may explain the irregularities at
about 5-in. size.

TABLE 20.—*Screen Analysis of Ball Load at Golden Cycle Mill*

| Diameter, In. | Weight, Lb. | Actual Per Cent. Wt. | Computed Per Cent. Wt.[a] |
|---|---|---|---|
| 5.5 to 5.3 | 814 | 12.83 | 12.32 |
| 5.3 to 5.1 | 795 | 12.54 | 11.43 |
| 5.1 to 4.9 | 836 | 13.18 | 10.57 |
| 4.9 to 4.7 | 542 | 8.55 | 9.74 |
| 4.7 to 4.5 | 545 | 8.59 | 8.94 |
| 4.5 to 4.3 | 535 | 8.44 | 8.18 |
| 4.3 to 4.1 | 430 | 6.78 | 7.46 |
| 4.1 to 3.9 | 352 | ·5.55 | 6.76 |
| 3.9 to 3.7 | 268 | 4.23 | 6.10 |
| 3.7 to 3.5 | 340 | 5.36 | 5.48 |
| 3.5 to 3.3 | 333 | 5.25 | 4.89 |
| 3.3 to 3.1 | ·286 | 4.52 | 4.33 |
| 3.1 to 2.9 | 265 | 4.18 | 3.80 |
| Total | 6341 | 100.00 | 100.00 |

[a] Computed by formula, $\% \text{Wt.} = \dfrac{Da^3 - Db^2}{Dm^3 - Do^3}$.

From the two tests reported, the agreement between the actual per cent. weight and the computed per cent. weight is as close as could be expected. One was on a wet-crushing conical mill, and the other was on a dry-crushing cylindrical mill, and as far as the available data are concerned, the law of ball wear seems to be proved. It may develop however, when more data are collected, that the wear, instead of being proportional to the cube of the diameter, will be proportional to some slightly higher or lower power.

### CONCLUSIONS AS TO BALL WEAR

1. In any mill, the rate at which the weight of any ball decrease is directly proportional to its weight.

2. In any mill, the rate at which the diameter of a ball decreases is directly proportional to its diameter.

3. In any mill, the rate at which the surface of any ball decreases is proportional to its surface.

4. Since the rate at which a ball loses weight varies as the work done upon it in the mill, it follows that the work done in wearing (or crushing) the ball varies as the weight of the ball. This is seen to be *Kick's* law.

5. It then appears that *Kick's* law holds true for the ball wear in a rotating mill.

6. The natural tendency is for the small balls to accumulate in the mill charge.

TABLE 21.—*Screen Analysis of the Ball Charge Under Various Conditions*

Per Cent. Weight

| Charged at, Inches → Removed at, Inches: | 5 / 0 | 5 / 1 | 5 / 2 | 5 / 3 | 5 / 4 | 4 / 0 | 4 / 1 | 4 / 2 | 4 / 3 | 3 / 0 | 3 / 1 | 3 / 2 | 3 / 2.5 | 2 / 0 | 2 / 0.50 | 2 / 0.75 | 2 / 1.00 | 2 / 1.50 |
|---|---|---|---|---|---|---|---|---|---|---|---|---|---|---|---|---|---|---|
| **Size, Inches** | | | | | | | | | | | | | | | | | | |
| 5.00 to 4.75 | 14.26 | 14.38 | 15.24 | 18.19 | 29.23 | | | | | | | | | | | | | |
| 4.75 to 4.50 | 12.83 | 12.93 | 13.70 | 16.36 | 26.28 | | | | | | | | | | | | | |
| 4.50 to 4.25 | 11.49 | 11.58 | 12.27 | 14.65 | 23.54 | | | | | | | | | | | | | |
| 4.25 to 4.00 | 10.22 | 10.30 | 10.92 | 13.04 | 20.95 | | | | | | | | | | | | | |
| 4.00 to 3.75 | 9.02 | 9.09 | 9.64 | 11.51 | | 17.61 | 17.89 | 20.13 | 30.47 | | | | | | | | | |
| 3.75 to 3.50 | 7.88 | 7.94 | 8.42 | 10.05 | | 15.39 | 15.63 | 17.59 | 26.62 | | | | | | | | | |
| 3.50 to 3.25 | 6.84 | 6.90 | 7.31 | 8.73 | | 13.36 | 13.57 | 15.27 | 23.11 | | | | | | | | | |
| 3.25 to 3.00 | 5.86 | 5.91 | 6.26 | 7.47 | | 11.45 | 11.63 | 13.08 | 19.80 | | | | | | | | | |
| 3.00 to 2.75 | 4.96 | 5.00 | 5.30 | | | 9.69 | 9.84 | 11.07 | | 22.96 | 23.84 | 32.63 | | | | | | |
| 2.75 to 2.50 | 4.14 | 4.17 | 4.42 | | | 8.08 | 8.21 | 9.24 | | 19.15 | 19.89 | 27.22 | 54.52 | | | | | |
| 2.50 to 2.25 | 3.39 | 3.42 | 3.62 | | | 6.63 | 6.73 | 7.57 | | 15.70 | 16.30 | 22.31 | 45.48 | | | | | |
| 2.25 to 2.00 | 2.71 | 2.73 | 2.90 | | | 5.30 | 5.38 | 6.05 | | 12.56 | 13.04 | 17.84 | | | | | | |
| 2.00 to 1.75 | 2.11 | 2.13 | | | | 4.13 | 4.20 | | | 9.78 | 10.16 | | | 33.00 | 33.54 | 34.82 | 37.72 | 57.14 |
| 1.75 to 1.50 | 1.58 | 1.59 | | | | 3.09 | 3.14 | | | 7.33 | 7.61 | | | 24.75 | 25.16 | 26.12 | 28.29 | 42.86 |
| 1.50 to 1.25 | 1.14 | 1.16 | | | | 2.23 | 2.27 | | | 5.30 | 5.50 | | | 17.87 | 18.16 | 18.86 | 20.42 | |
| 1.25 to 1.00 | 0.76 | 0.77 | | | | 1.48 | 1.51 | | | 3.52 | 3.66 | | | 11.87 | 12.07 | 12.53 | 187 | |
| 1.00 to 0.75 | 0.46 | | | | | 0.91 | | | | 2.15 | | | | 7.25 | 7.38 | 7.67 | | |
| 0.75 to 0.50 | 0.24 | | | | | 0.45 | | | | 1.07 | | | | 3.63 | 3.69 | | | |
| 0.50 to 0.25 | 0.09 | | | | | 0.17 | | | | 0.41 | | | | 1.38 | | | | |
| 0.25 to 0.00 | 0.02 | | | | | 0.03 | | | | 0.07 | | | | 0.25 | | | | |
| | 100.00 | 100.00 | 100.00 | 100.00 | 100.00 | 100.00 | 100.00 | 100.00 | 100.00 | 100.00 | 100.00 | 100.00 | 100.00 | 100.00 | 100.00 | 100.00 | 100.00 | 100.00 |

7. Since these small balls do very little crushing, and exclude ore and larger balls from the mill, if allowed to accumulate too long, a marked decrease in crushing efficiency will result.

8. Since the large ball is just as likely to strike the small pieces of ore and the small ball is just as likely to strike the large pieces of ore as the reverse, it would seem that all of the balls should be of a size to crush any of the particles of ore.

9. This means that the balls should be as nearly as possible of a uniform size.

10. Since spheres of uniform size provide the greatest amount of interstitial space, a mill charge composed of balls of uniform size will allow freer migration of the ore particles than a charge containing balls of different sizes.

### RECAPITULATION OF BALL-WEAR FORMULAS

$N_w$ = number of balls added in time $T$ to compensate for the ball wear.

$N_t$ = total number of balls in the mill.

$N'_t$ = total number of balls in any interval.

$N_1$, $N_2$, $N_3$, etc. = number of balls in the first, second, third, etc., intervals.

$D$ = diameter of any ball under consideration.

$D_m$ = diameter of balls added to compensate for wear.

$D'_a$ = mean diameter of balls in any interval.

$D_a$ = diameter at beginning of interval.

$D_b$ = diameter at end of interval.

$D_1$, $D_2$, $D_3$, etc. = diameter of balls at end of first, second, third, etc., intervals.

$W_t$ = total weight of balls in the mill.

$W'_t$ = total weight of balls in any interval.

$w$ = weight of any ball.

$R$ = rate of ball wear.

$R_t$ = loss in weight of the mill charge in time $T$. It is equal to $\frac{\pi}{6} D_m{}^3 S N_w$.

$T$ = time required for any ball to pass through any interval.

$S$ = weight of material from which balls are made.

(23)     $R = KW$.   Rate of wear for any ball of weight $W$.

(25)     $T = \dfrac{6.9}{K} \log_{10} \dfrac{D_a}{D_b}$.   Time to wear from $D_a$ to $D_b$.

(26)     $N'_t = \dfrac{6.9N}{K} \log_{10} \dfrac{D_a}{D_b}$.   Number of balls between $D_a$ and $D_b$.

(29)     $K = \dfrac{\pi S N_w}{6 W_t} D_m{}^3$.   A constant for any ball charge.

(30)     $W'_t = \dfrac{W_t (D_a{}^3 - D_b{}^3)}{D_m{}^3}$.   Weight of balls between $D_a$ and $D_b$, when all balls are allowed to wear out in the mill.

(31)     $\%W'_t = \dfrac{D_a{}^3 - D_b{}^3}{D_m{}^3} \times 100$.

$$(32) \quad W \text{ avg.} = \frac{\pi s(D_a{}^3 - D_b{}^3)}{41.4 \log_{10} \dfrac{D_a}{D_b}}.$$ Mean weight of balls between $D_a$ and $D_b$.

$$(33) \quad D'_a = \sqrt[3]{\frac{D_a{}^3 - D_b{}^3}{6.9 \log_{10} \dfrac{D_a}{D_b}}}.$$ Mean diameter of balls between $D_a$ and $D_b$.

$$(34) \quad \%W'_t = \frac{D_a{}^3 - D_b{}^3}{D_m{}^3 - D_o{}^3}.$$ Per cent. of total ball charge between $D_a$ and $D_b$ when all balls smaller than $D_o$ are removed from the mill.

### PRACTICAL APPLICATION OF THEORETICAL CONCLUSIONS

The question finally arises as to how fine-crushing practice can be improved by the application of any of the principles that have been set forth. It is felt that the chief benefit to the mill operator will be derived from the fact that he may be better acquainted with exactly what is going on inside the ball-mill under various conditions. He should have a mental picture of the action of the charge and know better how to correct the difficulties encountered. He should also have a better idea how to proceed in order to produce any desired result.

While it is possible mathematically, as has been shown, to calculate the proper mill speed for any definite volume of charge, the size of the balls to be used must be determined experimentally. The size of balls is a most important factor in crushing, and each different condition requires balls of a different size. Whether the ore is hard or soft, coarse or fine, does not affect the proper mill speed or the volume of the charge; these depend almost entirely upon the size, and possibly to some small extent on the characteristics of the mill. But the proper size of the balls can be determined only by a careful study of the existing conditions. Experimental data must here be resorted to and the following method is recommended as a good means for determining the exact size of the balls that should be used.

Charge the mill with large balls, say of 5-in. (127 mm.) diameter. A smaller size might be better, but balls should be used that are known to be too large. An ammeter or wattmeter should be connected in the circuit of the driving motor so that the operator may keep the ball load constant by observing the power required by the mill. For maintaining this constant ball load, only balls of the same diameter as are already in the mill should be added. That is, if the test is started with all 5-in. balls, at the end of 24 hr. these balls will all be, say 4¾ in.; the ball load should then be restored to its original weight by adding only 4¾-in. balls. Thus, each time balls are added a different size must be used.

In this manner the mill will be filled with balls of approximately uniform diameter at all times. Then by keeping the records of each day's run, it will be possible for the operator to determine just which size produced the best results under the conditions at hand. The ball charge should then be composed of balls as near this size as is practical and economical.

There are two methods of determining the proper size of the balls to be added at the end of each 24 hr. One method is to sample the ball charge and actually measure the balls. In some types of mills a sampler can be inserted through the discharge trunnion. If this is not possible the size of the balls can be computed, but samples should be obtained at certain intervals in order to check the computations. The method of computing the ball size comes directly from the ball-wear formula and is as follows:

First determine the ball wear for the previous 24 hr. This may be done by obtaining a rough calibration of the power meter in the motor circuit so that any definite decrease in power indicates a given decrease in the weight of the charge. This is, then, the ball wear in pounds per day.

Then in ball-wear formula (25), $T = \dfrac{6.9}{K} \mathrm{Log}_{10} \dfrac{D_a}{D_b}$; but from (29), $K = \dfrac{R_t}{W_t}$. Then $T = \dfrac{6.9 W_t}{R_t} \mathrm{Log}_{10} \dfrac{D_a}{D_b}$. $T$ is 1 day, $W_t$ is the original weight of the ball charge, and $R_t$ is the ball wear for one day. Then $\mathrm{Log}_{10} \dfrac{D_a}{D_b} = \dfrac{R_t}{6.9 W_t}$. In this formula, $R_t$, $W_t$ and $D_a$ are all known, and it is only necessary to solve for $D_b$, the diameter of the balls to be added. If only $R_t$ lb. of these balls are added, then any error in computing $R_t$ will not accumulate but will be corrected on the following day.

The above method for determining the proper size of the balls will, of course, require the careful attention of some one outside the ordinary mill crew. It also calls for a large assortment of balls of various diameters. As compensation for the trouble and expense necessary for the proper carrying out of this experiment, the operator stands a good chance of increasing the ball-mill capacity by a very considerable amount.

Once the proper size of balls is determined, the charge should be maintained so that it is composed of balls as little larger and as little smaller than the average diameter ball as is possible. Just how closely the proper ball charge can be maintained depends upon the facilities and economic conditions at the plant. The removal of the small balls, which is the main difficulty, is not so serious as it may seem at first; proper equipment makes this easy and inexpensive. It is to be hoped that the makers of ball-mills will succeed in producing a mill that will automatically discard balls of any desired diameter as rapidly as they are formed.

Another point which will bear investigation is the classifier capacity. As has been pointed out, large circulating loads seem essential for best efficiency. Classification is cheap compared with fine crushing and a classification capacity in excess of the capacity of the ball-mill is very much to be desired.

Acknowledgment is made for permission to publish data to the managers of the Mesabi Syndicate and allied interests under whose direction this work was done; to B. B. Gottsberger, General Manager of the Miami Copper Co., Miami, Ariz.; to A. L. Blomfield, General Manager of the Golden Cycle Mining and Reduction Co., Colorado Springs, Colo., who furnished valuable data; and to W. G. Swart, Fred A. Jordan and T. B. Counselman, at Duluth, who assisted in the experimental work and in the compilation of results.

## DISCUSSION

A. L. BLOMFIELD,* Colorado Springs, Colo. (written discussion†).— I congratulate the author on bringing out a paper of real service to the profession. His contention of uniform size in balls is borne out by my own experience; in coarse crushing at the Golden Cycle in 6 by 6-ft. (1.8 by 1.8-m.) ball-mills we unquestionably gain by screening out the small balls once or twice a month. The point of gaging the ball size to be used by the uniformity of screen sizing is of particular interest.

The necessity of sufficient return feed from classifier to the ball-mill is sound and clearly shown. In connection with this, I wish two further points had been gone into as thoroughly: (1) The effect of the quality of classification. In general, it is true that the smaller the per cent. of undersize in the feed the more effective is the mill's work. This is true on a bucking board, in tube-mills, ball-mills, and grinding pans. Given the possibility of returning a full-feed load to any mill, it has been my experience that the effective work in the grinder is almost proportional to the quality of the return feed. (2) Again speaking generally, the shorter the tube-mill the greater is the quantity of return feed necessary to keep it loaded. The classification thus keeps the oversize in the mill more free from finished product and thus the crushing more efficient. It is very easy to overload any long mill with too much return feed. The author's tests were made on an 8-ft. by 22-in. Hardinge. This type is capable of handling very large return feeds. I would like to hear Mr. Davis' views on the best length of mill, given the diameter.

Dealing with feeds to fine grinders: At the mill of the Great Fingall Cons. M. Co., in 1906, we found that the 5-ft. grinding pan gave the greatest tonnage in closed circuit, grinding a feed − 8+20 mesh to − 20 mesh when the feed was slightly over twice the effective work done, and that the effective work was almost directly proportional to the efficiency of classification.

* Manager, Mill Dept., Golden Cycle Mining and Reduction Co.
† Received Feb. 19, 1919.

I note the large number of classifiers in his flow sheet and agree that it is clearly demonstrated that sufficient classifier capacity should be installed, as they cost very little to run. Two 6-ft. classifiers to the ball-mill could probably be replaced by one 12-ft. quadruplex machine, though possibly the only gain would be a saving in first cost.

H. A. WHITE, P. O. Dersley, Transvaal (written discussion*).— The fact that the author makes no reference to the work of others on the same subject, I think, must be attributed to the neglect of the usual distinction between ball and tube mills, which is founded upon the presence of special lifting devices in the ball mill. The paper really deals with tube-mill theory, upon which quite a lot of work has been done. The most accessible reference will probably be Richards' "Ore Dressing," (Vol. III, p. 1336), but I will specially refer to my own papers in the *Journal* of the Chemical, Metallurgical & Mining Society of South Africa (Vol. V, p. 290 and Vol. XV, p. 176), where very much the same ground is covered.

It may be interesting to compare the results obtained by an independent handling of the same problem. The critical speed obtained is very slightly different owing to the variation in the value of $g$, which, of course, is not everywhere the same. But our author has omitted to note the fact that his "diameter of mill" must be reduced by the average diameter of balls used in order to make the theoretical particle correspond with the actual balls and, of course, only the inside diameter of the lined tube is meant.

The directions for drawing the parabolic curve of flight might be very much simplified by using the relation between the angle of impact and the angle of departure, which I discovered in 1905 and the author has proved on p. 269. The author notes that the locus of points of departure is a semicircle but does not observe that the locus of points of impact is a trisectrix. The maximum of $x^1$, when only one layer of balls is considered (the outside layer), is obviously $r$ and the angle of departure is 30° from the vertical.

In order to calculate the maximum value of the blow struck, the author uses the second order effects beside the principal one due to the height of fall of the balls from the vertex of the parabola. The author thus derives a value of 0.5775 for cos $a$, this value is identical with that found for the maximum height of fall, as might have been anticipated, but the derivation of the latter is very much simpler. He does not notice that the path in this case is through the center of the mill.

The author deviates from exact methods and uses the idea of center of gyration in place of summation of various layers of balls and no doubt the approximation is sufficiently close if $K$ were independently known;

* Received June 2, 1919.

but to ascertain $K$ the assumption is made that the average of the whole charge passes through the same cycle as the particle at the radius of gyration, and as the inner layers have a much shorter cyclic time than the outer and there are also many fewer balls, this assumption is much less accurate.   In fact, the author's figure of average number of cycles per revolution (1.444) may vary from more than 1.5 to less than 1.2, in accordance with the speed and loading of the tube.   However, if correction is made of tube diameter, as previously mentioned, the final results are not very far out.

Of importance is the question of maximum efficiency to be got out of any tube and the load for that efficiency; but though the author gives the best speed for any load he does not indicate which load and speed give maximum efficiency.   A further important question in actual running plants is the maximum capacity that may be gotten out of any tube, and as the efficiency is not widely affected this latter maximum may be the most important.   These points may be well illustrated from the Witwatersrand practice with standard tube mills 5 ft. 6 in. by 22 ft.   The effective diameter inside linings, allowing 2 in. for pebbles used is put at 59 in. and the speed for maximum efficiency is 27.96 r.p.m. with a load about 45 per cent., while the maximum capacity is attained at 32.2 r.p.m. with a load 3 in. above the center.

These theoretical deductions and some others of interest were confirmed by observation of a working model of 7 ft. diameter, the full account of which will be found in the papers mentioned.   I have to thank the author for the exceedingly clear manner in which his paper is arranged and for the able paragraphs on ball wear which should be of great practical use.

## Problems Involved in Concentration and Utilization of Domestic Low-grade Manganese Ore*

BY EDMUND NEWTON,† E. M., NEW YORK, N. Y.

(New York Meeting, February, 1919)

### INTRODUCTION

THE steel industry of the United States has depended in the past almost wholly upon imports for its supplies of manganese. Many of the important domestic sources yield ores leaner in their natural condition than the foreign ores the steel industry has been accustomed to use. To make them available, therefore, either the ores must be concentrated or the practice of the steel industry modified.

Roughly 25,000 T. annually of high-grade manganese ores are used for dry batteries, for chemical purposes, and in other ways; while approximately 750,000 T. are required for making steel.

By present practice, every ton of steel takes an average of about 14 lb. (1.8 kg.) of metallic manganese. This is generally added to the steel in the form of an alloy, the standard alloys being the 80-per cent. ferromanganese and the 20-per cent. spiegeleisen. During the year 1917, 286,000 T. of ferromanganese and 193,291 T. of spiegeleisen were made in this country, the former largely from imported ores; and 45,381 T. of ferromanganese was imported. The metallic manganese represented by these alloys was 304,000 T., the product of roughly 800,000 T. of high-grade ore and 345,000 T. of low-grade ore.

### Manganese Deposits in the United States

Before the war manganese ore was mined in relatively small quantities in the Appalachian region, including Virginia, Tennessee and Georgia, and in Arkansas, but owing to the increase in prices during the past three years manganese mining has been undertaken also in Montana, California, Arizona, New Mexico, Nevada, Utah, and Minnesota.

From data now available it appears that in this country deposits of high-grade manganese ores are usually small, while materials lower in manganese and higher in iron occur in appreciable quantities. Our total quantity of manganese-bearing material is relatively large, but on

* Published by permission of the Director of the U. S. Bureau of Mines.

† Formerly U. S. Bureau of Mines, Washington, D. C.

account of the difficulty of mining small deposits and the apparent undesirability of the low-manganese material in the steel industry many difficulties present themselves.

Manganese-bearing materials of the United States may be roughly classified as follows: (1) manganese ore proper, (2) manganiferous iron ore, (3) miscellaneous material: (a) manganiferous silver and lead ore, (b) zinc residuum from manganiferous zinc ore.

Manganese ore as now defined by the trade is material containing over 35 per cent. manganese and suitable for the manufacture of 70 per cent. ferromanganese. Manganiferous iron ore contains less manganese and more iron. Usually iron predominates but there is no hard and fast line of demarcation between manganese ore and manganiferous iron ore. Manganese and iron are usually closely associated in nature, and all gradations from very low manganese ores with high iron, to high manganese ores with low iron, may be met in various deposits or in the same deposit.

Manganiferous silver ore is similar to manganiferous iron ore, there being sufficient silver to make it valuable for that metal. It may be used for the production of manganese alloys and commercial considerations alone control the balance between the manganese or the silver value.

Zinc residuum is a byproduct from the smelting of certain zinc ores of Franklin Furnace, N. J., which contain considerable manganese. After the zinc is removed the remaining product, called residuum, is of nearly the same composition as natural manganiferous iron ore, and for a number of years has been used for making spiegeleisen.

### Metallurgical Requirements of the Steel Industry

Manganese is principally used in the steel industry in the form of manganese alloys. A less important use is for increasing the manganese content of certain pig irons to give them particular qualities, as for foundry purposes, or to assist in metallurgical operations of certain steel-making processes, as in the basic open-hearth process. The alloys of manganese generally used in this country are ferromanganese, formerly containing 80 per cent., but now 70 per cent. metallic manganese, and spiegeleisen, containing from 15 to 20 per cent. metallic manganese. The remaining content of these alloys is principally iron with small quantities of carbon, silicon and phosphorus.

For the past few years ferromanganese has been gaining in popularity with steel manufacturers, spiegeleisen declining proportionately. Until recently approximately nine-tenths of the metallic manganese used in the steel industry was in the form of the standard 80-per cent. alloy. "Ferro," as it is usually called, is easier to use than alloys containing smaller quantities of manganese, as the required quantity of that metal is contained in smaller bulk.

The difficulty of obtaining ores suitable for the production of "ferro" during the past few years has led to the consideration of using what may be called intermediate alloys with manganese contents varying between 20 and 80 per cent. In the electric furnace certain alloys can be made with a relatively large content of silicon in addition to the manganese and iron. The extent to which such alloys may satisfactorily be used in steel manufacture is not alone a technical or economic problem, but is largely controlled by the prejudice of steel masters against deviating from established practice.

Phosphorus is an undesirable element in finished steel. In the manufacture of manganese alloys all the phosphorus contained in the ore will enter the alloy and will be introduced into the steel when the alloy is added. It is permissible, however, for an alloy high in manganese to contain more phosphorus than one low in manganese, for in the former case less alloy is needed to introduce a given quantity of manganese.

For many years manganese alloys have been principally manufactured in the blast furnace, although recently certain plants have produced them in the electric furnace. The operation of a blast furnace on manganese alloys is in general similar to ordinary pig-iron practice, but there are high metal losses, principally in the slag and in the stack. The manganese content of the slag may be partly controlled by furnace manipulation, but it is evident that the total loss in this manner is directly proportional to the slag volume.

The results of increased slag volume are cumulative and serious. The accompanying increased loss of manganese requires that more ore be used per ton of alloy produced. The additional ore introduces more slag-forming constituents, requiring more coke to melt it, which in turn tends to produce more slag, while increased slag volume cuts down the daily output of alloy. The greater manganese loss decreases the ratio of manganese to iron in the alloy and unless proper allowance is made the alloy will be below the standard grade and therefore subject to penalty by the purchaser. Not only will the alloy sell for less, but the decreased daily output will lessen the total profits.

The alloy manufacturer endeavors to protect himself against these decreased profits by adjusting schedules for ore purchase. Although the endeavor is made to equalize the effects of poor ores in furnace practice, the alloy producer would prefer to buy better ores and pay correspondingly more, whenever they are available.

### Concentration of Domestic Low-grade Manganese Ores

The comprehensive term, concentration, as here used, is intended to include the improvement of low-grade material by any suitable means preliminary to smelting. The requirements of metallurgical practice control the classification of manganese materials into low-grade and high-grade.  In some cases, the term low-grade may refer to a low manganese content with respect to iron or to large quantities of non-metallic impurities.  The detrimental effect on metallurgical practice and the resulting penalties are incentives for attempts to improve the material or raise the grade before smelting.

### Factors Controlling the Possibilities of Concentration

In order to properly interpret the possibility of commercially concentrating any type of manganese-bearing material, it is necessary to consider many technical and economic factors.  For a particular property, district, or class of material it is necessary to obtain data on the following factors: (1) character and size of the deposits; (2) conditions affecting mining and marketing; (3) character of ore material as affecting the possible improvement of grade; (4) metallurgical value of crude ore and possible concentrate; (5) commercial considerations.

### Size and Character of Deposit

Obviously a deposit containing a large quantity of low-grade ore would warrant considerable experimental work in order to determine methods of treatment.  Conversely, if a particular type of material occurred in only one deposit and contained but a few thousand tons, it is evident that the value of the product to the industry at large would be relatively small, even if it were possible to concentrate it.  Therefore proper perspective should be obtained in order that no undue proportion of time be devoted to a concentrating problem which may be of considerable technical and individual interest but which would assist little toward furnishing any considerable portion of the industry's needs.

In other words, the mineralized mass must be of such size and character as to justify the expenditure of money in its development and beneficiation and return interest on the investment proportional to the risk taken.  This factor is of vital importance and it is feared that, under the stimulus of production incident to national needs during the war, sound business principles have at times been overlooked.

### Conditions Affecting Mining and Marketing

In addition to the classification of deposits on the basis of quantity and character, it is necessary to determine the natural factors controlling mining methods, transportation facilities and marketing.

The manganese deposits of the United States, while widely scattered and comparatively small, may nevertheless be mined by relatively simple and therefore cheap methods.  The mines are for the most part shallow, so that extensive, non-productive development and elaborate equipment are not necessary. Intricate problems of ventilation and drainage have not to 'be solved, and if all operations are competently directed, common mine labor will generally suffice. Limited tonnage means short life, and temporary support of excavations following more or less crude mining practice prevails. The cost of mining, however, will be more or less governed by the necessity of selective mining which in turn is determined by the variability in character of the ore, the feasibility of economic concentration, the transportation facilities and the distance from a consuming center. All these factors must be properly coördinated and their combined influence studied before intensive production from individual properties is started. By the elimination of waste, concentration may yield a product desired by the steel industry, but the cost may be prohibitive. The reduction of weight resulting from waste discard may enable the producer to offset excessive freight rates, but geographic isolation will invariably handicap an enterprise. Foreign ores will always find a market in the United States since the deposits from which they come are larger and more uniform in character, while the wage scale is low and railroad transportation can never compete with ocean freight.

## CHARACTERISTICS OF ORE AFFECTING BENEFICIATION

### Character of Manganese Minerals

There are a great number of minerals containing manganese, but relatively few that are commercially important. Usually it is rather difficult to identify accurately the manganese minerals contained in domestic oxidized ore. Several minerals may occur in more or less intimate association, and in some cases one has been formed by alteration of another. The hardness of the individual minerals varies widely. Pyrolusite is soft and may be readily pulverized between the fingers. Difficulty might be expected, therefore, in attempts to recover this mineral by the common processes of wet concentration. The other minerals are harder but usually brittle. While the character of the individual minerals is important, the association of the several manganese minerals with the gangue materials and the relation thereto often has a more important bearing upon the problem of concentration.

### Impurities Associated with Manganese Minerals

Manganese ore mined on a commercial scale always contains impurities. The presence of some of these will be obvious by simple inspection, while others may require chemical analysis for their determination. The impurities associated with manganese minerals may be classified as: (1) those derived from associated rocks or rocks partially replaced by manganese-bearing solutions, (2) those associated with the manganese in solution and deposited simultaneously, and (3) those chemically combined with manganese in the mineral. From the metallurgical viewpoint all are impurities and must be removed either before metallurgical treatment or by it.

For convenience, the common impurities in manganese ores may be classified according to certain general physical and chemical principles as follows:

1. Metallic: Iron, lead, zinc, silver, and in some cases, nickel, copper and tungsten.

2. Gangue: "Basic" lime, magnesia, baryta; "acid" silica and alumina.

3. Volatile: Water (atmospheric moisture and molecular water), carbon dioxide, organic matter.

4. Miscellaneous: Phosphorus and sulfur.

The chemical behavior of these impurities affects metallurgical operations, while the physical form in which they occur controls the possibility of removal previous to smelting, and the choice of methods of removal.

The proportion of manganese to useless or harmful constituents of the ore determines its value and desirability. The presence of appreciable quantities of any impurity means that more ore must be mined and smelted to produce a given weight of alloy. Some impurities, however, are more detrimental than others.

Metallic impurities, of which iron is the most common, will usually be reduced and retained by the alloy. The quantity present controls the desirability of the alloy for use in steel manufacture. Metallic impurities other than iron occur usually in such small quantities that they are not detrimental to the resulting alloy. Zinc is an exception; it is largely volatilized in the smelting and, if present in appreciable quantities, the fume will condense as oxide in the hot-blast stoves, which may interfere with furnace operations. Unless the furnace top gases are washed the stoves must frequently be cleaned, with consequent loss of time. When the price of zinc is high, the zinc oxide recovered from the stoves yields a substantial sum. Silver is neither detrimental to manganese alloys, from the standpoint of steel manufacture, nor advantageous. The silver content of a manganese alloy has no value, consequently no credit is allowed the miner for silver contained in an ore when it is to be used for manganese-alloy manufacture. In some cases the quantity of silver in a magniferous ore is such that it has greater value for the lead smelter. The manganese then acts as a flux.

The gangue impurities, classed above as basic and acid, may also be called slag-forming impurities. In smelting these impurities must be fluxed to form slag. Slag is usually considered a waste product of a smelting operation, but it has important metallurgical functions, and just sufficient slag must be present to perform those functions properly and economically. An excess of slag must be avoided. In manganese-alloy manufacture the slag contains more or less manganese which does not enter the alloy. The quantity of manganese thereby lost is dependent upon the basicity of the slag, the temperature, and the slag volume. The first two factors control the quantity of manganese in a given weight of slag, while it is obvious that a greater slag volume will result in a greater loss of manganese.

Silica is usually the predominating slag-forming constituent in domestic manganese ores. A certain quantity of silica is reduced to the metallic state in the smelting operation and is recovered in the alloy as silicon, but the larger part must be fluxed with lime, magnesia or other bases to form slag. Manganese-alloy slags should be basic, hence a larger quantity of slag will be produced from an ore with acid gangue than in normal iron blast-furnace practice. Alumina is a slag-forming constituent, and while usually classed with silica, it acts somewhat differently in blast-furnace operations. Brazilian ores are notably high in alumina but most domestic ores contain relatively small quantities.

Lime, magnesia and baryta in an ore are also slag-forming constituents, but they combine with the silica and alumina present and thereby reduce the quantities of these bases necessary in the form of limestone or dolomite for the furnace charge. Baryta is not common as a gangue mineral. It is not so strong a base as either lime or magnesia. While these constituents offset the metallurgical effects of silica or alumina, from the standpoint of evaluating an ore they represent weight; and if the ore must be transported a considerable distance to be smelted, it is doubtful whether their value as bases will equal the additional freight charge. Limestone can generally be obtained at low cost close to the smelter.

Volatile impurities are removed from the top of a blast furnace largely by the surplus heat. It is desirable, however, in order to reduce the loss of manganese from this cause to keep the top of a manganese-alloy blast furnace cool. Volatile compounds are not particularly detrimental to smelting. When carbonate ores are being treated the case is somewhat different, some metallurgists claiming that in treating rhodochrosite ores the ratio of carbon monoxide to dioxide in the furnace gases is disturbed, which has a detrimental effect on the reduction of the oxides of manganese in the upper part of the furnace. It has also been suggested that the carbon dioxide driven off combines with carbon of the coke, forming carbon monoxide in the upper part of the furnace, and thus increases the coke consumption. Definite data are not available on these points.

From the practical standpoint all the phosphorus in the ore mixture, together with that contained in the coke and limestone, is recovered in the resulting alloy. The permissible quantity of phosphorus in an alloy, such that it does not produce detrimental effects when added to steel, has not been definitely determined. The higher the manganese content of an alloy, the larger is the quantity of phosphorus that may safely be contained. Ordinarily steel makers desire as large a margin of safety as possible and therefore have specified that phosphorus shall not be above a certain percentage in an ore.

Sulfur usually exists in relatively small quantities in oxidized manganese ores, but in the case of the primary rhodochrosite ores of Butte. and other parts of the West there may be present considerable quantities of sulfides of iron and zinc. Sulfur is not a serious factor, as the conditions of blast-furnace operation, when making manganese alloys, are such that sulfur combines with manganese or lime and is readily retained by the slag, only traces entering the alloy.

Knowing the effect of impurities in manganese ores on blast-furnace practice, the methods of eliminating them may be considered. Ore-dressing deals with the separation of deleterious or useless materials from the more valuable minerals, thereby raising the grade and reducing the quantity of the concentrated product. To accomplish this it is essential that the physical and chemical characteristics of the ore be determined. These factors are governed largely by the type of deposit from which the ore is mined. As types of ore, entirely disregarding genesis, we recognize:

1. Rhodochrosite and rhodonite; carbonate and silicate ores, deposited in fissure veins or replacing original rocks.

2. Nodular ores: accretions of manganese oxide in soft plastic clays.

3. Manganese oxides deposited in small fissures or fracture planes, as breccia fillings, or as more or less impure beds.

4. Manganese oxides occurring as infiltrations, deposited in minute pore spaces, as particle replacements, or otherwise intimately mixed with the rock or gangue.

In the first class of deposits the principal gangue impurity is silica, although sulfides of silver, lead, zinc and iron are often found in appreciable quantities. The silica occurs both as quartz and chemically combined in rhodonite ores. In the carbonate ore, the carbon dioxide may be removed by calcination, thus effecting a concentration, but rhodochrosite decrepitates strongly when heated to a temperature where the oxide is formed, tending to produce an excessive quantity of fine material which is undesirable in practice. The breaking up of the particles by calcination will isolate some of the free silica, which on account of its larger sized particles may be screened out. The sulfide minerals may occur in such quantity that it is desirable to remove them by gravity methods of separation.

In the second class of deposits the nodules are of variable size and usually high in manganese. They do not appear to be contaminated internally with the inclosing material. The clays are soft, whereas the nodules are generally hard. This type is common in the Appalachian region of the United States. The clay may be separated from the manganese nodules by means of log washers, followed, where necessary and where the size of the deposits warrants the installation, by picking-belts, crushers, screens, and jigs.

With deposits of the third class the manganese minerals, although closely associated with the inclosing rock, are not generally contaminated by it and may be relatively pure. The method of treatment will vary with the size of the manganese particles and the hardness of the rock but will not differ essentially from the treatment of the second class. If there is little or no clay, the log washer will be omitted, while crushing, screening, jigging, and possibly tabling will make up the concentrating process.

It should be noted that if the mineral is largely pyrolusite, and therefore friable and soft, crushing may produce an excessive quantity of fine material, composed of manganese minerals which are difficult to recover by means of gravity or water methods of separation. If, however, the manganese mineral is of a hard, dense, massive variety, and the inclosing rock more friable, the problem is simpler. When the specific gravities of the minerals and the gangue approximate each other wet concentration is difficult unless there be a marked difference in the size of particles.

From the standpoint of concentration it is obvious that the association of gangue materials with the desired mineral in ores of the fourth class is so intimate that the finest crushing imaginable would not permit of separation by mechanical means. To this type the siliceous manganese ores of the Western States may be assigned. Ore-dressing experimentation has conclusively shown that where the silica is chemically combined with the manganese or where colloidal silica envelopes the manganiferous particles, wet-process or gravity concentration will not give the desired results.

### Concentration of Manganese Ores

It is not within the scope of this paper to describe in any detail actual ore-dressing or concentration practice. It is axiomatic that small deposits or mines of questionable life do not warrant elaborate plans or the adoption of intricate beneficiation processes. A general classification of methods applicable to the manganese industry is given below. These are all preliminary to the greater and final concentration of the desirable elements in the blast furnace from which the ferro-alloy is produced.

*Simple Methods of Concentration*

Selective mining.                    Water classification.
Hand picking.                        Roughing table treatment.
Jigging.                             Slime table or vanner treatment.
Screening.                           Pneumatic separation.
Log washing.                         Combination of two or more of the above.

*Complex Methods*

Magnetic separation:
   1. Without preliminary thermal treatment.
   2. With preliminary thermal treatment.
Electrostatic separation.
Hydrometallurgical processes:
   1. Leaching with various acids, precipitation by chemical substances.
   2. Leaching with various acids, precipitation by electrolysis.
   3. Leaching with various acids, evaporation of solution and heat treatment
      in rotary kiln.

Preliminary thermal processes:
   1. Drying, to remove hygroscopic moisture.
   2. Calcining, to remove carbon dioxide or combined water.
   3. Agglomerating fine concentrates to make them desirable for blast-furnace
      use.
   4. Volatilizing manganese at high temperatures in the presence of certain con-
      stituents which form readily volatile compounds.
   5. Direct reduction of oxides by carbon, under temperature control.
Miscellaneous processes:
   1. Flotation.
   2. Use of heavy solutions.

There are many standard machines for the concentration of ores
based on certain principles or combinations of principles, but it is unwise
and usually unsatisfactory to begin with the idea that a certain machine
will accomplish the necessary result on manganese-bearing materials.
As the character of the manganese materials varies greatly in different
districts, it is more logical to determine first the detailed physical and
chemical characteristics of the material. When such preliminary study
has shown the nature of the impurity and its relationship to the manganese mineral, it is easier to outline the general methods of treatment
which might reasonably be expected to accomplish the desired result.
The flow sheet, however, must be determined by experimental work.

*Commercial Considerations*

If the technical possibilities of beneficiating an ore have been favorably
determined, it is then necessary to ascertain whether such an operation
could be conducted on a commercial scale and a reasonable profit made.
The cost of the plant and its installation must be justified either by the
available ore, or upon the length of time during which the profit could be

made. The amortization of capital and interest on the investment must be included in the estimation of cost.

The effect of concentrating an ore is not always clearly appreciated. Concentration implies that an improvement of metallic content is made by the intentional elimination of impurities, but by so doing there is always a loss of the valuable mineral itself. When the grade of material is increased, the weight is decreased. In other words, in some cases from 2 to 25 T. of crude material may be necessary to produce 1 T. of high-grade concentrate. The income results from the sale of the smaller quantity of concentrate, but chargeable against this, will be the mining cost of the several tons of crude ore necessary to make that concentrate, the actual cost of treating the ore, the freight to market, and special overhead charges. Concentration may be necessary, however, to make the material marketable at all.

A detailed study of each project is necessary to determine the cost of mining, preparation of the ore, selling price of both crude ore and concentrates on the existing schedules, and the possible resulting profit or loss in either case.

## DISCUSSION

C. W. Goodale,* Butte, Mont.—I notice Mr. Newton refers very briefly to the carbonate ores of manganese, rhodochrosite, but he does not go into any special description of the treatment of that material. In some of the veins in Butte, Mont., there is a large amount of rhodochrosite, and a large tonnage of that material has been shipped to the East for making ferromanganese. At the works of the Anaconda company at Great Falls they treated several thousand tons of this manganese ore, running about 37 to 38 per cent. manganese, in electric furnaces, giving a product of about 78 per cent. of manganese. But just at that time the demand for ferromanganese from the West was discontinued and the material, about 1000 tons of 78 per cent. manganese, is on hand at the works at Great Falls. With the higher-grade manganese ore, running, we will say, 37 or 38 per cent., there are large bodies of ore perhaps 20 or 25 per cent., but containing too much silica to be valuable in their present condition. Some efforts have been made for water concentration of that material, but owing to there being only a difference of about 1 in the specific gravity of the quartz gangue and the manganese mineral, water concentration has offered some difficulties. I have been told, however, that some very satisfactory experiments have been made with magnetic concentration of this material. I think it is a new idea that this manganese ore can be treated successfully by magnetic concentration.

* Manager, B. & M. Dept., Anaconda Copper Mining Co.

KIRBY THOMAS, New York, N. Y.—I want to point out the possibilities in the business of manganese mining which have been made clear by the war. Before the war the manganese production in this country was from 3000 to 5000 tons, practically insignificant; although it had been larger some years before. However, under the stimulus of the war, we have reached a production of about 300,000 tons of manganese ore and at even the pre-war prices that represents a sum that is important even in our vast mineral production. With the termination of the war and the withdrawal of the stimulus of high prices through government encouragement, the manganese mining industry is apparently about to collapse and relapse into its former insignificant position. The war has shown us that there are possibilities in manganese mining in the United States. If the production of 3000 tons annually could be increased to 300,000 tons in 3 years, certainly there is the raw material, the base of an important industry, existing in this country. It must be admitted that it exists mostly under unfavorable conditions, as compared to the competing deposits of Brazil, India, and Russia. The raw material is here, and the engineering talent and the mining industry of this country should display the ability and the courage to tackle that problem. Cannot the engineers turn their special technical skill to realizing this national resource, notwithstanding the relatively unfavorable conditions under which it exists? It looks as if this whole business of manganese mining is likely to lapse until we have another war, or until there is a stimulus of protective tariff, etc., which shall throw us back on our own resources as regards manganese. The problem resolved itself not into one of mining but into a question of metallurgy, and this paper by Mr. Newton touches upon the metallurgical points, but very generally; it does not give enough definite results of what has been accomplished or what may be accomplished.

In the South the treatment of manganese ores was entirely crude, untechnical, unpractical, up to the last 2 or 3 years, but Mr. Newton and others connected with the Bureau of Mines have worked out a number of very important things in regard to the treatment of those southern manganese ores. They have studied and diagnosed the old log washer, which was hardly supposed to be a metallurgical instrument, and have found that the washers should be of a certain length and have a certain revolution and particular shaped blades. All that work is important. There has also been work done with regard to the chemical separation of manganese out of the silicate and carbonate ores. This is a problem that should be followed up.

In Maine there is a deposit of silicate ore, a whole mountain of it, that runs 22 per cent. manganese. It is located favorably and if someone can work out a process of getting the manganese in the form of steel or chemical manganese, it will add a large value to the country's mineral production. The same is true of the rhodochrosite ores of the West and the high-silica ores of Arizona and the Southwest. I urge that further consideration be given to this subject, with a view to continuing the investigation of processes of realizing the potential wealth that lies in the manganese deposits of this country. We are tackling a difficult problem and one in which the competition of the foreign supplies is against us, but it is a patriotic and selfish duty to try to develop our own industry and, through public or private stimulation, the work should be continued.

## Distribution of Coal Under U. S. Fuel Administration

BY J. D. A. MORROW,* WASHINGTON, D. C.

(New York Meeting, February, 1919)

THIS discussion relates to the distribution of coal under the direction of the U. S. Fuel Administration beginning Apr. 1, 1918. At that time a definite method of controlling and directing distribution was put into effect. Prior thereto, although some of the machinery utilized was in operation, especially in the few weeks immediately preceding Apr. 1, distribution had not been effectively and comprehensively organized.

The writer was called to the Fuel Administration by Dr. Garfield about the first of February, 1918, to assume general direction of the distribution of coal for the Fuel Administration, to develop the general plans, and to obtain practical men to assist in this work and in carrying through the program finally devised. A study of the situation and conferences with the Fuel Administrator and leading men in the business made it plain that it would be necessary to begin at the bottom, measure the undertaking, develop plans to meet the need thus shown, and build an organization to effect the distribution required. This work had to be done under war pressure in the midst of a bad coal shortage.

The successful performance of such a task is always primarily a matter of putting competent men in the right places and giving them the power and opportunity to function properly together. In the Distribution Division of the U. S. Fuel Administration, the men who were chiefly responsible for whipping plans into practical form and then for carrying out these plans were A. W. Calloway, of Baltimore, President of the Davis Coal & Coke Co., who served as the Director of Bituminous Coal Distribution; S. Lovell Yerkes, of Birmingham, Alabama, Secretary of the Grider Coal Sales Agency, who assumed general charge of distribution while the new plans and organization were being perfected and then became the Assistant Director of Bituminous Coal Distribution; Warren S. Blauvelt, of Detroit, who was Director of Coke Distribution; A. S. Learoyd of New York, Director of Anthracite Distribution, in connection with the Anthracite Committee, consisting of Joseph P. Dickson, President of Dickson & Eddy, Chairman; S. D. Warriner, President of the Lehigh Coal & Navigation Co., and W. H. Richards,

---

* General Director of Distribution.

President of the Philadelphia & Reading Coal & Iron Co.; A. M. Ogle, of Terre Haute, President of the Vandalia Coal Co., Director of State Distribution; and C. E. Lesher, Coal Statistician of the U. S. Geological Survey, who was kindly loaned by Director Smith to become Director of Statistics for our Distribution Division.

One of the first requisites in obtaining tangible results was to know how much coal would be required during the year, where it would be needed, and for what purposes. Without this information, it was impossible to make any definite plans for supplying the war needs of the nation. The first estimates of the Bureau of Statistics called for 736,000,000 short tons of coal. Subsequently, through known conservation of coal and through effective control of distribution, which made it possible to lower some estimates of necessary winter reserves, the figure was reduced to 723,400,000 tons. Of this amount 100,000,000 short tons was the estimated maximum possible output of anthracite, and the balance was the tonnage of bituminous coal required to fill all remaining coal needs.

The bituminous tonnage was classified to show the requirements for railroads, exports, ship bunkers, domestic consumers, public utilities, and industrial concerns. These latter were still further subdivided to give the requirements of particular industries, such as byproduct coke plants, iron and steel plants, etc. The tonnage needed for domestic consumers, public utilities, and industrial concerns was estimated by states, but that for railroads, exports, and bunkers was not so divided because in the distribution of coal for these purposes state lines were ignored. Distribution for the other consumers was placed on a state basis to insure equality of treatment among consumers in different parts of the country.

Anthracite requirements for domestic use were estimated with even greater care. The Anthracite Committee ascertained the shipments of domestic anthracite, in 1916, into some 22,000 communities. On the basis of these shipments, estimates of the needs of these communities for 1918 were prepared by the State Fuel Administrators and the Anthracite Committee and definite allotments of anthracite were made by states and communities. In order to supply the estimated needs of the northern and eastern states in which war activities had led to increases of population in many localities, it was necessary to withdraw anthracite from the western and southern states entirely and to decrease the shipments into various middle western states and parts of Canada.

### Comparison of Schedules and Shipments

With the requirements thus known, schedules were set up covering the movement of coal from the mines to the various consuming areas week by week throughout the year, allowing for differences in transportation conditions in summer and winter and making provision for sufficient winter reserves in the more remote consuming districts. These schedules were worked out in coöperation with the U. S. Railroad Administration and the shipments of coal were constantly checked against them. The more important movements, such as shipments of Navy and transport bunker coal were under daily check.

To show how these schedules were observed and how the movement approached and eventually surpassed the figures scheduled as production increased and as war activity stopped with the signing of the armistice, let me say that on July 6, for example, the rail movement of bituminous coal to New England totaled 3,058,000 tons since Apr. 1, against a schedule calling for 3,150,000 to this date, or 98 per cent. performance. On Sept. 28, rail shipments totaled 6,164,000 tons against a schedule of 5,849,000, or 105.4 per cent. performance. On Dec. 21, rail shipments totaled 7,763,000 tons against a scheduled total of 7,459-000, or 104.1 per cent. performance. On July 1, scheduled shipments to tidewater at ports from Hampton Roads north were 11,916,000 tons; actual shipments were 11,557,000, or 97 per cent. of the schedule. On Sept. 1, we had scheduled for those ports 19,860,000 tons and actual shipments totaled 20,013,000, or 100.8 per cent. performance. On Oct. 21, the figures were 23,963,000 tons scheduled and 23,843,000 shipped. On Dec. 21, thanks to the armistice, the shipments had run 2,688,000 tons, or 9 per cent., ahead of the schedule. A total of 28,000,000 tons was scheduled to go up the Great Lakes for the north-western states and Canada; we sent 28,153,000 tons. With similar precision and certainty, munition factories, arsenals, powder works, byproduct plants, etc., were kept running while stocks were accumulated insuring uninterrupted operation throughout the winter. In the same manner, retail dealers were given supplies for their domestic trade. Such results were only possible because of the complete control of shipments and the full information on which to proceed.

### SUPERVISION OF DISTRIBUTION

The bituminous producing fields were divided into 28 districts, with a District Representative in charge of the operators in each of these fields. These men were under the direct supervision of Messrs. Calloway and Yerkes. They had full information of the production, the obligations and shipments of producers in their respective territories, and executed the orders from Washington for the movement of coal from the mines to carry out the plans for the distribution of bituminous coal. A similar arrangement prevailed for coke. Each of these District Representatives obtained daily reports of the shipments from every mine under his jurisdiction. Each was a man in whom the producers had confidence, and they enjoyed the fullest and most loyal help and coöperation of the producers. Thus the coal operators themselves were organized into a part of the distribution machinery of the Fuel Administration.

Distribution among consumers in each state was supervised by a State Fuel Administrator, who had no interest in the coal business but whose duty it was to see that all consumers were treated impartially, and to call upon the District Representatives to assist any who might be in special need of coal. The distribution work of these state administrators was supervised, so far as was necessary, in Washington, by Mr. Ogle, the Director of State Distribution. He was constantly in close touch with the heads of the bituminous, anthracite and coke distribution bureaus.

### REPORTS AND PRIORITIES

To control and direct the movement of 2,000,000 tons of coal daily to 100,000,000 consumers, it was necessary to have not only current reports from the shippers at the 6000 mines to the District Representatives, but also weekly reports from some 90,000 industrial consumers and retail dealers. Duplicate reports went to each State Administrator. The Bureau of Statistics daily, under Mr. Lesher, checked, tabulated, and summarized reports from approximately 15,000 consumers and the information thus assembled went to the various executives who were supervising shipments.

To insure a distribution of coal that would contribute most directly to the nation's war program, particularly of the special grades of coal, it was necessary to have some degree of preference established among the various classes of consumers. That order of preference was fixed by Bernard M. Baruch, Chairman of the War Industries Board, working through the Priorities Board, of which Judge E. B. Parker was chairman. The entire distribution program of the Fuel Administration accorded with the preference as laid down by the War Industries Board. Consumers in Classes 1 and 2 were supplied first and were given ample stocks of coal before consumers in the lower classes were allowed to accumulate reserves.

### COÖPERATION WITH RAILROAD ADMINISTRATION

As a part of the effort to coöperate with the United States Railroad Administration most effectively, and as a practical means of insuring the movement of coal with a definite saving in transportation, a zone system of distribution from the mines to the consumers was applied in the central and southern part of the United States between the Rocky Mountains and the general line of Erie, Pittsburgh, and Baltimore. Such a system was under consideration by the Railroad Administration and the Fuel Administration prior to my connection with the latter. The plans for this system were pushed to completion and it was put into effect on Apr. 1.   The northeastern part of the country was not zoned because of the complexity of supplying that section; on the other hand, the western part of the country was not zoned because it divides itself geographically and so zones itself.   The zone system was not administered rigidly, but was modified as fluctuations in production required and as was necessary to permit the requisite movement of coals for special industrial uses, such as gas, byproducts, metallurgical, railroad fuel, etc.   The system saved 160,000,000 car miles in hauling coal from April to December and obtained a quicker return of empty cars to the mines.   It also helped to equalize the demand in different sections of the country, which, in turn, permitted the mines in all fields to run more nearly to capacity than ever before.

## Work of National Production Committee of U. S. Fuel Administration

BY JAMES B. NEALE,\* B. A., MINERSVILLE, PA.

(New York Meeting, February, 1919)

FROM the beginning of its activities, the members of the National Production Committee have felt that the following points were essential to the success of its work: The operators must feel that their operations were not 100 per cent. efficient; that they should be courteous and fair in their treatment of their employees; and that they should set an example of hard work and patriotic interest in increased production.

Every move made must appeal to the sense of fairness of all parties concerned.

The men must feel that in working more faithfully, and consequently producing more coal, they are rendering a distinct service to their Government in the time of its great need. As the workers feel that greater efficiency on their part results only in greater profits to the operators, the advantage to the operator should be lost sight of, as far as possible, and the advantage to the Government in fighting the war emphasized.

Both operators and workers should know that the Government considered increased production an obligation, and opportunity for service, resting on both parties and not on the operators or the workers alone, and that both parties could very much better the efficiency of the pre-war period. Had the committee intimated that it thought the coal shortage was due entirely to the idleness and inefficiency of the workers, its campaign would have failed, for the men would have resented the charge because only too frequently their efforts to produce more coal were thwarted by bad management; besides, the operators would have lacked stimulus to better effort. On the other hand, had the impression been given that the committee thought that the coal shortage was due entirely to bad management, the operators would have resented the charge because only too frequently their efforts to produce more coal were thwarted by the idleness and inefficiency of the workers; and the workers, having been thus indirectly pronounced 100 per cent. efficient and patriotic, would have lacked stimulus to better effort.

---

\* Director of Production, U. S. Fuel Administration.

### METHOD ADOPTED FOR INCREASING PRODUCTION

With these points in mind, the committee devised the following plan: A production manager was chosen for each of the twenty-eight large producing districts. For the most part these managers were nominated by the operators in each district and they became the centers of the production activities. At a large percentage of the mines production committees were formed, one at each mine. These committees consisted of six men—of whom three were appointed by the management and three were chosen by the workers—and worked under the direction of the production manager, who served as the umpire in case of division in the committee. The members of these committees were given certificates of appointment and badges and every effort was made to have them feel that their position was one of dignity and responsibility; that they represented the Government and were fortunate in having an opportunity to render special service to the Government during the war. Their duties were carefully defined in a letter which was broadly circulated. In the main these duties consisted in discussing ways to better the efficiency of the mine and its equipment and of the mine workers.

The members of the National Production Committee, in Washington, made every effort to protect mine labor from undue inroads by the selective draft, voluntary enlistment, munition plants, and other war activities. They looked carefully after the needs of the operators as to mine supplies and obtained proper priorities for such materials. They dealt with the railroads in reference to car supply and rendered valuable assistance to power plants selling electricity to the mines. Soldiers who had seen active service and could bring home to the workers the life in the trenches and their relation to it visited hundreds of mining camps and addressed the workmen. A great amount of material, such as posters, poems, etc., that it was thought would appeal to the men and arouse their patriotism, was distributed and personal letters were written to workers who made unusual records.

In many districts, the production managers established the daily tonnage necessary from each mine and had the local production committee accept it as the Government's requirement. This was one of the most stimulating features as it gave the men at the mine a definite goal and they felt that the standard was set by the Government and not by the operator, so the unwillingness to work hard for the operator's benefit was more or less lost. The net result was a very large increase in production.

During the five very active months the National Production Committee spent in Washington, the members came in close contact with hundreds of operators, miners' leaders, and mine workers. All were keenly interested in increased production and frankly expressed their views as to the causes that impeded production and as to the steps that should be taken to increase it. We discussed what were called methods for increased production, but which were really methods of increasing the harmony and coöperation between employer and employee, which is the true basis on which increased production must be built.

## NEED OF STIMULATING AMBITION IN WORKMEN

There is undoubtedly throughout the world, as is evidenced by the rise and spread of Bolshevism, a feeling that the good things have not been evenly divided; that some persons have had too much and many others too little. This is true in theory but it is difficult of correction. Doubtless the unequal distribution of wealth is due to the unequal distribution of worth, and until worth is equally distributed there can be no permanent equal distribution of wealth. It seems essential, therefore, to make an earnest persistent effort to emphasize the need of worth and to do everything possible to stimulate it. The Bolsheviki are trying to distribute wealth on the basis of might and disaster can be the only result of the effort. Is not the promotion of worth the proper counter-irritant to Bolshevism?

We, operators, as the pace-setters in the coal industry, must do our part. Our workmen must be, much more than ever, a matter of deep concern to us. We must plan well and unselfishly for their general well being and must use every effort to stimulate them to greater efficiency as to labor and to higher ideals as to the proper use of their lives. There must be created an ambition for more comfortable homes, and better food and clothing for themselves, their wives, and their children. The education of their children also must become one of their main desires. Now, their dominating desire is more leisure; fewer hours at work is their goal and increased wages help them to attain it, but at the expense of production. This ambition is disadvantageous both to the men and to the operators and the only way to relieve the conditions it produces is by creating new ambitions and aims. A man who needs $30 a week to live in his accustomed manner and who can earn $5 per day, will work six days a week. If his earning power is increased to $7.50 a day, he can earn the $30 a week necessary by working only four days, so there will be a tendency on his part to decrease his working days from six to four. This is uneconomical for all parties. By stimulating in the man higher aims and ideals, which will require more money for their satisfying, will not only increase his desire to work but will also tend to change his work into service. These aims and ideals will make him feel that his work is not merely dull toil but that it is an honorable means of enabling him to gratify his higher ambitions. The war has paved the way for this work on the part of the operators, for during the past 18 months, as never before, the work of the American workmen has been service in a great cause and now that the thought of work being service is in their minds, every effort should be made to intensify the thought and make it permanent. Surely war conditions cannot be a greater stimulant along this line than can peace conditions.

It is the duty of someone to promptly take this whole matter in hand; and if it is the duty of someone, surely that someone is the employer. He is the one who, by birth, environment, or some special virtue, is the one to whom this should appeal and he is the one best equipped for the leadership. The average workman is not just as we would want him to be, but that is no fault of his. He is the product of his opportunity. If we want him to be better, we must make for him a better opportunity. We get nowhere by complaining or by destructive criticism, but we can accomplish much by putting our shoulders under the wheel and lifting with a true desire to help for unselfish reasons.

If we are to assume this leadership we must first make firm our standing with the men along industrial lines. As yet the men do not feel that we actually owe them anything along the lines mentioned but they do feel that we have certain industrial obligations. These we must meet and satisfy the men that we are fair employers before we can expect to exert an influence along higher lines. The men now look upon us with suspicion. They feel that we are their enemies, that their best interests conflict with our best interests, and that we constantly try to take advantage of them. In days gone by, and to some extent today, many employers do try to take advantage of their workmen in an unfair way, so that the workmen are suspicious of all employers.

### Industrial Obligations of Employers to Men

The men feel that there are two industrial obligations on the part of the employer toward them. The first one is that the wages and working conditions, as defined in the scale contracts, must be met. We employers acknowledge this obligation, which is virtually a legal obligation and cannot be shaken off. Twenty-five years ago we did not acknowledge it and the workman who felt that he was paid less than a fair wage or less than the accustomed wage had scant chance to have his grievance heard, to say nothing of having it adjusted. In those days it was a curse and a blow from the boss to the worker. The worker took what was given him and if he was not satisfied he had to make the best of a bad bargain.

Twenty-five years have seen a great change along this line. Not only do the operators for the most part make an honest effort to abide by the scale agreements but they have set up courts of arbitration so that disputes under the agreements may be adjudicated by an impartial tribunal. The one fly left in the ointment, in this regard, is that very many men feel that the operator bitterly resents being taken before the board and that he will seek revenge against the man who brings the complaint. This feeling does away with much of the real virtue of arbitration. Every operator should make the men feel perfectly free to discuss their grievances with him and, failing to agree, to refer the matter to the board without any prejudice against the workman in days to come. If the operator is conducting his business fairly, he need have no fear of having his work reviewed by an impartial board.

The second industrial obligation that the employee feels should rest on the employer is that of making it possible for the employee to do a day's work when he reports for duty. This obligation the employer has not formally recognized in any way and it is the next step to be taken. We have recognized the first obligation and are very much better off for having done so. We must now formally recognize the second obligation and will be very much better off for doing that also. A new board should be set up or else the powers of the existing board should be broadened so that judgment may be passed as to whether or not the employer has failed in his duty of providing an opportunity to do a day's work. Before this board will come up questions as to whether or not the operator is to blame because a man is gassed out, because his place contains water, because there is no timber or cars, and various other obstructions to good working conditions. A man feeling aggrieved along this line should be encouraged to bring the matter to the attention of the board for the effect of having fair judgment passed would be valuable. No mine foreman wants to have an impartial board say that he is conducting the affairs in his mine in an inefficient way, so the very existence of such a board would be a tremendous stimulant to the mine management.

The board, also, should decide whether or not a workman failed in his duty by being absent from his work. There is just as much of a moral obligation on the employee to work, unless prevented by unavoidable causes, when there is an opportunity to do a day's work, as there is on the employer to furnish the opportunity. No good can result if the obligation is placed only on the employer. Both parties interested must share it. With these two obligations acknowledged, with means provided for adjudicating differences of opinion regarding them, and with the men feeling that they are welcome to bring their complaints, we will have our industrial skirts clean and the decks cleared for active leadership along lines looking toward higher ideals and better manhood.

## Obtaining an Equal Distribution of Worth

We all know the main influences that can be made powerful factors toward a more equal distribution of worth. We must be keenly interested in the home, the school, the church, and all social activities, in fact, in everything going on in the community that has an influence on the lives of the people.    Until now our thoughts have centered on material things, on mechanical devices, and along these lines we have made great progress; but we have paid little attention to the social, physical, and moral betterment of our employees.    For the most part we have not tried to see that the boys entering our employ each year were better than the boys who entered the year before.    We have given little thought to our workmen, excepting as workmen, and even then we have made no effort to teach them what we should.    We have complained bitterly of their actions many times and yet have done little or nothing to better their environment and opportunity and to stimulate higher ambitions and ideals.    We have engaged a young man as a fireboss, or a pitboss, and have said to ourselves "That young fellow is a comer.    I'll make a real man of him."    We should consider each one of our workmen a comer and try to make a man of him.    All of this may sound theoretical and impracticable, but it is not.    It cannot be done in a day or in a year, but it can and must be done if the good things of this world are to be gradually and fairly divided on the basis of worth, and are not to be suddenly and unfairly divided on the basis of might.    We, operators, are face to face with a condition that is exceedingly important and serious, and I believe we are strong enough to meet it in the proper way.

During the past 18 months the President of the United States, in his various addresses and messages, has set up a very high standard. He has given to the world high ideals and every effort must be made to live up to them.    The great mass of the people feel, in a hazy way perhaps, that peace is going to bring to the world better conditions; that there will be a more equal division of this world's goods; and that brotherhood will become much more a fact than before.    It will be a keen and dangerous disappointment if the President's ideals do not prevail in large measure.

## DISCUSSION

ROBERT PEELE,* New York, N. Y.—I should like to ask Mr. Neale how the members of the production committees at the individual mines, who came from the mine workers themselves, were chosen or appointed; also from what classes of employees they usually came?

J. B. NEALE.—In the union fields, they were chosen by a local.    In the non-union fields, they were chosen either by the men in mass meeting or by the company officials.    The last way was very unsatisfactory because it gave the men an opportunity to say that they were going to be passed upon as to whether they were slackers or not, by men chosen by the company.    When they were chosen by the locals or at mass meetings of the workers, the appointments have been very much more satisfactory.

---

* Professor of Mining, School of Mines, Columbia University.

R. D. HALL,* New York, N. Y.—The matter of inefficiency on the part of the operator is one that I think we may well take into account, because there is a very large inefficiency on the part of the coal operator; it has arisen from reasons almost beyond his control and will continue for these same reasons. It is no use arguing a moral responsibility on the part of the operator, when you make efficiency very largely a loss to the operator who puts it in force. If you ask a man working on a day wage to work with more efficiency he says, "Well, I may do it or I may not," it all depends how he feels, but he does not lose anything financially by not working hard. If you ask the man who is working by the ton to put out more tons, he makes a profit by putting out more tons. How is it with the operator? If he gives a man all the cars he needs and all the power he needs, if he runs after his working men and waits on them, he does not get his coal out any cheaper than before. He gives the man a better chance and there is more efficiency, but at the same time he has increased his expenses all along the line. At the end of the year he finds that his profits have been less than if he had had less efficiency. The efficiency is really against him in many ways. Because he gets a larger tonnage, he reduces the overhead to a certain extent; but his tonnage may not be any greater because he may work fewer days. As Mr. Taylor remarked some time ago, supposing he puts in a machine that is very much more effective than any he had before, he has the same scale; the miner gets out more coal, but he himself does not get any benefit out of it because he has a scale fixed, so much per ton.

So long as we have piece work of that kind, it will hardly pay the operator to go out of his way to help the working man; his interest is not with the working man, as it should be, and it never will be unless he is paid by the day. There is a great deal of efficiency in the transportation system in mines, in the dumping system, in the cleaning of the coal, and in all those various features, because it is on day labor; but when a foreman sees a miner doing nothing and learns that he is waiting for cars, he says, "They will be along in a short while." It does not worry the operator at all that that condition exists. He realizes that if he were to put on more drivers he would have some drivers waiting, and drivers are paid by the day and miners are paid by the ton, so he really feels it is more profitable for him to have fewer drivers and more miners. Not only that, but it has been found that when the miner is given cars whenever he needs them, he is liable to quit the mine early, with the result that in the afternoon things drag along and the work is very inefficiently done on the part of the day-hands; and it is their efficiency in which the operator is interested.

---

*Associate Editor, *Coal Age.*

So long as coal mining is conducted largely on the piece-work plan we will find a lack of efficiency on the part of the operator. There is no inducement to the operator for any efficiency that he may exhibit, and to my mind the only hope is that we will some time get a combined system in which the man will be paid partly by the day and partly by his work. When that is done, we will find that the operator will have an interest in reducing the number of men employed in his mine, and the miner will have the advantage, in that the more coal he produces the larger will be his income. But so long as the miner works entirely on the piece-work plan the operator will have no incentive. In fact, he will have an incentive entirely in the other direction, because the more efficient he is, the more he tries to give the miner a fair show, the more he will find his costs will mount up, and he cannot meet those costs under present conditions of competition because these costs do not bring him a return. I think that the whole trouble is the system on which the mines are worked, and that it will never be corrected until we have changed the system.

R. V. Norris, Wilkes-Barre, Pa.—I think some of the operators will wish to answer Mr. Hall. I do not agree at all with this proposition that the piece work of the miner tends to inefficiency; as a matter of fact, the piece work of the miner has been the custom in practically all coal-mining operations, for the reason that there is no means of supervising the miner's work and that piece work has been found to be the best method to encourage him and make it to his advantage to work. It is also an unquestioned fact that a very considerable part of the expense is overhead, which is reduced by increased tonnage, and I feel very strongly that the piece work of the miner is the best and most advantageous method for the mining work.

## Anthracite Mining Costs

BY R. V. NORRIS,* E. M., M. SC., WILKES-BARRE, PA.

(New York Meeting, February, 1919)

IT was stated in a former paper[1] that an intensive study of anthracite costs was being made by the engineers of the United States Fuel Administration. The results of this study are now available and are offered as a supplement to the former paper.

Anthracite costs as reported for the 6-months period, December, 1917, to May, 1918, inclusive, as compiled by the Federal Trade Commission, were used as a base, and charts similar to those described in the former paper were made for the standard white ash anthracite and for red ash and Lykens valley coals.

### ADJUSTMENTS

The adjustments of cost from a reported to a price-fixing basis, as described for the bituminous methods, were applied but showed only minor adjustments as necessary.

The great spread in anthracite prices on the varying sizes, which for the 6-mo. period under review ranged in average from $5.244 for nut to $2.074 for barley coal, makes the question of the percentage of sizes produced at the different collieries a vital one. The realization with the same prices for each size must be within very wide limits, when it is considered that the percentage of prepared coal reported from different collieries varied from over 80 per cent. to below 30 per cent. for fresh-mined coal. Hence, as the spread in prices for the various sizes must be predicated on some percentage, it is essential to find some method of adjustment to allow for this variation. The logical method of adjustment is to calculate actual costs to costs as of the standard percentage of sizes, so that the margin between the adjusted costs and the average realization shall be the actual margin for each colliery between its actual costs and actual realization due to its particular percentage of sizes. As a base for realization the actual percentage of sizes for fresh-mined coal for the 6-mo. period was adopted. This percentage is given in Table 1.

* Engineer, U. S. Fuel Administration.
[1] Cyrus Garnsey, Jr., R. V. Norris and J. H. Allport: Method of Fixing Prices of Bituminous Coal Adopted by the United States Fuel Administration. This volume.

TABLE 1.—*Percentage of Sizes of Fresh-mined Coal*

| Size of Coal | Mesh, in Inches | | | | Percentage of Sizes | | |
| --- | --- | --- | --- | --- | --- | --- | --- |
| | Through | | Over | | Fresh Mined | Washery | Fresh Mined and Washery |
| | Round | | Round | | | | |
| Broken............. | | 4½ | 3⅜ | 3¼ | 6.8 | 0.4 | 6.2 |
| Egg................ | 3⅜ | 3¼ | 2⁵⁄₁₆ | 2¼ | 14.6 | 1.2 | 13.5 |
| Stove.............. | 2⁵⁄₁₆ | 2¼ | 1⅝ | 1½ | 19.6 | 2.3 | 18.2 |
| Nut................ | 1⅝ | 1½ | | ¾ | 24.7 | 10.1 | 23.5 |
| Pea................ | | ¾ | | ½ | 9.1 | 10.0 | 9.2 |
| Buckwheat.......... | | ½ | ⁵⁄₁₆ | ¼ | 11.6 | 21.4 | 12.4 |
| Rice............... | ⁵⁄₁₆ | ¼ | ³⁄₁₆ | ⁵⁄₃₂ | 3.2 | 14.9 | 4.2 |
| Barley............. | ³⁄₁₆ | ⁵⁄₃₂ | ³⁄₃₂ | ¹⁄₁₆ | 4.9 | 27.5 | 6.8 |
| Boiler....·........ | ⁵⁄₁₆ | ¼ | ³⁄₃₂ | ¹⁄₁₆ | 3.9 | 8.8 | 4.3 |
| Screenings.... .... | ³⁄₃₂ | ¹⁄₁₆ | | | 1.6 | 3.4 | 1.7 |

For adjustment as a base for fixing a spread of prices the percentages used were, taken at even figures: prepared, 65 per cent.; pea, 9 per cent.; buckwheat, 12 per cent.; and smaller, 14 per cent.

The adjustment finally arrived at after long study was tested on actual reports from collieries having percentages that varied from over 80 per cent. to under 30 per cent. prepared coal and was found to be correct within a maximum variation of less than 1½ per cent. It was as follows:

| For Each 1 Per Cent. Variation | Above Standard, Per Cent. Deduction | Below Standard, Per Cent. Addition |
| --- | --- | --- |
| Prepared ............................... | 1.20 | 1.20 |
| Pea ................................ | 0.85 | 0.85 |
| Buckwheat.......... ..... ............ | 0.75 | 0.75 |
| Smaller.................. ............. | 0.50 | 0.50 |

As examples of the working of this adjustment with prices assumed at about the average for the 6 mo. and taking mines well away from average percentage of sizes.

| Size | Base Per Cent. Sizes | Base Price | Realiza- tion | Mine A, Per Cent. Sizes | Correc- tion, Per Cent. | Actual Realiza- tion | Mine B, Per Cent. Sizes | Correc- tion, Per Cent. | Realiza- tion |
| --- | --- | --- | --- | --- | --- | --- | --- | --- | --- |
| Prepared......... | 65 | $5.10 | $3.315 | 73.1 | −9.72 | $3.730 | 55.1 | +11.880 | $2.810 |
| Pea.............. | 9 | 3.70 | 0.333 | 6.4 | +2.21 | 0.237 | 15.3 | − 5.355 | 0.566 |
| Buckwheat....... | 12 | 3.20 | 0.384 | 10.4 | +1.20 | 0.333 | 13.7 | − 1.275 | 0.438 |
| Smaller.......... | 14 | 2.20 | 0.308 | 10.1 | +1.95 | 0.222 | 15.9 | − 0.950 | 0.350 |
| Totals......... | 100 | | $4.340 | 100.0 | −4.36 | $4.522 | 100.0 | + 4.30 | $4.164 |

| | | |
| --- | --- | --- |
| Assume cost for each mine ......................... | | $4.000 | $4.000 |
| Actual margin.................................... | | $0.522 | $0.164 |
| Standard realization............................. | | $4.340 | $4.340 |
| Calculated cost as of standard per cent. sizes | | | |
| $4.00 × 0.9564 % = | | $3.826 | $4 00 × 104.30 % = $4.172 |
| Calculated margin ............................... | | $0.514 | $0.168 |

The correction for Mine A is then −4.36 per cent. and the adjusted cost $3.826, showing 51.4 c. margin on the $4.34 standard realization against 52.2 c. actual margin. Similarly for Mine B, the correction is +4.30 per cent., giving an adjusted cost of $4.172 and a margin of 16.8 c. as compared with the actual margin of 16.4 c. Thus the adjusted costs on the chart bear a true relation to the realization received from a scale of prices for the various sizes based on the standard or average percentage of sizes adopted as a base, regardless of the actual percentage of sizes produced by each operation, and prices can be fixed from the chart line of adjusted costs which will result in giving each mine its intended margin. The correction, of course, is an allocation based on realization from the different sizes and could be made more accurately by taking into account each size produced, but at the cost of more time than was available for the work. With a material variation in price, different factors of correction should be calculated.

## ROYALTIES

A large percentage of the anthracite coal is owned in fee by operators, who also lease tracts contiguous to their fee holdings. As all report royalties on the basis of tonnage produced, the general average 15.5 c. per T. reported is misleading. The actual average royalty reported by operators mining generally from leased lands was 33.25 c., and by those generally mining from fee lands, 5.5 c. As relatively few operators mine exclusively from either class of lands, no data is available to show the actual average royalties paid, but it is believed that the present average would be approximately 40 c. per ton.

A few leases, notably those made by the trustees of the Girard Estate, owned by the City of Philadelphia, base the royalty payments on a percentage of the sale price of the coal at the mines instead of requiring fixed royalties; this percentage varies from 15 per cent. to as high as 28 per cent. of the price. As the labor war bonuses materially add to the sale price, these have resulted in excessive royalties and serious embarrassment to the operators, who were not allowed to increase the price of coal sufficiently to even fully absorb this additional labor cost and by whom the extra royalties must be paid out of already narrow margins. Many of the lessors have patriotically foregone the extra royalties due under these sliding scales, and as it is certainly improper that the public should be asked to pay these additional royalties, not contemplated in the original contract though enforceable at law, the U. S. Fuel Administrator has expressed himself as unwilling to consider such excess payments in any fixing of anthracite prices.

## COST CHARTS

Cost charts were made from averages of the 6 mo., showing both the reported and the adjusted costs for standard fresh-mined white ash anthracite, both by collieries and by operating companies. As, in the prices fixed by the President, Aug. 23, 1917, a differential of 75 c. per T. on pea size and above, equivalent to 52.95 c. per T. on all sizes, was established for the independent operators over certain companies with railroad affiliation, generally known as the "Companies." Charts were made by collieries for both the independent operators and the Companies, as well as a combined chart of the entire output. Similar charts were also made for the entire output of fresh-mined coal and for the total output

TABLE 2—*Average and Bulk Line Costs of White Ash Coal*

| Chart Number | Description | Costs, Averages Returned | Costs, Adjusted | Cost, 90 Per Cent. Bulk Line |
|---|---|---|---|---|
| | EXCLUDING WASHERY COAL | | | |
| 81 | All operations, each colliery separate... | $3.85 | $3.91 | $4.80 |
| 82 | All company operations, each colliery separate.......................... | 3.71 | 3.79 | 4.65 |
| 83 | All independent operations, each colliery separate.......................... | 4.37 | 4.36 | 4.97 |
| 86 | All operations, each company operating two or more collieries consolidated..... | 3.85 | 3.91 | 4.38 |
| | INCLUDING WASHERY COAL | | | |
| 91 | All operations, each company operating two or more collieries consolidated... | 3.57 | 3.77 | 4.36 |

TABLE 3.—*Average Prices Received for White Ash Coal*

| Size | Fresh Mined Coal | | Bank Coal | | Total Including Banks | |
|---|---|---|---|---|---|---|
| | Per Cent. | Average Price | Per Cent. | Average Price | Per Cent. | Average Price |
| Broken............... | 6.8 | $4.889 | 0.4 | $4.416 | 6.2 | $4.886 |
| Egg.................. | 14.6 | 5.028 | 1.2 | 4.815 | 13.5 | 5.027 |
| Stove................. | 19.6 | 5.161 | 2.3 | 5.060 | 18.2 | 5.160 |
| Nut................... | 24.7 | 5.244 | 10.1 | 5.246 | 23.5 | 5.244 |
| Pea................... | 9.1 | 3.687 | 10.0 | 3.696 | 9.2 | 3.698 |
| Total and weighted av'g prepared and pea...... | 74.8 | 4.959 | 24.0 | 4.544 | 70.6 | 4.947 |
| Buckwheat............ | 11.6 | 3.342 | 21.4 | 3.213 | 12.4 | 3.324 |
| Rice.................. | 3.2 | 2.482 | 14.9 | 2.452 | 4.2 | 2.473 |
| Barley................ | 4.9 | 2.231 | 27.5 | 1.767 | 6.8 | 2.074 |
| Boiler................ | 3.9 | 2.341 | 8.8 | 2.123 | 4.3 | 2.304 |
| Screenings............ | 1.6 | 2.202 | 3.4 | 1.555 | 1.7 | 2.162 |
| Total and weighted av'g small sizes........... | 25.2 | 2.795 | 76.0 | 2.339 | 29.4 | 2.697 |
| Grand total........... | 100.0 | 4.414 | 100.0 | 2.868 | 100.0 | 4.285 |

including washery coal by operating companies, all collieries operated by
a single company being consolidated. Similar charts were made for
red ash and for Lykens valley coal. These charts were interesting princi-

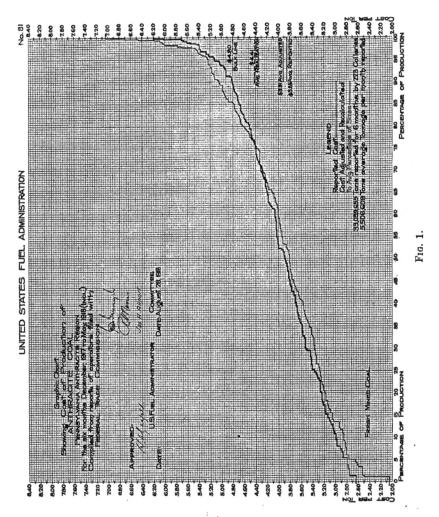

Fig. 1.

pally in that they showed that the higher prices charged for these special-
ties were justified by their higher costs of production. The charts given
are believed to fairly show the general costs of production.

These charts show the averages and bulk line costs for standard white ash coal given in Table 2. The average prices received are given in Table 3.

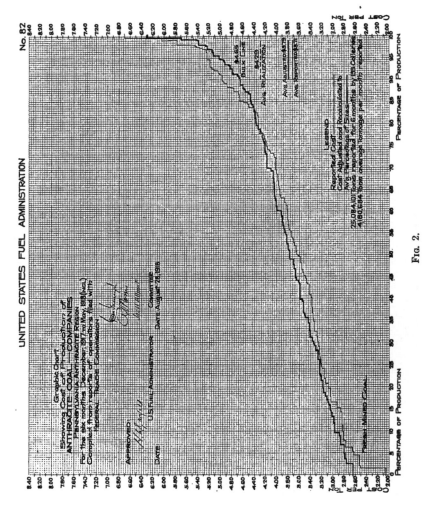

FIG. 2.

The prices received by the Companies and Independents have not been separately averaged, but calculating on the differential and assuming the percentages the same for Companies and Independents, which is only approximately the case, the selling price of fresh mined coal would

average for Companies, $4.287 and for Independents, $4.817.   Margins
over reported costs of Companies would be 58 c. and for Independents 45 c.
with a general average margin for all fresh-mined coal of 56 c. and for all
coal including washery of 71 c. per T., and under "bulk line" costs

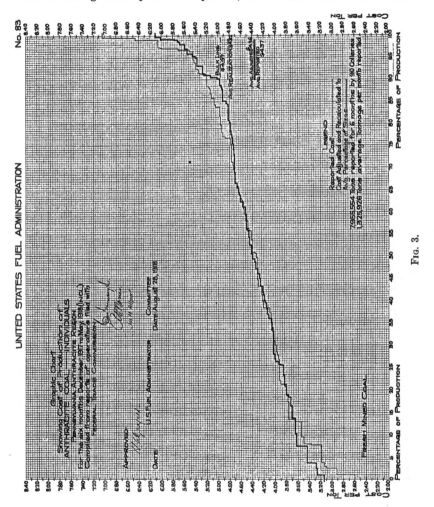

FIG. 3.

fresh-mined Companies, 36 c.; Independents, 15 c.; total, 39 c.; including
washeries consolidated sheets total of 7.5 c.

These margins include all expenditures for Federal income and
excess profit taxes, selling expenses, interest charges, expenditures for

improvements and developments to increase output, excess of capital expenditures over normal cost, and all profit on the investment of about $8 per ton annual output.

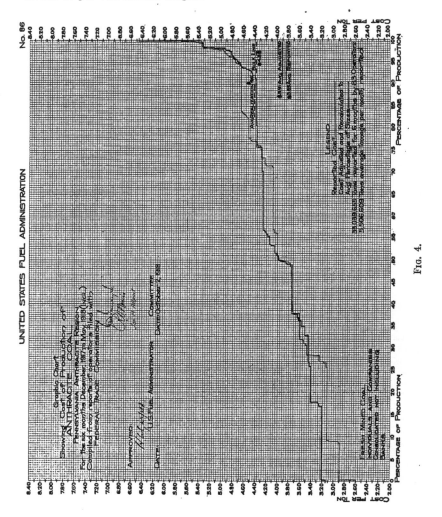

FIG. 4.

## LABOR INCREASES

Effective Dec. 1, 1917, a labor war bonus, ranging from 60 c. to $1.10 per day for labor and 25 per cent. for contract miners was granted over

and above the wage scales effective by agreement Apr. 1, 1916, expiring
Apr. 1, 1920, and the prices fixed Aug. 23, 1917 and modified Oct. 1,
1917, by reducing pea coal 60 c. per T., were increased by 35 c. per T.

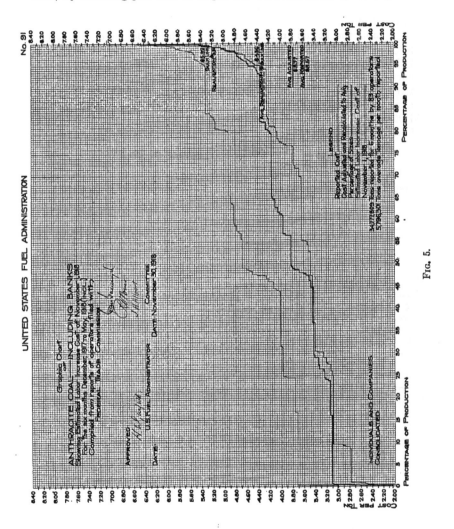

Fig. 5.

to compensate for this labor increase. The actual reported increase in
labor cost due to this advance was figured by the Federal Trade Commission from the operators' reports to be 60.3 c. From the actual payroll

figures later obtained by the U. S. Fuel Administration, this increase
was found to be 76.3 c. per T.

Effective Nov. 1, 1918, a second labor war bonus was granted.   The
calculated increase in cost due to this is shown in Fig. 5, on which the
increases for each operator are found by figuring from the payrolls, for
the 6 mo., the actual increase in pay which would have been given
applying the Nov. 1, 1918, increases, and dividing by the 6 mo. tonnage
of the colliery.   This line, adjusted to per cent. of sizes, plotted on the
chart shown in Fig. 5 and compared with the adjusted cost, shows an
increase in cost of 74.1 c.   As this was necessarily applied to the prepared and pea sizes 70.6 c. of the total the increase on these sizes was
$1.05 per T., which increase was allowed to balance the increased cost
of labor.

## PRICE FIXING

Except for the two increases to compensate for labor increases just
noted and the reduction, Oct. 1, 1917, of the pea coal price, the anthracite

TABLE 4.—*Prices Fixed by the President, Aug.* 23, 1917

| | White Ash | | Red Ash | | Lykens Valley | |
|---|---|---|---|---|---|---|
| | Company | Independent | Company | Independent | Company | Independent |
| Broken............... | $4.55 | $5.30 | $4.75 | $5.50 | $5.00 | $5.75 |
| Egg.................. | 4.45 | 5.20 | 4.65 | 5.40 | 4.90 | 5.65 |
| Stove................ | 4.70 | 5.45 | 4.90 | 5.65 | 5.30 | 6.05 |
| Chestnut............. | 4.80 | 5.55 | 4.90 | 5.65 | 5.30 | 6.05 |
| Pea.................. | 4.00 | 4.75 | 4.10 | 4.85 | 4.35 | 5.10 |

prices are as fixed by the President on Aug. 23, 1917.   The present realization, all companies and all sizes, including washery coal and both the
labor increases, is calculated to average $5.13 per T., while the bulk line

TABLE 5.—*Fixed Prices, Dec.* 31, 1918

| | White Ash | | Red Ash | | Lykens Valley | |
|---|---|---|---|---|---|---|
| | Company | Independent | Company | Independent | Company | Independent |
| Broken............... | $5.95 | $6.70 | $6.15 | $6.90 | $6.40 | $7.15 |
| Egg.................. | 5.85 | 6.60 | 6.05 | 6.80 | 6.30 | 7.05 |
| Stove................ | 6.10 | 6.85 | 6.30 | 7.05 | 6.70 | 7.45 |
| Chestnut............. | 6.20 | 6.95 | 6.30 | 7.05 | 6.70 | 7.45 |
| Pea.................. | 4.80 | 5.55 | 4.90 | 5.75 | 5.15 | 5.90 |

of the chart shown in Fig. 5, plus the November, 1918, labor increase would be $5.32.

CAPITAL INVESTMENT

The capital invested per ton output in the larger and better equipped collieries ranges from $5 to $11, with an average investment from $7.50 to $8.

PRICES FIXED

The prices fixed by the President Aug. 23, 1917, are given in Table 4. No price was fixed on sizes smaller than pea, which was decreased 60 c. per T. Oct. 1, 1917. There was a general increase of 35 c. per T. Dec. 1, 1917 and one of $1.05 per T. Nov. 1, 1918. Sizes smaller than pea were limited to a maximum 50 c. per T. below pea coal by order of Nov. 15, 1918.

TABLE 6.—*Average Cost, December, 1917, to May, 1918*

|  | Fresh Mined Coal, 35,256,550 T. Cost per Ton | Washery Operations, 3,431,916 T. Cost per Ton | Total Including Washeries, 38,688,466 T. Cost per Ton |
|---|---|---|---|
| Labor................................... | $2.593 | $0.687 | $2.423 |
| Supplies................................ | 0.616 | 0.260 | 0.584 |
| Transportation, mine to breaker............ | 0.004 | 0.007 | 0.004 |
| Royalty, current......................... | 0.153 | 0.102 | 0.148 |
| Royalty, advance........................ | 0.002 |  | 0.002 |
| Depletion............................... | 0.099 | 0.077 | 0.097 |
| Amortization of cost of leasehold........... | 0.014 | 0.024 | 0.016 |
| Depreciation............................ | 0.091 | 0.086 | 0.090 |
| Pro rata suspended cost of stripping........ | 0.023 |  | 0.021 |
| Contract stripping and loading............. | 0.009 |  | 0.009 |
| Taxes, local............................ | 0.054 | 0.034 | 0.052 |
| Insurance, current ...................... | 0.016 | 0.014 | 0.016 |
| Insurance, liability....................... | 0.058 | 0.018 | 0.055 |
| Officers' salaries and expenses............. | 0.030 | 0.019 | 0.029 |
| Office salaries and expenses................ | 0.048 | 0.024 | 0.045 |
| Legal expenses.......................... | 0.005 | 0.003 | 0.005 |
| Miscellaneous........................... | 0.026 | 0.023 | 0.026 |
| Total................................ | $3.841 | $1.378 | $3.622 |
| Increase over May to November, 1917..... | $0.764 | $0.365 | $0.719 |

The present fixed prices, Dec. 31, 1918, per ton of 2240 lb., f.o.b. mines, are given in Table 5. Smaller than pea is not to be sold within 50 c. of maximum pea-coal price. Thus the selling price of anthracite has been increased but 30.5 per cent. over the pre-war price, while the cost of production has gone up 52 per cent., the difference having been absorbed by the operators.

The average cost as reported for the 6 mo., December, 1917, to May, 1918, inclusive, prior to the increase of Nov. 1, 1918, but including that of Dec. 1, 1917, is given in Table 6.

## ACCOUNTING SUGGESTIONS

*Depletion of Lands.*—The proper charge for depletion is admittedly a difficult question. Under the present (1917) income tax law lands purchased since Mar. 1, 1913, may be depleted to return the cost price on exhaustion and lands purchased before that date may be depleted on their value as of Mar. 1, 1913. The second class may be divided into those operated on royalty by others than the owners, where the depletion accrues to the owners, and to those lands directly operated by the fee owners.

In the case of royalty lands the value as of Mar. 1, 1913, is clearly the present value as a varying annuity of the annual amount of the royalties for the term of the life of the property and as this sum is not dependent upon the vicissitudes of mining the discount is properly at the going value of money in most cases not exceeding 6 per cent.

In the case of lands operated by the fee owners the value as of Mar. 1, 1913, should properly be calculated as the present value of the profits year by year to exhaustion. Unfortunately, it is essentially impracticable to determine such a value, as the profits are a widely varying quantity and impossible to forecast far into the future. We would suggest substituting for the profits the present normal royalty value of the property, as indicated by recent leases of similar lands, and determining the present value on the same basis as royalty coal. In this case the question of profits is minimized and it would seem that the discount rate might properly be taken at 6 per cent., leaving excess earnings to take care of the mining risk. As in the case of a sale which would fix the depletion charge, the purchaser would expect to be able to earn more than 6 per cent. on any investment involving a mining hazard, and a 6 per cent. discount based on royalty values would probably fairly represent a fair sale value.

In determining the probable life of a property consideration should be given to its past history as to output, its state of development, probable future output with normal conditions, and to the quantity of coal available on the property or property tributary to its development. As the present value of an annuity for 50 years is within 5 per cent. of the present value of a perpetual annuity, it is probably wise to limit the present value of a property to the present value of the probable output of say 50 years regardless of the total tonnage or life of the property.

Acreage which is not estimated to be mined within the 50-year period should be carried in an investment account not subject to depletion and, when opened, treated as a new property.

The present value of property with less than the maximum life above indicated should be based on its probable life resulting, of course, in a higher depletion charge for short-lived properties.

As an example, the present value per ton assuming regular output to exhaustion would be for 10 years life 74 per cent. of the royalty rate, for 25 years 51 per cent., and for 50 years, 31½ per cent., the suggested limit.

Then, assuming 50 years' reserves and 25 c. per ton royalty rate, the present value would be 7.9 c. per ton, and with a 10 c. royalty rate but 3.15 c., while with a rate of 40 c., a fair present average anthracite royalty rate, the present value per ton would be 12.6 c. per ton for a fair average for depletion where reserves are at least equal to 50 years probable mining and increasing to 20.4 c. for 25 years' reserves.

*Depletion and Capital Charges.*—In general, it is our opinion that all plant and all opening and development charges less the value of coal produced during this period up to the time when the mine is sufficiently opened and developed to be able to maintain a normal production should be charged to capital and be subject to depreciation, and all charges for extensions including plant and development work necessary to maintain capacity should be charged to operating expense.

In the case of new development or plant designed to increase capacity or to materially reduce operating expense, a new development account should be opened and the cost of providing such increased output be charged to capital account.

The depreciation account should be so arranged as to extinguish the capital charge against any item of plant or development when such item is either worn out or ceases to be useful to the operation, and it might be well to classify into groups the various divisions of capital charges assigning to each group its probable average life and proper depreciation charge.

Repairs and renewals of parts should be charged to operation, but the life of plant should be figured considering such repairs and minor replacements. For instance, repairs and renewals in a breaker either to machinery or structure should be charged to operating expenses, but on its extension or replacement by a new structure the excess value considering depreciation of the new structure over the old which it replaces should be capitalized. Similarly, the retimbering of a shaft or main tunnel should be charged to operating cost.

The original capital charge against any improvement should be depreciated so as to be wiped off the books on its abandonment whether due to age and deterioration, to lack of further usefulness due to the exhaustion or abandonment of the mine, the expiration of the leasehold, or to other causes.

*Extraordinary Expense Account.*—Large items of expense properly belonging in operating cost but which would so affect monthly costs as to invalidate comparisons may be put into an extraordinary expense account and this distributed annually into the operating expenses.

*Insurance and Taxes.*—Fire insurance and workmen's compensation should be pro-rated with the monthly costs, as should local taxes. An insurance fund should also be maintained to provide against mining accidents such as fires, explosions, irruptions of large bodies of water with resulting flooding, extensive squeezes and other accidents which cannot properly be absorbed into operating expense.

Income and excess profit taxes are chargeable against profits and are not proper items of operating expense.

*Stripping Costs.*—The cost of stripping overburden should be kept in a suspense account and charged against the coal as produced. This requires a reasonably accurate estimate of the coal to be produced by stripping made from proper borings or in the case of worked-over areas, from mine data, and as such an estimate is an essential preliminary to any properly conducted stripping operation, there should be no difficulties in properly handling this account.

*General.*—It should be understood that the above accounting suggestions are the personal opinions of the author and have not been approved by any government department.

*Conclusion.*—The anthracite cost investigation supplements the bituminous price fixing previously described and, as it shows, probably for the first time, the actual range of anthracite costs, we believe that its inclusion in the *Transactions* of the Institute is warranted. The entire work is based on the costs reported to the Federal Trade Commission by the anthracite operators. The analysis and study was made by the Engineers Committee of the U. S. Fuel Administration, Messrs. Cyrus Garnsey, Jr., J. H. Allport, and the author, with the able assistance of Capt. R. H. Vorfelt, U. S. A., detailed temporarily to the Fuel Administration as Director of the Bureau of Investigation.

## DISCUSSION

EDWARD W. PARKER,* Philadelphia, Pa. (written discussion†).—The two papers on coal-mine costs and price fixing that Mr. Norris has contributed to the *Transactions* possess as much general interest and are of as much value to the coal-mining industry as any papers that have been presented in recent years. Until recently one of the troubles with the coal business, and especially the bituminous branch of it, has been that because of the lack of a uniform system of cost accounting (many operators, I regret to say, not having had any system at all), the industry as a

* Director, Anthracite Bureau of Information.    † Received Feb. 24, 1919.

whole may be said not to have known where it was at.  The system adopted by the Federal Trade Commission may not be absolutely perfect but it has accomplished something in that it has compelled a uniform method of reporting and tabulating costs giving results of these compilations which are comparable.

The Institute and the coal-mining industry owe a vote of thanks to, and of confidence in, Mr. Norris and his two associates on the Engineers Committee of the Fuel Administration, Messrs. Garnsey and Allport, for the great work they have done in interpreting the results compiled by the Federal Trade Commission, not to speak of the personal sacrifices made by them in giving their services to the government.

Those who have read Mr. Norris' paper will probably join me in wishing he had brought out a little more fully the real meaning of the charts so far as they indicate the difference between the cost of producing a ton of anthracite and the price the operators receive for their product. Mr. Norris could have done this more satisfactorily than anyone else. It seems to me that particular attention should be drawn to the fact that, according to these charts, and the tables accompanying them, not less than 25 per cent. of the total output of anthracite, whether "company" or "individual," during the period covered by Mr. Norris' paper, was produced at an actual loss on the operating costs alone; for, as I understand it, these figures do not include certain administrative salaries, interest on bonds or borrowed money, losses from bad debts, income or other Federal taxes, and some other overhead charges.  According to Chart 1, which gives the production and costs of all the companies reporting for a period of 6 months, about 10 per cent. of the output was mined at a loss of 40 c. a ton or over.  A study of Charts 1, 2, and 3 shows somewhat of a paradox.  Charts 2 and 3 show separately the operating costs of the "companies' and "individuals."  In each case 20 per cent. of the output was produced either at no profit or at a loss, whereas when we consider the total of the two, as shown on Chart 1, 25 per cent. of the total output was sold at or below cost.

The explanation, as I see it (and Mr. Norris will correct me if I am wrong), lies in the fact that the individual companies, on the government price-fixing arrangement were allowed a differential of 75 c. per ton on their output of prepared sizes above that permitted to the "company" operators.  The average realization for the "company" coal was $4.29, and the average realization on the "individual" output was $4.82.  Five million tons of "company" coal were mined at a cost in excess of $4.29, the average realization for that coal, while 1,590,000 tons of "individual" coal were mined at a cost above the average realization of $4.82.  The fact that the larger portion of the coal mined at a loss was "company" coal accounts for the apparent inconsistency.

The United States Fuel Administrator had before him the original of these charts and knew exactly what the situation was when the prices were fixed. As Mr. Norris has stated, while the bituminous prices were fixed upon the basis of cost and reasonable profits were allowed to the operators, the prices for anthracite were fixed apparently without any relation to the cost. When the advance in wages was granted by the supplemental agreement of November, 1917,[1] to take effect the first of the following month, it was estimated that the additional cost would amount to about 45 c. a ton. The Fuel Administration, however, seemed to be impressed with the idea that in preparing this estimate the anthracite operators had included a sufficient factor of safety to cover any possible chances, so instead of permitting an advance of 45 c. per ton to cover the estimated increase in cost, it permitted only an advance of 35 c. That the operators were out in their estimates by approximately 100 per cent. is shown by the fact that the actual results of the wage advance in November, 1917, was an average increase in cost of 76.4 c. per ton, or something more than twice the advance in price permitted by the Administration.

When, in November, 1918, it was found necessary again to increase the wages of mine workers, these facts were placed before the Fuel Administration and they were verified by the Administration's own Committee of Engineers and based upon the reports made to the Federal Trade Commission. As I have stated, the Fuel Administrator had before him these charts which are now presented by Mr. Norris in his paper, so that Mr. Garfield knew at that time that the operating cost of one-fourth of the anthracite output exceeded the prices obtained for it. He knew, moreover, that the average margin on the entire output of anthracite was not a safe one and that it did not represent what was specifically called for in the Act of Congress providing for the control of prices by the Fuel Administration, namely, that the prices fixed by the government should guarantee a fair and reasonable profit. The Fuel Administration on the first of February, when it practically laid down the scepter of office, confessed the injustice of its behavior toward the anthracite industry and admitted that the prices of anthracite should have been at least 50 c. more per ton than the Administration had allowed. The significant part of the statement which accompanies the order freeing anthracite from further governmental restrictions is as follows:

"For the purpose of arriving at a fair increase in price to cover the increase in wages recommended by the War Labor Board last October, an examination was made to determine the costs of the various anthracite producing companies. The result of this examination showed that the general increases in the price of materials and labor have raised the cost of mining anthracite to such an extent that many of the

---

[1] Supplemental to the agreement of April, 1916.

companies were not receiving a fair return and that some producers of necessary coal were actually sustaining a loss on the sale of coal at government prices, in spite of the two increases allowed on account of the advances to labor.

"The above statement is made \* \* \* \* out of fairness to those companies who have patriotically kept up their production to war needs, even at a cost which resulted in many instances in a loss, not only by individuals, but also by some of the railroad companies \* \* \* \*.

"Had the Fuel Administration's active control over maximum prices on anthracite coal been continued, the cost examination above referred to shows that it would have been necessary, on the basis of the present wage scale, to raise these maximum prices possibly as much as 50 c. a ton \* \* \* \* to prevent financial embarrassment and perhaps the closing of companies producing a substantial per cent. of the necessary anthracite output."

It is to be noted that in the opening sentence of this statement Dr. Garfield states that examination was made for the purpose of "arriving at a fair increase in price" and then in the final paragraph admits that a fair increase would have been at least 50 c. more than was permitted.

In the course of recent investigations into the anthracite industry by the Sub-Committee on Manufactures of the United States Senate, it was testified by the chief engineer of the oldest anthracite mining company in existence—and one that is now producing from 4,000,000 to 5,000,000 tons per year—that the margin between the selling price at the mine and the cost of production for the entire year of 1918 was 6.89 c. per ton, a little less than 7 c. Out of this 7 c., it should be remembered, must be paid the administrative salaries, interest on bonds, Federal taxes, and dividends.

That conditions were not improving toward the close of the calendar year, but on the other hand were going from bad to worse, was shown by the fact that this margin for November was 5.6 c., while in December there was a loss of 2 c. per ton. I understand the January statement shows even worse "in the red."

R. V. NORRIS.—I wish to say, in regard to Mr. Parker's discussion, that the prices were not fixed by the Fuel Administrator. He only allowed an addition for an increase in December; the prices were fixed by the President and, except in the case of pea coal, have only been adjusted for labor changes. The adjustments were based on averages and not on the labor increase of the higher cost companies. Chart 5 shows the increase for each company, and you will note a varying increase from a relatively small increase for the low-cost companies to a considerable increase, amounting to over a dollar a ton for all sizes, for the high-cost companies. This is perfectly natural, as the labor increase was not a percentage but a flat one, except in the case of the miners, and necessarily the high-cost companies had a higher labor cost and therefore a larger actual cash increase and a larger increase per ton.

S. D. WARRINER,* Philadelphia, Pa.—The anthracite industry is much indebted to Mr. Norris and the engineers of the U. S. Fuel Administration for the careful study they have made of the complicated subject of anthracite cost and realization. It is unfortunate, however, for the industry at present, and ultimately for those who desire, and require, an abundant supply of anthracite, that the Fuel Administration did not apply to anthracite the principles of price fixing recommended by its engineers and actually adopted in the case of bituminous coal. The application of these principles to anthracite would have stimulated productivity when greater production was a war necessity, and would have left the industry with its natural markets unimpaired and in better financial condition to meet the difficulties of the readjustment period. The statement of the Fuel Administrator on lifting restrictions is an acknowledgment that the policy of price fixing followed by the Administration with respect to anthracite was not based upon the costs as determined by his engineers, and that readjustments are necessary to prevent financial embarrassment and to avoid an ultimate shortage in supply with its consequent increased cost to the consumer.

The charts presented by Mr. Norris are illuminating but do not fully represent the present situation, which had arisen from the Fuel Administration lifting its restrictions and leaving the results of its price fixing to be paid for, not by the government as in the case of wheat, but by the industry upon which the regulation was imposed. These charts show costs that do not include the cost of marketing (at least 10 c. per ton) or interest on investment, which at 5 per cent. on the average capitalization found by Mr. Norris would be 40 c. per ton. We must, therefore, deduct from his realization 50 c. per ton before the operator realizes any recompense for the risk of his business as compared with an investment in a well secured bond.

The pressing demand during the war for anthracite throughout New England and the Atlantic seaboard, caused by the increased population of war workers and the activities of war industries, as well as the requirements of the government itself, necessitated a policy of zoning anthracite out of certain states and territories and prohibiting its use for many purposes for which it was normally used.

Following the signing of the armistice, the sudden let-down in activity, combined with the effects of a warm winter, speedily resulted in a surplus of coal which has made it impossible for the individual operator to secure the maximum price allowed by the Fuel Administration and has naturally brought all maximum prices down to company prices. Even at these

* Prseident, Lehigh Coal & Navigation Co.

prices coal cannot now be moved in volume for the reason that the market in the territory apportioned to the industry has been fully supplied, and the market from which it has been excluded has been supplied with adequate fuel substitutes for its normal requirements.

Chart 2 shows company realization allowed on the adjusted standard $4.29, or if we deduct 50 c. for selling and interest, $3.79. As against these figures the average cost was found to be $3.91, consequently of the total fresh-mined tonnage at least 35 per cent. shows an actual operating loss, and 60 per cent. a financial loss on a 5 per cent. basis for capital.

These figures (Chart 2) more nearly reflect the present relation between cost and realization than the figures in Chart 5, which show the increase in cost and realization due to the adjustment of Nov. 1, 1918. Chart 5 shows the combined cost of fresh-mined and bank coal, but as the market for bank coal largely ceased with the war, fresh-mined coal alone should now be considered. Furthermore, the realization shown on Chart 5 is the average of company and individual prices; and as only the company price is now obtainable the prices shown on Chart 2 should be taken as the basis to which the adjustment of Nov. 1, 1918, should be added.

As the price allowance of November merely covered the average increase in cost due to wages, each ton of coal mined above the average cost was not fully compensated by the increased price allowed. Furthermore, as the consumption of steam sizes (averaging 35 per cent. of the entire output) has shrunk with the slackening of industry, there has been a reduction in the prices of these sizes below the maximum heretofore obtained, thus reducing further the average realization. The real situation in the anthracite industry at the date of the lifting of the Fuel Administrator's price restrictions was very much worse than has been depicted, and it is probable that more than 60 per cent. of the output is now being mined at a loss.

The work of the engineers of the Fuel Administration is of great technical value in its pioneer study of the complicated subject of relative cost and realization, and, it is hoped, may go far to correct the popular misapprehension of great profits derived from the industry. From an economic standpoint, the price regulations actually enforced by the Fuel Administration are of interest as demonstrating the fact that any interference with economic laws is expensive exactly in proportion to the detail with which the interference is carried out. In abnormal times regulation is necessary, but it should be broadly constructive and not restrictive, and of such nature that when the necessity for such regulation no longer exists the laws of supply and demand may again assert themselves, without financial embarrassment to the industry of unnecessary expense to the consumer. The regulations of anthracite prices represented an extreme case of price fixing. The Fuel Administration fixed two prices for the same commodity, which in itself was rather penalizing to the producer than beneficial to the consumer, as local retail regulations became difficult.   Neither price was based on cost, as both company and individual mines were in both high-cost and low-cost group.'  Not only this, but arbitrary prices were fixed on practically each of the nine different sizes, the prices of which under commercial conditions have always relatively fluctuated according to demand.   With the reëstablishment of commercial conditions, the prices arbitrarily fixed disappear, the result being financial operating loss to more than 60 per cent. of the industry.   The consumer would not be especially interested in this phase of the situation were it not that with the indefinite closing down of high-cost operations, which is now taking place, the capacity of the anthracite mines is reduced below the average annual production normally needed.   This will inevitably tend to create a shortage, unless coal can be stored by the companies for future use.   Storage, however, is only economically practicable when there is stability of market conditions, otherwise the expense and risk makes it prohibitive.   For this reason the trade policy of the industry has been, so far as possible under existing laws, to promote stable conditions by the summer discount in price, thus encouraging the consumer to store a part of his winter supply, the operator storing the remainder.   By this method steady operation of the mines was insured and the annual supply required by the market obtained.   For a number of years the anthracite industry would otherwise have been unable to supply the market requirements.

Today the anthracite industry is left by the Fuel Administration under the handicap of inadequate prices, a wage structure still under government control, and a tendency toward declining prices in all commodities, with the problem of recreating the stable commercial condition necessary to the prosperity of the industry as well as to the comfort of those who are dependent on anthracite for fuel.

PAUL STERLING,* Wilkes-Barre, Pa.—There is an impression that the inspection of the coal has been let down during the strenuous times of the last 2 years, but in the company that I represent the inspection department maintained its standard during the entire period of the war and we maintained the same inspection in production that we always had prior to that time.

W. V. DeCAMP, Edgewater, N. J.—I realize that the discussion on the paper is entirely along the line of anthracite mining costs and that no effort has been made to state what factors tended to increase the cost of coal, other than increase of supplies and the increased wage, and I would like to ask if Mr. Norris has any data in regard to what possible decrease

* Mechanical Engineer, Lehigh Valley Coal Co.

in efficiency occurred during the period mentioned, especially the efficiency of labor? Also, if 25 per cent. of this coal was produced at a loss, did the companies that produced said coal at a loss make a reasonable profit in pre-war times, when the conditions were different, or were they companies that were beginning operations or increasing operations and working under some other than normal, high-production conditions? In addition, what general changes, if any, were made in the general mining methods, or in general organization methods, to meet the rapidly increasing costs, since that is a question in which the engineer is always interested, whether it is metal mining or coal mining?

R. V. NORRIS.—I will try to answer the three questions. In the first place, there was a notable decrease in efficiency, caused largely by the fact that the younger men went into the National Service, and that their work was necessarily taken up by older and less active men. The decrease was particularly notable in the transportation end of the service, as the older men did not have the snap of the younger men they replaced. There has been a very great decrease, from about 180,000 to about 142,000, in the number of employees in the anthracite region, with about the same actual production, but production was maintained not by increased efficiency, as would seem to be the case, but by more working days and longer hours, which the men in most cases willingly gave.

Second, the high-cost companies are all old, well-established and presumably profitable companies. In the case of the large high-cost companies, they have been paying dividends for many years. The excess cost was caused by their efforts to hold an increased production with a lessened supply of labor and that more poorly trained.

Third, there could be no sweeping changes made during the war; the effort was intense to hold the production. It was almost out of the question to make any serious changes in methods at that time. The production was, of course, largely held by reducing development work, and I look forward to a long siege of extra development work, to put the mines back to where they were at the beginning of the war. I am confident that development work was necessarily very much neglected and that the coal which had been put in sight by previous development work was largely mined to hold the production for war purposes.

EDWIN LUDLOW,* Lansford, Pa.—The question of decreased efficiency of the men working in the coal mines, which has just been raised, I am able to answer, as it is a point that is followed very closely by the Lehigh Coal & Navigation Co., and the tons per man hour is worked out in tabulated form. During the year 1917, the tonnage per man-hour was 0.026 and for the year 1918 it was only 0.022. The production for the years

---

* Second Vice-president, Lehigh Coal & Navigation Co.

1917 and 1918 was slightly in excess of 5,000,000 tons, with more working days in 1918 than in 1917. If the same production per man-hour had been maintained in 1918 as in 1917 the tonnage would have been 860,000 tons greater.

This loss in efficiency cannot be attributed entirely to the high wages and indifference of the men to keeping up a full production, as this was true of only a small proportion; while on the other hand our foremen and their assistants worked to the physical limit of their ability in long hours and without holidays in their endeavor to keep everything in operating condition. Our loss is more attributable to the 910 of our men who went into the service of the U. S. Government, of whom only 336 were drafted, the remainder enlisting. The class of work from which these men were taken was that requiring youth and quickness in order to produce efficient results, and the older men with whom we were obliged to fill their places could not be expected to do as much work as the younger trained force that had gone into the service of the government. There was also difficulty in keeping up the discipline, as the shortage of men made it impossible to replace anyone that we wanted to discharge and our efforts were all concentrated in trying to produce the best results with the material that we had.

In regard to the advances granted by the Fuel Administration in the selling prices of coal as compared to the advances granted in wages: A concrete example of one company may possibly bring this matter more clearly to those who are not interested technically in coal mining. The pay roll for the Lehigh Coal & Navigation Co. for the year 1918, on the basis of the wages established by the contract between the miners and the company, which was supposed to remain in effect for 4 years, expiring Apr. 1, 1920, would have amounted to $8,000,000. The war allowances that were granted to cover the increased cost of living and which were approved by the Fuel Administration, who endorsed the agreement between the men and the company, amounted to $3,000,000, and the increased selling price on the prepared sizes of coal, granted by the Fuel Administration, for the year 1918 increased the revenue from the coal sold by $1,500,000. This loss of $1,500,000 was partly made up by the increased prices obtained for the steam coals, but the prices of the steam sizes are dependent on the prices of bituminous coal, and if bituminous coal becomes cheaper, the prices of the steam sizes of anthracite will undoubtedly fall in proportion. No allowance was made by the Fuel Administration for the increased cost of material. This increase for the year 1918 over the year 1917, with the same tonnage produced, amounted to $906,000. The scale of wages as now fixed, including these war allowances, is practically on the basis of the wages paid to the railroad employees by the government.

The cost of living is still high, and it does not appear to be in any way practical or advisable to attempt a reduction of wages at this time. The anthracite companies and the consumers of anthracite are, therefore, faced with the fact that there will have to be an advance made in the selling price of the prepared sizes of anthracite coal in order to enable the companies to come out even. If there should be no advance, it would mean that a large proportion of the higher cost collieries would have to be closed, bringing on a decreased tonnage of anthracite, which would not enable the anthracite companies to produce enough coal next year to meet the demand; and with such a shortage the prices of anthracite would jump, as they have frequently done in the past when there has been a shortage of that kind and no restrictions as to the maximum prices which might be asked.

It would, therefore, appear to be the wisest for both the consumers and the companies that a gradual increase in prices should be made to cover the actual costs as determined by the engineers of the Fuel Administration as being the necessary prices to enable the companies to live, so that they may be able to maintain their tonnage and be prepared to meet the demands of the country for fuel next fall when coal will be needed as it always has been.

## Method of Fixing Prices of Bituminous Coal Adopted by the United States Fuel Administration*

BY CYRUS GARNSEY, JR., R. V. NORRIS, AND J. H. ALLPORT†

[[(Colorado Meeting, September, 1918)

### NECESSITY FOR PRICE FIXING

DURING the latter part of 1916 and the early months of 1917, due to war activities, there was a threatened shortage of coal which resulted in panic among consumers and a rush to obtain coal at once at any price. As a result of this insistent demand for immediate delivery, prices were bid up by the consumers to unprecedented heights; spot coal which had previously been selling at from $1.50 to $2 per ton was bid up to $5, $6, and, in exceptional cases, as high as $7.50 or more per ton.    Then when the April, 1917, contract period arrived, contracts could be made only at prices ranging from $3 up to $5 and $6 per ton for the year's delivery.

This condition caused such a demoralization of the business, and so much complaint, that some action to regulate prices was considered essential by the National Administration.

### LANE-PEABODY AGREEMENT

In May, 1917, a committee under the chairmanship of F. S. Peabody, of Chicago, was appointed by the Council of National Defense, through Mr. Lane, to consider the whole question of bituminous coal. This committee, with the Secretary of the Interior, Mr. Lane, after numerous meetings and long negotiations with the operators throughout the country, announced on June 29 an agreement between the Committee and the producers, fixing a tentative maximum price for bituminous coal throughout the country at $3 per net ton f.o.b. mines, to which was added 25 c. for selling commission to wholesalers.

This plan was based on the idea of fixing a maximum price, high enough to greatly stimulate production, with the expectation that the laws of supply and demand would, with ample production, operate to maintain fair and just prices for coal throughout the country.

---

* Read before the Anthracite Section, Wilkes-Barre, Pa., Aug. 10, 1918.
† Engineers to United States Fuel Administration.

## The "Lever Act"

With the country plunged into the greatest war of history, it became evident that distinct power should be given the Administration to control efficiently war necessities, food and fuel, needed in ever-increasing amounts, not only by our own country but by our allies, and for our armies in this country and abroad.    With this in view, the Sixty-fifth Congress passed House Bill No. 4961, generally known as the "Lever Act," entitled "An Act to provide further for the national security and defense by encouraging the production, conserving the supply, and controlling the distribution of food products and fuels."    It was approved Aug. 10, 1917.

Section 5 of the Lever Act authorizes the licensing of "the importation, manufacture, storage, mining or distribution of any necessities;" and Section 25, "to fix the price of coal and coke, whenever and wherever sold, either by producer or dealer."    It is further provided in this section "in fixing maximum prices for producers, the commission shall allow the cost of production, including the expense of operation, maintenance, depreciation and depletion, and shall add thereto a just and reasonable profit."    "In fixing such prices for dealers the commission shall allow the cost to the dealer and shall add thereto a just and reasonable sum for his profit in the transaction."

[The powers of the Federal Trade Commission as to coal were, with certain minor exceptions, transferred to the United States Fuel Administration by order of the President on July 3, 1918.]

## The "President's Prices"

On Aug. 21, 1917, the President announced prices for bituminous coal throughout the United States, specifying prices for run-of-mine, prepared sizes, and slack or screenings, by States and, in a few instances, by districts or by seams.    These prices for run-of-mine coal varied from $1.90 to $3.25, and were, in general, a very great reduction from the prices fixed by the Lane-Peabody Commission.

These prices were based on average figures on about 100,000,000 tons production, prepared by the Federal Trade Commission, from the very meager data in its possession, generally costs from the larger and lower-cost operations of each district.

## The Fuel Administration

On Aug. 23, 1917, Mr. Harry A. Garfield was appointed United States Fuel Administrator by the President, and to him was delegated the powers as to fuel, conferred by said act on the President.    On the same day, by Presidential proclamation, prices were fixed on Pennsylvania anthracite coal. From this date until early in January, 1918, numerous revisions and adjustments of "The President's Prices" were made by the Fuel Administrator, but no general verification or revision was attempted.

Early in January, the Engineers' Committee was constituted, and to this Committee was entrusted the making of a general review of costs, and the submission to the United States Fuel Administrator of the results of careful studies of the costs of producing coal throughout the United States. The Committee was not then, and never has been, authorized to fix prices on coal; the limit of its duties has been to study and report on methods of price fixing and to determine and furnish costs, leaving to the Fuel Administrator the personal duty of determining the amount of margin to be allowed.

The Committee's first work was a study of price-fixing methods which were, or might be, applicable to coal-producing conditions. In arriving at a logical and scientific plan for fixing the price of fuel the following methods were considered:

*Straight Cost Plus Method.*—The actual cost at each colliery plus a fixed sum or percentage of profit.

*Modified Cost Plus Method.*—The actual cost at each colliery plus a graduated profit decreasing as costs increase.

*Average Cost Methods.*—Prices fixed on the average cost in each district.

*Pooling Methods.*—All coal sold at the average cost of each district plus a profit, and the returns to each colliery adjusted through a clearing house at a price proportioned to its cost of production.

## DISCUSSION OF ADVANTAGES AND DISADVANTAGES OF THESE METHODS

### Straight Cost Plus Method

*Advantages.*—(a) All producers would receive the same profit, and no one would have an advantage over another in this respect.

(b) Apparently simple in plan and execution.

*Disadvantages.*—Impracticable of application, by reason of:

(a) Resultant multiplicity of prices, with grave disturbance of markets.

(b) Continual changing of prices due to inevitable variations in each producer's costs.

(c) Instability of the industry, due to the natural disposition of consumers to purchase the lowest-price coal.

(d) Inefficiency in operation always resulting from lack of incentive in cost plus operations.

(e) Material reduction in output and reduction in quality due to the natural tendency to mine the poorer and more expensive coal with a guaranteed profit, and to leave the better and cheaper coal in reserve to be mined on the return of normal conditions.

(*f*) Continual increase in all costs incident to extravagant methods encouraged by guaranteed profits.

(*g*) Labor unrest and constant demands for increases due to the knowledge of a guaranteed profit regardless of cost.

(*h*) Practical impossibility of arriving with technical accuracy at the costs of each separate operation.

(*i*) Impracticability of the Government's policing the mines and securing the same efficient operation and production attained by the individual producer under the stimulus of increased profits.

(*j*) Illogical, in that the better planned and managed operations are discouraged, as compared with poor and inefficiently managed properties.

### Modified Cost Plus Method

This is but a modification of the preceding, and the same discussion applies, modified only by the inclusion of a somewhat greater incentive to the better and more economical operations.

### Average Cost Methods

*Advantages.*—A minimum uniform price for each district or, if desired, for the entire country.

*Disadvantages.*—(*a*) The average cost is necessarily less than the cost of about half the total tonnage. Hence, a reasonable profit put on the average cost would not produce the necessary tonnage.

(*b*) The tonnage below and up to the average cost is actually produced by less than 30 per cent. of the operators of the country. Hence, the great majority of the operators producing at above average cost would be put out of business by a price based on the average.

### Pooling Methods

Pooling may be done on either cost plus, modified cost plus, or on the prices established by the United States Fuel Administration.

*Advantages.*—(*a*) A uniform price to consumers for sections and, if desired, for the entire country.

(*b*) A *present* lower price to consumers based on weighted average cost.

(*c*) A simplification of all present pooling arrangements, as all coal to each pool would have, or could be arranged to have, the same price.

(*d*) A return to the consideration of quality instead of cost, as, with all coal at the same price to consumers, the higher qualities would naturally be preferred.

*Disadvantages of Pooling Cost Plus or Modified Cost Plus Methods.—*

(a) Continual variation in pool prices, due to inevitable variations in producers' costs.

(b) Unfair and illogical, in that the better located and managed operations are made to pay tribute to the poor and badly managed ones.

(c) A general and considerable increase in cost inevitably resulting from any method involving guaranteed profits with a disregard of economy.

(d) A material reduction in output, due to lack of incentive and resulting inefficient methods, the employment of unnecessary labor, the mining of the more expensive and less desirable qualities of coal for the ultimate benefit of the mines, and the execution of development not immediately needed.

(e) A slacking of the efforts of employees, which is the usual result of a lack of incentive to the producer, with the resulting lack of interest.

(f) The installation of an unsound policy tending to encourage the inefficient and discourage the efficient producer.

(g) The ever-present temptation to allow costs to increase with the hope of readjustment of prices.

(h) Dissatisfaction to both labor and to producers from the knowledge that other and less efficient operations have higher limits of price.

The disadvantage of pooling on the prices fixed by the United States Fuel administration are the same as suggested above, without some of the special disadvantages of cost plus methods.

*Disadvantage of Pooling in General.* —(a) A very large capital required to handle such stupendous operations.

(b) Enormous and extended credits required to finance the producers.

(c) Lack of organization to handle this new business.

(d) Undesirability of creating such an organization with its army of additional employees at the present time.

(e) Inadvisability of putting a new and untried plan into operation at the present time.

(f) Impossibility of obtaining, with sufficient promptness, the costs necessary to fix pooling prices with the necessary accuracy.

(g) Interference with present established methods of handling coal, with serious risk of crippling its distribution and unnecessarily creating a shortage.

None of these suggested methods seemed to fill the peculiar conditions incident to price fixing of coal at the mines, and it devolved upon the Engineers' Committee to develop some method better suited to the conditions of the problem.

## PRICE FIXING METHOD ADOPTED

The study of the conditions indicated the necessity of finding a method of price fixing which would fill as nearly as. practicable the following requirements:

1. Result in a price fair to the public.
2. Prevent excessive prices or profiteering.
3. Prevent a multiplicity of prices in any district.
4. Encourage legitimate production.
5. Discourage production from inefficient and unduly costly operations.
6. Insure to the producer "the cost of production, including the expense of operation, maintenance, depreciation, and depletion, with a just and reasonable profit," as required by the Lever Act.

In arriving at a method promising to accomplish these results as nearly as practicable, the following system was developed. Costs obtained from the individual sheets filed by each operator with the Federal Trade Commission were studied, listed, and adjusted for price fixing. These figures, with the percentages of each cost in the total production of each district, were plotted on diagrams, showing graphically the range and extent of variation in each district. On these diagrams a line indicating the sources of indispensable tonnage, christened "the bulk line," is drawn as a base to which the Fuel Administrator personally adds a margin in his judgment necessary for each district.

### Advantages of System

The method of fixing prices by the "bulk line" principle recognizes the economic syllogism that "the price of any article necessary to a community will be fixed by the cost of producing that necessary portion of such article involving the greatest expense."

(a) This assures to all producers profits dependent upon their ability and exertions, only limited by the establishment of a reasonable price to the consumer.

(b) It does not unduly increase the price of coal to the consumer over the minimum price possible under other methods.

(c) It tends to encourage maximum production and necessary development by allowing to the producer the benefit of reduced costs due to greater production.

(d) It avoids bad feeling among the producers and among the workmen by allowing a fixed price in each district and not apparently showing favoritism to special producers.

(e) It tends to encourage the fit and discourage the unfit.

(f) The method is susceptible of refinement and extension, making it possible to eliminate undue profits to the producer and adjust prices from time to time to the ultimate advantage of the consumer.

## Disadvantages of System

(a) Considerable profits to the lowest-cost operators.

(b) A price for coal greater than one based on the average cost, by the amount by which the "bulk line" exceeds such average.

This method appeared to be better suited to the conditions than any of the others suggested, and after a careful study by the U. S. Fuel Administrator, it was adopted.

## COST DATA AVAILABLE

The Federal Commission had, by authority of the Act of Congress creating the Commission, the power to investigate costs, and to require, under penalty, reports of costs of operation. Under this authority, the Commission sent out, to each coal producer listed in the United States, blanks requiring a rather complete and detailed statement of costs of operation, and the realization obtained from the sale of this product.

These reports were generally available for the months of August and September, 1917, at the time the revision of prices was started, and reports for these two months, studied in connection with later reports, were generally used as a basis of costs for the first studies. It further developed that these two months were, in most instances, fairly representative, as to output, of the average of the year.

## ANALYSIS AND ADJUSTMENT OF COST SHEETS

Without desiring to impugn either the honesty or the accuracy of the cost sheets as presented, it was found essential to study and adjust them for use as a basis of scientific and accurate cost finding. Besides correcting slips and palpable mathematical errors, a considerable amount of revision was necessary. Many, especially of the small operators, were inexperienced in bookkeeping and submitted cost sheets which, while accurate in totals, were grievously mixed in details.

## Supplies

The item of supplies was found to vary so widely in the same mines in different months that the returns for single months were practically abandoned, and the figures were replaced by averages from all reports available, resulting in increases or deductions from the monthly costs as reported.

## Reserves

The item of maintenance was frequently misunderstood, in some instances all supplies and much labor being charged to this account; in others a fixed sum, and in still other cases, nothing at all was charged.

Depreciation was often put in as a guess; in some cases it was frankly stated that this seemed a good time to charge off improvements, and such were charged to the limit, and beyond.

Depletion of lands was also an item which appeared greatly to trouble some of the accountants. While generally understood, many very wild guesses, even up to the market price of the product, were found; also, many instances of depletion charges for lands operated on royalty or lease and not the property of the operator.

Contingent funds noted on the blank were generally omitted but, in a few cases, especially when the need of such funds had recently been felt, most ample allowances were made. After being considered, it was decided to apply in each district amounts obtained by studying the claims of the better operators of such district, after obtaining, from the best available sources, reliable figures as to the cost of lands and amount and value of improvements characteristic of the district.

The question of contingent reserves is a serious one. From a strict cost-accounting standpoint, no cost can be permitted until incurred. Nevertheless, such reserves are essential to an industry involving the great risk incident to coal mining, and with the full knowledge that such reserves are used only for major accidents or calamities, and that ordinary losses regularly incurred are charged to the costs of operation, it was decided to include a small amount for contingent reserve in the general allowance.

## Salaries

Executive and even superintendent's salaries were frequently omitted. In many cases of personal ownership, undoubtedly none were paid. It was considered only just to add to such returns reasonable allowances for salaries to place such reports on a parity with the majority of the operations which paid for such necessary service. On the other hand, occasional instances were found of reported salaries so excessive as to require adjustment downward to a reasonable parity with the general practice of the district.

A sliding scale of salaries adjusted, within broad limits, to the monthly tonnage of an operation was finally devised. Any salaries missing or below the minimum allowed were raised to the minimum, and those above the maximum were lowered to the maximum.

# FEDERAL TRADE COMMISSION

FILE No.........

## BITUMINOUS COAL REPORT ON COST, INCOME, AND TONNAGE FOR........, 191

(Period, Month or Year.)

LOCATION OF MINES

District............
State............
County............

COMPANY MAKING THIS REPORT { FULL NAME............
ADDRESS............

Mail to FEDERAL TRADE COMMISSION, WASHINGTON, D. C., on or before.....

......In the information required on this form.

The information required by this report is ordered to be furnished pursuant to the power of the Commission under subdivision b of section 6 of "An Act to create a Federal Trade Commission, to define its powers and duties, and for other purposes," and under paragraph 12, section 25, of "An Act to provide further for the national security and defense by encouraging the production, conserving the supply, and controlling the distribution of food products and fuel," which is as follows:

"The books, correspondence, records, and papers in any way referring to means of any kind relating to the mining, sale, or distribution of all mine operators or their persons whose coal and coke have or may be kept to this sale, and the books, correspondence, records, and papers of any person applying for the purchase of coal and coke from the United States shall at all times be subject to inspection by the said agency."

### PENALTIES

Failure to mail this report within the time required will subject the corporation to a forfeiture of the sum of $100 for each and every day of the continuance of such failure. Section 10, Federal Trade Commission Act.

Any person who shall wilfully make or cause to be made any false entry or statement of fact in this report shall be subject to a fine of not less than $1,000 nor more than $5,000, or to imprisonment for a term of not more than three years or to both such fine and imprisonment. Section 10, Federal Trade Commission Act.

## COST

| ITEM | LABOR | SUPPLIES | TOTAL | |
| --- | --- | --- | --- | --- |
| | | | AMOUNT | PER TON |
| **I. MINE EXPENSE** | $ | $ | $ | $ |
| (1) OPERATING EXPENSE: | | | | |
| (2) Mining, | | | | |
| (3) Yardage, | | | | |
| (4) Transportation, | | | | |
| (5) Ventilation, | | | | |
| (6) Drainage, | | | | |
| (7) Dead work, | | | | |
| (8) Tipple or breaker, | | | | |
| (9) Washery, | | | | |
| (10) Miscellaneous, | | | | |
| (11) | | | | |
| (12) Total, | | | | |
| (13) Power,      tons, | | | | |
| (14) Power-house fuel, | | | | |
| (15) Superintendence and engineering, | | | | |
| (16) Mine office, | | | | |
| (17) | | | | |
| (18) Total. | | | | |
| (19) TOTAL OPERATING EXPENSE, | | | | |

## INCOME

| | AMOUNT |
| --- | --- |
| | $ |
| (63) COAL SALES: | |
| (64) Local sales, | at per ton, |
| (65) Shipped, | at per ton, |
| (66) Railroad, | at per ton, |
| (67) Coke, | at per ton, |
| (68) Power-house fuel, | at per ton, |
| (69) | |
| (70) TOTAL SALES, | |
| (71) COST OF SALES: | |
| (72) Total mine expense (Item 35), | |
| (73) Coal purchased, | |
| (74) Total, | |
| (75) Inventory 1st of month, | |
| (76) Inventory end of month, | |
| (77) Increase or decrease in inventories, | |
| (78) TOTAL COSTS (Add 74 and 77), | |
| (79) GROSS PROFIT ON COAL SOLD (Subtract 78 from 70), | |

(20) GENERAL EXPENSE AND FIXED CHARGES:
(21) Royalty,
(22) Depleti n reserve,
(23) Depreciation  ...e,
(24) Maintenance  ...e,
(25) Contingent  ...e,
(26) ...s (exclude income and excess profit),  %
(27) Insurance—General,
(28) ...y, or Compensation,
(29) Officers' salaries and ...se,
(30) ...er general ffice salari s  nd expense,
(31) Legal,
(32) Miscellaneous,
(33)
(34) TOTAL GEN. EXP. AND FIXED CHARGES,

II. SELLING EXPENSE

(35) TOTAL MINE EXPENSE,
(36) Commissions,
(37) Advertising,
(38) Bad accounts,
(39) Allowances,
(40) Salesmen's salaries and expenses,
(41) Officers' salaries and expenses,
(42) Other office salaries and expenses,  %
(43) Miscellaneous,
(44)
(45) TOTAL SELLING EXPENSE,

III. OTHER EXPENSE

(46) Taxes—Income and excess profit,
(47) Interest,
(48) Sinking fund,
(49)
(50) TOTAL OTHER EXPENSE,

CAPITAL AND MAINT. RESERVE CHARGES:

(51)
(52) Improvements,
(53) Developments,
(54) Maintenance,
(55)
(56) TOTAL CAPITAL AND MAINT. RESERVE CHARGES,

SUPPLIES:

(57)
(58) Supplies on hand 1st of month,
(59) Supplies purchased during month,
(60) Total,
(61) Supplies used during month,
(62) Supplies on hand end of month,

MISCELLANEOUS PROFITS:

(80) Powder,
(81) ...t on pay roll advanced,
(82) Smithing,
(83) Heat, light, and power,
(84) Dwellings and farms,
(85) Stores,
(86) Standard-gauge railroad  ...ipment,
(87) Floating equipment,  ...,
(88) Coke and  ...,
(89) ...er  ore, ( ...ps,
(90)
(91) Total gross profit,
(92) Selling expense,
(93) Other expense,
(94) Total,
(95) TOTAL NET INCOME,

COAL TONNAGE. (2,000 Pounds)

| | PREPARED | RUN-OF-MINE | SLACK | TOTAL |
|---|---|---|---|---|
| (96) SALES: | | | | |
| (97) Local, | | | | |
| (98) Shipped. | | | | |
| (99) Railroad. | | | | |
| (100) Coke. | | | | |
| (101) Power-house fuel, | | | | |
| (102) | | | | |
| (103) TOTAL SALES, | | | | |
| (104) Add Inventory, end of month, | | | | |
| (105) | | | | |
| (106) TOTAL, | | | | |
| (107) Deduct Inventory 1st of month, | | | | |
| (108) Purchased coal, | | | | |
| (109) Total deductions, | | | | |
| (110) | | | | |
| (111) PRODUCTION, | | | | |

(112) Kind of coal,
(113) Name or number of seam,
(114) Thickness of seam,
(115) Number of mines,
(116) Remarks:

Number days worked,
Cause of idle days,

APPROVED AND CERTIFIED CORRECT: ............(Name.) ..........(Title.)
This Report Must be Signed by an Officer of the Company.

## Special Charges

Special charges were generally treated in detail, often spread over a reasonable time rather than allowed in a single month. In treating these a careful study of all reports available was made.

## Special Records

The Federal Trade Commission had required special explanations of all charges out of the ordinary, and all these records were available and were carefully studied and had great influence in deciding doubtful points.

## Outside Profits

The profits from farms, dwellings or stores are not properly mining profits, and accounts of these should be kept separate from mining expense. Where it is found that such accounts are separated, no deduction for such profits should be made, but where dwellings, particularly, are so intimately connected with the mining that no separation is possible, it is proper to include their operation with mining accounts.

## Fuel for Power

In general, charges for fuel for colliery power were allowed, and the tonnage divisor was made to include such fuel. In the opinion of the Committee it would, however, be advisable to eliminate colliery fuel from both sides of the account, and merely to keep a record of the amount used; by this method, the tonnage divisor represents the amount shipped and sold and is susceptible of accurate determination, while the fuel used is approximated, or even guessed at, too often to make the general records containing this item reliable.

Other items requiring occasional adjustment were the inclusion of washing costs, for which an extra charge is allowed, in the mining cost, and the inclusion of labor and supplies used in coking operations conducted by the same operators. In a few instances, the coke tonnage, or a mixed tonnage of coal and coke, was reported and used as a divisor to obtain costs per ton, resulting, of course, in a notably excessive cost.

In general, each and every cost sheet was studied carefully by at least one member of the Committee, all adjustments were considered by at least two, and only such adjustments were made as were warranted by the conditions and the necessity of placing all costs on the same basis, so as to find a just basis of cost for fixing the prices for each district.

## REPORTING COSTS

The great mass of cost figures, obtained from the above analysis of the cost of operators mining over 95 per cent. of the entire production of bituminous coal in the United States, would be merely confusing and of but little practical value if presented in tables of figures, and it was considered necessary to devise some plan to present these graphically so that they might be studied and compared with a minimum of effort and

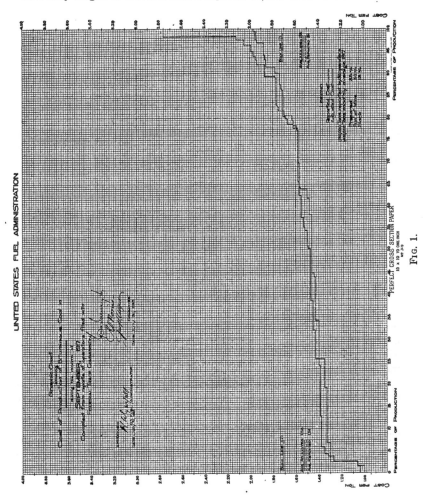

FIG. 1.

with maximum efficiency. After many trials, a chart was evolved which appears to have satisfactorily accomplished the ends sought.

The costs for each district, both exactly as reported and as adjusted, were arranged in order by 1-c. increments, beginning at the lowest cost, with the tonnage at each separate cost, whether from one or more operations; the percentage of total tonnage at each cost was calculated, and the cumulative percentage beginning at the lowest-cost tonnage was obtained.

### Charts

The percentages thus obtained were plotted on 10 by 10-space cross-section paper, resulting in a diagram like Fig. 1.

The dotted line shows the costs reported, and the full line the adjusted costs. The percentage of the total output between, or up to, any limits of cost can be determined by simple inspection. The "bulk line," or line of indispensable coal which must be assured of a minimum profit, after study of the conditions and necessities of any district, can be properly located, and from this a minimum profit necessary for the district can be determined.

### ADJUSTING THE "BULK LINE"

The "bulk line" is a matter requiring very careful study. Its location must be such as to conserve and encourage all necessary operations and thus assure the maximum coal supply from each district.

It is almost invariably found that at the high-cost end of the diagram are collected most of the doubtful enterprises. These include: mines which have failed under normal competitive conditions and have been reopened under the stimulus of the high prices preceding Government control; mines abandoned as exhausted and reopened for the few remaining pillars; new enterprises in the development stage; mines opened on beds so thin or of such poor quality that they could not operate under normal conditions; small mines on outcrop coal, often of poor quality, which have neither capital nor equipment for economical working; mines which have encountered faults or in which the coal has thinned or split, or the quality has so deteriorated as to prevent working at a reasonable cost; and, not the least of this group, mines so badly managed as to show unwarrantable costs of operation.

All these classes of mines are unjustifiable under war conditions. They use labor inefficiently. Often their records show less than half the tonnage per employee usually obtained in their district, and their elimination is an economical advantage to a district in releasing labor to more efficient mines.

In this high-cost group occasionally are found mines which have a coal of unusually high quality or fitted for special use, for which a market at prices above those of the district has always existed. Such mines, on proving their special conditions, may receive consideration for special prices sufficient to allow a fair profit on their higher costs.

After a study of all conditions, the "bulk line" is located as far as possible to exclude the classes of operations above mentioned, and to include all mines operating economically and efficiently. The margin above the ".bulk line" is sufficient to allow all but a very small percentage of the tonnage to be produced without actual loss, but with less than the minimum profit applied to all mines up to the "bulk line."

The charts have the further advantage that they show all the costs of any district without divulging the costs of any operation, yet by a very simple system of confidential keys the position of any separate operation can be almost instantly found and its cost sheets located.

The charts have the further advantage that almost any desired information as to costs or tonnage, averages, totals within desired limits, margins, excessive and subnormal costs with the tonnage involved, and other items of information often required, can be obtained very rapidly and with a minimum expenditure of time or labor. On one occasion, two members of the Committee calculated in a single evening the weighted average costs, both reported and adjusted, for over three-fourths of the entire bituminous coal output of the country.

### DISTRICTING

It is inevitable that mines should show wide variations in cost, due to the varying thickness and character of the beds worked, and to apply a single price to all the mines of a State would result in either allowing an unreasonable profit to those working the better and thicker beds, or absolutely put out of business the higher-cost districts.

Where an area examined shows wide variations of cost, it becomes necessary to employ some plan of separation, and to segregate into groups those mines operating under similar conditions. Such districting may be based on difference in beds, on thickness of coal, or by geographical and geological districts.

*Districting by beds* is only occasionally practicable, for the following reasons:

(a) Variation in thickness and quality in the same bed.

(b) Difficulty of identification of beds.

(c) Splitting of beds, changing one thick bed to two or more thin ones.

(d) Changes in mining conditions in the same bed, making radical differences in cost of mining.

*Districting by thickness* of beds seems at first glance the most logical method, but it has the fatal objection that, as nearly all beds become thin in places, two or more costs will be found in contiguous mines and often in the same mine. Further, this districting leads to gross profiteering, by attempting to classify mines by the thinnest coal, not by the average.

The terms "thick" and "thin" beds are particularly dangerous, as what would be considered thick in one region may be classed as thin in another, and the reverse; it is therefore manifestly undesirable for the United States Fuel Administration officially to designate any particular thickness as the dividing line between thick and thin.

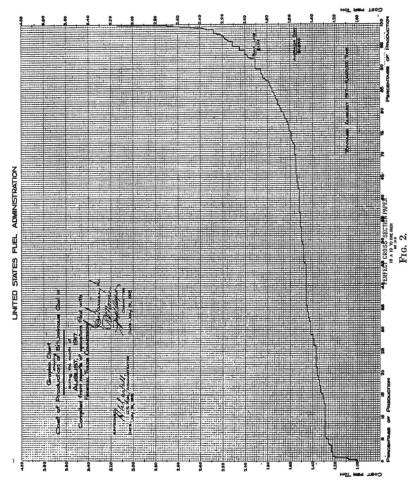

FIG. 2.

*Districting geographically* has the great advantage of making divisions susceptible of accurate description and eliminating all questions as to the proper price applicable to any colliery. It generally puts together mines having the same conditions and normally competing, avoids varying prices for coal of the same quality and character, and simplifies distribution and marketing. The difficulties in applying this method are greatest in fields where numerous beds of varying thickness and character are worked, resulting in considerable variations in price. It is also frequently difficult properly to classify operations near

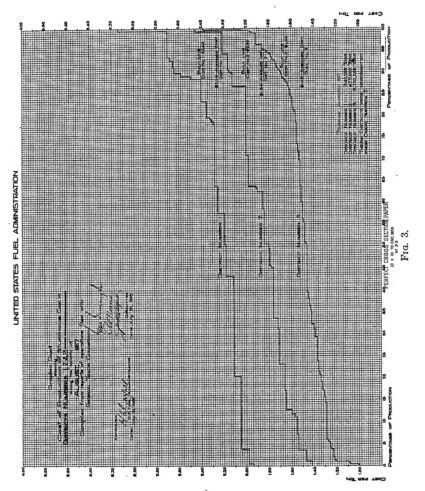

FIG. 3.

the borders of adjoining districts, and geographical districts are hence necessarily subject to some adjustment of boundaries.

Further, in designating districts, labor conditions are necessarily carefully considered. Neither miners nor operators wish to have the scale of wages changed in any mine by the throwing of such mine from one wage district to another, and before deciding on the boundaries of districts, it has been found essential to obtain the wage-scale and the boundaries of wage-scale districts. Further, it is found that, in making up the wage-scales, some very accurate districting has been done, and maps showing these wage districts are of great assistance in the final determination of proper boundaries.

It is the practice of the Committee to classify mines under study, first by counties or fields, and then to separate or combine them, as the case may be, to obtain districts containing, as far as practicable, mines operating under the same general conditions.

Diagrams 2 and 3 show the result of districting in an important territory producing about 75,000,000 tons of coal per year. Diagram No. 2 shows the costs for the entire area, and the very wide variations in cost due to different mining conditions are apparent. Diagram No. 3 shows the costs in the three districts finally segregated. These all permit price fixing without giving excessive profit, or putting the high-cost districts out of business, thus assuring the mining of the required tonnage of coal.

It will be noted in the three-district diagram that the low cost of No. 1 corresponds with the high cost of No. 2, and the low of No. 2 with the high of No. 3. If prices had been fixed on Diagram No. 2, for the entire State, the "bulk line" would have been placed at about $1.90. This would have put the whole of District No. 1 and 36 per cent. of District No. 2, or a production of about 500,000 tons per month, above the bulk line, giving these regions an insufficient margin, and checking production, if not stopping it; at the same time, it would have given to District No. 3, producing 5,700,000 tons per month, 30 c. per ton or over $1,700,000 per month unnecessary margin. Similar conditions throughout the country have been handled in this manner.

It will also be noted that the variation between high and low costs increases with the higher-cost districts. The angle of slope of the cost line increases from No. 3 to No. 1. This results in a somewhat greater margin between the average cost and the "bulk line," which is logical and necessary. The high-cost mines, having greater expenses, need a larger margin to attain the same percentage of profit. As an example of the districting necessary in exceptional cases, Fig. 4 is a map of the districts in West Virginia. On fixing prices for these, only seven different prices were found necessary for 13 districts.

## PRICE FIXING

The "bulk line" of the chart, with the margin added by the Fuel Administrator, gives the necessary realization for a district, but it does not completely fix prices. The price for run-of-mine is usually fixed at

the realization price, but where screening is desirable it is necessary to
fix a spread of price so that the operator receives as much as 1 c. or 2 c.
more for the screened product than for run-of-mine.

The tonnage of run-of-mine, prepared, and screenings for each district
is obtained from the cost sheets; then the average division of the screened
coal in percentage is computed and from this margins are determined
which will permit screening but not too greatly stimulate the practice.
For instance, if a coal will produce 55 per cent. prepared, and 45 per cent.

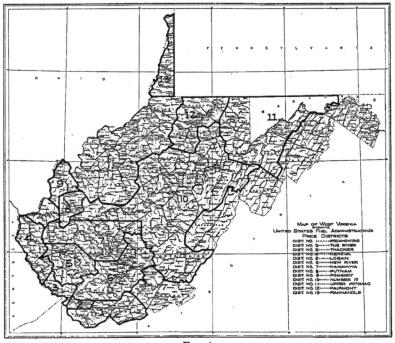

FIG. 4.

screenings, an equal margin above and below the run-of-mine price
would be indicated; but a coal which would produce 30 per cent. prepared
and 70 per cent. screenings would need a much larger margin for prepared
above run-of-mine and a very small margin below for screenings.

If the "spread" is not correctly figured, the result is to make some
combination unduly profitable, with the result that only that combination is found to be made. In some instances, run-of-mine from particular collieries shows a few cents margin over screened coal, and these
collieries will produce run-of-mine exclusively. In other cases, prepared
and screenings are more profitable, with the reverse result.

## Unusual Sizes

In many parts of the country, it was found that special sizes were customarily made. In Illinois, for instance, a considerable amount of coal is sized, about in accordance with anthracite practice, into egg, stove, nut, pea, and buckwheat, and in certain states so-called "modified run-of-mine," passing through 2, 3, 4, 5, 6, and 8-in. bars is a standard product. These specially prepared and modified run-of-mine sizes cost the operators something extra, and will only be made if the prices received yield a profit over the regular procedure. The Fuel Administration has met this condition by allowing special prices for specially sized coal, and for modified run-of-mine sizes, but all such prices are so calculated as to allow only enough profit on any combination to permit its existence and not enough to encourage the forcing of such size on consumers.

In price fixing of this sort, it is essential to obtain, from several independent and reliable sources, the percentages of the various sizes produced by screening from each coal likely to be used in this way, and carefully balance the costs, losses, and percentages of each size produced, to arrive at a proper price.

It is also necessary to evolve methods of preventing profiteering on special prices. For instance, it was found that after making sizes down to buckwheat from bituminous coal, in some cases the fine screenings, far below the standard mesh of the district, were run into the regular run-of-mine and sold at the run-of-mine price. This was handled by an order allowing a maximum of 30 c. below *screenings price* for any mixture of the fine coal below ½-in. mesh with any other coal. This is easy to police, as the mere report of sizes below the standard screenings mesh of any district involves the report of special fine screenings or "carbon" at the price 30 c. below screenings, and if such is not found, it is assumed to be mixed with commercial sizes.

## Margin

The difference between the mine costs, arrived at as above described, and the price is the "margin." This is far from being the profit, as many items of expense necessarily incurred are not included in the mine price. Such are: (*a*) Selling expense. (*b*) Improvements. (*c*) Developments to increase output. (*d*) Excess of capital expenditures over normal costs. (*e*) Contracts at lower than "Government Prices." (*f*) Interest on bonded indebtedness. (*g*) Income taxes. (*h*) Excess-profit taxes. (*i*) Profit on investment.

None of these items is properly included in "cost of production" under normal conditions, but in a war situation it is practically impossible to obtain money to capitalize expenditures for excess improvements, and developments, which would normally be capitalized and properly included in the permitted depreciation, particularly as all such expenditures are made at from two to three times their normal costs; it is a serious question whether the "margin" allowed should not be made large enough to include at least this class of expenditures.

## RESULTS OF PRICE FIXING

As our Government has been forced into this untried realm of price control by war conditions, it may be interesting to know the results. These, in general, are available only as applied to the latter months of 1917, before the labor increase, compensated for by the 45-c. general advance in coal prices above referred to. Diagram No. 5 shows the average costs, "bulk lines," and prices fixed for practically all districts in the country, as of August and September, 1917, and covers about 84 per cent. of the total output of bituminous coal for the period stated.

The costs for each district, in the proportion of its output to the total tonnage studied, are shown in full lines; the "bulk lines" are shown by dotted lines; and the prices fixed are indicated by dot-dash lines. The diagram also shows the weighted average costs, "bulk lines," and prices fixed for the tonnage included, and effectively disposes of the widely circulated aspersions of profiteering, of which industry has been so freely accused by people having no knowledge of the facts or willfully misrepresenting them.

The weighted average margin between costs and fixed prices for practically the entire bituminous coal production of the country is but 45.6 c., and between the "bulk line," which represents the higher-cost necessary coal, and the price fixed by the Fuel Administrator of but 26 c. When it is known that the capital invested per ton of yearly output ranges from $2 to nearly $8, and that the items above noted, which amount to a considerable sum per ton mined, must come out of this very narrow margin, it is evident that the coal business of the country is not only not on a profiteering basis, but is still on a very narrow margin of profit.

The average cost of the 84 per cent. of the total coal represented for the two months of August and September, 1917, was reported to be $1.696. The adjustments heretofore described raised this reported cost to $1.706, a very strong endorsement of the honesty of the reports made by the operators.

The average "bulk line" was fixed at $1.902, or 19.6 c. above the average adjusted cost. This represents the margin required to assure the mining of the necessary coal, as compared with the average cost, which, of course, involves the mining of only coal up to or below the average cost; in other words, half the available output.

The weighted average of all prices fixed is $2.162 per ton and the average margin above the "bulk line" is 26 c., representing all the above mentioned charges and all profit for the higher-cost necessary mines; the margin above the average weighted cost for the whole country is 45.6 c. per ton, which, compared with profits in other

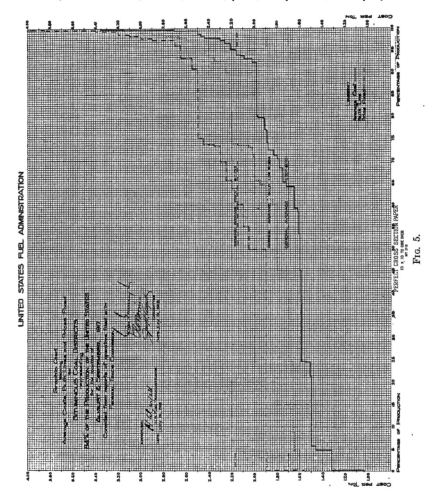

Fig. 5.

businesses, certainly does not show any signs of profiteering in the coal business as a whole. The prices fixed are also sufficient, on the basis of the reported costs, to permit the mining of 98.4 per cent. of all available coal, without loss.

The prices fixed from this complete investigation of costs have shown, in many cases, a remarkable compliance with economic laws. For instance, **in** Illinois the cost of coal from the different price districts, delivered in Chicago, is found to be practically identical, showing that

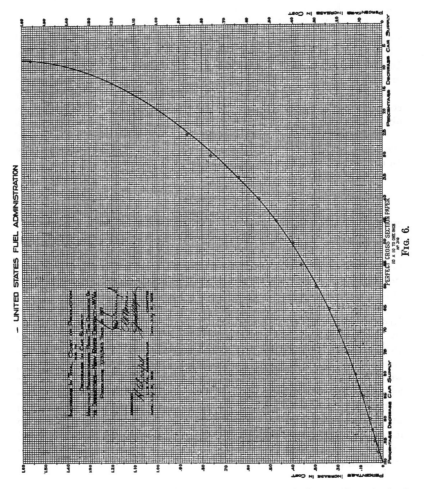

FIG. 6.

the mining of the higher-cost coal is due to its proximity to the principal market and the lower resulting transportation costs. High-grade coal shipped by lake and rail to Minneapolis was found to cost precisely the same per heat unit as a lower-grade coal shipped a much less distance all rail.

## Special Prices

The price-fixing program adopted is expected to take care of all normal mining conditions and to permit the operation, with a reasonable profit, of all mines necessary to the country under war conditions. There are cases, however, of old established mines which are producing coal of a quality specially suited to certain processes or requirements, which coal, while produced at a cost materially above the regional "bulk line," is necessary and has always commanded a special price. In these cases, a higher price, sufficient to permit operation, is usually granted. Also, in some cases, groups of small mines, not required to report, are found to be serving certain communities at prices below other coal available, considering the transportation charges, but with costs above the regional "bulk line." These also receive special prices.

In general, it is the policy of the Fuel Administration to encourage the operators to produce all coal needed and to place restrictions only on coal mined under conditions notably uneconomic.

## EFFECT OF SHORT TIME

While it is a syllogism in mining regions that "short time means increased costs," but little actual information as to the quantitative effect of lost time on the cost of coal mining is available. In the discussion as to the advantages or disadvantages of an even car supply to all mines, as compared with 100 per cent. supply to some and the remainder to the others, it devolved upon the Committee to determine, at least approximately, the effect of lost time on the cost of mining.

Fortunately, reports were available from 73 operators in the New River District of West Virginia, which had been made out and submitted by an eminent firm of expert accountants for each month of the year 1917. Each of these was carefully analyzed, and the percentage increase of cost for each of the 830 observations thus obtained was platted; weighted averages were then taken at each 2.5 per cent. from 70 to 100 per cent. working time, and for each 5 per cent. below 80 per cent. The result of this study is submitted on diagram No. 6 which has been checked by numerous observations from practically every field, and has been found, within reasonable limits, to be correct. This diagram can be and has been used in reducing to normal cost the reported costs of collieries shut down during parts of months.

## ANTHRACITE PRICES

Anthracite prices were fixed by the President on Aug. 23, 1917, and have not since been revised. The matter is now a subject of intensive study, and it is expected that after sufficient data have been accumulated a revision and scientific price fixing will be attempted.

The problem of anthracite price fixing presents all the difficulties encountered in the bituminous fields, complicated beyond measure by the varying percentages of sizes produced by different mines in the same region, and the still more widely varying percentage of sizes produced by the different regions.

## CONCLUSION

It should be generally known that the United States Fuel Administration exists only for war conditions. It expires by limitation of the Lever Act.

The Administration is endeavoring most earnestly to give both to the miners and to the consuming public a fair deal and no favor. It has accomplished incredible results in conservation of fuel and stimulation of output, but such results have only been possible by the earnest, whole hearted, and patriotic support freely given by operators, miners, and by the consuming public.

## DISCUSSION

EUGENE McAULIFFE,* St. Louis, Mo.—The method employed by the Engineer's Committee in arriving at a proper selling price for coal and coke represents hard painstaking effort based on a thoroughly scientific foundation; and the United States Fuel Administration can well point with pride to the work of this committee. The results so obtained placed the whole price-fixing issue above and beyond the sweeping criticism which was directed against the so-called Lane-Peabody prices. The meager cost data available, and the brief time allowed for study of costs, presented an insurmountable difficulty which the Peabody Committee was unable to overcome; its path was in no sense made easy by the attitude of certain producers who looked on the opportunity as one for the exercise of unbridled license, while consumers who hitherto had enjoyed, except for short seasons, an extremely low fuel cost, became not only hysterical in their actions but extremely pointed in their charges. The fact that the Engineer's Committee was able, within a few weeks, to quiet all criticism, winning the endorsement and commendation of both producer and consumer, indicates work of a character and scope deserving of the highest approval.

In the hurry and fervor incident to transforming a nation with but a limited naval force, and with only the nucleus of an army, into a great militant power, little serious consideration is being given to what might be called the readjustment period that will follow the world's war, and the question of making permanent provision for the intelligent control of the coal industry after the war is over is one that may well merit our attention. The experience of the past three years has demonstrated the vital relation that coal as a commodity bears to our governmental and economic structure, and with a war-inflated wage rate in effect when the pre-war competitive condition returns—perhaps suddenly—the very existence of many properties may be jeopardized by an excessive production cost, largely due to the application of uniform unit wage increases made without regard to the physical and market conditions surrounding certain properties.

The doctrine of the survival of the fittest is not a safe one to depend on, inasmuch as the elimination of the unfit might reduce production to a point that would put the remaining producers, temporarily at least, in a commanding position that would prove unsound and perhaps prejudicial to the industry as a whole. While attempting to guard against extraordinary conditions, such as may follow the war, the theory of individualism which gives every man incentive to put forth his best effort must not be lost sight of; it is not only the right but the duty of the State to apply to business a measure of regulation, sufficient to meet the requirements of the well-being of the country as a whole. The rapid unionization of our industrial population bespeaks a further tendency toward what might be termed "limited Socialism" which can, without doubt, be best administered through the medium of intelligent Government control, the regulatory body acting as an intermediary between the producer and his labor on the one hand, and the consumer on the other.

No greater safeguard could be provided for producer and user than the continuation, after the war, of a Federal Fuel Administration, endowed with plenary regulatory powers. The coal industry transcends in importance the street railway, water, gas, and electric power facilities, and should forthwith be classed a public utility of prime importance. Such regulation must be made broad enough to comprehend, not only the stability of the industry, including prices, but also the conservation of our coal resources; the safety of mine employees; all questions of social, intellectual and hygienic character which surround the mine workers; a definite and consistent plan for expansion commensurate with, but not in excess of our needs; the development of such additional mechanical devices as will admit of maintaining production in the face of a decreasing immigration of the men of that class who have in recent years entered the mines. The Engineering Committee has proved all these questions as possible of solution.

## Coal Mining in Washington

BY F. A. HILL,* SEDRO-WOOLLEY, WASH.

(Colorado Meeting, September, 1918)

Coal mining in the State of Washington offers many interesting problems for the mining engineer, due to the varied physical conditions occurring in different fields, and often in the same mine. The different coal seams worked in the same mine commonly require different mining and transportation methods for economical operation. The coal measures have inclinations ranging from 5° to vertical, and in the same mine the dip of the coal may vary from 30° to 50°. In all fields, the seams are extensively faulted, folded, and sheared.

The quality of the coal produced ranges from brown lignite to bituminous coking coal, the following being typical analyses:

|  | Lignite | Sub-bituminous | Coking Coal |
|---|---|---|---|
| Moisture............................... | 23.5 | 14.30 | 0.57 |
| Volatile hydrocarbon...................... | 31.0 | 33.03 | 23.57 |
| Fixed carbon............................ | 37.4 | 41.30 | 65.55 |
| Ash.................................... | 8.1 | 11.37 | 10.31 |
|  | 100.0 | 100.00 | 100.00 |

Very little of the coal contains as much as 0.75 per cent. sulfur.

The brown lignites occur mostly in Lewis and Thurston Counties, and as a rule are very cheaply mined. This coal has been successfully used in locomotives of the Oregon & Washington Railroad, after equipping them with specially designed fireboxes.

The sub-bituminous coals occur in all the coal-mining counties. Their cost of production varies widely, according to the character of the formations, the pitch, and the wall conditions. This makes a fairly good steam coal, but the lump is used mostly for domestic trade, after grading into lump, nut, and pea.

The coking coal is used for making coke, and for locomotive and other boiler fuel. It yields a strong metallurgical coke, quite high in ash, though seldom exceeding 18 per cent. ash. Most of it is susceptible to

---

* Manager, Cokedale Coal Co.

better cleaning than it now receives, but it will be difficult, from any seams now known, to produce a coke containing less than 16 per cent. ash.

Anthracite occurs in Whatcom County. The seams cannot be traced continuously for any great distance, although they outcrop here and there. These deposits lie on the edge of a Carboniferous field, but no seams have yet been found in that locality that would make commercial mines. West of that locality two seams have been worked; one has been closed for many years, and the other is producing a small amount of coal. In the Cowlitz Pass section of Lewis County, anthracite has also been found, but the field has not been thoroughly prospected, partly because it is remote from transportation.

With few exceptions, all the coal under 2 in. in size needs washing, and all over that size requires handpicking to eliminate the rock and heavy bone. Many of the mines yield as much as 30 per cent. waste, the heavy pitch of the seams precluding any separation of waste underground.

The coal has heretofore been graded as follows: lump, over a 2-in. screen; nut, 2 in. to ⅝ or ¾ in.; pea, ⅝ or ¾ in. to ⅛ in.; mine-run includes all but the lump. The coal under ⅛ in. is called sludge, and its market has been irregular; it will form a satisfactory powdered fuel provided it is mechanically dried before leaving the mine. If wet, it will not run, but will stand vertically in bunkers or cars.

The output of coal per employee per shift ranges from 1.55 to 5.60 tons, in profitably operating mines, the brown lignites producing the highest and the coking coals the lowest amount per employee.

Methods of mining are varied; room and pillar, chute and cross-cut, and an angle system which originated in Washington, are all used. In some cases planes are driven up the pitch, with room and pillar workings off the planes. Elsewhere, rooms are driven up the pitch in pairs, with gravity planes to handle the cars. Methods do not always follow the cheapest plan, where high percentage of lump coal is desired. Puncher machines are successfully used in a number of mines, producing a greater amount of lump and also costing less than shooting the coal.

Lump coal has been selling at $3 to $5.50 per ton f.o.b. cars; nut from $2 to $3.75; pea from $0.75 to $1.10; and sludge for any price that could be obtained, some of the coking sludge, however, selling at $1 per ton.

Bailey Willis estimates the thickness of the coal measures in the Wilkeson field at 10,000 ft. There are many coal seams in that field; in one locality 10 different seams are being worked, and there are three more that could be profitably mined. One geologist estimates 90,000,000 tons of coal in one 640-acre section.

The main coal field of the state is 56 miles long from north to south, and six miles wide; a large part of it has never been properly prospected, owing to its inaccessibility, the heavy growth of timber, and insufficient demand for the coal.

The opening and equipping of a coal mine in Washington will cost from $2 to $10 per ton of annual output, depending largely upon the cost of development to produce the desired output. This cost is hard to determine, even by those best acquainted with the field, unless the outcrop can be located by surface prospecting.

During the past 10 years fuel oil has been selling at coast points from $0.65 to $0.85 per bbl. Oil has been supplemented by coal to the extent of about 2,000,000 tons annually; hence coal operators, except those producing coke and the better grades of domestic coal, have fared poorly. With fuel oil at $1.50 per bbl., no doubt more coal properties will be opened.

## DISCUSSION

MILNOR ROBERTS,* Seattle, Wash. (written discussion†).—The coal fields of Washington, on which Mr. F. A. Hill's paper gives much detailed information, lie on the western slope of the Cascade Range, which extends north and south through the state just west of its center. The mines are on the lower half of the slope, the lignite fields of Lewis and Thurston Counties extending into valleys and low hills west of the mountains. An exception is to be noted in the case of the Roslyn field, a small but important area which lies on the eastern slope.

The degree of alteration which the coal of a particular field in Washington has undergone may be gaged roughly by the position of the field with reference to the Cascade Mountains. The lignites of Tenino, Tono and Castle Rock, on the railway line connecting Seattle and Portland, all occur in a region of low relief, in which the Eocene coal meaures have suffered only minor disturbances. Sub-bituminous coals, which are most typical of the state, occur in the foothills, as at Renton and Newcastle. Coals high in fixed carbon are found farther in the mountains, where the measures have been sharply tilted and folded, for example at Black Diamond and Carbonado. The two anthracite fields are located still higher, in rugged mountains where outcrops appear both in deep gorges and on ridges at elevations approaching 5000 feet.

The costs of mining appear erratic when compared with those in other states. The lignites, as a rule, are mined cheaply, as stated in the paper, but so also are certain other beds which lie at convenient angles of dip, have good roof, and contain gravity coal. Naturally, such beds already discovered have been attacked first.

Washington coals are being burned successfully in pulverized form in a copper-smelting reverberatory, in the heating plants of several buildings, and in a large central heating station in Seattle. It is impor-

---

* Dean, College of Mines, University of Washington.   † Received Sept. 13, 1918.

tant for the coal-mining industry that uses should be found for the fine sizes of coal; these products are considerable in amount, but scarcely any demand exists for them at present.  Briquets have been in use for domestic purposes for several years, and now the use of pulverized coal seems likely to offer a market for the fine sizes in the lignite and sub-bituminous grades.

The use of pulverized coal in boiler plants, to be located at the mines, is a possible source of power to compete in the cities of the Puget Sound region with hydro-electric power.  Washington has greater water-power resources than any other state in the Union, but it seems probable that coals of the cheaper grades and sizes can produce power in competition with hydro-electric plants, excepting those that enjoy unusually low costs. In future this comparison should prove still more favorable to coal, for the reason that water wheels have already reached a very high degree of efficiency, while power derived from steam falls far short of the ideal. Both the Puget Sound Traction Light and Power Co. and the City of Seattle have steam-power auxiliaries to their large hydro-electric power systems.

The coal fields of Washington are the only extensive ones in the Pacific Coast States; they supply a large territory and the coal is used for a wide variety of purposes, from gas making to bunkering steamers. Under such conditions it is unusually important that the coal be mined with the least waste, prepared as carefully as possible, and used to the best advantage.  The Pacific Northwest Station of the United States Bureau of Mines, located at the College of Mines, University of Washington, and working in coöperation with the College, is giving especial attention to these three phases of the industry.

## Use of Coal in Pulverized Form

BY H. R. COLLINS,* ALLENTOWN, PA.

(Colorado Meeting, September, 1918)

THE purpose of pulverizing coal before burning it is to make available every heat unit it contains. Machinery has been developed which will pulverize coal in one operation, delivering it to bins in front of the furnaces at an expenditure of about 17 hp.-hr. per ton, in a medium-sized plant. The cost of the operation depends upon the amount of moisture that must be expelled before pulverizing, the cost of labor, and the cost of coal delivered at the plant. At a small plant, requiring a pulverizer with a capacity of only $\frac{1}{2}$ ton per hour, the cost per ton pulverized will naturally be greater than at a plant requiring the largest pulverizer, with a capacity of 7 tons per hour.

The first step is to reduce large lumps to a size suitable for drying uniformly, before passing to the pulverizing mills; this is done in rolls, at a single pass. The second step is the elimination of moisture, in order to facilitate pulverizing to great fineness, while also increasing the heating effect and the temperature attainable when the coal is burned. There are other mechanical advantages in the handling of dried coal.

Driers are now manufactured which are able to eliminate moisture without distilling any of the volatile combustible matter in the coal; they are fired by hand or with pulverized fuel. The heat first surrounds the shell of the drier, being confined within a chamber where complete combustion takes place; the heated gases are then led through a duct to the discharge end of the drier and enter the inside of the shell at a temperature not exceeding 300° F. This temperature is maintained by the operator and is indicated by a pyrometer. Volatile combustible matter is not likely to be distilled until the temperature rises above 400° F.

On discharging from the drier, the coal is usually passed over a magnetic separator in order to prevent pieces of iron from going to the pulverizer. Two types are used: a magnetic pulley which automatically discharges its collection of iron, and a lifting type, from which the iron is removed by hand when convenient.

In the operation of pulverizing, the coal should preferably be reduced until 95 per cent. will pass through a 100-mesh and 70 per cent. through a 300-mesh sieve. Such a product is obviously an almost impalpable powder.

---

* Mechanical Engineer, Fuller Engineering Co.

After pulverizing, the fuel is conveyed by one of several methods to the point where it is to be used.  In several installations the pulverized coal is conveyed a distance of over 900 ft.  Where possible, a bin should always be installed at the furnace, in order to guard against interruption of supply.

Feeders are practically indispensable for regulating the passage of the fuel from the bin to the burner.  They are now made quite simple and highly efficient.  They deliver the pulverized coal in definite quantities into an air current of fixed volume, where the air disseminates the pulverized fuel, surrounding every particle and putting it into condition to develop all its energy.  The first to ignite are the volatile gases; these raise the temperature to the ignition point of the solid carbon, and before leaving the zone of heated air every particle has released its last heat unit.  It is entirely possible to obtain temperatures ranging between 1900° and 3500° F.; the highest temperature (3500°) I have observed was in an open hearth, when the average temperature of the furnace itself at the same time ranged from 3100° to 3200° F.

To justify the expense of erecting a special building, and installing special machinery to pulverize coal, the following advantages in its use may be enumerated:

1. Conservation of the country's fuel, by utilizing every heat unit in the coal, made possible by this method of consumption.

2. Reduction of labor for handling coal to the point of consumption, handling by the fireman, and the removal of ash and unconsumed fuel from the ashpits; practically all of this expense is avoided when fuel is burned in pulverized form.  All the coal is received at one point, and thereafter it is handled entirely by automatic machinery, the human element being thereby eliminated, except for supervision, adjustment, and repairs.

3. From actual experience with many grades of coal we believe that every carbonaceous fuel in solid form, from lignites to the graphitic anthracites of Rhode Island, will yield its maximum measure of heat, if burned in a truly pulverized condition.

4. Coal in pulverized form can be injected into a furnace on a column of air, at very low velocity, thus allowing the expanding gases to liberate their heat without erosion of the refractories.

5. Pulverized fuel permits the maintenance of a constant temperature in a furnace, when the relative amounts of fuel and air have once been set, and the body of the furnace has been brought up to the desired temperature.  It will continue thereafter under what is known as a test condition.  Furnaces can be operated in this manner hour after hour, as shown by charts of recording pyrometers.  The correct relationship between the amount of pulverized fuel and the volume of air, for any desired temperature, can be controlled automatically, after adjustment

to the particular grade of coal in use, thus using a minimum of excess air. We have obtained gas analyses as high as 17 per cent. of $CO_2$.

Many questions are asked on the subject of pulverized coal, and I have arranged the answers to them in the following order.

1. *Grades of Coal Used Successfully.*—For kilns, boilers, or metallurgical furnaces, coals of about 35 per cent. volatile, 50 per cent. fixed carbon, 8 per cent. or less ash, and 2 per cent. or less sulfur, are preferable.

2. *Experience with High-ash and High-sulfur Coal.*—Coals analyzing 25 per cent. ash, and 5 per cent. sulfur, have caused no trouble in kilns, boilers or metallurgical furnaces. Experience shows that sulfur is entirely consumed by burning in suspension, none of it being absorbed by the metal or other liquid bath, as does occur in the usual copper reverberatory furnaces used for melting electrolytic copper.

3. *Provisions for Storage.*—Storage bins for pulverized coal should be dust-tight and have steep hoppers, enabling old coal to leave the bin completely; accumulations of old coal are liable to fire, smoulder, and coke, causing more or less annoyance.

4. *Why Coal Should be Dried before Pulverizing.*—(a) To facilitate the pulverizing operation, giving the finest product with the least power consumption. (b) To permit high temperatures with the least consumption of fuel. Drying of the coal also promotes uniformity of temperature. (c) Dried coal will flow more easily from bins and through the feeders and burners. Coal should be dried to 1 per cent. of moisture, or less when possible, except that lignites can be readily handled with 5 to 8 per cent. combined moisture.

5. *Fineness of Grinding.*—The finer the coal the more rapid its combustion, with relatively higher efficiency. It is commercially and economically possible to grind coal so that at least 95 per cent. will pass a 100-mesh sieve and 70 per cent. will pass a 300-mesh sieve.

6. *Cost of Handling, Grinding, and Upkeep.*—The cost of preparing pulverized coal depends largely on the price, and on the moisture content of the coal to be used. On the basis of 200 tons per day, of coal containing 7.5 per cent. moisture, at present rates of wages and supplies, and with coal ranging in price from $1 to $8 per ton, the cost of pulverizing will be between 30 and 40 c. per ton, not including overhead charges, interest, or depreciation.

7. *Danger from Dust Particles Floating in the Air.*—A mixture of coal-dust particles in air will not ignite until it reaches a certain density; on the other hand, a mixture that is too rich in coal dust has a tendency to smother flame. Dust clouds should naturally be avoided outside the furnace chambers, and all sparks or flames should be kept away. Pulverized coal should be conveyed from mills to bins in as compact condition as possible; air currents should not be used to convey pulverized coal

if any other method can be devised. Leakages should be stopped, to prevent uncleanness and accumulations in inaccessible places. Carelessness in the handling of pulverized coal, and poorly designed plants, are the only causes of so-called explosions.

8. *Essentials of a Good Feeder.*—It must absolutely control the flow of pulverized coal to the burner, and prevent any rush or flooding of the fuel. This is essential for positive control of predetermined temperatures.

9. *Essentials of a Burner.*—A good mixing projector, or burner, should be so designed that it will receive the pulverized coal in regulated quantities, break up the stream of fuel, and so distribute it that each particle is surrounded by the correct proportion of air. It must also project the fuel into the furnace at the velocity required by the operation, and must be so proportioned as to deliver the necessary volume of air at the proper velocity. Four types of burners are employed:

(a) Induction type, in which a high-velocity jet induces and entrains the necessary additional air, and projects it into the furnace at low velocity; this type has the high-velocity air under control as well as the induced air.

(b) Positive type, in which the high-velocity air induces and entrains the fuel and projects it into a positive, larger column of low-velocity air, thereby breaking up the fuel stream evenly, and disseminating it through the larger column of low-velocity air, before it enters the furnace. The larger column of low-velocity air is usually preheated, in stoves located in a chamber through which the waste gases from the furnace pass; temperatures of preheating range from 100° to 600° F. in the better designed system of stoves. Both columns of air are positive, being generated by fans or pressure blowers, while gates regulate the quantity.

(c) Single type, in which the high-velocity air first induces and entrains the fuel stream, after which a high-pressure jet of air, applied usually in the center of the stream, gives a sharp projection of flame and quick distribution of the fuel through a larger volume of preheated air at quite low velocity. This type of burner is usually adjustable in direction. The heated air ranges in temperature from 2200° to 1300° F., as in open-hearth practice; usually 10 to 15 per cent. of the air enters with the fuel, and 85 to 90 per cent. from regenerators. The stack draft through the regenerative chambers is regulated by a valve.

(d) Single type, in which the high-velocity air induces and entrains the fuel and projects it into the furnace, as in rotary-kiln practice, under usually 5 to 6 oz. pressure from a fan. The additional air required for combustion is induced by stack draft, and enters around the hood and through the kiln discharge opening.

10. *Air Pressure and Effect of Stack Draft.*—Air pressures of ⅜ oz. entering the combustion chambers of some types of furnaces, from air

and fuel mixing burners, up to 2 lb. in pressure jets of other types, have been in successful, constant use for years. Stack draft should be of only sufficient intensity to create a partial vacuum in the furnace, thereby helping the fuel and air into and not out of the chamber; its strength must be enough, however, to extract all the products of combustion.

11. *Design of Furnace.*—Fuels low in volatiles but high in fixed carbon, as anthracite and coke braize, require a special furnace, in which the incoming fuel and air pass through the flame and the products of combustion, in a water-cooled, arched firebrick chamber, on their way to the furnace or boiler. When the volatile constituents of the fuel range from 1 to 3.5 per cent., it is difficult to support combustion unless a temperature above 900° to 1000° F., the flash-point of carbon, is maintained. The water-cooled arch, rear wall, and side walls are made of a special form of firebrick, which slips in place over the water tubes. After circulating through the tubes, the water passes to the hotwell or heater at approximately 190° F., entailing no appreciable loss of heat.

Lignites and bituminous coal require no special furnace. As the volatiles ignite between 600° and 700° F., from the radiant heat of the walls, the flame is self-supporting, and every heat unit in the fuel is liberated before coming into contact with any cold surface.

12. *Ash or Furnace Slag.*—Anthracite, coke braize and lignite ash do not slag. Bituminous coal ash will slag on the bottom of the furnace chamber, if not blanketed with cooler air, properly admitted, and if allowed to remain too long in the furnace. Most of the ash from pulverized coal passes away through the breeching to the cyclone, where the ash is separated from the gases. That portion which settles to the floor of the furnace should be removed from time to time; the quantity is very small and it is quite light.

13. *Furnace Temperatures and Slag Formation.*—Temperatures between 1800° and 3500° F. can be maintained in the flame. Slag forms more readily at high temperatures, necessitating proper blanketing with cooler air, always remembering the advisability of obtaining the maximum percentage of $CO_2$. We frequently obtain 16.5 to 17 per cent. $CO_2$ under operating conditions.

14. *Checkerwork in Metallurgical Furnaces.*—Experience seems to point to the necessity for vertical baffle walls where the waste gases enter the regenerative chambers. Turning the direction of the gases up and down several times tends to discharge the dust tangentially, allowing the major portion to settle in the bottom of the passages, whence it is easily removed, through proper cleaning doors at the sides, not interfering with the operation of the furnace. The gases then filter through checkerwork properly spaced and installed. The narrow side of the brick tile should be laid vertical and on rider walls, to permit

the use of longitudinal scrapers to remove the ash which may have passed by the vertical baffle walls. This arrangement will undoubtedly give the regenerating chambers a life equal to the best record ever attained, as the narrow edge of a vertical tile presents very little surface for the flocculent ash to rest and close the gas passages.

15. *Furnace Life.*—The life of furnaces in which pulverized fuel is used is equal to that of hand-fired, stoker, oil, or gas-fired furnaces. By absolute control of the quantity of coal and air, the velocity of the expanded gases can be reduced until erosion of refractories becomes hardly discernible.

16. *Economy of Pulverized Coal.*—In this connection, all the benefits of pulverized coal should be taken into account: labor saving, increased fuel efficiency, ability for closer adjustment, absence of smoke, etc.

The efficiency of hand firing depends upon the skill and reliability of the fireman. With the best of attention, a loss of 20 per cent. heating value is frequent, and it often reaches as high as 40 per cent., taking into consideration the analysis of the ash and of the flue gas.

Stoker firing is relatively more efficient and more regular than hand firing, but the feeding of moist coal wastes part of the heat in the most undesirable place. Losses also occur in breaking and removing the clinker, in the discharge of unburned fuel, and in the flue gas.

Producer-gas firing: Referring to W. H. Blauvelt's results,[1] we find that 131,280 cu. ft. of gas was produced from 1 ton of coal, and contained 20,311,162 B.t.u., or 155 B.t.u. per cubic foot, or 2270 B.t.u. per pound of gas. The composition of the coal from which this gas was made was as follows: water, 1.26 per cent.; volatile matter, 36.22 per cent.; fixed carbon, 57.98 per cent.; sulfur, 0.70 per cent.; ash, 3.78 per cent. One ton contains 1159.6 lb. carbon and 724.4 lb. volatile combustible, the energy of which is 31,302,200 B.t.u. Hence, in the process of gasification and purification, there was a loss of 35.2 per cent. of the energy of the coal. Producers are built today doing slightly better than this.

Oil and natural-gas firing: Coal, properly pulverized and burned, is on exactly the same basis as far as thermal capacity is concerned; and the price of the coal prepared and delivered into the furnace is directly comparable, on the heat unit basis, with the cost of fuel oil or gas delivered into the furnace, plus the slight additional cost for ash removal.

Savings by the adoption of pulverized fuel in the operation of various types of furnaces have been attained as follows: heating and bushelling furnaces, 20 to 25 per cent.; puddling furnaces, 30 to 50 per cent.; open-hearth furnaces, compared with gas producers, 30 to 40 per cent.; copper reverberatory, smelting ore, 30 to 45 per cent.

---

[1] *Trans.* (1889) **18,** 614 and quoted in Kent, page 819.

In other furnaces, the consumption has been reduced to the following figures: continuous billet heating, 160 lb. of coal per ton of billets; desulfurizing iron ore in rotary kilns, 296 lb. of coal per ton of ore; drying and nodulizing iron ore in rotary kilns, on basis of 30 per cent. free moisture and 11 per cent. combined moisture, 477 lb. of coal per gross ton of ore.

The above figures are from actual operations over extended periods, and confirm my contention that coal burned in true pulverized form is the only method by which every heat unit in the fuel will develop its full value.

## DISCUSSION

E. A. HOLBROOK,* Urbana, Ill.—To those who have followed the development of powdered coal, two questions often occur. First, as to the moisture in the coal. In Illinois we recognize that the bituminous coals contain about 12 to 15 per cent. of what we call "contained moisture," that is, the moisture is present in the coal although this may appear absolutely dry and dusty. In addition, the surface moisture may be present in any amount. I would like to ask what is the relative effect of these two forms of moisture upon the work of pulverizing. Does Mr. Collins refer to the contained moisture or to the surface moisture in estimating the degree to which the coal must be dried?

H. R. COLLINS.—In the lignite coals, which this country possesses in great amount and is beginning to develop to considerable extent, the "contained" or "combined" moisture does not trouble us; we have no difficulty in reducing lignites to a high degree of fineness. But when dealing with bituminous, sub-bituminous, or semi-bituminous, we prefer to reduce the combined moisture as well as the hydroscopic moisture as far as possible.

We have recently put into operation, at a plant, furnaces and boilers consuming the lignite from northern New Mexico, which contains between 4 and 5 per cent. moisture as we pulverize and burn it.

The evaporating efficiencies under the boilers have been so remarkable that we wish to check the results half a dozen times before quoting them. We have also just started two boilers at the Garfield smelter, and have obtained the same high efficiencies.

E. A. HOLBROOK.—My second question relates to the objectionable slagging that sometimes occurs when pulverized bituminous coals are used as boiler fuel. The first pulverized-coal installation I saw in this country was of Bettington boilers at one of the Nova Scotia coal mines; there their great difficulty arose from the slagging of the ash due to the exceedingly high temperatures.

* Supervising Mining Engineer and Metallurgist, U. S. Bureau of Mines.

H. R. COLLINS.—Slag, as a rule, does not seriously affect metallurgical work, but in boiler practice it has given a lot of trouble and required a great deal of study and experiment. Temperature we were all familiar with, but failed in the beginning to take into account the expansion of gases. If the air goes into the furnace preheated, its volume at that temperature must be taken into account. For many years, furnaces were made too small, and when we began to develop these high temperatures we were continually burning out our brick-work, especially when it became necessary, as in electric generating plants, to drive the boilers at 200 to 225 per cent. of their rated capacity. The driving of the gases through restricted passages, at high temperature and high velocity, caused almost complete erosion of firebricks in from 2 to 3 weeks.

After a great deal of experiment, we found that if the velocity of the heated gases were reduced, erosion of the firebrick ceased. In boiler installations, in order to complete the combustion of the pulverized fuel in suspension, we increase the volume of the combustion chambers.

The disadvantage of slagging has been greatly reduced by the practice of burning all the carbon in the fuel while in suspension, so that partly burned carbon is not projected against the refractories; and the passing of a column of cooler air from the outside through the bottom of the firebox diminishes slagging. To increase our $CO_2$ to 16 or 17 per cent., we diminish the air going through the burners and let it go in at the bottom. This seems to eliminate slagging, and the ash deposited is fine and flocculent, resembling discolored flour.

WALTER GRAHAM,* Washington, D. C.—Regarding the application of pulverized coal to the open-hearth steel furnace, Mr. Collins states that the sulfur in the coal will not detract from the quality of the steel, and that the ash does not interfere in the regenerators.

At present it is my special task to stimulate the production of steel and improve its quality. I am trying to help increase the quantity of steel by 10 per cent., which will be about 3,200,000 tons, or equivalent to the output of several large steel works.

The use of gas coal has not been very successful lately in steel mills. I found one works in which they were making only about 20 per cent. of their potential production, the trouble being mostly with the gas coal, which was unscreened. They were taking the gas from one furnace 10 hr. to get another one out in 24 hr., when the time should have been 10 or 12 hr.; they were making 9 heats instead of 21. Two of those were high in sulfur, and were taken for nut steel; of the seven others we rejected 65 per cent., leaving less than 20 per cent. of the potential output. The cause was mostly the poor preparation of the coal.

*Captain, Ordnance Department, U. S. Army.

While I was in the Fuel Administration, I was made a special investigator to report on the cleaning of coal. On going through the records of the Bureau of Mines and the Geological Survey, I estimated that there is 5 per cent. more ash and sulfur in the coal this year than before the war; this means 12.5 per cent. less efficiency. It also means 650,000 extra carloads of ash and sulfur to be carried this year, requiring 50,000 coal cars; a loss of 12.5 per cent. efficiency is equivalent to 85,000,000 tons of coal, which is greater than our shortage.

The importance of cleaner coal is most pressing. I see the necessity for it everywhere, and particularly in the steel works, where it affects the quality of pig iron, the quality of coke, and the quality and quantity of steel. It increases the consumption of manganese in the steel furnace and of limestone in the blast furnace, and is cutting down production.

H. R. COLLINS.—Copper is one of the most active absorbers of sulfur that we know. A year and a half ago, we made a trial of pulverized coal on one of the American Smelting and Refining Company's copper furnaces in New Jersey, melting electrolytic copper. The tests lasted a week, and the analysis of the resulting copper showed no increase in sulfur. The same seems to hold true in steel. The act of burning the coal in suspension, in such a finely divided state, seems to consume the sulfides, although sulfates are not effected.

BRADLEY STOUGHTON, New York.—It is embarrassing for a steel man, in a large company of coal men, to say that we blame the sulfur in our steel partly on the coal, but nevertheless we do. The statement that Mr. Collins has just made is contrary to all the experience of steel men, and I shall be very happy if he will show us that steel does not absorb a part of the sulfur that goes into the furnace in pulverized fuel, or in oil, or in gas. Any evidence that Mr. Collins has on this point ought to be presented, because it is very important. I do not think that the copper experiment is quite conclusive as to the behavior of steel.

H. R. COLLINS.—I am of the opinion that Mr. Stoughton has taken a wrong view of my remarks. I do not claim that steel does not absorb part of the sulfur that goes into the furnace, if it is present to be absorbed through not being properly burned out, and I suggest the following as a probable reason for the results that we seem to have obtained.

Sulfate is a volatile salt of sulfuric acid. Sulfide is a volatile compound of sulfur with a metal or other element. In the finely divided and thoroughly mixed particles of pulverized coal, combustion is so rapid and the temperature in the flame is so high that volatilization of the sulfide evidently takes place, and its percentage of sulfur is probably so thoroughly mixed in the large volume of waste gases that it becomes almost negligible and passes out of the furnace before it can enter the bath through the covering of slag.    This probably does not occur when burning fuel oil, as perfect atomization is very hard to obtain with a consequent slower combustion rate and retarded volatilization, allowing more time for sulfur, if present, to enter the bath.

With producer gas, the gas passes from the producer to the furnace at a low temperature (not sufficiently high to volatilize the sulfur) where it mixes with the air from the regenerators.    The heat slowly builds up and probably passes one-third of the way across the bath before the high temperature point is reached, which probably allows some of the sulfur to enter the bath before complete volatilization takes place, and the sulfur can leave the furnace with the waste gases.

MILNOR ROBERTS,[*] Seattle, Wash.—Coal is being burned in pulverized form in the State of Washington in large-scale operations under the general conditions described by Mr. Collins.    The coals that are being used are described in F. A. Hill's paper on "Coal Mining in Washington."[1]    The powdered coal is being used for heating, smelting and power generation.

The Puget Sound Traction, Light and Power Company's Western Avenue plant in Seattle, a central heating station containing 10 boilers aggregating 4100 hp., has replaced oil fuel with pulverized coal.    The pulverizing plant is located directly across the Avenue, the product being conveyed under ground and delivered to feeding bins above the boilers. Coal prepared at this plant is also distributed by truck to a number of large buildings having their own heating plants.    The Western Avenue station is believed by its operators to be the largest of its kind that is using powdered coal today.

Fifteen tests of Washington coals, in lots of about 10 tons, have recently been made at the Western Avenue station in a 300-hp. Babcock & Wilcox boiler.    The results, as reported by the chief engineer, are in part as follows: Evaporation per pound of coal, from and at 212° F., 8 to 9.3 lb.    Carbon dioxide in flue gas, 13 to 17 per cent.    B.t.u. per pound of powdered coal, 9688 to 12,734.    Efficiency, 65 to 77 per cent. In operation, an evaporation of 8.5 lb. is expected.    By way of comparison, it may be noted that the same company's Post Street power plant, where chain-grate stokers are used, with similar coals, under boilers aggregating 3000 hp., shows an evaporation of about 6 lb. per pound of coal.

Steam power plants in which coal is burned in pulverized form are in

---

[*] Dean, College of Mines, University of Washington.    [1] This volume: p. 371.

operation at the Black Diamond coal mines, where nine boilers are in commission, and at Newcastle, where the plant has been partly converted to the method. At the Tacoma Smelter pulverized coal has replaced oil in a reverberatory smelting furnace which treats copper ores.

It is feasible to use any of the Washington coals in powdered form, but the lignite and the sub-bituminous types have proved especially suitable. Coals of these grades, in fine sizes, may be bought on contract at low prices because they form practically a waste product.

ERSKINE RAMSAY, Birmingham, Ala.—Captain Graham's interesting remarks about the country's great necessity for clean coal bring to my mind certain advances made along this line in Alabama.

There is a strong demand today for some comprehensive means of adequately and properly determining the daily quality of the particular coal sent out by each and every individual miner, it being important that the means adopted shall show the percentages of slack, lump, and slate contained in the cars sampled, and it is essential that the cars thus sampled shall constitute a large proportion of the total cars dumped daily. Heretofore there has been no way to accomplish this.

The system in use at many Alabama mines consists simply of selecting at random from time to time, during each day, as many as, say, a dozen mine cars from various miners, the entire contents of each being laboriously hand-picked in order to ascertain just how much slate is in each one. The amount of slate found in each car picked is regarded as the fair average for all of the cars coming from the particular miner loading the sampled car.

One or two slate pickers are daily employed at the average mine. They pick by hand an average of about 1 per cent. of the total mine output, but, of course, a varying and indefinite proportion of the cars loaded by each miner. Fixed penalties, known as dockages, are made for given percentages of slate. At a few mines, however, as an inducement for the loading of clean coal, where the car picked is found to contain less slate than a fixed amount, a bonus or premium is given, which offsets in some degree the amount docked. As most of the mines do not give any bonus or in any way reward the loading of clean coal, there is not any special incentive for the loading of coal free from slate. With the daily picking of such small proportions of the total mine cars dumped, some of the miners may feel that they can safely ignore such examinations, and that dirty coal would run a good chance of not being detected.

The old system is uncertain and inadequate, for the reason that cars selected at such long intervals cannot possibly give the true average quality of all coal loaded by each individual miner. Sample cars from the same miner, selected at different times during the day, show varying percentages of both slate and slack, as cars loaded during the early part of the day's work, and especially from rooms, contain relatively high percentages of lump and low percentages of slate, as compared with those loaded later in the day when the places are being cleaned up.

Slate pickers are often paid so much per can of slate found in the cars they pick. This leads to confusion and discontent. The miners protest that the present haphazard system, as between man and man, is unfair, and it results in frequent and troublesome disputes between the management and the men.

The Ramsay mine-run sampler, which we are using, avoids such annoying disputes and disagreements, and secures without friction or trouble the loading of a much more satisfactory and uniform grade of clean coal. Improved conditions are brought about not only without additional expense, but even at less total cost. Our experience during the past 3 years has proved advantageous to both the operator and the clean coal miner. The miner who is compelled to improve the character of his work or go elsewhere is the one who habitually loads dirty coal. If he objects to this new plan of taking a fair sample from each of a large number of his cars, it is because it plainly and positively shows the character of coal he loads every day. One of the best points, however, is that, under our new system, a man loading clean coal is not subjected to a penalty on account of the dirty coal loaded by his fellow miner.

In order that every miner may know every day what grade of coal each miner is loading, the sample record as to both slate and slack is entered daily on a public slate tally sheet kept at the mine, and at the end of the month each man's monthly average is also shown. The moral effect of making this information public is good, even without penalties, as no man wants to be found at the foot of his class. When a plan entailing a penalty or the giving of a premium is in effect, the averages are used in determining the amounts.

In using our new sampler system, relatively small samples, running from 100 to 200 lb., are taken from a large portion of the total mine cars dumped daily. The apparatus we have in operation takes these small fair samples from the whole contents of a mine car. These samples are screened, so that the slate is quickly and thoroughly picked by hand, using the screen as a picking belt. When it is borne in mind that the time and labor required to hand-pick one of our 3-ton cars of coal, under the old way of sampling, is sufficient with the new plan to screen, pick, and record the 100-lb. samples from at least 75 different mine cars, the advantage of the sampler is apparent.

At our Banner mine, it formerly required three men to pick the entire coal contained in a limited number of mine cars, while with our present system three men take and pick the great number of relatively small samples. Formerly eight men were required at the picking belt, but now four men easily do the work. The efficiency and economy of our new arrangement is found in the surprising fact that the total amount of refuse going to the waste dump fell off 50 per cent. shortly after the sampler was installed. To put it differently, the refuse going to the waste dump, before the installation of the sampler, ran over 20 per cent. of the total output of the mine and this is now less than 10 per cent. On the mine's total monthly output, 40,000 tons, we get a saving of more than 4000 tons of coal, f.o.b. railroad cars, which means that 4000 tons of slate are now being kept in the rooms and in its place 4000 tons of coal are sent out.

The sampler screen speeds up the actual treatment of the samples, and gives a record of the percentages of slack and lump, thereby showing what miners are shooting their coal to pieces. This feature is especially valuable at mines loading domestic coal.

The sampler has a series of pockets or compartments arranged transversely in the bottom of the tipple chute. The covers permit any sample pocket to be opened at will, thus varying the portion of the car from which the sample is taken, as the pocket can be opened at any time during the flow of the stream of coal from the car. In this way samples are taken from any part of the car. Five samples are taken by the New Castle machine just as fast as the cars can be dumped and without interfering in any way with the speed of the dumping operation. The samples are taken, screened, picked, weighed, and recorded at the rate of one every 2 minutes. With the data furnished by the sampler, the operator knows what parts of the mine, if any, are giving the best product, and he can therefore make changes of correction where necessary. Where the coal is loaded in the railroad cars as run-of-mine, the tally sheet record not only shows the quality of coal loaded in every railroad car but it actually shows who loaded the coal.

H. N. EAVENSON, Gary, W. Va.—Mr. Collins states that he can burn any fuel having from 1.5 to 3 per cent. of volatile matter, but that, for almost all purposes, he prefers fuel having 35 per cent. or more. The highest-grade coal that we use is the semi-bituminous coal, which is low-volatile (about 15 per cent. to 20 per cent.) and low-sulfur, but high-heat content coal, running as high as 15,500 B.t.u. per pound. I would like to know what his experience with such coals' has been; whether they are less desirable than the higher-volatile coals on account of trouble with the brick work, or in operation, or on account of the higher price usually charged for those coals.

H. R. COLLINS.—The low-volatile anthracites can be burned satisfactorily only in a special furnace, for the simple reason that the temperature must be brought high enough to burn the carbon as it enters. As to the higher-volatile coals, there is a point between the low-volatile and those having the percentage of volatiles mentioned at which some slight trouble is encountered, but there is no danger of not being able to burn any of it. The only annoyance occurs when designing the proper fire-boxes for fuels ranging from 1.5 up to 25 per cent. volatiles. The size of the returning arch must be reduced as the volatiles increase. With coal having 25 per cent. volatile, there is no trouble in burning it in an open chamber.

A. V. ADAMSON,* New York, N. Y. (written discussion).—The experience of users of pulverized fuel in metallurgical work, particularly for open-hearth furnaces, has demonstrated that high ash and sulfur in pulverized fuel are a detriment, the sulfur in the coal entering into the finished product; hence, in this class of work it is extremely important, when pulverized coal is to be used, that the coal be selected with care, and only those having low percentages of sulfur be used, in combination with proper means for combustion.

I have found that no general statement can be made as to the cost of pulverized coal, each installation being an individual engineering problem and necessitating a varying capital expenditure; as each installation has its own power cost, the actual cost of pulverizing varies between wide margins.

It has been found that no general rule can be laid down as to either fineness or dryness; for some uses a wide range is permissible in both factors, while some classes of fuel require different treatment from others.

Anthracite, coke breeze, and lignite ash will, under certain conditions, form slag, the melting point of the ash from fuel of this character being practically the same as of that from bituminous coal; furnace temperatures above the melting point of ash must therefore cause the formation of slag. However, it should be noted that furnace temperatures attainable with this class of fuel are not usually so high as those from bituminous coal, which probably explains the statement that the former do not slag. In a well designed furnace, temperatures above 2250° F. (1235° C.) will be obtained with that class of fuel, and slags will be formed.

Too much attention apparently has been centered on high $CO_2$ contents of stack gases. A high percentage of $CO_2$ is frequently obtained at the expense of the brickwork, the furnace temperature being increased by diminishing the excess air; a high percentage of $CO_2$ thus simply means a small proportion of $O_2$.

I cannot agree that the life of a pulverized-coal furnace is as long as that of other forms of furnaces. If the statement be modified to the effect that the life will be the same as that of other forms of furnaces

---

* Construction Engineer, Locomotive Pulverized Fuel Co.

operating under the same capacity rating, this may be true. My experience is that with most pulverized-fuel systems the life of the furnace is less. This expense, however, is more than offset by the advantages of using coal in pulverized form.

In general, the papers on this subject are inclined to lay too much stress on the advantages of pulverized coal, regardless of its manner of burning, and overlook the operating difficulties. A wide range of choice is open as to the equipment for drying and pulverizing, since the development of this machinery has been in progress for many years in the cement industry. Hence, the furnace problem is reduced to proper handling, feeding, and burning.

Ash, slag, and clinker are a real problem; to prevent their formation on tubes, checker-work, and furnace walls, is absolutely essential. Methods of combustion, and furnace design are of vital importance. Scientific analysis of the difference in heat transmission due to radiant fuel and ash in suspension has offered a real opportunity for research. None of these items can be disregarded; yet the beaten highway of cement practice has offered no answer because in long-flame rotary kilns, such as are used in cement work, these questions are of no importance.

For four years the organization with which I am connected has labored with these subjects, and I am frank to say that the ground has not yet been fully covered. Such progress as has been made is due to progressive development based on practical operating experience on locomotives, heating furnaces, and boilers. The facts that have been definitely settled can be summarized as follows:

1. Pulverized coal is not suitable for all purposes, and the character of the fuel available is one of the factors causing the necessity for separately considering each proposed installation.

2. Honeycomb and ash on tubes, and slag on walls and furnace floor, can be controlled.

3. Flame velocity must be reduced to a minimum.

4. No continuous operating efficiency or continuous good results can be obtained with pressure air for combustion, even if only ⅜ oz. pressure is used; and the more nearly a furnace approximates a balanced-draft condition, the better the operating results.

5. The incoming fuel must be mechanically mixed with the air-conveying medium to prevent a pulsating flame and the formation of coke due to lumps of pulverized coal in the feed.

6. In general, the future for the use of coal in pulverized form for steam generating, metallurgical, and chemical work, offers a real opportunity for fuel economy and increased production; it is the one way by which shortage of oil, gas and coal, together with transportation and labor difficulties can be overcome. Due consideration, however, must be given to the difficulties inherent in and the troubles incident to improper

design and installation. In other words, an installation for the economic utilization of fuel in pulverized form is not one which can be purchased over the counter, and special experience or development is essential for each individual installation.

THE CHAIRMAN (J. W. RICHARDS, South Bethlehem, Pa.).—I understand that Mr. Adamson has had experience with powdered coal in locomotives. Are you free to tell us of that?

A. V. ADAMSON.—My experience with locomotives was in Brazil for the Government-owned railroads. Twelve of these locomotives were built and equipped for pulverized fuel in America and shipped to Brazil in the early part of 1917, where they were put into fast-passenger service. The purpose was to burn native Brazilian coal, a means of using which had not been found. On the regular run of 110 miles for the round trip and several hours standby, 6 metric tons of Brazilian coal were consumed. Compared with the same amount on the hand-fired locomotives using American coal at approximately $40 per ton, this effected a saving for each trip of approximately $120. After 6-weeks' service test on one of these locomotives, without opening the firebox except for the dumping of the ash and slag, the flue sheet and tubes were found to be absolutely clean.

These Brazilian coals run very high in sulfur, ranging from 4.5 to 6 per cent. However, no difficulties were experienced due to the sulfur in the coal, neither did it have any deleterious effect on the boiler. These coals run high in moisture and ash, the ash running from 22 to 30 per cent. The moisture was reduced before pulverizing to 3 per cent., and less if possible. A low grade of Brazilian lignite was successfully burned on several trips, although the moisture after pulverizing ran as high as 12 per cent.

CHAIRMAN RICHARDS.—I would like to ask Mr. Collins if the coal must necessarily be dried to 1 per cent. moisture in every case. If it is dried to that extent and is not used at once, coal will absorb 1 or 2 per cent. of moisture from the air. I have been told that coal has been pulverized satisfactorily though containing much more than 1 per cent. moisture.

In regard to the efficiencies attained by pulverized fuel, I believe that the heating arrangement of the furnace itself is a most important factor. Thus, when burning lump coal a grate is necessary, entailing losses through the bars and sides of the grate, and radiation through the roof; whereas, when burning powdered coal, the heat losses from the firebox end of the furnace are largely eliminated.

Mr. Collins states that the costs of different coals prepared and delivered in the furnace are directly comparable on the heat-unit basis.

I wish to urge caution as to the comparing of different fuels on the heat-unit basis, as usually determined. The calorimeter gives a thermal value which includes the heat of condensation of any water produced by the combustion; in practice, this heat of condensation is never utilized and the calorimetric results ought to be corrected to that extent before comparing the practical calorific powers of different fuels. That fuel which has the largest amount of combustible hydrogen and yields the most water, will show a disproportionately high calorific power in the calorimeter, because it will be credited with the heat of condensation of all the water that is formed, which, in practice, always goes up the chimney, as vapor. I should recommend using the practical or metallurgical calorific power of the fuels, and not the calorimetric value, as ordinarily used.

H. H. STOEK, Urbana, Ill.—I should like to ask how much of the ash is lost by passing up the stack, and would this not be troublesome in settled communities?

H. R. COLLINS.—From a boiler of 250 hp. rating, and running at about 150 per cent. capacity, we obtained about a barrel of ash out of the firebox in 24 hr. Some ash settled in the back chamber underneath the heating surface of the boiler, and the rest passed out of the stack. The fine ash is so light that it floats high in the air, and it lands probably miles away.

CHAIRMAN RICHARDS.—One disadvantage of pulverized coal as a matter of national economy is that it offers no possibility of saving any byproducts. Hence the carbocoal process, followed by pulverization of the semi-carbocoal, from which some of the volatiles have been extracted, would seem to offer great economies.

H. R. COLLINS.—The Chairman's remark that one disadvantage of pulverized coal is that it offers no possibility of saving any byproducts, as a matter of national economy, is quite true, but when we consider the great quantity of fuel burned today without any thought of saving byproducts, it would seem that the saving possible from burning fuel in pulverized form alone is an economy it was hardly possible to obtain under the older methods.

While it is quite true the carbocoal process permits the saving of byproducts and as these byproducts are undoubtedly of great value, there is no reason for, or objection to, pulverizing the semi-carbocoal and placing it in a condition to utilize all the heat units remaining, on heat work for which it can be adapted.

R. F. HARRINGTON,* Boston, Mass.—I should like to ask about the application of pulverized coal to malleable and gray iron. At

---

* Hunt Spiller Mfg. Co.

one time, considerable difficulty was experienced, due to the deposition of the ash on the metal bath.

H. R. COLLINS.—As a general rule, the molten metal is drawn first, and the ash, which forms slag on the top of the bath, leaves the furnace last, the same as in open-hearth practice; hence the deposit of ash ought not to be objectionable.

In heat-treatment furnaces, particularly those for malleable iron, there is no trouble at all if the flues are correctly proportioned to the amount of fuel burned, so that the velocity of the spent gases will be sufficient to carry off the ash.

THOS. A. MARSH,* Chicago, Ill.—I would like to ask what success has been attained in the burning of pulverized coke, a fuel containing less than 3 per cent. volatile.

H. R. COLLINS.—In order to burn pulverized coke, or other fuel low in volatiles, the temperatures must be higher than usual, since carbon ignites at around 900° to 1000° F. Hence we have found it necessary to pass both anthracite and pulverized coke underneath an arch, returning the ignited fuel around the incoming fuel so as to raise its temperature above the ignition point; after which the flame passes into the furnace itself. That is the only way in which we have been able to burn the low-volatile fuels satisfactorily.

CHAIRMAN RICHARDS.—Is it not possible to mix a certain proportion · of bituminous coal with anthracite or coke, in order to gain the advantages of both?

H. R. COLLINS.—This can readily be done; about 32 per cent. bituminous coal, ranging above 30 per cent. in volatile matter, can be mixed with anthracite, and the combustion made nearly perfect.

---

* Green Engineering Co.

# Carbocoal

BY CHARLES T. MALCOLMSON,* CHICAGO, ILL.

(Colorado Meeting, September, 1918)

An elaborate series of experiments has been conducted during the past three years at Irvington, N. J., which has resulted in the perfection of a process for the manufacture of smokeless fuel from high-volatile coals, and for the recovery and refinement of the coal-tar products derived therefrom. These experiments have been financed by Messrs. Blair & Co., of New York, and were conducted under the direction of Charles H. Smith, a member of this Institute, and the inventor of the process.

The low-temperature distillation of coal has interested investigators for many years. Sporadic attempts have been made to solve the mechanical problems, but until the Smith process was developed, they were not carried to conclusions of economic value. The present coal shortage and the increasing demand for smokeless fuels make this subject one of timely interest.

## Description of Plant

The following equipment was installed and operated during the experimental period: Four horizontal and two vertical units of commercial size for the low-temperature distillation of the coal; two vertical, two horizontal, and two inclined benches for distillation of the briquets at medium and higher temperatures; presses and auxiliary equipment necessary for making briquets; and a complete byproduct recovery and tar-refining plant. This commercial equipment is provided with gas and electric meters, pyrometers and other apparatus for recording accurately the results of all experiments.

There is, in addition to the commercial equipment, a complete chemical laboratory with distillation and recovery apparatus, including facilities for refining and cracking the tar and measuring the yields and calorific value of the gas. This apparatus makes possible a study on a small scale, of the various problems involved in the process.

## Description of Process

Mr. Smith's experiments have resulted in the production on a commercial scale of:

1. A fuel, called Carbocoal, which, for convenience in handling, is prepared in briquet form.

* President, Malcolmson Briquet Engineering Co.

2. A yield of tar more than double that obtained·in the ordinary by-product coking process.

3. Ammonium sulfate in excess of that normally recovered in the ordinary byproduct coking process.

4. ˙Gas,·in amount approximately 9000 cu. ft. per ton of coål carbonized, which is at present used in the process.

The essential features of the Smith process are the two distillations carried on at different temperatures, first of the raw coal and second of the raw briquets. The raw coal, after being crushed, is distilled at a relatively low temperature, 850° to 900° F., and the volatile contents are thereby reduced to the desired point. The result of this first distillation is a large yield of gas and tar, and a product rich in carbon, termed semi-carbocoal. The semi-carbocoal is next mixed with a certain proportion of pitch obtained from the tar produced in the process, and this mixture is briquetted. The briquets are then subjected to an additional distillation at a higher temperature, approximately 1800° F., resulting in the production of carbocoal, the recovery of additional tar and gas, and a substantial yield of ammonium sulfate.

The characteristic feature of the primary distillation is that it is continuous and that the coal is constantly agitated and mixed during the entire operation. This is accomplished by a twin set of paddles which also advance the charge through the retort. By this means, all portions of the charge are uniformly distilled, and by controlling the speed at which the charge moves through the retort, the distillation may be arrested at any desired stage. As only a partial carbonization is permitted in the primary distillation, the hard metallic cells characteristic of coke are avoided. The period of distillation is 1 to 2 hr., and the continuous retort has a carbonizing capacity of one ton of coal per hour.

In the subsequent distillation of the briquets, all evidence of the pitch as a separate ingredient disappears. There is a marked shrinkage in the volume of the briquet, with a corresponding increase in density, but no distortion of its shape. ˙This distillation requires 4 to 5 hr., and is performed in an inclined, self-charging and self-discharging bench.

The carbocoal represents more than 72 per cent. of the weight of the raw coal, the exact percentage depending upon the volatile content of the latter.

## CHARACTERISTICS OF CARBOCOAL

Carbocoal is dense, dustless, clean, uniform in size and quality, and can be readily handled and transported long distances without disintegration. It is grayish black in color, slightly resembling coke, but its density more nearly approaches that of anthracite coal. It is manufactured in briquet form and can be made in any size, from 1 oz. to 5 oz. The larger sizes are better suited to locomotive purposes, and the smaller sizes for domestic use.

Heretofore, devolatilized fuels, such as coke, have not attained the high rates of combustion desired for locomotive, marine and general steam purposes; and their greater displacement has operated against their general use where transportation cost or stowage space has been an important factor. Carbocoal overcomes these objections. It is actually a relatively soft but tough form of carbon, readily attacked by oxygen in combustion; and for this reason, requires much less draft than other high-carbon fuel.

### CARBOCOAL FOR STEAM PURPOSES

Carbocoal has been tested by the Long Island Railroad; by the Pennsylvania Railroad at its testing plant at Altoona; by the Carolina, Clinchfield & Ohio Railroad; and by the United States Navy.

These tests have demonstrated that the fuel is smokeless; that it will evaporate from 8.5 lb. of water at a combustion rate of 100 lb. per square foot of grate surface per hour, to 12.8 lb. of water at a combustion rate of 27 lb. per square foot of grate surface per hour, from and at 212° F. per pound of fuel fired; and that it requires no greater draft than bituminous coal. A maximum combustion rate of 166 lb. per square foot of grate surface per hour has been reached for a short period.

Carbocoal has been found particularly suitable for the following purposes:

1. Marine and locomotive service, where limited grate area and restricted boiler capacity demand efficient fuel; where smoke is objectionable or dangerous, as in the case of ships in time of war.

2. Stationary boilers, where smoke pollution of the air is offensive and dangerous to health.

3. Domestic uses, including furnaces, stoves, ranges and open grates, where cleanliness is a desirable factor.

4. Kilns, drying and roasting ovens, and all purposes where an intense and uniform heat is desired.

5. In metallurgical furnaces, as a substitute for coke.

6. Gas producers.

### CARBOCOAL AS A DOMESTIC FUEL

Carbocoal has been subjected to practical tests in household use for over a year. It fulfils all requirements of a domestic fuel. It can be burned satisfactorily without change of furnace or grates, responding readily to changes in draft. The uniformity of combustion, absence of fines, even distribution of ash, and absence of clinker as compared with the coal from which it is made, are additional characteristics in favor of this fuel.

Tests have demonstrated that carbocoal can be banked satisfactorily over night, requiring no more attention, and with no greater consumption, than anthracite.

### COMPARISON WITH OTHER BRIQUETS

Carbocoal is compressed into briquet form to obtain maximum density, to minimize transportation costs and the losses incident to handling; and to secure the efficiency of combustion resulting from uniformity of size.

Briquets of bituminous and anthracite coals have been manufactured for many years. Such briquets are made from the smaller sizes or screenings of coal, with the addition of a binder, such as coal-tar pitch. In carbocoal, however, an entirely new product is obtained, differing from the original coal in chemical and physical properties. The briquets contain no binding material to soften or disintegrate in the fire. Carbocoal must therefore be recognized not as a mixture, but as a new product, the result of a process of manufacture.

### ANALYSIS OF CARBOCOAL

The amount of ash and sulfur in the carbocoal depends upon the characteristics of the coal from which it is made. The summarized proximate analyses of carbocoal, manufactured from 25 different coals at the Irvington plant, are shown in Table 1.

TABLE 1.—*Analyses of Carbocoal*

|  | From Run of Mine, Per Cent. | From Washed Coal, Per Cent. |
|---|---|---|
| Moisture | 1.00 to 3.00 | 1.00 to 3.00 |
| Volatile matter | 0.75 to 3.50 | 0.75 to 3.50 |
| Fixed carbon | 82.00 to 88.00 | 85.00 to 90.00 |
| Ash | 8.50 to 12.00 | 7.00 to 10.00 |
| Sulfur | 0.5 to 1.50 | 0.6 to 1.50 |

The percentage of byproducts recovered from clean coal is greater than that recovered from high-ash coals; therefore the careful preparation of the raw coal by washing or other means is profitable.

### TAR AND ITS PRODUCTS

The total yield of tar by the carbocoal process is large. Coal containing 35 per cent. volatile combustible produces more than 30 gal. of waterfree tar per short ton.

The tar products recovered from the distillation of the coal, at the low temperature used in this process, are different in nature from those obtained in other processes where high temperatures are used. At the lower temperature, there is an abundance of tar vapors, and a relatively small yield of gas of high illuminating value. At the higher temperature these primary products are split up, with a consequent increase in the gas yield and a corresponding decrease in its illuminating value and in the amount of tar vapors recovered. There is also an increase in the percentage of residuals, the pitch increasing from 30 per cent., in the low-temperature distillation, to 64 per cent. or more when high temperatures are used.

The tar obtained in the primary distillation of the coal has a specific gravity of 1.00 to 1.06. It contains a large percentage of light solvent oils, tar acids, and cresols, but very little carbolic acid and no naphthalene or anthracene. The free-carbon content of this tar is low. The light oils contain appreciable quantities of naphthenes, pentane, hexane, hexahydro-benzenes, and also hydrocarbons of the paraffine series, which make these oils valuable as a substitute for gasoline.

A satisfactory method of removing the paraffine and aromatic portions of the light oil has been developed, so that c. p. benzol and toluol can now be obtained by this process. During the low-temperature distillation period, 20 to 28 gal. of tar, including the light oil obtained from the stripping of the gas, are recovered, the exact amount depending upon the volatile content of the coal. This low-temperature tar contains approximately 30 per cent. of pitch and 70 per cent. of tar oils, as compared with 50 to 60 per cent. of pitch and 40 to 50 per cent. of oil products contained in ordinary gas-house and coke-oven tar.

In the second or high-temperature distillation, 5 to 6 gal. of tar are added to the above yield. This tar is heavier than that obtained from the first distillation, and is similar to coke-oven tar.

Table 2 compares the tars and light oils obtained in the production of carbocoal with those obtained in the ordinary byproduct coking processes.

TABLE 2.—*Recovery of Liquid Products per Ton of Raw Coal*

| | Distillation Temperature °C. | Byproduct Coke Oven | | Carbocoal, First Distillation | | Carbocoal, Second Distillation | |
|---|---|---|---|---|---|---|---|
| | | Gal. | Per Cent. | Gal. | Per Cent. | Gal. | Per Cent. |
| Light oil........ | 0–170 | 0.27 | 3.47 | 1.58 | 6.60 | 0.003 | 0.05 |
| Middle oil...... | 170–230 | 0.44 | 5.85 | 3.29 | 13.70 | 0.036 | 0.60 |
| Creosote oil.... | 230–270 | 0.78 | 10.37 | 3.11 | 12.95 | 0.126 | 2.10 |
| Heavy oil....... | 270–360 | 1.26 | 16.81 | 8.88 | 37.00 | 2.485 | 41.42 |
| Pitch........... | ........ | 4.66 | 62.18 | 6.90 | 28.75 | 3.290 | 54.83 |
| Loss........... | ........ | 0.09 | 1.32 | 0.24 | 1.00 | 0.060 | 1.00 |
| Totals........ | ........ | 7.50 | 100.00 | 24.00 | 100.00 | 6.000 | 100.00 |

In addition to the above yield of tar, there is obtained, in both the byproduct coke oven and the carbocoal process, by stripping the gas, from 2 to 3 gal. of light oil. This yield depends upon the characteristics of the coal carbonized.

Approximately 30 per cent. of the fractions from 170° to 360° C. in the carbocoal process are tar acids; the remainder of the fractions are neutral oils.

The value of the products from the distillation of tar depends, of course, on the extent to which the tar is refined. The fractionation and subsequent treatment of the tar oils, which is a part of this process, give the products shown in Table 3, in carbonizing 1000 tons of coal; the figures are based upon data obtained from carbonizing run-of-mine coal from Clinchfield, Va.

## TABLE 3

1. Carbocoal............................................ . . 725 tons.

*Analysis*

| | Raw Coal | Carbocoal |
|---|---|---|
| Moisture.................... | 0.72 | 1.84 |
| Volatile matter.............. | 35.01 | 2.75 |
| Fixed carbon................ | 57.23 | 85.64 |
| Ash ................ ......... | 7.04 | 9.77 |
| | 100.00 | 100.00 |
| Sulfur..................... | 0.63 | 0.52 |

2. Sulfate of ammonia...........................20,000 to 25,000 lb.
3. Other nitrogen products, principally pyridine bases  2,000 to  4,000 oz.
4. Motor spirits...................................1,800 to 2,200 gal.
   (or c. p. benzol, 250 gal., c. p. toluol, 500 gal., motor
   spirits, 1000 gal.)
5. Crude tar acids, principally cresylic acids...............  4,040 gal.
6. Water-white naphthas...............................  3,500 gal.
7. Creosote oil....................................  5,450 gal.
8. Heavy creosote oil ................................  4,660 gal.
   Other products, used in process:
9. Pitch..............................................  10,000 gal.
10. Gas, of 530 B.t.u., approximately....................9,000,000 cu. ft.

Pitch is always an element of questionable value in tar distillation. The Smith process, however, utilizes all of its pitch for briquetting the semi-carbocoal produced by the first distillation. Moreover, the valuable portions of this pitch are recovered in the tar and gas resulting from the second distillation. It is therefore noteworthy that all the tar products recovered by this process are oil derivatives.

### AMMONIUM SULFATE

The primary distillation of the raw coal gives only a small yield of ammonium sulfate. The secondary distillation of the raw briquets, however, brings the amount up to approximately 21 lb. per short ton of raw coal carbonized.

## GAS

In the primary distillation of the raw coal, from 5000 to 6000 cu. ft. of gas per short ton of coal is recovered. This has a heating value of 650 to 700 B.t.u. per cubic foot. The distillation of the briquets yields also about 4000 cu. ft. of gas of 350 to 400 B.t.u. per cubic foot. The process in its present stage of development uses all of the gas recovered from both distillations.

## AVAILABLE COALS

The carbocoal process has been applied to both coking and non-coking coals. It has been found to work satisfactorily with the non-coking coals of Utah, Washington, and Illinois, and the coking coals of Pennsylvania, Virginia, West Virginia, Tennessee, and British Columbia. Through the application of this process, many of the black lignite or sub-bituminous coals of our Western States may be converted into a fuel of higher economic value.

## BYPRODUCT GAS PRODUCERS

Another application of the Smith process equal in importance to the manufacture of carbocoal is the adaptability of a certain part of the process to the production of electric power. The carbon residue, or semi-carbocoal, the residue of the primary distillation, is a non-caking fuel, practically tar free. Although representing 72 per cent. of the originally coal by weight, it contains nearly all the nitrogen originally in the coal. It therefore provides an ideal fuel of high nitrogen content for the byproduct producer. A combination of this kind would recover the maximum percentage of tar and ammonia products, with a large yield of low B.t.u. gas, which can be burned satisfactorily under steam boilers, and thus produce cheap power.

Byproduct gas producers have been used to a considerable extent in Europe, where fuel is expensive and where non-caking coals have been available; but in America the development of this industry has been retarded by difficulties arising from the use of caking coals.

## DISCUSSION

W. ROWLAND COX,* New York, N. Y. (written discussion†).—The process described by Mr. Malcolmson undoubtedly represents a great stride toward conservation of our natural resources. Without discussing the subject from an economic standpoint, I wish further to emphasize the fundamental principles of the Smith process in order that its economic value may be fully appreciated.

---

\* Mining Engineer.          † Received Aug. 7, 1918.

The process is essentially a means of producing an ideal fuel, as regards its combustion qualities, by the removal of the so-called volatile matter from the coal, in the form of byproducts, which have value for other purposes far in excess of their fuel value. In other words, coals which contain large amounts of gas, tar, and ammonia, and are not suitable for domestic or industrial fuel, or for treatment in byproduct coke ovens or gas retorts, because they possess poor coking qualities, can be prepared for commercial use by the Smith process with a very high degree of efficiency.

Much work has been done along this line, both in this country and in Europe, in developing a means of preparing coal, as the economic advantages have long been recognized. To the best of my knowledge, the Smith process is the only one that has proved its commercial adaptability.

Distillation of coal at 900° to 1000° F. (480° to 540° C.) yields a large amount of tar and oils. The residue of the distillation, however, is a very friable semi-coke, wholly unsuitable for use as a fuel, unless it can be charged directly into a byproduct gas producer. The problem which confronted the inventor was then to convert this carbon residue into a commercial product, and it is in this step that the Smith process has overcome the obstacles which prevented a number of other methods from becoming commercially practicable.

It was found that by pulverizing the carbon residue it could be briquetted with pitch as a binder. Such briquets, however, were not smokeless, and possessed all the other disadvantages, such as softening during combustion, smell, and other characteristics of briquets made with a pitch binder.

The inventor found that by distilling the briquets made from the carbon residue, at a temperature between 1400° and 2000° F. (760° and 1095° C.), the material was completely devolatilized, and a briquet of high specific gravity was produced.

It was also proved that by making the first distillation incomplete, the final briquets burned very freely as compared with similar briquets made from anthracite or coke, and it is this rapid combustion that so distinctly differentiates this fuel from other low-volatile fuels now on the market.

As to the byproducts of the process, I would emphasize one feature. It is well known that low-temperature carbonization produces a high yield of light oils of the benzol series, but with such large amounts of paraffines present that the toluol will not be accepted by the manufacturers of explosives. The inventor and his associates, I am advised, have developed a means of removing the paraffines, and are producing higher yields of benzol and toluol than in ordinary coke-oven practice.

J. M. FITZGERALD,* Rochester, N. Y. (written discussion†).—Until the end of the war at least, carbocoal will help to satisfy the enormous requirements of the Government for a smokeless fuel for the use of vessels engaged in crossing the Atlantic; just now, the Government is practically commandeering the output of the principal mines in West Virginia, Maryland, and Pennsylvania, producing the so-called "smokeless coals," for their use. The need for smokeless coal increases as more ships are launched, and it is entirely probable that within a year bunker requirements will have grown to such proportions as to demand practically the entire output of most of the mines that produce low-volatile coals in Pennsylvania, and for a large portion of the product of the Pocahontas and New River fields. About a year ago, ships were required to carry a "sufficient quantity of smokeless coal to carry them for two days through the submarine zone;" as the submarines now have no limits to their zone, it has become necessary to use smokeless coal for the entire trip between America and Europe.

This new fuel has been given a most thorough test by the United States Navy, and it has been shown to be not only smokeless, but particularly suitable for marine and locomotive service where a high-grade fuel is demanded because of restricted grate and boiler capacity.

As to the yield of oils by this process, Dr. Mollwo Perkins, in a recent paper[1] on the importance of the oil requirements of the Navy, spoke of recovering oil from coal by a low-distillation process similar, as to temperature, to the primary operation of the Smith process. Mr. Perkins stated that to produce 300,000 gal. of oil daily, only 15,000 tons of raw coal would need to be treated, and that the erection of suitable plants in the coal districts would be simple. The adoption of such measures would release, for other use, 100 tank steamers now employed in carrying oil to England from Mexico and other remote places.

Dr. Perkins' figures are sensational when one stops to figure that to build 100 tank steamers, averaging 6000 tons dead weight, would involve an expenditure of around $120,000,000. To build plants that will treat 15,000 tons of coal per day, according to estimates, would cost only $25,000,000. While America will not be able to effect so large a saving in ocean transportation, the building of these plants in the important coal fields will introduce economies in railroad transportation, and reduce the cost of power and fuel for domestic use.

An important subject in every home is the supply of domestic coal. In the east, anthracite coal is used almost exclusively for domestic purposes; little coke is used, because, owing to its bulk and the corresponding

* President, Davis Machine Tool Co.
† Received Aug. 13, 1918.
[1] *Journal,* Institution of Petroleum Technologists (Apr., 1918) 4, 121–125.

increase in freight charges as compared with coal, it can not readily move to points far from its place of manufacture.

Only a few months ago, Professor Breckenbridge, of New Haven, estimated that the anthracite coal measures would be exhausted in about 100 years. One of the dominating features of the new process is that it eliminates from further consideration the possible exhaustion of anthracite; carbocoal is smokeless, is said to be even more cleanly for domestic use, and it suffers no breakage. As bituminous coals in nearly all of the fields of the United States are susceptible to this treatment, this smokeless fuel can be made available for use in many sections so distant from the anthracite fields that the cost of hard coal is prohibitive. This, it is argued, will go far toward the saving of transportation and of cars.

Briefly, the Smith process gives promise of supplying to nearly every bituminous producing section of the country the means of manufacturing its raw coal without loss of the valuable byproducts. What has retarded the more general installation of byproduct coke plants in this country has been the fact that, to dispose of some of the most abundant byproducts, the plants would have to be located near large centers of industry and population. The Smith process makes it practicable to treat, by the low-distillation method, coals of proper quality wherever they may be mined. The valuable byproducts can be shipped to any point. If there is not sufficient local demand to consume the full output of carbocoal made from the residue, this fuel can be shipped in competition with run-of-mine or prepared coal. Having the same weight as anthracite or bituminous coal, it will move at the same freight rate, and not bear the handicap of extra charges that are maintained on coke by reason of its bulk.

NEWELL W. ROBERTS,* New York, N. Y. (written discussion†).—One of the most valuable points brought out in Mr. Malcolmson's paper is the quantity and quality of the byproducts.

In the early days of coal distillation, the carbonization of coal was carried on primarily for the yield of gas, and the tar resulting therefrom was considered a necessary evil. The yield of tar, however, although considered large in those days, was relatively small as compared with present-day results. Even yet we find that the yield of byproducts does not fully meet our requirements. The coke-oven plants are producing an average of little more than 7 gal. (26.5 l.) of tar per ton of coal carbonized, although some are producing as high as 9 gal. (34 l.) per ton. In addition to the tar, by stripping the gas, some 2 to 3 gal. of light oil is obtained, the average yield being slightly under 2.5 gal. (9.4 l.), of which approximately 0.3 gal. (1.13 l.) is toluol.

* Vice-president, International Coal Products Corporation.
† Received Aug. 7, 1918.

The Smith process produces an average of 30 gal. (113.5 l.) of tar from both distillations, from coal having volatile contents of 35 per cent.; approximately 24 gal. (91 l.) are obtained from the first, and 6 gal. (22.7 l.) from the second distillation. This tar is unusually light in gravity, ranging from 1 to 1.06, while in some cases the gravity has fallen below 1. The difference between Smith tar and coke-oven tar is very well brought out in Mr. Malcolmson's paper. Coke-oven tar contains 62 per cent. of pitch, and the tar from the Smith process less than 40 per cent. This, however, is to be expected. In the first place, the primary distillation is carried out at a low temperature, and there is no cracking of the oils and vapors. That the difference in yields is due to the difference in temperature at which the two processes are operated is very well illustrated by a simple cracking tube. Take almost any oil and pass it through a tube at 800 to 900° F. (430° to 480° C.) under atmospheric pressure, and there is practically no cracking. Increase the temperature to 1500° F. (815° C.) and large quantities of gas and tar are formed, the latter containing 40 to 60 per cent. of pitch. This is what occurs in coke ovens; the oils come in contact with the hot coke, or the heated walls of the retort, and are cracked, producing gas and reducing the yield of tar.

The value of the tar from the Smith process can only be estimated, but it will undoubtedly be more valuable than any coal tar now on the market. Of the 30 gal. of tar produced by both distillations, 15 to 20 gal. (57 to 76 l.), dependent on the raw coal used, can be recovered in the form of marketable oils.

The light oil from the tar, as well as the light oil from the gas, contains, as might be expected, a certain quantity of paraffin hydrocarbons. These vary from 4 to 20 per cent., depending upon the nature of the coal used. Ordinarily, light oil containing paraffin hydrocarbons in the amount just mentioned would be valueless for use in the manufacture of c. p. benzol and toluol, but with the Smith process it has been found possible entirely to eliminate the objectionable paraffin hydrocarbons. After the demand for c. p. toluol has ceased, these light oils, as has already been demonstrated by actual tests, can be used as an excellent motor fuel. The middle and heavy oils, while at present not attracting so much attention as the lighter fractions, should be in great demand in the future.

At present there is a good market for creosote and flotation oils, and all the oils produced at the Irvington plant have been sold at a very good price for these purposes. No attempt has as yet been made to separate the tar acids from the oils, and the crude cuts, just as they are taken from the still, are used as such. The higher fractions are rich in tar acids, containing as high as 40 per cent., and averaging well above 30 per cent. acidity.

The middle fraction, distilling between 170 to 230° C., is rich in cresylic acid, and laboratory tests have shown that as high as 0.5 gal. of acid

can be obtained from each ton of coal carbonized. A small quantity
of phenol is also present in the middle oil fractions but it is doubtful
whether it would pay to extract it. The light oil fraction from the tar,
distilling below 170° C., also contains tar acids, but in much smaller
proportions. The analysis of this fraction shows it to contain about
10 per cent. of tar acid, the bulk of which is cresylic. With a domestic
production of cresylic acid, the American markets will be well supplied,
and its importation in large quantities will no longer be necessary. This
should result in a lower cost, and as most forms of commercial cresylic
acid darken on storage, which in numerous cases is found objectionable,
it should be possible to put on the market a product more satisfactory
to the consumer.

As to the oil obtained by stripping the gas, this is obtained in the
same quantity as from coke ovens. It is richer in toluol, yielding approximately 0.35 gal. (1.32 l.) per short ton (907 kg.) of coal carbonized.
In addition to the tuluol recovered from the gas, 0.13 gal. (0.49 l.)
is also obtained from the tar, making a total of nearly half a gallon from
each ton of coal, or practically double the average yield from coke ovens.

The Smith process also produces a good yield of ammonia. The first
or low-temperature distillation yields 4 or 5 lb. (2 kg.), and the second
or high-temperature distillation 15 to 16 lb. (7 kg.) of sulfate, a total
which compares very favorably with coke-oven and gas-house practice.

CHARLES M. BARNETT,*New York, N. Y. (written discussion†).—During the last few years, much thought has been given by coke-oven engineers
to the distillation of coal at low temperatures, to obtain a smokeless fuel
and a larger yield of tar, benzol, etc., than is obtainable in existing byproduct coking processes. The intermittent process, as used in England,
has many disadvantages: Restriction to the use of certain coals; charging
and discharging of retorts is costly; fuel obtained is bulky, low in specific
gravity, and difficult to handle without excessive breakage in transportation; in carbonization, coals nearest the sides of the retort are coked more
than those in the center, making a uniform yield impossible; yield of
ammonia is low.

The Smith continuous retort appears to have overcome these difficulties. The higher the volatile content, the higher the yield of oils. No
special coal is required for the process. Coal which has a tendency to
swell in carbonization can be easily handled in this retort, which would be
impossible in the intermittent process. The interlapping paddles agitate
and mix the material thoroughly, and each particle of coal is brought into
intimate contact with the sides of the hot retort, giving a uniform product.

* Of the Clinchfield Coal Corporation.   Formerly Pres. and Gen'l Mgr., Chesa-
peake & Ohio Coal and Coke Co.
   † Received Aug. 19, 1918.

The question of feeding coal into the retort and discharging the carbon residue therefrom, which has been found troublesome in all other continuous processes, has been solved satisfactorily in the Smith process. The heating system of the retort enables the operator to obtain any desired temperature at any part of the retort; access to the gas burners is obtained underneath, and control of the air for combustion is easily accessible, being adjacent to the gas burners, so that adjustments to the gas and air can be made at the same time. The dampers connecting the heating flues to the waste-heat flue are operated from the top of the retort. Air used for combustion is pre-heated in the recuperator underneath the retort, cold air and waste gases flowing countercurrent to each other.

The carbon residue obtained from the primary carbonization is a fuel of uniform quality, and averages 8 to 10 per cent. volatile matter. It is easily crushed and would make ideal powdered fuel, with a considerable saving of national resources through recovery of the valuable byproducts, as compared with present methods of using raw powdered bituminous coal. The carbon residue could also be stored with less danger of spontaneous combustion than raw bituminous coal. This opens up a large field for its economical storage in the tropics and at bunkering stations around the world.

As to the briquetting of the carbon residue, the pitch employed (only a small percentage) is obtained from the tar distilled in the plant, and it is found that more than enough for the requirements of the briquetting plant is available. This is an ideal way of disposing of the pitch, as in normal times pitch does not find a ready market in large volume. In the secondary carbonization, some of the pitch is recovered, some gasifies, and the remainder combines as carbon with the finished product.

The process is also valuable for carbonizing coals of high sulfur content, such as certain Illinois coals, which are not suitable for metallurgical coke, but make a good domestic fuel in the form of carbocoal.

CHARLES CATLETT,* Staunton, Va. (written discussion†).—I witnessed a test of carbocoal on a Mallet locomotive on the Clinchfield road in September, 1917. The locomotive was pulling approximately 3000 tons up a ½ per cent. grade at a speed of between 9½ and 11 miles an hour. A comparison was presented between carbocoal made from the Upper Banner coal, mined by the Clinchfield Coal Corporation, and a former test made under very similar conditions when using straight Upper Banner coal. There was no difficulty in keeping the steam gage steady between 195 and 200 lb., and the results indicated by the two tests were approximately the same. But taking into consideration all the conditions surrounding the tests of the carbocoal, the impression

---

* Chemist and geologist.  † Received Aug. 19, 1918.

made upon me was that the latter would have afforded even better results if the conditions of the tests had been fixed with reference to that particular fuel.

The engines of the Clinchfield road, after many experiments, had been drafted and arranged with special reference to Upper Banner coal, and the fireman was familiar with what was necessary to get the best result from it. This coal, which is considered one of the best locomotive fuels in the East, carries about 32 per cent. of volatile matter, and is strongly coking; yet in this test it was replaced by carbocoal, carrying not over 4 per cent. of volatile matter, and having no coking qualities, without any change in the drafting or arrangement of the fire-box. It was evident that the fire-box could have contained and consumed a very much larger amount of carbocoal than was fed to it. A small matter bearing upon the test was the fact that the briquets were of domestic size, and being of round shape would not pile up on the shovel as would run-of-mine coal; thus, with the same exertion the fireman could not put the same weight of fuel into the fire-box when firing carbocoal as when firing run-of-mine coal. This could easily be obviated by a change in the form of the scoop, or in the shape of the briquet.

One great advantage of the carbocoal, as compared with other forms of fuel, is that brought about by the formation of the semi-carbocoal at a low temperature. When coal has been coked at a low temperature and allowed to cool, it no longer softens and continues the process of coking when again heated. The character of coke so formed is entirely different from that of coke made at a high temperature. It probably occludes more oxygen, and is more readily acted on by the oxygen of the air, and probably burns with less excess of air than is required for burning other fuel.

One interesting factor is the marked reduction in the formation of clinkers, as compared with the coal from which the carbocoal is made; this is hard to understand. The carbocoal will naturally carry somewhat more ash than the coal from which it is made, but owing to the uniform grinding and mixing of the coal, the ash is distributed uniformly through the material, and probably every particle of the ash is covered with a layer of soft coke in the initial coking. This may have some effect in reducing the tendency of the ash to clinker, and may enable it to drift away from the zone of highest temperature before it becomes so fused as to give trouble.

In addition to its value, as described, for ordinary gas producers or Mond gas producers, I imagine the carbocoal would have special value in connection with suction gas plants. My impression has been that the wider adoption of such plants has been retarded by the difficulty of securing adequate supplies of suitable fuel.

While carbocoal undoubtedly can be used for steam purposes, and its use under certain conditions is desirable, it looks to me as if it would have much greater value in the domestic trade, where it could take the place of anthracite in those sections which are at present remote from anthracite supply.

I believe one of the difficulties in connection with the production of large amounts of ammonia from coke-oven gases has been the apparent limit to the demand, in many sections, for coke; but the demand for such a material as carbocoal will be far more extensive and universal, and it is not unreasonable to expect that in time a large amount of nitrogen compounds, so greatly needed in agriculture, will be derived from this source.

F. R. WADLEIGH,* Washington, D. C. (written discussion†).—The carbocoal process, as patented and developed by Mr. Smith, would seem to have greater possibilities than any other fuel development of late years, or indeed for many years. Low-temperature distillation of coal is not new, of course, but the experimental results obtained by the Smith process go far beyond and differ from all previous work of the kind, both in methods of operation and in the products obtained. The process is a long step toward the conservation of fuel resources; indeed, its development in foreign countries, especially in England, France, Italy and Spain, or wherever suitable coals are available, will aid in solving many important fuel problems, as well as others not strictly pertaining to fuel, as such.

In Great Britain, for instance, where all oils must either be imported or made from coal or shales, but where the best grades of high-volatile coal are available, a carbocoal plant would be highly profitable; the carbocoal itself would fill the demand for a smokeless house and steam fuel, while the yield of byproducts, especially the oils, would help to reduce the quantity of oil to be imported, while assuring a definite and steady supply and no uncertainty as to losses of ships. A plant carbonizing 2000 tons of coal per day would produce about 15,000,000 gal. per year of various oils; to import the same amount from the Mexican oil fields would take about five tanks, each making five trips per year.

Regarding carbocoal, the fuel, it is understood that the railroads have allowed the same freight rates for its transportation as for coal. In this respect, it will have a considerable advantage over coke, which is always charged a higher rate than coal, owing to its bulk. As regards transportation, experiments have shown that carbocoal has decided advantages in loading and unloading cars, as well as in freedom from breakage, the latter being practically nil.

The fact that the process is entirely self-contained is a distinct asset—everything it needs, it makes; only one raw material—the coal—has to be supplied.

* Mining Engineer, Emergency Fleet Corp.     † Received Aug. 17, 1918.

As a means of conserving our supplies of anthracite coal, carbocoal presents a most promising future; not only has it been proved to be a better domestic fuel than anthracite, but the fact that carbocoal can be made from any high-volatile coal solves the serious question of distributing anthracite. The fact that carbocoal, as now manufactured commercially, contains less volatile than anthracite insures its smokeless combustion; its uniform shape and size also make for improved combustion.

The writer has seen carbocoal burned on a locomotive, running with 60 per cent. cut-off, at a rate of 166 lb. per sq. ft. of grate per hour, and with 16-in. draft in the smokebox, with absolutely not a vestige of smoke; burning with a long yellowish-white flame, yet every part of the surface of the fire entirely visible, even to the unaccustomed observer—a performance impossible with either coke or anthracite.

On the same locomotive (a standard "Mikado" as used in freight service) at a rate of 80 r.p.m. and 50 per cent. cut-off, over 11 lb. of water were evaporated per pound of dry fuel, from and at 212° F.; yet the carbocoal contained only 12,291 B.t.u., showing a boiler and grate efficiency of 84 per cent. It would seem from such performances that carbocoal is the answer to the smoke question on locomotives—a truly large field, when it is remembered that one railroad alone buys some 3,000,000 tons of anthracite, coke, and so-called "smokeless" coals, solely to reduce the smoke nuisance.

Other possibilities in the use of carbocoal would seem to lie in its substitution for coke in various industrial processes, such as beet-sugar manufacture, smelting, and various chemical processes, where a smokeless fuel is required, and possibly also for foundry use. This would be especially the case west of the Mississippi River, where low-volatile coals and good coking coals are comparatively scarce. The high-volatile, high-nitrogen coals in Washington should be especially well adapted for the carbocoal process, and there would be a ready market, both for the carbocoal and for the byproducts.

One rather unusual feature of the process, as applied to fuel, is the fact that it greatly reduces the percentage of oxygen in the carbocoal as compared with that in the original coal. In the case of one well known Eastern coal, which normally contains from 7 to 10 per cent. oxygen, the carbocoal produced from it contained slightly less than 1 per cent. As a given amount of oxygen has about the same effect on coal combustion as a similar percentage of ash would have, or, in other words, the B.t.u. in dry coal vary directly with the percentage of oxygen, this may explain some of the extraordinary results that have been obtained with carbocoal in boiler and locomotive furnaces.

Another promising feature is the entire suitability of the semi-coke, or coke residue, obtained after the first distillation, and before briquetting, for use in byproduct gas producers; here its entire freedom from caking, or tar, its uniform quality and ease of conveying, in addition to the fact that the first distillation does not remove any of the original nitrogen in the raw coal, all make the semi-coke an ideal fuel for such use.

This semi-coke would also seem to offer possibilities for use under boilers, either stoker or hand-fired, where smokeless combustion could be easily obtained. Its use under such conditions would, however, be dependent upon location, as the semi-coke could not well be transported for any distance.

As a smokeless fuel for the Navy and Merchant Marine, especially in war time, carbocoal should be satisfactory. Indeed, it is understood that tests made by the U. S. Navy at its Annapolis testing plant have given quite remarkable results, both in evaporation, quick response to draft changes, and entire absence of smoke. As a result of these tests, the Emergency Fleet Corporation has taken up the question of using carbocoal on ships passing through the submarine zones.

F. W. SPERR, JR.,* Pittsburgh, Pa.—Mr. Malcolmson states that the carbocoal process produces ammonium sulfate in excess of that normally recovered in the ordinary byproduct coke process. Table 3 indicates that 20,000 to 25,000 lb. of sulfate of ammonia are obtained in carbonizing 1000 tons of Clinchfield coal. This is a high-grade coal and yields from 26 to 28 lb. of ammonium sulfate per ton in a modern byproduct coke oven. On the next page, 21 lb. is stated as the yield of ammonium sulfate per short ton of raw coal; this is less than the average production from any well operated byproduct coke-oven plant treating coal of the character usually coked in Pennsylvania and Ohio.

The statement as to the comparative yields of light oil is somewhat misleading in view of the results obtained during the past year or two by the more modern byproduct coke plants. The plants put in operation by the H. Koppers Co., during the past two years, are producing about 3.5 gal. of light oil per short ton of coal, and several of these plants are producing more than 4 gal. per ton. The yield of pure toluene runs from 0.45 gal. to 0.55 gal. per ton.

The quality of the light oil obtained from the carbocoal process is of much greater importance than would be inferred by one not closely familiar with the subject. The presence of paraffins in the benzol and toluol fractions seriously detracts from their commercial value; the benzol and toluol fractions derived from the carbocoal process contain about 50 per cent. of paraffins. It should be thoroughly understood what these paraffins are. They are simply hydrocarbons that go to make up ordinary petroleum, and if a successful method for the removal of paraffins from benzol and toluol has been developed, this can at once be applied

* Chief Chemist, H. Koppers Co.

to several well known processes by which benzol and toluol can be made directly from petroleum, which have as yet been total failures on account of the contamination of these products by paraffins.

It is very doubtful whether the method for conducting this operation, of which Mr. Malcolmson speaks, has been made commercially practicable. This separation of paraffins from benzol and toluol presents the utmost technical difficulties, and although a great number of investigators have worked on this subject, and although millions of dollars have been spent by various chemical and munition concerns in attempting to solve the problem, no successful commercial process has yet been put into operation, and the great number of failures makes one skeptical as to any new and untried suggestion of this sort. The presence of paraffins in excess of a few per cent. in either benzol or toluol is fatal to the successful use of either of these materials in the manufacture of high explosives and other important chemicals for which they are so much in demand. The Government specifications for pure toluol stipulate that not more than 2 per cent. of paraffins shall be present. It is very important to keep the distinction between the light oil obtained by the carbocoal process and that obtained from coke-oven gas well in mind. The former is of no value except as a source of motor fuel, while the latter is practically indispensable as a source of our most important high explosives.

CHARLES H. SMITH,* New York, N. Y. (written discussion†).—I should like to clear up certain points raised by Mr. Sperr, as he has apparently misunderstood some of the comparisons made by Mr. Malcolmson between the carbocoal process and that of the byproduct coke-oven industry, and more particularly his reference to our statement that the carbocoal process produces "ammonium sulfate in excess of that normally recovered in the byproduct coke process."

Mr. Sperr calls attention to the fact that Table 3 of Mr. Malcolmson's paper states that 20,000 to 25,000 lb. of ammonium sulfate is recovered from the carbonization of 1000 tons of Clinchfield coal, and that later in the paper the statement is made that 21 lb. of sulfate of ammonia is recovered per ton of coal carbonized. In the first case, this paper is dealing specifically with the results from Clinchfield coal, and in the second case it is dealing with an average sample of high-volatile coal, running from, say, 33 to 35 per cent. volatile matter.

Mr. Sperr states that if this same coal were carbonized in a modern byproduct coke oven, the yield would be from 26 to 28 lb. of sulfate of ammonia. A modern byproduct coke oven, in first-class operating condition, might be made to yield the results outlined by Mr. Sperr. On

---

* President, International Coal Products Corporation. Inventor of the carbocoal process.

† Received Oct. 2, 1918.

the other hand, I have in my files the official report of a series of tests by the H. Koppers Co. on this same Clinchfield coal, under date of June 3, 1914, in which the practical yield on an 18-hr. coking time is given as follows: Ammonium sulfate, 20 lb.; coke, 69 per cent.; tar, 7,4 gal. The coke manufactured from this coal was soft and spongy in structure, and the H. Koppers Co. reported that in order to obtain a satisfactory metallurgical coke it would be necessary to admix approximately 25 per cent. of low-volatile coal. A further test was made on June 5, 1914, using 25 per cent. of Pocahontas and 75 per cent. of Clinchfield coal, from which the following practical yields were reported: Ammonium sulfate, 19 lb.; coke, 76 per cent.; tar, 5 gal.

The necessity of an admixture in order to obtain coke with a satisfactory structure is the point we particularly desire to emphasize in comparing the two processes, rather than a direct comparison of the results that could be obtained by carbonizing the same coal under the two methods. Modern byproduct coke practice is confined almost entirely to the carbonization of mixtures of coal, ranging from 26 to 31 per cent. in volatile, in order to get coke having a satisfactory structure. The average yield of sulfate of ammonia in carbonizing such mixtures will approximate 19 lb. per ton of coal carbonized. The carbocoal process, on the other hand, is especially adapted to carbonizing coals of higher volatile content, and particularly those coals ranging from 33 to 38 per cent. in volatile, without the necessity of admixtures. This process, therefore, is utilizing a fuel that would naturally yield a considerably larger quantity of sulfate of ammonia and tar. The higher yields in the carbocoal process, stated in the Malcolmson paper, are therefore to be expected, and it was not intended to make a comparison with what could be accomplished in byproduct coke ovens if the structure of the coke were disregarded and high-volatile coal were used for carbonization. This explanation may serve to clear up the point raised by Mr. Sperr.

In regard to the yields of light oil given in the paper, there was no intention to belittle in any way the results of the byproduct coke process. We have no doubt that the statement of Mr. Sperr in regard to the yield of light oil in some of the latest H. Koppers Co. plants is accurate. The figures given in the Malcolmson paper, however, do not go into the results obtained under the very best conditions in the carbocoal process, but were intended to represent average conditions. In testing fuels in any carbonization process, the range of results varies considerably with different coals, and average results should be taken, rather than specific results obtained under the most favorable conditions. Certain coals tested in the carbocoal process yielded as much as 5 gal. of light oil per ton of coal carbonized, from the primary distillation alone. The particular fuel in question was a cannel coal and could, therefore, not be considered a representative fuel.

In regard to the quality of the light oil obtained by the carbocoal process, and the value of the tar oils in general, this is a matter which can only be definitely proved by experience in marketing carbocoal oils on a large scale. If we estimate the average yield of light oil from by-product coke ovens as $2\frac{1}{2}$ gal. and of crude tar as $6\frac{1}{2}$ gal., we would have the following market values under present conditions:

$2\frac{1}{2}$ gal. light oil at 50 c. per gal............................... .1.25
$6\frac{1}{2}$ gal. of tar at 4 c. per gal.............................. 0.26 ·
                                                                          ———
Total.............................................. $1.51

The carbocoal process will yield 20 gal. of tar distillates, with an average value of 24 c. per gallon, or a total value of $4.80, and by referring to prices prevailing prior to the war it will be found that this type of oil never sold below 15 to 17 c. per gallon. The present market value of the two products can therefore be easily compared.

In regard to Mr. Sperr's assumption that benzol and toluol fractions derived from the process contain about 50 per cent. of paraffin, we think Mr. Sperr's assumption inaccurate. We permitted Mr. Sperr's representatives to make certain tests on a small apparatus, in order to confirm the statements made by us regarding the yields. These representatives informed us that the total yields obtained in these tests were greater than the average yields represented by us from the same coal. The apparatus used for these tests was not a commercial retort, and we have learned by experience that this laboratory apparatus yields a light oil containing approximately 30 to 50 per cent. of paraffins, as compared with 4 to 20 per cent. of paraffins in the light oil derived from the commercial retort. Coke-oven engineers will recognize that the quality of the results obtained from distilling coal in a metal retort, where the charge is limited to 30 lb., will be quite different from those obtained in a commercial retort operating with a capacity of 1 ton per hour, although we have found the quantity of byproducts obtained per ton to be practically the same. We might also note the particular object of the "cracking" referred to in the Malcolmson paper. By this method we have completely eliminated all the paraffins from our light oil, and by adding some of our higher-tar oil fractions we have obtained total yields of toluol per ton of coal several times the total yield per ton obtained in the coking process.

The present shortage of toluol and its high price will warrant such measures, but in peace times it is universally admitted that the light oil from both processes will be largely marketed as motor spirits, in which case the paraffin hydrocarbons present in these oils will not be objectionable, and on account of lower freezing points and ease of ignition may probably be beneficial.

In regard to the practicability of "cracking" petroleum oils and producing c.p. toluol therefrom, I might add that, notwithstanding the early failures of certain processes, this is now being done on a very large scale, and that the results therefrom, up to the present time, are entirely satisfactory.

In making a comparison of the two processes, we do not believe it was Mr. Malcolmson's intention to compare the financial benefits or the economic status of the carbocoal process with that of the established byproduct coke industry. The two processes represent different types of carbonization, yielding substantially different quantities and qualities of various products. The carbocoal process yields a smokeless fuel of great density, available for many uses for which coke is not suitable, while coke, on the other hand, has well-established markets for metallurgical use. The byproduct coke process gives a substantial yield of surplus gas, and is therefore located near where the consumption of gas is great, where it will yield its greatest value. The carbocoal process uses all of its gas in the distillation of the coal and in the refining of the oils. Carbocoal plants will therefore be located at the mines and the carbocoal shipped to the consumer by direct route, requiring transportation facilities for only 72 per cent. of the raw coal. In the coking process, the entire coal required for distillation must be shipped to the ovens, usually located at a considerable distance from the mines, and the finished product, coke, representing 70 per cent. of the original coal, re-shipped to the consumer, requiring double transportation facilities. Byproduct coke ovens produce light oil and tar, while the carbocoal process produces oil distillates. The sulfate of ammonia from the same coal would be approximately the same in both processes, but on account of the highervolatile coal ordinarily used in the carbocoal process, as compared with that utilized in the byproduct coke process, the yield is greater in the carbocoal process.

The railroads have established coal rates for the movement of carbocoal, on account of its density, as compared with coal. The rates on coke, on account of its lack of density, are approximately 15 to 20 per cent. higher than coal rates between the same points. This alone is of enormous economic benefit, as compared with coke used for fuel purposes.

The comparisons between the two processes are not made for competitive purposes, but to define clearly just what products are derived from the carbocoal process as compared with the already well-known products from byproduct coke ovens, and their general suitability for market.

A. W. CALLOWAY,* Washington, D. C. (written discussion†).—In a discussion of Mr. Malcolmson's paper which describes the method of

* Director of Bituminous Distribution, U. S. Fuel Administration; President Davis Coal and Coke Co., and Pittsburg Terminal R. R. and Coal Co.

† Received Oct. 7, 1918.

coal distillation developed by Mr. C. H. Smith, attention is first directed
to the fact that apparently Mr. Smith has accomplished by mechanical
agitation what byproduct coke-oven builders have been working many
years to accomplish by other means; namely, the complete distillation
of coal in minimum length of time with a maximum of byproduct
recovery. The coke-oven builders are not changing the fundamental
design of their ovens, but are steadily making them narrower. Two
benefits are derived from the narrower ovens; first, a low wall temperature, which increases the byproduct recovery, and second, a faster coking
time.

In the usual byproduct coke oven, the coal charge is stationary, and
the heat necessary for the distillation must be transmitted or conducted
from the walls of the oven slowly through the charge, driving off the volatile matter and coking the charge from the sides to the center of the oven,
the result being a comparatively slow distillation, with the temperature
of the walls higher than necessary to coke the coal, because the heat has to
be conducted through that part of the charge already coked; and while
the charge in coking shrinks away from the walls there is a gap left which
retards the transfer of heat from the walls and also allows the gases given
off to pass up the sides of the hot wall, where some of the hydrocarbons
are broken down, causing loss in byproduct recovery.

An interesting development in this connection is the discovery that
with the narrow ovens and short coking time it is possible to coke coals
that could not be satisfactorily coked in wide ovens; if it is granted that
quick application of heat to coal increases its tendency to coke, then one
would expect, in a mechanically agitated retort where all particles of
coal in the charge come directly in contact with the hot walls, to find it
possible to coke coals that could not be coked in the ordinary byproduct
coke ovens. This is just what Mr. Smith has demonstrated can be done,
in his primary retorts at his experimental plant at Irvington.

If the mechanical features of this operation have been successfully
worked out at Irvington, as indicated in the paper, then Mr. Smith
has indeed made a tremendous stride in the direction of utilizing large
fields of bituminous coal throughout the country to the best advantage.

In the primary retorts in the carbocoal process, where the charge of
coal is in motion, and each particle of coal comes in contact with the
heated wall of the retort, it is possible to operate with a much
lower wall temperature, and the coking effect is obtained without the
breaking down of the hydrocarbons to such an extent as they are
broken down in the other byproduct coke ovens. This has long been
recognized as the ideal condition for coal distillation, and if Mr. Smith
has perfected the method we should then expect his results in the
recovery of byproducts to show a very high production of tar and light
oils, with probably a considerably reduced yield of gas. We would not

expect a maximum yield of ammonia at this low temperature, but Mr. Smith in his secondary distillation appears to recover approximately as much ammonia as ordinarily is recovered in the byproduct coke oven distillation.

*Tar.*—Mr. Malcolmson has stated that a yield of approximately 30 gal. of tar per ton of coal is obtained in the combined primary and secondary distillations in the carbocoal manufacture. Judged by the standards of the best practice in byproduct coke-oven work, this yield appears excessive, but with Mr. Smith's extremely low-temperature distillation I believe that such recovery may be obtained, although it is at the expense of a certain amount of gas. On account of its value, however, the increase in tar should offset the value of the gas that would be obtained from a higher-temperature distillation.

*Sulfate of Ammonia.*—Mr. Malcolmson claims a recovery of approximately 20 to 25 lb. of ammonium sulfate per ton of coal. depending upon the volatile matter of the coal. This compares favorably with byproduct coke-oven practice.

*Light Oils.*—Mr. Malcolmson reports light oil recovery of 2 to 3 gal. per ton of coal. This, I should say, is slightly below the best byproduct coke-oven practice on coals of the same volatile contents. It is my understanding, however, that his percentage of toluol in this light oil is high, as would have been expected from his low-temperature distillation.

*Gas.*—The production of gas is reported to be from 5000 to 6000 cu. ft. of very rich gas in the primary distillation, and 4000 cu. ft. of lean gas in the secondary distillation, making a total of 9000 to 10,000 cu. ft. of gas per ton of coal, the total heating value of which is about 85 per cent. of the heating value of the gas produced in the byproduct coke-oven operation, and, I should say, compares favorably, considering the low temperature of the primary distillation. Mr. Malcolmson states that at the present stage of development all of the gas recovered is required for the primary and secondary retorts; considering the low temperature required for the operation, and the fact that only 45 per cent. of the gas made in byproduct ovens is required for this operation, it appears to the writer that there is yet room for improvement, either in the design or construction of the retorts, whereby it should be possible to get better economy out of the gas used, and probably have surplus gas sufficient to furnish power for the plant and for the distillation of the pitch required for a binder in making the briquets.

Mr. Malcolmson called attention to the possibilities of the application of the Smith process in the production of a fuel suitable for use in the byproduct gas producers, in which the gas would be scrubbed and the ammonia extracted, his idea being to use the semi-carbocoal; that is, the residue of the distillation in the primary retorts, and which contains considerable ammonia, for a fuel for byproduct gas producers in connection with steam plants. It is the writer's opinion that this might be practical, but better economy would be effected by the use of this fuel in byproduct gas producers furnishing gas for internal-combustion engines, in which a greater efficiency would be derived from the gas than if it were fired under boilers for steam generation.

## Low-temperature Distillation of Illinois and Indiana Coals

BY G. W. TRAER, CHICAGO, ILL.

(Milwaukee Meeting, October, 1918)

THE distillation of bituminous coals at what is commonly termed low temperature, and the quantities, nature and adaptabilities of the products have been the subject of considerable experimentation, during recent years. Fortunately, the earlier work in this country was done by men whose scientific training qualified them properly to record and interpret the results of their experiments. The work of Prof. Parr, of the University of Illinois, is especially notable in this respect.

Prof. Parr's work, added to that of English experimenters, demonstrated certain things to a degree which is generally regarded as convincing. They have proved that a caking bituminous coal, when subjected to a temperature of not more than 1000° F. (preferably somewhat less) will yield light tar having low specific gravity (about 1.06) and high value; also, that this yield, with a given coal, will depend upon the percentage of hydrocarbons in the coal, the maximum temperature used, and the promptness with which the distillate gases are removed. It is assumed that air is excluded from the retort, and that the coal is subjected to heat for a sufficient time to drive off all of the tarry hydrocarbons.

The results attained in the first, or low-temperature stage of the carbocoal process, recently described by C. T. Malcolmson,[1] corroborate these general principles. It is the purpose of the writer to describe other work of a practical nature, directed toward the commercial adaptation of these principles, which was begun nearly two years ago; this investigation was assisted by A. J. Sayers, of the Link-Belt Co., and Carl Scholz.

The first step was to draft tentative designs for the handling of the coal and coke so as to produce a coke having a suitable structure for sizing, storage, shipment, and handling, without the necessity of briquetting. Prof. Parr's results had indicated that the rather open-pored volatile coke produced by him had a low ignition point, which would contribute toward easy control of the fire in stoves, house furnaces, fire-boxes, and boilers. It is recognized by those experienced in the marketing of fuels that the hard non-volatile coke from the high-temperature process,

---

[1] This volume, p. 393.

while an excellent fuel in the hands of those qualified to handle it, nevertheless falls far short of being an ideal domestic fuel, and is still less well adapted to other uses, to which it was believed a high-volatile coke would be more suited. The high ignition point of hard non-volatile coke makes it difficult to burn satisfactorily in a small or thin fire. With a strong fire the drafts have to be watched closely because of the tendency to over-intensity of combustion. In spite of its high heat value and cleanliness in handling, it has therefore fallen short of being a popular domestic fuel. Another obstacle to the general use of hard coke for domestic purposes is the fact that, during active periods in the blast-furnace industry, the furnaces will pay such high prices for the coke as to divert it from domestic use. Domestic consumers feel, therefore, that it is not a reliable source of supply at all times, and when the hard-coke makers wish to return to the domestic industry, they have to build up their business all over again, against a handicap.

Another important factor is that, while it has proved quite impracticable to use non-volatile hard coke in locomotives previously using high-volatile bituminous coals, without radical changes in the fire-boxes, the indications were that high-volatile coke could be employed for this important purpose, without material changes in the fire-boxes.

### EXPERIMENTAL PLANT

C. M. Garland was retained to revise the designs and supervise the construction and operation of an experimental plant, which was installed at Chicago in 1917. The operation of the plant began early in October and, with some stoppages for corrections indicated by its use, was operated until Dec. 24. Altogether about 140 tons of coal were passed through the plant, about 100 tons of which was Franklin County mine-run crushed to pass through a 2-in. round-hole shaker screen. The other coals tested, in lots ranging from 4 to 12 tons, comprised Harrisburg and Eldorado in Saline County, Ill.; Indiana 4th Vein from the J. K. Dering mine, Clinton, Ind.; Indiana No. 5 from Grant mine No. 4; and Indiana No. 5 from Knox County.

The plant comprised one oven or retort, together with the usual condensing and scrubbing apparatus for the distillate gas, and other necessary machinery and appliances.

The oven was horizontal and rectangular, about 60 ft. long, 4 ft. high and 20 in. wide. At each end was a double-door lock, one serving for the admission and the other for the withdrawal of the containers carrying the coal, without admitting any substantial amount of air. These locks were of suitable length to hold one container; they were not heated, nor was a space of the same length as the lock immediately inside of it. The remainder of the retort was supplied with down-draft flues, each of which led to a large off-take flue in the base of the oven, and thence to the stack.  The flues were about 1 ft. square in their inner dimensions and were laid with ordinary rectangular silica brick, as tongued and grooved brick in interlocking shapes could not be obtained for many months.  As the volume of clean gas yielded by the low-temperature process is insufficient for heating the oven, provision was made to supply city gas for the additional heat necessary; in the commercial plant, producer gas will be supplied for this purpose.

A track was laid through the oven with 16-lb. steel rail, 11-in. gage, to carry the containers.  These were introduced at intervals into the charging end, each successive container pushing those which had preceded it; when the retort was filled, a finished container would be discharged whenever a fresh one was inserted.  The train of containers was pushed by a hydraulic ram, and they were charged with coal and the coke was discharged from them manually.  In the commercial plant, the pushing will be done by an electric ram, and the charging and discharging will be mechanical.

The design of these containers was the central feature of our process.  Those used in our experimental plant were about 4 ft. long, 4 ft. high and 12 in. wide.  Partitions, 12 in. apart, divided each box into four sections, each 12 in. square.  Each box was placed on a cast-iron truck mounted on four cast-iron wheels.  When the truck and box were removed from the oven, the box was lifted off, the coke discharged, the box replaced on the truck, and returned to the other end of the oven for recharging.

One of the principal purposes of this form of container was to afford means by which the heat could be quickly conveyed through the mass of coal.  Experience with the high-temperature byproduct oven has shown that when heat is applied only to the side of the charge, the progress of the coking is very slow, probably only $\frac{1}{2}$ in. per hour.

It is impracticable to make a semi-coke of substantially uniform volatile content, unless special means are employed for conveying the heat to the center of the charge more quickly than is possible by means of a heated side-wall only, because if the side-wall temperature is not raised above 1000° F. the progress of the coking toward the center will be prohibitively slow; indeed, it has proved impracticable to convey the heat, at that temperature, through 10 in. of coal.  If the heat is raised above 1000°, by the time the tarry volatiles are all driven off at the center of the charge, the volatile contents of the coke on the outer side will be greatly reduced, thus giving a coke of widely varying character.  Furthermore, the higher temperature will cause a proportionate decrease in the valuable light oils and a corresponding increase in pitch; the percentage of coke also will be decreased.  There will be an increase in gas, but it will fall far short of compensating for the losses mentioned.

The first container boxes were made of sheet steel, and were divided into four compartments each 12 in. square.  It quickly developed that this column was too large for short-time coking, with a reasonably uniform volatile contents. Additional compartments were then put in some of the boxes; while, in others, pipes of varying diameters were placed in the centers of the large compartments. Both of these expedients produced usable coke.

A cast-iron box also was used; the shell and partitions were ¾ in. thick, and the compartments varied from 4 to 8 in. in width. This cast-iron box showed great superiority over all other forms, and proved eminently satisfactory for the purpose. The compartments were tapered slightly, so that by the time the boxes reached the open air, the temperature had been reduced enough to cause a slight shrinkage and the coke slipped out of the compartments without difficulty. It showed no tendency whatever to cling to the metal.

### CHARACTERISTICS OF SEMI-COKE

The semi-coke thus produced proved a most satisfactory fuel for all general purposes. Although our retort had only side-wall heat, we were able to produce a coke practically homogeneous in structure and volatile contents. By keeping the charge in the heat long enough, we could produce a hard coke with practically no volatile contents, or by withdrawing the charge as soon as it was fused to the center, we could change the coal into the form of coke while but little of the original volatile contents had been removed. Our experience indicates that the ideal fuel for general use is obtained by leaving about 18 per cent. of volatile matter in the coke; this gives a satisfactory yield of coke, while, at the same time, practically all the volatile-producing oil or tar has been removed, and the coke thereby rendered smokeless. Using compartments of suitable width, the coke is uniform in structure and has sufficient strength to permit sizing and handling. The specific gravity is somewhat less than that of hard coke, but not enough so to be a practical objection. The breeze and fines burn satisfactorily, either by themselves or with the coarse sizes; in fact, the behavior in the firebox, in that respect, is quite similar to that of bituminous coal. It ignites as readily as bituminous coal. The fire is easily controlled, and was successfully banked for 8 to 10 hr. in house furnaces during severe winter weather. It burns with a short, bright, clear blaze, and is clean to handle and fire.

An 8-hr. test of the semi-coke made from three different kinds of coal was made on a switch engine in one of the large railroad yards in Chicago. The coals were Franklin County, Ill., Indiana Vein No. 5 from north of Terre Haute, and Indiana Vein No. 5 from Knox County. In clearing up the pile from which this tender load was taken, all of the fine stuff was gathered up with the coarse, so that there was a large percentage of it in the tender load, which, in proportion of coarse and fine, resembled mine-run coal.

There was a light coal fire in the locomotive and on this the semi-coke was started. It steamed rapidly from the start and made no smoke throughout the test. The amount of cinders thrown from the stack with a heavy exhaust was but a small proportion of the amount coming from bituminous coal. The semi-coke steamed rapidly enough to make it unnecessary to use the steam blower when the exhaust was off, as is done to lessen the smoke when bituminous coal is used; this effects a substantial saving of coal. The engine did its regular yard work for about 4 hr., the crew being changed at the third hour. By this time a thin clinker had formed which, however, was readily removed during the noon hour. Another thin clinker was formed and removed during the afternoon, but some coal, of which there was some at the bottom of the tender, had been fired on top of the semi-coke before this happened. The Engineer of Tests of the railroad stated that this clinkering probably resulted from the mixing of semi-coke made from three different kinds of coal, and from the excessive amount of fine stuff. Mr. Scholz, who was on the engine during a considerable part of the test, was of the same opinion; and they agreed that the semi-coke had satisfactorily shown its adaptability for locomotive use without material change in the firebox. The semi-coke was entirely new to both firemen on this run.

In regular practice, a semi-coke made from one kind of coal will clinker less than the coal from which it is made. This is because the semi-coke burns up without permitting the lumps to break down or melt; there is, therefore, a better circulation of air through the fire-bed than with coal. In regular practice there also would be a great deal less fine stuff than was in this tender load. In a plant equipped entirely with casting-boxes, the plant run of semi-coke would contain a very small percentage of fine stuff; from his observation in our experimental plant, Mr. Garland estimates this at less than 10 per cent.

### Volatile Products of Distillation

In the operation of our experimental oven, it was demonstrated that it would not be possible to use a brick wall between the side flues and the containers, and at the same time make a practically complete recovery of the oil and gas in the coal. This was partly due to the small amount of gas, $\frac{1}{2}$ to 1 cu. ft. per pound of coal, obtained in the low-temperature process, as compared with 5 to 6 cu. ft. in the high-temperature process. A small air leakage, such as exists in the high-temperature ovens, would therefore produce disastrous results; even in high-temperature ovens, there is more or less trouble in keeping down the percentage of carbon dioxide in the gas. In our experimental oven, the use of plain silica brick instead of special interlocking brick increased the leakage materially, although special precaution was taken in laying up the brick and in coating the interior of the oven with special high-temperature cement.

The quantitative determination of gas and tar was therefore impossible with this oven. These factors, however, have been demonstrated repeatedly, not only by Prof. Parr, for practically every coal mined in the State of Illinois, but also by numerous others. The curves of Fig. 1 show the relation between the coking temperature and the tar‑produced, in gallons per ton, for a coal containing approximately 30 per cent. of volatile matter. These curves are based on experiments made by Lewes, and correspond with the results obtained by Prof. Parr. The Illinois and Indiana coals have 35 per cent. or more of volatile hydrocarbons and will therefore produce correspondingly more tar than the 30 per cent. coal illustrated in Fig. 1.

For our commercial oven, a cast-iron retort will be used, which will be lined on the flue side with silica brick. This retort is ribbed to prevent warping, and is fixed only at the center, leaving the ends free to expand.

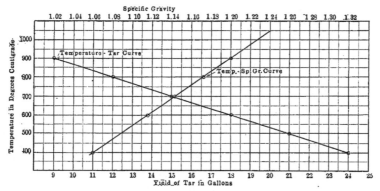

FIG. 1.—RELATIONS BETWEEN RETORT TEMPERATURE, AND SPECIFIC GRAVITY AND YIELD OF TAR, PER TON OF COAL CONTAINING 30 PER CENT. VOLATILES.

Since the temperature within the oven does not exceed 1000° F., this will not only insure a gas-tight construction, but the retort will be considerably more permanent than a brick structure, and will cost less for maintenance than brick retorts of this form.

The light tar produced by the low-temperature process is a liquid having the consistency of heavy cylinder oil. Its significant characteristics, as disclosed by Prof. Parr's work, are: first, the large percentage of the light fraction cut at 210° C. (about 18 per cent.); second, the large percentage of the middle fraction (about 52 per cent.) and the high percentage of the tar acids therein; third, the very low proportion of pitch (about 30 per cent. as compared with 65 per cent. or more in the heavy high-temperature tar), the low melting point of this pitch, and its low percentage of free carbon.

Taking Franklin County, Ill., coal as a basis, the results of operation at our experimental plant indicated that the products of low-temperature distillation will be about 77½ per cent. of semi-coke having 18 per cent. volatile combustible matter, and about 25 gal. of light tar and oil per ton of coal. There will also be a small amount of ammonia, but since the nitrogen in the coal, which is the source of the ammonia, does not leave the coal freely until the temperature is considerably above 1000° F., the greater part of the nitrogen remains in the semi-coke; if this product is then made into producer gas in a byproduct plant, a very large recovery of the ammonia may be made. Treated in this way, the recovery from Franklin County coal would be about 100 lb. of sulfate of ammonia per ton of coal. This assertion is based upon data furnished by C. W. Tozer of London, England, regarding his experiments with a coal closely resembling that of Franklin County. In the high-temperature non-volatile coke plants, the nitrogen is all driven off the coal, but only the lesser part of it is recoverable as ammonia, because the temperature of the oven is so high that the greater part of the nitrogen does not combine with hydrogen, or, if such a combination is made at first, it is later broken up before it can be recovered.

Based upon our experience at our experimental plant, the heating period required with the casting containers will not exceed 8 hr., and it is probable that this may be reduced. The length of time required depends upon the greatest distance which it is necessary for the heat to travel through the coal. A shorter period could be attained by reducing this distance, but if it is too greatly shortened, the result might not be compensatory, by reason of structural and other considerations.

### HIGH- VS. LOW-TEMPERATURE DISTILLATION

Ample data were secured from the operation of our plant upon which to base enlarged and perfected designs for a commercial plant, and to furnish a sound basis for estimating the cost of installing and operating it. This has been carefully worked out by Mr. Garland and the writer, and the results are very attractive as to earning power. Our plant will cost considerably less than a high-temperature plant of the same capacity, and our products will yield a larger return per ton of coal. Franklin County coal, which has a natural moisture content of about 8½ per cent. will yield 12 to 15 per cent. more semi-coke, by the low-temperature process, than can be derived from it as hard coke by the high-temperature process. There will be at least double the amount of tar and oil, and commercial investigations indicate that its value per gallon will be several times as great. The high-temperature process will yield several times as much gas as the low-temperature process, but the surplus gas is always sold at a very low price. The high-temperature process also has some advantage in the greater amount of ammonia produced, but this also is more than offset by other considerations. The opportunities for commercial development of the low-temperature process seem to be very great; it is not visionary to predict the substitution of semi-coke, in the Central and Western states, for a large part of the anthracite and so-called smokeless coals which now come into this territory, by rail or lake.

Members of the Institute are doubtless well informed as to the large number of valuable sub-products which may be derived from light tar and oil, and the rapidity with which this chemical industry is now developing in the United States. Dr. David T. Day is authority for the statement that the cracking process for deriving motor fuel from mineral oils has reached a stage of development indicating that it is possible, in this way, to secure nearly 20 gal. of gasoline, or motor fuel of the same value, per ton of Franklin County or similar coal.

## DISCUSSION

S. W. PARR,* Urbana, Ill. (written · discussion†).—Multiplication of argument is unnecessary to establish the desirability of coking coals at low temperatures, that is to say, below 1200° F. The value of the semi-coke thus produced would doubtless go far toward solving the problem of smoke prevention, and suggests the possibility of the substitution of a smokeless fuel for anthracite and so-called smokeless coals. Whether such coke would have any value as metallurgical coke is a question which cannot be answered for lack of any experimental data in the use of such material. Of course interest is accentuated, at the present time, in the amount and character of the tars, which promise greatly to exceed in values the products obtainable from the high-temperature process.

The difficulties encountered in adapting the principles of low-temperature carbonization to industrial methods are well nigh insurmountable. Briefly stated, there is involved the heating of the center of a mass of non-conducting material by the external application of heat. It is a well established fact that the center of the mass of coal in a standard byproduct oven does not reach the stage of decomposition until after approximately 14 hr. have elapsed, the complete process requiring approximately 18 hr., and all this at a high temperature, approaching the limit of safety for the refractory material used in construction. Up to the present time, the common method of overcoming this difficulty has been by narrowing the cross-section of the retort; in Mr. Traer's experiments a width of from 4 to 8 in. has been adopted. This raises a serious question as to the possibility of economic operation under industrial conditions where the putting through of a large tonnage is essential.

* Professor of Applied Chemistry, University of Illinois.     † Received Sept. 23, 1918.

However, the results as indicated in this paper are exceedingly valuable in that they constitute an additional verification of the character and value of all the byproducts obtained. I would call attention, however, to one seeming inconsistency in his discussion relating to the character of the tars. In the descriptive matter of the paper, the condensible products are referred to as light tars, for example, "the light tar produced by the low-temperature process is a liquid having the consistency of heavy cylinder oil," and again the yield of this material is referred to as "about 25 gal. of light tar and oil per ton of coal." The question might therefore be raised as to what is meant by "light tar." In the curve, Fig. 1, the yield of tar in gallons is directly proportional to the specific gravity, which would seem to indicate that the higher the yield of tar the heavier these condensible products. In our work the reverse is true; in general, the lower the temperature at which the decomposition is effected the lower the specific gravity and, in a general way, the higher the yield of tar. It is indeed quite possible to produce a tar, or rather an oil, under these conditions, which has a specific gravity slightly under 1. With an increase of temperature, on the other hand, by reason of a different type of decomposition, the tars have a specific gravity which may run as high as 1.2. It would seem, therefore, that Fig. 1 would more accurately indicate the results if the specific-gravity figures were reversed. At least, if the results were thus tabulated it would be more consistent with the values obtained in our own experiments.

C. M. GARLAND,[*] Chicago, Ill. (written discussion[†]).—In the development of Mr. Traer's method for the low-temperature coking of Illinois coals, after the principles laid down by Prof. Parr, it was soon demonstrated that, in order to yield a product of uniform volatile content, the coal must be coked in thin layers, and consequently must be handled in comparatively small volumes. This means, in the first place, that great care must be observed in the design of the coking chambers to prevent the loss of the products of distillation, and, in the second place, that labor-saving devices must be utilized to the greatest possible extent in order to reduce the cost of handling.

There is, however, one advantage in the coking of the material in small volumes, and that is the decrease in the time required for coking. Notwithstanding that the temperatures used are in the neighborhood of 1000° F. (538° C.), it would be safe to predict that the coking time will ultimately be reduced to 3 or 4 hours.

In Mr. Traer's process of carrying out low-temperature distillation, considerable attention has been given to the matter of gas-tight retorts for the distillation; the use of low temperatures has made possible the

---

* Consulting Engineer.     † Received Oct. 5, 1918.

employment of cast-iron retorts, which can be made practically gas tight, thereby reducing the loss of distillate to a negligible amount.

The reduction of the labor item has been accomplished by the development of a system for the handling of the retorts and containers, involving electrically operated pushers and pullers for the moving of the containers in the retorts, and hydraulically operated doors for the quick opening and closing of the retorts.

The retorts, which may be made of any practical length from 50 to 150 ft. (15 to 45 m.) are operated in pairs, the containers traveling in opposite directions in the two retorts of each pair. A container discharged from one retort is unloaded onto a conveyor, and is then refilled and charged into the next retort of the pair. By this arrangement almost as much coke can be handled per man as in the high-temperature process.

In calling attention to this phase of the process, it should be pointed out that coke is formed in containers having a capacity of approximately 1500 lb. (680 kg.) of coal. The coke is formed in slabs approximately 6 ft. (1.8 m.) long, 1 ft. (0.3 m.) broad, and 4 in. (10 cm.) thick. Since the material is handled in units of 1500 lb., the process does not compare in any way with the briquetting process in which the cost of handling, due largely to the small unit volume, has been all but prohibitive under the conditions existing in this country. Due to the comparatively low cost of handling the material, as developed for this process, and to the high value of the byproducts, the erection of plants having a capacity as low as 125 tons of coal per 24 hr. should be profitable. From plants having a capacity of 300 tons of coal per 24 hr. an excellent return is assured.

CARL SCHOLZ,* Chicago, Ill. (written discussion†).—Mr. Traer's paper on low-temperature distillation of coals is particularly timely, because it deals with two phases of special interest to coal miners and coal consumers.

The action of the Fuel Administration in prohibiting the shipment of Eastern coals to the Western markets presents a problem to the coal producers of the Central West, to prepare their fuel in such manner as to make it compete with the Eastern high-grade bituminous and anthracite coals. This matter has had the serious consideration of many engineers for a number of years, but the low ebb of the coal industry has made investments, even in experimental plants, an impossibility, and it is only in the last 2 or 3 years that some of the Western carriers have been willing to aid in the development of the tests conducted by Mr. Traer.

The writer's interest in the low-temperature distillation of coal dates back to some experiments conducted in the Kanawha District of West

* Consulting Mining Engineer, Chicago, Burlington & Quincy Railroad Co.
† Received Oct. 8, 1918.

Virginia by the process known as "Charite," which is nothing more than the burning of egg coal in long low piles, covered with slack or coke breeze to prevent contact with the atmosphere, thus driving off the higher volatile gases without fusing the coal. Later on, the question of a continuous coking process was investigated, in connection with L. L. Summers, which resulted in the construction of an experimental plant at Haney, Ill.; here some interesting observations were made, the most important of which was the fact that nearly all of the Illinois coals would coke if finely ground and coked under compression.

Since the beginning of the war in 1914, the demand for byproducts has been revived, and the writer recognizes the need for comprehensive and extensive study of this problem because of the increased opportunities to market byproducts at higher prices than were ever dreamed of before, and the need for replacing Eastern coals, which were required for other purposes. The transportation of West Virginia coals to the West by rail has always been an uneconomic procedure, which was emphasized by the car shortage and the need of railway facilities for war purposes. With this in view, the tests conducted by Mr. Traer were very carefully followed.

It is only in recent years that hard coke has been used for domestic purposes, due to the difficulties in obtaining sufficient draft under ordinary conditions. The coke produced by the low-temperature process overcomes all of these difficulties; and, aside from possessing all of the advantages of the best domestic coals, this fuel will store for an indefinite period without any deterioration or slacking. Undoubtedly, when it becomes a commercial commodity, many new uses will be found for it.

G. W. TRAER (author's reply to discussion).—Prof. Parr's discussion develops two points, upon which it seems desirable to comment. First, as to putting through a large enough tonnage to secure economic operation. At each manipulation of each oven 1500 lb. (680 kg.) of coal will be charged and the resultant coke discharged from a like amount of coal. Sixteen slabs will be discharged at each manipulation, each approximately $5\frac{1}{2}$ in. by 11 in. by 6 ft. (14 cm. by 28 cm. by 1.8 m.). The operating schedule of a commercial plant, based upon our observations and calculations, indicates a "through put" of at least 12 tons of coal per day for each oven. We have made these figures conservatively and feel confident of materially exceeding them in a fully developed operating system. Our oven plant, with containers, will not cost as much as high-temperature plants of the same capacity. They do not require container equipment; but this is much more than offset by the cost of the great basal heat regenerative flues. Our operating requirements are quite susceptible to the application of mechanical devices and we believe our labor cost will be well within the necessary limit.

Second, as to Fig. 1 in my paper, showing the specific gravity of the tar. It was not intended to show that the specific gravity of the tar rises with decreasing heat, since of course the contrary is the fact. One of the curves shows the relation between temperature and specific gravity and the other the relation between temperature and yield. Our consulting engineer prepared the chart according to his customary method for showing two dimensions.

## Development of Coke Industry in Colorado, Utah, and New Mexico

BY F. C. MILLER,* TRINIDAD, COLO.

(Colorado Meeting, September, 1918)

THE metallurgical fuel of Colorado, Utah, and New Mexico has been a very tardy member in the caravan of western industrial progress. The history of western coke has naturally been closely related to the development of the metallurgy of iron and steel and the precious metals, in the State of Colorado.

It was at Leadville, with a population of 300 in 1877, that the *Sentinel* was published by Richard S. Allen. The *Sentinel* gave very full accounts of the various mineral discoveries at the new town of Leadville, which found their way to the more widely circulated journals of the East, and attracted the attention of many people. At the close of 1878, a census showed a total of 5040 persons.

The first smelter was established in 1877, at Malta, two miles from Leadville, by A. R. Meyer, but was not successful. It was succeeded by the Harrison Reduction Works, in 1878.

The La Plata began with one furnace in June, 1878, and in 1879 had four furnaces in active operation. These furnaces used charcoal for fuel, and, because of hot tops, suffered a tremendous loss in lead and silver. This difficulty was overcome by the erection of 50 single-bank 10-ft. beehive ovens at Crested Butte, Colo., in 1880. The slack coal sent to these ovens was very pure, and yielded a coke carrying as little as 4.5 per cent. ash.

Among the pioneers in the Leadville district were Grant, Eddy, James, Billings, Eiler, and Dickson. To these men the West owes the present expansion of its mining industry, and the erection of metallurgical processes for the treatment of precious metals.

On Mar. 1, 1876, Pueblo was connected to the Atchison, Topeka & Sante Fe by the Pueblo & Arkansas Valley branch. The first smelter was erected in Pueblo by Mather and Geist. This modest plant grew into the immense Pueblo Smelting and Refining Co., now a part of the American Smelting and Refining Co. Coke for these small furnaces was made from unwashed Engleville coal, in 186 single-bank 12-ft. ovens at El Moro, in the Trinidad district, and it was not uncommon for the percentage of ash to exceed the percentage of fixed carbon in the coke.

On Jan. 23, 1880, the Colorado Improvement Co. consolidated

---

* Chief Chemist, Colorado Fuel and Iron Co.

with other companies to form the Colorado Coal and Iron Co., with Gen. William J. Palmer at the head.   In 1881, this company erected a small iron and steel works at Bessemer, which has grown into the present Minnequa Plant of the Colorado Fuel and Iron Co.   The early plant employed 900 men; the present plant employs over 7000 men.   ·The Colorado Coal and Iron Co. never made a success with its steel venture at Bessemer, the principal cause of its failure being inferior coke, which contained 40 per cent. ash, had no strength to resist pressure, and was so fine in size that it filled the voids between the limestone and iron ore.

The reorganization of the Colorado Coal and Iron Co. in August, 1896, to the present Colorado Fuel and Iron Co. resulted in very active progress toward the improvement of its coke.   Washeries, at that time, were not a staple on the market, and consisted chiefly in very costly experiments.   Sopris was the first coke plant to have a fairly efficient washery.   In 1899, the Colorado Fuel and Iron Co. purchased the patent right to the Forrester plunger jig, which became the standard unit for all of the washeries—Sopris, El Moro, Starksville, Segundo No. 1, Segundo No. 2, Tabasco, and Tercio—built in rapid succession during the years 1900 to 1902.

During this period, a great many standard batteries of double-bank 13-ft. ovens were constructed by the Colorado Fuel and Iron Co. and by coal companies operating in New Mexico and Utah.   Previous to 1898, there were approximately 1600 coke ovens operating in Colorado, Utah, and New Mexico, without a single coal washery.   At the end of 1902, there were 5900 coke ovens in operation, with 12 coal washeries. That short period will undoubtedly go down in history as the Coke Age of Colorado, Utah and New Mexico.

The Utah Fuel Co. erected ovens at Castle Gate, in 1888, to take care of the Montana trade, but after a brief period of operation, it was found that the coking property of the coal was rapidly deteriorating as the mine developed.   An exhaustive examination of the entire property controlled by the Utah Fuel Co. resulted in the discovery of coking coal at Sunnyside.   This coal does not possess uniform coking qualities, and it is necessary to disintegrate the slack to a fineness of $\frac{1}{8}$ in. before the best results can be obtained.   Slack from this mine is low in impurity, yielding coke carrying about 12 per cent. ash.

In 1900, the Dawson Fuel Co., operated by the Phelps-Dodge corporation, at Dawson, New Mexico, erected 95 double-bank 13-ft. ovens and a washery, which gave the usual inefficient results.   After a few years of experimenting with jigs and other coal-washing appliances, this company constructed an all-steel washery and assembled units of double-compartment jigs, concentrating tables, centrifugal driers, and appliances which make this washery second to none in the United States.   In 1905, 480 under-flue ovens were erected.   These ovens are 11 ft. in diameter, of the beehive type, but the spent gases are carried through flues under the oven floor before going to a central spent-gas main leading to tall stacks. The hot gases are passed under boilers, but the system is so arranged that they can by-pass the boilers on their way to the stacks, or the boilers can be fired by hand if occasion should demand. A great saving in boiler fuel is accomplished by this form of oven construction, and the coke has a structure suitable for copper smelting. The product of these ovens is extracted with the Covington machine.

The Colorado Fuel and Iron Co. experimented with a similar oven, called a stack oven. This was of the beehive type, but the gases passed through flues under the floor and then up through a stack. The object was to coke the charge from the bottom up and from the top down in one operation, and reduce the time of coking from 48 to 24 hr. After several years of practice, the system was abandoned because there was not sufficient heat under the floor to make a dense coke; there was more sponge on the bottom of the prisms than on the top.

Mr. E. H. Weitzel was appointed Manager of the Fuel Department of the Colorado Fuel and Iron Co., in 1908, and began an active development in oven construction. Coke machines were installed at Segundo and Tabasco, and experiments were made on oven building. Previous to that date, oven retaining walls had been built of stone, laid with a batter of 2 or 3 in. The results of experiments show conclusively that reinforced concrete walls, without batter, are superior to stone. The retaining walls of the ovens constructed at Cokedale by the Carbon Coal and Coke Co., operated by the American Smelting and Refining Co., are reinforced concrete; they have been operating about 10 years with practically no repairs to the retaining wall.

The Covington coke extractor was introduced to offset the rapid decrease in available oven labor. It met this condition with a margin of profit in favor of the machine and, at the same time, the structure of the product was improved by making straight 72-hr. coke. The Koehler ovens, operated by the St. Louis, Rocky Mountain & Pacific, at Koehler, New Mexico, installed machines in 1915.

In 1909, the Colorado Fuel and Iron Co. shipped samples of coal to Joliet for a test in the Koppers byproduct ovens; the results were not satisfactory. In 1910, samples were shipped to Glasport, Pa., to be tested in the Otto Hoffman byproduct ovens; these tests were not all that could be desired, but they showed that a part of our western coal would make satisfactory coke in a byproduct oven. In 1916, another series of tests was made in the Koppers byproduct ovens at St. Louis, and the results proved satisfactory.

The H. Koppers Co., of Pittsburgh, has erected 120 ovens at Minnequa, Colo. They are divided into two batteries, and have a coking capacity of 46,000 tons per month; each oven is 40 ft. long, 10 ft. high and 19 in. wide.  It is presumed that the coking period will be between 16 and 18 hr.  There is no definite decision as to the ultimate recovery of byproducts.  Provision has been made at this time for the use of the excess gas, recovery of ammonium sulfate, recovery of tar and oils, and a special plant for benzol recovery.  An approximate cost of $5,000,000 will cover the ovens, coal washery, benzol plant, coke-screening plant, and various appliances.  The plant began the manufacture of coke on July 8, 1918.

### The Physical Properties of Western Coke

| Coke | Number of Ovens | Apparent Specific Gravity | Per Cent. Coke | Per Cent. Cells | Weight per Cu. In., Grams | Crushing Strength, Pounds |
|---|---|---|---|---|---|---|
| Colorado Fuel & Iron Co., Colo.: | | | | | | |
| Tabasco............ | 302 | 0.842 | 47.31 | 52.60 | 13.80 | 1,620 |
| Starkville........... | 190 | 0.814 | 44.76 | 55.24 | 13.34 | 1,500 |
| Sopris.............. | 272 | 0.988 | 54.06 | 45.94 | 16.11 | 2,500 |
| Segundo............ | 800 | 1.12 | 59.69 | 40.31 | 18.46 | 2,500 |
| Tercio.............. | 600 | 0.818 | 45.54 | 54.46 | 13.40 | 1,560 |
| Cardiff............. | 165 | 0.785 | 43.23 | 56.78 | 12.84 | 1,553 |
| Crested Butte....... | 154 | 0.805 | 44.20 | 55.80 | 13.17 | 1,476 |
| Victor American, Colo.: | | | | | | |
| Delagua............ | 160 | 0.722 | 38.34 | 61.66 | 11.63 | 1,120 |
| Hastings............ | 197 | 0.757 | 40.96 | 59.04 | 12.40 | 1,250 |
| Gray Creek......... | 99 | 0.793 | 42.12 | 57.88 | 12.70 | 1,300 |
| Dawson Fuel, N. Mex.: | | | | | | |
| Dawson............. | 575 | 0.738 | 40.72 | 59.28 | 12.05 | 900 |
| St. Louis, Rocky Mountain & Pacific, N. Mex.: | | | | | | |
| Gardiner............ | 184 | 0.752 | 40.41 | 59.59 | 12.29 | 1,163 |
| Koehler............. | 210 | | | | | |
| American Smelting & Refining Co., Colo.: | | | | | | |
| Cokedale........... | 350 | 1.05 | 55.92 | 44.08 | 17.20 | 2,485 |
| Durango............ | 58 | 0.784 | 42.18 | 57.82 | 12.84 | 1,275 |
| Utah Fuel Co., Utah: | | | | | | |
| Sunnyside.......... | 566 | 0.783 | 44.53 | 55.46 | 12.80 | 1,546 |

Machine-drawn ovens: Segundo, 500; Koehler, 210; Tabasco, 302; Dawson, 480.
The average ash in Colorado and New Mexico coke is between 16 and 17 per cent.; in Utah coke, between 12 and 16 per cent.

The foregoing table shows the physical properties of the coke produced in the ovens of Colorado, Utah, and New Mexico.

## DISCUSSION

C. H. GIBBS,* Salt Lake City, Utah (written discussion†).—The development of the coke industry in Utah had a somewhat checkered career for the first 50 years of its existence. About 1851 the iron-ore deposits of the Iron Springs and Iron Mountain mining districts, Iron County, Utah, were discovered. Obtaining of iron for any purpose at the time involved its shipment from the Mississippi River to Utah, by ox-team. The result was that the pioneers endeavored to establish an iron-ore smelting plant at what was known as Iron City, near Iron Mountain, in Iron County. Bishop Taylor, who headed this enterprise, endeavored to obtain coal suitable for coking from the Cretaceous coal measures immediately east of Cedar City. A few small beehive coke ovens were built and lump coal was coked in these ovens; or, more properly speaking, the volatile matter of the lump coal was driven off, leaving a carbonized material that retained the lump shape, although it was extremely friable. This coke was promptly found to be unsuitable for iron smelting, not only due to its extreme weakness under smelting load, but also because of the fact that it carried about 5 per cent. sulfur. Bishop Taylor made a very thorough and exhaustive effort, covering a considerable period of years, to locate suitable coking coal in Southwestern Utah. His efforts failed, and in 1902 the iron-ore property acquired by him and associates was sold to the Colorado Fuel and Iron Co. It is of interest to note in this connection that the iron-ore deposits of Iron County at present are the largest known undeveloped iron-ore fields on the North American continent. It is safe to say that there is at least one hundred million tons of 55-per cent. ore actually in sight.

In 1878, coal-mining operations were begun at Connellsville, in Huntington Canyon, Emery County. Ten 8-ft. beehive ovens were constructed, and lump coal was carbonized for use in the Salt Lake valley smelters. The product was somewhat better than that obtained in Iron County, but was still of very inferior quality. It had to be hauled a distance of 140 miles by wagon to the two small smelters then operating in Salt Lake valley. These ovens operated only a very short time.

The next move toward obtaining a coking coal was in 1889 when the Pleasant Valley Coal Co. began operations in the Castle Gate district. In 1890, eighty 8-ft. beehive ovens were constructed, and the product was used until 1900 in all the smelters in Utah. During this period, 124 additional ovens were constructed, making a total of 204 beehive ovens operating at Castle Gate.

The Castle Valley coal field was thoroughly prospected during the '90's for a better quality of coking coal, and in 1898 natural coke was

---

* Geologist, Utah Fuel Co　　　† Received Sept. 14, 1918.
VOL. LXI.—28.

discovered at the present location of Sunnyside. This natural coke was formed by the burning and carbonizing of outcrop coal. Due to its hardness it was uncovered by weathering, which lead to prospecting and the opening of the Sunnyside mines.

The Sunnyside district carries the only genuine coking còal found in Utah. There are several areas in which semi-coking coal occurs, but none that in any way approaches the Sunnyside product either for strength, quality or low ash content. Mining operations were begun at Sunnyside in 1900, and the product was shipped to the ovens at Castle Gate. As soon as the Utah smelters obtained coke made from Sunnyside coal the smelting results were so much more satisfactory than those obtained from the Castle Gate product that the Utah Fuel Co., as rapidly as possible, has increased the Sunnyside output to such an extent that it now takes care of the entire coke requirements of this territory.

The Sunnyside coal ranges in thickness from 5 to 14 ft.; it is hard and blocky; stands weathering well, and is the premier steam fuel coal of Utah. Very little of it, however, has been used for steaming purposes, due to the fact that the entire production of this district is used either for coke making or in byproduct gas plants located in Utah, Idaho, Montana, and Nevada. Throughout the coking area the coal shows evidence of considerable movement due to local bending, which is undoubtedly responsible for the fact that the Sunnyside district coal is of coking quality. Wherever the synclinal bending in the Sunnyside district ends, the coking quality of the coal practically disappears. In order to prepare the coal for coke ovens, the entire product is crushed so that it passes a $\frac{1}{4}$-in. screen. This crushing serves two purposes: first, that of producing a decidedly more uniform coke, and second, ringwall and breeze losses are materially reduced.

The following table gives the annual production of coke in Utah from 1890 to date:

| Year | Tons | Year | Tons | Year | Tons |
|------|------|------|------|------|------|
| 1890 | 8,395 | 1900 | 32,730 | 1909 | 184,745 |
| 1891 | 7,947 | 1901 | 39,860 | 1910 | 150,677 |
| 1892 | 7,242 | 1902 | 73,230 | 1911 | 174,000 |
| 1893 | 16,007 | 1903 | 158,099 | 1912 | 302,457 |
| 1894 | 16,057 | 1904 | 156,337 | 1913 | 332,396 |
| 1895 | 22,517 | 1905 | 220,706 | 1914 | 349,898 |
| 1896 | 20,449 | 1906 | 259,924 | 1915 | 301,420 |
| 1897 | 23,619 | 1907 | 317,925 | 1916 | 424,828 |
| 1898 | 28,327 | 1908 | 180,074 | 1917 | 374,775 |
| 1899 | 26,882 | | | | |

As soon as the good quality of Sunnyside coke became generally known it was shipped not only to Utah, but north to Anaconda and East Helena, Montana; and to Nevada, Idaho, and parts of California.

In 1907 the Castle Gate beehive plant was abandoned, having been replaced by new ovens at Sunnyside. At present the Sunnyside coke-oven plant is the largest single beehive operation in the United States, consisting of 819 ovens of 12- and 13-ft. size. The coke produced carries a very low, uniform ash content, and is practically free from sulfur and phosphorus. Average analyses of this coal and of the coke made from it are as follows:

|  | COAL | COKE |
|---|---|---|
| Moisture | 1.20 | 0.81 |
| Volatile matter | 39.13 | 0.83 |
| Fixed carbon | 53.69 | 87.36 |
| Ash | 5.98 | 11.00 |
|  | 100.00 | 100.00 |

Sulfur in the coke will run from 0.7 to 1 per cent., and phosphorus from 0.03 to 0.05 per cent. An increasing tonnage of this coke is being used in Salt Lake valley and also on the West Coast, for steel making.

## The Byproduct Coke Oven and Its Products

BY WILLIAM HUTTON BLAUVELT,* E. M., M. S., SYRACUSE, N. Y.

(Colorado Meeting, September, 1918)

THE technical and engineering problems in the manufacture of coke are today the problems of the byproduct oven. Except in a few special localities, practically no beehive ovens have been built in the United States for the last 5 or 6 years except as renewals of old plants, and during these years the total number of beehive ovens in existence has been steadily falling. On the other hand, the number of byproduct ovens has increased from 4624 in 1911 to a total of about 7660 in operation at the end of 1917; with 2800 building, this makes a total of about 10,460 which will be in operation in 1918, or shortly thereafter.

Preliminary Government estimates place the total production of coke for 1917 at 56,600,000 tons, the largest tonnage in the history of the industry. Of this 34,000,000 tons, or 60 per cent., was beehive, and 22,-600,000 tons, or 40 per cent., was byproduct coke. Some time in 1918 the production of byproduct coke should pass the beehive output, and when the byproduct ovens now building are completed the total byproduct capacity will be about 40,000,000 tons of coke per annum, which is over 70 per cent. of the record-breaking total coke production of 1917. The reasons for these changes in the coking industry are familiar to all of us, and need not be discussed here.

### IMPROVEMENT IN OVEN DESIGN

Since the byproduct oven was brought to this country from Europe, in 1892, it has been radically improved, and from the point of view of American metallurgical practice it is safe to say that the American ovens are superior to those of Europe. The points of superiority are mainly those of larger units and larger output per unit, and the greater extent to which labor-saving machinery has been introduced. The increased output is largely due to the use of silica refractory material, which permits higher heats and shorter coking time than are employed in Europe. The modern American oven will carbonize commercially more than 20 tons of coal per day, and contrary to the European idea that slower

---

*Consulting Engineer, Semet Solvay Co.

operation is necessary to conserve the plant, this rate of operation, with only current repairs, can be maintained for an indefinite time. The by-product oven is not old enough in the United States to make it possible to state from experience how long an oven can be operated at maximum efficiency. There are a number of plants in different parts of the country which were built 16 or 18 years ago, which are more efficient today than in the first year of operation, and which show costs of operation that compete well with those of the most modern plants.

The accompanying illustrations, Figs. 1 and 2, show the earliest and the latest development of one of the prominent types of ovens in use in this

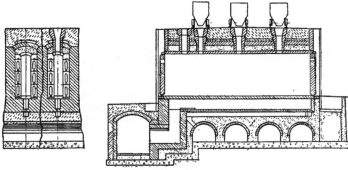

FIG. 1.

country. It is not the purpose of this paper to discuss mechanical details of oven development, and the illustrations show sufficiently well the principal changes in the design. The first ovens had a capacity for carbonizing 4.4 tons of coal per day, as compared with the present capacity of over 20 tons mentioned above.

The early ovens were economical in heat consumption because the heat in the waste gases was utilized efficiently in raising steam for the operation of the plant. On account of the growing demand for oven gas for metallurgical and other uses, the change to the regenerative type of oven has been general. While this type of oven is not so economical of the total heat produced by the combustion of the gas as is the combined oven and boiler plant, it makes available for use half again as much of the surplus gas as did the older type of oven. Modern ovens require for carbonization of the coal less than 40 per cent. of the total heat in the gas produced.

### Description of Oven

Fig. 2 illustrates one of the principal types of the modern oven, and a few words of description may be appropriate here. The byproduct oven is essentially a closed chamber, heated from the outside, and in the coking process the volatile matter of the coal is distilled off with careful exclusion of air.    This point distinguishes it in principle from a beehive oven, where the heat is generated by combustion within the oven itself, with the resulting destruction of everything except the coke.    The oven shown consists of a chamber about 36 ft. (11 m.) long, 12 ft. (3.66 m.) high, and a width depending upon the coal that is to be carbonized.    In American practice the average width varies from 16.5 to 21 in. (41.9 to 53.3 cm.). Modern ovens usually have about 2 in. (5 cm.) of taper toward the discharge end to facilitate the pushing of the coke.

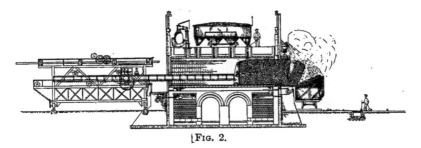

[Fig. 2.

The ovens are usually heated by a portion of the gas recovered from the distillation of the coal, although sometimes producer gas is substituted. In all the successful ovens this gas is burned in a series of flues which control the travel of the burning gases and distribute the heat over the entire side of the oven.    The success of any oven design depends very largely upon its ability to distribute the heat evenly over the entire surface which forms the wall of the oven, and it is by no means an easy problem to control this distribution accurately over an area of 432 sq. ft. (40 sq. m.).    In some ovens the flues composing the heating system are vertical and in others horizontal.    The oven shown is the chief exponent of the latter arrangement.

Both systems have their advocates, and the details of both have been worked out so that they can be relied upon to give the uniformity of heating necessary for satisfactory operation.    Naturally the representatives of each system prefer their own type.    The advocates of the horizontal flues find them more accessible for supervision, easier to control, and easier to maintain.    The somewhat lower temperature that can be maintained in the upper flues under all conditions of operation is a desirable feature for the best results.    One feature of the oven shown is the strong middle wall between the ovens extending from the foundation to the top, which gives the structure great stability and permanence and acts as a heat reservoir to help maintain uniform temperatures, which is important, because when a new charge of cold wet coal is dropped in the oven there is a great demand for heat to start the coking process quickly. These middle walls also permit any oven to be repaired or entirely relined with a minimum interference with adjacent ovens.

### Heating the Oven

Fig. 3 shows the flow of the air and gases through the system. From this illustration it will be noted that one pair of dampers near the stack does all the reversing of the air and gas. The cold air enters the system from a fan located at a point near these dampers, flows through one of the

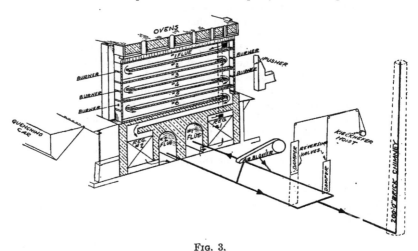

Fig. 3.

two main flues, and is distributed to the regenerators by a system of firebrick slides. The air is heated in the regenerators to about 1000 to 1100° C. before entering the flue system of the oven and meeting the gas to be consumed.

It will be noted that this gas is supplied at several points in the flue system, and that it flows steadily through the supply pipes, its direction in the oven being reversed by the reversal of the current of air, so that no attention need be given to this point by the operator. That is, when the currents are flowing upward through the flue system each gas stream is bent upward and burns in the flue above the gas pipe, which is opposite the partition wall. When the flow is downward, the gas streams are bent downward. By means of these several supplies of gas the heat is supplemented as needed and all the air supplied is consumed by the last admission of gas. The products of combustion pass out to the other regenerator and thence to the reversing dampers and chimney in the same manner as in an open-hearth furnace.

In the earlier types of ovens, when luminous gas was used, much of the heating was done by radiation, but in modern practice, where the benzol is removed from the gas, the combustion is not luminous and the heating is done mainly by conduction.   To this end, in the system described it has been found quite advantageous to maintain a reasonably high velocity of the hot gases in the flues, making use of the principle which has been developed in some types of hot-blast furnace stoves and other furnaces, where a high velocity of the hot gases has been shown to be most effective in the transfer of heat to the adjacent surfaces because it sweeps away the sluggish layers of gas lying against the surfaces, which, if not removed, would form efficient non-conductors of the heat which we desire to transmit.

The total amount of heat delivered to the combustion flues is usually regulated by the pressure on the supply of air and gas; and the distribution of the heat among the various flues is controlled by a simple regulation of the relative amount of gas to each flue.   Each flue may be conveniently inspected through a peephole accessible from the working platform outside; and it is found that the convenience of this inspection and control is an important factor in obtaining the best heating conditions.

### Operation of Byproduct Oven

The coal is charged into the oven from a hoppered larry car running on the top of the oven structure, the fuel usually being delivered through four charging-holes.   The fifth hole, shown in the illustration, serves to carry off the volatile products of the distillation.   The coal is leveled after charging by an auxiliary ram on the coke pusher, the space being left between the top of the coal and the top of the oven for the outgoing gases to reach the point of discharge.

Upon completion of the coking process the doors are raised, and the coke is pushed out by an electrically operated ram.   The doors are then lowered and sealed and the oven is ready for another charge.   Many attempts have been made to develop a mechanically sealed door, but the conditions of operation make this mechanical sealing a difficult problem, and nearly all plants still use the mud joint.   Careful designing of the door and its seat make this mudding a much less serious matter than might be supposed.

### Use of Silica Brick

As stated above, silica brick is largely used in modern byproduct oven construction.   There seemed at first many obstacles to its successful use for such construction, the principal objections being the large amount of expansion of the brick when heated and the tendency to spall, due to sudden changes in temperature. When silica brick was first used in the crowns of beehive ovens it was thought that it would be an absolute failure, but it was unexpectedly successful. It was first employed in byproduct ovens at the plant of the Cambria Steel Co. and was soon after generally adopted.

It has been found that, since most of the expansion takes place below a red heat, the variations in temperature during operation do not give any trouble, and the superior resistance of silica brick to high temperatures and its satisfactory conductivity under operating heats make it a superior material. Notwithstanding its success in American practice, I understand it has not been employed to any extent in Europe, and only a few years ago some prominent English brick makers stated that they did not believe it could be employed, notwithstanding statements that came from America.

### Size and Width of Ovens

The cubic capacity of a byproduct oven has increased nearly fourfold since its first introduction into this country. The larger capacity has the advantage of reducing the cost of operation per unit of product, and the smaller number of units also reduces cost of construction and repairs. Increases in length and height of the oven chamber have no effect on the product and are limited only by structural and heating conditions. The width of the oven, however, is a more important factor.

In the early days of the byproduct oven in Europe it was thought that the lean coals of Belgium, running as low as 15 per cent. volatile, required a very narrow oven, perhaps not over 14 in. (35.5 cm.) wide, while for high-volatile coals a width of 20 in. (50.8 cm.) was not infrequent. American practice has not demonstrated any such simple rule as this, and perhaps the reason may be found partly in the fact that oven plants are generally located at the point of consumption of the coke, making it easier to obtain mixtures of coals, and for this reason most plants operate on about the same average percentage of volatile matter. It has not been definitely shown in American practice that a particular width of oven is best suited to coal of a certain composition. Narrower ovens have certain advantages:

First, a faster coking time per inch of width with a given flue temperature.

Second, a somewhat higher yield of byproducts.

Third, a somewhat more uniform cell structure as between the portion of the coke nearest the wall and that nearest the center of the oven.

On the other hand, the larger number of operations of the oven crew when handling the narrower ovens increase the labor cost and the repairs.

It has yet to be developed what is the best oven width to average these conditions.   The widest ovens operating in this country have an average width of 21 in. (53.3 cm.) and the narrowest 12 in. (30.6 cm.).

It will probably take much further investigation to determine where between these limits can be obtained the best equalizing of conditions, and whether the mixture of coals naturally available for a given plant will be a factor in determining the best average width.   Some coals require higher temperatures than others to produce the best coke, but it is not clear that this fact will lead to the selection of wider ovens for such coals.   It is not clear that we can say today that any width is definitely the best for a given set of conditions, but much study is being given to this subject, and we may before long have some better data.

## QUALITY OF COKE

In the discussion of this subject I shall not go into the chemical composition of coke, since the permissible amount of any impurity for a given use is well understood, and the different elements are easily determinable. The physical structure of the coke is quite as important as its chemical composition.   It is the physical structure which gives coke its advantage for metallurgical work over other forms of solid fuel, and it is important that the structure should be adapted to the conditions under which the coke is to be used.

The blast furnace is the great coke consumer.   In the days of the beehive oven one kind of coal gave a coke with a certain physical structure, and another coal gave another structure.   Furnaces either adapted their practice to the coke, or changed their coke supply.   Coke was recognized as hard or soft, porous or dense, and that was about all that was known regarding physical structure.

Mr. Brassert says in his paper on "Modern American Blast Furnace Practice," read in 1914, that "the early coke produced in our byproduct ovens, even from the same coals as were successful in the beehive oven, burned too slowly and made our furnace operations exceedingly difficult, by preventing rapid and continuous movement of the stock.   The lack of knowledge and experience along these lines was responsible for the slow progress attending the introduction of byproduct ovens in this country."   The economy of the byproduct oven practically forced its adoption by the furnace operators, and for several years, as Mr. Brassert states, "at a number of American plants byproduct coke has been made which rivals in quality our best beehive product."

The byproduct oven, with its variable mixtures of coals, variable heats, coking time, width of oven, fineness of coal charged, and other controlling factors, permits a control of coke structure formerly impossible. The problem is to determine, first, what is the structure best adapted to standard furnace practice, while recognizing that special practice requires modifications of structure; second, what conditions are necessary to produce it.

Notwithstanding the general acceptance of Grüner's theory of ideal combustion in the furnace, the production of a high thermal head at the tuyeres is of the first importance, and the best coke is that which reaches the tuyeres in proper condition to produce the highest temperatures in the tuyere area, and in just sufficient quantity to do the amount of work required there under the conditions produced at this maximum temperature.

The ideal coke is one that will descend through a furnace shaft to the combustion zone in front of the tuyeres with the least loss from attrition and oxidation, and when it arrives there will burn at the highest possible rate. Of course, these are paradoxical qualities. However, Mr. Walther Mathesius points out in his interesting paper on "Chemical Reactions of Iron Smelting" that "modern American coke-oven practice has made enormous strides toward approaching this apparently paradoxical ideal." He stated that this is accomplished by producing coke with an open-cell structure, in which the cell walls themselves are amply strong and well protected by a graphitic coating.

The time of contact of the blast with the coke in the tuyere area can be only a few seconds and the speed of any chemical reaction decreases as the relative quantities of reacting and resulting substances approach equilibrium. Therefore, the farther these relative quantities remain from the status of equilibrium, the higher the rate of resultant combustion. With the facts now before us, I am disposed to believe that we, in seeking to produce the best blast-furnace coke, should aim to produce an open-cell structure, with cell walls strong and hard. Later experience may, however, show that there are other requirements that are not now known to us. It is not necessarily true that the open-cell structure is the same thing as a high percentage of cell space. The advantage of an open-cell structure is that it gives the oxygen of the air easy access to the carbon. It is entirely possible that a coke of very fine cell structure, having say 50 per cent. of cell space, might offer less surface for prompt combustion under practical conditions than another coke containing larger cells but having the same percentage of total cell space.

The composition of the cell wall, which it is agreed should be hard, thin and strong, and, according to Mr. Mathesius, covered with a graphitic coating that is smooth and bright, is a much more complicated matter. What are the conditions of coal mixture and coking which produce this kind of wall? I think we have not yet found the answer to this question, although we know some of the conditions that are favorable to this result. The coal mixture, the degree of fineness of grinding, the coking time, and the heats are probably all factors.

Our search for the best coke structure to meet a given set of furnace conditions is not an easy one, but we know better which path to start on than we did even a few years ago. Are we not agreed on the following points at least?

1. The coke must be hard.

2. It must have an open-cell structure; that is, cells of good size and approximately 50 per cent. of cell space.

3. It must have a high rate of combustibility.

Can we add anything more to this list? Some investigators have concentrated their comparisons on the rate of combustibility, but I cannot believe that this test alone is sufficient to determine the best coke structure, because it ignores one of the sides of the paradox. While the best coke must burn rapidly at the tuyeres, it must also resist attrition and oxidation during its descent in the furnace. Good-sized cells and a good percentage of cell space, coupled with a hard structure, would seem to give a coke corresponding to Mr. Mathesius' definition. Testing the rate of combustion has been a help to us, and I hope we will find a test for hardness of structure better than the crushing of the 1-in. cube specimens, over which so much time used to be spent in the days when John Fulton wrote his book on "Coke."

### Foundry and Domestic Coke

In addition to the iron blast furnace, foundry cupolas afford a steady market for coke, and in recent years the domestic market has grown to importance. Before the war, sales of coke for domestic use amounted to over 1½ million tons per annum, although these sales really rather demonstrated the value of coke as a domestic fuel than showed what were the ultimate possibilities of the trade. Foundry coke is usually made on a somewhat slower coking time than furnace coke, as cupola practice calls for larger and tougher pieces.

The production of domestic coke is now a greatly different industry from what it was in earlier days, when most of the domestic coke was produced in gas works. At that time it was often regarded as something to be got rid of with the least trouble, rather than as a regular and important source of income to the works. Great care is now taken in the sizing and preparation of domestic coke and the complete removal of breeze and dirt. The following table shows approximately the usual sizes produced:

| | Size of Square Through, In. | Opening of Screen Over, In. |
|---|---|---|
| Egg coke | 3.5 | 1.5 |
| Nut coke | 1.5 | 1.0 |
| Pea coke | 1.0 | 0.5 |
| Breeze | 0.5 | |

These screens are varied locally to meet market requirements. It would be desirable to have a standard for each domestic size throughout the country, but where the demand for one size is large relative to another, it is natural for the operator to modify his screen openings accordingly.

In American practice, where domestic coke competes mainly with anthracite, a hard structure is generally regarded as desirable. This is contrary to English practice, where familiarity with bituminous coal as a domestic fuel makes quick inflammability a desirable quality, so that coke containing a considerable amount of volatile matter is often sold successfully. Even the partially coked coal, known as "Coalite," which is produced by a low-temperature distillation, is preferred to the hard clean coke which is becoming familiar to the American consumer.

## PREPARATION OF COAL

The practice of locating American ovens at the point of consumption of the coke has made it easy to bring together several coals of different qualities, and this has in many cases permitted the production of a better coke by admixture than could be made from any one of the coals used alone. For example, a coal of excellent coking qualities but high in ash can be mixed with a low-ash coal of poorer cokability, or an excess or deficiency in the coking quality of the main coal supply can be corrected by mixture with another coal.

Except in cases where the source of coal supply is fixed, it seems desirable to arrange the plant for the preparation and mixing of as many as three or four kinds of coal. Means permitting the accurate proportioning of the different coals are, of course, essential in such a plant. The mixture of coals generally calls for their being ground quite finely, in order to bring the particles of the different qualities of coal closer together. The general practice [is to grind so that from 70 to 80 per cent. or more will pass through a ⅛-in. screen. This fine grinding is also often advantageous where there are many pieces of slate of considerable size in the coal, since each large piece of slate in the coke is usually a center of cracks that tend to shatter the structure.

Various degrees of fineness in grinding are often adopted to improve the structure of the coke, but in some cases, especially where only one coal is used, a better structure is produced from considerably coarser coal. Fine grinding reduces the weight of the coal mass per cubic foot, and consequently the capacity of the oven. With ordinary coals the weight of charge is generally computed on the basis of 40 to 43 cu. ft. (1.13 to 1.21 cu. m.) per net ton, depending somewhat on the coals, and largely on the degree of fineness.

### Use of Coals Formerly Considered Unsuitable

The byproduct oven, with its control of heats, coking time, mixture of coals and fineness in grinding, has permitted the use of many coals for making metallurgical coke which would not have been suitable with the beehive oven. As the best coals are being progressively exhausted, this point becomes more important. Owing to the extraordinary conditions caused by the war, some coals hitherto considered entirely unsuitable for making metallurgical coke have been successfully used; for example, the unwashed coals of southern Illinois.

To use these hitherto unsuitable coals successfully it is necessary to make a careful study of the coking qualities of each one in order that any deficiencies it possesses may be supplied by the proper selection of the coals forming the mixture. Very satisfactory furnace results have been obtained when using important percentages of coals heretofore considered unsuitable, and these results are encouraging on account of the progressive disappearance of the best of the coking coals now available.

## Recovery of Byproducts

The distillation or coking of coal in closed retorts, now generally known as byproduct ovens, was first introduced for the purpose of saving the several valuable products contained in the volatile matter of the coal, which were all destroyed in the beehive process. The advantages in the production of coke outlined above were not then realized. In fact, it was often thought that a sacrifice of coke quality must be accepted in order that the byproducts might be recovered. Fortunately it has developed that the best oven conditions for the production of coke are, in general, the best for the production of byproducts also.

The composition of the volatile matter in bituminous coal is very complex and the changes through which these compounds pass during the process of distillation are also complicated. It is not appropriate here to attempt to trace out these changes or to discuss the method of formation of the ammonia, benzol or other byproducts. It is well known that only about one-fifth of the nitrogen in the coal combines with hydrogen to form ammonia, and that the remainder is found in the coke and gas.

Attempts have been made to obtain higher yields of ammonia by blowing steam into the coke, somewhat after the manner followed with the Mond producer, which recovers from three to four times as much ammonia, but none of these processes has been developed commercially. Higher coking heats tend to break down the combinations to their ultimate elements. For example, the gas produced from high-temperature coking will contain more hydrogen and less marsh gas, and a larger volume will be produced of lower calorific value. When coking at very low temperatures, as in the production of "Coalite," practically no benzols are produced and the hydrocarbons recovered are of the aliphatic series. A small quantity of very rich gas is obtained and the ammonia yield is low. At high coking temperatures the tendency is for the toluol, the more complex hydrocarbon, to break down into benzol, and at very high temperatures the yields of both are reduced. The temperature conditions to which any particular particle of hydrocarbon is submitted in its travel from the original piece of coal up to the hydraulic main are so complicated and so varied that a theoretical discussion of these reactions is of little value.

### Cooling the Gas

The gases passing off from the oven are heavily laden with various hydrocarbon vapors and with water vapor and consequently carry a large amount of latent heat. So the cooling of the gas and the condensation of the vapors require efficient condensing apparatus and large amounts of water. The gas-works manager of early days used to deal gently with his gases in removing the tarry vapors and cooled them very slowly in order to prevent their taking up the benzols from the gas, with injury to its illuminating power, but in coke-oven practice this precaution is not considered necessary.

Much improvement has been effected in the efficiency of the condensing apparatus since the early days of the tubular cooler, in which the gas circulated around the tubes of an apparatus similar to a vertical boiler, while the cooling water flowed through the tubes on the countercurrent principle. I have in mind one single-stage condensing apparatus 10 ft. (3 m.) in diameter and 45 ft. (13.7 m.) in height of effective cooling surface in which 12,000,000 cu. ft. (340,000 cu. m.) of gas per day is cooled on the average from 80° to 20° C.; in which the cooling liquor enters the apparatus at a temperature 5° below the exit temperature of the gas, and leaves it at not more than 10° below the entering gas temperature. In such an apparatus very little condensate separates from the gas at a low temperature, and as a result the benzols remain in the gas and the separation of tarry matter from the ammonia liquor is sharp and easily effec'ed.

### Removal of Tar

A small quantity of the lighter tar vapors remain in the gas with the benzols. These may be removed by some efficient form of the old Pelouse and Anzin tar extractor, or where the gas is washed with water to remove the ammonia in the indirect process of ammonia recovery, these light tars are efficiently washed out. An efficient system of removing these light tars in a high-tension electric field has been developed by F. W. Steere, the plan adopted being similar to that used in the well-known Cottrell system, except that the special conditions permit the use of an alternating current of the necessary high voltage, since the problem is merely the agglomeration of the minute globules of tar to permit their effective removal by tar extractors of the usual type. This process has been worked out commercially and is simple and effective.

*Tar as Fuel*

One of the interesting developments in the operation of byproduct ovens in connection with steel plants is the growing use of tar as a fuel for metallurgical furnaces. I have in mind one plant producing over 20,000,000 gal. of tar per year, in which the entire amount has for years been utilized as fuel in open-hearth and heating furnaces, where its value as fuel is much higher than its value in the ordinary tar market. Generally speaking, the tar used in steel furnaces is handled in the same manner as ordinary petroleum fuel oil, having about the same viscosity, and its use presents no difficulties after the details of its application have been worked out. It has a higher fuel value per gallon than ordinary petroleum fuel oil.

*Recovery of Ammonia*

It was formerly the universal practice to extract the ammonia from coke-oven gas by scrubbing with water, which has a large capacity for absorbing ammonia at moderate temperatures. This was the old gas-works method, and the processes as used in oven plants are the same except as modified to meet the conditions of large volumes and efficient recovery. The ammonia is recovered by these absorbing processes in the form of a crude liquor, usually containing from 0.5 to 1 per cent. of ammonia.

This liquor usually contains practically all of its ammonia in the free state, whereas the condensate from the coolers carries considerable amounts of fixed ammonia, sulfate, carbonate, chloride, and sulfide. After removal of the tar by decantation the combined liquors are distilled with steam, after setting free the fixed ammonia by means of lime, the product being the crude ammonia liquor of commerce. In other plants the vapors from the still are passed through a sulfuric-acid bath and sulfate of ammonia is formed.

The so-called direct and semi-direct processes for the manufacture of ammonium sulfate have been very generally applied to oven plants in recent years. The direct process requires only the removal of the tar from the gas without any more cooling than is necessary for this purpose, after which the gas is passed through a bath of sulfuric acid and the sulfate recovered. The direct process has been applied to a considerable extent in Europe, where the percentage of chlorides in the coal is quite small, but where the amount of chlorides is relatively large it is necessary to remove them before the acid bath is reached.

This introduces the "semi-direct" process, in which the gas is cooled to remove the chlorides and then reheated and passed into the acid. The fixed ammonia condensed out in the cooling process is distilled and generally the ammonia from the stills is returned to the gas before it reaches the acid bath. The percentage of ammonia that has to be distilled in the "semi-direct" process depends upon the amount of fixed ammonia in the gas and varies from 15 to 40 per cent. of the total ammonia.

Sulfate of ammonia was formerly the only source from which pure ammonia in the form of *aqua ammonia* or anhydrous ammonia was produced. This process consists in the distillation of the sulfate of ammonia using lime to free the ammonia from the sulfuric acid, the small amount of hydrocarbons remaining in the sulfate being removed by charcoal filters or other similar means.

If anhydrous gas is to be produced the gases in the top of the still are dried by means of quicklime or other desiccating material, and the pure gas is then compressed and pumped into the steel cylinders so familiar to the trade. *Aqua ammonia* is produced by absorption of the pure gas in distilled water. This process is essentially a simple one, but the purification of the ammonia gas and the necessity for providing against the blocking of the apparatus by accumulations of calcium sulfate scale somewhat complicate the work.

In recent years, by refinements in the distillation process it has been made possible to produce pure ammonia direct from the crude weak liquor and this product is quite equal in quality to that produced from ammonium sulfate as above described. These processes have the advantage of avoiding the waste of sulfuric acid inherent in the ammonium-sulfate process.

## Effect of the War on the Ammonia Market

The war has introduced many unprecedented conditions to the ammonia producer. Before the war the price of sulfate of ammonia varied, in general, between 2½ and 3 c. per pound. Official quotations are now running from 7.25 to 7.35 c. While these high figures do not represent the average of actual transactions, any more than the $55 per ton frequently quoted represented prices at which the pig iron of the country was moving last summer, yet they illustrate the unsettled conditions of the market. Recently the Government has asked for very large amounts of ammonia for war purposes, and this call for ammonia has introduced much discussion as to the best methods for producing it.

In times of peace the great bulk of the ammonia goes into sulfate for fertilizer purposes, and most oven plants have been built for supplying this product. The Government's needs have caused many inquiries regarding the changing of these plants to produce other forms of ammonia, but this is usually practically out of the question on account of the great time and expense involved, since large parts of the recovery plants would have to be entirely rebuilt in order to change the product. Fortunately there are several practical methods for the conversion of ammonium sulfate into ammonium nitrate, and the synthetic ammonia plants will soon have a large production directly available for this purpose.

The total ammonia from all oven plants now built or building is estimated at somewhat over 130,000 tons of $NH_3$ per annum.

### COKE-OVEN GAS

In the earliest European byproduct ovens the gas distilled from the coal in the coking process was not sufficient in amount to heat the ovens properly and had to be supplemented with solid fuel. As the efficiency of the heating increased, the surplus, which was rather unreliable, was utilized for raising steam. As the supply became still larger it began to be introduced as one of the regular sources of fuel in metallurgical and manufacturing plants.

In America, oven gas has long been used for domestic purposes, and now over forty important cities and towns are wholly or partly supplied with oven gas, and a number of years ago the daily sales for domestic purposes had reached 50,000,000 cu. ft. Europe has lagged far behind America in this application of oven gas.

Since the first introduction of byproduct ovens into America many experiments have been made in the use of oven gas for metallurgical furnaces, and several experiments have been recorded describing unsuccessful efforts to apply it to open-hearth steel furnaces, although it has long been used in heating furnaces, soaking-pits, etc. During the last few years, however, the whole subject of using oven gas in open-hearth furnaces has been successfully worked out, and it is now in regular operation in several important steel plants. Since the benzol is now generally extracted from oven gas, it is the practice to burn a certain amount of tar along with the gas to give it luminosity and higher radiating power.

From 12 to 16 gal. (45 to 60 l.) of tar, together with from 8000 to 8200 cu. ft. (226 to 232 cu. m.) of gas, the quantity depending upon its calorific value, are required per ton of ingots produced, and it is found that furnaces fired in this way can be depended upon to produce from 10 to 15 per cent. more steel than when heated with producer gas. These figures are based on work at stationary furnaces. On the basis of these figures it may be stated that the surplus gas and the tar obtained in the production of a ton of coke will furnish the necessary fuel to produce a ton of ingots from an open-hearth furnace. Without considering the value of the other byproducts recovered, this statement illustrates the importance of the byproduct oven as a fuel supply to a blast furnace and steel plant.

### Separation of Gas

Early in the development of the byproduct oven in America the principle of the separation of the richer portion of the gas, which comes off during the early part of the coking period, from the leaner portion, was developed in a practical way. This has enabled the oven plants to produce for sale gas having a calorific value much higher than could possibly be produced in ordinary gas works from the same coal. The accompanying charts (Figs. 4 and 5) show the composition of a gas from

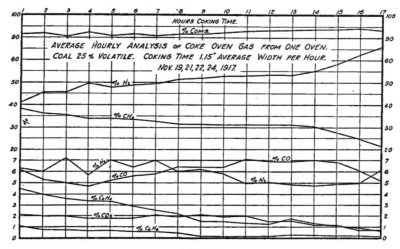

FIG. 4.

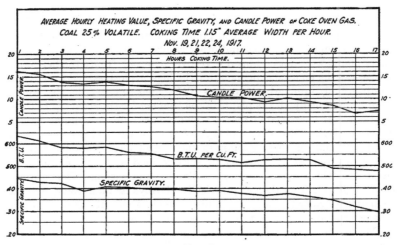

FIG. 5.

an oven running on a coal of about 25 per cent. volatile matter, analyses being taken each hour during the coking period. At the time these analyses were made the oven was coking at the rate of about 1.15 in.

(2.9 cm.) of width per hour.  This chart shows plainly the progressive decrease in the percentages of methane and the progressive increase in hydrogen and carbon monoxide, as the coking proceeds.

The separation of the rich from the lean portion of the gas is probably not advantageous when the surplus is to be used for fuel purposes, as, of course, separation introduces some complications into the apparatus, but where a higher calorific value than the average is desired, the separation method is of great value and can be accurately controlled.

### Recovery of Benzol

Before the general use of the byproduct oven, benzol and its homologues, toluol and xylol, or solvent naphtha, were produced from the distillation of tar, which usually contains about 3 per cent. of these compounds.  Now the main source of supply is coke-oven gas, from which the benzols are scrubbed by bringing the gas into contact with a heavy oil of petroleum or tar.  This process is usually carried on after the removal of the ammonia.  The benzolized heavy oil is distilled to drive off the benzols and the heavy oil is cooled and returned to the heavy-oil scrubbers.  The benzols are recovered in the form of a "light oil," containing usually about 60 per cent. of benzol, 12 per cent. of toluol, and 6 per cent. of solvent naphtha, although these percentages vary considerably with the details of the coking process and the light-oil recovery.

Before the war the light oil produced was used largely for enriching illuminating gas, and the refined products were employed in the manufacture of paints, varnishes and other similar industries.  At that time many plants were not equipped to recover light oil.  Since the beginning of the war the enormous demand for the high explosives, picric acid and trinitrotoluol, familiarly known as TNT., has resulted in the equipment of practically all the plants in the country for the recovery of these products.  It is estimated that the plants now built or building have a capacity of 120,000,000 gal. (454,200,000 l.) of light oil per year.  For the production of these explosives the benzol and toluol have to be thoroughly refined by washing in sulfuric acid and by accurate fractionation, which processes require skillful supervision.  The removal of the light oils from the gas practically destroys its luminosity, but the reduction in calorific value is relatively small, about 10 B.t.u. per cubic foot per gallon of light oil recovered per ton of coal.

When the demand for high explosives ceases after the war a large amount of the benzol now produced will probably be sold as motor benzol.  This product is usually a mixture of benzol, toluol and solvent naphtha in the proportions in which they occur in the light oil, and it is obtained by a comparatively simple process of refinement from the light oil.  Motor benzol has a fuel value from 18 to 20 per cent. higher than gasoline, and may be used alone with the ordinary carbureter if a somewhat larger proportion of air is used, or it may be mixed with gasoline in any proportion. Pure benzol freezes at 5.18° C., and toluol at −92° C. No trouble is experienced from freezing, even in the coldest weather, when proper percentages of toluol or gasoline are mixed with the benzol.

The dye and color industry in the United States has greatly expanded during the war, and it will flourish after peace has returned if given proper Government protection. This will furnish a large market for benzol, which is the raw material for aniline. Also it seems not improbable that trinitrotoluol will be widely utilized as an industrial explosive. It is claimed that it can be manufactured more cheaply than nitroglycerine, and as a commercial explosive it has many advantages over nitroglycerine compounds. Trinitrotoluol is very stable and ignites only at high temperatures, and even then it does not explode. It does not freeze, as does nitroglycerine, and on many counts it has advantages on the side of safety and convenience. When used alone it produces large amounts of carbon monoxide; but this can be entirely prevented by combination with other bodies, such as ammonium nitrate for example, if these are added in proper proportion to produce complete combustion.

## Conclusion

From the above brief discussion of the development of what has grown to be one of the important American industries, it will be seen that the byproduct oven has given this country something which is much more than merely a cheaper method of producing coke. It gives the coke user a material which can be accurately adapted to his special needs by an intelligent application of the several variables that may be introduced into the coking process. It renders available many coal deposits not heretofore considered suitable for making coke. It furnishes a new and large supply of gaseous and liquid fuel for metallurgical, industrial, and domestic uses. It adds very greatly to the supply of available fertilizers which our overworked and underfed lands are coming to need so sadly, and it furnishes an abundant supply of raw materials for a number of our chemical industries, especially the great dye and color industry which our war experience has shown is vital to a large number of our manufacturing operations. It is hard to suggest any other industry of comparatively recent introduction which does more to conserve the natural resources of our country.

## DISCUSSION

A. K. McCosh, Coatbridge, Scotland (written discussion*).—It is well known in Great Britain that oven operators in the United States

___

* Received Sept. 3, 1918.

have been able to reduce the coking time much below European standards by the employment of higher temperatures.    Our linings have always been the factor that has prevented our doing the same, and it would be of much interest to coke-oven operators here if Mr. Blauvelt would give a typical analysis of the bricks his company uses for the linings of its ovens, and if he could see his way to send me a sample so that I could have the texture examined, I should personally be very much obliged.

The lining brick which my company has found most·successful is of a ganister variety, and has the following analysis: silica, 80.8 per cent.; alumina and iron oxide, 17.8 per cent.; lime, 0.8 per. cent.; magnesia, 0.07 per cent.; but with this brick the least carelessness on the gas regulator's part may destroy the lining.

It would also be of interest if Mr. Blauvelt could give the temperatures of the air leaving these generators and also of the oven flues.

I should be glad to know the life of the linings operated at these high temperatures.    I have known oven linings to last as long as 10 years, and, in our case,if any lining fails before the end of seven years we consider that it has done badly.    This is with washed coal containing about 12 to 15 per cent. of free moisture, and charged by compressing machines.

As regards the size of ovens, the width and lengths given by Mr. Blauvelt are much the same as in this country, and the additional capacity is obtained by increasing the height.    At a great number of the plants on this side, the coal is compressed before charging, and the maximum height of the charge is about 7 feet.

It would be interesting to know whether compressed charges are ever used in America and whether you find it feasible to compress a charge for an oven 12 ft. high.

Has the construction of ovens longer than 36 ft. ever been considered, and what is thought to be the limiting factor in this dimension?

GRAHAM BRIGHT,* East Pittsburgh, Pa.—Beehive coke ovens are usually located at the mines, where the gases from the ovens are not strongly objectionable because the communities are not thickly built up.    The byproduct oven will naturally be located near centers of industry, to which the coal can be brought from different directions and desirable mixtures can be made; where also the byproducts can be readily marketed, while the gas can be utilized for industrial and domestic purposes.

Not much thought seems to have been given to the gases escaping from the byproduct coke oven, which are more or less objectionable in thickly populated sections.    When the byproduct oven shall have been more widely installed, this question is likely to become serious, and it will be necessary to apply remedies.    May I ask what has been done, or is going to be done, in this direction?

* Westinghouse Elec. and Mfg. Co.

S. A. Moss,* Lynn, Mass.—One cause of escaping gas is that the gas exhausters do not supply the proper suction. If a satisfactory regulating governor is applied and the correct kind of gas exhauster is used, the trouble will be eliminated.

J. I. THOMPSON,† Pittsburgh, Pa.—The Koppers Co. has made a careful study of the question of smoke prevention. The smoke caused by the operation of the byproduct coke ovens is infinitely less than that caused by beehive ovens.

The difficulties in preventing the last traces of smoke in byproduct ovens are largely mechanical but are serious difficulties, considering that it is necessary to charge the 10 to 15 tons of coal into a heated oven in such a manner that no gas shall escape. The principal loss of gas occurs during this operation of charging the ovens when it is not possible to conduct all of the gas into the mains leading to the byproduct apparatus.

A great deal of study is being given to this matter, and the Koppers Co. has constructed a plant in St. Paul which is equipped with certain devices for correcting this nuisance. The method employed is to carry a portion of the escaping gases into the stack and discharge them high enough above the ground so that they will not be objectionable, while the remainder will be recovered and led into the byproduct apparatus.

W. H. BLAUVELT (author's reply to discussion‡).—The prevention of smoke from byproduct-oven plants is not as simple as would appear from Mr. Moss's statement. All well designed byproduct-oven plants are equipped with ample exhauster capacity, and the suction on the ovens is controlled by highly sensitive pressure governors that control the pressure within a fraction of a millimeter. Of course, this apparatus can get out of order, causing the batteries to smoke, but this does not often happen.

As stated by Mr. Thompson, the smoke produced is made almost entirely during the time of charging the ovens, when they are cut off from the hydraulic main. The amount of smoke that escapes at this time is extremely small, compared with the amount produced by a beehive-oven plant, but the mechanical difficulties in preventing any escape of smoke at charging time are considerable. Several devices are in operation at a number of plants which collect most of the smoke made during charging and several devices are being worked on for its complete elimination. These will probably be employed only where the smoke is objectionable on account of the location of the plants, as the value of the byproducts in this smoke is very small.

* Mechanical Engineer.  † Chief Engineer, H. Koppers Co.  ‡ Received Nov. 2, 1918.

## Theory of Volcanic Origin of Salt Domes

BY E. DE GOLYER, NEW YORK, N. Y.

(Colorado Meeting, September, 1918)

### INTRODUCTION

VOLCANIC origin was among the first of the theories advanced to account for the occurrence of the salt domes of the Gulf coastal plain, northern Louisiana, and eastern Texas, and it is still being re-stated in various forms by the most recent contributors to our knowledge of the geology of such deposits.

Argument has been largely by analogy and attention has repeatedly been called to the Mexican oil fields, particularly those of the Tampico-Tuxpam region, because of the occurrence of oil in close proximity to volcanic plugs in those fields, an occurrence supposed to have some resemblance to the common occurrence of oil in the salt domes of the United States.

The author, having spent several months of the past year in a general study of salt domes and in a detailed study of certain domes in the United States, and having been acquainted for some time with the exploration of similar domes in the Isthmus of Tehuantepec region, Mexico, and with the general geology of the Tampico-Tuxpam region, Mexico, has reviewed the various theories of volcanic origin for salt domes which have been advanced to the present time. The purpose of this paper is to present such review and to discuss the supposed Mexican analogy.

### REVIEW OF VOLCANIC THEORY

R. Thomassy[1] who had visited the Five Islands of Louisiana between 1857 and 1859, in 1860 described them with some care, noting their distinct character, alignment, and similarity, and ascribing to them a volcanic origin. He regarded them as being akin to the mud lumps in the delta of the Mississippi and was impressed by their symmetry and by the ponds or lakes which occur on several of the islands and which he regarded

---

[1] R. Thomassy: "Géologie pratique de la Louisiane." Paris, 1860. See particularly Chapter 8 on the intervention of hydrothermal and volcanic forces in the formation of lower Louisiana.

as being of extinct crater types. He believed that he detected evidence of corrosion by thermal and acid waters on certain rocks found near Petite Anse and supposed that these rocks had been ejected from the depths of the Gulf through explosive action. He apparently suspected the presence of rock salt and regarded it as resulting from the evaporation of sea water by volcanic heat.

In May, 1862, the main mass of salt was discovered at Petite Anse by a negro who was cleaning out and deepening one of the old brine wells under the direction of John Marsh Avery. This was the first discovery of the salt nucleus so characteristic of the domes, and observers first began to learn something of their true structure.

Thomassy[2] again visited the Island and first brought the salt to the attention of the scientific world, recalling his theory of volcanic origin "in the sense that it (Petite Anse) comes from a volcano of water, mud, and gas," and repeating from his former work the theory that the rock salt was formed from the evaporation of sea water by volcanic heat.

Richard Owen,[3] in 1865, while serving in the Federal Army and stationed at New Iberia, studied the Petite Anse deposit cursorily and concluded that the Island was not of volcanic origin but that the salt was produced by the evaporation of sea water in lagoons which were cut off from the sea by low barriers but were continually being filled with sea water at occasional high tides.

Goesmann,[4] in 1867, reported on the Petite Anse deposit. He rejected Owen's salt-pan theory and suggested that the salt resulted from the evaporation of brine springs originating from beds of rock salt in some older geological formation; a theory akin to the later and more elaborate ones of Hill and Harris.

Eugene W. Hilgard[5] examined this same deposit in 1867, at the request of Joseph Henry of the Smithsonian Institution, and concluded that the island was not of volcanic origin, but was a Cretaceous outlier, with cappings of drift and other alluvial material, the salt itself being of Cretaceous age. He suggested a genetic relationship between the salt deposit and the Calcasieu sulfur deposit.

---

[2] R. Thomassy: Supplément à la géologie pratique de la Louisiane. Ile Petite Anse. Société Géologique de France, *Bulletin*, 2d series (1863) **20**, 542–544, plate.

[3] Richard Owen: Report, on Quaternary Rock Salt Deposits in Louisiana. St. Louis Academy of Science, *Trans.* (1868) **2**, 250–252.

[4] C. E. Buck and C. A. Goesmann: On the Rock Salt Deposit of Petite Anse, Louisiana Rock Salt Co. Report, American Bureau of Mines; 35 pages, 2 plates, New York, 1867. Original source not consulted. Goesmann is quoted by H. C. Bolton, New York Academy of Science, *Trans.* (1888) **7**, 124–125.

[5] E. W. Hilgard: On the Geology of Lower Louisiana and the Salt Deposits of Petite Anse Island (1872). Smithsonian Contributions, *Separate* No. 248, 32–34, Washington.

Otto Lerch, in 1893, published two papers setting forth the results of his observation of geological features including the salines in north Louisiana. In the first of these papers[6] he concludes that no volcanic eruptions and no violent contortions have disturbed the geological formations of north Louisiana; and in the second paper,[7] after noting that the existence of the salines can only be traced along the summit of a Cretaceous ridge, which traverses the state diagonally from its northwest corner to the Island of Petite Anse, he arrives at a conclusion which combines to some extent both Hilgard's outlier theory and a theory similar to volcanic origin. It was his idea that at the close of Mesozoic time enormous plutonic forces had thrown up mountain chains of vast extent and that the isolated peaks in the mountain chains thus formed had served as Cretaceous outliers in the Tertiary sea. He was further of the opinion that the Balcones fault and the basaltic outbreaks along it are of contemporaneous origin.

T. Wayland Vaughan,[8] in 1895, expressed opinions very similar to those of Hilgard with regard to the origin of the north Louisiana salt domes.

Clendenin,[9] as a result of investigations during the summers of 1894 and 1895, in 1896 expressed the opinion with regard to Thomassy's theory that he could not interpret any evidence seen as proof of powerful volcanic convulsions. He rather followed Hilgard, whom he quotes extensively, but suggested, somewhat after Lerch, that the Cretaceous outliers are the result of differential elevation rather than differential erosion. He recognized clearly the earth movement during comparatively late geological time.

A. C. Veatch[10] early in 1899 left off his study of the salines of north Louisiana to investigate the Five Islands. He states that only two of the islands at the time furnished definite data on the method and date of their formation, Belle Isle showing a very distinct dome-shape fold and Petite Anse seeming to represent a fault block. He thought that the faulting might possibly be due to solution of the underlying salt but that

---

[6] Otto Lerch: A Preliminary Report upon the Hills of Louisiana, North of the Vicksburg, Shreveport and Pacific Railroad. Louisiana State Experiment Stations, *Geology and Agriculture* (1893) 1, 27.

[7] Otto Lerch: A Preliminary Report upon the Hills of Louisiana, South of the Vicksburg, Shreveport and Pacific Railroad, to Alexandria, La. Louisiana State Experiment Stations, *Geology and Agriculture* (1893) 2, 53–109.

[8] T. W. Vaughan: Stratigraphy of Northwestern Louisiana. *American Geology* (1895) 15, 208–229.

[9] W. W. Clendenin: A Preliminary Report upon the Florida Parishes of East Louisiana and the Bluff, Prairie, and Hill Lands of Southwest Louisiana. Louisiana State Experiment Stations, *Geology and Agriculture* (1896) 3, 236–240.

[10] A. C. Veatch: The Five Islands, Louisiana. Louisiana State Experiment Stations, *Geology and Agriculture* (1899) 5, 259–260, pls. 19–31.

it was probably due to orographic movements and concluded that the formation of the island began with a possible initial movement (evidences of which had thus far been seen only on Petite Anse) in probably late Tertiary time, that the main folding and faulting which occurred in the Pleistocene was followed by the depression of the whole coastal region and deposition of the upper yellow clays, that during the succeeding high level period the deep channels of the coastal rivers were excavated and the lake valleys formed on the Island and that the subsidence which followed had continued to the present day. These views by Veatch are particularly interesting in view of his subsequent independent promulgation of a theory of volcanic origin.

In 1896, A. F. Lucas found salt under Jefferson Island, or Cote Caline, and thus initiated the most fruitful part of his long period of study and prospecting of the domes of the Gulf coastal plain; a work which culminated so successfully and happily in his bringing in of the discovery well of the Spindletop pool and of the coastal oil fields in January, 1901. This discovery gave impulse to the intensive exploration by the drill of the various salt domes of the coastal plain, operations from which we have gathered most of our present knowledge regarding the occurrence of the salt domes and without which we should still be but little removed from the speculative days of Thomassy and Hilgard.

Lucas, in 1896[11] and 1898,[12] published two papers on Louisiana salt and in 1899[13] published a third and more complete one. He was more concerned with occurrence than origin, merely noting a belief that the geological formation of the Five Islands is Quaternary, while the salt deposits belong to the Tertiary period and are supposed to rest on the Cretaceous and that the salt formation at Jefferson Island does not lie in undisturbed stratification but seems to have been folded and contorted while still in a plastic condition. Lucas clearly recognized the existence of the dome structure of Spindletop but was inclined to think that it was due to gas pressure.[14]

Harris[15] at first regarded Spindletop as a Cretaceous fold.

Robert T. Hill,[16] in 1902, published his theory that the salt of the domes is one of the resultant products of columns of hot saline waters ascending under hydrostatic pressure.

---

[11] A. F. Lucas: The Avery Island Salt Mine and the Joseph Jefferson Salt Deposit, Louisiana. *Engineering and Mining Journal* (Nov. 14, 1896) **62**, 463–464.

[12] A. F. Lucas: Louisiana Salt Resources. *American Manufacturer* (Dec. 23, 1898) **63**, 910–911.

[13] A. F. Lucas: Rock Salt in Louisiana: *Trans.* (1899) **29**, 465–466.

[14] Robert T. Hill: The Coast Prairie of Texas, *Science* N. S. (1901) **14**, 327.

[15] *New Orleans Picayune,* Mar. 27, 1901, requoted from Harris, 1907.

[16] Robert T. Hill: The Beaumont Oil Field, with Notes on the Other Oil Fields of the Texas Region. *Journal,* Franklin Institute (1902) **154**, 273–274.

Veatch,[17] reported on the salines of north Louisiana in detail, in 1902, but did not express any modification of his views as expressed for the Five Islands.

Eugene Coste,[18] in 1903, advanced the theory that the salt domes are the result of "volcanic emanations bringing the water, salt, sulfur, oil and gas from the interior in the state of vapors and gases, which condensed more or less near the surface" etc.  He regarded the domes and salt masses as "the dying distant echo of that tremendous volcanic energy which, a little farther south, in Mexico, Central America and in the islands and along the southern coast of the Caribbean Sea, is to this day so powerfully active."

Lee Hager[19] presented, in 1904, an able paper summarizing the existing knowledge regarding the southern salt domes and proposed, in greater detail, a theory somewhat similar to that proposed by Coste.  Hager's theory is essentially as follows:

By contact with the molten intrusives, vast quantities of gas were generated from the reduction of metallic sulphides and the distillation of lignites and organic substances.  These gases, accompanied by steam under tremendous pressure, forced their way to the surface through the unconsolidated sands and clays of the overlying Tertiary material, perhaps giving rise to mud volcanoes, such as occur in many of the world's great oil fields at the present day.  Heated waters from great depths found vent along the same channels, carrying in solution carbonates of lime and magnesium, gypsum and salt.  By ebullition and evaporation these solutions became concentrated until, saturation resulting, precipitation commenced, forming the neck-like masses of salt, gypsum and dolomite now encountered.  With the cooling of the intrusive masses and the choking of the vents, the process practically ceased.  A period of subsidence followed, during which the Coastal Quaternary beds, which at present cap the mounds, were laid down, followed by a secondary movement along the old lines of weakness, resulting in the present elevation of the mounds above he surrounding prairie.

Hager admits the weakness of his hypothesis to he in "the assumption of intrusives concerning the existence of which we have no evidence whatever," but finds some support for it in the occurrence of intrusive rocks and the upward displacement by them of overlying strata in the Austin, Texas, area as well as in the common occurrence of some form of vulcanism in most of the world's Tertiary oil fields.[20]

---

[17] A. C. Veatch: The Salines of North Louisiana.  Geological Survey of Louisiana, *Report* (1902) 47–100.

[18] E. Coste: The Volcanic Origin of Natural Gas and Petroleum.  *Journal*, Canadian Mining Institute (1904) 6, 89, 104.

[19] Lee Hager: The Mounds of the Southern Oil Fields.  *Engineering and Mining Journal* (1904) 78, 137–139, 180–183.

[20] From his numerous citations of the occurrence of mud volcanoes in various oil fields, one must conclude that Hager regarded them as true volcanic phenomena.  While some mud volcanoes are undoubtedly of true volcanic origin, many are known to be only gas seepages through very wet and plastic clays and to give no indications whatever of volcanic origin.

Veatch,[21] in 1906, after describing the igneous intrusions of southern Arkansas, concluded with regard to the origin of the salt domes of northern Louisiana and eastern Texas as follows:

Whether the forces producing these unique domes were in any way associated with those producing the intrusions just mentioned it is as yet impossible to say; but the irregularity of their distribution, the great symmetry of all the domes which have been carefully studied, the difficulty of explaining this symmetry by any manner of folding not associated with igneous intrusions, and the suggestion which this symmetry carries of force applied at one point from below, just as a sharp-pointed little dome might be formed in a sheet of dough by pushing upward with a blunt pencil, indicate similar igneous intrusions beneath these great thicknesses of relatively plastic, recently deposited cretaceous sediments as the cause of these domes.

This conclusion, so similar to that previously put forward by Hager, was arrived at independently by Veatch, who had not yet become acquainted with Hager's theory.

Harris,[22] in 1907, suggested that the domes were formed by "concretion-forming forces as well as the power of crystallization," and concluded that though the domes might be "located along lines of weakness, faults, or fractured anticlines, they are not to any great extent due to tangential, mountain-making forces, nor volcanic upheavals, nor igneous plugs." He further notes that igneous plugs and dykes usually cause great irregularities in the earth's magnetic fields and notes that the area under consideration, which contains a number of saline domes, shows irregularities entirely too slight to indicate igneous activity in the region. Harris restated and developed his theory further in 1908, 1909 and 1910. In 1909, he[23] called attention to the fact that certain nearby Arkansas volcanic rocks are magnetic.

In 1909, Coste[24] elaborated upon his former declaration of volcanic origin for salt domes and noted that,

In the Mexican oil fields the volcanic action has been a little more intense and instead of only the hot gases, vapors and waters piercing up more or less through the horizontal strata to form the salines, as in Texas and Louisiana, we see the volcanic lava cones themselves piercing up boldly through the plains. There is no doubt that these lava cores in Mexico surrounded with petroleum and other solfataric emanations are one and the same volcanic phenomenon as the vertical chimneys of salts, hot waters and hot petroleums of the Texas-Louisiana salines.

Clapp,[25] in 1912, concluded with regard to the theory suggested

[21] A. C. Veatch: Geology and Underground Water Resources of Northern Louisiana and Southern Arkansas. U. S. Geological Survey, *Prof. Paper* 46 (1906) 29.

[22] G. D. Harris: Notes on the Geology of the Winnfield Sheet. Geological Survey of Louisiana, *Bulletin* 5 (1907).

[23] G. D. Harris: Magnetic Rocks. *Science*, new ser. (Mar. 5, 1909) **29**, 384.

[24] Eugene Coste: Petroleum and Coals. *Journal*, Canadian Mining Institute (1909) **12**, 290.

[25] F. G. Clapp: The Occurrence of Oil and Gas Deposits Associated with Quaquaversal Structure. *Economic Geology* (1912) **7**, 377.

by Hager that "his theory would seem to be supported by the existence of volcanic plugs in the Coastal Plain of Mexico, accompanied by many of the Texas phenomena." He states that "these plugs are arranged in straight lines in a manner similar to the sahne domes."

Lucas,[26] in 1914, states that he has "long held the opinion that igneous rock in the form of laccoliths, batholiths, or sills may underlie the salt domes of Louisiana, Texas, and elsewhere." He further states that he has "held the opinion with Mr. Coste that emanations of gas through these domes constitute evidences of volcanic origin." He concluded, however, that "the salt was not deposited by evaporation, but must have been deposited by saline waters ascending from great depths."

In a still more recent paper, Coste[27] continues,

That these are solfataric volcanic emanations is made plain by the occurrences of oil a little farther south, along the same coastal plain, in Mexico. There one can actually observe numerous volcanic necks of olivine basalt, scattered at wide intervals, and also distributed along fault lines similar to the lines connecting the salines of Texas and Louisiana."

Washburne,[28] in 1914, seems to favor the volcanic origin of salt domes, regarding the salt as having been precipitated from saturated brines of sedimentary origin by hydrochloric acid from igneous intrusions.

Norton,[29] in 1915, suggested still another somewhat complex variation of the theory of volcanic origin, regarding the initiation of the salt deposits as resulting from "the intrusion of molten rocks into the underlying Paleozoic sediments along lines of structural weakness."

Dumble[30] regards Norton's paper as one which "seems a step in advance of former ones."

Quite recently, Oliver B. Hopkins[31] has expressed a view that the structural conditions found at salt domes are such as "are most commonly produced by igneous intrusion, of which no direct evidence has been found in any of the numerous salt domes of the Gulf Coast," and that it is quite probable the forces forming the salt domes "were connected with igneous activity."

---

[26] A. F. Lucas: In discussion of Hoefer's paper on the origin of petroleum. *Trans.* (1915) **48**, 494.

[27] E. Coste: Rock Disturbances Theory of Petroleum Emanations *vs.* the Anticlinal or Structural Theory of Petroleum Accumulations. *Trans.* (1915) **48**, 510.

[28] C. W. Washburne: Chlorides in Oil-field Waters. *Trans.* (1915) **48**, 687–694.

[29] E. G. Norton: The Origin of the Louisiana and East Texas Salines. *Trans.* (1915) **51**, 508.

[30] E. T. Dumble: The Occurrences of Petroleum in Eastern Mexico as Contrasted with those in Texas and Louisiana. *Trans.* (1916) **52**, 263.

[31] O. B. Hopkins: The Palestine Salt Dome, Anderson County, Texas. U. S. Geological Survey, *Bulletin* 661 G (1917) 264.

Shaw[32] has also recently restated the various factors which make the theory of volcanic origin attractive and has discussed briefly the bearing of the Mexican occurrences of oil on the theory itself.

In discussing the weakness of the volcanic theory because of the absence of igneous rock, Rogers,[33] who is at present engaged in a study of salt domes, has noted "the discovery, during the past summer, of volcanic ash in the sediment around one or two of the domes; of a rock resembling a porphyry in a well at Damon Mound; and of an undoubted igneous plug about 50 miles north of the salt-dome belt."

### Evidence of Volcanic Activity in Salt-dome Regions

The strength of the volcanic theory of origin of salt domes lies in the fact that the intrusion of a volcanic plug or neck is the only geologic agency now known to us which affects structurally the contiguous sedimentary rocks in the same manner as they are affected by the intrusion of the salt masses. There is considerable variation of form both in salt and igneous rock masses, but the general contour, symmetry, and size of the explored portion of many of the salt masses show a remarkable similarity to like phenomena connected with many igneous rock intrusions.

The weakness of the theory, for the salt dome regions of Texas and Louisiana particularly, lies in the absence of igneous rocks and in the absence of evidence of metamorphism of the sedimentary rocks. The absence of igneous rock may be explained as being the result of deep-seated igneous intrusions which do not nearly reach the surface, but it is hard to conceive of igneous intrusion accompanied by such giving off of hot gases, vapors, and waters as are usually postulated, which would not have resulted in at least a partial metamorphism of the sedimentary rocks forming the walls of the ducts or fissures through which such emanations escaped, presumably the present locus of the salt.

Furthermore, the salt domes give indisputable evidence of movement in quite recent geologic time. If this movement was the direct result of the force of igneous intrusion, one would expect to find some regional evidence of dying vulcanicity such as seismic disturbances, hot springs, etc.

The only evidence of this type which has come to the attention of the author is the high temperature of many of the crude petroleums and underlying waters of the Gulf Coast oil fields. Temperatures ranging from 80° F. or 90° F. to 180° F. have been observed. The only other oil-field temperatures known to the author which are generally higher than these are oil and water temperatures from the Tampico-Tuxpam region,

---

[32] E. W. Shaw: Possibility of Using Gravity Anomalies in the Search for Salt-dome Oil and Gas Pools. *Science*, new ser. (Dec. 7, 1917) 46, 553–556.

[33] In discussion of Matteson's paper, "The Principles and Problems of Oil Prospecting in the Gulf Coast Country." *Trans.* (1918) 59, 480.

Mexico.[34]  The Mexican temperatures are very evidently due to latent volcanic heat and it is possible that the Texas and Louisiana temperatures are also the result of volcanic heat.

Harris[35] apparently believes the Gulf Coast temperatures can be explained by the normal increase of the temperature of the earth's crust with depth, generally accepted as 1° C. with each 30 m. of depth, or 1° F. per 54.6 ft.  The author does not believe that this explanation is satisfactory, but suggests the probability that much of the heat may be the result of chemical reactions.  Sulfurous waters are of common occurrence, pyrites of iron, other metallic sulfides and free sulfur are not uncommon, and sulfuric acid has even been found in the domes.  It is suggested that a systematic study of temperatures in the Gulf Coast region might throw light on this very interesting subject.

The volcanic theory can also be made to account for the enormous amounts of gypsum, free sulfur, hydrogen sulfide, sulfurous waters and small amounts of metallic sulfides which occur around various of the salt domes.  Instead of the sulfur being derived from the gypsum, as is generally suggested, is it not possible that the gypsum is derived from the alteration of limestone by sulfuric acid, as has been suggested by Kennedy,[36] who notes that sulfuric acid has been found in many of the domes.  It has also been found in volcanic waters.

Sulfur dioxide is one of the commonest of volcanic emanations. By hydration, it readily forms sulfurous acid, which may be further oxidized to sulfuric acid.  It also has the power, in the presence of water, of converting chloride into hydrochloric acid: $SO_2 + Cl_2 + 2H_2O = 2HCl + H_2SO_4$.  The common occurrence of hydrochloric acid in many volcanic emanations suggests that such reaction may occur. The free sulfur was possibly produced by this reaction of hydrogen sulfide, which is also a common volcanic emanation, with sulfur dioxide in the same manner as suggested for solfataric sulfur.

The high sulfur content of the oils from the salt-dome fields of Texas and Louisiana is a characteristic which they have in common with the oils of the Isthmus of Tehuantepec salt-dome region and the Tampico-Tuxpam region in Mexico, and in which they differ from most other oils. This condition suggests that oils from each of the regions may have been subjected to the action of sulfurous waters and gases of volcanic origin.

Salt itself is one of the commonest of volcanic sublimates but, so far

---

[34] The author has discussed the Mexican temperatures in some detail and given certain Gulf Coast temperatures in an article in *Economic Geology* (1918) **13**, 275.

[35] G. D. Harris: Oil and Gas in Louisiana.  U. S. Geological Survey, *Bulletin* 429 (1910) 8.

[36] Coastal Salt Domes.  *Bulletin*, Southwestern Association of Petroleum Geologists (1917) **1**, 34–59.  See particularly pp. 48 and 55.

as is known to the author, no deposits of salt at all comparable in magnitude with those of the domes is known to occur in any volcanic region.

These various mineral occurrences and suggested reactions do not prove the theory of volcanic origin but they do indicate that it accounts for many of the minerals found in the domes as well as do the various other suggested theories.

The actual occurrences of igneous rock in the vicinity of the Texas and Louisiana salt domes consist of certain igneous rocks in southern Arkansas; certain other igneous rocks in the Austin, Texas, area; the igneous mass which forms the oil-producing rock in the Thrall field; and the various occurrences mentioned by Rogers, as already cited.

The volcanic ash mentioned by Rogers as having been found in the sediments around one or two of the domes is of no importance as an indication of volcanic activity in the salt-dome region, since the ash may have been air-transported hundreds or even thousands of miles, but the rock resembling porphyry from Damon Mound and the igneous plug 50 miles north of the salt-dome belt are of considerable importance, and further information as to their occurrence will be of great interest in this connection.

In considering occurrences of igneous rock associated with salt domes, it is worthy of note that Hahn[37] states that close connection with propylitic, doleritic, or basaltic extrusions may exist for Algerian salt domes, that relation of salt domes with ophiolitic magma may be probable in the Pyrenees, and that in Germany, dykes of basalt with included fragments of rock salt are observed piercing some of the salt deposits.

## Proposed Mexican Analogy

Up to the present time, oil in quantity has been encountered in two geologically and geographically distinct regions in Mexico; the Tampico-Tuxpam or northern region, and the Isthmus of Tehuantepec, or southern region.

The Tampico-Tuxpam region consists of that portion of the Gulf coastal plain extending generally from the Soto la Marina River in central Tamaulipas to the Misantla River in north-central Vera Cruz.

The region is made up of a thick series of limestones of Cretaceous age, the equivalent of the Comanche, overlain unconformably by 600 to 800 ft. of alternating thin-bedded argillaceous limestones and shales, which grade imperceptibly into an overlying series of 2000 to 4000 ft. of shales and marls containing occasional thin sandstones and limestones. This shale series, known generally as the Mendez shale, is of Eocene age and is overlain by various series of limestones, conglomerates, sandstones, and shales of Oligocene age, so far as has been determined.

---

[37] F. F. Hahn: The Form of Salt Deposits. *Economic Geology* (1912) **7**, 129.

This entire sedimentary series has been folded in Oligocene or post-Oligocene time by the same forces that formed the Sierra Madre Oriental. The rocks of the coastal plain lie in a series of folds more or less parallel to those of the Sierra Madre, the general strike changing from a N.–S. direction in the northern part of the region to a NNW.–SSE. direction in its central part, and a NW.–SE. direction in its southern part, the general dip being toward the Gulf of Mexico. The folds are generally sharp along the mountain front and parallel to it but gradually become less intense as the distance from it increases.

At periods apparently subsequent to the major folding, the region was subjected to considerable volcanic activity and intrusions of volcanic plug, laccolith, dyke, and sill forms occurred. The igneous rock is of basic type, for the most part olivine basalt, but including occasional dolerites, gabbros, etc.

Considerable brecciation and metamorphism of the contiguous sedimentary rocks resulted from the intrusions. Metamorphism by hot waters and gases ascending from the igneous rock was so extensive that in many instances the pipe of metamorphosed rock reached the surface, though the igneous rock did not reach it. This condition exists at the Cerro Tampule in Aguada, Cerro Chapopotal in Cerro Viejo, and at Furbero, where the igneous mass has been reached by the drill at a depth of some 1500 ft. below the surface.

Evidence of any doming around the igneous plugs of this region similar to doming around the salt plugs of Texas and Louisiana is so scarce that it does not seem probable that such structure exists generally. Evidence on this point is very meager, because the geology in the vicinity of the plugs is often masked by talus débris, soil, rank vegetation, etc., but in cases where wells have been drilled near the igneous plugs, there has been little or no evidence of doming.

The oil deposits are not found around these plugs in domed strata, as has often been suggested and as is held by Coste; but in structures formed by the major folding and faulting. Oil is found around the volcanic plugs Cerro de la Pez and Cerro de la Dicha in the Ebano field and conditions appear on the surface to resemble the salt-dome fields more nearly than at any other point in Mexico. Ordonez,[38] the only person who has studied this deposit, states regarding this general form of intrusion that "the volcanic plug or pipe of ashy and compact basaltic lava, of which the neck is the upper end, has not produced any considerable disturbance of the shale around it."[39] In the Casiano-Tepetate field, where the sedimentary rocks have suffered considerable intrusion, folding and faulting

---

[38] E. Ordoñez: The Oil-fields of Mexico. *Trans.* (1915) **50**, 860.

[39] The author has given a number of examples of wells drilled near plugs in another paper, The Effect of Igneous Intrusions on the Accumulation of Oil in the Tampico-Tuxpam Region, Mexico, *Economic Geology* (1915) **10**, 660.

seem to control the occurrence of oil, and instead of the productive pool being developed on top of or around a dome, it seems to be confined to a fold striking N.–S. and bounded by a parallel fault on the west side. The productive area as outlined at present is some 1½ miles long by ¼ mile wide.

Some doming by igneous intrusion probably exists. Laccoliths which have domed the overlying strata are common in the San Jose de las Rusias region, Tamaulipas. The oil of the Furbero field occurs in the altered igneous rock of a laccolith or sill, which has been intruded along a plane of stratification of the sedimentary rocks folded previously into a gentle anticline, as well as in the metamorphosed shales which envelop the intrusion. The intrusion of this igneous rock accentuated the pre-existing fold in the overlying beds.[40] Accentuation to a considerable degree of the normal regional dip on the eastern side of the volcanic plugs Cupelado, Huiltepec, and Dos Hermanos, all lying a short distance southeast of Furbero, has also recently been called to the author's attention.[41] Straight lines suggesting faults are often determined by three plugs in this region but there is no general parallelism between the lines so determined nor with the general structural features of the region. The field evidence does not justify any such conclusion as that expressed by Clapp in stating that the plugs are arranged in straight lines in a manner similar to the salt domes of Texas and Louisiana.

No deposits of salt of the salt-dome or other types are known to occur in the Tampico-Tuxpam region. The suggestion by Shaw that gradation phases may be found between salt domes and intrusions and that the salt domes may be absent from the Tampico-Tuxpam region, due perchance to more consolidated rock, would not seem to hold, since salt domes occur in east central Texas in a series of sediments which are quite as consolidated as are the sediments of the Tampico-Tuxpam region.

The Isthmus of Tehuantepec region consists of a portion of the northern coastal plain of the Isthmus of Tehuantepec in the vicinity of Minititlan. It lies almost wholly in the cantons of Acayucan and Minititlan, State of Vera Cruz, and is some 200 miles southeast of the southernmost part of the Tampico-Tuxpam region. Generally speaking, the Isthmus of Tehuantepec is a great E.–W. striking anticlinorum, the folding probably of post-Oligocene date, involving Mesozoic and Tertiary sediments. The salt-dome region lies on the north flank of this structure. The sediments of the salt-dome region are of Tertiary and Quaternary age, except for small patches of Jurassic or Cretaceous limestone which have been brought up by a salt dome near Chinameca.

---

[40] The author has described this occurrence in more detail in "The Furbero Oil Field, Mexico." *Trans.* (1916) **52**, 286–280.

[41] Personal communication from E. B. Hopkins.

The oldest known rocks outcropping or reached by the drill in the salt-dome region consist of a series of massive blue to gray marls, clays, and shales, containing some sandstones. They probably have a thickness of several thousand feet and are of Miocene-Pliocene age. They are overlain by a series of thin-bedded sandstones interstratified with still thinner shales. This formation is of Pliocene age and is overlain in turn by several hundred feet of white to cream colored clays, reddish sands, and thick beds of rounded quartz gravels believed to be of Pleistocene age.

This entire series of rocks has been domed and folded by the intrusion of salt masses, though the unconformable Pleistocene shows so little deformation that it is not believed to have been deposited until after the intrusion of the salt masses was more or less complete.

The salt domes of the Isthmus are very numerous and in places they are closely grouped together. Several of the domes, such as the San Cristobal-Capoacan and the Concepcion, are fairly symmetrical flat-topped domes very similar to those of the Texas-Louisiana region, but others are elongated domes or sharp anticlines. The Soledad and Tecuanapa domes are of this type.

With the domes are associated impure limestones, gypsum, and generally such minerals as are found in the Texas-Louisiana region. The only known occurrences of igneous rock consist of two isolated outcroppings near Paraje Solo, some 15 miles east of Minititlan, and the volcanic rocks in the San Andres Tuxtla district, adjoining the region on the north and west.

No general alignment of the salt domes of this region similar to that observed in the domes of the Texas-Louisiana region can be made. The structures produced by the salt masses generally resemble folds rather than symmetrical domes occurring at the intersections of faults. These folds may often be traced from one dome to another, the surface geology being much better exposed than in the Gulf Coast region of the United States.

### Conclusions

The lack of evidence of any considerable volcanic activity in the salt-dome regions of Texas and Louisiana or in the salt-dome region of the Isthmus of Tehuantepec, Mexico, as well as the entire absence of any salt masses of the nucleus type in the Tampico-Tuxpam region, Mexico, a region of considerable volcanic activity, would seem to argue most strongly against the acceptance of the volcanic theory of salt-dome origin.

The volcanic theory affords a plausible origin for the sulfur, sulfides, certain of the gases, the mineralization of waters, and indirectly the alteration of limestone into gypsum, but it is of little more service in this respect than other theories. It affords also an acceptable source.

of heat in accounting for the high temperatures of oil and water in the Texas and Louisiana dome region, but in view of the various minerals present in the domes, heat from chemical action would seem to afford quite as satisfactory an explanation.

Finally, since the attractiveness of the theory of volcanic origin of salt domes lies in the similarity of structural effect produced by the intrusion of the salt in many domes to that produced by the intrusion of the igneous rock of a volcanic neck or plug, it follows that we are interested in the conditions and forces which caused the salt to act in a manner so like to that of an igneous magma. To argue that because the structural effects produced by salt and igneous intrusions are identical, the former is a result of the latter, is little more acceptable than to argue that the latter is a result of the former, and only seems less reasonable perhaps because igneous plugs and the forces forming them are of more common occurrence and are better known than the rarer salt-dome types.

It is difficult to conceive of the formation of a tip of salt over 3000 ft. thick, as at Humble, on the end of an igneous plug during the course of or preparatory to an upward thrust which would result in the considerable displacement of contiguous rocks, as at the Palestine, Texas, salt dome where the displacement was 3000 to 3500 ft. Furthermore, the salt may take other forms resembling igneous intrusions, such as the probable laccolithic form of the salt in the Tecuanapa dome, Isthmus of Tehuantepec region; a form which does not lend itself to any theory that it is underlain by a similar igneous intrusion.

Rather, the structure of the domes would seem to make it evident that the salt itself must have flowed. Such flowage was most likely due to great pressure exerted upon buried salt masses by the weight of overlying rocks, and at considerable temperatures and probably accompanied, as has been suggested by van der Gracht,[42] by recrystallization of the salt. This flowage theory has long been held by European geologists and seems to have been fairly well established by them for certain salt plugs where the geology is well exposed both at the outcrop and by underground workings.[43]

The marked resemblance between certain of the European salt domes and those of the Texas-Louisiana and Isthmus of Tehuantepec regions, and the difference of the salt domes from other known forms of salt deposits, suggest most strongly certain genetic similarities. The author believes that the theory of intrusion of salt by flowage must receive in future more consideration from American geologists than it has in the past.

---

[42] The Saline Domes of Northwestern Europe, *Bulletin,* Southwestern Association of Petroleum Geologists (1917) **1**, 85–92.

[43] F. F. Hahn: The Form of Salt Deposits. *Economic Geology* (1912) **7**, 120–135.

## DISCUSSION

J. A. UDDEN,[*] Austin, Tex. (written discussion[†]).—I have read Mr. De Golyer's paper with a great deal of interest. It seems to me a very able summary and discussion of the various theories set forth to account for the origin of our salt domes on the coast. It has always appeared to me that what we need is more observations on the facts involved and I regret very much that so far we have not been able to devote much time to the study of our salt-dome fields on the coast in this state. One feature that I do not find mentioned in Mr. De Golyer's paper is the so-called "cap-rock" overlying the salt. The nature of this rock seems to vary from a coarsely crystalline limestone to a fine-grained limestone containing sand and in which the limestone seems to be interstitially introduced material. We should be prepared to accept, if necessary, more than one theory. To my mind, the igneous theory and the artesian theory, if I may so call it, may very well both be true. The "squeeze" theory and the artesian theory may, it seems to me also, jointly explain the phenomena. Neither the volcanic nor the salt "squeeze" theory explains the presence of the cap-rock. The latter certainly does not explain the extensive sulfur deposits which occur in association with the cap-rock. Whichever additional theory will be found best to explain the salt-dome structures, when we shall know them better, it seems to me that the theory first advanced by R. T. Hill and later elaborated by G. D. Harris is not likely to be entirely replaced. Whatever cause has first set into action an upward circulation of gases and liquids in our salt domes, it seems to me that to a gentle circulation of the kind postulated by their theory is to be ascribed the uppermost and the later phenomena in our salt domes on the coast, with which we are most familiar; such as the cap-rock, the sulfur, and the gypsum and anhydrite deposits.

E. T. DUMBLE,[‡] Houston, Tex. (written discussion[§]).—The domes of the Texas coastal plain are structural features, consisting of bosses or stocks of salt, gypsum or anhydrite, or of combinations of these, intruding into and occurring in connection with broken and uplifted sedimentary beds of Cretaceous and Tertiary age. Some of these domes show on the surface as mounds of greater or less elevation, while others are only known from drilling records. They may be divided into two classes: interior domes and coastal domes.

The interior domes include those which occur in a zone 40 to 50 miles in width lying immediately east of the Cretaceous-Tertiary contact,

* Director, Bureau of Economic Geology and Technology, University of Texas.
† Received May 27, 1918.
‡ Consulting Geologist, Southern Pacific Co.
§ Printed in *Bulletin* 143 (Oct., 1918) under title of "Origin of the Texas Domes."

and extending from the Sabine to the Colorado. With a single exception, these domes, so far as now known, occur in approximate alignment with the contact, but at irregular intervals. East of the Brazos, the domes are entirely surrounded by Eocene sediments.

The coastal domes are found nearer the Gulf, lying entirely within Neocene territory and stretching from the Rio Grande to the Sabine. They are much more numerous than the interior domes, and, apparently, are ranged along several lines having a general northeast-southwest direction. No domes are known in the Oligocene belt.

## INTERIOR DOMES

The interior domes so far identified are Butler's, in Leon county, Palestine and Keechi in Anderson county, Brooks and Steen in Smith county, and Grand Saline in Van Zandt. In connection with these interior domes we must also consider the Sabine Peninsula, of Harris, which is, without question, genetically connected with them.

The Palestine and Keechi domes have been described recently by O. B. Hopkins,[1] who gives a clear idea of their composition and structure. Others have been described at different times, and an excellent résumé of the literature is given by De Golyer.

The Palestine dome is 6 miles west of Palestine. Here we find a depression of irregular shape, with a maximum diameter not exceeding $\frac{3}{4}$ mile (1.2 km.). The bottom of the depression is occupied by a shallow lake with its surface 50 ft. (15 m.) below the general level. The banks on the eastern and northern sides slope upward gradually, but that on the west is more abrupt. The lowest rock exposed is a sandstone, which, as proved by its fossils, is of Woodbine age. Between this and the underlying body of salt, 140 ft. (43 m.) below, there is 85 ft. (26 m.) of gray to yellow water sand, 40 ft. (12 m.) of dark gray sandy clay, under which there is in places a cap-rock of hard limestone of varying thickness. Apparently, therefore, the Woodbine rests directly upon the salt mass.

The Woodbine, at its exposure, shows a dip of 46° to the northwest. It is overlain by the Eagle Ford, Austin, Taylor(?) and Navarro beds, all of which dip northwest at angles varying from 40° to 50°. No beds were found which can be referred either to the Midway or the Lower Wilcox, the lowest Tertiary beds being sands, clays and lignites belonging to the Middle or Upper Wilcox. According to Hopkins, these Wilcox beds in the vicinity of the dome show southeast dips of 38° to 57°, which decrease within $1\frac{1}{2}$ miles to 20° to 30°, and within 3 miles become normal. On the northeast and southeast the Claiborne beds reach

---

[1] U. S. G. S. *Bull.* 661 (1917) 253. The Brenham dome, also described in Hopkins' paper, either belongs to same class as interior domes, or is intermediate between them and the coastal domes. It certainly is in no way related to the coastal domes.

within 3 to 5 miles of the Dome, but show little, if any, change from normal dip.

Six miles (10 km.) northeast of this locality, the Keechi dome shows the Austin Chalk at the surface, surrounded by the Navarro beds, and these are in turn encircled by the Wilcox, which dips away from the dome at angles varying between 20° and 30° (Hopkins). As in the case of the Palestine dome, the Claiborne is 3 to 4 miles northeast and southeast of the Keechi dome.

The thickness of the beds as interpreted from logs of the wells would be approximately 500 ft. (152 m.) for the Navarro and Taylor, 800 ft. (244 m.) for the Austin and Eagle Ford, and 400 ft. for the Woodbine. The salt mass reached at 2200 ft. (671 m.) was drilled into for 900 ft. (274 m.), a 30-ft. (9 m.) bed of water sand being encountered in it at a depth of 2900 ft. (884 m.).

The Butler dome, which is 6 miles (10 km.) southwest of the Palestine dome, has not been so carefully examined, but appears to resemble the Palestine very closely and gives the same geological section. Brooks and Steen domes, in Smith county, are also Cretaceous islands.

Grand Saline, on the contrary, while undeniably of similar origin to these five Cretaceous islands, and a dome in structure, is entirely overlain by the Wilcox and shows no Cretaceous at the surface. Furthermore, the body of salt is more limited and probably not more than ½ mile square. The workable salt here appears as a lens or boss and not as a stock. The salt occurs at a little over 200 ft. (61 m.), and in places attains a thickness of 300 ft. (91 m.), thinning out at the edges. It is immediately overlain by gypsum, or by alternations of limestone and salt. The beds above these are heavy limestones, shales, and clays, and dips of 30° or more have been observed.

On the surface, the Sabine Peninsula is a belt of the Wilcox, in places more than 30 miles (48 km.) in width, extending along the Texas-Louisiana line from Vivian to Sabinetown. It is flanked east, west, and south by bodies of Lower Claiborne. From north to south, the average surface slope of this peninsula is less than 1 ft. per mile (0.2 m. per km.).

Well sections show the Upper Cretaceous beds dipping southward from Red River at about 50 ft. per mile (9.5 m. per km.) to Vivian. Here the Sabine uplift begins, which brings the Cretaceous up again within 500 to 700 ft. (152 to 213 m.) of the surface. Basing the estimate on the top of the Annona Chalk, the dip of the Cretaceous between Vivian and Sabinetown is 300 ft. or about 3 ft. per mile (0.6 m. per km.). Above the Cretaceous, the sections show 250 ft. (76 m.) of Midway and up to 450 ft. (137 m.) of Wilcox. The well records show that the Cretaceous rocks were folded first along northeast-southwest lines and later at right angles to this.

The Cretaceous of the Peninsula, therefore, instead of continuing on its normal dip, forms a mesa or table-land, and seemingly had this structure during lower Eocene time. Harris suggests that it formed an island during the Claiborne.

## Coastal Domes

Beginning on the south we find, some 30 miles (48 km.) north of the Rio Grande and a few miles west of Raymondville, on the St. Louis & Brownsville Railway, an elliptical-shaped lake about 5 miles (8 km.) in circumference with water nowhere exceeding 3 or 4 ft. in depth and a bottom of pure rock salt. It occupies quite a depression in the widespread plain of gray sands, and is known as the Sal del Rey. It is apparently a salt dome, and probably the only one in which the salt occurs at the surface.

The next important dome to the northwest is Loma Blanca in Brooks county, 6 miles southeast of Falfurrias. A lake, 3 miles in length by ½ mile in breadth, lies north of it, the water of which is only a foot or two in depth. Loma Blanca itself covers an area of probably 50 acres (20 ha.), and its top has an elevation of 75 to 100 ft. (23 to 30 m.) above the lake. Its northern side rises somewhat abruptly from the margin of the lake and is covered with a soft, gypseous sand containing blocks and boulders of selenite. The sand thins out toward top of the mound, and near the summit clear, transparent selenite is seen to cover an area of several acres. The selenite is in layers 2 to 6 in. (5 to 15 cm.) in thickness and is perfectly transparent. Wells drilled on this mound show that the selenite stock has a thickness of more than 1000 ft. (305 m.). Around the foot of the mound the gypseous sand is gradually covered by gray, siliceous sands. The deposit of gypsum evidently underlies a large area, for the lake is bordered on the north for more than a mile by a bluff 25 to 30 ft. high of gypsum sand with blocks of selenite.

Between the Colorado and the Brazos we find some large domes, among which are Big Hill near Matagorda, Bryan Heights, at the mouth of the Brazos, and Damon Mound about 30 miles northwest of the latter. These mounds are prominent features in the Coastal Plain.

Big Hill rises gently from the shores of Matagorda Bay to a height of 36 ft. (11 m.) but dips sharply on its northern side to the level of the prairie. Its surface material consists of yellow Port Hudson clays. It first attracted attention as a probable oil field, and both gas and oil were found, but its production soon failed. The drilling, however, discovered considerable bodies of sulfur along the eastern side of the mound, filling caverns in porous limestone beds at about 1000 ft. (305 m.) in depth. So far as our records show, although salt waters were encountered in drilling, no body of salt has yet been found.

Bryan Heights has about the same elevation as Big Hill, w ith a dia n eter of about 1 mile and rather gentle slopes on all sides. The surface is composed of Port Hudson clays, which are apparently 500 ft. (152 m.) in thickness, underlain by 200 ft. (61 m.) of Lafayette and other materials which rest upon a dolomitic cap-rock. Below this there occurs some 300 ft. (91 m.) of gypsum and sulfur with some sandstone. This lies directly upon the salt stock, which is found in the dome at about 1100 ft. (335 m.). In this mound the sulfur is associated with gypsum and anhydrite. The only limestone is the cap-rock.

Wells to the northeast and to the south have been drilled to considerable depths, one of them to more than 3000 ft. (914 m.), entirely in stratified sediments, with no signs of dome materials. In the few fossil forms found in drilling outside the mound, nothing occurs to indicate that the beds are earlier than Miocene.

Damon Mound, in the northern corner of Brazoria county, is much larger than either of the others, covering between 2000 and 2500 acres (809 to 1012 ha.), and rising 80 ft. (24 m.) above the prairie. Its surface is composed of Lafayette sands and gravel, and it thus forms a Pliocene inlier in the Port Hudson.

One of the earlier wells struck gypsum at 171 ft. (52 m.) and continued in it for 400 ft. (122 m.), the bottom 30 ft. (9 m.) being a mixture of gypsum and sulfur. Immediately below this was loose sand and then 500 ft. (152 m.) of rock salt with bottom not found. Subsequent drilling shows that the "gypsum" body is a mixture of anhydrite, gypsum and selenite. Some limestones are found on the sides of the mound and in the wells. Kennedy gives this section of the region near the original well:

Red and blue clays with some heavy beds of sand.... 360 ft. (110 m.)
Limestone and gypsum........................... 190 ft. (58  m.)
Gypsum and sulfur............................. 40 ft. (12  m.)
Rock salt...................................... 500 ft. (152 m.)

Drilling to the south and southwest of the mound has developed commercial oil deposits in sedimentary beds which seem to dip away from the mound at angles of 30° to 50°. Wells off the mound, but near it, have been drilled to over 5100 ft. (1554 m.) without finding any trace of mound materials.

Stratton Ridge lies 8 to 10 miles (14 km.) northeast of Bryan Heights. Three wells along a line running north and south found gypsum and anhydrite at depths varying from 862 to 1840 ft. (263 to 561 m.), but so far no well has gone through it. No salt, oil or sulfur has yet been found.

Hoskins Mound lies 10 miles northeast of Stratton Ridge. A good production of oil was obtained from the shallow sands on top of the dome. One well drilled to the south encountered the cap-rock at about 800 ft. (244 m.), below which they found over 500 ft. (152 m.) of gypsum and anhydrite with some sulfur. The only fossils found indicate the age of the sedimentary beds to be Lower Pliocene or Upper Miocene.

The Humble dome, which shows on the surface an extreme elevation not exceeding 20 ft. (6 m.), is probably 1½ mile (2.4 km.) in diameter. On top of the dome, gusher oil was found at 1000 to 1200 ft. (304 to 366 m.). The cap-rock, as it is called, is dolomitic and is in places accompanied by gypsum, the two having a maximum thickness of about 200 ft. (61 m.). These rest directly upon the mass of rock salt.

Upon the decline in production of gusher oil from the dome proper, wells were drilled at various distances from it on all sides. The Esperson wells, 1 mile to the south, found light oil in shales, and later, similar oil was found on the north and in larger quantities to the east of the dome. The series of beds in which the oil was found consists of shales and gumbo with some sand, and dips away from the dome on all sides. Between the producing area on the dome and that of the shale-oil belts on its flanks, there is a strip ½ mile or more in width in which oil is not found in any quantity, and in this the belts are apparently much broken. The salt mass which occurs at 1400 to 1600 ft. (427 to 488 m.) on the dome, was struck at 2320 ft. (707 m.) at the western edge of the shale-oil belt lying east of the dome. This well was drilled to a depth of 5410 ft. (1649 m.) without getting through the salt. To the east of this, wells considerably more than 3000 ft. (914 m.) deep found only stratified materials.

These conditions indicate clearly that the domes comprise an intrusive mass of salt, gypsum, or anhydrite, coming up through the sedimentary beds, which are broken and tilted. The intrusive action is also shown by the sills of salt and gypsum which are found in some of the domes.

No fossiliferous beds have been found in connection with these Coastal Domes that are earlier in age than the Miocene, except in one well at Sour Lake, where Harris found Jackson fossils, and in another at Saratoga in which forms of similar age were reported. No oil was found in either of these wells, and other drillings near them failed to find any extension of these deposits, which are, therefore, seemingly restricted to very narrow limits.

### Origin of Intrusives

The association of the gypsum, salt, and anhydrite suggest their derivation from sea water by evaporation. Attention has been called to the fact that the occurrence of the gypsum above the salt in these domes indicates that these substances may have had some other origin, but none of the hypotheses suggested seems to have any better grounds than that of the familiar one of evaporation of brines from sea water, especially as the geological facts indicate the probability of favorable conditions for such formation in this region.

The salt masses of the interior domes lie below the Woodbine. Therefore, their deposition must have taken place prior to the beginning of Upper Cretaceous sedimentation. There were two periods antedating this which were especially favorable for the deposition of gypsum and salt.

Prior to the Comanchean Cretaceous, East Texas was part of a land mass, a region of erosion and not of deposition, with drainage into the epi-continental sea of North Texas. The encroachment of the Comanchean sea was from the west and south, and the character of its sediments, as seen in the extreme eastern exposures in Arkansas and southward into Texas, are those of shallow water, showing that the old land area was but slightly submerged. The Trinity in Arkansas carries considerable beds of gypsum, a condition which was duplicated in West Texas, where, in the Malone mountains, we have hundreds of feet of gypsum of Lower Cretaceous age. There is, therefore, no reason why salt and gypsum deposits of this age may not be expected in the area of northeast Texas occupied by the interior domes.

A second period favorable for such deposits is found in the interval between the Comanchean and the Upper Cretaceous. While we have no such positive evidence of the accumulation of such deposits of sea salts at this period as those already mentioned, the fact that for hundreds of miles the contact between the Buda limestone, which marked the close of Comanchean deposition, and the Eagle Ford (which overlies the Woodbine when the latter is present) shows no signs whatever of erosion, proves that during the long period that elapsed between them the top of the Comanchean must have remained at or near sea-level, and in such relation to it that no terrigenous sediments could be laid down on it. In the more littoral zone of Northeast Texas, the Buda is represented by clays, and the conditions would be even more favorable for the formation of salt basins and the accumulation of gypsum and salt prior to the beginning of Upper Cretaceous sedimentation. There is every reason to believe, therefore, that the gypsum and salt found in connection with the interior domes may have been deposited during the Lower Cretaceous or in the Mid-Cretaceous interval.

While sediments of Jackson age have been found in connection with the salt of the Coastal Domes, in no place has salt been found beneath these deposits. It is found, however, underlying the Neocene beds; what we know, then, is that they are pre-Neocene.

In the Coastal area, so far as is evidenced by the beds at the surface, marine sedimentation closed with the Eocene, and during the Oligocene and Neocene the deposits are largely those of land and fresh water.

That the withdrawal of the sea at the close of the Eocene was accompanied by the deposition of beds of massive gypsum is clearly shown at the southern end of the belt of Gulf Coast Eocene on the Conchos River, in Mexico. Here the Frio clays, which are the uppermost Eocene beds and are probably of Jackson age, form a large portion of the Pomeranes Mountains. They carry in their upper portion heavy beds of massive gypsum, alabaster, and selenite, interbedded with clays. The pre-Oligocene age of the gypsum beds is clearly shown by their relation to the yellow clays of the San Rafael formation lying to the east.

While similar conditions are not positively known to have occurred in East Texas, it is probable that they did, and that the salt and gypsum which occur in connection with the coastal domes was deposited at the time of this emergence and prior to the deposition of the Corrigan sands.

## ORIGIN OF DOMES

The character and structure of the domes show conclusively that they are the result of orogenic action and that this activity has been manifested at different times and in different degrees.

The interior domes and the Sabine Peninsula were begun by crustal movements at the end of the Upper Cretaceous, and were involved in later movements at the end of the lower Eocene and during the Claiborne or middle Eocene, resulting in total elevations of 2500 ft. (762 m.) or more above their normal horizon.[2]

Similarly, the coastal domes were elevated by orogenic forces acting after the deposition of the Lafayette and prior to that of the Port Hudson clays. The elevations in this case are fully as great as in those of the interior domes, as is proved by drilling. In the case of Damon Mound, the elevation does not appear to have entirely ceased, since the San Bernard River, flowing on its western side, seems to be still cutting its canyon where it crosses it, although to the north and south its banks are low and flat. The movements are most probably the result of isostasy.

## QUESTION OF VULCANICITY

According to the writer's point of view, the question of vulcanicity cannot enter into the origin of the materials of which the domes are composed, since, although salt may be formed by volcanic emanations, we have here no need to invoke such a theory of its origin, since a more natural one is probable. Neither is it believed to be the actual dome producer, but rather an accompanying factor or feature of the forces which caused them.

Both salt and anhydrite become liquid at far lower temperatures than any lava, and existing in considerable bodies, as they probably did here, could be readily forced into any opening caused by flexures or faulting of the crust above them more easily than a plug of basalt. Apparently this is what has occurred, and since the term vulcanism is used by Chamberlin and Salisbury[3] "to embrace not only volcanic phenomena in the narrower sense, but all outward forcing of molten material, whether strictly extrusive or merely ascensive," it is thought to cover the phenomena of the intrusive salt, gypsum, and anhydrite stocks in these domes, as fully as if the stocks were basalt or other rock material.

---

[2] O. B. Hopkins: U. S. Geological Survey, *Bulletin* 661 (1917).

[3] T. C. Chamberlin and R. D. Salisbury: "Geology," 2d Ed., 590. New York, Henry Holt & Co., 1906.

## Lithology of Berea Sand in Southeastern Ohio, and Its Effect on Production

BY L. S. PANYITY,* COLUMBUS, OHIO

(Colorado Meeting, September, 1918)

THE State of Ohio is among the pioneers in the production of oil and gas. Numerous anticlinals, such as the Macksburg, Cow Run and Newport, have been thoroughly developed, and the pools found in connection with them are quite in accord with the structural theory. Numerous oil and gas pools found in the Berea sand, however, demand a different explanation.

The study of these pools belongs to the sub-surface geologist, as it is possible to understand the factors controlling "off-structure" accumulations only after a careful study of well records. By carefully analyzing the various conditions over a large territory, it is possible to understand the otherwise meaningless mass of data that one may have at hand, and put it into tangible shape, from which various conclusions may be drawn.

The territory here discussed includes Belmont, Guernsey, Muskingum, Noble, Monroe, Morgan, Washington, Athens and Meigs counties, containing the majority of the prolific oil and gas pools of the Carboniferous system in southeastern Ohio.

The general dip of the strata is to the southeast, the entire territory lying on the northwestern flank of the Appalachian geosyncline. The best known oil- and gas-producing horizon in the territory is the Berea sand, which lies near the lower part of the Mississippian, with the Sunburry shales (the coffee-shales of the driller) above, and the Bedford shales below it. The Bedford is underlain by the thick Ohio shales. The Ohio and Bedford contact is considered to be the dividing line between the Carboniferous and Devonian systems.

The Berea is one of the most persistent of recognizable oil and gas horizons; its outcrop has always been found in its expected place, and it has always been penetrated whenever the drill went deep enough. It is not expected, however, even of the most uniform formations, that no lithological variations should occur, and the Berea is no exception. Nevertheless, there appears to be a certain regularity in its variation, the main features of which directly affect the accumulation of petroleum.

* Geologist, Ohio Fuel Supply Co.

It is the purpose of this paper to point out this characteristic condition in the sand, and to attempt to show its observed effects on production.

The first noticeable feature of the Berea sandstone is that the formation is thickest at its outcrop, being between 50 and 60 ft. (15 and 18 m.) in average thickness. This thickness is maintained down the dip for a great distance, when suddenly the sand becomes thin, and remains so. This thinning seems to appear in a well-defined zone. From a study of well records near McConnelsville, Guysville, and Bedford, it is noticed that the Berea appears to be in three strata. First, a top layer of sand about 35 ft. (10 m.) thick, which is invariably full of water. Below this is a shale "break" of about 25 ft., then a lower and thinner sand, from

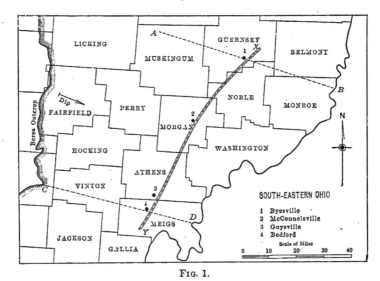

Fig. 1.

1 to 15 ft. thick; this lowest, or thin stratum, is the "pay sand" (Fig. 2 and 3).

In the vicinity of Byesville, north of the places above mentioned, this double structure of the sand is altered, and the thick and thin strata do not overlap. The change is shown by a sudden thinning of the thick water-bearing formation; then to the east follows a conspicuous territory of "no-sand," and further east the thin sand makes its appearance (Fig. 4).

Connecting the points where the change in the sand is noticeable, a hypothetical line ($X - Y$ in Fig. 1) is obtained, running a little east of north. In the territory under discussion, the line is quite well defined and its position may be located from well records. North of this territory, its course is not so easily followed; the direction of the line appears to change and at Byesville it seems to bear to the east. In this district the drill generally stops when the water in the thick sand is found, so the presence or absence of the thin lower sand is not known.

### THIN-SAND TERRITORY

In the territory east of the line, the sand everywhere is found to be thin, and quite lenticular. The position of the thin sand near the line is up the dip, and the best gas fields in the Berea are found near this (western) limit of the sand; east, and parallel with the line, are such gas

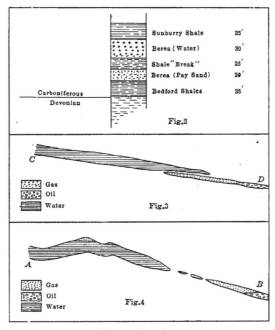

FIGS. 2, 3 and 4.

fields as Barnesville, Temperanceville, Summerfield, Macksburg, Dudley, McConnelsville, and Rutland. East of the gas, down the dip, the best Berea oil pools are found. Little water is found in this stratum below the oil. In places where anticlinal structure is present, the gas, oil and water are found in relative positions in accordance with the structural theory, but the controlling factor in most of these pools is the western limit of the thin sand, in proximity to the hypothetical line. The oil found in this sand is of the highest, or Pennsylvania, grade.

## THICK-SAND TERRITORY

In the territory west of the line, a thick water-bearing sand is present throughout. The pools are few and very small, the only one of any magnitude being the Corning. The oil in this field is of a lower grade than the thin-sand oil, and is known as Corning grade. Two small pools, the Oakgrove and Otsego, lying northeast and northwest of Byesville, are also in the thick-sand territory, the former being on a strong anticlinal "nose," the latter on the flank of an anticlinal fold, with gas at the top. In almost every instance, when the drill reaches this thick sand, a showing of dark oil is found, and the few scattered small pools in this sand are no doubt associated with "structure." However, the territory west of the line appears to be free of prominent structural disturbances, which is of the greatest importance in a district where connate water is found in great quantity in the sand.

## CONCLUSION

It will be noticed that the thin-sand territory is by far the better oil and gas producer of the two, and the size and staying qualities of the wells are much better than of those in the thick-sand district.

It appears possible that these two horizons are entirely separate strata; the up-dip limit of the lower, thin sand, and the lower limit of the upper, thick sand, however, appear to be too closely related to permit such a conclusion. A close study of numerous well records in proximity with the hypothetical line would no doubt throw more light on this question. A better understanding of the lithology of the Berea sandstone could be obtained, I believe, by the study of that portion of its outcrop which parallels Lake Erie and passes eastward into Pennsylvania.

It has long been the belief of a number of practical oil and gas men, based on personal experience, that a thick Berea sand is not a good producing stratum in Ohio. One cannot but agree with them, and it is hoped that the conditions here described account for this belief.

## Interpretation of So-called Paraffin Dirt of Gulf Coast Oil Fields

BY ALBERT D. BROKAW, PH. D., CHICAGO, ILL.

(Colorado Meeting, September, 1918)

THE so-called "paraffin dirt" of the Gulf Coast oil fields has been considered an indication of the possible presence of oil and gas, and not a few wells have been brought in solely on the basis of such evidence. In a sense, it may be empirically justified as evidence of the presence of oil or gas, though it cannot be said to be infallible. If the writer is correctly informed, its association with oil and gas was pointed out by Lee Hager some years ago, and while it is commonly mentioned by geologists and operators in conversation, no discussion seems to have found its way into print. The present interpretation is based on laboratory study rather than on field investigation, and is put forth with the hope of stimulating discussion which shall lead to more complete understanding.

The term "paraffin dirt" has been applied to soils with a peculiar texture, which has been described as "curdy" or "rubbery." When moist, the material breaks much after the fashion of "green" cheese. It is rubbery under compression, but does not resemble rubber in tenacity or cohesion. When dry, the material ranges from hard clods to a horny mass of high tensile strength. The shrinkage in drying is very great, especially in the hornlike matter.

The moist material ranges in color from dark brown in the specimens rich in organic matter to grayish in specimens containing more inorganic matter. All of the specimens studied were readily attacked by molds, which in some cases grew in great profusion on the samples. Samples exhibited a rather characteristic "swampy" or "mucky" odor when wet, and a characteristic odor of humic soils when moist.

· If the material is brought into suspension in water, very fine sand settles out and may be recovered by successive washing and decantation. The sand is almost entirely quartz, but contains occasional fragments of feldspar—more or less kaolinized and iron-stained. Practically all of the sand passes through a 100-mesh screen and the grains are extremely sharp —possibly below the limit of rounding. A flocculent mud, of clay and organic matter, remains in suspension for days. Some material is taken into solution in the water, imparting a slight tendency to froth when vigorously shaken.

The results of partial analyses of four samples of "paraffin dirt," and, for comparison, one of a coastal plain soil are given in Table 1.

TABLE 1

| No. | As Received, Loss at 100°–110° | Dried Ignition Loss | Composition after Ignition | | | | |
|-----|-----|-----|-----|-----|-----|-----|-----|
| | | | SiO₂ | Al₂O₃ | Fe₂O₃ | CaO | MgO |
| 1 | 22.5 | · 12.1 | 73.8 | 15.7 | 2.3 | 0.8 | 0.9 |
| 2 | 46.5 | 36.6 | 68.4 | 21.0 | 4.9 | 2.6 | Tr. |
| 3 | 56.5 | 63.7 | 64.9 | 25.6 | 2.7 | 1.6 | Tr. |
| 4 | 41.0 | 51.0 | 73.5 | 22.9 | | 1.5 | N.d. |
| 5 | N.d. | 11.7 | 71.9 | 17.6 | 5.4 | 0.6 | N.d. |

1. One mile north of Spanish Lake, S . Martin's Parish, La., near test well.
2. Near discovery well, New Iberia Oil Field, Iberia Parish, La.
3. Lake Dauterive, St. Martin's Parish, La.
4. Near discovery well, St. Martin's Parish.
5. Soil taken near paraffin-dirt exposure.

From inspection of these analyses, it is apparent that the range in proportions of inorganic substances is considerable, and in no way characteristic of the paraffin dirt as differentiated from ordinary soil. With the exception of No. 1, all the paraffin-dirt specimens show very high moisture and ignition loss, which go hand in hand with organic matter. All but No. 1 burn rather readily when thoroughly dry, and the residue from ignition may be considered ash. Unfortunately, there is no accurate analytical method for separating colloidal silica from quartz, as it has been suggested that colloidal silicic acid might be responsible for the characteristic texture of paraffin dirt. Fortunately, the results of later experiments rendered further investigation along this line unnecessary.

Inasmuch as nothing characteristic was found in the inorganic analysis, attention was directed to the organic constituents. A 50-gm. sample of paraffin dirt was boiled with water for several hours, then filtered. This extract froths on boiling. When mixed with alcohol, it yields a precipitate resembling soluble starch, dextrine and carbohydrate gums. On evaporation, the extract becomes viscous and dries to a pale brownish-yellow coat over the porcelain dish. When thoroughly dry, this material readily scales off, and can be collected without difficulty. When moist, it is gelatinous, but neither markedly adhesive nor plastic.[1]

A second 50-gm. portion was boiled in very dilute sulfuric acid (3 gm. in 1000 c.c.) for 8 hr. Ammonium hydroxide was added to precipitate iron and aluminum and the solution was boiled to drive off excess ammonia. The calculated amount of barium hydroxide was then added to precipitate the sulfates and decompose any ammonium salts. After further boiling, the material was allowed to settle, then was decanted and transferred to an automatic filter. The colloidal nature of the solution made

[1] See description of humic matter. E. W. Hilgard: *Soils* (1906), 124. New York, MacMillan Co.

filtration extremely slow, and 15 days were required for the liter of solution obtained.   A second filtration removed practically all turbidity and the solution was evaporated to dryness in small lots in porcelain evaporating dishes.   The resulting material was similar in every way to the hot-water extract, but was somewhat darker in color.   From 50 gm. of moist paraffin dirt, 4.75 gm. of this extract was obtained, though no great effort was made to make either the extraction or the recovery quantitative.   No attempt to identify this material chemically was made because of the well-known complexity of humus.[2]

A sample of soil (No. 5 in Table 1) was worked to a paste with a solution of this extract, and the texture of the mud was noted during slow drying.   As soon as the paste began to stiffen, it showed in a mild degree the peculiar "curdy" fracture of paraffin dirt, to a degree depending on the concentration of gum used in making the paste and upon the stage of drying.   In a word, this organic extract is believed to be the material that is responsible for the characteristic texture of paraffin dirt, which is not appreciably different from that of certain well decomposed peats, especially those containing fine sand, as in the Cambrian area of Wisconsin.[3]   The lumps of mud mentioned were attacked by mold when kept in moist air, exactly after the fashion of paraffin dirt.

The writer concludes, therefore, that the so-called paraffin dirt is a form of peat, or peaty soil.   This is further evidenced by the high moisture, ignition loss, and shrinkage on drying, mentioned in connection with the analyses.   Most of the material available to the writer is neutral peat, showing no action toward litmus and yielding no brown color to ammonium hydroxide solution.

Since paraffin dirt is not so restricted topographically as ordinary peat, some further explanation is required.   The failure of peat substance to accumulate in ordinary soils is due to its oxidation to black peat or humic acids by atmospheric oxygen or to its elimination through complete oxidation.   The suggestion here made is that in certain cases where the soil atmosphere is largely not air, but gas seepage from below, peaty matter may accumulate and be preserved from oxidation; indeed, conceivably the work of the anaerobic organisms involved in the structural disintegration of plant matter may continue, even though not protected by more or less complete submergence as is the case with ordinary peat formation.

Rumor has it that hydrocarbons have been detected in paraffin dirt. Paraffin has been identified in ordinary peat.   The paraffin of paraffin dirt (if any is present) need have no relation to petroleum deposits.   It is

----

[2] E. C. Shorey: Some Constituents of Humus.  *8th International Congress of Applied Chemistry, Sec.* 7 (1912) **15**, 247.

[3] H. L. Walster, oral communication.

highly improbable that any of the low-boiling hydrocarbons, *e.g.*, hexane, etc., are present, as they are highly toxic to a wide variety of organisms. Methane, however, is not toxic to many anaerobes, and if the soil atmosphere contains large amounts of methane and correspondingly small amounts of oxygen, the oxidation of the products of anaerobic decay would doubtless be arrested or inhibited, and peaty accumulations would result.

In a sense, then, the place of paraffin dirt among the evidences of oil and gas rests on the possibility that it may indicate gas-saturated soils—and obviously such soils are present in the vicinity of gas seepages—though it does not necessarily follow that every gas seepage finds it accompanying paraffin dirt; the presence of any one of many possible toxic substances in the gas might prevent peat formation.

In conclusion, the writer wishes to acknowledge his obligation to various geologists for material and for descriptions of the occurrence of "paraffin dirt," notably Messrs. W. Van der Gracht, R. A. Conkling, S. Wells, W. E. Wrather, C. W. Hamilton, and especially Mr. F. B. Plummer, who collected the material used in the greater part of this study. It should also be noted that detailed description of the material in its field relation is left to others, as this paper is based solely on a laboratory study.

## DISCUSSION

W. E. WRATHER, Wichita Falls, Tex. (written discussion*).—The appearance of Mr. Brokaw's paper dealing with the chemical composition of "paraffin dirt" will be welcomed by oil geologists who have worked in the Gulf Coast district. It is to be hoped that it will lead to a discussion of the subject which will throw some light on the relation this peculiar substance bears to the occurrence of oil deposits.

The problem of locating oil pools in advance of drilling is at the best a very intangible matter in the district along the Gulf Coast, and it is therefore desirable to make the most of any clew which may be of the least assistance to this end. Geologists have been familiar with "paraffin dirt" for a number of years, and have informally discussed its value as a surface indication of oil. But doubtless owing largely to the fact that they have been unable to define its exact composition, very little information has appeared in print on the subject.

The presence of this baffling and somewhat indeterminate substance is now quite generally accepted as one of the most reliable indications of a gas seepage. The question is whether paraffin dirt necessarily indicates that the escaping gas is an emanation from an oil deposit, or whether it merely indicates that marsh gas (methane) has here escaped from its

---

* Received June 1, 1918.

source in the abundant decomposing vegetation entombed in the sediments of the old Mississippi Embayment, and the Gulf Coastal Plain.

The presence of this material has been relied upon to a great extent, both by oil producers and geologists, in the location of wildcat wells, since the discovery that it was associated with gas seepages. It is quite true that other surface indications were often present at such localities, but if, in addition, paraffin dirt could be found in the vicinity, this fact was accepted as the final justification for drilling a test well, despite the fact that no one seemed to understand its chemical composition or to know why it should be associated with oil. In fact, for a long time it has generally been accepted that the name is a misnomer, that very little, if any, paraffin is present in it, or, for that matter, any other known fraction of crude oil.

The name first appeared following the early development of Batson. It was applied to a gummy, reddish-colored bed of this substance found on Batson Prairie, which had a fancied resemblance to true paraffin such as accumulated in oil lines and well tubing in the older Eastern fields. It was found that the material would burn, which further confirmed the belief that it was paraffin. Indeed, it is not strange that on the above criteria, this name should have been chosen, for there is often an undoubted resemblance in general appearance to paraffin. All trace of this deposit at Batson has long since disappeared.

The writer's personal opinion has been that the peculiar "rubbery" characteristic of paraffin dirt might possibly be due to colloidal silica, brought up in solution from below through the agency of hydrogen sulfide gas. This gas is commonly present in the Coastal oil pools, and is precipitated at the surface as an impregnation of the soil, perhaps owing to the loss of gas, and to the precipitating properties of humic acids derived from decomposing vegetable matter. Whether or not this was a tenable premise on strictly technical grounds has been a question, as no attempt was made to work out a chemical formula to fit the case. The idea merely took form as the result of numerous observations of the conditions under which the deposits were found.

These conditions seem to necessitate a moist, humid climate, and the presence of vegetable matter. The writer has examined gas escapements in the semi-arid region of west and southwest Texas, several of which have since been found to be associated with oil or gas, but nowhere in this area has he ever seen anything to resemble paraffin dirt. The character of the escaping gas in those regions did not apparently differ from that found in East Texas and Louisiana seepages which were accompanied by such deposits. It seems to be true also that "paraffin" dirt will not accumulate where gas escapes through water, provided the water covers the gas escapement most of the time. Repeated attempts to find traces of it in the bottom of pools through which gas was boiling incessantly were invariably unsuccessful. Sulfur gases, escaping through permanent pools of water, however, often leave a black or bluish colored "sour" mud around the gas escapement, which has no resemblance whatever to paraffin dirt. Typical occurrences of this class include the well known original gas escapements at Markham and Anse la Butte. While the writer never saw the hole at Spindletop through which hydrogen sulfide gas escaped before the pool was developed, it is his opinion, from descriptions, that this probably corresponded quite closely to the two above-mentioned localities.

Apparently the most favorable locality for the formation of paraffin dirt beds is a damp, seepy spot, over or around which vegetation flourishes, though the spot need not necessarily be marshy. It is quite as often true that the deposits are found on the open prairie as in timbered regions.

The range over which paraffin dirt has been found by the writer lies in the Gulf Coastal Plain, bounded on the West by the Colorado River, the boundary swinging northward north of Humble and including east Texas as far north as Shreveport. The dirt is quite common in western and southern Louisiana, but none has been seen in northeastern Louisiana, Arkansas or Mississippi, although the climate and vegetation in these localities does not seem to differ in any important particulars from those in east Texas and Louisiana. There seems to be no good reason for its absence in parts of northern Louisiana, and it may very likely have been found there by others more familiar with this section. Geologists who are familiar with conditions in Cuba think they have identified the dirt there also.

It will be observed that the territory over which paraffin dirt is common lies in the region of 40 in. or more of annual rainfall, and is so situated geographically as to permit of only slight differences in conditions of vegetation. In other words, over this section, conditions are such as might well permit the formation of peat deposits. It will also be noted, however, that paraffin dirt is found over only such portions of this territory as afford known gas seepages. It seems more than likely that if climatic conditions were suitable to support a satisfactory mantle of vegetation, with attendant conditions of moisture, paraffin dirt might well be expected to accompany gas seepages in sections where it does not occur.

Returning to the question of the character of the gas accompanying deposits of this material, it has been the writer's observation that sulfur gas was nearly always present. Hydrogen sulfide gas is quite commonly, if not universally, present where oil is found associated with salt domes. In Shelby and Sabine Counties, Texas, and in Sabine Parish, La., paraffin dirt is quite abundant, particularly along Flat Fork and Sabine River, near old Sabinetown. Near Negreet, La., a short distance east of Sabinetown, the gas seepages are decidedly sulfurous. The gas near Negreet burns with a pale blue, almost invisible flame, and in crevices and joints in the clays, crystalline sulfur is found.  This region, however, belongs more properly·in a geologic sense to the Sabine Uplift than to the Coastal Plain, and the deep gas associated with the Sabine Uplift is "sweet" gas quite free from sulfur.  It may therefore be true that the seepages of sulfur gas are not emanations from the deep gas sands, but have their origin in the sediments above the oil and gas horizons. The territory in and around the Caddo field is surely an ideal place to expect paraffin dirt beds, unless they are in some way dependent on the presence of sulfurous gases, but the writer has never personally encountered them there.

In Terrebonne and adjoining parishes of Louisiana, gas escapements and paraffin dirt are so abundant as to create a suspicion that not all the gas is petroleum gas.  Deep drilling has demonstrated the presence of an abundance of deeply buried vegetation, notably tree trunks, some partially carbonized, others partly decomposed to a pithy texture of the lightness of cork.  In fact, if this proof of buried organic matter were not at hand, one might safely assume its presence, owing to the semi-alluvial character of the sediments in that region.  It is the writer's belief that more or less of the surface gas is derived from this source. It is no doubt true also that a part of it comes from deep gas or oil sands. Evidence of this fact is found in the discovery of gas in the Knapp and Gulf Refining Co. wells near Montegut, below 1600 ft., and in the McCormick well 15 miles southeast of Houma, at over 2700 ft.  In none of these wells is salt-dome structure indicated, and we must assume that if a salt dome is present, these wells are considerably offside, or else the salt core is so deeply buried as to be beyond the reach of the drill.  If oil is found in this region, it would naturally be expected around salt domes. This section lies in a portion of the Mississippi Embayment where recent sediments are undoubtedly very thick, and it is likely that if salt cores are present, they are very deeply buried.

The writer is prone to accept Mr. Brokaw's suggestion that the peculiarities of paraffin dirt are due to unoxidized peaty material which has been deposited in a gas-saturated soil, since it fits in with all the conditions enumerated above.  He is inclined to believe, however, that the gas saturating the soil must necessarily be a sulfurous gas, but is open to conviction on this phase of the subject.  He believes also·that this material, once formed, is not subject to ready oxidation or decay, as paraffin dirt is occasionally found where it appears that the gas escapement has ceased.

The idea of the vegetable or peaty character of paraffin dirt is not altogether new.  Dr. J. A. Udden, in a personal communication dated March 13, 1913, with reference to specimens sent him by the writer from localities in Louisiana, said: "I find that most of it will burn, leaving about

30 per cent. ash. Only a minor part of this ash appears to be of mineral origin, such as sand. I find that it is nearly all soluble in potassium hydrate. Its elasticity and other physical qualities make me believe that it is identical with the mineral which has been described as "dopplerite," in Germany; or the related mineral known as "phytocollite." This mineral is supposed to be a precipitation of vegetable matter which has combined with soda or potassium, or some other base. The precipitation has taken place in such waters as those known as "black waters," containing much humic material in solution."

To what extent, then, may paraffin dirt be relied upon as an indication of oil? The writer's attention was first attracted during 1909 to the probable association of this substance with oil and gas at the above-mentioned locality near Negreet. His associates, Messrs. L. P. Garrett and B. S. Sorelle of Beaumont, were also of the opinion that paraffin earth was a valuable surface indication and in our subsequent joint investigation of probable oil prospects, particular attention was paid to this substance. Reëxamination of certain localities proved that paraffin dirt was present where we had previously failed to find it, but had later concluded conditions were such that it should have been found. In looking over some of the gas seepages around several of the developed pools which had been relied upon in the early prospecting, paraffin dirt was discovered, notably on the Slaughter tract at Humble, and at Goose Creek. Although we were sometimes unable to find it, we concluded that it had no doubt been present at Sour Lake, Jennings, Anse la Butte, and elsewhere, but that all traces of it had been obliterated around the older pools, before our investigation, by the incessant and intensive changes incident to oil-field development when confined to such small areas. Instances where it was relied upon in advance of the discovery of oil include Edgerly, La., Cow Bayou, Orange County, Tex., and the New Iberia, La., pool. Instances where its presence is known but where prospecting has not yet proved the presence of oil are numerous and well known, and need not be enumerated here.

The writer's opinion is that this substance almost invariably indicates the presence of a gas seepage, and probably one in which sulfur gas is present. These seepages along the Gulf Coastal Plain are quite likely to be of deep-seated origin, the gas under high pressure having found its way to the surface along the face of the salt core or through the numerous intercommunicating water sands. But it may also, where climatic conditions are suitable, indicate gas seepages which are not associated with oil. It may therefore be classed as one of the most important indications of oil along the Gulf Coastal Plain, inasmuch as the gas which causes its deposition is quite apt to originate in an oil pool. It would perhaps be a better statement of the case to say that in this region paraffin dirt more than likely indicates the presence of a submerged salt core.

This statement may be premature, however, since recent investigation, particularly in southern Louisiana, has brought to light a large number of new deposits which have not as yet been drilled.    One is loath to admit the possibility of the existence of such a large number of new salt domᴏs as seem thus to be indicated, until further drilling has been done.

E.  G.  Woodruff,* Tulsa, Okla. (written discussion†).—In thi- paper on paraffin dirt, Dr. Brokaw has given us valuable data on a subs ject which is of both scientific and financial interest to the Gulf Coast oil operators.    It is regrettable that he has been forced to confine his attention to a laboratory examination of the earth, rather than to include a field study also.    My own experiences have been limited to field observations supplemented by reports from chemists.    These chemists find no paraffin in the "earth."    It is called paraffin earth by the field scouts because, as described by Dr. Brokaw, it looks like paraffin-impregnated clay.

Some of the Gulf Coast operators think that the presence of the paraffin earth is associated with oil fields.    This supposition is perfectly rational since in most, if not all, of the low clay-coated oil fields, this pe- culiar spongy earth is found on the surface.    It is not found where the surface strata are sand or very sandy clay.    As far as I am aware, it is found only where there are gas escapes.    I had an opportunity to study this earth, especially in Terrebonne Parish, La., where there are gas wells of immense capacity.    Here the surface is flat and swampy, and both the gas escapes and the paraffin earth are common.

A paraffin earth locality may embrace a considerable area, possibly one hundred acres, with the paraffin earth spotted over the area; each spot irregular in shape and covering a few square feet or yards.    The ground is largely a clay soil carrying a large percentage of vegetable matter, and is saturated with water, or at least very moist.

Where soil cracks occur, the pure paraffin earth occurs as an incrusta- tion along the walls of the cracks.    When a small soil block from two to four inches across, for example, is examined, it is found that this coating quickly merges into an intermixture of paraffin earth and native soil which grades into unaltered soil in the interior of the soil blocks.    Stated another way, there seems to be a coating of paraffin earth on the sides of the blocks of soil behind which there is a partial alteration of the soil, while in the interior of the blocks there is no alteration.

In the field it appears as if gas escaping along weather cracks in a moist humus soil has altered the soil to a rubbery clay.    I am inclined to think that most commonly this gas is marsh gas produced by the altera- tion of the vegetation, probably near the surface.    In some places, it may

* Chief Geol., Sperry Oil & Gas Co        † Received June 4, 1918.

be that petroleum gas has produced the effect, but no doubt this is rare. I believe, therefore, that the presence of paraffin earth does not necessarily indicate an oil or gas reservoir of commercial importance.

The process by which this earth is produced can only be suggested. It may be due to a sort of plucking action of gas-agitated water on the clay soil, or it may be due to chemical action. I am inclined to think it is the latter.

In private correspondence, Prof. Heinrich Ries speaks of the paraffin earth as a hydrous aluminum silicate and mentions its occurrence at various places in the Southern States.

LEE HAGER,[*] Houston, Tex. (written discussion[†]).—In no other oil region of the world, perhaps, have the geological principles, of such high value in many fields, found so little practical application as in the Gulf coastal belt of Louisiana and Texas. The ever-present mantle of recent beds everywhere obscures the relations of the formations involved, and renders all questions of structure—faulting, folding, deformation— in large measure matters of conjecture.

The discovery of oil on a defined dome at Spindletop led at once to the belief that similar results would follow the exploitation of similar well known mounds. Prospecting was at once begun at High Island, Big Hill in Jefferson County, Barber's Hill, Damon Mound and other like localities. At shallow depths all encountered heavy masses of limestone, gypsum, and rock salt—conditions practically identical with those existent at the Five Salt Islands of Louisiana. In each case the porous, dolomitic cap-rock so prolific at Spindletop was found to be either wanting or barren.

From that time, the attention of oil men was directed more and more to the secondary phenomena which general observation had shown to be present in close association with the known domes and prospective fields: seepages of oil,[1] escapes of gas, acid, and sulfureted waters, brine springs and saline flats. To the escapes of gas, being most universal, attention was chiefly attracted. These gas escapes exist in every field thus far

---

[*] President, Tidewater Oil Co.     [†] Received June 14, 1918.

[1] The existence of oil seepages in the coast country has been questioned. As a matter of fact, such seepages were present in a shallow dug well on the Porcio place at Anse La Butte; at the oil spring at Sulfur Mine, where drilling had developed a little oil as early as 1872; at Vinton, where the seepage oil from wells 18 to 20 ft. in depth was sold to local sawmills; at Sour Lake, where the oil appeared in shallow dug wells at the edge of the lake, and where some oil was actually produced prior to the discovery of Spindletop; at Saratoga, where an oil spring and beds of asphalt had led to shallow drilling as early as 1862; at a spot examined by William Kennedy and by myself on the bank of Pine Island Bayou, at the north edge of the Batson field; at Hockley, where a shallow dug hole has exposed oil-saturated rock; on the Rachal ranch at White Point, where asphaltic matter occurs with much free sulfur, at the "Sulfur Hole;" and finally at Piedras Pintas, where a bed of asphalt exuding oil led to the drilling of the discovery well.

developed in the Gulf region, with one possible exception—Hoskins Mound.

At these spots, where it is probable that gas has escaped for long ages, it was early noticed by the prospectors that certain deposits were left in the ground. Where the gases were heavily charged with "sulfur gas'—$SO_2$ or $H_2S$—streaks of free sulfur were left in the ground. Where sulfates and a little free acid were present, "sour dirt" resulted, as at Sour Lake, Hockley, Damon Mound and White's Point, among other localities.

But more common even than the above named indications was the so-called "paraffin dirt." This was probably discovered at about the same time by various parties who were digging surface holes for the purpose of burning the gas. My attention was called to this "curiosity" half a dozen times in the first few months after the discovery of Spindletop. I saw it in quick succession at Spindletop, Vinton, Hackberry Island, Anse La Butte, Bayou Bouillon and Jennings. The fact that it was always associated with escaping petroleum gas early favored the conclusion that it had an intimate connection with, and was probably a deposit from, the gas itself. I at first thought it to be a form of ozokerite. An analysis showed this to be an error.

The name "paraffin dirt" was given to it by Judge Douglas, of Beaumont, who dug up some of this material at a spot in Batson Prairie when gas was escaping He organized the Paraffine Oil Co. and drilled the discovery well at this locality. These circumstances did much to attract the attention of oil men to this substance.

The characteristics of this material have been well described by Dr. Brokaw. It occurs in spots or beds from 2 to 20 ft. across. For the most part, these spots are bare of vegetation—mere wallow-like holes or depressions in the ground. The soil is resilient to the step, like rubber. When dug, the material falls apart in small squares. Intimately mixed with the soil itself, and between the cracks and joint-planes, is a yellow, jelly-like substance much resembling paraffin, or soft yellow beeswax. Where sulfureted gases are present, the stuff is blackened along the joint-planes, or shows streaks of free sulfur. In such cases, it is often slimy, and has a vile odor. When dry, a great shrinkage occurs in the mass; it lies loose in its bed, becomes a dark brown in color, and assumes a horn-like hardness.

Professor Fenneman makes mention of this substance in his report: "On the east side of the marsh (at Anse La Butte) an asphaltic substance is found, at places, about 8 in. beneath the surface of the ground. This substance so strongly impregnates the sub-soil as to give to it something of the consistency of rubber." He speaks further of an "asphaltic substance sometimes found impregnating the soil at shallow depths. . . . . It is well illustrated at Anse La Butte, Sour Lake and the Tar Springs of

Jasper County, Texas." It is apparent that he regards the paraffin dirt as an oil residue. The occurrence at Tar Springs is a typical asphalt bed. Both asphalt and paraffin dirt occur at Sour Lake.

For the most part, only a few small beds of paraffin dirt are found at any locality. Escaping gas is invariably present at the spots where these beds occur. That this gas is always petroleum gas, and no other, I believe the facts detailed below will go far to establish.

Matteson[2] says: "Concerning the paraffin dirt which Rogers mentions, that dirt has been analyzed, and we have not been able to make much out of it, except that it doesn't contain any paraffin; but where drilling has taken place in the vicinity of this paraffin dirt, the results have been a failure."

What are the facts?

It occurs in many spots at Bayou Bouillon and at New Iberia, both of which are small fields. The largest deposit of this material known in the coast country occurs at Anse La Butte. The best well in the field—5000 bbl. per day—was struck within 100 ft. of this deposit.

It occurs, again, on the Clement and Arnaudet tracts in the heart of the Jennings field. It is found in several spots in the south part of the Edgerley field surrounded by wells.

It occurs in two places on the Vincent land at Vinton, on the flat, and in the heart of the field.

It is found again in three places in the Terry field.

It occurs on the flat southwest of Spindletop—at one spot about 600 ft. from the Lucas gusher.

It occurs along the west margin of the lake at Sour Lake—again in the very heart of the field.

It is found in two or three spots in the center of the Batson field. It received its name from these occurrences.

It occurs at North Dayton. The discovery well was drilled by Judge Douglas, who opened up the Batson field.

It is present at Humble, on the Slaughter farm, on the Long and Moore tracts, and close to the discovery well drilled by C. E. Barrett—all in the oil field.

It occurs at Goose Creek, on the Minnie Gaillard, Tabb and Smith tracts, and on land owned by the writer and associates—all in the heart of the field.

It occurs in several spots on the Hogg tract in the Columbia field.

It is found at Barber's Hill in the close vicinity of all the wells which have produced any oil.

It occurs on the southwest side of the Big Hill field in Matagorda County.

---

[2] *Trans.* (1918) **59**, 486.

It occurs, finally, at Hackberry Island, at High Island and at Pierce Junction.

None of the three last-named localities has yet produced oil, but all are well known saline domes, and have shown strong evidences of oil in a number of wells. .

These localities comprise all the known occurrences of this material in the developed oil belt of the coast country. Fifteen of these localities, including practically all of the greatest fields, have produced oil in some quantity. The other three occur in connection with saline domes regarded by all oil men as possible fields. In view of these facts, Matteson's statement that, "where drilling has taken place in the vicinity of this paraffin dirt, the results have been a failure," would seem to be rather overdrawn.

It is true that in the lower Louisiana region—the Napoleonville-Houma district—this material occurs in a number of localities. In every instance, the occurrence is connected with escaping gas. The Lirette field below Houma has yielded heavy gas wells. At three other localities wells have been drilled without results—the deepest to a depth of 2600 ft. This evidence is of a neutral character since these wells are located near the trough of a great geosyncline. If saline domes exist in this region, they are probably buried to great depth, possibly 6000 or 7000 feet.

At the Markham, White's Point, and Damon Mound fields, as well as at Hockley and Johnson's Bayou, the escaping gases are heavily charged with $SO_2$ and $H_2S$. Sulfates are formed, resulting in the presence of the so-called "sour," or "copperas" dirt. At none of these localities is any paraffin dirt found in connection with the escaping gas. If the theory of Dr. Brokaw holds good, may not this absence be due to the fact that the anaerobic organisms involved in the formation of peat are in these instances destroyed by the ferrous sulfate present? The germicidal properties of this agent are well known.

Dr. Brokaw's theory seems to offer a valid explanation of the manner in which these deposits are formed. A gas-saturated soil is requisite to prevent oxidation. Everything points to the conclusion that gas has been escaping at these paraffin spots for ages. The supply is constant and inexhaustible. Methane, where found, occurs in limited quantity and with no concentrated point of escape. Air is fatal to the necessary organisms, as is the free acid present with heavily sulfureted gases. May not all these conditions go to account for the fact that where paraffin dirt occurs petroleum gas is always present?

Several questions in connection with this theory of origin remain obscure: In what manner is the silica reduced to its hydrated colloidal condition? Experiments conducted by the staff of the Roxana Petroleum Co. indicate that no fluorine is present in the escaping gases. What, then, is the form of the alkaline agents probably requisite? Some age-long process involving the paraffin tar, the acetic acid and ammonia yielded by the decomposition of peat matter, or the potassium acetate present in the sap of many plants?

Again, these paraffin beds vary from 2 to 7 ft. in depth. They seldom exceed 10 ft. in diameter. The enclosing clays carry little organic material. If this substance is a form of peat, in what manner has it come to impregnate the surface clays to a depth of 6 or 7 feet?

W. G. Matteson,* Fort Worth, Tex. (written discussion†).—Dr. Brokaw has evidently given a great deal of time and study to this phenomenon and his conclusions seem sound and logical.

Mr. Lee Hager was probably the first geologist of repute to attach any great importance to the occurrence of paraffin dirt in connection with the Gulf Coast oil-producing salt domes. The writer has been reliably informed that through his discussion the idea was promulgated that the presence of this so-called paraffin dirt was one of the most important and reliable indications of the presence of an oil-producing salt dome. Despite very strong evidence to the contrary, Hager evidently still has considerable confidence in this theory, as brought forth in his discussion of Dr. Brokaw's paper. Hager here gives all the paraffin-dirt occurrences known to him, and adds, "Fifteen of these localities, including practically all of the greatest fields, have produced oil in some quantity."

The acceptance of the list of occurrences as submitted by Hager depends upon what material we are justified in including under the term "paraffin dirt." If we agree to a broad classification, whereby paraffin dirt is to include all spongy, earthy, decomposed, vegetable material, or spongy soil impregnated with the same, then Hager's list will have to be not only accepted but perhaps enlarged. On the other hand, if we confine ourselves to a strict interpretation of the substance known as paraffin dirt, based upon typical occurrences, general understanding, and application, several of Hager's occurrences will be seriously questioned by many geologists. The typical material is excellently described by Hager as a dirt or soil-like substance "resilient to the step, like rubber, which falls apart in small squares when dug. Intimately mixed with the soil itself and between the cracks and joint planes is a yellow, jelly-like substance resembling paraffin or soft, yellow beeswax." The material is generally brownish to dark brown in color.

Many able geologists, who have carefully investigated the territory at Anse La Butte, have failed to observe any such material as above described within "100 ft. of the best well in the field." A deposit of spongy, vegetable matter was found near the heart of the field, but it

* Consulting Petroleum Geologist and Engineer.    † Received Sept. 24, 1916.

in no way resembled paraffin dirt. In a personal communication to the writer, William Kennedy states that he could not find any of the dirt in the proved part of the field, but observed some peat-like material considerably to the east, where several dry holes had been drilled in the vicinity of the occurrence. In the Jennings field, some peculiar clays were observed, and again at Terry, some slimy, stinking vegetable material and soil were found which in no way resembled paraffin dirt, even in the altered form described by Hager. Here again several dry holes were drilled in close proximity to this substance. William Kennedy, one of the ablest and most experienced Gulf Coast investigators, has stated that he was never able to find any evidence of paraffin dirt near or around Spindle Top, although he has examined that area carefully several times.

As to the statement that this dirt is found close to the discovery well at Humble, C. E. Barrett did not drill the discovery well; his well went into gypsum and was abandoned. D. R. Beatty, in 1905, brought in the first commercial producer at Humble. Some paraffin dirt has been found at Humble, however, although the occurrence had nothing to do with the location and discovery of the dome. Some quantity of typical material occurs at Hackberry, yet five or six wells, 1600 to 2200 ft. in depth, failed to obtain even gas.

Hager claims that Batson was discovered through presence of this paraffin dirt, the discovery well being located close to its outcrop. It is doubtful if Judge Douglass would have made his location in the vicinity of the dirt, were it not for the quantity of gas escaping near there. In view of many subsequent drilling failures on evidence of this paraffin dirt, too much weight should not be given the Batson occurrence.

It is true that small quantities of paraffin dirt are found in the vicinity of several oil-producing salt domes, but it is most significant that, with possibly the exception of Batson, cited by Hager, none of these domes was located and drilled by reason of the presence of this dirt. The dome or mound-like character of the area, or other topographic features, the presence of sour springs, salt springs, asphalt seeps, gas seepages, etc., this was the evidence by which 95 per cent. of the oil-producing salt domes were located and drilled.

What has been the result when well locations have been made according to the presence of paraffin dirt only, and without some of the more reliable and conclusive evidence such as has just been cited? In Chambers County and at Bayou Caster, wells drilled near paraffin dirt failed to find oil in commercial quantity and were abandoned. Wells drilled at the eastern vicinity of Anse La Butte, where the dirt is said to occur, were abandoned as failures. The largest and most characteristic deposit of this dirt is found in Sabine County, south of Low Creek, in the vicinity of Sabinetown. Wells drilled on the evidence of this dirt here found only some gas, but no oil in commercial quantity. These are a few of the instances the writer had in mind when making the statement to which Hager takes exception, that "where drilling has take place in the vicinity of this paraffin dirt, the results have been a failure." And even Hager's confidence in this material must have undergone some change when he has been reliably reported as turning down the Columbia field.

As to Hager's concluding arguments, he bases the importance of the paraffin dirt as an indication of the nearby presence of oil on his contention that the gas so intimately concerned in the formation of this dirt is *petroleum gas*. Why petroleum gas? Would not methane or marsh gas serve equally well? Owing to its widespread occurrence in the Gulf Coastal plain, would not marsh gas be a more common and plausible source? Does not the fact that this paraffin dirt is found widely scattered, and in places where drilling has failed to yield oil in anything like commercial quantity, strengthen the evidence that marsh gas is more concerned in the formation of the dirt than petroleum gas? In escaping, marsh gas may be diffused over a small area at the surface instead of issuing from one particular spot. Would not this method of escape explain the areal extent of these paraffin dirt deposits better than if the escaping gas were confined to one vent? As for quantity of gas, the writer has observed places in the Gulf Coastal plain where marsh gas has been escaping in considerable and undiminished volume for many years. Moreover, have we any direct evidence to prove that the formation of this dirt is the product of ages? And if petroleum gas is the gas concerned in the process of formation, why do we not find this paraffin dirt extending to considerable depth along the passage of escape taken by the petroleum gas?

In concluding this discussion, the writer wishes to emphasize the following points:

1. Certain spongy soils, highly impregnated with vegetable matter, should not be mistaken for paraffin dirt.

2. Some reported important occurrences of paraffin dirt are not the typical dirt, but the material referred to in 1.

3. Small quantities of paraffin dirt have been observed at a few of the important oil-producing salt domes along the Gulf coast, but none of these domes, with the possible exception of Batson, was discovered and drilled on the evidence of this paraffin dirt.

4. Drilling, based solely on the presence of paraffin dirt and often where this dirt has been observed most characteristically and in unusual quantity, has almost invariably failed to open up new commercial fields, or to develop oil in anything like paying quantities. Thousands of dollars have already been lost on such propositions.

5. Paraffin dirt as an indicator of the presence of domes and the

commercial accumulation of oil, is so erratic and unreliable as to be of little concrete value to the careful, conservative investigator.

EUGENE WESLEY SHAW,* Washington, D. C. (written discussion†).— Paraffin earth is of especial interest because of the apparent difficulty of determining its chemical nature. I have submitted specimens to three chemists at different times with suggestions as to special tests that might be made. The results are indicated in the following reports:

"The material contains a small amount of a faintly yellowish, waxy substance apparently of paraffin order. It also contains much anhydrous silica. From the dried material only 0.8 per cent. of soluble silica could be extracted."[3]

"The amount of material was very small and the waxy material extracted from it, only a minute trace, was too small for a very satisfactory examination. It proves, however, to be soluble in boiling absolute alcohol, but to precipitate out on cooling. On treatment with sulfuric acid only part of it was destroyed, the remaining particles having the appearance of paraffin wax. You are justified, therefore, in the conclusion that this is paraffin wax and not the solid fatty acid. The material is entirely insoluble in caustic alkalis."[4]

The analysis[5] of the third set of samples is as follows:

| | Sample No. 1 | Sample No. 2 |
|---|---|---|
| Water at 130°................... | 55.46 | 56.71 |
| Loss on ignition above 130°........ | 13.38 | 12.73 |
| $SiO_2$............................ | 21.33 | 21.20 |
| $Fe_2O_3$.......................... | 1.73 | 1.67 |
| $Al_2O_3$, etc. ..................... | 6.09 | 5.86 |
| CaO.............................. | 0.36 | 0.43 |
| MgO.............................. | 0.31 | 0.47 |
| $K_2O$............................. | 0.77 | |
| $Na_2O$............................ | 0.20 | |
| Combined carbon.................. | 3.13 | |
| $SiO_2$ soluble in 5 per cent. NaOH... | 0.92 | Does not form suspension in water. |
| Organic matter soluble in gasolene.. | 0.01 | |

Sample No. 1 was collected by H. E. Minor, who suggests a reaction involving hydrogen sulfide and yielding colloidal silica.

Sample No. 2 was collected by Alexander Deussen, who suspects that the substance is mainly colloidal silica but contains a hydrocarbon akin to ozokerite.

* U. S. Geological Survey.        † Received Nov. 8, 1918.
[3] Chase Palmer: U. S. Geological Survey, July 10, 1916.
[4] Report of the Director of the Bureau of Mines to the Director of the U. S. Geological Survey, Sept. 23, 1916.
[5] Made by R. C. Wells, of the U. S. Geological Survey. Reported Mar. 21, 1917.

According to the reports of David T. Day, who made the examination for the Bureau of Mines, and of Chase Palmer, paraffin earth really contains paraffin; on the other hand, Wells and Minor report none. None of them seems to have found much colloidal silica. Ries, according to Woodruff,[6] reports that the paraffin earth is a hydrous aluminum silicate. Udden, according to Wrather,[7] concludes that it is dopplerite or phytocollite which contains neither aluminum nor silica. Considerable quantities of the natural jelly are obtainable and the difficulty in its identification or even of determining its general chemical character is indeed remarkable. Our most convincing line of argument seems to be that any substance whose chemical constitution cannot be determined from good-sized specimens must be humus.

That the substance is real admits of no argument even if there is disagreement in the reports of chemists and apparent difficulty in its isolation. That it is characteristic of the low Coastal Plain salt-dome country, if not confined to it, seems also satisfactorily established. That it is not dopplerite (composed of carbon, hydrogen, and oxygen in the proportions of about $10 : 12 : 5$) or any other recognized hydrocarbon compound, seems evident from the chemical tests. That specimens from different localities are similar is true so far as known. Whether or not it consists largely or in part of colloidal silica produced by the action of gas, as suggested by Alexander Deussen,[8] seems to be still a question, as does also the suggestion by H. E. Minor[9] that $H_2S$ may be an essential factor. Minor[10] reports as follows concerning one specimen: "It is practically insoluble in benzine, chloroform, alcohol, or benzol; however, it is almost wholly soluble in nitric acid and yields a residue of carbon with sulfuric acid. Upon distillation a thick reddish-brown fluid of offensive odor is obtained. In a slide made from the purest of the "paraffin" (a colorless, transparent, jelly-like mass) magnified 500 times, small rod-like objects are seen resembling bacteria."

In a later letter he writes, concerning a sample from New Iberia: "I found the sample to be literally alive with very small worms resembling earth worms, but much smaller in size. These worms died immediately upon exposure to the air, showing that in all probability they are anaerobic in nature. Upon puncturing the skin, it is seen that they secrete a large amount of sticky gelatinous substance. I do not wish to advance the idea that these worms are responsible for the formation of paraffin earth, as this may be only a local occurrence, but nevertheless it is of interest."

Paraffin, as Brokaw remarks, is said to have been found in peat; but even if paraffin is, under certain conditions, formed at the surface,

---

[6] E. G. Woodruff, p. 491.　　[7] W. E. Wrather, p. 488.　　[8] Letter to writer.
[9] Oral communication.　　[10] Letter to writer.

it is safe to say that much of that which has been observed at the surface has migrated from considerable depth.    Hence the occurrence of any paraffin, except marsh gas, at the surface is not valueless in oil prospecting.    On the other hand, neither true paraffin nor paraffin earth can be regarded as a strong indication of oil.

As Kennedy says, paraffin earth occurs both in oil fields and elsewhere. At least it is found in many places where oil has not yet been discovered, as, for example, in parts of Terrebonne Parish, La.    Wrather notes that sulfurous gases and considerable exposure to the air (lack of prolonged submergence) seem to be requisite and Woodruff says that the substance "is not found where the surface strata are sand or very sandy clay." It seems to me that it will be well worth the cost to publish both laboratory and field observations such as these by Brokaw, Wrather and others, for more advancement has resulted in understanding this mysterious substance since last April (the date of Brokaw's paper) than in all the preceding years that paraffin earth has been known.    Further, it seems to me probable, from the data now available, that it will be found that the substance develops under certain peculiar conditions only and that it is of more definite or restricted chemical composition than humus or peat, or "peat substance."

## Possible Existence of Deep-seated Oil Deposits on Gulf Coast

BY ANTHONY F. LUCAS, WASHINGTON, D. C.

(Colorado Meeting, September, 1918)

THE discovery of oil in 1901 on the Spindletop dome, Texas, inaugurated a new industry on the Gulf Coast, an industry which has grown with the discovery of successive fields, until today it engages the services of thousands of workers and employs enormous capital. New fields are being discovered from time to time and doubtless some still remain to be found, though of late years discoveries have become more infrequent. Nowadays several hundred dry holes are drilled each year in a fruitless and blind effort to discover new fields, for as yet geologic science has developed no effectual method of locating the coastal oil deposits in advance of drilling.

Moreover, despite the occasional discovery of new fields, the total production of the Gulf Coast is today no greater than it was in 1906, for the added production of the new fields has been offset by the rapid decline and more or less complete exhaustion of some of the older ones. Careful geologic work within the fields has in some cases increased the production temporarily, but has developed no really new supplies. The Gulf Coast oil industry seems to have passed its period of greatest expansion and to be declining at a fairly steady rate, and this condition is naturally viewed with alarm by the more farsighted operators.

In my opinion, the time has come for the adoption of radical and aggressive methods of prospecting, and a fraction of the money wasted yearly in drilling shallow wells in hopeless locations might well be devoted to this purpose. Many facts lead me to believe that all the salt-dome oil has had a common origin; that it has migrated up from considerable depth along lines of structural weakness; and that a deep well, properly located, stands an excellent chance of discovering the parent reservoir and thus of developing new and probably great supplies. This paper is presented as a discussion—necessarily hypothetical and based largely on personal opinion—of the possibility of encountering deep-seated oil deposits beneath the salt domes.

### SALT-DOME STRUCTURE

All of the oil produced in the coastal region of Texas and Louisiana is probably associated with salt domes, though in Goose Creek, Edgerly, and one or two other fields no salt has yet been actually penetrated. As

a result of the innumerable wells that have been drilled on the various domes, it is now known that a typical salt dome consists of a very thick mass of pretty pure rock salt, generally almost flat-topped, but sloping abruptly away from the rim on every side. The flat top of the salt is generally covered by rock 25 ft. to several hundred feet thick, consisting chiefly of limestone, dolomite, anhydrite, or gypsum, with generally more or less sulfur. The sediments above the salt are slightly domed and those on the sides of the salt mass generally slope at angles of 30° to 60°, and in some fields at even greater angles. These sediments consist of sand, gravel, shale, and gumbo, arranged in beds so lenticular and irregular that they can seldom be correlated from one well to another.

Of the many salt domes already discovered on the Coastal Plain, there are no two whose structures are identical. Each has its individual peculiarities of size, height, steepness, character of cap-rock, and wealth or absence of oil. Some of the domes, like the phenomenally rich Spindletop, are only a few hundred acres in extent; others, like Humble, Damon Mound, and some of the salt islands of Louisiana, cover several square miles. Most of these domes are roughly circular in outline, but some, like South Dayton and Blue Ridge, are greatly elongated. Many of the domes form more or less distinct mounds at the surface, the highest rising 100 ft. or so above the surrounding plain, but others are overlain by sunken areas or lakes, and still others by perfectly level country in which no clue to the structure can be obtained. The cap-rock of some mounds, like Sulphur and Bryan Heights, contains commercial quantities of sulfur, whereas that of others contains only traces. Gypsum and anhydrite are almost the sole constituents of some cap-rocks, whereas in others limestone and dolomite greatly predominate. Pyrite is common in nearly all, and in one or two, galena, sphalerite, chalcopyrite, etc., have been found in small quantities.

Finally, the salt itself is physically different in different mounds; in most of them it is hard and well crystallized, but in others it is soft, granular, and almost incoherent. As may be seen in the salt mines at Petite Anse and Grand Cote Islands, La., the salt is characterized by peculiar gray streaks or markings. These do not seem to have any regular trend, and are often contorted and twisted into the most fantastic shapes (see Fig. 1).

## OCCURRENCE OF OIL ON SALT DOMES

The salt-dome oil occurs under conditions just as irregular and impossible to determine in advance of drilling as the structure of the dome itself. The oil at Spindletop occurs chiefly in a hard cavernous limestone layer which forms the cap-rock of the dome. At Welsh, Saratoga, and many other fields, the oil is found in the loose sands above the cap-rock.   At Anse la Butte, Damon Mound, and some other fields, it occurs partly or wholly in the deeper and sharply dipping sands which lie on the flanks of the domes and below the level of the cap, while in the Humble field it is found under all three conditions.   In only one dome, Belle

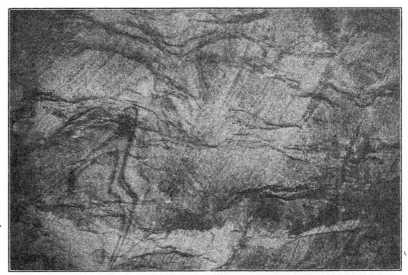

FIG. 1.—GRAY MARKINGS ON SALT, AS FOUND AT PETITE ANSE AND GRAND COTE ISLANDS.

Isle, La., has oil been found within the salt itself, and in this locality the salt contains much gas under enormous pressure and also a small quantity of high-grade paraffin oil.

As few of the domes are more than a mile across, it is evident that the productive area is very limited in extent, though where oil occurs on the flanks of the salt mass the field is somewhat larger.   The largest field, Humble, is, however, less than 2 miles square.   Salt-dome oil thus occurs under conditions very different from those in the Appalachian fields, where the oil is found in well defined gently folded sands traceable over large areas.   In the Gulf Coast fields the wells are more extensive and less likely to find oil, and though their production may be enormous for a short period their decline is usually rapid.

Of the four or five dozen salt domes now known in Texas and Louisiana, less than two dozen have produced oil in commercial quantities and less than one dozen have become really important fields.   Many are perfect blanks as far as production is concerned, and though apparently similar to richly productive domes and characterized by apparently identical surface indications, they may yield only a puff of gas while drilling and a showing of oil so slight that it may have found its way into the well from the surface machinery.   These conditions suffice to emphasize the practical importance of arriving at a solution of the origin of the domes and their associated minerals, and of determining, if possible, the ultimate source of the oil and the conditions which have controlled its migration and accumulation.

## PRESENT METHODS OF EXPLORATION AND DEVELOPMENT

The search for oil on the Gulf Coast at present consists of two phases, the search for new and hidden salt domes, and the drilling of old and plainly visible domes which have yet produced no oil.   As the occurrence of oil on the dome is very irregular, two dozen or more wells may be drilled before commercial supplies are found.   As already stated, these supplies may be phenomenally rich considering the area involved, but they soon become exhausted and the search must then be resumed.

Owing to the flat and monotonous topography of the Coastal Plain and to the fact that the oil-bearing rocks are overlain by a thick and structureless mantle of Pleistocene deposits, ordinary methods of geologic field work are of no avail.   Some domes are marked by prominent and unmistakable hills, but others appear only as low and inconspicuous mounds.   The so-called gas mounds, and even large ant hills, have sometimes been confused with true domes, and wells have been drilled on them. Other domes have no topographic reflection whatsoever and the search for such domes therefore resolves itself into a search for surface indications of oil; gas seeps, the so-called paraffin dirt, oil or asphalt exudations, sulfur or salt springs, and mud volcanoes are commonly regarded as the best indications.   Even if they lead to the discovery of a new salt dome, however, there is, of course, no assurance that the dome will prove productive of oil.

Quite recently several different and as yet untried methods for locating salt domes have been proposed.   Shaw's suggestion,[1] that the difference in specific gravity between the ordinary sediments and the salt, limestone, etc., composing the dome may be large enough to cause perceptible gravity anomalies, appears plausible and has received much favorable mention.   Rogers' work[2] on California oil-field waters has shown that the water associated with the oil is very different chemically from the ordinary ground water of the region, and it may be that the analysis

[1] E. W. Shaw: Possibility of Using Gravity Anomalies in the Search for Salt-dome Oil and Gas Pools. *Science*, N. S. (Dec. 7, 1917) **46**, 553–556.

[2] G. S. Rogers: Chemical Relations of the Oil-field Waters in San Joaquin Valley, California.   U. S. Geological Survey, *Bulletin* 653 (1917).

of water from shallow wells on the Gulf Coast will prove an aid in the discovery of salt domes.

Another method for locating hidden domes has been suggested by a high authority in geologic and physical science, whose name I am not as yet permitted to mention. This method is based on the fact that the domes are composed chiefly of salt, gypsum, and limestone, while the surrounding material is clay, silt, and gravel. Such difference in rock character is likely to be marked by at least a small deflection in the magnetic currents, and if this deflection exists, the presence of a dome might be detected by a sufficiently delicate instrument. The scientist I refer to has perfected an extremely delicate dipping needle, which, it is claimed, will easily detect magnetic differences only one-twentieth as strong as the ordinary dipping needle can detect. Knowing the capabilities of the gentleman, I am sure that as soon as he is ready to release his discovery he will receive every consideration, not only by leading scientists but also at the hands of the oil companies who are so vitally interested in the discovery of new domes.

### WIDE DISTRIBUTION OF OIL IN MEXICAN GULF REGION

Before discussing the source and origin of the salt-dome oil, a brief review of the known occurrences of oil in regions adjacent to the Gulf of Mexico, and especially under the Gulf, may be of interest. Showings of oil are found in many localities along the littoral of the Gulf of Mexico and the Caribbean Sea, including points in Venezuela, Colombia, Panama, Costa Rica, Honduras, Salvador, Guatemala, the Isthmus of Tehuantepec and the Tampico region of Mexico, and the coast of Texas and Louisiana. Indications of oil are also found in Cuba, the Isle of Pines, Panama, Haiti, and Santo Domingo, the Lesser Antilles, the Windward and Leeward Islands, and Trinidad.

In the Isthmus of Tehuantepec, oil in considerable quantity is produced from salt domes apparently similar to those in Texas and Louisiana. In the tremendously rich Tampico region, the oil is associated with igneous plugs and dikes, and in Cuba and the Isle of Pines, it occurs either in serpentine (or decomposed igneous rock) or near the contact between serpentine and highly folded sedimentary rocks. Evidences of igneous action are plentiful throughout the Mexican Gulf region except in that portion belonging to the United States.

Perhaps the most interesting indications of oil in the whole region are those which reveal its presence beneath the Gulf of Mexico itself. Lieut. John C. Soley, U. S. N.,[3] in his paper on "The Oil Fields of the Gulf of Mexico," gives a chart showing the location of all reported occurrences, and discusses their origin and geologic significance. This chart is re-

---

[3] *Scientific American Supplement* (Apr. 9, 1910) **69**, 229.

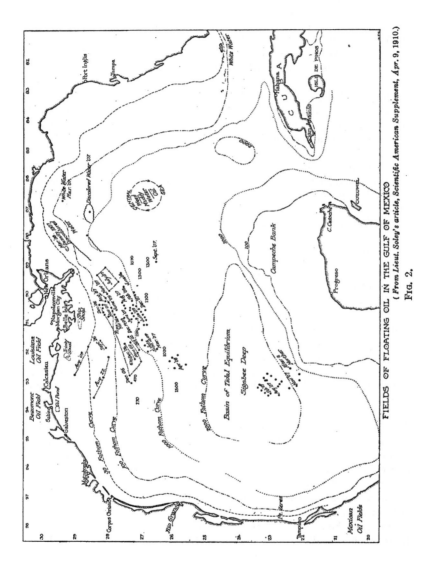

FIELDS OF FLOATING OIL IN THE GULF OF MEXICO

(*From Lieut. Soley's article, Scientific American Supplement, Apr. 9, 1910.*)

FIG. 2.

produced herewith (Fig. 2), together with extracts from his interesting paper.

*Location.*—The accompanying chart shows the position and extent of the oil field as located by the reports received during the last five years.[4] The area within which floating oil has been found extends from latitude 26 deg. N. to 28 deg. 30 min. N., and from longitude 89 deg. W. to 93 deg. W. As the location extends in the direction of the flow of the main current of the Gulf Stream, it is probable that the area shown is much larger than the actual area from which the oil escapes. The most active part of the field is about 120 miles south of Trinity Shoal off the Louisiana coast.

*History.*—The appearance of bitumens in various shapes in the Gulf of Mexico dates back to the earliest history of this country. . . . . The floating oil has been noticed from time to time and has been reported in the newspapers occasionally, but in later years records have been kept carefully so that it is possible to locate the origin of the field and to determine its extent. . . . .

Near the shore at Sabine Pass, west of the west jetty, there is a deposit of a viscid tarry liquid of bituminous odor, which impregnates the mud, and when the mud is stirred by wave action or mechanically, the globules of oil rise to the surface and spread out so that, in storms, there is no break, and the location has become celebrated as an oil pond. The appearance is irregular and the origin is uncertain; the oil probably occurs in a porous bed under the water, whose cover is sufficiently pervious to permit a constant though very slow seepage, sufficient to allow the overlying mud to become saturated.

A peculiar substance called sea wax is frequently found on the beaches between Sabine Pass and Matagorda. This is found in large cakes as large as 6 or 8 ft. long and 1 or 2 in. thick. It is undoubtedly a petroleum or asphaltum residuum and its presence points to the existence of springs of liquid bitumen somewhere in the Gulf.

The chart shows a number of oil spots south of Trinity Shoal from Ship Shoal to Sabine Bank, but these have only been reported in later years and are undoubtedly the refuse from the oil ships and barges when they clean tanks before going into Sabine Pass.

Along the Mexican coast and near Campeche Bank, floating oil is reported occasionally, but the first record appears in 1907, and it may come from either of two sources; from the northern field drifting before a gale or the overflow from the Mexican oil fields.

During the last 7 years, the reports from vessels that have passed through the oil field in the Gulf have been frequent, principally because attention has been especially directed to it; the positions where it has been reported are plotted on the chart, so that its limits have been determined with considerable accuracy, and the point of origin has been located almost exactly as being at latitude 27 deg. 30 min. N. and longitude 91 deg. 30 min. W. A number of vessels have reported from this position that the oil was seen bubbling on the surface, while the report on September 10, from the steamship "Comedian," described it particularly as coming up in three jets. It is generally described as dark yellow, sometimes so thick that a vessel passing through will hardly make a ripple on the water. The oil floats away from the source in large fields, but it absorbs oxygen from the air and evaporates quickly. On evaporating the oily residue of hydrocarbons disappears and the emulsion, mixing with the water, first has the appearance of slime which is generally reported as discolored water, and later it turns the water milky white, which appearance is often reported in the eastern part of the Gulf current and as far south as the Florida Reefs.

---

[4] Note that Lieut. Soley's paper was published in 1910.

*The Source.*—The location at 27 deg. 30 min. N. 91 deg. 30 min. W. is nearly south of the Trinity Shoal from which it is distant about 120 miles, and the chart shows depths of 600 fathoms. The Continental Shelf, with a general southeastward dip, is 90 miles in width from the Coastal Plain to the edge of the Shelf, at a depth of 100 fathoms (600 ft.); in 25 miles from the edge, the bottom drops off to 600 fathoms (3600 ft.), and in 20 miles more to 1000 fathoms (6000 ft). The source may be at any one of these depths between 100 fathoms and 1000 fathoms. . . .

*Genesis.*—The oils which are found on the borders of the Gulf and those which appear floating on the deep sea are so different in their characteristics that they seem to have been separated by nature into two classes, one having an organic and the other an inorganic origin.

The bituminous deposits along the shores, the methane of the mud lumps, the sea wax, the oil ponds, are probably all due to organic agencies. The vegetable and animal remains which have given rise to local deposits, the presence of the fucoids and diatoms in enormous quantities, which slowly decompose and sink into the oozes, the climatic conditions, and the temperature of the water, all contribute material for the change into hydrocarbons. But the oils in these deposits contain no soluble salts and no free sulfur. In some cases plants predominate in the formations, in others animal remains, and the resulting products vary as petroleum varies in different fields.

The floating oil in the deep water is probably of azoic origin. . . .

As the flow of oil at or near the same spot has been more or less continuous ever since the days of which we have any record, it is probable that there is an enormous oil pool underneath the Coastal Plain and the Continental Shelf of Louisiana and Texas, which lies at a depth of more than 2000 ft. At the time of the last subsidence of the crust, the impervious cover of the pool was probably borne down by the great weight of the superposed deposits until the cap-rock was broken, allowing the continuous escape of some oil which had accumulated below that stratum. Under any circumstances, it would be easy and natural for the oil to pass up through the water to the surface and be dissipated, but when we find the oil rising through a great depth of water with such violence as to displace the water in jets, it is evident that the pressure of gas behind the porous rocks is enormous.

The oil deposits of Louisiana which are known to be under the Coastal Plain, and those which are apparently under the extension of the Coastal Plain into the Continental Shelf, either immediately preceded or immediately followed the salt and sulfur solutions; they are all sulfur oils and, below the oil horizons, the sand is generally filled with salt water. Sulfur deposits are due to the action of volcanic gases; hydrocarbons are derived largely from the igneous rocks; chemical reactions in volcanoes need the coöperation of water and, in fact, steam, charged with carbon dioxid and sodium chlorid, is the mainspring of all volcanic phenomena; the gases of an expiring volcano consist of vapor, sulfurous acid and sulfureted hydrogen, hydrochloric acid and chlorid of ammonium. All these conditions are favorable to the deposits of the hydrocarbons which, associated as they are with volcanic and intrusive phenomena, are believed to be due to inorganic agencies.

It will be noted that Lieut. Soley in part ascribes the appearance of bodies of oil off the Louisiana coast to the escape of oil under great pressure along lines of structural weakness from a main reservoir below. This excellent paper was called to my attention only recently, though I have long been aware of the reports of sea captains concerning oil bodies on the Gulf and have long considered them evidence of the presence of oil and gas activities or disturbances beneath the sea bottom. In my opinion, the escape of these bodies of oil periodically must necessarily be ascribed to excessive gas pressure accumulating on a line of structural weakness in a main reservoir, so that when the pressure rises above or in excess of its overburden, it opens the reservoir and ejects a certain quantity of oil and gas until, automatically, it is closed again by hydrostatic pressure, to be again reopened at a later period, somewhat on the order of the intermittent geysers of Yellowstone Park. Similarly we know of oil wells flowing "by heads" at intervals of 5, 10, or more minutes, simply because of the fluctuating balance between hydrostatic pressure and gas pressure.

## Source and Origin of Salt-dome Oil

Owing to the fact that all our prospecting on the Gulf Coast has so far been confined to the upper surface of the salt domes, we know little of their complete structure. We have innumerable and conflicting theories as to the origin of the domes and of the oil found on them, but little positive knowledge. On one point, however, all geologists now agree, viz., that, as I contended in 1901, the salt domes are of secondary origin, and the oil was not formed in the beds from which it is now produced, but has migrated into them. Some believe that the oil originated fairly close to where it is now found, but others that it has ascended from a considerable depth. Many facts lead me to concur in the latter view.

When I began boring on Spindletop, in 1901, I never for a moment expected to find oil under the conditions which prevailed in Corsicana or the other stratified oil fields, for my exploratory work carried out previously on the barren salt domes of Louisiana had taught me that conditions were very different. I have always believed that the salt, gypsum, dolomite, etc., composing the dome are of secondary origin and that the oil and gas have migrated to their present position. In this connection, I had a long series of arguments[5] with Prof. R. T. Hill in 1901, and it was only, with difficulty that I convinced him of the existence of dome structure and of the secondary character of the dome-forming materials, facts which Hill and all other geologists now admit. In the present paper, therefore, I am advancing no new idea but am simply discussing and developing the views which I have held from the beginning.[6]

As already stated, when the oil reservoir on a dome is tapped, a large production is generally obtained, but this production in no case has

---

[5] My discussions with Hill have been fairly set forth by Harris. See G. D. Harris: Rock Salt, Its Origin, Geological Occurrences and Economic Importance in the State of Louisiana. Louisiana Geological Survey, *Bulletin* No. 7 (1908) 65–68.

[6] *Science*, N. S. (Aug. 30, 1901) **14**, 326–328.

proved of a permanent nature, but begins to decline after a certain time, be it a few days, a week, or longer.    The reason for this is very apparent. The gas and oil being confined under pressure in cavernous dolomite, on being released comes out with much force, but as soon as the pressure abates the flow of oil decreases in volume and ultimately stops.[7]    Putting the well under the beam may again increase production, but as there is no real oil sand, in the sense of the great sand encountered in West Viriginia or Pennsylvania, no permanent results can be obtained or expected    Furthermore, the area of a true dome is very small, say between ½ mile and 1½ miles in diameter, and though in some instances considerable oil has been produced on the slope of a dome, as at Humble, most of the Gulf Coast oil has been derived from the cap-rock or the irregular sands above it.

The gas pressures in many of the coastal fields are stupendous and have resulted in the destruction of many drilling rigs.    At Saratoga, a gas pocket blew the tools from the well and formed a crater several rods in diameter, in which was swallowed up the wreck of the derrick and machinery; at Spindletop, one wrecked the machinery.

Disastrous blow-outs have also occurred at Humble, and in Belle Isle a gas pressure of over 1000 lb. was encountered in the salt itself.    In the neighboring parish of Terrebonne, where no salt domes are known, enormous volumes of gas under high pressure have been discovered, permanent enough to warrant the laying of pipe lines to convey them to New Orleans, a project which is now in contemplation.    In the typical salt-dome fields, however, the blow-outs are due simply to pockets of gas, and they continue only until these pockets are exhausted.    With the exhaustion of the gas in the pocket, it is safe to conjecture that an equal volume of oil is permitted to creep upward and distribute itself along the easiest way of travel.

The well that encountered the great gas pressure in the salt at Belle Isle was drilled in 1907 by I. N. Knapp, through my instrumentality. This well, a partial record of which is given in *Bulletin* 429 of the United States Geological Survey, reached the depth of 3171 ft.    From numerous horizons large quantities of rock-salt cuttings were brought out with the bailer, and these, when dumped on the derrick floor, would jump like firecrackers or popcorn, leaving the floor covered with a scum of oil and paraffin.    Fig. 3 is an enlarged photograph of a specimen of this salt, showing the inclusions of oil and gas.    Indeed, the surface of the adjacent slush pond, 100 ft. in diameter, where the returns of the well were dumped, was covered during a cold spell with about an inch of solid

[7] There are records of a number of wells showing that after a lapse of a certain period they were found to be again producing, for a time, one or more hundred of barrels per day, thus proving that the oil traveled upward, refilling the exhausted locality.

paraffin. At about 2700 ft., the drill apparently passed out of salt into a calcareous rock like anhydrite and magnetite, and at 3171 ft. about 3 bbl. of reddish oil of about 40° Bé. gravity were secured while an effort was being made to bail the well (a change in color from canary yellow as at the upper surface was noted) when the casing collapsed and the hole had to be abandoned. The presence of gas under such great pressure in the salt itself down to 2700 ft., and of oil below, proves to my mind that these hydrocarbons must have been forced up from a considerable depth.

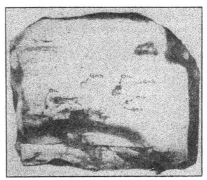

FIG. 3.—OIL AND GAS INCLUSIONS IN SALT CRYSTAL.

The initial source of the petroleum cannot be definitely determined at the present time, though in my opinion there are many facts, in addition to those already cited, which indicate that it is partly or largely of inorganic origin and that the domes themselves are certainly due to igneous forces. The sulfur present in the cap-rock of the domes, occasionally in great quantity, cannot be satisfactorily explained except as the result of dying vulcanism or solfatara fumarole action. At Belle Isle, a crevice in the salt mass 590 ft. deep is filled with sulfur-impregnated rock[8] and at Sulphur, Bryan Heights, and Damon Mound, several hundred feet of limestone or gypsum, richly impregnated with sulfur, are found. The suggestion that the sulfur is derived through the reduction of gypsum at moderate temperatures cannot be substantiated in the chemical laboratory, where a temperature of many hundred degrees Centigrade is required for the reduction of gypsum.

The drilling of almost every dome has also disclosed hollow tubular, or worm-shaped, concretions composed of pyrite. The presence of the small openings in the center of these tubes suggests that they were formed by hot sulfur gases forcing their way upward, which in cooling

[8] A. F. Lucas: Review of the Exploration at Belle Isle, Louisiana. *Trans.* (1917) **57,** 1034.

condensed and expanded and left the sulfur and iron in capricious forms resembling worms. The presence of sulfides of other metals—galena, sphalerite, chalcopyrite, etc.—also indicates connection with igneous forces.

Dr. Eugene M. Coste,[9] the exponent of the volcanic theory of the origin of oil, boldly takes the ground that petroleum is of inorganic origin and is the result of solfataric volcanic emanations. Dr. Coste reasons that animal organisms entombed in rock could not produce oil by distillation and that vegetable remains decompose into carboniferous matter, such as peat, lignite, coal, marsh gas, etc. Moreover, the distillation of organic remains takes place in nature soon after their deposition and they are thus lost largely as methane and carbon dioxide. The irregularity of gas pressures is also cited as evidence of the deep-seated origin of oil, for if gas pressure is really hydrostatic in character it should be fairly regular. On the other hand, in volcanic regions, gaseous, liquid and solid hydrocarbons, carbon dioxide, salt, and other chlorides, hydrogen sulfide, sulfur, gypsum, and hot calcic and siliceous waters are found. It is well established that the salt domes themselves are made up of secondary material which has been brought up along lines of weakness from great depth, and as the oil and water are generally quite hot it is entirely reasonable to suppose that they were also brought up from below. The high gas pressure encountered, far greater than hydrostatic pressure, also indicates deep-seated origin.

It is a well known fact, conceded by the best authorities, that petroleum occurs in rocks of all ages and nearly all kinds. I recently had occasion to examine, through the courtesy of Dr. Coste and Dr. Ledoux, a quartz crystal containing inclusions of oil and gas (Fig. 4). This occurrence was described in 1898 by Reese,[10] who states that Sir Humphrey Davy,[11] in a study of the fluid contents of cavities in rocks, mentions a single instance in which naphtha was found in a quartz cavity. Davy states that "the quartz crystals containing the oil and gas inclusions were found at Diamond Post Office, near Gunterville, Marshall County, Ala., near the Tennessee line. The inclusion presents the appearance of petroleum, in that it has the yellow-green fluorescence. A crystal from the same locality was crushed on filter paper which, having absorbed the oil, showed grease spots and gave the characteristic odor of petroleum and burned with a smoky flame. Another evidence of the nature of the

---

[9] E. M. Coste: Volcanic Origin of Natural Gas and Petroleum. *Journal*, Canadian Mining Institute (1903) 6, 73–128.

[10] C. R. Reese: Petroleum Inclusion in Quartz Crystals. *Journal*, American Chemical Society (Oct., 1898) 20, 795–797.

[11] Sir Humphrey Davy: On the State of Water and Aëriform Matter in Cavities Found in Certain Crystals. Royal Society of London, *Philosophical Transactions* (1822) 367–736.

liquid is that petroleum occurs in the neighborhood where the crystal was found."

G. S. Rogers,[12] in his discussion of Matteson's paper on oil prospecting in the Gulf Coast country, mentions "the discovery during the past summer of volcanic ash in the sediments around one or two of the domes; of a rock resembling a porphyry in a well at Damon Mound; and an undoubted igneous plug about 50 miles north of the salt-dome belt." These new discoveries lend added weight to the igneous theory, and the close association of oil with igneous rock in the neighboring regions of Mexico and Cuba should not be overlooked.

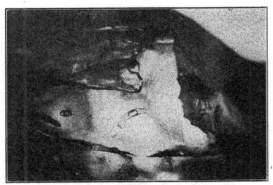

FIG. 4.—OIL AND GAS INCLUSIONS IN QUARTZ CRYSTAL. × 25.

The probability that the salt-dome oils, as well as the salt domes themselves, are a product of volcanic forces leads me to suggest a hypothesis involving the existence of laccoliths under the domes. According to the late Dr. G. K. Gilbert, laccoliths in the Henry Mountains of southern Utah have produced similar dome-like elevations at the surface. The molten rock has risen through a vertical pipe or fissure, but being unable to burst across the superincumbent bed has insinuated itself between the strata and by lifting them up has caused a doming visible at the surface. This explains the known structure of domes more adequately than would a sill or plug of igneous rock, such as is found in the Mexican fields.

As we are totally in the dark as to the position or age of the laccoliths, their form, composition or depth, that may have given rise to the domes, it is impossible to present in detail their relations with the salt, though I have attempted to do this in the accompanying hypothetical sketch (Fig. 5). This represents a cross-section of the Belle Isle dome which is very evidently located on a prominent line of structural weakness. This

[12] *Trans.* (1918) **59**, 480.

line is marked by the sulfur-filled channel in the salt above and by the presence of oil and gas under great pressure in the salt nearer the base of the dome.   Although the exact character and age of the strata beneath the salt but above the supposed laccolith are unknown, it is probable that they are domed in conformity with the known upper surface of the salt mass.

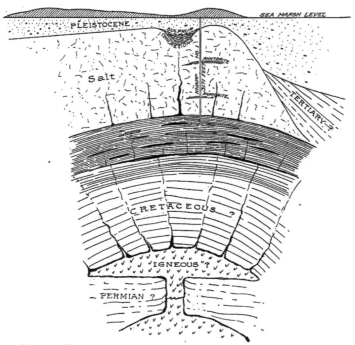

FIG. 5.—HYPOTHETICAL SKETCH SHOWING POSSIBLE RELATIONS OF
BELLE ISLE, LA., DOME.

## METHOD OF EXPLORATION PROPOSED

Whether the foregoing hypothesis of the origin of the domes is correct or not, the fact remains that the domes are there and it is for us to make the best of it.   When I drilled on Spindletop in 1900, I was told by some of the best geologists in the United States that this slight elevation and accompanying phenomena did not prove it a dome and had no signifi-cance, and that I was wasting my time and money; and similarly today many geologists doubtless disagree with my views that further supplies of oil are to be found in the domes at greater depth.   However, the vari-ous conflicting theories and opinions that have been offered by leading geologists since the problem sprang into existence in 1901 embolden me to suggest that theories be ignored and that a thoroughly practical test be made. We know positively only that the Coastal Plain is covered with Pleistocene sediments and that at points beneath it there exist salt domes carrying oil. As to how the domes originated or where the oil came from, we know nothing. I will therefore put my suggestions on a practical basis in the hope that if they are acted upon some definite knowledge of practical value will result.

Although several deep wells have been drilled on salt domes, none of them was located with reference to its value as a deep test. A well at Humble penetrated salt to 5300 ft., but it was intended to find oil at moderate depth and was not located with reference to lines of structural weakness, as a deep test should be. Similarly, many deep wells have been drilled into the sediments off the sides of various domes and have found no oil, but this, of course, proves nothing. At Damon Mound, for example, a well has just been drilled to 4620 ft. in order to test the theory that the big pool would be located off the dome at a deep level, but this well has found only some oil seeps and water. The presence of water in the sediments on the flanks of the domes is to be expected, while the chance of finding water in a well drilled in a central location on the dome proper would be either greatly minimized or disappear altogether.

Of all the domes with which I am familiar, Belle Isle seems to offer the greatest promise of rewarding a deep test. This dome rises prominently out of the marshes that surround it for many miles and is the farthest southeast of the series. As already stated, there is a pronounced line of weakness crossing this dome, as evidenced by a sulfur-filled channel in the salt, some 590 ft. deep. Far down in the salt itself, oil and gas under great pressure are found and the largest quantity was encountered at the greatest depth reached. It is very unfortunate that Mr. Knapp's well could not have been drilled deeper. At the beginning of operations, Mr. Knapp expected to go only about 1800 ft. and used a small rotary and standard rig. During the drilling, he found reason to become interested in reaching a greater depth and by great skill and careful management succeeded in reaching 3171 ft., though, owing to the heavy gas pressure, he was unable to secure good samples of the rock penetrated below about 2700 feet. At 3171 ft., he bailed out 3 bbl. of reddish paraffin oil of about 40° Bé. gravity, when the casing collapsed and the well had to be abandoned.

Another favorable indication at Belle Isle is the presence of crystals of sphalerite, galena, barite, native sulfur, etc. These crystals are sharp and well developed and have evidently been formed in place, for they show no signs of erosion. Although sulfur also occurs in great quantity at Bryan Heights, Sulphur, and Damon Mound, which presumably indicates that these domes are also located on partly open fissures, the actual presence of oil and gas in the salt itself at Belle Isle leads me to consider this dome most favorable for a deep research well.

It is, of course, impossible to state even approximately how much deeper a well would have to go at Belle Isle in order to tap the main reservoir of oil. Oil in paying quantity might be found only a few feet below where Knapp stopped and where the signs were most promising, though on the other hand it may lie several hundred or a thousand feet deeper, or more.

It has never been possible to drill on the Gulf Coast with the standard tools so extensively used in many oil fields, owing to the drift and quick-sand encountered and to the probability of blow-outs. At the time I drilled on Spindletop, the rotary method was in its infancy and during the past 15 years enormous improvements have been made. Extremely heavy tools are now manufactured and many extra appliances have been devised. The rotary or hydraulic machine of the early days of my pioneer work compares with the rotary of today as the early steam-engine compares with our modern triple-expansion engines.

In order to make as complete a test as possible, one should be prepared to drill a deep record well, even though on Belle Isle oil may be struck a short distance below where Knapp stopped. The project of drilling, if necessary, a 7000-ft. well is today perfectly feasible, for the Hope Natural Gas Co. of Pittsburgh is now drilling a well in West Virginia which on Feb. 19, 1918, had already reached a depth of 7363 ft. For a deep well on Belle Isle, I would recommend a specially constructed hydraulic rotary machine for the first 4000 or 5000 ft., with extra heavy cable tools to proceed farther, if indications warrant it.

The drilling of a well of this character should be carried on in co-öperation with a competent geologist of the United States Geological Survey, who could examine samples of the rock, oil, gas, and water encountered, and test the temperature at various depths.

In conclusion, I may state that I have no financial interest in exploit-ing this enterprise on Belle Isle or elsewhere, and have simply set forth my opinion in the hope that my early discoveries on this island and on the Coastal Plain may be still further extended for the benefit of any who care to accept the hazards involved, and for the advancement of science.

## DISCUSSION

G. Sherburne Rogers,* Washington, D. C. (written discussion†).— We are indebted to Captain Lucas for an interesting contribution to the literature of the salt-dome oil fields, and especially for his suggestion that a more aggressive method of prospecting be adopted. The occa-

---

* Geologist, U. S. Geol. Survey.          † Received Sept 3, 1918

sional discovery of new fields on the Gulf Coast is in itself sufficient to show that the oil possibilities of the region are by no means exhausted, and those who have studied the area most thoroughly are the most optimistic regarding its future. Nevertheless, the search for new fields, according to present methods, involves the expenditure of great sums of money annually, for the proportion of dry holes to producing wells drilled yearly in the Gulf Coast region is much greater than in most areas, even if all the wells drilled in what is commonly regarded as proved territory are included. Not only are the present methods of locating new domes very uncertain, but when a dome is discovered there is no assurance that it will be oil-bearing; or, if so, whether the oil occurs in the cap-rock, in the sands above it, or in some small isolated lens of sand far down on the flanks of the salt mass. As Captain Lucas points out, about all we positively know is that the domes are there and that some of them are oil-bearing; their origin is a matter of controversy and the original source of the oil is conjectural. Under these conditions, it is not surprising that the search for oil is conducted more or less blindly and that dry holes are common.

Many wells are drilled each year in an endeavor to find oil at depths of 2000 or 3000 ft. (609 to 914 m.) or less, in locations which most geologists would condemn without hesitation. As Captain Lucas suggests, the money spent in this way might be used to far better advantage in drilling one deep well in a carefully selected locality. Opinions may differ as to whether such a well would find oil, and many geologists will probably believe that the chances are against it, yet all will agree that the boring, whether it finds oil or not, will probably throw light on the origin of the domes and on the source and mode of accumulation of the oil, and will therefore be of great interest, not only to the scientist, but to the oil operator.

Our opinion of the chance that a deep well would have of finding new reservoirs of oil depends on our conception of the origin and course of migration of the oil. I cannot concur in Captain Lucas' belief that the domes are due directly to volcanic activity, and that the oil is of inorganic origin. In my opinion, the facts indicate that the salt has been squeezed up under the influence of pressure from deep-seated salt beds and that the domes are themselves salt plugs or laccoliths rather than the product of igneous plugs or laccoliths below.[1] According to either view, however, or to any of the numerous other theories extant, it is evident that a line of communication, or at least of weakness, exists between the general horizon at which the oil is now found and very much deeper and older beds.

Most geologists agree that the oil did not originate in the beds in which it now occurs, and believe that it has migrated up from the older strata.

---

[1] I have discussed this view in a paper entitled, "Intrusive Origin of the Gulf Coast Salt Domes," *Economic Geology* (1919) 14, 178.

Many, however, consider that the migration has been a short one, and at Humble, for example, there is a good reason to believe that the oil originated in the lower Tertiary Yegua formation which immediately underlies the deep oil sands.  On the other hand, there are several horizons in the Cretaceous which are petroliferous in northern Texas and Louisiana, and the Carboniferous horizons of the Mid-continent field are, of course, extremely rich.  Cretaceous rocks are known to underlie the salt-dome region, and there is every reason to suppose that Carboniferous formations are also present; whether or not these formations contain oil is, of course, conjectural.  If they ever were petroliferous in this region the oil must either have ascended along the salt core and collected on the salt domes or in some convenient reservoir en route, or it must still be present in the deeper formations today.

It might be argued that the pressures and temperatures to which the lower Cretaceous and Carboniferous rocks are now subjected would have resulted in the cracking and destruction of whatever oil they may have contained.  It appears, however, that unless the gaseous products of cracking are permitted to escape the process goes on only until an equilibrium is reached and that complete conversion of oil into gas and residuum would be attained only at far higher temperatures and pressures than would prevail.  In the Ventura County fields (California), oil is now produced from the Topa-Topa formation, which apparently was at one time buried under more than 20,000 ft. (6096 m.) of sediments; and in some of the Wyoming fields, also, oil is produced from horizons that were formerly deeply buried.  We do not know the greatest depths at which oil can exist, but they are certainly below levels yet reached by the drill.

The enormous quantity of oil per acre that has been produced at Spindletop suggests that either salt has a mysterious faculty of attracting oil from the sediments, as some European geologists believe, or that the gathering area is greater than would appear, or that some of the oil has ascended from unknown reservoirs below.

The occurrence of gas and oil in the salt at Belle Isle, as described by Captain Lucas, is an indication of deep oil deposits in that locality.  Belle Isle is flanked by a great thickness of Pleistocene sediments and the oil encountered at 3000 ft. is very possibly Tertiary oil that has migrated into the salt, though, on the other hand, it may have been driven from older beds and may have migrated some distance vertically.

Nothing is known regarding the shape of the stem or lower portion of the salt cores, or of its structural relations with the surrounding sediments.  Whether oil (if present) could be retained in porous strata abutting against the salt, or whether conditions are such that it must have migrated upward, is a matter for speculation.  In the writer's opinion, most of the salt cores probably incline up to a laccolithic or mushroom shape, and if so, a well starting on the dome and penetrating salt might pass out of the salt and into sediments lying against the stem of the salt core in such a way as to form excellent reservoirs. These matters can, of course, be settled only by drilling, and estimates of the chance of finding oil in a deep well must, in view of our present lack of knowledge, be based largely on personal opinion or conjecture. As more accessible possibilities are becoming exhausted, the question of deep drilling is becoming increasingly attractive. If a deep well finds oil at all, it is likely to discover rich reservoirs and will pay for itself indirectly through the light it will shed on the unsolved problems of the Gulf Coast oil fields.

## Geology of Oil Fields of North Central Texas

BY DORSEY HAGER,* TULSA, OKLA.

(Colorado Meeting, September, 1918)

NORTH CENTRAL Texas has recently become a center of interest for the oil men of America. The bringing in of the McClosky well at Ranger, Eastland County, and the shallow pool at Brownwood, Brown County, in 1917, has stimulated interest in this area to fever pitch. Oil men from all over the United States are now investing there. The area of present interest is shown by the accompanying map (Fig. 1).

The money spent in leases runs into millions of dollars. It is no exaggeration to say that a strip of country 200 miles (321 km.) long and 125 miles (201 km.) wide, comprising some 15,000,000 acres (6,070,310 ha.), has been leased practically solid at a cost for rentals and bonuses of at least $1 per acre. The test holes contracted for will certainly number 400, at a cost of at least $5,000,000. From what has been done in the past six months, $20,000,000 at least will be spent. To pay returns on this amount of money, new production to the extent of at least 12,000,000 bbl. must be obtained. At present, the production from new fields will not average over 5000 bbl. per day; three wells at Ranger are producing 3000 bbl.; 250 wells at Brownwood produce 1000 bbl.; and the Gray well, Coleman County, is as yet an unknown factor.

However, at Ranger there is every indication of developing a good pool covering from 1500 to 2000 acres (607 to 809 ha.), more or less, but the wells are deep, 3400 to 3800 ft. (1036 to 1158 m.), and cost $35,000 to $40,000 to drill. Large wells are necessary to pay for such expensive holes. As new wells are drilled, the gas pressure will be lowered rapidly, and large production need not be expected. For those oil men who expect a second Cushing or an Eldorado, Ranger holds little of promise.

At Brownwood, Brown County, there are some 250 shallow wells (depths from 200 to 350 ft.) averaging 4 to 7 bbl. per day. There is a chance of an extensive producing area for these shallow sands to the southwest, and the opening of several thousand acres of shallow oil territory, and also some promise of deeper oil horizons in the Ranger horizon, but probably all under 2500 ft.

At present, the lease brokers and speculators, and only a handful of oil men, have made any money. More fields must be developed, and it is more particularly with these possibilities that this paper deals. Lack of water for drilling purposes has undoubtedly held back development so far this year; this part of Texas has had a drouth for two years and there is an actual scarcity of water. Geological investigations so far undertaken cover nearly all the area outlined, and while part of the

---

* Petroleum Geologist and Engineer.

area has not been closely worked, it is not too early to make some definite statements as to the probable prospective pools.

## GENERAL GEOLOGY

The present oil pools lie in the Carboniferous belt of rocks (Fig. 1) consisting of sandstones, limestones, and shales belonging to the divisions shown in the accompanying outline (Table 1). The names are peculiar to central Texas, and are derived from towns where these beds were first studied in detail by early investigators.

TABLE 1.—*Stratigraphy of Central Texas*

| | | | | Thickness feet |
|---|---|---|---|---|
| Mesozoic | Cretaceous | Comanche beds | Sandstones, shales, limestones | 200–500 |
| Carboniferous | Permian | Albany—Wichita | Principally limestones, shales | 1200 |
| | Pennsylvanian | Cisco Canyon Strawn | Limestones, shales Shales, limestones Principally shales, sandstones | 900 600–800 2500–4000 |
| | Mississippian | Bend series, } 800–1000 ft. } | Smithwick shale Marble Falls limestone Lower Bend shale | 400 450 50 |

For the relation of these beds to one another see Fig. 2.

The Bend series carries the Lower Bend shale, the Marble Falls limestone, locally called the Bend limestone, and the Smithwick shales.

The Bend limestone is probably the producing horizon at Ranger, Eastland County; at Parks, near Breckenridge, Stephens County; at Caddo, Stephens County; at Coleman, Coleman County, in the Magnolia deep test; and of the gas in the Magee well southwest of Bangs, in Brown County. The productive horizon in the Bend limestone may be at the top, at the bottom, or in the middle of the limestone series, depending upon the porosity of several beds; the Bend limestone is not homogeneous, but comprises limestones which vary greatly in porosity and hardness. In places, the limestones are parted by thin shale bodies.

In the writer's opinion, the "sands" or "reservoirs" are most probably thin beds of cherty conglomerate. At the outcrop of the Bend south of San Saba and also Richlands Springs, thin beds of very porous chert are found at the base of the Bend, as well as through the lower part of the formation.

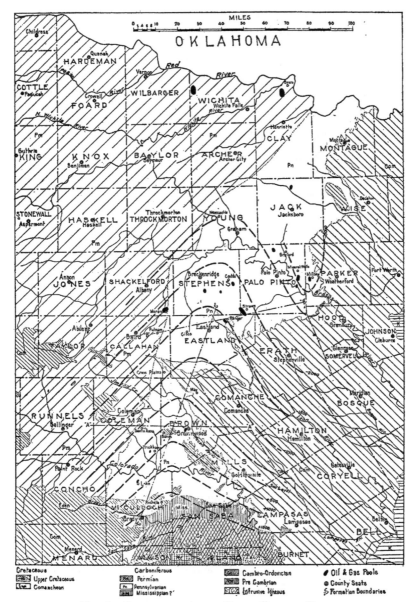

Fig. 1.—Geological map of North Central Texas.
(Hager, Bates & Kemp, Tulsa, Okla.)

The Smithwick shale series, or upper Bend, produces gas in the Chestnut well, 9 miles (14 km.) south of Mineral Wells, Palo Pinto County, the oil at Trickham, Coleman County, and the oil in the Gray well northeast of Coleman, Coleman County. The sandy members in the shale series are the productive horizons.

The Strawn horizon produces at Strawn, in Palo Pinto County; at Millsap, Parker County; and at Moran in Shackleford County. This horizon carries good sands, or, more properly speaking, sandstones.

The Canyon series produces oil at Brownwood, Brown County; at Lohn, McCulloch County; and a little oil near Abilene, in Taylor County.

The Cisco horizon produces the deep wells at Petrolia and at Electra in Clay and Wichita Counties, and the shallow oil at Moran, in Shackleford County. This oil occurs in sandstones.

The Albany-Wichita beds produce the shallow oil at Electra and Petrolia. The producing horizons here are sandstones.

TABLE 2.—*Specific Gravities of Some Texas Oils*

| FIELDS | DEGREES BÉ | REMARKS |
|---|---|---|
| Strawn | 40.0 | |
| Caddo | 39.5 | |
| Ranger | 39.0 | Dark green color. |
| Brownwood | 40.0 | Dark amber color |
| Millsap | 44.0 | Pure amber color; 50 per cent. gasoline. |

From this review, there are at least six distinct oil horizons, each containing separate sands, from one to three in number, in the Carboniferous beds, notably the Bend limestone, the Smithwick shales, the Strawn, the Canyon, the Cisco, and the Albany-Wichita beds, which are all producers at different places.

TABLE 3.—*Producing Sands*

| OIL FIELD AND COUNTY | PRODUCING HORIZONS | DEPTHS TO PAY SAND, FEET |
|---|---|---|
| Abilene, Taylor | Cisco (Penn.) | 2100 |
| Brownwood, Brown | Canyon (Penn.) | 90–350 |
| Caddo, Stephens | Bend (Miss.) | 3150 |
| Coleman, Coleman | Smithwick (Miss.) | 2350 |
| | Bend (Miss.) | 3400 |
| Electra, Wichita | Albany (Penn.) | 200–1000 |
| Gray, Coleman | Smithwick (Miss.) | 2412 |
| Millsap, Parker | Strawn (Penn.) | 2150 |
| Moran, Shackleford | Cisco (Penn.) | 350–2150 |
| Parks, Stephens | Bend (Miss.) | 3150 |
| Petrolia, Clay | Albany (Permian) | 200 |
| | Cisco (Penn.) | 1800 |
| Ranger, Eastland | Bend (Miss.) | 3428 |
| Strawn, Palo Pinto | Strawn (Penn.) | 700–1100 |
| Santa Anna, Coleman | Smithwick (Miss.) | 1560 |
| Trickham, Coleman | Smithwick (Miss.) | 1020–1200 |

*Probable Stratigraphic Range of Oil Fields*

There is no logical reason why the basal Permian beds should not produce oil; below it are the Upper Pennsylvanian beds, which, 75 miles (120 km.) west of Santa Anna, should be reached at a depth of 3000 ft. (914 m.).   In fact, there seems to be no reason for contesting the possibilities of oil over an area at least 125 miles in width and 200 miles long.

The only argument against production in these upper beds is the scarcity of sandstones in the basal Permian and in the Cisco, upper Pennsylvanian.   Some sands are present, however; their persistence is the only question.   However, the Permian horizon does produce in the northern part of the area described; notably at Electra and Petrolia, and there is no good reason for assuming that it will not produce in the southern part of the area.

*The Cretaceous Unconformity*

A feature of much importance is the Cretaceous unconformity, so pronounced in this area in Texas.   Fig. 2 shows the relation of the over-

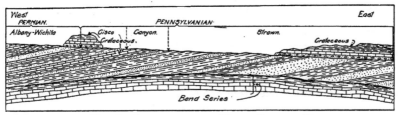

FIG. 2.—STRATIGRAPHY OF NORTH CENTRAL TEXAS.

lying Cretaceous beds to the underlying Pennsylvanian and Mississippian.   The gentle regional dip of the Cretaceous beds is eastward, and of the Pennsylvanian westward.   The Mississippian corresponds with neither series of beds.

The Cretaceous caps a large area of the Strawn beds in the eastern part of this area.   One can see no reason why production may not be found in the Strawn beds hidden under the Cretaceous cover. In several places, folds carry right into the Cretaceous beds, notably the Bull Creek and the Ebony anticlines in Mills County.   Other folds probably lie hidden under the Cretaceous, but the only way to discover such folds is to prospect for them with the drill; a set of shallow tests might uncover something that might later justify deep testing.

### GEOLOGICAL STRUCTURE

In no other areas are ordinary haphazard wildcatting methods so excusable as here, since the geologist can be of little assistance in guiding initial prospecting.

The most significant feature of all these oil pools is their relation to structural folding of the beds; every pool so far discovered occurs on a fold. This folding is of three general types: (1) closed domes (Fig. 3); (2) plunging anticlines, called noses (Fig. 4); (3) terraces (Fig. 5). True closed structure is scarce, although some closed folds have been discovered. Electra and Petrolia may be classed as of true dome types; Moran, Strawn, Ranger, and Coleman are terraces on the surface. Other folds of the terrace type are found near Brownwood. In Mills and San Saba Counties, true closed domes are known. These folds have full closures of 50 to 80 ft., and are unusually large for Texas.

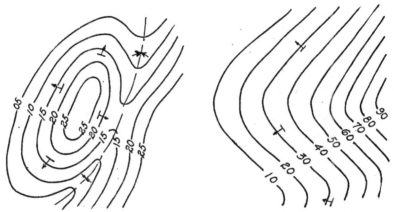

FIG. 3.—CLOSED DOME.                 FIG. 4.—PLUNGING ANTICLINE.

From all the evidence at hand, the folds that may reasonably be expected to develop the pay fields will be mainly of the types shown in Figs. 4 and 5. The true dome type, which is the usual producing structure in South Texas, Oklahoma, Kansas, Wyoming, and California, is uncommon, though present, in northern Texas. The Saline dome, as found in the Gulf Coast area, is absent here. The folds thus far discovered seem to have been formed by lateral pressures, not by the peculiar upward bulging in connection with salt cores, so common further south.

### Size of Structures

The folds vary in size from small wrinkles of no commercial value up to folds several thousand acres in size; however, the known prospective folds are mostly small, ranging from 320 to 2000 acres of favorable land. At least 50 folds having prospective value have already been discovered and leased in this broad area.

### Direction of Folding

Three sets of folds exist in Texas, having axes trending respectively: (1) Northeast–southwest; (2) northwest–southeast; (3) north–south. Most of the folds extend in a northeast-southwest direction, while the northwest-southeast are next in number. The former, roughly parallel to the strike of the beds, might be called strike folds, while the latter can be called dip folds. As the normal or regional dip of the surface beds is northwest throughout the Pennsylvanian and Permian areas, the important reversal on all folds is to the southeast and the northeast.

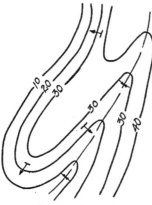

Fig. 5.—Terrace.

### Persistency of Folding with Depth

It is interesting to note that on most of the folds already drilled through the Smithwick shales and the Bend limestone, commercial oil or gas has been found. This is notably the case at Ranger, at Breckenridge, at Caddo, at Coleman, at Trickham, and the Magee Field south of Bangs. This seems to show that folding in the Upper Pennsylvanian persists into the Mississippian horizon; probably following a line of earlier folding.

### Relation of Smithwick and Bend Series to Overlying Beds

One of the most interesting and important subjects at present is the relation of the underlying Bend horizon to the overlying beds. This oil horizon is the main deep prospective measure in the minds of the oil operators, and is perhaps too much emphasized at present.

The Pennsylvanian beds overlying the Bend series dip west at a rate of 60 to 90 ft. per mile (11.4 to 17.1 m. per km.), the dips diminishing from east to west; but the underlying Bend does not dip accordingly, as shown by the records of the deep wells thus far drilled to the Bend; also, from conditions near the outcrop of the Bend (see cross-sections *AA*, *BB* and *CC*, Figs. 6, 7 and 8 respectively).

### Bend Arch

An important structural feature apparently underlies the Pennsylvanian beds, according to evidence derived from outcrop and well records. This feature is what the writer designates the Bend Arch, a large plunging fold which extends northeastward from the outcropping beds in San Saba and McCulloch Counties (see contours on map). The cross-section (Figs. 6, 7, and 8) shows the general relationship of the Bend to the overlying beds.

The importance of this fold lies in the fact that one looks for local oil accumulation in the minor folds upon it, and also that it materially affects estimates of depths. The contour lines on the map are drawn with 500-ft. (152-m.) intervals, and will unquestionably be modified by additional data. They are based on the top of the Bend

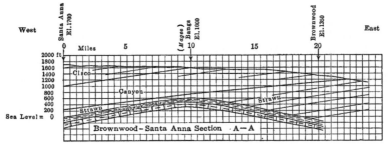

Fig. 6.

limestone as a datum, and are corrected for mean sea level. The writer expects faulting and perhaps other large folds to appear with further detailed study. The average north dip in the Bend is 32 ft. per mile (6 m. per km.).

Section *AA* indicates a cross-section east and west across the fold. The Strawn beds outcrop 3½ miles (5.6 km.) east of Brownwood. At

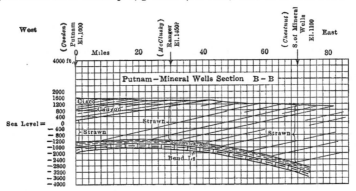

Fig. 7.

Brownwood, there are 1100 ft. (335 m.) of Strawn shown by well records; near Bangs, 200 ft. (61 m.); and at Santa Anna, 200 ft. Wells at Coleman, northwest of Santa Anna, show about 700 ft. (213 m.) of Strawn. There is a marked convergence between the Strawn from Brownwood to Bangs, and then the Strawn runs nearly parallel to the Bend and does not thicken appreciably until west of Santa Anna.

It will be noticed, in the cross-sections, that very little Strawn is present on top of the big arch in the Bend, especially in the area from Bangs south to Rochelle. Along this line, production, if any, will come from the Canyon, the Smithwick shales, and the Bend horizons. East of this line, the Strawn may produce. Wells so far tested west of this fold have produced oil and gas in the Smithwick shales at Santa Anna and at Coleman, and below the Bend limestone at Coleman.

Section *BB* shows the condition from the Chestnut well, 9½ miles (15 km.) south of Mineral Wells, across Ranger to Putnam. The Chestnut well shows 3600 ft. (1097 m.) of Strawn; the Ranger well 2200 ft. (670 m.); and the well at Putnam, 1400 ft. (427 m.) more or less.

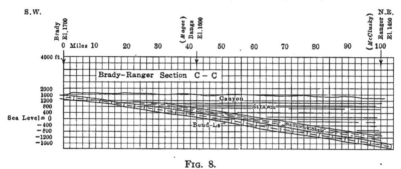

FIG. 8.

Section *CC* shows the relation of the beds from Brady, in McCulloch County, to Ranger. At Brady, the Canyon lies unconformably on the Bend. Further north, near Bangs, 200 ft. of Strawn is shown, while at Ranger 2200 ft. of Strawn is present, showing a marked thickening of the Strawn beds northward along the axis of the fold.

A study of the cross-sections *AA*, Fig. 6; *BB*, Fig. 7; and *CC*, Fig. 8, is of great interest. One is struck at once with the marked fold in the Bend showing in *AA* and *BB* and the changes in the thickness of the Strawn. The Strawn normally dips northwest. Its strike varies from N. 30° E. to N. 50° E. The strike of the Canyon and Cisco beds is N. 20° E. with a northwesterly dip.

### Possibility of Oil Migration from Smithwick Shales to Strawn Horizons

Another point that may be of interest is the possibility that the Smithwick shale, a black carbonaceous shale, is the source of oil in the Strawn. The Strawn beds lie unconformably on the Smithwick shales, and it would be a simple matter for oil to leave the Smithwick and enter the Strawn at the point of contact, without crossing bedding planes (see Fig. 2). Folds in the Strawn beds will form traps or reservoirs for this migrating oil, and are worth testing, not only in the Smithwick and Bend horizons but also in the Strawn. In the writer's judgment, new pools will be opened in the Strawn series.

## Summary

1. At least six producing horizons, from Mississippian to Permian, are found in this area of Texas. Each general horizon carries one or more pay sands.

2. The areal extent of the probable pools comprises a strip 200 miles (321 km.) long and 125 miles (201 km.) wide.

3. The Cretaceous beds cover some good prospective land which is worth prospecting. The geologist can be of but little assistance there.

4. All pools so far discovered are on structures such as domes, terraces, and plunging anticlines.

5. The folds are not large, 300 to 2000 acres (121 to 809 ha.).

6. At least 50 untested folds are known to geologists.

7. The direction of folding is mainly northeast-southwest. Some northwest-southeast and north-south folding is found.

8. The Pennsylvanian beds lie unconformably on the Mississippian series.

9. The Bend Arch is an important new structural feature, deduced from drill records and outcrops. It may assist in governing accumulation, and influences estimates of depths to the Bend producing horizon.

10. The migration of oil from Smithwick shales to Strawn beds is possibly due to the angular unconformity between the two series. This may largely account for oil in the Strawn series.

11. The above review indicates a possibility of numerous small pools in north central Texas, varying in depth from 250 to 4000 ft. (76 to 1219 m.). Stratigraphic and structural conditions are favorable, and most of the folds so far tested have proved productive.

## Acknowledgments

In presenting this paper, the writer wishes to thank Messrs. Leon Pepperberg, C. T. Griswold, Robt. T. Hill, Lee Hager, and Dilworth Hager, for valuable suggestions and information, and also to thank Robt. Jordan of Mineral Wells and Mr. Robinson of Santa Anna, for courtesy in furnishing well records. The writer has drawn freely from the *Texas State Survey Bulletin* for nomenclature and description of the geologic formations.

Much of the matter is new, however, and some of the correlations and conclusions may be open to question, so the writer invites discussion to clear up the doubtful points.

## DISCUSSION

WALLACE E. PRATT, Wichita Falls, Tex. (written discussion*).—
Mr. Hager has touched upon interesting problems related to the broad
subject with which he has entitled his paper.

As to the existence of the important structural feature in the buried
Bend series, which Mr. Hager constructs from well logs, and designates
as the "Bend Arch;" there seems to be general agreement among a
number of observers. Indeed, M. G. Cheney[1] and Robt. T. Hill[2] have
each published structural maps of the Bend Arch that do not differ
essentially from the map shown in Mr. Hager's Fig. 1.

The reflection, in the overlying, unconformable Cretaceous, of
certain folds in the Pennsylvanian, with which the petroleum accumulation has to do, is an important observation on the part of Mr. Hager,
since, as he says, much critical territory is covered by Cretaceous, and it
is highly desirable that the underlying Pennsylvanian structure be
mapped, if possible.

If the folds in the Pennsylvanian are of post-Cretaceous age, then it
should be possible, in spite of the intervening unconformity, to determine approximately the position of buried folds beneath the Cretaceous
overlap. The idea of post-Cretaceous folding would not be out of accord
with the suggestion by Carroll H. Wegemann that the folding of the
Permian beds in southern Oklahoma was still in progress as late as the
Tertiary.[3] But most observers have related the folds in the Pennsylvanian and Permian beds at the surface, both in North Texas and southern Oklahoma, to the process of emergence of the land area at the end of
the Permian.

It seems more probable that the reflection of Pennsylvanian folds in
the unconformable Cretaceous which Mr. Hager observed is due to long
continued persistence of stresses which were largely relieved at the close
of the Permian, than that the principal period of folding was post-Cretaceous; this being the case, it seems likely that only folds of greatest
intensity in the Pennsylvanian will be decipherable in the Cretaceous
overburden, and that many promising structural features are so concealed by the Cretaceous as to escape detection except by drilling.

Mr. Hager concludes that the direction of the axes of most of the
folds of the North Central Texas region is northeast-southwest, a smaller
number having a northwest-southeast alignment. He touches here
upon a subject intimately related to former shore-lines, old land areas,
etc., which may well be made the theme of other specific contributions

* Received June 17, 1918.
[1] *Oil Trade Journal* (May, 1918) 9, No. 5, 75.    [2] *Ibid.* (June, 1918) 9, No. 6, 89.
[3] U. S. Geological Survey, *Bulletin* 602 (1915) 34.

to the geology of North Central Texas. Exception may be taken to the general statement as to the relative number of folds on each of the two axial lines, it being admitted, however, that folds do occur along both sets of axes. It is my observation that a large majority of the folds in this area trend northwest, or west-northwest; in this category, also, may be placed most of the important folds, both from the viewpoint of size and of past production. These are the "dip-folds" of Hager, and their alignment follows the Wichita-Arbuckle trend so clearly displayed in the producing folds of southern Oklahoma, adjacent to North Texas.

It may be remarked, however, that many of the folds, the axes of which do trend northeast-southwest, are found in that part of the general area of which Mr. Hager writes, where he seems to have made most of his observations, namely, the southern and eastern parts of North Central Texas as a whole. In this connection, the suggestion arises that in the southern part of the general area, near the Llano-Burnett uplift, the tectonic lines which have been found to be most conspicuous in this old land mass, that is, northeast-southwest lines, govern also the folding in the Pennsylvanian rocks; whereas, farther north, and over a greater part of the whole area, the Wichita-Arbuckle alignment (west-northwest) is predominant.

## Oil in Southern Tamaulipas, Mexico

BY EZEQUIEL ORDOÑEZ, MEXICO CITY, MEXICO

(Colorado Meeting, September, 1918)

THE great activity with which the oil resources of the northern Cantons of the State of Veracruz have been developed has largely resulted from the great success obtained by the important explorations carried out since 1902 in the Ebano district in the North, and several years later near Tuxpam, in the South.  It must not be thought that these successes were quickly obtained, since the preliminary studies by good experts, the acquisition of what were thought to be the most promising coast lands, the extensive clearing thereof, and finally the number of unsuccessful drillings at great depth, signified an original outlay of some tens of millions of dollars before finding commercial oil.

The discovery of oil in industrial quantities at Cerro de la Pez, Ebano district, and the famous Dos Bocas gusher some years later, which was burned, once and for all made famous the Veracruz coast, where today is concentrated the Mexican output of mineral oil.  In 1917, this production was nearly 61,000,000 bbl.  It is well l nown that this output represents only a fraction of what the wells in actual production can furnish, because with adequate means of transportation and storage, the present extraction could be somewhat over 300,000,000 bbl. a year.

These enormous potential oil resources of the Veracruz coast proceed from a relatively small number of wells, scattered over a few oil fields, separated from each other by large unexplored areas, wherein may be found other favorable fields which, in time, will undoubtedly become just as great centers of oil production.

Experts who know our Gulf Coast believe that lands with commercial oil resources lie not only between the Pánuco and Tuxpan Rivers, but that indications are that the oil zone of Mexico, with more or less interruption, takes in the coast zone from Northern Tamaulipas to the foot of the Sierra Madre of Chiapas, and the banks of the Usumacinta River.  This generalization is not merely the outcome of optimism, but is the result of the persistency with which one finds, throughout this Gulf Coast, the most usual characteristics upon which we depend for recognizing oil-bearing lands.

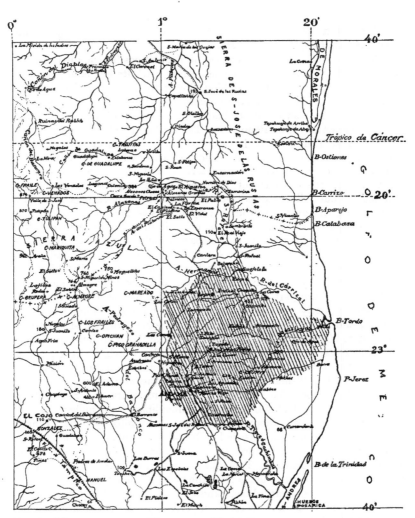

FIG. 1.—MAP OF SOUTHERN TAMAULIPAS, MEX. SHADED AREA REPRESENTS BASALT

In this short paper, we can discuss only the oil possibilities which, in our opinion, exist in that part of the State of Tamaulipas extending southward from the Soto la Marina River and its tributaries, that is to say, south of the San José de las Rusias property, which everyone is watching at present, because the Corona company is making certain very important explorations there.

The wide coastal belt and littoral of Tamaulipas, in the district near Tampico, to the north of the Tamesi River, with its low lands, lagoons and belt of sandstone hills reaching almost to the San Andres lagoon, sheltered on the sea side by sandy dunes, greatly resembles the sea shore and coast of the Canton of Ozuluama to the South of Tampico, in the State of Veracruz. The Pánuco and Tamesi Rivers, at their joint outlet to the sea at Tampico bar, have cut through a chain of hills which is merely an extensive strip of Neozoic formation which has largely disappeared over a great part of the interior. Before the Pánuco opening existed, from the outskirts of Tampico to the sea, the waters of the Pánuco undoubtedly emptied into the Gulf through various channels, representing the overflow from a very extensive lagoon stretching for many kilometers inland, which is now drained. This explains the estuary material covering the entire coast for 50 km. (31 mi.) inland, including oyster banks and other sea shells as far as Topila, and those estuary deposits and aluvions which have adhered to the first high undulations to the west and at a considerable distance from Tampico.

In the southern part of the State of Tamaulipas, there is a vast area of low-lying gently undulating coast, measuring over 150 km. from east to west, and about the same from north to south, with an elevation of from 100 to 200 m. (328 to 656 ft.) above sea level. This low-lying coast continues toward the northwest by a wide valley which, rising gently, is used by the railway between Tampico and Ciudad Victoria. This flat southern coast of Tamaulipas is bounded on the west by the Sierra del Abra, or Tanchipa. To the north, the Buena Vista and Sierra Azul ranges, forming part of the Sierra de Tamaulipas, form higher valleys. Important units of the Sierra Azul are joined by hills to a coast range, not very high, called San José de las Rusias. The flat coast, mentioned above, extends south to Veracruz territory beyond the Pánuco and Tamesi Rivers. On this coast, in Tamaulipas, there are only two important breaks which are of importance from the point of view of oil possibilities; these are: (1) The beautiful peak known as Bernal de Horcasitas, which resembles a sharp and narrow crest supported by a cone with a very wide base, the top of the rocky crest being over 1000 m. above sea level. (2) The group of mountains situated near the village of Presas Aldama, dominated by the Cerro Cautivo, the top of which is some 650 m. above sea level. The slopes of this mountainous mass approach the sea shore between Tordo bar and the sandy Point of Jerez.

Both the Bernal and Cautivo mountains, separated by a distance of 70 km., are volcanic and their bases are formed by extensive flows of basaltic lava; both mountains are very recent. Bernal has a relatively short formation history, implying lava flows in close succession from one or few craters, after which eruptions a lava cap formed over the principal crater and filled it; the crater has disappeared by erosion and the lava neck has a graceful appearance as seen from the distance. The Cautivo mountain mass is very different. Here we find a group of volcanoes having a much longer period of activity, during which numerous lava flows overlapped one another, building the whole mass higher and higher. On the summits can be identified various craters, some of considerable dimensions. Among the most recent ones we may mention one of explosive type situated on the southern slopes of the Zapotal crest. This crater has a diameter of about 1.5 km. and is about 100 m. deep at its lowest edge, with an extensive flat surface in the bottom. The walls of the crater are of basalt, basaltic breccia and tufa, with limestone, sandstone and marl coming from the adjoining formation underlying the eruptive rock.

The importance of the eruptive center of the Cerro Cautivo, in the neighborhood of the village of Presas Aldama, is not to be measured by the number and height of the summits and craters which compose it, but rather by the area which the lava from these volcanoes has covered in the form of a thin and uniform mesa. The lava from the large and small volcanoes, spreading to the north and northwest of Presas Aldama, cover an area of not less than 2500 sq. km. (965 sq. mi.). Several overlapping flows sustain the crests and craters of Cautivo, but the greatest spread of the lava field stretches in the form of a slightly elevated and uniform plateau, abruptly cut at its edges by erosion. To the northwest and west the lava crust almost reaches the southern slopes of the Sierra Azul, and to the north the San Rafael River washed away part of it, thus forming a wide valley whence this river reaches the sea bordering the southern end of the San Jose de las Rusias range.

As regards the secondary hills and small volcanoes, subordinate in origin to the eruptive center of Cautivo, they usually occur at the edge of the lava crust, but are also to be found on top of it. Among the small hills scattered throughout this vast low plateau, only a few may be mentioned, such as the cone with an open crater, known as Cerrito del Maiz, to the south near Presas Aldama; the three consecutive hills, Los Tres Hermanos, to the west of the Lagarto ranch; the cones with craters or dome-shaped hills at the edges of the lava at the foot of the Sierra Azul; and lastly, the remarkable little cones on the ranches of Bejarano, Real Viejo, El Sombrerito, La Muralla, etc., etc. The majority of these basaltic cones are important as suggesting the possibility of encountering oil near them, as usually occurs in the Veracruz oil districts.

It may be of interest to state briefly some data furnished us by the study of these basaltic cones in relation to the Tertiary land formation underlying them.

The geological formation of the southern Tamaulipas çoast is very uniform, both in physical constitution and in structure. The entire formation is Tertiary (Eocene). The top layers are of limestone alternating with phosphatic lime, sandstone, and thin beds of shale. This formation may be studied very well near Rancho del Cojo, on the flanks of the hills cut by erosion, forming that broad and low depression which extends between the heights of Cojo and those of Barranco ranch east of Gonzalez station of the Tampico-Monterey R. R. In various other parts of that area the formation of El Cojo is found with good exposures. The strata, as is everywhere the case on the southern Tamaulipas plain, are almost horizontal or in successive undulations and domes, invariably having a monoclinal structure slightly dipping to the east. Below the limestone strata of the top of the formation, thin limestone beds, with shales, are found overlying the very thick shale formation, known as the Mendez shales, which has a thickness sometimes exceeding 3000 ft. (914 m.) and rests upon the San Felipe formation or upon Cretaceous limestone.

From the village of Presas Aldama to the north the country is flat over the basaltic lava or malpais for nearly 40 km. passing the ranches of Zanapa, Lagarto, and others, to near Rancho Bejarano, where the basaltic plateau ends and Eocene rocks are again exposed by the wide erosion valley of the San Rafael River.

A well-aligned series of almost isolated high mountains, with peaks broken into fantastic forms, arises from the eastern foot of Tamaulipas range, forming an advanced ridge the base of which is touched by the malpais. This row of high peaks limits the Coastal plain on the west. Prominent among these mountains are Jerez, Platano, Plateros, and Cerro Gordo or Alazanes, all having elevations exceeding 1500 ft. (460 m.) above sea level. A little further on from the Bejarano ranch, toward the Real Viejo farmstead, the river flows between hills at the foot of the high mountains to the west and the coast range of San José de las Rusias to the east. At the junction of the eastern hills with the low ramifications of Las Rusias range, at Lomas de la Encarnación, a broad gap divides the waters of the San Rafael river from those of the important Soto la Marina river.

Leaving the northern edge of the basaltic plateau between La Coma, Guajolote, and vicinity of the Bejarano ranches, the Tertiary shales appear throughout the valley of the San Rafael river with here and there small basaltic cones. Near certain of these cones there are some oil seepages, which suggest that in this San Rafael river valley there is a wide area of great prospective value. Even more, it is probable that the malpais of Presas Aldama, in its eastern and southern portions, covers a large area of excellent oil ground, a prolongation of that of the San Rafael valley.

Toward the bank bordering the San Rafael river bed, or at the edge of the basaltic plateau which also bounds the valley of this river, there are excellent cuts where not only the Tertiary rocks may be observed, but also the basaltic necks in their relation to those sedimentary rocks. Among the most instructive cuts carved by the river, I would mention a low hill near Real Viejo ranch, 10 km. to the north of Bejarano ranch, and on the right bank of the San Rafael river. A wall has here been cut 20 m. high in the Tertiary shales, at the same time cutting the basaltic hill rising from the river bed to a height of 50 m. This wall plainly shows the contact between the lava and the shales, clearly marking the line of separation between these rocks. The Mendez beds, made of yellow and red shales intercalated with fine-grained reddish sandstone, in slight undulations, show a gentle uplift at the junction with the basaltic rock. Near

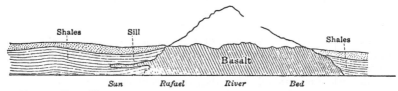

FIG. 2.—REAL VIEJO LAVA PLUG AND SURROUNDING SHALES SHOWING SLIGHT MOVE-MENT OF THE SHALES UPWARD AT THE CONTACT WITH THE LAVA, AND SOME META-MORPHISM. A SMALL SILL IS NOTICED IN THE ILLUSTRATION.

this lava are some layers of shale metamorphosed into a hard schistose rock resembling flint. This metamorphism, produced by heat upon contact with the lava, is commonly found in the Veracruz oil fields where there are basaltic cones. We have seen this at Juan Casiano, La Pez, Cerro Azul and at many other places on that coast. A sill about 4 ft. (1.2 m.) thick of lava from the Real Viejo hill enters into the shales, and though this sill is not very long in the cut, it must extend much farther in places which are not visible. These lava intrusions in the form of sills in the shale formation of our oil lands are more frequent than is generally believed, and many drillers are familiar with the appearance of these lava strata in certain oil districts, even at considerable depths and at great distances from basaltic plugs.

This cut of the San Rafael river in the Real Viejo ranch illustrates various important facts to which I would draw attention. First, the sedimentary strata, upon contact with a basaltic neck, are slightly moved upward with the lava plug. Second, a slight metamorphism of the sedimentary rock occurs at its contact with the lava, due to heat. Third, the lava intrudes as sills in sedimentary rocks.

On various occasions, I have already stated that plugs of lava traversing the Tertiary strata of the Gulf Coast oil fields do not perceptibly
move those strata, and we take as our reason for this belief the observation
of what happens in the explosive craters or *xalapascos* to which type
many of the diminutive volcanoes of our eastern coast belong.  Our
observation is opposed to the statements of Garfias and Hawley in a
recent article[1] in which the authors, on the strength of observations by
Geikie and other writers on certain volcanic districts, maintain that a
synclinal movement of the sedimentary rocks takes place around the
necks in our oil lands, and that this synclinal curve, together with other
phenomena which they try to explain, facilitates accumulations of petroleum in the vicinity of the necks.  This phenomenon might occur, but it
is not fully proved in our oil fields.  Whereas we continue to assert the
almost immovable position or a gentle uplift of sedimentary rocks around
the neck; the formation of broken zones at the contact of the lava and
the sedimentary rocks, greater at some places than at others; and lastly,
the absorption of sedimentary material by the lavas and hot waters, gases,
and vapors, all determining the formation of extensive hollow or porous
spaces very recently filled with liquid hydrocarbons.  The roots of the
plugs of lava, ramified like those of a plant, have served as channels to
bring the mineral oil to the hollow or porous spaces prepared during the
appearance of the lava plugs.  It also frequently happens that large accumulations of mineral oil, when they come in limestone subjacent to the
Mendez shales and San Felipe formation, fill cavities in these limestones
which have been caused exclusively by circulating waters without the
help of any volcanic phenomena.  In the Pánuco district it has happened
that, while drilling a hard limestone, suddenly the drill has been lost in
a big cave.

If oil has sometimes accumulated in old hollow spaces made by water
circulation in the limestone, there is room for the belief that the slight
upheaval and breaking caused in limestone by the effect of explosions,
or by the sudden passage of lava, has produced fractures which, even at
a great distance, may facilitate the communication between the drill
holes and an oil seam, so the connection between oil and volcanic
phenomenon is purely mechanical.

In front of the Real Viejo ranch, the San Rafael river bed passes
at the foot of a low hill, called Sombrerito, formed at its base of shales
and sandstones, a continuation of the Real Viejo Eocene shales.  Higher
up, shales with a greater number of thin limestone layers are seen, then
the cavernous limestone of the Cojo formation.  Erosion has respected
there these top formations, due to a thick lava capping with a ruined
crater which forms a small indent.  Between the basaltic lava and the

---

[1] Funnel and Anticlinal Ring Structure Associated with Igneous Intrusions in
the Mexican Oil Fields.  *Trans.* (1917) **57**, 1071.

limestone, on the northwest slope of this hill, there is a small oil seepage which is carried away by waters of the arroyo during the rainy season.

I have already said that the ample valley of the San Rafael east of Real Viejo and Bejarano is a most promising field for the occurrence of mineral oil. The existence of oil seepages at various points in the wide lava field of Presas Aldama indicates that this oil region is very extensive, and that undoubtedly the malpais conceals good ground which, but for the lava crust, would show oil seepages.

At Jobo ranch in the eastern part of the malpais, at no great distance from the main mass of Cautivo Mountain, a *chapopote* seepage appears through thick lava. To the southeast of Jobo, toward the edge of the lava field, are other seepages at Sabino ranch.

Undoubtedly, even the good Mexican oil lands, far from the centers of actual exploitation, to become productive require assiduous and detailed geological studies, drilling and exploration, implying considerable investments.

## DISCUSSION

V. R. Garfias,* Palo Alto, Cal. (written discussion†).—Regarding the statement of Mr. Ordoñez, on page 538, concerning the synclinal curving of sedimentary beds caused by the extrusion of volcanic necks, that "this phenomenon might occur, but it is not fully proved in our fields," I beg to reply that the funnel and anticlinal-ring structure *is actually associated* with igneous intrusions in the Mexican oil fields, as we were able to ascertain by plotting an underground contour map and making a structural model of the oil-bearing beds of the Ebano field.

E. De Golyer, New York, N. Y. (written discussion‡).—The paper under discussion is not only of interest in connection with its subject but it is a contribution of considerable importance to our knowledge of the possible effect of igneous rocks upon the accumulation of petroleum in the Mexican fields.

Señor Ordoñez[1] has long held that the intrusion of igneous rocks in the Mexican fields has had no important structural effect in the accumulation of petroleum and does not seem to have found it necessary to change his opinion because of his studies in Tamaulipas.

Clapp,[2] as a result of his most recent consideration of the Mexican fields, has concluded that:

* Geologist and Civil Engineer.                    † Received Aug. 19, 1918.
‡ Received Sept. 6, 1918.
[1] Sobre Algunos Ejemplos Probables de Tubos de Erupcion. *Memorias* Sociedad Cientifica "Antonio Alzate" (1904–05) **22**, 141–150. The Oil Fields of Mexico. *Trans.* (1914) **50**, 859–863.
[2] Revision of the Structural Classification of Petroleum and Natural Gas Fields. *Bulletin*, Geological Society of America (1917) **28**, 586.

At the base of the upheavals (around the plugs) and surrounding them in close proximity the Tamasopo limestone and overlying formations form pockets or places of catchment where large deposits of oil have accumulated. In the Tamasopo limestone and the San Felipe beds these oil deposits were presumably concentrated from surrounding portions of the same strata, owing to the upheavals mentioned; possibly with the assistance of heat.

The presence of the oil accumulations surrounding the plugs is sometimes, although not always, evinced by large seepages of oil. Some cases are known where the lower beds actually reach the surface and a true quaquaversal structure exists. Whether this is common has been doubted, but it is certain that definite doming does exist surrounding some of the plugs. At any rate, it is a fact that where the plugs exist pockets of oil have accumulated, and the conical plugs themselves may be considered as quaquaversal structures.

Garfias and Hawley,[3] following and developing the theory advanced by Garfias[4] in 1912, apparently largely as a result of his observation of wells in northern Alazan and a consideration of the analogy suggested by driving a nail through a book, have tried to find in the action of the intrusion a reason for the common occurrence of oil in the vicinity of igneous plugs in Mexico. They propose a theory of a circular anticlinal ring and funnel structure surrounding igneous necks and argue by analogy, citing numerous sections and examples of such structure, though they fail to cite any Mexican occurrences, merely concluding that there is no "question of the existence of these conditions in the Mexican fields."

I have long held with Ordoñez that important occurrences of petroleum in the Mexican fields, under the conditions described by Garfias and Hawley, and by Clapp, have not yet been proved to exist and, although I am not yet prepared to accept wholly the Ordoñez theory that the "roots of the plugs of lava, ramified like those of a plant, have served as channels to bring the mineral oil to the hollow or porous spaces prepared during the appearance of the lava plugs," I agree that one important effect of the intrusions has been the formation of pore space by brecciation and metamorphism. I have further suggested that the greatest importance of the porous zones thus formed is that they cut across the sedimentary strata and so provide for transverse migration of petroleum.

The accompanying sketch of the Tepetate-Casiano pool, showing the surface relation of the outcrop of igneous rock[5] to producing wells and dry holes, is instructive in this connection, since the region is one of considerable igneous activity. No evidence of doming similar to that suggested by Clapp, nor of anticlinal ring and funnel structure similar to that suggested by Garfias and Hawley, has yet been seen in this field. In the sketch, only those wells which have been completed to the Tama-

---

[3] *Trans.* (1917) **57**, 1071–1082.

[4] The Effect of Igneous Intrusions on the Accumulation of Oil in Northeastern Mexico. *Journal of Geology* (1912) **20**, 666–672.

[5] Igneous rocks mapped by C. W. Hamilton.

sopo limestone are shown. The occurrence of the deeper dry holes or salt-water wells on both flanks, and of shallower producing wells along a narrow, unbroken, N.-S. striking zone are the distinctive features of this field as thus far developed by exploration. The surface geology does not show a marked fold corresponding to the ridge in the Tamasopo limestone that is shown by drilling operations. The structure has been ten-

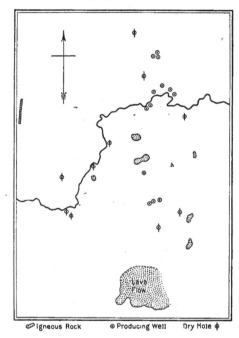

*Igneous Rock*    ⊚ *Producing Well*    *Dry Hole* φ

SKETCH MAP OF TEPETATE-CASIANO OIL FIELD.

tatively explained as being the result of folding which started in the Tamasopo before the deposition of younger rocks. The strike of the field parallel to that of the dyke to the west and parallel to the line of weakness shown by the three domes or dyke outcroppings to the east of the field indicate that faults, the existence of which cannot be determined at the surface because of the lack of good rock exposures, may have been the controlling factor in determining the structure of this field, one of the most important in Mexico.

A few kilometers to the south is another important field, the Los Naranjos, in which the discovery well was located by the writer. This field is a broad dome; the producing wells are located on its crest, and the only known igneous rocks consist of a dyke on the south flank and another on the eastern flank.

The oil industry is to be congratulated in having a clear recognition, by so distinguished a Mexican as Señor Ordoñez, of the preliminary preparation and the risks incident to exploratory drilling. Attention was early attracted to the petroleum seepages of Tamaulipas, and particularly to those of the San José de las Rusias region. In 1864, permission to exploit petroliferous substances at San José de las Rusias and at Chapopote, near Aldama, were granted to a certain Ildefonse Lopez, and similar permission was granted in 1865, to one Parades to exploit *chapopote* at Carancitos.

In 1873, Alajandro Prieto in his book, "Historia, Geografia, y Estadistica del Estado de Tamaulipas," describes the profusion of *chapopote* seepages in southern Tamaulipas and concludes (p. 264): "This class of product, which is encountered in such abundance in Tamaulipas, should form some day, by itself alone, a branch of industry which will doubtless offer greatly to the prosperity of these communities."

In 1889, a well was drilled to a depth 40 m. near San José de las Rusias by a certain Manuel Flores and is said to have gushed oil.

The oil field department of S. Pearson & Son, in 1907, commenced drilling near Los Esteros in southern Tamaulipas and completed four wells before abandoning the enterprise as unproductive. Subsequently the American International Fuel and Petroleum Co. drilled 8 wells in the same region. In 1909–1911, the Tamesi Asphalt and Petroleum Co. drilled two wells near the same place. All of these wells were unsuccessful. In 1913, Tampico Oil, S. A., drilled a dry hole near Jopoy, on the Tampico River.

In 1910, the Texas Mexican Asphalt and Petroleum Co. moved a small drilling machine to the San José de las Rusias property and drilled a number of shallow wells, none of which was successful. Finally, a standard rig was brought in and a shallow well was drilled at a seepage rising along a dyke near Encarnacion. The rig was moved a short distance and a well was drilled to a depth of 2274 ft. (690 m.) and abandoned. The Corona Petroleum Co. (Royal Dutch-Shell group) then became interested, took over the lease and, after careful surveys by competent geologists, commenced drilling in 1914. To date, they have completed and abandoned as dry holes five wells having depths of 3450, 4000, 3496, 3030 and 3370 ft. (1052, 1219, 1066, 924, 1027 m.) respectively, and are now drilling two other wells.

During the past year, the Mexican Gulf Oil Co. completed and abandoned as a dry hole, at a depth of 3280 ft. (999 m.), a well on the Sabino Gordo property near Chapopote, just south of Aldama. This well was drilled near a seepage at a small hill of rock metamorphosed by igneous intrusions, and resulted in a salt-water well.

It is quite probable that in the purchase of leasehold, drilling of unproductive wells, and for various incidentals in an unsuccessful attempt to develop a productive oil field in the State of Tamaulipas, as much as a million dollars has already been expended, most of it under competent technical direction.    The region has been recognized as probable oil territory since the earliest time and yet not a single commercially ˙productive well has been encountered.    Nor has it been proved that the state will not be an important oil producer at some future date.    The conclusion reached in the paper under discussion that "even the good Mexican oil lands, far from the centers of actual exploitation, to become productive require assiduous and detailed geological studies, drilling and exploration, implying a considerable investment" would seem to be irresistible.

Economic and Geologic Conditions Pertaining to Occurrence of Oil in
North Argentine-Bolivian Field of South America

BY STANLEY C. HEROLD,* PITTSBURGH, PA.

(New York Meeting, February, 1919)

CONSIDERABLE interest has been shown, during recent years, in the
possibilities of developing oil fields in the South American Republics,
now that the exhaustion of our present fields can be seen in the not very
distant future. The demand for fuel oil and its products has been increasing far more rapidly than our increment of production during the
last few years. Increase of consumption has been largely curtailed by
the lack of sufficient supply.

The main factors that have contributed to our previous neglect of
the southern fields have been as follows:

(a) Lack of sufficient demand for the product.

(b) Competition with producing fields, having large production,
low cost, and better geographical position, such as Tampico, Mexico.

(c) Greater interest in home fields with ready market at hand.

(d) Our ignorance of southern economic and geologic conditions,
due to the distance between those areas and our financial centers.

(e) Economic conditions in the southern countries, necessitating
high costs of exploration and exploitation of the fields.

(f) Our lack of understanding of the business methods of the Latin-
American.

It must be borne in mind that the oil-development work done to the
present time in South America has, with one exception,[1] been undertaken
upon lands which possessed direct evidence of oil in seepages or asphaltic
deposits, occurring either immediately upon the lands or in their close
vicinity. The development of oil fields in South America is therefore
in the pioneer stage. It is not at all improbable that "hidden fields"
will be uncovered in future work, as has occurred notably in the mid-
continental field of this country within recent years.

## LOCATION

The areas grouped as a unit under the title of North Argentine-
Bolivian oil field extend in a narrow belt from the north-central part of

---

* Consulting Oil Geologist.

[1] The accidental discovery of oil in boring for water at Comodoro Rivadavia,
Argentina.

the Province of Salta, Argentina, northward into the central part of Bolivia. This belt lies in a direction a few degrees east of north, slightly curved with its concave flexure to the west, passing through the border

FIG. 1.

town of Yacuiva at latitude 22° South and longitude 63° 30′ West of Greenwich. It extends from 18° South to 23° South latitude.

To the west of the field lie the frontal ranges of the Andes System, beyond which extend high rugged mountains, deep narrow gorges and high plateaus, culminating in the crest of the continental range.    Toward the east extends the great expansive plain, reaching over northeast Argentine, southeast Bolivia, northern Paraguay, and into southern Brazil.

Notable seepages of oil occur in the following political subdivisions of the two countries:

Argentina:
Department of Orán, Province of Salta.
Bolivia:
Department of Tarija.
Department of Sucre (Chuquisaca).
Department of Santa Cruz.

The location of this field has given rise to the name of "Sub-Andean Petroliferous Zone," a term used by some writers.

### LITERATURE

The literature on the oil regions of South America is very limited, particularly that in the English language.    A résumé has recently been compiled in a valuable paper by Frederick G. Clapp.[2]

As to the literature in the Spanish language, among the most important publications are the reports of Dr. Guido Bonarelli,[3] of Buenos Aires, who, as government geologist, has made diligent study of many petroliferous areas in his country and in southeastern Bolivia.    Chemical analyses of the Sub-Andean petroleum have been made by Dr. Ernesto Longobardi.[4]    Reports of Bodenbender[5] and Brackebusch[6] give valuable information regarding the geology of adjoining regions in the provinces of Salta and Jujuy.    Steinmann[7] has made examination of the Sierra de Aguaragüe region, particularly in southeastern Bolivia.    All the above reports have been freely consulted in the preparation of this paper.

---

[2] *Trans.* (1917) **57**, 914.

[3] *a*, G. Bonarelli: Las Sierras Subandinas del Alto y Aguaragüe y Los Yacimientos Petroliferos del Distrito Minero de Tartagal, Provincia de Salta. *Republica Argentina. Anales del Ministerio de Agricultura Sección Geologia, Mineralogia y Minera* (1913) **8**, Núm. 4.

*b*, La Estructura Geológica y Los Yacimientos Petroliferos del Distrito Minero de Orán, Provincia de Salta. *Ministerio de Agricultura Dirección General de Minas* (1914). Núm. 9. Ser. B.

[4] Los Petroleos Subandinos y sus Relaciones Geo-quimicas. *Anales,* Sociedad Quimica Argentina (1915) **3**, 423.

[5] Informe sobre una exploración geológica en la región de Orán (Provincia de Salta). *Ministerio de Agricultura Boletin* (1906) **4**, Núm. 4, 5.

[6] Brackebusch: Estudios sobre la formación petrolífera de Jujuy. *Boletin*, Academia Nacional de Ciencia de Córdoba (1882) **5**, 137–252.

[7] *a*, G. Steinmann, H. Hoek and A. V. Bistram. Zur Geol. des sudöstl. Boliviens. Tentralbl. für M. G. u P., Stuttgart, 1904.

*b*, H. Hoek and G. Steinmann: Erläuterung zur Routenkarte der Exped. Steinm. Hoek, V. Bistram in d. Anden v. Bolivien. *Petermans Mittheilungen* (1906) **1**.

## Nomenclature

Some ambiguity apparently exists among references to the North Argentine-Bolivian region, due to varied spelling and multiplicity of names for the same locality. The following corrections and correlations are given with the hope of clarifying such ambiguities:

*Capiazuti Wells.*—Situated 4 miles (6.4 km.) northwest of the village of Aguaray. They are in Argentina, on Capiazuti Creek, 25 miles (40.2 km.) south of the Bolivian border. These wells are sometimes referred to as "Las Minas de Aguaray." Not to be confused with Cuarazuti.

*Iquira Springs.*—Seepages 2 miles (3.1 km.) south of Capiazuti, on Iquira Creek. Often included in the term "Las Minas de Aguaray."

*Tartagal Wells.*—Situated 3½ miles (5.6 km.) southwest of the village of Tartagal, Department of Orán, on Galarza Creek. Sometimes termed "Las Minas de Tobar," or "Las Minas de Tartagal."

*Ipaguazu Springs.*—Seepages of petroleum on Agua Salada Creek, 12 miles (19.3 km.) east-northeast of Yacuiva. Synonymous with "Lomas de Ipaguazu." Included in the "Gran Chaco" seepages. Incorrectly written Ipaguaciu or Ipahuaso.

*Yacuiva.*—The Bolivian frontier town, though 2 miles south of the border recognized between the two countries; viz., 22° south latitude. Also correctly written Yacuiba, but incorrectly Ayacuiba.

*Aguaragüe.*—The frontal range of mountains commencing a few miles southwest of Tartagal, extending northward into Bolivia. Sections of the range are sometimes designated, "Cerros de Tartagal," "Cordon de Itaque," etc. The northern extension comprises the "Sierras de Santa Cruz." Improperly written Aguaraygus, Aguaraygua, etc.

*Cuarazuti.*—A creek in the Department of Tarija, Bolivia, north of Yacuiva. Incorrectly written Guarazuti or Kuarazuti.

*Gran Chaco.*—The great plains extending over northern Paraguay, southeastern Bolivia and northern Argentina. The Province of the Gran Chaco of Bolivia includes the eastern parts of the departments of Tarija and Sucre (Chuquisaca). Geographically divided into Chaco Oriental or Chaco Paraguayo, Chaco Central and Chaco Austral, the latter two largely included within the boundary of Argentina.

## Topography

The North Argentine-Bolivian oil region lies along a boundary line between rough mountainous country to the west and extensive plains to the east. A series of mountain ranges, with parallel trend from north to south, stretches westward with increasing altitude until the crest of the great Andes Range is reached. Between these ranges are long narrow river valleys with occasional river terraces, which frequently serve as locations for villages at altitudes varying from 10,000 to 12,500 ft. (3048 to 3810 m.) above sea level.

The entire mountainous relief has been produced by the same dynamic forces which caused the uplifting of the main Andes Range, modified by subsequent weathering and stream erosion. The easternmost range, known as the Sierra de Aguaragüe, with its northern extension in the Sierras de Santa Cruz, constitutes the frontal range, extending from north to south for a distance of approximately 300 miles (483 km.), with altitudes between 1000 and 3000 ft. (305 and 914 m.) above the level of the adjoining plains.

Extending eastward from the frontal range lie the great plains (El Gran Chaco) of northern Argentine, southeastern Bolivia, northern Paraguay and adjoining parts of Brazil. With a slight inclination to the southeast, they stand at elevations varying from 1600 ft. (488 m.) in the south to 2400 ft. (732 m.) above sea level in the north. Slight vertical displacement has apparently taken place in various sections of the plains, as is shown by the flexures in the stratified deposits of Quaternary age. This movement has caused a distinct topography of low rolling hills, partially eroded by subsequent torrential stream water. Typical of this topography are the hills 6 miles (9.6 km.) to the northeast of Aguaray, and also the hills of Ipaguazu northeast of Yacuiva. They stand between 200 and 300 ft. (61 and 91 m.) above the surface of the plains.

Numerous small streams flow from the deep narrow gorges in the mountains and enter upon the plains with slackened gradient. Most of these proceed for only a few miles, as they lose themselves in the deep loose sand deposits so frequent and extensive in the flat country. Where the water has the opportunity to collect in volume before leaving the rocky beds in the mountains, more persistent streams flow with a southeasterly course and find their way eventually to the Paraguay River. Thus it is with the Pilcomayo and Bermejo Rivers.

The extreme northern end of the region is within the water-shed of the Amazon River, through the successive tributaries Río Grande, Río Mamore, and Río Madeira.

## Climate

The climate of this region is subtropical, with temperature variation between 40° F. in winter to 90° F. in summer. Occasionally more extreme temperatures of 30° and 115° are experienced, though generally for short duration only. The humidity of the summer atmosphere is rather oppressive, and rainfall at this season is heavy. The winter months are rainless, to such an extent that resident natives suffer from lack of good water for domestic purposes. As the region is south of the equator, the seasons are the reverse of those in the northern hemisphere.

It is during the summer months that the great migration of insects and animals takes place from the tropical jungles in the north. The most conspicuous insects are the mosquitoes, flies, and grasshoppers. There are several varieties of flies that are inconvenient, but their bite, fortunately, is not serious. Some malaria exists, though it is not at all so common as in the tropical countries. On the whole, very little sickness prevails, so the region may be classified as a healthy one.

The great pest of the country is the grasshopper, *longosta*, flying from the north in hordes of millions, destroying all vegetation that may happen to lie within their path. No satisfactory method of combating them has yet been discovered. Several plans are now under investigation and experimentation.

### INHABITANTS AND INDUSTRIES

The inhabitants may be divided into three groups: Saltanean, Bolivian, and Indian. In addition to these are found a very small proportion of foreigners, such as Spanish, English, Italian, Syrian, French, and North American.

The Saltanean of high class is of Spanish descent, proud, aristocratic, courteous, alert and prosperous, possessing these qualities perhaps to a higher degree than most other natives encountered by the writer in his travels south of the Canal. He has distinguishing characteristics in customs and language that are readily recognized by Argentinos from other regions. The laboring or peon class has a considerable admixture of Indian blood, and furnishes faithful and trustworthy workers. Wages are standardized at one peso (44 c. gold) per day.

The high-class Bolivian differs little from the Saltanean. He shows, however, a greater proportion of Indian blood as a rule, in having a darker skin. Peon classes of both countries have wandered across the boundaries to some extent; this cannot be said of the higher class.

Here is the home of the Chiriguano and Mataco Indian. Both retain most of their tribal customs, though the former has undoubtedly advanced by his contact with civilization. The Chiriguano lives in houses equal to those of the civilized peon and he himself is quite trustworthy. The Mataco, however, is more deceitful and dishonest. He will kill to steal if he feels assured that he will not be apprehended. As he has no permanent abode he wanders by groups or families, and camps by the roadside where the close of day may chance to find him. Both tribes, men, women and children, are employed as cane cutters during the sugar harvest, their employment being arranged for in advance by the planter's traveling agents.

The raising of sugar, corn and cattle constitutes the chief occupation of the people. While most of the mountainous regions are uninhabited, the valleys and plains are well utilized in cultivation or grazing. As a rule, the plantation and cattle owners live in the towns and cities and leave the property in charge of trusted employees. Some enterprises are run on an extensive scale, notably the sugar plantations with large refineries under the same management. Thousands of acres of land are devoted to this product.

The custom of living in small villages is very popular with the poor native. Every few miles are groups of houses ranging in size from 3 to 50 habitations. By so living the social tendency is satisfied and the native need not be far from his work.

## TIMBER

A large portion of this section of the country is covered with timber, especially in the mountains. True tropical jungle does not exist in the south, due to the low winter temperatures. As one travels toward the north, however, a gradual change in the nature of the forest growth is noted, the softer woods of the south being replaced by the more valuable hard woods. Palms also become more conspicuous toward the north.

## APPROACH AND ACCESSIBILITY

The terminal at the town of Embarcación, in Argentina, is the nearest approach made by a railroad to the oil region. This road is a narrow-gage line operated under the management of the "Ferrocarril Central Norte," connecting the cities of Tucumán, Salta, and Embarcación, with a junction point between the latter two at Güemes. The broad-gage line of the company "Ferrocarril Central Argentina" connects Tucumán with Buenos Aires.

Another approach to the field is by way of Asunción, the capital of Paraguay, which is directly connected by rail with Buenos Aires. The distance from Asunción to Yacuiva is approximately 455 miles (732 km.) overland, whereas Embarcación is but 114 miles (183 km.) Furthermore, Asunción is on the opposite side of the Río Paraguay from the field; the river at this point would be difficult to bridge.

Over the route to Embarcación, through passenger express service runs twice weekly between Buenos Aires and Salta, with direct connection at Tucumán. The trains carry sleeping and dining accommodations and the trip requires 36 hr. to Salta. Between Salta and Embarcación there is daily service by slow train, requiring 14 hours.

As for freight service, considerable time is consumed and expense incurred by the necessary transfer of material from cars of one gage to those of the other at Tucumán. Rates at present are high.

Rail distances between points are as follows:

| | KILOMETERS | MILES |
|---|---|---|
| Buenos Aires to Tucumán | 1344 | 835 |
| Tucumán to Güemes | 294 | 182.6 |
| Güemes to Salta | 46 | 28.6 |
| Güemes to Embarcación | 223 | 138.2 |
| Buenos Aires to Embarcación | 1861 | 1155.8 |

No Bolivian railroads reach the interior from the direction of La Paz; therefore this route is impossible as an approach for freight.

Away from the railroad lines, freight is carried by two-wheeled carts. There is considerable traffic between points in Argentina and Bolivia over the road leading northward from Embarcación. This road, in the dry season, is in bad condition, and worse in the wet season, when it becomes impassable. The river crossings are heavy, due to deep loose deposits of sand in some and large boulders in others. Weak bridges span the narrower gullies in some parts.

Sand areas are so numerous on the flat plains that the road must cross some of them. The clay sections of the road are very badly rutted by the heavily laden carts.

An extension of the railroad from Embarcación would represent a simple engineering problem.

The Pilcomayo and Bermejo Rivers have beds about 400 ft. (122 m.) wide from bank to bank. Between these banks the stream meanders from one side to the other, though on occasions of excessive rainfall, the entire bed may be covered. These streams can not be reckoned on for transportation.

The Paraguay River, with the Paraná, into which the former empties, is navigable throughout the year. River boats drawing from 4 to 6 ft. ply regularly between Asunción and Buenos Aires, a distance of approximately 745 miles (1200 km.). The river is a free trading route and would, therefore, possess that advantage over the rail to Buenos Aires. As above stated, the distance between the field and the river at Asunción is approximately 455 miles (732 km.).

## OIL SEEPAGES

Seepages of oil, and asphalt deposits, have been known to exist in this part of the continent for many years. The native Indians had used the oil for medicinal purposes and knew of its inflammability before the arrival of the first Spanish explorers. To this day, small quantities are collected by both Indians and civilized people for domestic use, as the oil is often of such high quality that it may be used in crude lamps in its natural state.

Several springs are very persistent in their flow, though they give only two or three quarts, individually, per day. Among the best known are the following, named in geographical order from south to north:

| NAME OF CREEK | LOCATION |
|---|---|
| Galarza | Department of Oran (Argentina) |
| Iquira | Department of Oran (Argentina) |
| Agua Salada (Ipaquazu) | Department of Tarija (Bolivia) |
| Los Monos (Villamontes) | Department of Tarija (Bolivia) |
| Caigua (Villamontes) | Department of Tarija (Bolivia) |
| Peima | Department of Tarija (Bolivia) |
| Oquita | Department of Sucre (Bolivia) |
| Mandiyuti | Department of Sucre (Bolivia) |
| Espejos (Santa Cruz) | Department of Santa Cruz (Bolivia) |

With the exception of the Agua Salada spring, the seepages lie along a north-south line bordering the eastern flank of the Sierra de Aguaragüe, between Galarza Creek, near Tartagal in Argentina, and Espejos Creek, near Santa Cruz in Bolivia. The springs occur in creek beds, where the flowing water keeps the rock surfaces clean and does not allow the heavier residue from evaporation to clog the pores.

FIG. 2.—SEEPAGE OF OIL ON THE BANK OF GALARZA CREEK.

In addition to the above springs, numerous other localities have been reported by competent natives to show signs of oil. These are also in creeks crossing the same line, and are said to show oil on the surface of quiet water before the drying-up of the streams. During the dry season the oil can sometimes be detected by odor from fresh soil at these places. Among such creeks are the following:

In Argentina: Zanja Honda, Yeriguarenda, Yacuy, Piquirenda, Ñacatimbay, Capiazuti, Carapari (Itiyuro). In Bolivia: San Roque (?), Caipitandé, Cuarazuti.

In order to collect the oil, the natives have sunk small pits, not over 1 cu. ft. in size, over the place where it comes to the surface. Some water also collects, but the oil is easily skimmed off the top.

## Quality of Oil

There is considerable range in the quality of oil coming from the various springs. In general, the springs may be divided into two groups according to the weight of the oil. The light oils are of light yellow or orange color, with greenish fluorescence, and vary in gravity between 41° and 46° Baumé. Heavier oils are of deeper colors, reddish-brown to almost black, and also fluorescent; their gravity varies between 20° and. 27° Baumé.

Dr. Longobardi gives the results of tests (Table 1) performed upon samples, the naphtha contents of which had been slightly reduced by evaporation at the springs.

Samples from the Agua Salada and Espejos springs, which have been sent to the United States, have sometimes been prematurely discredited on account of their apparent high gravity.

Most of the oils are of paraffin base, though some of the heavy ones contain some asphalt. Usually a small amount of sulfur is present.

TABLE 1.—*Results of Tests on Samples from Some of the Springs*

| Locality | Color | Gravity at 15° C. | Fractional Distillation | | |
|---|---|---|---|---|---|
| | | | Begin at | To 300° C., Per cent. | Over 300° C., Per cent. |
| Light Oils: | | | | | |
| Agua Salada......... | Amber.......... | 45° Bé | 153° C. | 91 | 9 |
| Los Monos........... | Reddish.. ....: | 42 | 160 | 78 | 22 |
| Caigua.............. | Orange.......... | 41 | 158 | 79 | 21 |
| Espejos............. | Orange.... ..... | 46 | 154 | 78 | 22 |
| ·Heavy Oils: | | | | | |
| Galarza............. | Dark brown..... | 23 | 195 | 15.5 | 84.5 |
| Iquira.............. | Brown.......... | 20 | 235 | 11 | 89 |
| Peima.............. | Brown........... | 26 | 152 | 31 | 69 |
| Oquita.............. | Dark brown.. . | 27 | 180 | 41 | 59 |
| Mandiyuti........... | Red-brown...... | 26 | 178 | 31 | 69 |

All show greenish fluorescence. Distillation test by Engler method.

## DEVELOPMENT

The results of the development attempted in the past have been unsatisfactory, principally due to the lack of sufficient knowledge on the part of the operators concerning the stratigraphic and structural conditions existing in the field. Wells have been placed in localities where there would seem to be no possibility of reaching the oil horizons which are in evidence on adjoining ground, and where local structural conditions are not particularly favorable for the accumulation of petroleum below the surface. Obviously, such wells are unable to give definite proof regarding the possibilities of the field at large.

Other enterprises have failed on account of financial difficulties, lack of proper organization, or mismanagement. A company intending to work in this field should have sufficient capital to tide it over excessive costs before expected returns. A well of 3000 ft. (914 m.) depth may be calculated to cost between $65,000 and $75,000. Transportation and imported labor are expensive, and delays are frequent notwithstanding the large quantity of spare parts necessarily required. Drillers must be carefully selected. Proficiency does not seem to be a more important qualification than personality and adaptability, for in camp the companionship must be agreeable and modern conveniences are generally absent.

Although thousands of claims have been legally granted by the governments of Argentina and Bolivia, only seven or eight honest attempts have been made to develop various sections of the field. The most conspicuous work has been that done on the Galarza, Capiazuti and Mandiyuti creeks. A brief discussion of these follows.

### Galarza Creek

The work on Galarza creek was started by Señor Francisco Tobar in 1907. The presence of two natural seepages of oil induced him to sink several pits by hand in the creek bed and on the banks above. He invariably encountered a soft porous sandstone, so saturated that the oil collected in the bottom of the pit. He secured two portable rigs of 1000-ft. capacity and sank four wells. Dr. Bonarelli gives the following data regarding these wells:

Well No. 1.—Depth reached 188 m. (617 ft.). Encountered two strata with petroleum; the first at 70 m. (230 ft.), the second at 140 m. (460 ft.).
Well No. 2.—Depth reached 241 m. (838 ft.), encountering five strata bearing oil. (Showings.)
Well No. 3.—Depth reached 75 m. (246 ft.); nothing encountered.
Well No. 4.—Depth reached 39.5 m. (130 ft.), encountering oil-bearing sand from 37 m. (121 ft.) downward, and at 39.5 m. a surging spring of gaseous water with taste and odor of petroleum.

Thus the drilling was unsuccessful. Nothing more than a show of oil was encountered below the original oil-bearing strata exposed in the creek bed below the well locations.

### Capiazuti Creek

Drilling at the side of Capiazuti Creek was encouraged by the presence of the oil springs on Iquira Creek, two miles to the south. With these springs the two wells drilled here were evidently located by "topographic correlation" for they are poorly located with respect to the stratigraphic and structural conditions of the immediate vicinity.

The first well was landed at 240 m. (787 ft.) with 4-in. casing, encountering nothing, and was abandoned on account of the small diameter at that depth. The rig was moved westward about 15 m. and landed at

FIG. 3.—STEEL DERRICK ON CAPIAŹUTI CREEK.

the same depth with 10-in. casing, when further work was postponed indefinitely.

These holes have been sunk since early 1912 under the management of the "Dirección General" of the Argentine Government. The rig was imported from Holland and has a capacity of 750 m. (2460 ft.). The derrick is entirely of steel and stands 85 ft.

This same rig had previously been used (1910–1911) in drilling two dry holes outside of the Capiazuti canyon.  One. was located 1 km. east of the Iquira springs, and the other 2 km. still farther east, in the village of Aguaray.  The first hole reached a depth of 133.6 m. (438 ft.), encountering only water at 128.2 m. (420 ft.).  The second hole was abandoned at 46.6 m. (153 ft.) on the advice of Dr. Bonarelli.

### Mandiyuti Creek

In 1911–12, the Farquhar Syndicate, organized for the apparent purpose of acquiring title to many of the largest Argentine and Bolivian enterprises, obtained petroleum concessions on Mandiyuti Creek, in Bolivia.  A complete outfit was imported at Buenos Aires, shipped to Embarcación and hauled by mules a distance of 225 miles.

One well was drilled to a depth of 157 m. (515 ft.), encountering a considerable showing of oil at 505 ft.  Drilling ceased, and soon after the syndicate went bankrupt.

It is interesting to note the difference between the analyses of samples taken from the springs and from the hole.

From Springs:
    Color: Reddish brown, with greenish fluorescence.
    Density: 26° Baumé.
    Distillation test (Engler method):

|  | Per Cent. |
|---|---|
| From 178° to 300° C. | 31 |
| Residue, over 300° C. | 69 |

From Drill-hole:
    Color: Amber, with light greenish fluorescence.
    Density: 38° Baumé.
    Distillation test (same method):

|  |  |
|---|---|
| From 70° to 150° C. | 18.0 |
| From 150° to 300° C. | 40.0 |
| Residue, over 300° C. | 42.0 |
| Paraffin residue, sulfur content | 0.129 |

In general, the oil encountered at depth may be expected to be of higher quality than that seeping at the surface in other parts of the field.

## Geological Conditions

### Stratigraphy

The surface of the oil region is entirely covered by sedimentaries. The only signs of igneous rocks are a few scattering fragments to be found in the creek beds, which have evidently been transported from the higher mountains of the Andes Range to the west.  Such transportation took place previous to the late diastrophism which gave rise to topographic features now eroded to their present condition in this "Sub-Andean" province.

The stratigraphy of this area has been a subject of discussion for some years past by the geologists who have worked here or in close vicinity. Fossil remains have been found in but one horizon sufficient to classify its age. Other strata covering the greater part of the area are apparently non-fossiliferous, for diligent search has failed to produce specimens. Consequently, classification has largely been based upon correlation of strata and comparative lithology

FIG. 4.—A VIEW OF THE SIERRA DE AGUARAGÜE, IN BOLIVIA. THE SUB-ANDEAN TERTIARY SERIES IS SHOWN ALONG THE CREST.

The recent work of Dr. Bonarelli has established the correctness of the stratigraphic interpretation previously given by Steinmann and Bodenbender, though Bonarelli has proceeded farther in differentiation by subdividing the groups formerly recognized.

Bonarelli's divisions of the Aguaragüe formations are as follows:

    (f) Recent alluvial.
    (e) Quaternary.
    (d) Sub-Andean Tertiary.
    (c) Upper sandstone (*Areniscas superiores*).
    (b) Calcite-dolomite horizon.
    (a) Petroliferous formation, or lower sandstone.

### The Petroliferous Formation

The term *Formación Petrolífera* was first used by Brackebusch in his early studies of the northern Argentine provinces. He later desig-nated the series "Salta Series" (*Sistema de Salta*) obviously with the same stratigraphic limits as signified by the first name. . This series is divided by Bonarelli into *a*, *b*, *c*, and *d* above, the term *Formación Petrolífera* being retained by him for the present recognized oil-bearing series. Steinmann, who had worked in the north around Santa Cruz, had designated the series of sands and shales underlying the calcite-dolomite horizon as the "Lower Sandstone" (*Areniscas Inferiores*).

This series is composed of a succession of medium-grained sandstones with beds of laminated or massive shale. The sandstone is generally massive, though sometimes thinly bedded, of medium harness, and gray in color. Crossbedding is shown in a few localities. The shale is hard, compact and of bluish-gray color; the thinly laminated beds often contain considerable sand. The color of both sandstone and shale shows alteration to yellow or brown on the surfaces exposed to the weather. Occasionally marl in thin bands, and also conglomerates are present.

Bonarelli estimates the total thickness of the series to be over 6000 m. (19,600 ft.) though it is possible that he has measured beds repeated by folding and faulting. There can be no doubt, however, of the first 2000 meters.

Respecting the age of the formation, the same author is of the opinion that it is Jurassic, and possibly older. Steinmann and Bodenbender considered it Cretaceous. The preponderance of evidence seems to support the latter view. Future work may reveal its age with precision, provided that fossil remains are found in the beds.

The *Formación Petrolífera* is not only the core of the entire Sierra de Aguaragüe range, but also underlies the young deposits of the plains stretching toward the east from the mountains. The oil seepages of this area, and also others beyond this belt to the west and southwest, not included within this paper, are intimately associated with this formation. The exposed strata from which the Agua Salada oil is flowing are probably of the same age and possibly of the same series. Correlation is impossible, however, until fossils can be found, for the intervening surface between Agua Salada and the Sierra de Aguaragüe is completely covered by Quaternary deposits.

This formation appears to be the basal series of the region. There has been, as yet, no recognition of an older series to the northeast of the province of Tucumán.

### The -Calcite-dolomite Horizon

Steinmann and Bonarelli have both recognized the series *Horizonte calcáreo-dolomítico* in the northern and southern divisions of this region, respectively, overlying conformably the petroliferous formation. It is described by the latter as follows: "A thin formation of calcitic and dolomitic strata with rare intercalations of other lithologic types; white or reddish, beds of variable thickness, very compact, with amorphous structure, conchoidal fracture, frequently with bands or nodules of jasper."

Steinmann places the age of this horizon as Cretaceous, on the evidence of fossils which he encountered within the formation. The thickness of the beds varies from 2 to 15 ft. (0.6 to 4.6 m.). While it is considered a persistent stratum, it is sometimes difficult to recognize, possibly because of local variations in character or on account of the ease with which it may be covered from view by débris.

## The Upper Sandstone

The calcite-dolomite horizon is overlain unconformably by the Upper Sandstone (*Areniscas superiores* of Steinmann and Bodenbender). This series is non-fossiliferous and is composed chiefly of sandstone and shale. Bonarelli describes it as follows: "This formation begins at the bottom with a series of strata mostly sandy and conglomeratic; overlain by a series of yellowish and whitish sandstones, by impure sandy marls, and above this a series constituted by repeated alternation of soft sandstone reddish, yellowish and gray with marls and sandy or limey shales, more or less compact." The reddish and yellowish hues are due to weathering. The thickness of the entire series is approximately 600 ft. (183 m.)

The Upper Sandstone probably belongs to the Tertiary Epoch. Steinmann did not differentiate the specific groups which he found associated with petroleum but he stated that he believed them to be Cretaceous, with Tertiary possibly in the upper part.

## Sub-Andean Tertiary

Dr. Bonarelli appears to have been the first to differentiate the *Terciario subandino* series which overlies the Upper Sandstone with apparent conformity throughout the Aguaragüe region. The series consists largely of thickly bedded, fine-grained, white, soft sandstone, interbedded with soft, light-colored shales. Some shale beds contain a small amount of sand or lime material. No fossils have been found in the series.

This series is mainly of interest with respect to the topography of the area. It forms the crest of the Sierra de Aguaragüe and is subject to rapid erosion by mountain streams which consequently run in steep, narrow gorges. Along the top of the ridge, throughout the length of the range, the series is exposed in almost vertical cliffs from 300 to 700 ft. (91 to 213 m.) high. The entire thickness reaches over 6000 ft. (1829 m.).

## Quaternary and Recent

The extensive *Chaco* plains are covered with loosely stratified, horizontal beds of sand, clay and gravel. This formation probably measures at least 400 ft. (122 m.) in thickness and has evidently been laid down upon previously peneplained surface of older rocks, including the *Formación Petrolífera*. In some localities the strata show slight flexures. A typical exposure can be seen at the Itiyuro Creek near the village of Campo Duran, to the northeast of Aguaray.

The recent alluvial deposits of loose clay, sand, and gravel, unstratified, have béen laid down in various sections of the plains, especially along the foot of the range, by streams which are capable of carrying considerable detritus during the rainy season.

## Structure

The regional structure, consisting of extensive thrust faults, anticlinal folds and synclinal folds, is not particularly difficult of interpretation and is, therefore, comparatively well known. The local features often present more difficult problems, owing to the obscurity of the older strata under younger deposits and to the lack of good horizon markers in the various groups.

The great thrusts originating below the Pacific Ocean off the west shore of the continent have pushed up the Andes Range, a series of parallel mountain ridges extending from north to south, producing folding and faulting of the sedimentaries previously laid horizontally over the area. The effects of the major thrusts have been transmitted toward the east, and perhaps on account of a greater rigidity of formations at this latitude, the resultant disturbances stand out so prominently at the great distance from the axis of the major ranges.

The folding of the Sierra de Aguaragüe may not have been synchronous with the upheaval of the main range. The structural features of the Aguaragüe were produced at the end of the Cretaceous period, and smaller movements have probably taken place periodically since that time. This is, no doubt, the cause of frequent earthquakes in the region today.

A sketch map by Dr. Bonarelli, showing the structural features of the Argentine portion of the Sierra de Aguaragüe and adjoining territory to the east and west, is reproduced in Fig. 5.

The strong thrusts have upturned the petroliferous formation along the Aguaragüe Range so that they are now standing at 75° or higher, dipping east or west, with a strike of N. 10° E. Stresses have been greatly relieved by a series of thrust faults between the ranges to the west and along the eastern flank of the Aguaragüe. The frontal range does not mark, however, the eastern limit of the action of the thrusts, for there is evidence of steeply dipping strata underlying the Quaternary deposits of the Chaco plains, at least within a certain radius of the range.

The Sierra de Aguaragüe fault extends along the eastern flank of the range for almost its entire length. It plays the important rôle of aiding

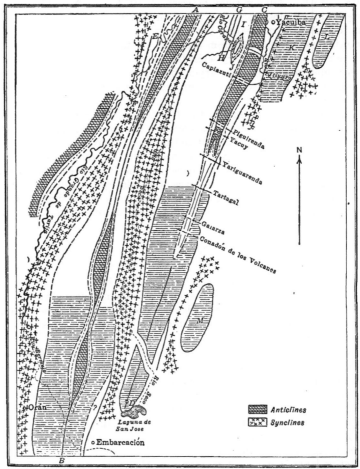

FIG. 5.—DR. BONARELLI'S SKETCH SHOWING STRUCTURE IN THE ARGENTINE SECTION OF THE NORTH ARGENTINE-BOLIVIAN FIELD.

the petroleum to the surface at the numerous springs. In Argentine, the fault plane has a strike of approximately N. 20° E. and swings to due north in Bolivia. The dip is to the west, varying between 40° and 60°. A thrust displacement of over 1500 ft. (457 m.) has taken place.

The calcite-dolomite horizon accompanied the underlying formation in its distortions, as it had previously been deposited conformably upon them.

The upper sandstone series presents a different aspect from the lower strata. With a strike varying between N. 10° E. and N. 26° E., it has a westerly dip ranging from 3° to 30° in the south and north respectively. This formation is oil-bearing on Galarza Creek at the "Tobar wells." Migration has taken place from the older strata below.

The Sub-Andean Tertiary series has a strike conforming to that of the upper sandstone, with a dip varying between 5° and 35°. This and the younger groups have, so far as is known at present, no association with petroleum.

During a period as late as the Pleistocene, the central part of the South-American continent was covered by an extensive continental sea. In northern Argentine and southeastern Bolivia, the surface had been previously well base-leveled with but few, if any, places remaining of sufficient height to escape immersion in the Pleistocene sea. Later, the interior of the continent underwent gentle uplifting to its present level. The beds are not absolutely horizontal, however, as slight flexures, anticlinal and synclinal, have been observed to cover extensive areas.

## SUMMARY

Extending from northern Argentine northward into central Bolivia is a belt of petroleum seepages. On account of the remoteness of the district it has, heretofore, been little considered by oil operators. The regional geology is comparatively well understood but the local features have not been carefully detailed.

Development work in the past has been done on an unscientific basis and has led to failures. At the present time, access to the region is somewhat difficult but no serious problem would be encountered in improving the conditions. The nearest railroad terminal is at Embarcación, 114 miles (183 km.) south of the Bolivian border, or 72 miles (116 km.) from the nearest manifestation of petroleum in natural springs.

The oil is of high quality and the seepages occur in creek beds along the Sierra de Aguaragüe fault, and at other isolated places.

Native labor is good and government policies are sympathetic toward foreign exploitation.

Though the structural features of the region, as a unit, have been worked out by reconnaissance surveys, there still remain many local sections upon which no detail study has been made.

Several small areas have been proved unfavorable for production, though the region as a whole cannot be condemned on this account.

## DISCUSSION

EUGENE COSTE,* Calgary, Alta —Evidently in Argentine they have a petroliferous province just east of the Andes where very much the same conditions exist as are found on the east side of the Rocky Mountains and in the plains adjacent. It strikes me, though, that the same mistake is being made in Argentine in the exploration for oil as was made in Alberta, Canada; viz., drilling the wells too near the seepages in the belt in the eastern part of the petroliferous province, too near the mountains. This same mistake was made in Alberta, with disastrous results; viz., only fissure production was found. The strata so near the mountains are tilted at a very strong angle, such as is described in this paper, and the dislocations are intense. On the other hand, we have found that when we get 100 or 200 mi. away from the mountains we get into oil and gas belts which are very prolific. So far our fields in Alberta have been gas, but in Wyoming, as you know, they get many thousands of barrels a day. I have no doubt that conditions in Argentine are exactly similar but that they have so far kept too near the mountains and that they want to explore the belts farther east to obtain better results.

MOWRY BATES,† Tulsa, Okla.—In Oklahoma we have a very similar condition; in the southeastern part of the state there is a decided uplift with highly folded and faulted country. In this we have the Fort Smith gas field of Arkansas where you get only gas. To the east of this field are the great areas of asphaltic sandstone; some distance to the west there are indications of oil and to the northwest are the oil fields. This would indicate a partial distillation in the ground. This is fairly close to the original oil field which bears out your statement.

EUGENE COSTE.—May I refute that idea of partial distillation? I think it can easily be proved that there is no such thing as partial distillation. Taking Alberta, for instance; in the deep syncline we have many regions absolutely undistilled but full of gas. There never was in these strata heat enough to cause the slightest distillation as shown by the coals and lignites in these strata which are entirely undistilled.

D. B. DOWLING,‡ Ottawa, Ont.—I do not propose to start any argument, but really there is a comeback to what Mr. Coste says. Geography is probably not geology, but geology is geography. You cannot regard structure as geography on the surface, but it is geology as soon as you get under the surface. In Canada, it is quite true that they went too close to the Rocky Mountains. When boring in the foothills, as

---

*President and Chief Engineer, Canadian Western Natural Gas, Light & Power Co., Ltd.
† Petroleum Geologist and Engineer.                    ‡ Geol. Survey.

Mr. Coste says, they were too close to the mountains where they got into a faulted area, and if there had been storage, it probably would have lost a good deal of the oil.

The situation in the Calgary field is one that might be interpreted in many ways, but in front of the foothills there is a very deep syncline. The gas and oil that is recovered on the edge of the foothills is of such a nature that we must assume that it is not the pure crude oil. It seems to be somewhat refined, or a condensation product. The solution, or at least the argument that appeals to me, is that as the deeper part of that basin is, say, 6000 ft. below sea level, there must have been some breaking up of the oil on account of pressure and of the heat, and we are getting gas, traveling up toward the surface in the foothills; and as it comes up, it probably condenses again, and is caught in folds on the edge of the syncline so that we have a natural condensing machine there. The oil is not found on the eastern side of the folds, it is found on the western side. On the eastern side and on the summit there is a little light oil, gasoline, we will say, blowing into the well once in a while; now that looks like a natural condensation of the gas and a storage behind the summit.

Following the further observation of Mr. Coste, we must allow that there is one place possible to get some oil. If we go farther out, we must cross the deeper part of that syncline and then we have another chance, which we are getting now on the eastern side of the deep syncline. There we have troubles, such as were discussed here with regard to water. The il is found in the lower part of the Cretaceous. Beneath that is some bed or beds of Devonian limestone. In this case we have the oil sand lying on the limestone. In the Peace river region, the oil is at about 100 ft. above sea level, on the Athabaska it is at 800 ft., so that we have on the surface of the limestone a very steep slope; and as there are, in places, broken beds of pebbles, there is some flow of water down that slope, probably from the north, as well as from the northeast, and when you strike the limestone, you have a very big chance of trouble with water. The question is how to prevent the driller from getting into that trouble.

\
## Petroleum Hydrology Applied to Mid-Continent Field*

BY ROY O. NEAL,† BARTLESVILLE, OKLA.

(New York Meeting, February, 1919)

THERE are two main sources of the water that floods productive oil or gas sands. The water may rise from the lower depths of the producing stratum, or it may come from beds above or below the oil-bearing formation. Usually the recovery of oil is decreased by water entering the oil sands, and most oil-field waters, especially those of deep wells, tend to foster the formation of an emulsion, which is expensive to treat. This paper deals with a method of distinguishing between waters that encroach upon oil-bearing beds from sources in the same stratum and waters that reach the oil sands from horizons above.

In order to remedy effectually water difficulties in oil and gas wells, it is absolutely necessary to determine the source of the invading water. Sometimes evidence such as the structural relations between the strata penetrated by neighboring wells and those found in the well under investigation is not adequate; also, data from drill logs and well records, from mechanical tests made on the wells by plugs, testers, drilling tools, etc., and from tests by chemical indicators such as eosin and Venetian red, may be unsatisfactory, for such data may fail to locate the source of the infiltrating waters. As a resort in such cases, the application of chemical analyses, that is, the comparison of an analysis of the water in question with that of typical waters from the various water horizons in that particular district, has in certain instances proved of value. This method has been used to advantage in the Westside Coalinga field of California, where the source of water may be fairly definitely determined from its composition.

The chief conclusions concerning the chemistry of the oil-field waters in California by G. S. Rogers[1] are: "Oil-field water is not necessarily salty, as is generally believed, and may not be even slightly salty to the taste. The degree of concentration of chloride in such water is governed primarily by local conditions and is not affected by the position of the water in relation to oil. Sulfate diminishes in amount as the

* Published by permission of the Director of the U. S. Bureau of Mines and of the Empire Gas & Fuel Co. The data for this article were obtained by the writer in an investigation for the benefit of the sub-surface division of the Geological Department of the Empire Gas & Fuel Co. while he was chemist for this company.

† Petroleum Experimental Station, U. S. Bureau of Mines.

[1] Chemical Relations of Oil-field Waters in San Joaquin Valley, California. U. S. Geological Survey, *Bulletin* 653 (1917) 6.

oil zone is approached and finally disappears.  The concentration of carbonate increases as the oil zone is approached, but depends largely on the concentration of chloride.  The horizon, with respect to the oil zone, at which these alterations take place, is different in each field."

The conditions existing in the Mid-Continent field are very different from those in California; in fact, each pool probably has its own peculiarities, wholly different from those of any other district.  In all pools in the Mid-Continent fields that have come to the writer's attention, the sulfates increase as the oil horizon is approached.  In fact, in most cases the sulfate content of top waters is practically nil whereas in the bottom

TABLE 1.—*Analyses of Bottom Water Below 2500-ft. Sand in Augusta Field*

| | Average of Twenty Samples Distributed Over Field | Northern Extremity of Field | Southern Extremity of Field | Center of Field, Depth 2545 Ft. |
|---|---|---|---|---|
| Primary salinity, per cent...... | 79.18 | 77.30 | 80.03 | 79.10 |
| Secondary salinity, per cent.... | 19.98 | 21.82 | 18.66 | 19.19 |
| Secondary alkalinity, per cent. | 0.84 | 0.89 | 1.32 | 1.71 |
| Chloride salinity, per cent..... | 92.08 | 88.68 | 90.88 | 91.84 |
| Sulfate salinity, per cent...... | 7.08 | 10.43 | 7.73 | 7.27 |
| Iron, parts per million........ | 22 | 32 | 38 | 56 |
| Calcium, parts per million..... | 1,779 | 1,987 | 1,580 | 1,808 |
| Magnesium, parts per million.. | 464 | 445 | 544 | 455 |
| Sodium and potassium, parts per million.................. | 11,146 | 10,879 | 11,592 | 11,317 |
| Bicarbonate, parts per million. | 313 | 234 | 353 | 456 |
| Sulfate, parts per million...... | 2,086 | 2,635 | 1,986 | 1,893 |
| Chloride, parts per million..... | 20,007 | 19,552 | 20,384 | 20,384 |
| Total solids, parts per million | 35,817 | 35,764 | 36,477 | 36,369 |

waters the sulfate salinity has reached as much as 10.5 per cent.  The sulfate content has proved a trustworthy indication of bottom waters, which is contrary to the results obtained by Rogers for the California fields.  The bicarbonates vary irregularly and are not consistent, none of the waters examined showing more than a trace of carbonates.  The total solids of the bottom waters always was considerably lower than that of any of the top waters, and is a fairly reliable property to study in differentiating between top and bottom waters.  More recently the method has been used extensively by the Empire Gas & Fuel Co. in the Butler County fields of Kansas and to a slight extent in the Blackwell field of Oklahoma.

In Table 1 are the analyses of bottom waters of the Augusta field.

The primary salinity[2] is due to the presence of alkaline salts of the strong acids; secondary salinity is due to the presence of calcium and magnesium salts of the strong acids. Secondary alkalinity is due to the presence of the salts of magnesium, calcium, and iron of weak acids. Chloride salinity is equal to the sum of the combined chlorides present and sulfate salinity is equal to the sum of the combined sulfates. The uniformity of the various bottom waters is conspicuously apparent from these data. From the results of more than 125 samples in the Augusta field, it was found that the sulfate varied from 7.15 per cent. to 10.52 per cent.; the chloride salinity ranges from 88.44 per cent. to 94.25 per cent. of all the properties, with practically all the results below 92 per cent.; and the primary salinity between 70.72 per cent. and 85.55 per cent. of all the properties.

In Table 2 are given the analyses of top waters, above the 2500-ft.

TABLE 2.—*Analyses of Waters Above 2500-ft. Sands in Augusta Field*

|  | From 2160-ft. Water Sand | From 1280-ft. Water Sand | From 1660-ft. Water Sand |
|---|---|---|---|
| Primary salinity, per cent......... | 80.24 | 80.92 | 74.41 |
| Secondary salinity, per cent....... | 19.74 | 19.02 | 25.55 |
| Secondary alkalinity, per cent..... | 0.02 | 0.06 | 0.04 |
| Chloride salinity, per cent......... | 99.62 | 99.94 | 99.96 |
| Sulfate salinity, per cent......... | 0.35 | none | none |
| Iron, parts per million........... | 117 | 48 | 52 |
| Calcium, parts per million........ | 4,950 | 6,710 | 7,440 |
| Magnesium, parts per million...... | 1,915 | 2,762 | 2,960 |
| Sodium and potassium, parts per million...................... | 34,452 | 49,298 | 37,026 |
| Bicarbonate, parts per million..... | 10 | 68 | 41 |
| Sulfate, parts per million......... | 252 | trace | trace |
| Chloride, parts per million........ | 67,300 | 95,800 | 78,800 |
| Total solids, parts per million... | 108,996 | 154,686 | 126,319 |

sand, from distinctly different water horizons in the Augusta field. The primary salinity of some 50 samples of top waters varied from 70.72 per cent. to 85.55 per cent. of the total properties; the secondary salinity from 14.12 per cent. to 29.25 per cent.; the chloride salinity from 87.90 per cent. to 99.98 per cent.; and the sulfate salinity from a trace to 0.35 per cent.

The chief distinction between the top and bottom waters is the percentage of total solids. The content of solids in the top waters averages

[2] Chase Palmer: Geochemical Interpretation of Water Analyses. U. S. Geological Survey, *Bulletin* 479 (1911).

four or five times as great as that of the bottom waters. The difference in the chloride salinity between top and bottom waters is a reliable index to use in differentiating the various waters. The distinctive character of the waters as regards the sulfate content can be used in classifying top and bottom waters.

From these data, it is obvious that the top and bottom waters are distinctly and uniformly unlike and that it is an easy matter to differentiate them; it is extremely difficult, however, to distinguish between the various top waters, inasmuch as there appears to be no uniformity in their composition. The top and bottom waters being so different in total solids, and hence in specific gravities, it has been suggested[3] that a

TABLE 3.—*Analyses of Water in Augusta Field*

| No. | Lease and Well No. | Section | Total Solids | Chlorides | Sulfates |
|---|---|---|---|---|---|
| 1 | Love No. 15................ | 29 | 49,932 | 29,120 | 2,424 |
| 2 | Love No. 6................. | 29 | 40,860 | 23,296 | 2,860 |
| 3 | Love No. 7................. | 20 | 41,028 | 23,296 | 2,992 |
| 4 | Long No. 2................. | 8 | 43,500 | 24,960 | 2,096 |
| 5 | E. Varner No. 9............ | 17 | 46,732 | 25,792 | 2,700 |
| 6 | E. Varner No. 7............ | 17 | 49,008 | 24,544 | 2,424 |
| 7 | Smith No. 11............... | 20 | 46,620 | 26,208 | 2,396 |
| 8 | Smith No. 14............... | 17 | 41,200 | 22,464 | 2,644 |
| 9 | Haskins No. 8.............. | 17 | 153,600 | 94,640 | 892 |
| 10 | Miller No. 6............... | 2 | 34,032 | 19,136 | 2,644 |
| 11 | Martin No. 5............... | 2 | 40,300 | 24,960 | 2,516 |
| 12 | F. Varner................. | 16 | 36,336 | 20,592 | 2,560 |
| 13 | E. Varner No. 9............ | 16 | 42,528 | 22,464 | 2,940 |
| 14 | F. Varner No. 16........... | 16 | 155,956 | 97,760 | 696 |
| 15 | Ralston No. 25............. | 9 | 35,588 | 22,464 | 2,328 |
| 16 | Scully No. 4............... | 9 | 33,320 | 18,720 | 2,444 |
| 17 | F. Varner No. 12........... | 16 | 37,428 | 21,636 | 2,748 |
| 18 | Ralston No. 18............. | 8 | 41,872 | 24,544 | 2,588 |
| 19 | Scully No. 10.............. | 9 | 39,152 | 22,464 | 2,876 |
| 20 | Curry No. 2................ | 16 | 37,428 | 25,584 | 2,404 |
| 21 | Ralston No. 6.............. | 9 | 41,756 | 24,028 | 2,752 |
| 22 | F. Varner No. 12........... | 16 | 47,496 | 26,208 | 2,920 |
| 23 | F. Varner No. 1............ | 16 | 36,148 | 21,216 | 2,860 |
| 24 | Ralston No. 5.............. | 17 | 43,808 | 25,792 | 2,896 |
| 25 | Curry No. 1................ | 16 | 36,000 | 22,048 | 2,484 |
| 26 | Feltham No. 15............. | 10 | 43,128 | 24,960 | 1,512 |
| 27 | Brown No. 10............... | 16 | 39,656 | 22,880 | 2,952 |
| 28 | Ralston No. 9.............. | 9 | 38,736 | 22,048 | 2,840 |
| 29 | Curry No. 3................ | 16 | 90,684 | 58,240 | 1,864 |
| 30 | Feltham No. 13............. | 10 | 37,364 | 21,632 | 2,436 |
| 31 | E. Varner No. 6............ | 16 | 38,320 | 20,592 | 2,588 |
| 32 | Cunningham No. 4.......... | 16 | 36,116 | 20,592 | 2,416 |

[3] J. O. Lewis, Petroleum Technologist, U. S. Bureau of Mines, in interview.

hydrometer be used to distinguish the waters. In most cases only three determinations—namely, total solids, sulfates, and chlorides—are necessary to detect mixed or top water that has infiltrated into the oil sand, as is shown by Table 3. It is apparent that samples Nos. 9, 14,

TABLE 4.—*Properties of Water in El Dorado Field*

|  | Bottom Water From 2450-ft. Sand, Per Cent. | Top Water From 1500-ft. Horizon, Per Cent. |
|---|---|---|
| Primary salinity.................... | 80.20 | 81.32 |
| Secondary salinity........ ........ | 17.90 | 18.66 |
| Secondary alkalinity............... | 1.90 | 0.02 |
| Chloride salinity................ .. | 92.72 | 99.80 |
| Sulfate salinity................ .... | 5.38 | 0.18 |

and 29 are top waters or of a mixed type. Complete analyses are recommended when interpreting results of a questionable mixed water. There is a close agreement among the various bottom waters, the results from the various leases in the field nearly coincide. The analyses of

TABLE 5.—*Analyses of Water in El Dorado Field*

|  | Composition, in Parts per Million | | Reacting Values, in Per Cent. | |
|---|---|---|---|---|
|  | Bottom Water From 2450-ft. Sand | Top Water From 1500-ft. Horizon | Bottom Water From 2450-ft. Sand | Top Water From 1500-ft. Horizon |
| Iron........................ | 42 | 63 | 0.15 | 0.04 |
| Calcium ................... | 1,350 | 5,663 | 6.68 | 5.02 |
| Magnesium................. | 377 | 2,935 | 3.07 | 4.28 |
| Sodium and potassium. .... | 9,318 | 52,652 | 40.10 | 40.66 |
| Bicarbonate................ | 587 | 40 | 0.95 | 0.01 |
| Sulfate..................... | 1,310 | 236 | 2.69 | 0.09 |
| Chloride.................... | 16,640 | 99,640 | 46.36 | 49.90 |
| Total solids.... . ...... | 29,624 | 161,229 |  |  |

the top waters, even from the same horizon, do not correspond even to a degree whereby they can be specifically identified.

Practically identical results were obtained in the El Dorado field, the analyses of the water being very similar to those of Augusta, the only difference being in the smaller content of total solids in bottom waters as is shown by Tables 4 and 5.

The following reaction coefficients were used in calculating the reactive values of the various constituents in these waters:[4]

| | | |
|---|---|---|
| Calcium....... 0.0499 | Sodium....... 0.0435 | Bicarbonates ... 0.0164 |
| Magnesium.... 0.0822 | Potassium .... 0.0256 | Sulfates....... 0.0208 |
| Iron.......... 0.0358 | Carbonates.... 0.0333 | Chlorides...... 0.0282 |
| Aluminum ...: 0.1107 | | |

Analyses of the waters from the typical water horizons in the Blackwell (Oklahoma) field are relatively similar to those of the Kansas field, excepting that the concentration of the various salts is somewhat higher. The analyses are given in Table 6.

TABLE 6.—*Analyses of Water in Blackwell Field*[5]

| | Water From Depth of 2226-ft. | Water From Depth of 2640-ft. | Water From Depth of 3412-ft. |
|---|---|---|---|
| Primary salinity, per cent................. | 87.97 | 93.06 | 78.24 |
| Secondary salinity, per cent.............. | 12.02 | 6.93 | 21.71 |
| Secondary alkalinity, per cent............ | 0.01 | 0.01 | 0.05 |
| Chloride salinity, per cent............... | 99.99 | 99.99 | 99.59 |
| Sulfate salinity, per cent.... ............ | None | None | 0.36 |
| Iron, parts per million................... | 52 | 185 | 174 |
| Calcium, parts per million............... | 5,950 | 3,412 | 2,407 |
| Magnesium, parts per million............. | 3,855 | 2,470 | 9,353 |
| Sodium and potassium, parts per million.... | 91,276 | 104,079 | 62,089 |
| Bicarbonate, parts per million............ | 25 | 21 | 58 |
| Sulfate, parts per million................ | None | None | 475 |
| Chlorides, parts per million.............. | 162,240 | 173,680 | 126,800 |
| Total solids, parts per million........... | 263,398 | 283,847 | 201,356 |

The general definition of a connate water is, according to Rogers:[6] "A sample of salt water may reasonably be called connate if it approximates ocean water in chemical composition and if it occurs in rocks of marine origin in which the circulation of the water is very slight. For practical purposes, therefore, connate water may be defined simply as fossil sea water." Although it could not be expected that fossil sea water would remain unchanged in its original chemical constitution when in contact with various rocks, it is remarkably striking how closely the composition of Augusta bottom water resembles that of ocean water. This is shown by the analyses given in Tables 7 and 8.

[4] Taken from Kansas State Board of Health Reports.
[5] Oil strata are at 1700 ft. and 3400 ft. (518 m. and 1036 m.).
[6] *Op. cit.*, 22.

Paramount in importance is the collection of representative samples of waters from the various water horizons and from wells distributed over the entire field. Reliable samples can be taken only at the time of drilling into water strata and it may prove profitable, especially in a newly developed field, to collect representative samples, label

TABLE 7.—*Properties of Sea Water and Augusta Bottom Water*

|  | Augusta Bottom Water,[7] Per Cent. | Sea Water,[8] Per Cent. |
|---|---|---|
| Primary salinity.... | 79.18 | 78.60 |
| Secondary salinity.... | 19.98 | 21.10 |
| Secondary alkalinity.... | 0.84 | 0.30 |
| Chloride salinity.... | 92.08 | 90.30 |
| Sulfate salinity.... | 7.08 | 9.24 |

carefully while drilling the well, and preserve for future reference in case water troubles arise later. After an adequate number of trustworthy samples of typical waters have been examined, it is possible to form an estimate on the probable location of the water in question after comparing with analyses of water from known horizons. It might be emphasized

TABLE 8.—*Comparison of Sea Water and Augusta Bottom Water*

|  | Composition, in Parts per Million | | Reacting Values, in Per Cent. | |
|---|---|---|---|---|
|  | Augusta Bottom Water[7] | Sea Water[8] | Augusta Bottom Water[7] | Sea Water[8] |
| Sodium and potassium......... | 11,146 | 11,100 | 39.59 | 39.31 |
| Calcium..................... | 1,779 | 420 | 7.23 | 1.77 |
| Magnesium.................. | 464 | 1,300 | 3.12 | 8.92 |
| Iron........................ | 22 | ..... | 0.06 | |
| Sulfates.................... | 2,086 | 2,700 | 3.54 | 4.62 |
| Chlorides and bromide......... | 20,007 | 19,410 | 46.04 | 45.22 |
| Bicarbonates................ | 313 | ..... | 0.42 | |
| Carbonates.................. | ..... | ..... | ..... | 0.16 |
| Total solids................ | 35,817 | 35,000 | | |

that too much care cannot be exercised in procuring representative samples, also that generalizations should not be drawn from a few specific cases—that is, too few analyses.

It would be a matter of speculation to say that the application of chemistry to water problems in other Mid-Continent fields would be

[7] Average of 20 analyses of Augusta bottom water.
[8] Mean of 77 analyses of sea water, by W. Dittman, given by Chase Palmer in U. S. Geological Survey *Bulletin* 479.

as beneficial as they have been in Butler County, but inasmuch as good results have been obtained in all the pools that have been examined, namely, El Dorado, Augusta, and Blackwell, as well as the California fields, it seems possible that the principle can be practically demonstrated in other fields, although the characteristics of the water in each pool will be distinctive.    This method is not recommended as foolproof or as a panacea, but should not be neglected in determining the source of waters intruding upon oil-bearing formation, because it has been so successfully applied in the Butler County fields and will undoubtedly prove of much value in other fields, especially where complicated water difficulties are encountered.

The writer wishes to express his appreciation of the assistance and coöperation rendered by A. J. Diescher, A. W. McCoy, L. E. Jackson, and W. A. Williams of the Empire Gas & Fuel Co., J. O. Lewis of the Bureau of Mines, and Dr. Chase Palmer and Dr. G. Sherburne Rogers of the U. S. Geological Survey.

## DISCUSSION

G. SHERBURNE ROGERS,* Washington, D. C. (written discussion†).— Mr. Neal's paper on the petroleum hydrology of the Mid-Continent district is a welcome contribution on a subject concerning which none too much is known; it is, I think, the first published description of the chemical character of the oil-field waters of the great Mid-Continent area When studying the oil-field waters of the California fields several years ago, the differences in chemical composition between waters occurring at various horizons seemed to be sufficient to enable the operators to use water analyses as a basis for determining the source of the water flooding a well; and it is gratifying to find that Mr. Neal has arrived at a similar conclusion.

As Mr. Neal points out, the Mid-Continent waters are very different in character from those of the California fields, and casual inspection would seem to indicate that the laws governing their variation are diametrically opposed to those that seem to have operated in California. In the Mid-Continent fields, the upper waters are highly concentrated solutions of sodium, calcium, and magnesium chlorides and are practically lacking in sulfates; whereas the waters underlying the main producing oil sands are much less concentrated, contain sulfates, and closely resemble ocean water.   In the California fields, the waters above the oil sands are moderately strong solutions of sodium, calcium, and magnesium sulfates, with carbonates and chlorides in minor proportion; whereas the waters close to the oil sands or below them are generally

* Geologist, U. S. Geological Survey.        † Received Jan. 27, 1919.

much more concentrated and much higher in chlorides, but are invariably lacking in sulfates. Broadly speaking, therefore, concentration increases with depth in California, but in the Mid-Continent field it diminishes; sulfates decrease and disappear as the oil sand is approached in California, but in the Mid-Continent field they are absent above the main oil sand and appear only beneath it.

The first point of difference, concentration, is evidently due to the great difference in geologic conditions that distinguishes the two areas. The fields of San Joaquin Valley, California, are located in the foothills about 500 ft. (152 m.) above the center of the valley, and as the formations dip steeply and outcrop close to the fields considerable surface water enters the rocks and drains down toward the valley. Drainage is impeded, especially in the deeper beds, by a series of anticlines which separate the fields from the main valley, and the troughs back of these anticlines constitute traps from which apparently the original connate water has never completely drained. The upper waters are therefore similar to the normal groundwater in any semi-arid region, their salts being derived from the rocks through which they pass; the deeper waters, however, are partly or largely connate in character and in some localities closely approach ocean water in density. In the Mid-Continent area, on the other hand, the fields are located far from the outcrop of the producing sands and the dips are very gentle; meteoric waters penetrate only to shallow depths and the underground circulation beneath is exceedingly sluggish. The very high concentration of these waters may be due to the fact that they have remained practically stagnant for long geologic periods and that the rocks with which they have come in contact contained disseminated salt and other chlorides. The theory, advanced by Mills and Wells, that ascending hydrocarbon gases exert an appreciable drying effect on underground waters, and thus increase their concentration, may also be applicable. The decidedly lower concentration of the waters beneath the deepest oil sand in the Augusta and Eldorado fields may be explained by supposing that these waters have not come into contact with salt; that their circulation at some remote time was freer than that of the waters above them has ever been; or that, since they underlie the deepest known oil sand, they have not been subjected to the drying effect of ascending hydrocarbon gases and therefore retain their original density.

The second point of difference between the Mid-Continent and California waters, variation in sulfate, is probably also due to difference in geologic conditions. In the California fields, the shallow waters are high in sulfate, those close above the oil zone contain less sulfate but generally some sulfide, those in and below the oil zone carry neither sulfate nor sulfide. The theory that the sulfates in this area were reduced to sulfides (including hydrogen sulfide) through the action of certain hydrocarbons seems plausible, therefore. In the Mid-Continent field, Mr. Neal states that the waters above the main oil sands, which in the fields described lie at about 2500 ft. (760 m.), are sulfate free, whereas those beneath them contain sulfate; but he does not take into consideration the presence of higher and commercially less important oil and gas horizons. Thus, in the Eldorado field, considerable oil is produced from a sand at about 600 ft. (182 m.) and there are prolific gas sands at about 900 and 1200 ft.; in the Augusta field, the 600-ft. horizon carries gas and there is another and more important zone of gas sands between 1400 and 1600 ft.; and in the Blackwell field, gas is encountered at numerous horizons between 225 and 3275 ft. and some oil in several sands between 2000 and 3350 ft. The waters lying above the main oil sands in these fields, therefore, cannot be considered top waters in the sense in which I used the term in describing California conditions—as waters unaffected by oil or gas—and the absence of sulfate in these upper waters is thus in accord with the theory that sulfates are eliminated by the reducing action of hydrocarbon materials. The absence of sulfate in the waters beneath the oil sands in California is thought to be due to the fact that the oil and the deeper waters migrated together from unconformably underlying formations. In the Mid-Continent field, there is no evidence of extensive upward migration and the presence of sulfate beneath the oil sands is not, therefore, an argument against the sulfate reduction theory. Nevertheless, despite the strong evidence in favor of this theory to be found in the California fields, I would not venture to apply it in the Mid-Continent district without further evidence. The oils of the two regions are very different in character and the reaction may be induced only by some constituent present in California oil and absent in the Mid-Continent variety. Moreover, the fact that the ratios of calcium to magnesium are distinctly smaller in the upper Mid-Continent waters than in the deeper ones suggests that most of the sulfate, in conjunction with calcium, has been precipitated out as gypsum. These problems cannot well be solved until more extensive studies of the Mid-Continent and Appalachian oil-field waters have been made.

It is interesting to note that the upper waters analyzed by Mr. Neal are not true bitterns, despite their concentration, and that except for the absence of sulfate they bear a general resemblance in properties to ocean water. In this respect they differ from most Appalachian oil-field waters and many other strong natural brines, which are characterized by high proportions of calcium and magnesium chlorides. The deeper water, which is almost identical in properties and in density with ocean water, is also highly interesting. The discovery of a Pennsylvanian fossil water so similar to the water of the ocean today raises a difficult problem for those who contend that connate water cannot long remain in the rocks without losing its original character. On the other hand, it certainly does not support the views of A. C. Lane and others who have attempted to trace variations in the composition of the ocean through geologic time, and who suggest that the age of a fossil water may be determined roughly from its composition.

The analyses presented by Mr. Neal appear to have a bearing also on the search for deeper oil sands in the Augusta and Eldorado fields. The fact that the waters within the general productive zone are several times as concentrated as those directly beneath it indicates a distinct change in conditions at the base of the deepest known oil sand. The geologic nature and cause of this change are not clear, but the mere fact that a change exists suggests that the true base of the oil-bearing zone has been reached and that no deeper sands will be found. The appearance of sulfate in these deeper waters may have a similar significance, though, for the reasons mentioned, I should not attach much weight to variation in sulfate until the waters of this region have been more extensively studied.

Mr. Neal's statement, that the waters of each pool probably have their own peculiarities and that no attempt to use water analyses in practical oil-field work should be made until the local variations had been studied, is thoroughly warranted. Conditions in the Mid-Continent field are very different from those in California, and the waters of both districts differ from those of the Appalachian and of the Gulf Coast fields. In all oil fields, the underground circulation is naturally restricted and the waters are usually salty; but except in this regard there is no more reason to assume that all oil-field waters are alike than to assume that all river waters are similar. However, as most of the world's oil is produced from Tertiary and Cretaceous formations under conditions resembling those in California, rather than those in the Paleozoic fields of the United States, it is probable that oil-field waters the world over have a general resemblance to those in California. This inference is supported by a study of the analyses of such oil-field waters from Russia, Roumania, Galicia, Egypt, India, etc., as I have been able to collect.

R. Van A. Mills,* Washington, D. C. (written discussion†).—Mr. Neal's paper describing the successful application of petroleum hydrology in the Mid-Continent fields marks an important advance in this new field of investigation, the far-reaching possibilities of which are only beginning to be recognized. That his data and conclusions, based upon studies in the Mid-Continent fields, should accord closely with the results of similar studies in the Appalachian fields is interesting. A few of the points of difference, as well as those of similarity between hydrology problems in the two localities, seem worthy of discussion.

The chloride salinity and predominance of sodium, calcium, and mag-

---

* Petroleum Technologist, U. S. Bureau of Mines.

† Received June 30, 1919. Published by permission of the Director, U. S. Bureau of Mines.

nesium in the waters overlying the main oil and gas horizons in the Mid-Continent fields studied by Mr. Neal, place these waters in the same general class as the deep-seated waters accompanying oil and gas in the Appalachian fields. Again, the bottom waters in these Mid-Continent fields resemble the Appalachian brines except for the generally lower concentration of dissolved constituents and the sulfate content of the Mid-Continent bottom waters. The respective percentages of primary and secondary salinity, predominant qualities of the deep waters in both localities, are practically the same.

The shallow top and intermediate waters of the Appalachian fields occurring in or closely underlying the coal measures contain bicarbonates, chlorides, and sulfates, but their geologic relationships and chemical qualities differ widely from any of the Mid-Continent waters described by Mr. Neal. Some of the Applachian top waters are primary alkaline but their predominating quality is secondary alkalinity. Analyses of shallow top waters in the Mid-Continent fields are needed for comparison with the Appalachian top waters.

The increase of sulfates as the main oil horizons are approached in the Mid-Continent fields furnishes a striking contrast to Appalachian conditions wherein sulfates, though present in shallow top waters, are rarely found in the deeper brines. The content of noteworthy proportions of sulfate in some Appalachian top waters is probably due to the oxidation of sulfides in the coal measures; it also seems probable that the acquisition of sulfates by some intermediate waters is due to down-dip migrations of shallow ground waters. In contrast to these conditions, the sulfate-bearing bottom water of the Augusta field appears to be fossil sea water that has remained practically inert with respect to the reservoir rocks and petroliferous hydrocarbons of that locality. The absence of sulfates in the deeper Appalachian waters, those of the Lower Pennsylvanian, Mississippian, and Devonian series is explained, in part at least, by their content of calcium, barium, and strontium salts, the high concentrations of dissolved constituents, and the increase of temperature with depth, which favors the deposition of calcium sulfate from solution. These simple geochemical relationships in certain Appalachian fields may be accepted in lieu of the theory of the reduction of sulfates by petroliferous hydrocarbons and are substantiated by the occurrence of precipitated sulfates in the wells and pay sands and by simple field and laboratory experiments.[1]

In passing this subject, attention is called to Wegemann's report upon the Salt Creek oil field, Wyoming,[2] wherein it is shown that deep-

---

[1] These experiments for the U. S. Bureau of Mines will be described in succeeding papers.

[2] Carroll H. Wegemann: The Salt Creek Oil Field, Wyoming. -U. S. Geological Survey, *Bulletin* 670 (1918).

seated waters accompanying oil and gas in Cretaceous sands carry noteworthy proportions of sulfates. How widely the absence of sulfates in waters associated with oil and gas holds true is problematic and should not, of course, be accepted as a generalization even in strata of the same geologic age.

Another comparison, that of relative concentrations of so-called top and bottom waters in the two localities, is made possible by Mr. Neal's paper. In the Appalachian fields, an increase of concentration with depth is the rule, though certain fields, especially those in southeastern Ohio, afford noteworthy exceptions. In that locality, the brines associated with oil and gas in lenticular parts of the Berea sand (early Mississippian age) have attained only one-half the concentration of brines in the Big Lime, Keener, and Big Injun sands (Late Mississippian Age), which lie several hundred feet higher in the stratigraphic column directly beneath the erosion surface that marks the contact between Pennsylvanian and Mississippian strata.

In explanation of this condition, it seems probable that during the period of pre-Pennsylvanian erosion a part at least of the gaseous content of these beds escaped through thin covers of limestone, sandstone, and shale. In the absence of an excessive influx of surface waters, such an escape of gases would have concentrated the associated waters. This theory of concentration of oil- and gas-field waters by evaporation into hydrocarbon gases has been previously advanced[3] and the conditions just cited are regarded as exceptionally favorable for its application. If there are similar relationships in the Mid-Continent fields, the theory is strengthened. The occurrence there of highly concentrated top waters overlying comparatively dilute bottom waters and the noteworthy absence of gas in the main oil sands of these same fields suggests the concentration of top waters through the escape of gases that should otherwise occur with the oil.

From a technologic point of view, it matters little what theories upon the origin of the waters in different fields are accepted, but to establish primary causes of water injury to wells and pay sands and to develop methods of preventing or remedying such injury is imperative. The corrosion of well casings and tubings through the hydrolysis of certain dissolved salts and the plugging of well cavities, tubing perforations, and even the interstices of pay sands by inorganic mineral deposits formed through the agency of oil-field waters are common causes of the under-

---

[3] C. W. Washburne: Chlorides in Oil-field Waters. *Trans.* (1914) **48**, 687–693.

R. Van A. Mills and R. C. Wells: Evaporation of Water at Depth by Natural Gases (abstract). *Journal,* Washington Academy of Science (1917) **7**, 309–310.

Evaporation and Concentration of Waters Associated with Petroleum and Natural Gas: U. S. Geological Survey *Bulletin* 693 (1919).

ground loss of gas and oil.   The Bureau of Mines is developing methods of preventing and remedying these common injuries to wells and pay sands through the use of cheap chemical reagents.[4]   This is an application of petroleum hydrology which, though only in the experimental stage, offers considerable promise in the conservation of gas and oil.

Analytical methods, very similar to those described by Mr. Neal, for detecting the infiltration of injurious waters into oil and gas wells and establishing the source of these waters before attempting to shut them off have also been outlined in Geological Survey *Bull.* 693.[5]   The different waters in each field are regarded as chemical reagents having definite properties and capacities to react with each other to form new types of solutions and precipitates.   Thus in addition to analytical comparisons of the waters of different horizons, we have as criteria to establish the source of infiltrating waters in some Appalachian fields precipitates of barium sulfate and strontium sulfate with minor proportions of calcium sulfate in the wells indicating the infiltration of sulfate-bearing top waters.   In certain fields, calcium carbonate on the bottoms and walls of wells or incrusting tubings and inner surfaces of casings is an indication of top-water infiltration.   Again, an abnormally low concentration of the dissolved constituents of a brine together with low proportions of chloride and high proportions of sulfate or bicarbonate in the dissolved constituents is a further indication of dilution by top waters.

The infiltration of bottom waters in the Appalachian fields is not easily detected because of the similarities between most deep waters and also because natural differences in concentration or differences in the relative proportions of dissolved constituents may be confused with differences brought about through induced concentration.[6]   In the Mid-Continent fields, studied by Mr. Neal, it appears that there probably is not enough gas with the oil and water to cause these effects, but in the Appalachian fields oil is practically always accompanied by gas, the movement and escape of which brings about a definite order of change in the associated waters.   Physical and chemical equilibrium is disturbed by the drilling and operation of wells.   Carbon dioxide is lost from solution together with other gases and carbonates of calcium, magnesium and iron are deposited in wells and pay sands through the breaking down of bicarbonates; induced concentration is brought about through the removal of moisture in expanding gases; and other dissolved constituents are lost from

---

[4] R. Van. A. Mills: Process of Excluding Water from Oil and Gas Wells.   U. S. Patent Application No. 258260.   Also the acid treatment of wells to remove precipitated carbonates.

[5] The manuscript for this bulletin was prepared in 1917, the field work having been done in 1914, 1915, and 1916.

[6] Concentration of deep-seated waters brought about through the evaporation and removal of moisture in gases passing to and issuing from oil and gas wells.

solution as the points of saturation for different salts are reached. Chemical reactions and temperature effects enter into these induced changes, but the subject, which has been outlined in Geological Survey *Bull.* 693, is too lengthy for discussion at this time. The fact is evident that in the Appalachian fields analytical comparisons of waters collected at different periods in the life of a well or field furnish only part of the criteria necessary to establish the source of infiltrating waters. In making analytical comparisons of waters in the Appalachian fields, the careful consideration of induced effects[7] is absolutely necessary.

In closing this discussion I quote Chase Palmer[8] as follows:

"Oil-field waters must be studied, not only in the usual analytical way, but also by reactions of certain constituents Moreover, quantities of water much larger than the quantities used for analysis must be available. The active substances present only in small amounts must be obtained from the waters and studied. If oils and waters are studied experimentally we can then learn the true relations between the oils and waters.

"The study of waters and oils experimentally, not analytically, is the study of petroleum hydrology.[9] It is a new study, a new line of attack laid out by the organic and water chemist."

Palmer's conception that ground waters are natural chemical reagents with definite properties and capacities to react is fundamental, and his classification of waters[10] based upon this conception which Mr. Neal and many other hydrologists have followed marks probably the greatest single advance in modern hydrology.

---

[7] The deep-seated effects of extracting gas, oil, and water from their reservoir rocks.

[8] Personal communication.

[9] The term "petroleum hydrology" originated with Chase Palmer.

[10] Chase Palmer: Geochemical Interpretation of Water Analyses. U. S. Geological Survey *Bulletin* 479 (1911).

## Water Troubles in Mid-Continent Oil Fields, and their Remedies

BY DORSEY HAGER* AND G. W. MCPHERSON,† TULSA, OKLA.

(New York Meeting, February, 1919)

THE rapid increase of water troubles in the Mid-Continent oil fields is causing much alarm. Troubles occur at Towanda, Eldorado, Augusta, Cushing, Blackwell, and Healdton, although they had not been acute in the Mid-Continent field until about two years ago, when the unusual conditions in the deeper oil fields were first encountered. California faced the same situation, but, thanks to aggressive measures, has largely overcome the dangers.

The following analysis of water trouble may throw some light on the subject and be of assistance in solving the problems involved.

### WATER TROUBLES CLASSIFIED

The presence of water in large quantities in oil sands has the following results:

1. Diminishes oil production.
2. Diminishes casing-head gasoline production:
   (a) By curtailing the gas flow.
   (b) By making the use of vacuum pumps unsuccessful.
3. Increases lifting costs:
   (a) By making it necessary to pump large quantities of water, which requires a fast motion and long stroke (third hole).
   (b) By requiring the use of compressors for air lift.
   (c) By causing break-downs and delays due to the high speed necessary to pump water.
   (d) By making it necessary to treat "cut" or emulsified oil.

Oil production is seriously curtailed by the presence of large quantities of water. Lease records show that wells are shut down 40 to 60 per cent. of the pumping time where serious water trouble occurs. While a small quantity of salt water may cut the paraffine and keep the oil moving, several hundred feet, or a hole full of water, effectually "kills" the oil and gas. The quick return of wells to production, once the water is shut off, shows how wells have been affected. The killing of gas sands

---

* Petroleum Geologist and Engineer.         † Production Expert.

naturally means a decrease in gas volumes. Also, where the hole is full of water, vacuum pumps are worthless.

When attempts are made to pump off the water, the wells must be pumped fast and with a long stroke (third hole). This results in rapid crystallization of the rods, in numerous breaks, and in much belt trouble, all causes of expensive delays. Some operators have installed air-lifts to pump off water in great quantities; this calls for expensive compressors, and $1 to $2 per barrel of oil for lifting expense is not unusual. The treatment of "cut" or emulsified oil also calls for considerable extra equipment and expense.

## Some Results Obtained by Shutting off Water

It is necessary only to give a few results obtained by shutting off water to show how important this procedure is.

The Ohio Cities Oil Co., on its Sina Crow lease in the North Cushing oil field, Oklahoma, cemented more than 12 wells. These wells made from 3 to 20 bbl. of oil apiece before shutting off water, and all the water that could be pumped, ranging from 100 to 150 bbl. per well. After cementing, these wells showed increases of 10 to 150 bbl. of oil, and yielded only 1 to 10 bbl. of water; in fact, some wells made none. The casinghead gasoline content increased from 400 gal. to over 1400 gal. per day. Lifting expenses were greatly reduced, as the well pumped 90 to 100 per cent. of the time instead of 40 to 60 per cent. By eliminating water, all the principal troubles ceased.

## Classification of Water

In nearly all oil fields, water is found to occur at some place in the oil sands. This water may lie immediately below the oil and in the same sand *throughout all* of the oil pool (bed D, Fig. 1), or it may occur around the edges of the pool (B, Fig. 1). It may also lie below the oil, but in a different stratum (C, Fig. 1). Water may also occur in sands above or between oil horizons (A and C, Fig. 1).

In California, Wyoming, and areas of steep dips, it is most unusual to find water in the oil sand near the center of the pool. However, in the Mid-Continent field, where low dips are the rule, water is often found in the same sand as the oil, even in the heart of the pool (bed D, Fig. 1). Water in the oil sand is found around the edges of all Wyoming and California oil pools; this is known as "edge" water. Water below the oil sand is generally called "bottom" water (beds C and D, Fig. 1); and water above is called "top" water. The writer prefers the terms "primary" and "secondary" waters.

Primary water may be defined as water that was present in the oil sand before drilling occurred; it includes "edge" water and "bottom" water. where there is no break in the sand.  All water entering the oil sands from above, between, or below, may be termed secondary water, and its entrance is made possible only by the drilling of oil-wells.  Top water, and also bottom water, in lower sands come under this heading.

As the oil and gas are drawn from the oil sands, primary water generally replaces them.  This is because the water is generally under high head, sometimes as much as 1500 lb. per sq. in.  This water cannot be exhausted from the sands, but its encroachment can often be checked and so directed that wells need not be abandoned, until most of the available oil is secured.  At present, only from 10 to 20 per cent. of our oil is secured, when 50 to 75 per cent. should be obtained.

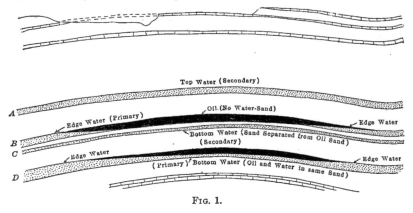

Fig. 1.

Fig. 1 illustrates a number of conditions that can and do occur in oil fields, where more than one oil sand is present.  These conditions existed before any wells were drilled.  $A$ is a water sand; $B$ carries oil and water; $C$ is a water sand; $D$ carries oil and water.  A condition of equilibrium exists in those sands; there is no travel nor migration of water from one sand to another and the only way to upset this stable condition is to drill wells.  Deeper sands often occur, but for purposes of illustration only two oil sands are shown; in some fields, five or six oil sands exist, and numerous water sands.

There is no possible way for primary water at $A$ to enter the lower sands without being admitted by drilling.  At $B$ it is noticed that water is around the edge of the sand.  On top of the fold, water does not exist. However, were a well drilled below the $B$ sand into the $C$ sand, bottom water would be found.  This would then be true secondary water.  Also, it may be noticed that sand $C$ would provide bottom water for $B$ and top water for $D$.  Sand $D$ shows edge water, which, however, is at the same time bottom water, a condition common to thick sands in regions of low dip like the Mid-Continent.   By drilling too deep into the sand this water would be encountered.

## SOURCE OF WATER FLOODING

The effect of water flooding may all be the same, but the source of the water may be different.

Well 5, Fig. 2, shows how bottom water may flood a well.

Well 3, Fig. 2, shows how top water may flood a well.

Well 5, Fig. 2, also shows how water between sands may flood wells, when improperly shut off.

Wells 4 and 5, Fig. 2, show how water in one well may flood a neighboring well.

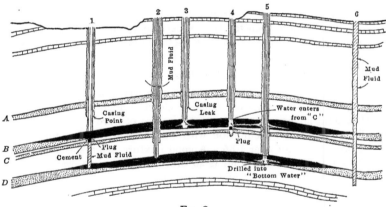

FIG. 2.

To remedy water troubles, it is necessary first to determine the source of the water, but this is by no means a simple matter.   In some cases it is quickly ascertained, but in others only careful experiment will decide. Chemical analysis may be successfully employed where the waters show marked differences in composition.   Aniline dyes may be used as tracers in some cases.

## REMEDIES FOR WATER TROUBLES

Once the source of the water has been ascertained, remedial operations must be employed.   These remedies may call for total abandonment of wells in some cases, for a change of casing points in others, for new casing, for the use of mud-laden fluid, or for cementing off water.   Remedies should be employed with the full coöperation of the oil operator not only of the immediate property but on the adjoining properties.

If a well is so hopelessly flooded that remedial means cannot restore it, and if it is a menace to neighboring wells, abandonment should be ordered.  Before abandoning, however, the well should be plugged in such a manner that there is no possible chance for water to use the well as a channel by which the flooding of adjacent wells may still go on. Before leaving a bad well, see that it is filled with mud fluid, of proper consistency, from top to bottom (see well 6, Fig. 2).  Mud fluid forms a more efficient seal than any other known means of plugging.

A change in casing points may be necessary where a gas sand or "dry" sand has become a water sand, due to flooding from adjacent wells.  In such case, the shutting off of the water may be best accomplished by underreaming the hole to form a new seat, and then using mud fluid or cement in back of the casing.

New casing, of course, is required if leaks have been found in the casing.  Casing may be eaten through by acid waters, a common cause, or the casing may have been faulty.  Where heavy heads of water occur, any tendency to weakness in the casing may develop into breaks.  Collapsed casing, leaks around couplings, or holes in the casing may result.

### Mud Fluid

Mud fluid is best used when abandoning wells, or for filling behind casing.  Limitations of its use must be clearly understood.

Mud fluid is the name given to a mixture of pure clay with water. The fine clay forms an emulsion with water when the specific gravity of the mixture is not over 1.35.  A good mixture has a specific gravity of 1.2 to 1.3.  This mud fluid is so much heavier than water or gas that it shuts off gas sands and keeps water back in the sands instead of moving in the drill hole.

Fig. 2 illustrates a variety of conditions.  Well 1 was drilled too deep.  The lower sand $D$ was non-productive, so it was mudded off, and a wood plug and cement seal were set at the bottom of sand $B$. This effectually protects $B$ from water at $C$, and the $C$ water cannot enter $D$.

Well 2 is producing from sand $D$, and the effectual shutting off of sand $B$ by mud fluid is shown.

Well 3 shows how a casing leak will allow water to enter a sand.  The remedy is to put in new casing.

Well 4 was drilled to sand $C$, but plugged off successfully.  Sand $B$ is, however, endangered by wells 3 and 5.  The remedy is to shut off water in 3 and 5.

Well 5 shows the improper shutting off of sand $C$, and the consequent flooding of sand $B$ from $C$.  Also, well 5 shows how, by drilling too deep into the sand, the upper part of sand $D$ is threatened.  The remedy is to shut off the offending sands.

Well 6, in the syncline, has been abandoned and filled with mud fluid.

## CEMENTING METHODS

For shutting off water under pressure, cementing is the best method. The old methods of using wooden or lead plugs, of bottom packers, and of seed bags, are obsolete. These methods work under some conditions, but safety-first requires the use of cement. Cementing methods, which comprise by far the largest part of remedial work, are not so simple as may first appear. Many factors enter into the question, and the average lease boss has neither the experience nor the training necessary to appreciate them all; yet failure in observing any one of them may result in loss. Observation of the following points is essential to success:

1. As little as possible of the pay oil sand must be covered. Many failures have resulted from cementing the pay sand; although water may be shut off, so also may the oil and the gas.

2. The method of introducing the cement must be the one best applicable to the conditions in the field.

3. Care should be exercised in selecting the brand of cement used for this work. In some cases it may be necessary to use accelerators to give rapid sets, but when these are used they must not appreciably weaken the cement.

4. The work must be done as quickly as conditions will warrant. Loss of production and expensive work on wells must be reduced to the minimum.

The system of cementing employed most successfully is the one that takes these factors most completely into account. The introduction of cement has given rise to several distinct systems, all of which seek to attain the same results. These methods are: The bailer dumping method, the tin-tube method, the tubing and packer method, the Perkins method, the McDonald method, the McPherson system.

The bailer method of emptying cement into the hole is the simplest, but least efficient, except where a plug of only limited depth is required. It fails, however, where it is necessary to put cement into a hole that has large gas pressures and moving water.

The tin-tube method, which consists in filling tin tubes with cement, lowering them in the hole, and later crushing the tubes by the drill, is very little used at present.

The four last mentioned methods are the best. They are more or less similar, with enough variation to have distinctive names. In all of them, the cement is mixed in big boxes on the surface and is pumped into the hole through casing or tubing, precaution having first been made to insure circulation of water. Rotary mud pumps are employed for pumping in the cement.

The Perkins system consists in placing a disk packer in the casing, which is driven down upon the cement and forces all of it from the hole and up behind the casing.

The tubing system consists in gravitating the cement into the hole through open tubing, and letting it settle to bottom. Cement is mixed at the surface and pumped into the tubing; gravity does the rest.

The McDonald system consists in dropping dry cement through the casing or tubing and adding water to it. It is a simple method, especially applicable to shutting off bottom water.

In using the McPherson system, a canvas packer is generally placed in the bottom of the tubing in such a way that, when expanded, it limits the level of the cement, which is pumped through the tubing, to the level of the bottom of the tubing. By using an expanding packer, cement may be pumped under pressure into a shot hole and some of it forced into the pores of the sand. A pressure of 200 to 300 lb. is employed.

The McPherson system differs from the other systems in the employment of special canvas packers in some cases where shot holes are the rule, and by a special device for the cleaning of the walls of shot holes by causing a strong rotative action of the water on the walls of the hole. This action insures a clean surface of the wall of the hole and results in a better contact for the cement.

The Perkins method has met with excellent success in California. The McDonald system has given favorable results in Illinois and eastern fields. Variations of these methods are also employed in the Mid-Continent fields. However, the McPherson system has thus far been the most successful in the Mid-Continent fields, as it has been developed to meet the special needs of that area.

### Conclusion

Water flooding in the Mid-Continent is a menace that must be met quickly. Only the active coöperation of the oil operators will save many wells. In California, the oil operators settled their difficulties by coöperating for shutting off water. Some action must be taken in the Mid-Continent, whether it be through private, state, or federal agencies. Common business sense and national conservation demand it.

## DISCUSSION

I. N. KNAPP, Ardmore, Pa. (written discussion*).—The writer would first call attention to the fact that the Mid-Continent field was credited from 1900 to 1915 with a production of about 641,000,000 bbl. of oil without any burdensome water troubles being reported. In fact, the authors say the conditions were not acute until about two years ago when the unusual conditions in the deeper oil fields were first encountered. The writer disclaims any personal knowledge of the district mentioned but,

---

* Received Mar. 18, 1919.

from the experience he has gained in directing drilling in various fields
by both the cable-tool and the rotary methods and also in plugging and
cementing wells, he fails to see how the analysis of water trouble given
in the paper under discussion could be of any practical assistance in
solving the problems so vaguely outlined. Possibly it might induce
academic discussion, if not limited to practical operating conditions.

A recent bulletin of the Bureau of Mines gives definite and practical
statements that can be easily understood by one skilled in the art of
drilling wells and heading off water troubles.

From an operating standpoint subsurface knowledge must be based
on logs of completed wells. Absolutely no one can have such knowledge
as that indicated in Fig. 1 of the paper in question without drilling wells,
and no one can question such facts when disclosed by the drill.

Under classification of water the paper states "This water (primary)
cannot be exhausted from the sands, but its encroachment can often
be checked and so directed that wells need not be abandoned, until
most of the available oil is secured." This is a live question to the
writer. A couple of years ago he had some wells drilled and found the
conditions comparable with sands A and D of Fig. 1 with B and C left
out; but nothing in the paper, so far as can be found, indicates how this
water encroachment is or can be checked or directed particularly under
the high heads named.

Fig. 2 indicates several wells presumably drilled with the cable tools
but not shot. The figure and description are so general, however, that
it is difficult to discuss them. Take well 3 for instance. No one would
drill much ahead of the casing without testing to find if it were perfectly
water tight. This being assured, drilling is continued and an oil well is
developed but not shot; it is put to producing. Water begins to show
with the oil. The paper states "the remedy is to put in new casing"
but does not state why. The figure indicates that the leak is at the
point of the casing; there is nothing to indicate that anything is wrong
with the casing itself. What would be the usual course of such an event?
The pumper would report the well as making water. If there was only
a small amount of water leaking the writer would wet a 25-lb. sack of
coarse corn meal and pour it in outside the casing; after an hour or so
he would scatter in a 10-lb. sack of bird shot, which would facilitate
the sinking of the corn meal to the bottom. If this treatment was not
successful, he would pull the rods and tubing and run a casing tester to
definitely locate the leak, which in this case would be found at the point
of the casing. An attempt to get a shutoff would be made by driving
the casing. If this was unsuccessful, he would plug the hole some 30
or 40 ft. below the casing, lift the casing, say, 10 ft. from its seat and
underream and make and test out a new seat.

Of course if it was thought, in the first place, that the water had broken through under the casing, the hole should be plugged at once to prevent flooding the sand.  But why pull out a perfectly good casing and put in a new one?  Would that "throw some light on the subject" or would it be "of assistance in solving the problems involved?"

The authors say "Casing may be eaten through by acid waters, a common cause."  It is quite possible surface waters may occasionally cause such acid corrosion, but deep-seated waters are usually alkaline.

The writer once had water trouble by new casing being eaten through in two or three months.  The cause proved to be electrolysis.  Inside of 300 ft. of 6¼-in. casing used to shut off the fresh water was placed 600 ft. of 5-in. for the oil string shutting out the salt water, which formed an electrolyte between the casings for 300 ft., resulting in two plates of a wet electric cell.  On measurement, this cell was found to generate current enough to eat through the casing in the time mentioned.  The damage occurred only at a few points along the casing, the rest being as good as new.  The cure in this case was to pull the outside casing as soon as the well was completed.  At first, this casing was left in from a fear that the salt water would contaminate the fresh surface water and spoil the land-owners' water wells.  This fear proved to be groundless.  Electrolysis may be the undiscovered cause of many casing failures.

The paper states that mud fluid is a mixture of pure clay with water, making a fluid of certain gravity.  The writer believes that a proper mud fluid depends on mixing with water a clayey material that will remain in suspension for a considerable period of time and have the proper gravity or weight.  A pure clay mixed in water might make a fluid that would precipitate so rapidly as to be useless for the purpose intended.  Laboratory experiments show that matter in suspension imparts a certain viscosity to fluids and the rate of settling is by no means dependent solely on the difference in specific gravities, or on the fineness of subdivision of the solids.

The paper states that "Mud fluid forms a more efficient seal than any other known means of plugging."  The writer begs to differ with this statement.  A cement fluid is much better for it will act by its superior weight in cases where a mud fluid fails and, in addition, has the property of setting and hardening, which are requisites in plugging.  He once saw in the Vivian, Louisiana, field a gas well showing an open flow measurement of 80,000,000 cu. ft. per 24 hr. through an 8-in. casing with a probable rock pressure of 450 lb. that had blown out from a depth of about 1100 ft.  It took about five days to get the well lubricated and killed so as to shut in the gas.  In the course of a couple of days the gas worked through outside the well casing that had been sealed with mud fluid only.  As soon as this leak was discovered a quantity of good thick mud was pumped into the well, but the mud soon appeared in spots at the surface over an area of 500 ft. around the well.  When it was seen that the mud was not able to kill the gas escaping outside the casing, about 100 sacks of Portland cement was mixed neat into as heavy a fluid with water as could be pumped and run into the well. This effectually plugged the well in about 2 hr. We were confident that this success was entirely due to the excess weight or gravity of the cement over mud fluid as there was not time for enough set to take place in the cement to cut any figure in the initial result.

Under cementing methods, the paper says "for shutting off water under pressure, cementing is the best method." The writer must dissent from this statement. It is absolutely necessary to equalize all water pressures and kill all gas to successfully use cement. I believe this is common knowledge supposing that a hydraulic Portland cement is to be so used.

It further says "The old method of using wooden or lead plugs, of bottom packers and of seed bags are obsolete," but H. R. Shidel's paper presented at this meeting mentions the recent trial by "plugging with wood, lead, and limit plugs" in the Augusta field. Perhaps the seed bag is obsolete but I have had use for one in the past year; the bag is all right when properly applied.

Six methods of cementing are named but the last four are said to be the best; these are as follows: the tubing and packer method, the Perkins, the McDonald, and the McPherson. The paper says "In all of them, the cement is mixed in big boxes on the surface and is pumped into the hole," but on the next page it says "The McDonald system consists in dropping dry cement through the casing or tubing and adding water to it." Which is right?

Five years ago the writer presented a paper to the Institute on cementing oil and gas wells.[1] This was applicable particularly to wells drilled by the rotary method, but the general remarks on the proper use of Portland cement as to its setting, hardening, and testing have general application. The writer has had many years' experience in testing and using various hydraulic cements but fails to discover how the paper under discussion can, on the whole, be considered as of any practical assistance in solving water troubles in any oil field.

---

[1] Cementing Oil and Gas Wells. *Trans.* (1914) **48**, 651; Notes on Cement Masonry. *Proceedings*, American Gas Light Assn. (1902) **19**, Appendix. Also, *American Gas Light Journal* (Nov. 10, 1902) **77**, 665.

## A Concrete Example of the Use of Well Logs

BY MOWRY BATES,* TULSA, OKLA.

(Colorado Meeting, September, 1918)

THE following example of the practical application of engineering geology is of interest in that it demonstrates the advantage of keeping accurate records of all wells, whether drilled by one's self or by others, together with the advantage gained by gathering such data during the process of development.   The fault described was found in the autumn of 1915, during the hurried development caused by the discovery of the famous Gusher Bend pool in Red River Parish, Louisiana.   The writer was at the time employed by one of the operating companies as geologist, and the example is a portion of the routine work.

### LOCATION

The oil fields of northern Louisiana and east Texas are located in Caddo, Bossier, De Soto and Red River Parishes, Louisiana, and Marion County, Texas.   A very narrow strip of the eastern portion of the latter is productive.   The producing oil pools are known as the Caddo Field, in the northern part, and the De Soto-Red River Field, in the southern. The two fields are 60 miles apart.   There are several gas fields between them, but no important supply of oil has been found, though there are several producing wells in the Anona chalk, locally called the Chalk Rock, in the Elm Grove gas field in Township 16 North, Range 11 West (see Plate 1).

### GENERAL GEOLOGIC AND STRUCTURAL FEATURES

The general geology of northern Louisiana has been excellently described by A. C Veatch in *Professional Paper* No. 46, U. S. Geological Survey; by G. D. Harris, *Bulletin* No. 429, U. S. Geological Survey; and later by George C. Matson and Oliver B. Hopkins in *Bulletins* No. 619 and 661C, U. S. Geological Survey.

---

* Petroleum Geologist and Engineer.

The portion of the section referred to in this paper is that lying below the Wilcox formation and the flood plain of the Red River, which drains the Red River oil field.

The Red River runs through approximately the center of the field and is parallelled by lateral bayous or old river channels. The river bottom, at this point about 12 miles wide, is filled with alternating sands, clays or gumbos, and gravel, having a depth as great as 120 ft. (36 m.). Both to the east and west of the bayous, rise low sandy hills, rarely attaining an elevation of more than 100 ft. (30 m.) above the river bottom. The only surface exposures are Wilcox sands and clays, of Eocene time. The formations are so soft and easily eroded that it is impossible to do any detailed surface work unless one encounters one of the thin seams of lignite which are found at from 60 to 200 ft. below the surface. It is seldom that these seams are exposed except along the banks of the bayous and it is difficult to correlate them or even find them. All work must therefore be done from the study of well records.

Below the Wilcox sands and clays are the Midway and Arkadelphia formations, which are supposed to be respectively the Lower Tertiary and the uppermost formation of the Cretaceous. As both consist of a few thin, soft sandstones in thick soft shale and clay deposits, it is impossible to differentiate them, and in the study of records one is compelled to disregard both.

Fortunately the next formation below is a hard sand, which often has a calcareous capping, making it easily distinguished in drilling. This is the Nacatoch sand or, as it is locally called, the Gas rock. In the Red River Field it has an average thickness of 130 ft., though this varies greatly to the east and the west, and in some of the wells in the south central part of the state the Nacatoch cannot be distinguished with certainty.

Below the Nacatoch is the Marlbrook marl, consisting of blue to white shales. We then come to the Anona chalk, which is about 100 ft. thick. Below the Anona is another thick bed of blue and white shales, with some irregular sand, known as the Brownstown marl.

Below the Brownstown is the Eagle Ford shale, which includes the Blossom sand member, called the sand rock, or Lower Gas Rock, by the drillers. The Eagle Ford is composed of varying sands and shales, with no regular order or thickness until near the bottom, where a soft limestone is found, as shown in the section. It is just below this lime that the oil-bearing sands are found. The exact age of the producing sands is at present somewhat uncertain. It was assigned to the Woodbine for a number of years, but a fossil found in a deep well, which assuredly came from the bottom of the well, has been assigned by Stephenson to the Eagle Ford, which would place the producing sand somewhere in the lower portion of this formation.

*Generalized Section of Formations Supposed to Underlie the De Soto-Red
River Oil Field, Louisiana*

| System | Series | Thickness, Feet | Group | Formation |
|--------|--------|-----------------|-------|-----------|
| Quaternary... | Recent | | | |
| | Pleistocene | 0–200 | | . |
| Tertiary..... | Eocene....... | 300–900<br>200–300 | ..............<br>.............. | Wilcox<br>Midway |
| Cretaceous... | Gulf (Upper Cretaceous) | 300–600<br>100–160<br>150–750 | ..............<br>..............<br>.............. | Arkadelphia clay<br>Nacatoch sand<br>Marlbrook marl |
| | | +100<br>150–500 | ..............<br>Austin........ | Anona chalk<br>Brownstown marl |
| | | 400–700 | .............. | Eagle Ford shale (including Blossom sand member)<br>Woodbine sand |
| | Comanche (Lower Cretaceous) | 0–400 | Washita........ | Denison<br>Fort Worth limestone<br>Preston |
| | | 25– 30 | Fredericksburg | Goodland limestone |
| | | 500–800 | .............. | Trinity sand |

The Red River oil field is in the southern part of the Sabine uplift,
the higher portion of which is outlined in the key map (Fig. 1). The
oil is found in a series of domes which are superimposed on a long narrow
fold running from the center of Twp. 12 North, Range 12 West, to the
center of Twp. 13 North, Range 10 West. The western portion of this
fold is known as the De Soto oil field, the eastern as the Red River-Crichton oil field.

### RED RIVER-CRICHTON OIL FIELD

The Red River-Crichton oil field is situated almost entirely in the
Red River bottom, though a small portion of the eastern end is in the
hills east of Coushata Bayou. It was opened in 1914 by the Gulf Refining
Co., who drilled a well in Section 14, Twp. 13 North, Range 11 West. A
short time later Wolfe & Keen drilled Weiss No. 1 in Section 18, Twp.
13 North, Range 10 West. The latter well had an initial production of

6500 bbl. per day and produced many thousand barrels before any offset wells were completed.

In the late summer of 1915, a well was drilled on a small lease in the northwest corner of Section 25-13-11, which started the famous Gusher

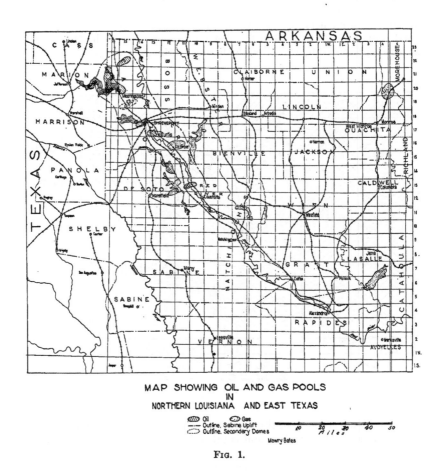

MAP SHOWING OIL AND GAS POOLS
IN
NORTHERN LOUISIANA AND EAST TEXAS

Oil    Gas
Outline, Sabine Uplift
Outline, Secondary Domes

Mowry Bates

Fig. 1.

Bend pool. The leases were all small, and a large number of wells were drilled very close together. Wells came in as large as 4000 bbl. per day, but the decline was very rapid, owing to the loose character of the sand and the closeness of the drilling.

## METHOD OF MAPPING STRUCTURE

A study of a large number of well records in the various Louisiana fields indicated a close relationship between the producing sand and the Nacatoch sand above. The interval was found to be fairly constant in the various pools, though it varied considerably from one pool to another. In the Crichton field the interval from the Nacatoch to the producing sand is about 1670 ft. (509 m.). The attitude or form of the Nacatoch sand is represented by contours, the figures on the contours (Fig. 2) being

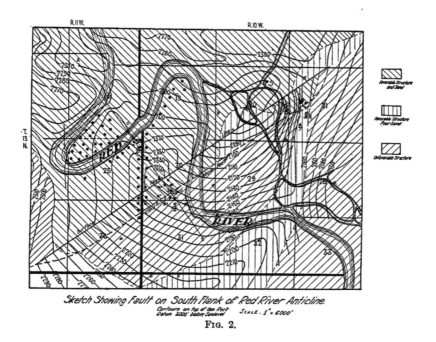

Sketch Showing Fault on South Flank of Red River Anticline.
Contours on Top of Sea Rock
Datum 3000' below Sealevel     Scale. 1" = 6000'

Fig. 2.

the elevation of the top of the Nacatoch sand above a datum plane 3000 ft. below sea level. This datum is taken in order to have positive contours, the larger numbers being higher. The Nacatoch sand is selected as the key rock because of the marked change in character from the Arkadelphia clays immediately above, whereby the top of the Nacatoch can be immediately detected and recorded. The interval to the oil sand is sufficiently constant to be useful, and the shallow depth enables one to keep ahead of development in close-in work and project structure in advance of the drill.

RED RIVER FAULT

The development of Sections 25, 30, 20 and 21 was proceeded with as rapidly as possible. The writer was at all times collecting records from various sources and plotting them in order to keep in advance of the drill. It was at first thought that the southern lines of Section 25 and 30 were in a syncline, as indicated by elevations of the Nacatoch from the

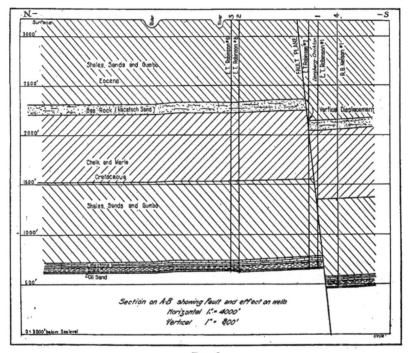

FIG. 3.

well in the center of Section 32. Then the well marked 1 in Section 30 came in, making 3000 bbl. where a dry hole or a small well was expected. This log was kept secret, and we did not have the depth of the Nacatoch until we secured the entire log. It was then found to be on the top of a dome the higher part of which had been thought to be farther to the northwest. A glance at the log showed a marked difference from any other in the field. The interval to the Nacatoch, instead of being 1682 ft. (512 m.) as in well No. 3 to the northwest, was only 1472 ft. (448 m.), or 210 ft. (64 m.) less than it should have been. It was at once apparent that faulting was present. The same day, wells No. 6 and 8, in Section 20 and 21 came in, showing intervals of 187 and 111 ft., respectively.

A study of all literature and records of normal faults in soft formations indicated that the majority had dips between 50° and 70°. A dip of 60° was assumed, and the fault plotted. Fig. 2 is the original contour map, which it has not been necessary to change since the first draft. Fig. 3 is Section *A-B* as marked on Fig. 2. It was estimated that a well drilled over 282 ft. south of well No. 1 would be beyond the limit of the producing sand, and would pass through the same horizon 210 ft. lower, which was sufficient displacement to throw the producing sand below the water level. This would cause dry holes in an area where the accumulation of oil was controlled by the condition of saturated sands, as in this field. This was found to be accurate within 14 ft. when well No. 4 was completed. The same condition was found to exist in the wells along the fault to the northeast, and it was feasible to predict a well or a dry hole from the location on the map. The distance south of the fault to which the producing sand might extend was, of course, controlled by the amount of displacement at each point.

After locating the fault underground, it was found on the surface between wells No. 8 and 9, where a sand rests against a clay. Matson and Hopkins, in *Bulletin* 661C, U. S. Geological Survey, extended the fault much farther to the southwest than I have done, but the evidence indicates that the extent mapped is correct.

The mapping was so accurate that the well immediately north of No. 1 was due to pass through the gap, or break, in the Nacatoch sand; which, in fact, it did, and it is the only well in the field in which no gas rock is encountered.

An interesting point discovered in the subsequent drilling of the wells to the south of the fault was found in well No. 4, which made some 30,000,000 cu. ft. of gas the first day; toward evening of the second day after drilling in, water began to show and in a few hours the well went entirely to water. This is explained by the theory that the dome was almost completely formed and the accumulation of gas had begun before the faulting took place. If the 2190-ft. contour south of the fault is extended to the southwest, and the higher contours are sketched in their proper relations, it will be found that a dome is shown with its high point a little south of the present high; and where the syncline is shown south of the fault a gradual southeasterly dip will be found. When the faulting occurred, the gas collected in the higher portion of the dome, and to the south of the fracture it was trapped and carried down in the sand. It was sealed in place by the water coming in from the south, and when tapped by the drill it made a large gas well for a short time, until the pressure was reduced and the water was allowed to come in from the higher structure to the south.

The Gusher Bend dome followed the usual rule of the minor structures on the Sabine uplift in having thinner and more compact sands on the southern flank, the thicker and more favorable sands being found on top and on the north flank. Hence the small favorable structures to the south of the fault were barren except for a limited amount of gas and the usual slight showing of oil which is found in so many wells in northern Louisiana.

The economic value of the above work was not appreciated at the time by the writer's superiors. The company for which the work was done was not fully convinced as to the reliability of this type of study and against his advice proceeded with the drilling. Every well south of the fault was dry, as predicted, and many thousands of dollars were spent without results.

## Cement Plugging for Exclusion of Bottom Water in the Augusta Field, Kansas

BY H. R. SHIDEL,* AUGUSTA, KANS.

(New York Meeting, February, 1919)

THIS paper summarizes the results obtained from the preliminary cementing of wells in an effort to cut off the bottom water.   The object of this work was two-fold:

(1)  To prevent the oil sand from becoming flooded.

(2) To plug off bottom water, thereby preserving the individual well and reclaiming production.

Valuable suggestions and help have been given by the following named: Messrs. Kyle and La Velle, of the U. S. Bureau of Mines; Magnolia Petroleum Co.; Freed Oil and Gas Co.   The cementing work was carried out under the personal supervision of L. J. Snyder.

In an unpublished paper on the "Water Problem in the Augusta Field," S. K. Clark reaches the following conclusions:

(1)  That the great amount of water present is bottom water, occurring in the Varner sand, the main producing or the 2500-ft. horizon.

(2)  That the only striking connection between structure and water is in the area of the marked fault on the Ralston, E. C. Varner, and F. Varner leases in sections 8, 9, 16 and 17.

(3)  That the oil occurs in porous streaks, generally separated by fine-grained, well cemented sand, which is barren.   Possibly two or three such pay streaks may be found.   That under a pay, fine-grained sand occurs, which is presumably barren at the time of drilling, but soon reveals water.

The writer takes partial exception to the last point, because well defined shale, slate, lime, or hard sand breaks have been encountered in a great many cases, separating the pay streaks under which water is often found.   This is not an invariable occurrence, as cases have been noted when the oil has been followed immediately by water in the same stratum.

In an effort to overcome the water menace, the following methods of plugging were tried:

(1)  Plugging with wood, lead, and limit plugs.

(2)  Plugging with sand pumpings.

---

*Resident Geologist, Empire Gas and Fuel Co.

(3) Mudding the sand, removing the packer, and driving a limit plug.

(4) Cementing.

A few variations of the last process will be discussed on the following pages. The success obtained by the above methods has been variable. In some places the plugging has been remarkably successful temporarily, as in the early efforts on wells of the Penley lease. In a few instances, as Brown No. 4, Sec. 16, the plugging was decidedly effective in improving both the cemented well and the surrounding ones.

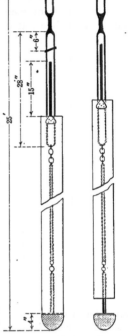

Before Dumping   After Dumping

FIG. 1.—DETAIL OF TRIP BAILER.

## MOYLE NO. 4, SEC. 10

This well produced a large percentage of water for some time. A limit plug was driven at the bottom of the well but proved unsatisfactory. In driving the plug through the shale-lime break much material was broken off, thus forming an imperfect seal. It was then decided to cement. A trip bailer was constructed (Fig. 1) and enough cement was put in the well to fill it to the top of the break. This cementing sealed the hole with material having practically the same character as the shale-lime formation.

The cement was allowed ample time (14 days) to set before the well was pumped. The results of this test were satisfactory, as can be seen from the relative percentages of oil, B. S., and water produced before and after cementing (Table 1).

TABLE 1.—*Moyle No. 4, Sec. 10*

| Before Cementing | | | | After Cementing | | | |
|---|---|---|---|---|---|---|---|
| Date | Oil | B. S. | Water | Date | Oil | B. S. | Water |
| Apr. 19, 1917..... | 33 | 7 | 60 | Jan. 9, 1918..... | 94 | 2 | 4 |
| May............. | 32 | 3 | 65 | Feb. 6.......... | 98 | 1 | 1 |
| Aug............. | 6 | 0 | 94 | Mar. 21......... | 98 | 1.5 | 0.5 |
| Sept............ | 17 | 1 | 82 | Apr. 18......... | 98 | 1 | 1 |
| Oct............. | 23 | 0 | 77 | May 16......... | 99 | 1 | |
| Nov............. | Well cemented. | | | June 7.......... | 100 | | |
| | | | | June 20......... | 99.5 | 0.5 | |

BRANT NO. 3, SEC. 2, TWP. 28 S., R. 4 E.

The cementing of this well proved so unsatisfactory that it was drilled out and re-cemented.    Drilling out and cementing was done three times during one week.    Other methods were tried, in an effort to shut off bottom water temporarily.    A packer was placed but failed to work, although sufficient mud had been forced in to shut off the water.    The cement was then put in on top of the mud and a limit plug was driven

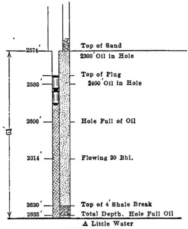

FIG. 2.—BRANT NO. 3.    CEMENT, LIMIT PLUGS AND CHARACTER OF SAND WHEN DRILLING IN.

into it; a second limit plug was then driven on top of the first.    This job has been successful, as shown by the behavior of the well before and after (Table 2).

TABLE 2.—*Brant No. 3, Sec. 2*

| Before Cementing | | | | After Cementing | | | |
|---|---|---|---|---|---|---|---|
| Date | Oil | B. S. | Water | Date | Oil | B. S. | Water |
| Apr., 1917........ | 18 | 2 | 80 | Feb. 21, 1918.... | 20 | .. | 80 |
| May............. | 19 | 4 | 77 | Feb. 23......... | 16 | .. | 84 |
| Sept............. | 9 | .. | 91 | Feb. 25......... | 8.1 | 1 | 91 |
| Oct............. | 15 | .. | 85 | Apr. 19......... | Well cemented again. | | |
| Nov............. | 11 | .. | 89 | May 17......... | 44 | 4 | 52 |
| Jan. 22, 1918..... | Plugged with cement. | | | May 23......... | 70 | 25 | 5.0* |
| | | | | June 20......... | 80 | 14 | 6 |

* Average of 10-min. tests.

## SCULLY No. 6, SEC. 28

This well was cemented to the top of a shale break and a limit plug was driven in the top of the green cement. The well was put to pumping, three days after cementing, with unsatisfactory results. The relative percentages of oil, B. S., and water were as shown in Table 3.

TABLE 3.—*Scully No. 6, Sec. 28*

| Before Cementing | | | | After Cementing | | | |
|---|---|---|---|---|---|---|---|
| Date | Oil | B. S. | Water | Date | Oil | B. S. | Water |
| Aug. 4, 1917...... | 3 | .. | 97 | Mar. 29......... | 50 | .. | 50 |
| Sept. 21.......... | 5 | .. | 95 | Apr. 4.......... | 24 | 3 | 73 |
| Oct. 15........... | .. | .. | 100 | Apr. 18......... | 10 | .. | 90 |
| Feb. 8, 1918...... | .. | .. | 100 | May 16.......... | 4 | .. | 96 |
| Mar. 25.......... | Well cemented. | | | May 31.......... | 3 | .. | 97 |
| | | | | June 20......... | 2 | .. | 98 |

The results point conclusively to the fact that 14 days should be allowed for the cement to set.

## SCULLY No. 9, SEC. 28

The same procedure was again followed in this well, which started to pump water before the flow of oil. At first the well flowed at the rate of 500 bbl. per day, but gradually diminished; the average production

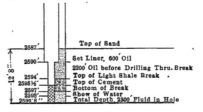

FIG. 3.—MOYLE No. 23. CEMENT AND CHARACTER OF SAND WHEN DRILLING IN.

during 60 days was 75 bbl. The well is now pumping about three times as much water as oil.

## MOYLE No. 23, SEC. 10

A good top pay was encountered (Fig. 3), followed by a shale break under which was water. The well was then cemented to 2595.5 ft., but this work proved unsuccessful, as the well has produced a high percentage of water since cementing. After testing, the cement was proved to be of a poor grade, so that this work has to be done over. Since cementing, the well has produced about 40 per cent. of oil.

### CUNNINGHAM No. 6, SEC. 16

This well was cemented April 1, 1918, the top of the cement being at 2491 ft. The results are shown in Table 4.

TABLE 4.—*Cunningham No. 6, Sec. 16*

| Before Cementing | | | | After Cementing | | | |
|---|---|---|---|---|---|---|---|
| Date | Oil | B. S. | Water | Date | Oil | B. S. | Water |
| Sept. 5, 1917.... | 3 | ... | 97 | Apr. 25, 1918.... | 75 | .... | 25 |
| Oct. 31.......... | 3 | ... | 97 | May 7.......... | 79.6 | 3.2 | 17.2 |
| Jan. 17, 1918.... | ... | ... | 100 | May 15.......... | 97 | 2 | 1 |
| Mar. 14........ | ·11 | ... | 89 | June 1.......... | 81 | 12.5 | 6.5 |
| | | | | June 13........ | 84.5 | 1.0 | 14.5 |

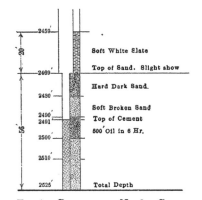

FIG. 4.—CUNNINGHAM No. 6. CEMENT AND CHARACTER OF SAND WHEN DRILLING IN.

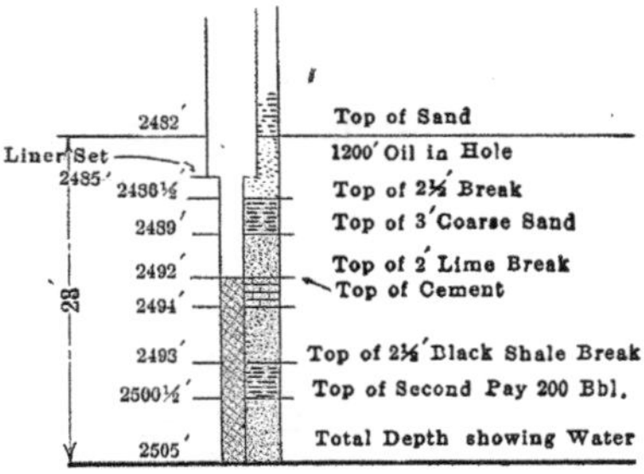

FIG. 5.—SCULLY No. 10. CEMENT AND CHARACTER OF SAND WHEN DRILLING IN.

### SCULLY No. 10, SEC. 9

Two pays were encountered, which were separated by distinct breaks (Fig. 5). The second pay showed a little water at the time of drilling. Six days later, one sample gave 100 per cent. water while an average of an 8-hr. sampling tested 72.4 per cent. water. This well was cemented to shut off the bottom pay, and since then has produced very satisfactorily, as shown in Table 5, yielding 3 to 5 bbl. of oil per hour.

TABLE 5.—*Scully No. 10, Sec. 9*

| Before Cementing | | | | After Cementing | | | |
|---|---|---|---|---|---|---|---|
| Date | Oil | B. S. | Water | Date | Oil | B. S. | Water |
| Mar. 26, 1918.... | 26.6 | 1.0 | 72.4* | May 1, 1918..... | 99 | 1 | |
| Mar. 25......... | ..... | ..... | 100 | May 2.......... | 98.5 | 0.75 | 0.75 |
| Apr. 10......... | Well cemented. | | | May 4.......... | 95.1 | 3.0 | 1.9 |
| | | | | May 6.......... | 98.3 | 0.9 | 0.8 |
| | | | | May 7.......... | 97.8 | 1.6 | 0.6 |
| | | | | June 12......... | 100 | | |
| | | | | June 19......... | 98.9 | ..... | 1.1 |
| | | | | July 1.......... | 98.5 | ..... | 1.5 |

*Average of 8-hr. testing.

## MILLER No. 7, SEC. 2

It was decided to mud the sand before cementing. A packer was placed and a pressure of 500 to 600 lb. was applied continuously for three

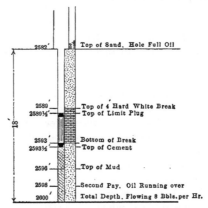

FIG. 6.—MILLER No. 7. CEMENT, LIMIT PLUG AND CHARACTER OF SAND WHILE DRILLING IN.

days. A limit plug was driven in, but did not hold. One-fourth sack of cement was then put in, and a limit plug was driven in it (Fig. 6). The well was allowed to stand for a while and was then pumped. The results before and after were as shown in Table 6.

The results of the cementing of the 13 wells were as follows:

1. Six were very successful, in that the water was shut off in the well itself and production was reclaimed.

2. Four were successful, in that the percentage of water produced was decreased. One of this number is to be cleaned out and re-cemented.

## TABLE 6.—*Miller No. 7. Sec. 2*

| Before Cementing | | | | After Cementing | | | |
|---|---|---|---|---|---|---|---|
| Date | Oil | B. S. | Water | Date | Oil | B. S. | Water |
| May, 1917....... | 31 | 16 | 53 | Mar. 11, 1918.... | 96 | 2 | 2 |
| Aug. 10......... | 10 | 1 | 89 | Mar. 14......... | 29 | 37 | 34 |
| Sept. 5.......... | 8 | ... | 92 | Mar. 18......... | 56 | 42 | 2 |
| Oct. 13.......... | 9 | ... | 91 | Mar. 20......... | 83 | 16 | 0.5 |
| Nov. 6.......... | 5 | ... | 95 | Mar. 21......... | 84 | 16 | 0 |
| Jan. 19, 1918.... | 7 | ... | 93 | Mar. 23......... | 91 | 8 | 1 |
| Feb. 8.......... | 13 | ... | 87 | Mar. 25......... | 82 | 27.5 | 0.5 |
| Mar. 11......... | Ce- | | | Mar. 26......... | 74 | 25 | 1 |
| | mented. | | | Mar. 30......... | 98.5 | 0 | 0.5 |
| | | | | May 2.......... | 62.0 | 32 | 6.0 |
| | | | | May 23......... | 54.0 | 46 | |

3. One was successful, in that, while the well continued to produce water, it also produced an average of 75 bbl. per day for two months. It was producing no oil before cementing.

4. In one well, oil and water were plugged off, and when an effort was made to drill out the cement a strong flow of water was encountered. The sand was so flooded that it was decided to abandon the well. The cementing of this well caused an increase of production in the nearby wells.

5. One was successful, in that the amount of water produced was diminished to a few barrels, and an offsetting well began producing oil.

6. Four of the wells cemented did not give satisfactory results, as the proportions of oil and water produced were practically unchanged. These wells are to be abandoned as non-productive, after it is proved that the sand is flooded.

7. One was a complete failure. The cement was drilled out and the second operation was very encouraging, as an increase of production was noted. This well is to have more work done on it.

8. One well was cemented so high that it has to be drilled out and tested.

9. One well was proved to be no longer productive, and is to be abandoned.

Nine wells have not been tested as yet.

Preliminary attempts to shut off the bottom water by cementing have been successful in most cases. There are certain disadvantages to be considered, although they are generally overbalanced by the advantage.

### DISADVANTAGES OF CEMENTING

The process may cement oil sand as well as the water.

The well has to be shut down 10 or 14 days.

The cement has to be carefully tested for setting qualities.

If but a few feet of sand are to be cemented, generally some means have to be taken to keep the cement from being agitated. Driving a temporary plug and mudding the sand have been tried.

## Advantages of Cementing

The cement assumes the same shape as hole; it does not leave cavities.

There is no pounding or jarring, thus preventing shattering of the sand.

The cement can be partially or entirely drilled out if unsatisfactory. Cement drills about the same as a lime formation.

When pressure is applied the cement will penetrate the sand and seal it.

The amount of cement to be placed can be carefully gaged. One sack of cement fills 7 ft. in a $5\frac{3}{16}$-in. hole and $5\frac{1}{2}$ ft. in a $6\frac{5}{8}$-in. hole.

It is estimated that to clean out and cement a well costs $600. In all of the successful cases, this cost was returned in a few days. The wells in which production was not reclaimed showed sufficient evidence to warrant abandoning; all casing and other equipment could then be removed.

## DISCUSSION

Mowry Bates, Tulsa, Okla.—In the first part of this paper the author says: "In an unpublished paper on Water Problem in the Augusta Field, S. K. Clark reaches the following conclusions: (1) That the great amount of water present is bottom water, occurring in the Varner sand, the main producing or the 2500-ft. horizon. (2) That the only striking connection between structure and water is in the area of the marked fault on the Ralston, E. C. Varner, and F. Varner leases in sections 8, 9, 16 and 17."

That paper must have been written a long time ago because practically all of the producing wells are in water now and were on structure, that is on the sides of the numerous domes found in the El Dorado and Augusta fields. They are now practically all making water and the water occurs in the lime where the oil is mostly found. The oil is forced out by a hydraulic flow, there is no gas in the producing zone at all. The lower wells strike water first; the ones highest up on the structure will make entirely oil. As the oil is taken out the water flows up the structure. The Shumway lease, of which I suppose you have all heard, has produced up to date enough oil to fill a tank $\frac{1}{2}$ mi. square and 7 ft. deep from the porous lime formation, which is more oil than the lime usually contains. That lease is practically all water today and I do not see how they can shut the water off by plugging.

EUGENE COSTE,* Calgary, Alta.—My experience is that in many cases you cannot avoid the water. If the water is in the sand itself, the only thing you can do is not to drill too deep into the sand. If you have drilled too deep, sometimes it is possible to plug the bottom water and in that way so minimize the trouble that the water becomes an advantage, for a little water is really an advantage.

There are two main things to avoid in these water troubles: First, the water from above, to avoid which one must be sure that the casings are tight and set at the right depth; they should be cemented if necessary before drilling into the oil sand. That is an easy thing to do with a little patience and care. Second, the most important thing is to avoid water trouble from the outside of the field. This outside water comes from wells beyond the pool, in the dip mostly, but sometimes not in the dip. The state should intervene on those wells that are just water wells, and should compel the operators to plug a well before abandoning it. The operators themselves inside the fields will easily overcome all the other water troubles which will not affect the total production of the field. Although certain particular leases may get too much water as the oil becomes exhausted, yet the field as a whole will not be affected.

MOWRY BATES.—About two weeks ago we asked some of the operators in the Mid-Continent field, who had been plugging the wells, what the effect had been. They said the production was increased for a very short time. But it seems to me that if this oil is produced on an anticline, and it is pretty well established now-that it is, and as the pressure is practically hydraulic pressure at the rate of 0.4 lb. per foot of depth except in a very few cases of synclinal oil, as fast as you extract the oil the water must ascend. As for plugging a well, you only plug about 7 in. and generally there are about 6 acres around each well through which water can pass. You cannot stop the ascent of the water.

When you take out the oil, the water is going to follow. You can plug an individual well or a dry hole that is making water and save it for a very short time, but only for a short time. That is absolutely proved in the Cushing field, where some of the wells were plugged. The production was increased about 50 per cent. at the time, but it is down again and all the wells make just as much water now as before, though some of them were cemented over. This is true in all cases of the Bartlesville sand but some of the wells that were making water from the Tucker sand have been successfully plugged as there is a break between the sands.

* President and Chief Engineer, Canadian Western Natural Gas, Light & Power Co., Ltd.

A. W. AMBROSE,* Washington, D. C.—In California they had very good success in the Coalinga District and elsewhere by coöperation between the companies. As a result, in certain wells that were considered hopelessly gone to water, it was possible not only to shut off the water but to increase the production. When you consider that a driller can never pick up a break of a few inches unless the break has some characteristic feature, I fail to see how it would be possible to pick up a 3-in. shale break at a depth of 2200 ft. in a well drilled by a rotary.

MOWRY BATES.—While working for the Gulf company, I was authorized to stop a well and test it. I washed it out for about 2 hr., then moved the bit for 5 or 10 min., and then washed it again. It cost the company a lot of money, but I was trying at that time to get some samples for the Geological Survey.

A. W. AMBROSE.—Another way is to put a core bit on the bottom of your rotary drill pipe; otherwise, I do not see how, when you wash, where you have 2000 or 3000 ft. of open hole, you are going to pick up such a small shale break, as the shale above may cave in and show in the washings. Take the Eldorado Augusta case: Many of the companies have no record but the thickness of the oil sands. They do not know, for instance, whether in the 50-ft. producing horizon there are one, two, or three breaks or one unbroken producing horizon. The companies there have everything to gain and nothing to lose by coöperation, which has proved very successful in California. In the Kern River field, some wells were producing as high as 15,000 bbl. of water per day by the aid of air compressors in order to get 40 or 50 bbl. of oil. Some of these compressors must have cost anywhere from $200,000 to $300,000. When the compressors were first used, operators lifted a small quantity of water; then the flow of water increased as the flow channels in the sand opened and soon the old compressors had to be replaced for newer and larger ones. By coöperating and spending about $5000 to trace the source of that water and connecting one well, it was possible to junk several thousand dollars worth of compressors.

I. N. KNAPP,† Ardmore, Pa.—My experience is that it is bad practice to wash out a well at any particular point to determine the formation. It is extremely difficult in rotary drilling to get reliable samples of the material as it is passed through. If clear or turbid water is used, the cuttings tend to dissolve or separate by the jigging action of the ascending circulation, and the heavy particles tend to lodge in the irregularities in the walls of the well between the end of the casing and the bottom of the hole. In such cases it is very difficult to find any

---

* Petroleum Technologist, U. S. Bureau of Mines.        † Mechanical Engineer.

material in the overflow from which to judge of the formation through which the drill is passing. If a heavy mud is used there is much less dissolving, separating and mixing of the drill cuttings, and on washing out a sample from the mud overflow, particles may be found that fairly indicate the material just drilled through.

If you wish to get an accurate well log with samples of an unconsolidated formation at any particular horizon with the rotary it is necessary to take a core. This is a pretty hard proposition but I have accomplished it. I have taken out as much as 12 ft. of core at a time. It is difficult to get a soft core out of the core barrel and lay it out on the ground without considerable breakage, but I did get some pieces 2 and 3 ft. long. The material was very friable and easily broken with the fingers.

I have been told of a device that has been in practical operation in the Baku, Russia, oil field for some time that will pick up an accurate sample of the formation from the walls of wells drilled in unconsolidated formations at any selected point between the end of the casing and the bottom of the well. Such a device is greatly needed in this country.

I have had top and bottom water troubles in fields where the oil sands themselves were saturated with oil only and this without a trace of water. If the top water in such cases was efficiently shut off in the first place there was no future trouble. The bottom water after once being definitely located could not cause trouble unless deliberately drilled into. We have heard a great deal about water in the same sand with the oil and of mudding and cementing this afternoon but no reference has been made to Texas and Louisiana conditions and practice. What has been said does not seem to agree with the practice down there.

Mowry Bates.—I would like to return to what Mr. Ambrose has said. I cannot see why plugging one well is going to shut off 6 or 7 acres of sand. It will shut off the water up to the level of the oil, but as the oil comes out the water comes up.

A. W. Ambrose.—We are arguing on different points. You are assuming that the water occurs in the oil sand while I am assuming that it may not be.

Mowry Bates.—It is in the Mid-Continent field.

A. W. Ambrose.—That same thing was contended by the oil operators in California until they were shown that there was a separation between the oil and the water sands. When they plugged off the water, they found there was a break that had never been considered, and the point is whether or not similar conditions hold in Augusta.

Mowry Bates.—I do not think they do.

I. N. KNAPP.—A bulletin recently issued by the Smithsonian Institution[1] contains a section on petroleum and a section on natural gas. The writer of the petroleum section claims that so long as the ownership of oil in the ground is determined by vertical property boundaries, arbitrarily dividing a geologic unit or reservoir into many parts, just so long will there be hurried production with all its train of waste and losses. The writer of the natural-gas part claims gas-field operating conditions should be regarded as a natural monopoly and that it is a primary need of the industry to have mandatory pooling of field operations, coupled with an adequate market price.

It would be a fine thing for the operator to individually control all the land in any one oil or gas pool or reservoir. It would ultimately result in the landowner getting more royalty and in the conservation of both oil and gas and also lower the cost of production. I am fearful that all the oil and gas in our country will be exhausted before such conditions can be brought about.

EUGENE COSTE.—The great trouble of water in the oil and gas fields is due to the careless man, the one who wants to pull out his casings and sell them, or the one who relies too much on the drillers and does not assure himself that the casings are properly set in the right place, and properly cemented when necessary. A careful operator will take the necessary precautions to have his casings right but when a careless man on another lease does not case off properly, he communicates the different water sands with the oil sand either from above or below. I think the sands from above are especially dangerous. When a man pulls out all his casings and does not have his well properly plugged, the greatest danger is from the fresh-water sands. The fresh water from above is the most dangerous because it gets the weight of the thousands of feet between the surface and the oil sand, which can often easily overcome the pressure in that sand.

I entirely agree with the gentleman who recommends coöperation among the operators, but would include also coöperation and regulation by the state—regulation by an authority that can make the operator give the proper information to the inspector so that he can do his work intelligently. If that is done, it will save millions of dollars worth of oil.

Even after a field has been used for many years, no strong hydraulic pressure is found to interfere with production. We know that absolutely because gas fields with original pressures of, say, 1000 lb. have had very little trouble with water, when the proper precautions were taken, even when the pressure had declined to 50 lb. in the field, and even down to nothing.

[1] See Smithsonian Institution, United States National Museum, *Bull.* 102, Pt. 6. Petroleum: A Resource Interpretation; Pt. 7. Natural Gas: Its Production, Service and Conservation.

THE CHAIRMAN (DAVID WHITE,* Washington, D. C.).—The use of the terms "primary" and "secondary" as incorporated in this paper tends to create confusion on account of the use of the terms in the discussion of the chemical composition of oil-field waters. However, in attempting to suggest substitutes, one encounters the difficulty of length and cumbersomeness of such substitute terms as indigenous or autochthonous, but he might use foreign and native waters.

I feel that the studies which have been made from the chemical standpoint of the oil-field waters—G. S. Rogers is carrying on such studies now—tend to show great value in the analysis of the waters with reference to detecting their origin. Top waters, such as those described, should be readily identified by their analyses.

---

* U. S. Geological Survey.

## Staggering Locations for Oil Wells

BY ROSWELL H. JOHNSON, M. S., PITTSBURGH, PA.

(Colorado Meeting, September, 1918)

THE prevailing system of locating wells on a rectangular basis, as shown in Fig. 1-A, has developed because of the exigencies of offsetting at boundary lines. When, however, a very large tract is being drilled, it is often possible to abandon this inferior method for the superior arrange-

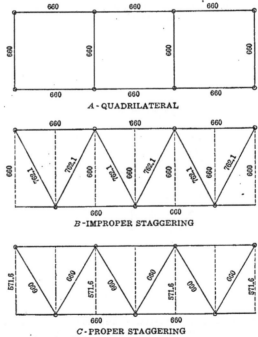

FIG. 1.—THREE SYSTEMS OF SPACING WELLS.

ment known as "staggering," which is illustrated in Fig. 1, B and C. The advantages of the triangular over the rectangular method are evident from Fig. 2, computed on the basis of one well to 10 acres (4 ha.). With the triangular system, no land is more than 409.4 ft. (124.8 m.) from a well, while with the rectangular system, 14,400 sq. ft. (0.33 acre or 0.13 ha.) lies more than that distance from a well.

The purpose of this contribution is to assist the producer in planning

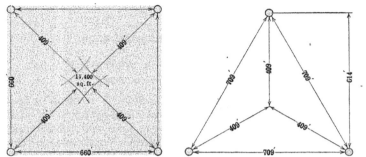

FIG. 2.—QUADRANGULAR AND QUINCUNX ARRANGEMENTS COMPARED, ON BASIS OF ONE WELL TO 10 ACRES.

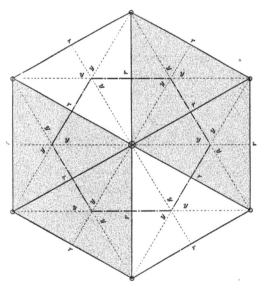

FIG. 3.—DIAGRAM FOR DEMONSTRATING FORMULAS FOR EQUILATERAL TRIANGULAR ARRANGEMENT OF WELLS, WHERE $y$ = DISTANCE BETWEEN ROWS, $r$ = DISTANCE BETWEEN WELLS.

the staggering arrangement of wells. Some companies, in staggering wells, have retained the same distance between the rows as between the wells in the row (Fig. 1-B), but to get equal distances between any two adjacent wells, which is desirable in order to yield maximum drainage, the triangles must be equilateral, which necessitates the shortening of the distance between the rows (Fig. 1-C), making the quincunx arrangement, as customary in arranging tank farms.

The formulas for computing the distance between wells thus arranged were derived from Fig. 3, representing a well surrounded by the six nearest wells; the dot-dash line midway between the central and the surrounding wells outlines the hexagon drained by this well. This hexagon consists of 12 right triangles, each of the triangles having as its altitude one-half the distance between the wells ($r$), and as its base one-third the

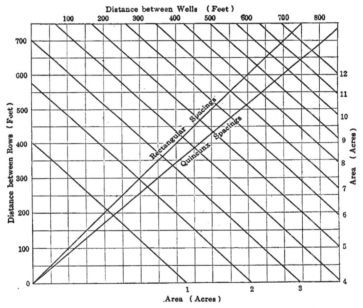

Fig. 4.—Graphic representation of relation of distance between rows, distance between wells, and area (in acres) tributary to each well.

distance between the rows ($y$). The hexagonal area drained by each well therefore reduces to $ry$.

To obtain the relationship of the distance between rows ($y$) to the distance between wells ($r$), it is only necessary to remember that the right-angled triangle having these two distances for its sides has angles of 30° and 60.° By the geometric properties of the equilateral triangle,

$$r : y : : 2 : \sqrt{3}$$
$$y = \frac{r}{2}\sqrt{3} = 0.866r$$

Correspondingly, $r = 1.154734y$

To obtain $r$ when the area tributary to one well is given: knowing that area $= ry$, and that $y = 0.866r$,

$$\text{Area} = 0.866r^2 \text{ or } 1.155y^2$$

Using these formulas, Table 1 is so arranged that for a given distance between wells, the appropriate distance between rows, and the area in square feet and in acres tributary to each well, may be read at once where the arrangement is by equilateral triangles. Similarly, Table 2, for a given distance between rows, shows the corresponding distance between wells, and the area tributary to each. Table 3 gives the proper distance between wells and between rows for a given area tributary to one well.

These relationships are shown graphically in Fig. 4, from which smaller, intermediate, and larger values may be obtained. The diagram may also be useful for arranging tanks and gas wells, and for other purposes. Beginning at the lower left-hand corner of Fig. 4, the upper line gives the appropriate distances where the arrangement is rectangular, and the lower line the distances where the arrangement is in equilateral triangles (quincunx). The diagonal lines crossing these show the area per well, in acres.

TABLE 1.—*Quincunx Arrangement, Distance between Wells Given*

| Distance between Wells, Ft. | Distance between Rows, Ft. | Area per Well | | Distance between Wells, Ft. | Distance between Rows, Ft. | Area per Well | |
|---|---|---|---|---|---|---|---|
| | | Sq. Ft. | Acres | | | Sq. Ft. | Acres |
| 300 | 259.8 | 77,940 | 1.79 | 500 | 433.0 | 216,500 | 4.97 |
| 310 | 268.5 | 83,235 | 1.91 | 510 | 441.7 | 225,267 | 5.17 |
| 320 | 277.1 | 88,672 | 2.04 | 520 | 450.3 | 234,156 | 5.38 |
| 330 | 285.8 | 94,314 | 2.17 | 530 | 459.0 | 243,270 | 5.58 |
| 340 | 294.4 | 100,096 | 2.30 | 540 | 467.6 | 252,504 | 5.80 |
| 350 | 303.1 | 106,085 | 2.44 | 550 | 476.3 | 261,965 | 6.01 |
| 360 | 311.8 | 112,248 | 2.58 | 560 | 485.0 | 271,600 | 6.23 |
| 370 | 320.4 | 110,548 | 2.72 | 570 | 493.6 | 281,352 | 6.46 |
| 380 | 329.1 | 125,058 | 2.87 | 580 | 502.3 | 291,334 | 6.69 |
| 390 | 337.7 | 131,703 | 3.02 | 590 | 510.9 | 301,431 | 6.92 |
| 400 | 346.4 | 138,560 | 3.18 | 600 | 519.6 | 311,760 | 7.16 |
| 410 | 355.1 | 145,591 | 3.34 | 610 | 528.3 | 322,263 | 7.40 |
| 420 | 363.7 | 152,754 | 3.51 | 620 | 536.9 | 332,878 | 7.64 |
| 430 | 372.4 | 160,132 | 3.68 | 630 | 545.6 | 343,728 | 7.89 |
| 440 | 381.0 | 167,640 | 3.85 | 640 | 554.2 | 354,688 | 8.14 |
| 450 | 389.7 | 175,365 | 4.03 | 650 | 562.9 | 365,885 | 8.40 |
| 460 | 398.4 | 183,264 | 4.21 | 660 | 571.6 | 377,256 | 8.66 |
| 470 | 407.0 | 191,290 | 4.39 | 670 | 580.2 | 388,734 | 8.92 |
| 480 | 415.7 | 199,536 | 4.58 | 680 | 588.9 | 400,452 | 9.19 |
| 490 | 424.3 | 207,907 | 4.77 | 690 | 597.5 | 412,275 | 9.46 |

TABLE 2.—*Quincunx Arrangement, Distance between Rows Given*

| Distance between Rows, Ft. | Distance between Wells, Ft. | Area per Well | | Distance between Rows, Ft. | Distance between Wells, Ft. | Area per Well | |
|---|---|---|---|---|---|---|---|
| | | Sq. Ft. | Acres | | | Sq. Ft. | Acres |
| 250 | 288.7 | 72,175 | 1.66 | 480 | 554.3 | 266,064 | 6.11 |
| 260 | 300.2 | 78,052 | 1.79 | 490 | 565.8 | 277,242 | 6.36 |
| 270 | 311.8 | 84,186 | 1.93 | 500 | 577.4 | 288,700 | 6.63 |
| 280 | 323.3 | 90,524 | 2.08 | 510 | 588.9 | 300,339 | 6.89 |
| 290 | 334.9 | 97,121 | 2.23 | 520 | 600.5 | 312,260 | 7.17 |
| 300 | 346.4 | 103,920 | 2.39 | 530 | 612.0 | 324,360 | 7.45 |
| 310 | 358.0 | 110,980 | 2.56 | 540 | 623.6 | 336,744 | 7.73 |
| 320 | 369.5 | 118,240 | 2.71 | 550 | 635.1 | 349,305 | 8.02 |
| 330 | 381.1 | 125,763 | 2.89 | 560 | 646.7 | 362,152 | 8.31 |
| 340 | 392.6 | 133,484 | 3.06 | 570 | 658.2 | 375,174 | 8.61 |
| 350 | 404.2 | 141,470 | 3.25 | 580 | 669.7 | 388,426 | 8.92 |
| 360 | 415.7 | 149,652 | 3.44 | 590 | 681.3 | 401,967 | 9.23 |
| 370 | 427.3 | 158,101 | 3.63 | 600 | 692.8 | 415,680 | 9.54 |
| 380 | 438.8 | 166,744 | 3.83 | 610 | 704.4 | 429,684 | 9.86 |
| 390 | 450.3 | 175,617 | 4.03 | 620 | 715.9 | 443,858 | 10.19 |
| 400 | 461.9 | 184,760 | 4.24 | 630 | 727.5 | 458,325 | 10.52 |
| 410 | 473.4 | 194,094 | 4.46 | 640 | 739.0 | 472,960 | 10.86 |
| 420 | 485.0 | 203,700 | 4.68 | 650 | 750.6 | 487,890 | 11.20 |
| 430 | 496.5 | 213,495 | 4.90 | 660 | 762.1 | 502,986 | 11.55 |
| 440 | 508.1 | 223,564 | 5.13 | 670 | 773.7 | 518,379 | 11.90 |
| 450 | 519.6 | 233,820 | 5.37 | 680 | 785.2 | 533,936 | 12.26 |
| 460 | 531.2 | 244,352 | 5.61 | 690 | 796.8 | 549,792 | 12.62 |
| 470 | 542.7 | 255,069 | 5.86 | 700 | 808.3 | 565,810 | 12.99 |

TABLE 3.—*Quincunx and Quadrangular Arrangements, Area per Well Given*

| Area per Well | | Quincunx | | Quadrangular |
|---|---|---|---|---|
| Acres | Square Feet | Distance between Wells, Ft. | Distance between Rows, Ft. | Distance between Wells, Ft. |
| 0.5 | 21,780 | 158.6 | 137.3 | 147.6 |
| 1.0 | 43,560 | 224.3 | 194.2 | 208.7 |
| 1.5 | 65,340 | 274.7 | 237.8 | 255.6 |
| 2.0 | 87,120 | 317.2 | 274.7 | 295.2 |
| 2.5 | 108,900 | 354.6 | 307.4 | 330.0 |
| 3.0 | 130,680 | 388.5 | 336.4 | 361.5 |
| 3.5 | 152,460 | 419.6 | 363.3 | 390.5 |
| 4.0 | 174,240 | 448.6 | 388.5 | 417.4 |
| 4.5 | 196,020 | 475.8 | 412.0 | 442.7 |
| 5.0 | 217,800 | 501.5 | 434.3 | 466.7 |
| 5.5 | 239,580 | 526.0 | 455.5 | 489.5 |
| 6.0 | 261,360 | 549.4 | 475.8 | 511.2 |
| 6.5 | 283,140 | 571.8 | 495.2 | 532.1 |
| 7.0 | 304,920 | 593.4 | 513.9 | 552.2 |
| 7.5 | 326,700 | 614.2 | 531.9 | 571.6 |
| 8.0 | 348,480 | 634.4 | 549.3 | 590.3 |
| 8.5 | 370,260 | 653.9 | 566.3 | 608.5 |
| 9.0 | 392.040 | 672.8 | 582.6 | 626.1 |
| 9.5 | 413,820 | 691.3 | 598.7 | 643.3 |
| 10.0 | 435,600 | 709.2 | 614.2 | 660.0 |
| 10.5 | 457.380 | 726.7 | 629.3 | 676.3 |
| 11.0 | 479,160 | 743.8 | 644.1 | 692.2 |
| 11.5 | 500,940 | 760.6 | 658.7 | 707.8 |
| 12.0 | 522,720 | 776.9 | 672.8 | 723.0 |

## Natural-gas Storage

BY L. S. PANYITY,* COLUMBUS, OHIO

(New York Meeting, February, 1919)

THE question of natural-gas supply is receiving careful consideration in many parts of the country, as in the winter months it is quite a problem to have on hand sufficient gas to satisfy the demand. Increasing the output of wells by the application of vacuum has been tried with various results and large companies have attempted to keep up the supply with gas compressors. The possibility of storing natural gas in the sands of exhausted gas pools has been tried in a few instances with satisfactory results. This method may prove of practical value in solving the problem, especially in the case of towns that formerly obtained gas from their immediate vicinity but now must search for new pools.

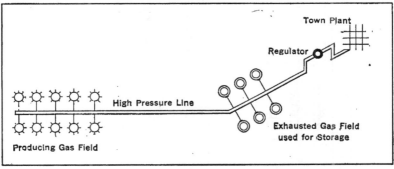

FIG. 1.

In all cases, only part of the available supply is utilized during the warm weather, so that many wells are shut-in, yet during the winter months, the supply is not sufficient even with all the wells on the line; in such cases it would be of great value if a large volume of stored gas were on hand, obtained through wells that would have been standing idle during the summer.

Idle producing wells having considerable "rock pressure" will force gas into the exhausted, or storage wells, and this gas will be used only when the regular supply falls short (Fig. 1). If two gas wells of different

* Geologist, Ohio Fuel Supply Co.

pressures are connected, the one having the greater pressure will feed the other, until the pressures are at an equilibrium. The same results will be obtained if an exhausted gas well is connected to a high-pressure gas line. High-pressure lines equipped with regulators near the town plant have considerable pressure, so that storage wells connected to such a line will receive gas from the line as long as the line pressure is greater than the well pressure. During a period of heavy consumption of gas, the pressure

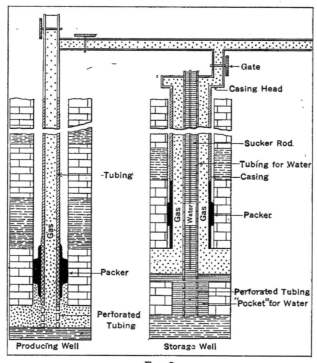

FIG. 2.

on the main line is greatly reduced, so that gas from the storage wells will flow into the lines. An arrangement of this kind will work automatically, the flow of gas into or from the storage wells depending on the pressure carried in the line.

It is advisable in most instances that the flow into and from the storage wells be regulated by means of gates, instead of automatically. In addition, the volume of gas should be metered as it is forced into or passes from the wells; the pressure also should be noted. By properly charting the meter and pressure records, the characteristics of the individual wells may be determined.

The geological conditions existing in the storage field, as well as in the producing field, must be taken into consideration. Best results are obtained where the storage reservoir is in a shallow sand and the producing horizon is deeper, so that the producing sand will have a higher pressure. This will have a tendency to reëstablish the original rock pressure in the artificial reservoir. It is unlikely that a rock pressure greater than that which existed originally can be obtained. A lenticular-shaped sand body is preferable, as a good control over the entire reservoir is necessary. The location of all wells that have been drilled must be known and put in such a condition that they may be used; if this is not possible the wells must be properly plugged.

In some instances the sand used for storage may make large quantities of water; a difficulty that may be overcome by pumping. Best results are obtained by drilling a pocket below the sand in which the water may accumulate and from which it may be pumped through the working barrel, in the same way as it is customary to pump oil, the storage and recovery of the gas being through the casing-head (Fig. 2).

This storage method may be used to advantage in many towns; for instance, Tiffin, Ohio, formerly the center of a large gas-producing area, but now dependent on outside sources, which in times of great demand are not sufficient for the needs of the town. It may be possible to find nearby an exhausted gas pool in such condition that this method may be tried. The system allows considerable latitude and may be installed to suit the requirements of the particular case.

## DISCUSSION

L. S. PANYITY.—I made inquiries from the Smith and Dunn people, who are the originators of the compressed air and gas method of increasing the production of oil wells, as to how much pressure the sand will stand; that is, relative to the original rock pressure. They tell me that they have not been able to show, up to date, any pressure in a sand greater than the original pressure. I have also made a statement to that effect, but I am really dubious as to whether that is right, as I am not sure that you cannot get a greater pressure. However, I think that a greater pressure would be dangerous because it might cause a loss of gas.

C. H. SHAW,* Lawton, Okla.—In regard to the gas that was put into the pool up in New York State, may it not be possible that, on account of errors, etc., in measuring or computing, there is a mistake in the apparent fact that more gas was gotten out than was put in?

---

* Attorney-at-Law and Petroleum Engineer.

L. S. PANYITY.—Unfortunately, I have not the papers I expected from those people to show the actual conditions, but I do not doubt that such an error might exist. It is possible that the renewed action and pressure in the well might have caused some sort of commotion there that produced a little more gas than was obtained before.

THE CHAIRMAN (DAVID WHITE,* Washington, D. C.).—The New York storage was limestone.

L. S. PANYITY.—It was limestone; they call it a flint rock, though, I believe. It is in the western part of New York and is on a dome structure called flint field. They have no other name for it.

CHAIRMAN WHITE.—Was the gas measured at the storage well both going "in" and "out?" Assuming that your well storage is in a different sand from that producing your gas, did you ever make an analysis of your incoming and, again, of the outgoing gas, to show what difference there might be?

L. S. PANYITY.—I have not obtained those figures.

EUGENE COSTE,† Calgary, Alta.—I am glad that this subject has been brought before the Institute, because it is a practical subject; I have used such a storage in exhausted sands for many years with success. The Welland field in Ontario is 30 years old, covers a large area, and the gas sands are close-grained sands. Parts of the field have exhausted because the gas was first obtained in these parts and the wells have been used constantly for 30 years. In the newer parts, where we got the gas later, we had a stronger pressure. For years, in the summer time, when we did not require high-pressure wells, we turned the gas into low-pressure wells and know that we have obtained a larger quantity of gas in the winter time by that method.

This method was suggested to me by the action of a well south of Buffalo, which we used to call Big Zoar Well. It belonged to the United Co. piping gas from Pennsylvania to Buffalo and was close to their main line. The gas was in a limestone, filled with flint of the Corniferous formation. It was a large well, 15,000,000 cu. ft., but it would get down to about 800,000 cu. ft. after a few days of use. As it was very much nearer Buffalo than the main supply in Pennsylvania, this well was kept closed in until we had a long cold spell causing gas pressure to drop too low. Then the Big Zoar Well was turned in and it would help the Buffalo Company to get over the cold spell nicely. Immediately the pressures picked up again the company would close the well in. In that way it

---

* U. S. Geological Survey.

† President and Chief Engineer, Canadian Western Natural Gas, Light & Power Co., Ltd.

was found that this well was a splendid storage and was worth hundreds and thousands of dollars to that company. That gave us the idea of doing the same thing, using the storage in the Canadian field of the more or less exhausted parts of the field. For years now, we have transferred gas from one part of the field, which was under pressures of say, 300 lb. to another part under pressure of 100 lb.; also from 3000-ft. wells where gas was obtained into another limestone, the Trenton, where we had 1000-ft. pressure. These we often kept feeding in the upper rock. We measured the wells and know that after thus feeding in them for a while the low-pressure wells would recover a great deal and that instead of measuring 200,000 to 400,000 cu. ft. they would measure up to 1,000,000 cu. ft. or more. In that way we were in better shape to meet the cold weather the following winter. It wouldn't be surprising if by that method one could obtain a little more gas from the low-pressure part of the field than would otherwise be possible. I mean of its own gas, of the gas belonging to these parts of the field, because in that way the gas has a chance to flow under higher pressure, or compression, for a longer period. Everybody knows that as the pressure goes down, the friction increases, so that where the gas couldn't travel at all and enter the well under the low pressure, by building the pressure around that well up to, say, 200 lb. the gas will come in, because the friction will be less at the higher pressure. In that way, not only will one recover all the gas put in, but some that could not be gotten otherwise.

L. S. PANYITY.—There may be some difficulty in getting the land needed for such purposes. There is no doubt that it will be a hard thing to do with three or four operators in a certain field, without their coöperation. Besides, it will be hard to acquire the land if the owners know what you want to use it for. I think, however, that difficulty can be overcome for a small fee, or by the promise to the owner of a certain amount of free gas.

I. N. KNAPP,* Ardmore, Pa.—I believe the free-gas conditions of a lease should have some definite limit. Unconditional "free gas" leads to abuses and great waste. For lease or other considerations, as rights of way, it should be limited to amounts to be determined by meter measurements or by specifying the number of gas lights allowed and if for fuel use by specifying the number and size of the mixers allowed.

The possibility of storing natural gas in sands of exhausted gas pools might be, in many cases, a good engineering proposition, but it cannot be considered a practical business proposition under the usual competitive conditions that exist in the natural-gas business. Under the usual conditions of diversified land ownership and leasing with all lack

---

* Mechanical Engineer.

of coöperation, various operators drill into the same pool or reservoir and drain out the gas under each other's land. To attempt to use for storage purposes the depleted sands of any one pool having such various competing ownerships would be highly impractical, considered from operating and business standpoints, unless there was coöperation with the interests owning adjacent wells and leases or the purchase of the same. Also, it would be very difficult to find out whether all old or abandoned wells in any pool had been properly plugged and thus prevent the waste of gas brought from a distance for storage.

Another objection to Mr. Panyity's proposition is the probability that no lease made in the past contemplated the use of a gas sand for storage purposes. Very likely, then, the operator would have a lawsuit on his hands for the unlawful use of a lease if he attempted the storage of such gas without first obtaining additional lease rights.

In some instances salt water could not be pumped from a gas well in the customary way of pumping oil wells on account of the deposition of salt in the pump valves or in the working barrel or tubing. Very ingenious compressed air jets have been devised and arranged with suitable piping to blow out such salt-water accumulations. In many pools it would be impracticable to drill a water pocket below a gas sand, as suggested, since such deeper drilling would probably intensify water troubles.

I know that it is perfectly practical to compress air and use a depleted gas sand for storing the air from the actual use of such a method for a period of seven years. In this case the compressed air was distributed by pipe lines for use in field pumps for gathering oil and in blowing oil out of wells, instead of producing by the usual oil pumping methods. The wells were about 720 ft. in depth. With this storage system, many times the capacity of the air compressor could be used for short intervals without any appreciable loss of air pressure and when no air was being used there was no appreciable gain in pressure. Two two-stage compressors were used with approximately 200 hp. gas-engine capacity; we run 24 hr. per day. I do not remember the compressor capacity in cubic feet of free air. The original rock pressure of the depleted sand was 305 lb. and the open flow capacity of the well as a gasser was around 2,000,000 cu. ft. per 24 hr. When the air was attached to the well, the gas had been used down to a rock pressure of 115 lb., the production was small, and the gas was wet and troublesome to use. In the course of a few months' use of the compressors, a rock pressure of 290 lb. of air was developed and there was no water trouble. The extent of the sand used for air storage was small, probably not over 20 or 25 acres, as shown by dry holes and exhausted wells.

The thickness of the porous gas sand probably did not exceed 7 ft. I imagine the occasional gas sands of this region were like isolated hillocks of sand resting on the salt-water sand that underlaid the whole country. The depth to the gas sand was around 800 ft. and it underlaid the oil sand. I felt that I could do no possible damage by using this small depleted gas sand for air storage. But to avoid possible trouble I purchased the fee of 80 acres of land so as to cover fully the area used for the purpose mentioned.

We hope we may soon have enforced collective coöperation in the natural-gas business together with a proper franchise price. In the past the price paid for gas delivered to the consumer has never been commensurate with the price of the fuel it has displaced. When the present governmental attitude, which tends to prevent unity of action in the gas fields, is abandoned we may hope to see Mr. Panyity's project for the storage of gas utilized to the full extent that it deserves.

### Losses of Crude Oil in Steel and Earthen Storage

BY O. U. BRADLEY,* E. M., MUSKOGEE, OKLA.

(Colorado Meeting, September, 1918)

THE extent of losses, due to evaporation, sediment, and water, in crude oil stored in steel tanks, is a very interesting question, and particularly so at this time, when every reasonable measure should be employed to eliminate all possible losses of this important natural product. Available information on this subject is incomplete; e.g., during the development of the Cushing Field, considerable surplus oil was stored in steel tanks, but from time to time, owing to changes in weather, some losses by evaporation and short storage room on the leases, these tanks were topped out, thus rendering inaccurate any system designed to determine the average rate of evaporation or other losses of the oil over a given period. Furthermore, the losses, as shown by the records at the time, as a result of hasty gages, failed to take into account the temperature and gravity of the oil. Losses may be classified roughly as occurring from evaporation, presence of sediment and water, and leakage. The coefficient of expansion and contraction of crude oil, in relation to temperature conditions, is of material importance; also, the rate of evaporation is dependent upon the gravity of the oil, as the escape of the lighter hydrocarbons in fresh Cushing oil, of from 40 to 42 gravity, on a warm day is considerable, and from available records of temporary storage, it is safe to assume that a loss of 1 to $1\frac{1}{2}$ per cent. in volume will easily occur in light crude oil of Cushing grade over a period of 6 months, including the summer season. The presence of sediment, water, and other impurities, in the oil will also cause more or less deterioration in quality, particularly when the fresh production is run direct from the gage tanks, or from the wells on the lease to steel storage, as is often the case when an oil field is being rapidly developed. The bad oil will settle to the bottom of the tanks and may be determined by thiefing and running a centrifuge test on the sample. The amount of water in the oil depends entirely on the conditions under which it is produced. Sometimes in large producing wells, with heavy gas pressure, some water may come in with the oil, particularly if the well has been drilled to the top of a water sand. Under these conditions, cut oil is often produced, which, if not settled out, is carried over into the steel tanks.

The losses by leakage in steel tanks are difficult to determine, for

---

*Oil and Gas Inspector, Department of the Interior.

the reason that some tanks are built better than others. Contractors are required to turn out work that will reduce the losses due to leakage to a minimum, but in hurried building, necessitated by emergency conditions in a rapidly developing oil field, the character of the work is often not up to standard.

A record of losses, due to sediment and water, on Cushing stored oil, covering a period of from 5 to 7 months, is shown in Table 1.

TABLE 1.—*Gage Showing Sediment and Water*

| Tank No. | Gage | | Sediment | | Water | | Total Per Cent. |
|---|---|---|---|---|---|---|---|
| | Date of Filling | Date of Test | Barrels | Per Cent. | Barrels | Per Cent. | |
| 1 | 55035.26 5/31/1914 | 54810.47 1/31/1915 | 970.29 | 1.7 | 513.18 | 0.9 | 2.6 |
| 2 | 55603.76 6/9/1914 | 54470.27 1/31/1915 | 665.04 | 1.2 | 513.39 | 0.9 | 2.1 |
| 3 | 54894.52 6/18/1914 | 54140.57 1/31/1915 | 1045.83 | 1.9 | 512.25 | 0.9 | 2.8 |
| 4 | 54907.10 6/28/1914 | 53816.85 1/31/1915 | 970.68 | 1.8 | 284.98 | 0.5 | 2.3 |
| 5 | 54478.39 7/15/1914 | 54102.88 1/31/1915 | 744.35 | 1.3 | 591.64 | 0.1 | 2.3 |
| 6 | 54982.84 8/9/1914 | 54643.34 1/31/1915 | 897.30 | 1.6 | 591.80 | 0.1 | 2.6 |
| 7 | 54653.97 6/15/1914 | 53937.81 1/31/1915 | 862.50 | 1.5 | 431.25 | 0.7 | 2.2 |
| 8 | 54931.53 6/30/1914 | 53724.89 1/31/1915 | 843.83 | 1.5 | 555.23 | 1.3 | 2.8 |
| 9 | 54766.65 7/4/1914 | 53812.81 1/31/1915 | 766.90 | 1.4 | 630.41 | 1.1 | 2.5 |
| 10 | 54784.53 7/17/1914 | 53767.56 1/31/1915 | 653.25 | 1.2 | 440.27 | 0.8 | 2.0 |

No record of temperature or gravity was kept, and therefore it is not possible to give the evaporation. The possible leakage was not considered.

In measuring the oil in steel tanks, in order to determine losses, the depth is taken at the two hatch holes by steel tape, and an average is computed from the results. This steel tape should be graduated to $\frac{1}{16}$ in. and a 5-lb. plumb-bob is preferable to one of lighter weight.

It is known that losses in earthen storage are high, but to the author's knowledge there was but one case in the Cushing Field where an attempt was made to determine such losses, and then no temperatures or gravities were recorded. The reservoir used was prepared as carefully as possible in the face of the emergency due to the necessity of providing for a large production on the property. The banks were tamped, the inner surface puddled and the whole reservoir roofed over shortly after filling. The oil was run in during the hot summer months and removed before cold weather set in.   The record on filling, removal and percentage of loss is as follows:

BARRELS

Time of filling and total number of gross barrels run in
May 6 to June 28, 1914, inclusive................ 35,039.54
Removal of merchantable oil, Aug. 9 to Sept. 2, 1914.. 28,651.10
Loss........................................... 6,388.44
Percentage..................................... 18.2

In the case of Healdton oil, which is of heavier gravity and somewhat different in character, the following is a record of a loss on three steel tanks, two of which were filled and one partly filled direct from the wells on the lease as the oil was produced.   The time of filling these tanks covers a period of 8 months, the months and the quantities of oil run into them being as follows:

| MONTH | TOTAL BARRELS |
|---|---|
| Aug., 1916..................................... | 10,463.34 |
| Sept., 1916.................................... | 17,571.74 |
| Oct., 1916..................................... | 22,964.93 |
| Nov., 1916..................................... | 20,803.99 |
| Dec., 1916..................................... | 20,441.20 |
| Jan., 1917..................................... | 15,432.25 |
| Feb., 1917..................................... | 9,627.32 |
| To Mar. 12, 1917.............................. | 3,220.15 |
| | 120,524.92 |

A portion of the last run of oil was used to top out the two full tanks, so that on March 12, the measurement and total gross barrels in each tank were as shown in Table 2.

TABLE 2

| Tank No. | Measurement | | Gross Barrels |
|---|---|---|---|
| | Ft. | In. | |
| 1 | 29 | 10 | 52,806.36 |
| 2 | 29 | 4⅝ | 52,014.52 |
| 3 | 8 | 11¼ | 15,704.04 |
| | | | 120,524.92 |
| Gravity of the oil, 30 | | | |

The quantity of oil checks exactly with the total quantity run, as shown above, because no gage was taken other than the total monthly amount run into each tank, which was always calculated from the difference between the previous month's measurements and those of the month for which the record was desired.

The record on the purchase of the oil, after tanks had been restrapped, with tests for sediment and water, is shown in Table 3.

TABLE 3

| Tank No. | Date of Test | Measurement | | Gross Barrels |
|---|---|---|---|---|
| | | Ft. | In. | |
| 1 | June 30, 1917 | 29 | 10½ | 52,880.24 |
| | | Sediment..... 1 | 3⅛ | 2,038.48 |
| | | Water........ .. | 2½ | 358.17 |
| | | Merchantable oil.................. | | 50,483.59 |
| 2 | June 17, 1917 | 29 | 4⅞ | 52,051.46 |
| | | Sediment..... 1 | .... | 1,754.33 |
| | | Water........ .. | 2¾ | 389.11 |
| | | Merchantable oil.................. | | 49,908.02 |
| 3 | June 13, 1917 | 8 | 11⅜ | 15,722.54 |
| | | Sediment..... 4½ | .... | 647.24 |
| | | Water........ , | 1⁵⁄₁₆ | 143.35 |
| | | Merchantable oil.................. | | 14,931.95 |

|  | PER CENT. | |
|---|---|---|
| Tank No. 1—Sediment......................... | 3.85 | |
| Water........................... | 0.67 | |
| | | 4.52 |
| Tank No. 2—Sediment......................... | 3.37 | |
| Water........................... | 0.74 | |
| | | 4.11 |
| Tank No. 3—Sediment......................... | 4.11 | |
| Water........................... | 0.91 | |
| | | 5.02 |

These losses are high, I believe, as undoubtedly some of the heavy oil in these tanks could have been sun cured and recovered; and, furthermore, it was flush production run to steel, which is not the usual practice in the storage of crude oil.

Table 4 gives a record on six 55,000-bbl. tanks in the Healdton Field, showing total losses, due to evaporation, leakage, and bad oil, over a period of 11 months:

TABLE 4

| Tank No. | Standard Temp. and Gvty. Correc. Vol., March 1, 1917 | Correc. Vol., Jan. 31, 1918 | Leakage and Evap., Barrels | Sediment and Water, Barrels | Total Loss, Per Cent. |
|---|---|---|---|---|---|
| 1 | 52,722.46 | 52,395.13 | 327.33 | 1,241.38 | 2.97 |
| 2 | 52,428.09 | 52,162.95 | 265.14 | 1,088.02 | 2.58 |
| 3 | 53,000.32 | 52,122.02 | 878.30 | 1,202.78 | 3.92 |
| 4 | 51,452.78 | 50,724.22 | 728.56 | 1,049.89 | 3.40 |
| 5 | 53,485.61 | 53,086.59 | 399.02 | 1,294.89 | 3.16 |
| 6 | 53,605.75 | 53,064.02 | 541.73 | 991.12 | 2.85 |

The above results do not.show any considerable uniformity as regards evaporation and leakage. I believe this is due, in large measure, to the variations in leakage. An examination of the conditions of these tanks in the field shows the familiar oil stains on their sides. This is particularly true of tanks 3, 4 and 6, while others are almost free from stains of this character. The quantity of the bad oil in the tanks seems to average up to about what would be expected in crude oil that has been in storage for a period of 11 months. Inasmuch as the water and sediment were not separately determined in the tests, it cannot be stated in what proportion these two materials were present. Under treatment or steaming, a certain percentage of what is termed bad oil could be recovered. The general average percentage of loss, due to deterioration in quality of oil, leakage, and evaporation in these tanks was, therefore, 3.15 per cent. for the period above specified.

## Gaging and Storage of Oil in the Mid-Continent Field

BY O. U. BRADLEY,* A. B., MUSKOGEE, OKLA.

(Colorado Meeting, September, 1918)

THE methods of handling the oil output of the Mid-Continent fields are not unlike those practised in other oil fields of the United States, and it is not expected that this paper will present any entirely novel ideas. Some features of the practice of handling this oil may be of interest, however.

### WOODEN GAGING TANKS

Practically all of the oil sold to pipe lines and other purchasing agencies throughout the Mid-Continent field is gaged from wooden tanks. Steel gaging tanks are slowly coming into use, and are preferable in many ways—leakage and evaporation are reduced, and the strapping and recalculation of capacities are unnecessary. The wooden tanks are slightly coned to facilitate tightening the hoops. The sizes in common use are 250, 500, 800 and 1600 bbl. The number and the capacity of the gaging tanks upon any producing oil property depend upon its production, and the facilities for disposing of the output. The 1600-bbl. size was the most common in the Cushing field, particularly during the period of flush production.

The tanks are distributed over the leased premises in groups of two, three, and four tanks each, and these are filled from flow tanks, or through direct pipe connections with the heads of the wells. Flow tanks are merely temporary reservoirs to facilitate the transfer of oil from the wells to the gaging tanks. They diminish fire risk, and are considered indispensable on small producing properties where water is pumped with the oil, as they permit the refuse and water to settle out before the production is run to the gaging tanks.

### Strapping and Calculations

After a gaging tank has been strapped (measured) and its contents computed, a deduction of 3 per cent. is made from its computed volume. The strapping of a tank requires four steps: (1) The outer circumferences

* Oil and Gas Inspector, Department of the Interior.

of the tank, at selected heights, are determined by steel tape; (2) the thickness of ten staves is averaged, in order to arrive at inside diameters; (3) the depth is measured; (4) the space occupied by vertical supports for the roof is figured. The following two examples show complete data of two strappings of 1600-bbl. tanks. It will be noted that the measurements, in the first case, were taken at the prescribed positions; while in the second, the actual measurements are converted to the prescribed positions.

TANK No. 139.—*Amy Ferguson Farm*, 1600-*Bbl.*

| Height above Bottom | Circumference, Feet | Difference in Circumferences |
|---|---|---|
| 0′ 6″ | 88.09 | |
| 2′ 6″ | 86.96 | 1.13 |
| 4′ 6″ | 85.95 | 1.01 |
| 6′ 6″ | 84.84 | 1.11 |
| 8′ 6″ | 83.60 | 1.24 |
| 10′ 6″ | 82.37 | 1.23 |
| 12′ 6″ | 80.89 | 1.48 |
| 14′ 6″ | 79.54 | 1.35 |
| | 672.24 | |

These measurements were all taken at the regular positions required by the accepted system of strapping.

Stave 2 in. thick. Deadwood, 23 pieces of 2 by 6-in. uprights. 110 ft. of 6 by 2 in. across top. 80 ft. of 1 by 6-in. braces at 6 to 7 ft. from bottom of tank.

TANK No. 420.—*James Durant Allotment*, 1600-*Bbl.*

| Actual Measurements | | Interpolated Measurements | | Difference in Circumferences |
|---|---|---|---|---|
| Height above Bottom | Circumference, Feet | Height above Bottom | Circumference, Feet | |
| 1′ 0″ | 87.86 | 0′ 6″ | 88.16 | |
| 2′ 6″ | 86.96 | 2′ 6″ | 86.96 | 1.20 |
| 5′ 0″ | 85.62 | 4′ 6″ | 85.89 | 1.07 |
| 6′ 6″ | 84.73 | 6′ 6″ | 84.73 | 1.16 |
| 9′ 0″ | 83.23 | 8′ 6″ | 83.53 | 1.20 |
| 10′ 6″ | 82.33 | 10′ 6″ | 82.33 | 1.20 |
| 12′ 6″ | 81.12 | 12′ 6″ | 81.12 | 1.21 |
| 14′ 6″ | 79.94 | 14′ 6″ | 79.94 | 1.18 |
| | | | 670.66 | |

Stave, 2 in. thick. Depth, 15 ft. 6½ in. Deadwood, 11 pieces 2 by 6 uprights. 90 ft. of 1 by 6 braces 6 to 7 ft. from bottom (see Fig. 1).

From the figures derived in this manner, tables are computed showing capacity in barrels at different depths, this method having been adopted universally by pipe lines and producers alike. Fig. 2 is a graphic illustration of the manner in which the calculations are made. The volume of the tank is calculated at different zones, and a computed constant is applied over a given vertical distance. Comparing a number of gage tables, it will be found that the zones are not selected with any regularity; in another table it would probably be found that the first constant of volume would change at 6 in., instead of 9 in., then at 1 ft., 1 ft. 3 in., 1 ft. 9 in., etc., instead of at the points shown in Fig. 2. It is observed, however, that changes occur at heights which are multiples of 3 in. These zones are all frustums of cones and are considered perfect in form. Their volumes are computed for depths of ¼ in., in the manner illustrated by the following calculation of an average ¼-in. frustum in the zone between 6 ft. 6 in. and 7 ft. 6 in. of Fig. 2.

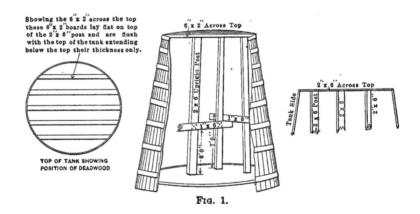

FIG. 1.

Circumference at 6 ft. 6 in.......................... 84.75 ft.
Circumference at 7 ft. 6 in.......................... 84.16 ft.

Mean circumference................................ 84.45 ft.
Mean outside radius............................... 13.44 ft. = 161.28 in.
Average thickness of staves........................ 2.00 in.
Mean inside radius................................ 159.28 in.
Volume of frustum = $\pi$ (159.28)² × ¼ ............ 19,925.68 cu. in.
Deadwood, 11 pieces, 2 by 6 by ¼ in................. 33.00 cu. in.

Net contents...................................... 19,892.68 cu. in.
Barrel = 42 gal. × 231 cu. in...................... 9,702.00 cu. in.
Capacity in barrels............................... 2.05 bbl.
Standard contents, deducting 3 per cent............ 1.988 bbl.

## STEEL TANKS

The steel tanks in common use in the Mid-Continent field are 35,000 and 55,000 bbl. Fig. 3 shows the ground plan of a 55,000-bbl. steel

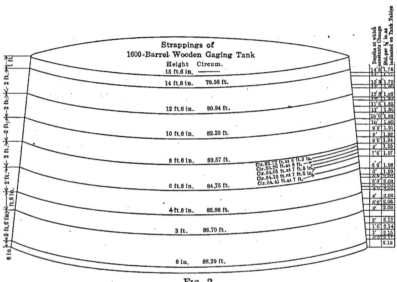

FIG. 2.

tank and fire wall; also the usual arrangement of steel storage tanks on a 160-acre (64-ha.) tank farm. The actual area covered by a 55,000-bbl. steel tank, with its fire guard and wall, is usually 1.55 acres. Six tanks may be located on 40 acres (16 ha.), at 600-ft. (182-m.) centers, 12 tanks on 80 acres, and 20 tanks on 160 acres, thus requiring an average area of 6.66 acres for each tank on an 80-acre farm, and 8 acres on a 160-acre farm. The dimensions for 55,000-bbl. tanks, are as follows:

|  | FEET |
| --- | --- |
| Diameter of shell tank | 114.5 |
| Width of fire guard, including wall | 90.0 |
| Height of wall | 6.0 |
| Base of wall | 20.0 |
| Crown of wall | 4.0 |

The contract price for grading tank sites includes the construction of the fire guard, and varies from $350 to $2000, according to the topography of the ground and the necessity for rock excavation. The proper construction of a fire guard requires about 1500 cu. yd. (1147 cu. m.) of dirt.

The support of the roof requires considerable timber, the cubical displacement of which is carefully figured and deducted. The following

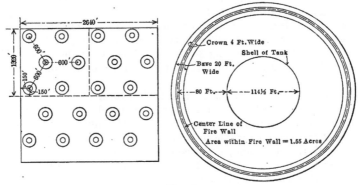

FIG. 3.

list covers the approximate amount of deadwood required in the construction of a 55,000-bbl. steel tank 30 ft. high.

| Posts | | Inch | No. of Pieces | Sills | Inch | Ft. |
|---|---|---|---|---|---|---|
| Outside circle | 24 | 6 by 6 | 24 | | 2 by 8 | 16 |
| Second circle | 20 | 6 by 6 | 20 | | 2 by 8 | 14 |
| Third circle | 16 | 6 by 6 | 16 | | 2 by 8 | 12 |
| Fourth circle | 8 | 6 by 6 | 8 | | 2 by 8 | 10 |
| Center post | 1 | 6 by 6 | 68 | | | |
| | 69 | | | | | |

The posts are tied together by two rings of 1 by 6. This takes 136 pieces of the same lengths as the sills. The rings are also tied together radially, requiring 128 pieces of 1 by 6, 15 ft. long. The plates on top of the posts, tying them together, are not included in the deadwood calculations; they are all 5 by 12, of the same number and length as the sills. The total volume of the above list of deadwood, excluding plates, is thus 779.52 cu. ft.

### VARIATION AND CALCULATION OF VOLUMES

When a wooden tank gets to leaking badly, the hoops must be driven and a restrap made, from which a new table is figured. The frequency of the strapping depends upon the kind of wood used, its condition, and the manner of constructing the tank; also upon the weather. A restrap of a 1600-bbl. tank will show a difference in capacity of from four to 20 bbl. or more, depending upon the condition of the tank when the hoops are driven. For example, at the end of 10 months the restrapping of a 1600-bbl. tank reduced its capacity, at 14-ft. depth, from 1314.12 to 1297.80 bbl.

Gage poles used to measure tanks are graduated to ¼ in., to which the smallest volume calculated in the tank tables corresponds.

Computation of oil runs is very simple. Having a table for a 1600-bbl. tank and one of the run tickets showing the measurements of depth, suppose that the top measurement or "turn-on" is 15 ft. 2 in.; this denotes a capacity of 1427.16 bbl. The bottom measurement, or "shut-off" may read 1 ft. 7 in., indicating 366.04 bbl. direct from the table. The difference is 1061.12 bbl., representing the quantity of oil run from that tank.

A comparison of a number of tank tables, those issued by pipe-line companies for the purchase of oil, and those computed by lessee companies for the gaging of their production, shows that the differences are slight, not exceeding those chargeable to the personal equation of the measurer. For example, the same tank, at the same depth, was calculated as 1239.24 bbl. by one party and 1241.40 bbl. by another, only a week earlier. In many other cases, likewise, the differences have been found to be practically negligible.

### LEAKAGE AND EVAPORATION

Leakage from tanks is frequent, and gagers always make deductions for it. The amount deducted is arbitrary, and varies according to their judgment, both as to the time required to empty the tank and the size of the leak. Producers have generally been perfectly willing to submit to such deductions in order to have their oil run, particularly when there is flush production in any district.

Evaporation losses vary according to the gravity of the oil, weather conditions, and time of exposure. Field-tests with gaging tanks, on light-gravity Cushing Field oil, have shown from 1 to 2 in. shrinkage, or 8 to 16 bbl. lost from a 1600-bbl. tank in 24 hr. Under varying conditions, with rapid handling of the oil, it was not possible to get accurate data, as temperatures were not observed, and as a part of the loss in volume was due to the escape of occluded gas. On tanks freshly filled from wells making gas along with the oil, under considerable pressure, pipe-line concerns will sometimes make an arbitrary deduction of 1 to 1½ in. to allow for subsequent shrinkage of volume. This is not done, however, except when it is necessary to turn on green tanks at once to maintain available storage capacity on the lease. The producer soon remedies such a condition by erecting more tankage so as to allow a reasonable time for full tanks to settle before turning them on the line.

### PRODUCTION ACCOUNTING

When large quantities of oil are being produced from a property it is usually desirable to keep a record on the daily movements of oil, showing

## FORM 701.—*Stock Report*

Date,........................Owner..........................Lessee,........................
Stock, 7 *A.M.*

| Battery No. 1 | | | Battery No. 2 | | | Battery No. 3 | | |
|---|---|---|---|---|---|---|---|---|
| Well Nos.................... | | | Well Nos.................... | | | Well Nos.................... | | |

| Tank No. | Ft. | In. | Tank No. | Ft. | In. | Tank No. | Ft. | In. |
|---|---|---|---|---|---|---|---|---|
| 416 | 12 | 10 | 402 | 14 | 3 | | | |
| 417 | 2 | 4½ | 403 | 7 | 1 | | | |
| 398 | 15 | 3½ | 341 | 1 | 7 | | | |
| 418 | 9 | 5 | 384 | 12 | 5 | | | |
| Total, | 39 | 11 | Total, | 35 | 4 | Total, | | |

| | Ft. | In. |
|---|---|---|
| Battery No. 1 | 39 | 11 |
| Battery No. 2 | 35 | 4 |
| Battery No. | | |
| Total stock, | 75 | 3 |

## *Oil Run*

| Company | Tank No. | On | | Off | | Amount Run | | Barrels |
|---|---|---|---|---|---|---|---|---|
| | | Ft. | In. | Ft. | In. | Ft. | In. | |
| | | | | | | | | |
| | | | | | | | | |
| | | | | | | | | |
| | | | | | | | | |
| Prairie, O & G.............. | 66,765 | 15 | 1 | 1 | 11 | 13 | 2 | 1451.12 |
| Kathleen.................... | 402 | 14 | 3 | 1 | 8 | 12 | 7 | 1169.52 |
| | | | | | Total | 25 | 9 | 2620.64 |

| Production lost, account: | Ft. | In. |
|---|---|---|
| Cuts. ......... | | |
| Evaporation.... | | |
| Cleaning tanks. | | 2½ |
| Total....... | | 2½ |

| | Ft. | In. |
|---|---|---|
| Stock today.................................. | 75 | 3 |
| Stock following day........................... | 81 | ½ |
| Increase (or decrease) in stock last 24 hr........... | 5 | 9½ |
| Total amount of oil run today................... | 25 | 11½ |
| Total day's production......................... | 31 | 9 |

........................
Gager.

losses by evaporation, pipe-line deductions for leakage, waste when cleaning tanks, etc., thus permitting a balance to be made each day between the total amount of oil produced, the amount on hand, the pipe-line runs, and the losses for the day. For this purpose, Form No. 701 may be used.

By the use of this form (in connection with No. 702, detailing production, not here reproduced for lack of space) the complete cycle of turning on, filling, shutting off, and running to pipe lines, may be kept with the measurement and amount in each tank in every battery on the lease. Any irregularity or losses will at once appear in the report. When the lease is running on short storage room and it is necessary to switch tanks frequently, the forms constitute a valuable record of the manner in which the oil is handled. On leases with flush production, when tanks are switched night and day, the only error that may occur is to allow a stream of oil to run into a tank from which the pipe line is taking oil; this would have to be done deliberately if the time of "shut-off" on the record of production going into each tank is kept carefully. Furthermore, a decided decrease in oil production on any day may be noted and its causes investigated. The forms may be made to show the time of "turn-on" and "shut-off" on all pipe-line runs when there is sufficient tankage on the lease to allow the oil to settle out, but when tanks are being filled and emptied in rapid succession the time of the "turn-on" of any tank to the pipe line company would be practically simultaneous with the "cut-off" of the incoming stream on the production record; and the "shut-off" of the pipe line would be the same as the "turn-on" of the stream to fill the tank again.

## GAGE TICKETS

Tickets for recording runs of oil from producing leases are issued in several slightly different forms, of which two are shown on the following page. (1) Ticket of lessee company with a pipe-line company; (2) ticket for runs of oil to storage by lessee company. The essential information on every ticket includes the name of the lessee, the farm owner, the pipe-line company, ticket number, size of the tank from which oil is run, "turn-on" and "shut-off" measurements, from which the number of barrels is calculated from the tank tables.

## GAGING OF OIL IN STEEL TANKS

Owing to the large size of steel tanks, it is not practicable to use them on producing properties; they are employed primarily for the storage of oil by pipe lines or other purchasing agencies. Measurements are usually made with a steel tape graduated to $\frac{1}{16}$ in. The gaging of steel tanks by different parties will frequently show slight differences, due to one or more of the following causes:

1. The tank may not be absolutely level, making the oil deeper on one side than on the other.

### ALMA OIL CO.

CHECK WITH·

### MAGNOLIA PIPE LINE COMPANY

Dist. No.................}    *Feb. 21* .............191 *6*.

Steamed.......... Degs.

Ticket No. *3320* ..........

Cold.................Deg s

................*Roxana Pet. Co.*.................... Owner

................*Rayson Wesley*....................Farm.

| Tank No. | | FEET | INCHES | Office Calculations |
|---|---|---|---|---|
| *3209* | OIL FIRST MEAS'T | 6 | 9 | *423.44* *65.34* |
| Size *500* | SECOND MEAS'T | 1 | ½ | *358.10* |
| Well Nos. *1 to 4* | B.S. 1st. M. | | | |
| | 2d M. | | | |
| | WATER 1st. M. | | | |
| | 2d M. | | | |

By    *Suction Steel Tank No. 17*

.................................................. Station

Remarks:
   *Duplicate Ticket*

   *No. 3320 Roxana Pet. Co.*

| TIME | 1st Meas.......... M. | Witness: | *J. F. McGuiger* |
|---|---|---|---|
| TIME | 2d Meas....... M. | Witness: | |

GAGE TICKET NO. 1

### GATES OIL COMPANY

*July 23,* ...........191 *6*.

Ticket No... *130* ..........

   *Gates Oil Co.* .............. Owner of

Wells No's. .. *14 to 17  19 to 21*

................*Rayson Wesley*................ Farm.

| Tank No. | | FEET | INCHES | Official Calculations |
|---|---|---|---|---|
| 7 | OIL First Meas't | 7 | 5½ | *468.02* *62.64* |
| Size *500* | Second Meas't | 1 | 0 | *405.38* |

By    *Old 1194*

Station

Remarks    *New 1237*

       Operator's check

| TIME | 1st Meas't.... *E. R. Mann* | Gager. |
|---|---|---|
| | *815* A. M. *J. W. Loughry* | Well owner's witness. |
| TIME | 2d Meas't.... *E. R. Mann* | Gager. |
| | *815* A. M. *J. W. Loughry* | Well owner's witness. |

GAGE TICKET NO. 2

2. The plates on the bottom of the tank may be sprung down or up at the points at which the measurements are taken, under the manholes.

3. The personal equation, that is, the manner of handling the steel type.

4. Sometimes the sludge, or B.S., at the bottom of a steel tank will vary in thickness or consistency, and unless the plumb bob is heavy, the bottom may not be accurately recognized and sagging of the tape may give high readings.

## LOSS OF OIL FROM STEEL TANKS

In the calculation of the average rate of shrinkage or evaporation and deterioration of crude oil over a period of time, it is necessary to take into account the temperature, the character and gravity of the oil, and the climatic conditions.

A number of grades of oil are produced in the Mid-Continent field, and, while practically all are of lighter gravities, some heavy oil has been stored in southern Oklahoma. No accurate data as to the exact loss during storage in Oklahoma are available, but all the circumstances point to the conclusion that deterioration and shrinkage of fresh oil run to storage will be at least 3 per cent. during the first 12 months, and possibly 2 per cent. during the following 12 months. It is well known that the sampling of a 55,000-bbl. tank of crude oil will show a variation in quality and gravity from the bottom to the top of the tank; this is easily demonstrated by thiefing the tank at successive zones, and making centrifugal tests of the samples. Sediment and other adulteration will appear most prominently toward the bottom of the tank.

There are no absolutely air-tight coverings over any steel tanks in the Mid-Continent field, and the escape of the lighter constituents of the higher-grade crude oil by evaporation is very rapid during the hot summer months. Losses from this cause are considerable, and the more serious because the lighter elements, which escape, are the more valuable, being very productive in the manufacture of gasoline. One company in this field operates a topping plant, taking off some of the lightest constituents of its stored oil and running the remainder back to steel storage. The company thus obtains a certain proportion of the gasoline in the crude oil which otherwise would evaporate, and the general value of the oil is not reduced to any great extent, particularly when it is remembered that the crude oil must stand for a period of one or two years in steel tanks before it undergoes the usual refining operations.

## Principles of Mining Taxation*

BY THOS. W. GIBSON,† TORONTO, ONT.

(New York Meeting, February, 1919)

THE object of taxation is the raising of a revenue. Unless a tax accomplishes this, it is a failure. · The right to take for public purposes a part of the moneys obtained from the carrying on of private enterprises is an attribute possessed only by a lawful government, and may indeed be called the supreme attribute of government. In free, democratic communities this power must of necessity rest upon the consent of the governed. Power implies obligation, hence it is obviously the duty of a government, in framing a system of taxation, to provide one which, while effective in producing a revenue, is also just and impartial.

It will be conceded that mines and mining property ought to bear a fair share of the public expenditure. The question is, upon what basis taxation should proceed. Two separate taxing principles are usually contended for: Taxation should be proportionate to the benefits received by the person or property taxed; taxation should be based on ability to pay. A third is sometimes added, taxation should have reference to the extent and value of the natural resources enjoyed.

As applied to mining taxation, it is pointed out by those who support the first mentioned basis that, mines being usually situated in rocky and hence sparsely peopled regions, mining companies are frequently obliged to construct roads, railways, telegraph and telephone lines, schools, water and sewerage systems, and many other amenities and conveniences of civilization the cost of which in regions better suited for agriculture, and consequently more thickly populated, has either been defrayed locally, or can be spread over a sufficiently large taxable group of people to materially lighten the financial burden on the mine. The mining company, it is argued, having made these outlays, should receive credit for them and be relieved of taxation to a corresponding extent.

There may be some force in this contention, but the situation is capable of relief without attempting to place taxation upon so unworkable a basis as the value of benefits received. Who is to be the judge of these benefits, and how are they to be valued? The very existence of a mining company and its ability to carry on business in peace and security depend on the prevalence of law and order, which is a result of good government. The state provides protection against violence, and courts of law for

* Presented at the joint session with the Canadian Mining Institute.
† Deputy Minister of Mines, Canada.

redress against wrong. The relationships among the social units of a modern well-governed state are so complex and interdependent as to make it impossible to fix, with any degree of accuracy, the money value of the services rendered by the whole, or in other words, the government, to any one unit.

Again, the state existing for the benefit of the individual units and extending its care and protection to all, it follows, in so far as property is concerned, that the greater the value the larger are the benefits which its owner derives from the state, and consequently the greater the tax he should pay. It would seem, therefore, that the first basis—benefits received—really occupies the same ground as the second basis—ability to pay.

The latter basis presupposes the possession of capital productively employed, for it is obvious that taxes could not long continue to be paid out of capital which remained unreplenished. Any tax upon legitimate industry which diminishes the amount of capital available or necessary for that industry is objectionable, for the reason that its effect must be to cripple, or in the end even to extinguish such industry. Moreover, the only source out of which an unremunerative enterprise can pay taxes is its capital. These considerations bring us at once to a recognition of the fact that the burden of taxation can properly be placed only on net earnings or profits.

Adverting to the third ground for taxation previously mentioned, namely, the value and extent of the natural resources owned or controlled, it is clear that this is a suggestion for the taxation of monopolies. Natural resources, though great, are limited and cannot be equally enjoyed by all. A water power easy of development and capable of supplying electric current for a considerable body of people becomes, let us say, the property of an individual or company. The owner, being without competitors, is able to raise his charges to the highest point, which would probably be the level at which energy could be generated by steam or brought in from elsewhere. Few will deny that such profits are a fair subject of taxation.

In a real, though not in so complete a sense, the possession of a valuable deposit of ore constitutes a monopoly. All land does not contain mineral values; indeed, the proportion of valuable mineral land to the entire surface of the earth is very small, and the owners of mineral deposits are monopolists in the sense that they enjoy this advantage over the great majority of their fellow-beings. Where such advantage enables them to make unusual profits, taxation is amply justified.

In the case of the third basis, what is really put forward for taxation is not so much the monopoly, real or fancied, as the profits which it ensures to the owner, and its application logically leads to the same conclusions as those drawn from the second basis or even the first.

Now, mining is a business more than ordinarily subject to uncertainty of result, and if the capital invested in a mine is to be in danger of serious depletion by means of taxes, because "Dame Fortune's fickle smile" has not been won, the effect will not only be disastrous to the particular mine but deterrent upon the engaging of capital in mining enterprises generally. On the other hand, it is true that mines are sometimes profitable on a scale far beyond that of an ordinary commercial business. I need not stop to multiply instances. The copper mines of Michigan, Montana, and Arizona; the silver bonanzas of Virginia City or Cobalt; the gold mines and diamond diggings of South Africa; and the nickel deposits of Sudbury will come to your minds. Returns such as accrue from deposits like these are the sustaining force of the mining industry and enable it to maintain a vigorous existence despite countless individual failures. Nevertheless, if deposits so valuable pass into private ownership, there is nothing unreasonable in regarding them as charged with a lien in the form of a demand that a fair share of the profits derived from working them shall enure to the public benefit.

In applying the principle of taxing profits or net earnings, it is, of course, implied that all the usual and necessary costs of operation shall be deducted from the gross proceeds. These include wages of labor, superintendence, power, explosives, timber, transportation, expenses of marketing, administration, and similar charges. Depreciation of mine buildings and plant should be allowed for on a basis that will amortize their cost during the lifetime of the mine. Mining machinery is valuable only so long as the mine is in operation; when the ore is exhausted it rarely has more than a scrap value.

Initial capital really invested, carefully distinguished from water, should be exempt from taxation, as until it is returned, either in dividends or some other form, profits in the strict sense of the term cannot begin. By initial capital is meant the money invested in purchasing the land from the government and in opening up and equipping the mine. It is apparent that should the property be subsequently sold for a larger sum, it would not be equitable, so far as the state is concerned, to treat the additional cost to the new owner in a similar way, for the increase in value simply represents a capitalizing of future profits, the express object of taxation.

Opinions may differ as to whether dividends to shareholders should be reckoned among the expenses of operation, and so go untaxed. It is not the practice of shareholders to refrain from dividing the profits of a paying mine until the entire capital investment is amortized, but rather to require a distribution to begin at the earliest possible moment. Prudent management, in most cases, will be satisfied if a yearly sinking fund is provided sufficient to wipe out capital before the exhaustion point arrives, and there is no unfairness in the state's request to be treated on the same basis as the shareholders themselves, and to require its tax to be paid out of the gross sum of net earnings, irrespective of dividends altogether. It may happen that the company instead of distributing its profits in dividends, will reinvest them wholly or partly in enlarging its operations, or in new or improved machinery and equipment, or even in the purchase of additional properties. The destination of profits or the purposes for which they are used, being wholly within the control of the company, should not affect the right of the state to its share.

A more fundamental question in dealing with mining taxation, whether on the net profits, ad valorem, or any other plan, is provided by the fact that the assets of a mine are, in the nature of things, vanishing assets. The first bucketful of ore raised from a mine marks the beginning of the end. The rich storehouse which it has taken nature ages to fill is ransacked by man in a few years. Fairness requires that this special feature of the mining industry should be given full weight in any system of taxation.

This would not be a difficult problem if the end could be foreseen from the beginning, and if the years of a mine's life could, like those of a man's life, be estimated by actuarial methods. Borings, shafts and cross-cuts, in the case of large homogeneous masses such as bodies of iron or copper ore, may give reliable data for such a calculation, but in many cases these methods besides being expensive are difficult of application and uncertain in results. Besides, while the drill is at work and reserves are being blocked out, ore is being raised and treated, shareholders are clamoring for dividends, and the state is asking for taxes. The most that can be done in many instances is to make a conservative guess and proceed accordingly, subject to adjustment as development goes on.

The practical problem, however, for taxation and dividend purposes, is not how long will the mine last, but how soon can the initial capital investment be recovered? After such recovery, the entire net proceeds of the mine may safely be regarded as profit, and treated accordingly.

Usually it will be the part of wisdom to make the period for capital replacement as short as possible. Pressure for returns, however, is imperious and not to be resisted, consequently it is not the usual practice of companies to formally amortize their capital out of net earnings, but rather to pay these out as dividends. The effect of large dividends on the market value of shares is usually very marked, hence the price of shares in large dividend-paying mining companies, is almost invariably too high, since investors overlook the fact that a mine, unlike a farm, cannot be worked forever.

It is sometimes urged that the taxation of profits should be equated over a period of years so as to avoid heavy fluctuations of revenue, which are inconvenient to the state or local body dependent upon mine tax receipts for financing. It should be remembered, however, that, especially in precious metal mines, the rise and decline of a property cover usually a comparatively brief period. A mine can best pay a large tax when it is earning large profits, but when the end is in sight and profits dwindle the ability to pay is lessened. The adoption of a five- or even a three-year term would undoubtedly tend to shift the high point of taxation to a later period in the mine's history, and might entail a burden on the closing years which they could not rightly bear. On the whole it seems advisable to close the account at the end of each year.

The net earnings or profits basis for taxation appeals to the sense of fairness. If the mine is a failure, the state receives nothing; if moderately successful, the state gets a moderate return; if it prove a bonanza, the state revenue benefits accordingly. The incidence of taxation is better adjusted than under any other form. Experience has shown it to be acceptable to the mining community itself, which is a strong recommendation. The taxing authorities must of necessity be clothed with sufficient powers of enquiry and examination, and adequate penalties provided for offenses. Given these, the difficulty of enforcement is reduced to a minimum.

The fact is that in any method of mining taxation, excluding those which are admittedly arbitrary in their nature, the underlying principle is taxation of profits. This is borne out by an examination of the ad valorem system in vogue in many states of the Union.

State constitutions almost invariably have imbedded in them a provision that all properties shall be assessed for taxation, and that taxation shall be uniform. This provision precludes any method of taxation avowedly based on output or profits, as well as a specific tax of any kind on mining property or products. In a few of the states the constitution permits specific taxes, but the prevalent basis of taxation is the assessed value of tangible property, to which a uniform rate of taxation is applied. Ordinarily the requirement is for assessment at the actual value, but in practice this provision is disregarded, and in most cases the assessment value is less than the real value. Some states have regularized this disregard, and provide that the assessment shall be for only a specified percentage of the actual value. The valuation is made by the local taxing authorities, and is the basis upon which is levied both state and local taxation.

Most states have now a State Tax Commission, also a Board of Equalization, which may or may not be the same body. The assessment of mining properties by local officials is naturally far from uniform, and the function of the Equalization Board is to adjust the valuation to a common standard. It is obvious that while within the area of a local taxing unit, be it town, city or county, if the same rule of assessment be applied to all properties, it is a matter of indifference whether the basis is 50 per cent., 75 per cent., or the full actual value, for the same rate is applied to all.

For purposes of state taxation, however, it would be inequitable to collect the same rate on properties in a town assessed at 50 per cent. or 75 per cent. of their value as upon properties in another town appraised at their full value; hence the necessity for the process of equalizing.

But even in estimating the amount at which a mine of any magnitude should be valued for taxation purposes, it is evident that a degree of skill and experience is required, little likely to be possessed by the local assessor. Mere guessing at the value produces endless confusion and irritation. In some of the states of first class mining importance, notably Michigan and Minnesota, the Tax Commissioners determined to adopt some more scientific method. The Michigan mines are of iron and copper. The orebodies are extensive and valuable. Local conditions, chiefly the fact that the mines are confined to the northern peninsula and the farming lands of the state to the lower peninsula, led to perennial dispute as to the share of taxation which the respective sections of the state should bear. In 1911, the State Tax Commission employed J. R. Finlay to survey and value the iron and copper mines for assessment and taxation purposes. In making this valuation Mr. Finlay applied five factors: (1) Tonnage of ore contained in mine; (2) estimated life of mine; (3) cost of operating; (4) annual receipts from sales of ore; (5) rate of interest for ascertaining present value of deferred production.

The diamond drill is the distinctive tool for exploring the iron mines of Michigan. Most of the mining companies that had been operating for any length of time had delimited their orebodies by borings, and had ascertained their approximate dimensions and consequently the reserve tonnage. Their average production during the four or five years previous, divided into their ore reserves, gave the probable life of the mine. Operating expenses deducted from receipts from ore sales gave the profit per ton; this figure, multiplied by the average output, gave the yearly profit, or in other words, constituted an annuity payable until the mine was worked out. The present value of this annuity, discounted at 6 per cent., Mr. Finlay set down as the valuation of the mine for taxation purposes. Taking a concrete case, for the sake of clearness, let us assume that Mr. Finlay was valuing the Ferrum mine. On the basis of the company's borings and allowing for probable undiscovered ore, the contents were 10,000,000 tons. The average production was 500,000 tons a year, hence the mine had a future lifetime of 20 yr. The average cost per ton for operating was $3, and the average price for ore sold $3.50, leaving a profit of 50 c. per ton; this, on 500,000 tons of ore would give an annual profit of $250,000 for 20 years, the present value of which was $2,867,475. This sum would be the valuation of the Ferrum mine.

Iron mining in Michigan has had its ups and downs. In 1914, when the great struggle of the nations broke out, the average loss per ton of ore to the mining companies was $0.07712, yet the taxes collected from them, based on Mr. Finlay's valuations, were equal to $0.12009 per ton. In 1915, there was an improvement. The output of the iron mines was 13,151,612 tons and the average selling price was $2.79102 per ton. Much of the ore is mined subject to royalty payable to the owner of the fee, the average in 1915 being $0.23136 per ton. Including the royalty, the average profit of iron ore was $0.52380 per ton. Out of this margin the royalty charge had to be met, and in addition, taxes equal to $0.13784 per ton. That is to say, the net profit remaining to the mining companies, after allowing for certain other smaller items, was $0.14674 per ton. Thus, after deducting expenses of production from the selling price, nearly one-half of what was left went as royalty to the fee owner, and the remainder was divided almost equally between the mining company and the tax collector. Before payment of the royalty the taxes amounted to 26.31 per cent. of the profits; after payment of the royalty, to 47.13 per cent. Judged by pre-war standards, this is pretty heavy taxation.

Mr. Finlay's method has the merit of being systematic, and it takes into account the essential factors of value, but it is evident that no allowance is made for the unexpected. A demand may arise for iron ore far in excess of anything previously known, with a corresponding increase in price and consequently in profits. On the other hand, stagnation may set in, prices may fall, and iron ore may become practically unsaleable. In the former case the valuation of the mine, based as it was on normal conditions, was too low; and in the latter, too high.

In Minnesota a similar method is in vogue, but a somewhat elaborate classification into productive and unproductive mines and prospects, with a varying standard of valuation in the several classes, has been worked out in the endeavor to arrive at equitable results.

The Finlay valuation of the Michigan iron mines gave them a total value of $129,000,000, which was a large increase over the total valuation under the old methods. The companies protested vigorously, and the Tax Commission reduced the valuation to about $90,000,000. At about this figure it practically remained stationary for a number of years, the amount of ore brought in sight year by year equaling the amount extracted.

In the case of the copper mines, the Finlay system had an entirely different result. The aggregate value reported was $69,000,000. This sum was so far below the total of the previous valuations, and also so far below the value of the mining properties as shown by the market value of their shares, that the companies, fearful of the bearish effects upon the stock market, demanded that their old assessments be restored. This request the Tax Commission also granted, and the copper mine valuations have remained at about their old figures.

Whatever may be urged in favor of the Finlay method of valuation as applied to the copper and iron mines of Michigan and the iron mines

of Minnesota, with their immense masses of ore capable of fairly accurate
delimitation and ascertainment of quantities, and individually fairly
uniform in quality, it is clearly inapplicable to deposits such as those
of the precious metals, where the veins are small and subject to great
irregularity in their size and direction, and the value of their contents.
The diamond drill cannot be depended upon to the same extent in disclos-
ing the value of the deposits.   There are other minerals of value notoriously
pockety and capricious in their habit; for all such the Finlay method is
unsuited.   In many cases it could not even be tried, for the essential
requirement in estimating tonnage, namely, diamond-drill borings on a
sufficient scale, is lacking.   Even where orebodies are large and of
fairly uniform value, unless there has been this preliminary examination
by the drill, the necessary data are absent.   Cases can be cited where
mining companies, even after years of working, were uncertain or even
ignorant of the size of their deposits.   To make a notable instance from
the Province of Ontario, the Canadian Copper Co. working the Creighton
nickel mine, were apprehensive that its productive limits were being
reached, and consequently prepared to exploit the Frood mine, another
large orebody, but lower in nickel and copper contents.   They built a
town, laid out streets, equipped the new mine with machinery and hoisting plant, and began to work it.   Concurrently they continued to explore
the Creighton by diamond drilling, and encountered at lower depths unexpected and very large reserves of ore.   It paid the company to cease
all work at the Frood and to continue operations at the Creighton.   Had
a valuation been made of the Creighton mine on the Finlay plan before
the reserves were disclosed, it would have fallen far short of the correct
amount.

   Undoubtedly one effect of the Finlay system in Michigan and
Minnesota has been to discourage exploration for ore.   As soon as the
drill brings ore into view, it is subject to valuation and taxation, notwithstanding that it may not be actually worked for years.   Every prudent
mining company desires to know its position regarding reserves of ore,
so that it may be justified in making adequate capital expenditures for
the winning and treatment of the same, but if to establish new orebodies
is to materially increase their taxation, the result is apt to be a slowing
down of the drill.

   It is apparent that the Finlay or any other method of mine valuation
in the end rests on profits.   Unless a deposit can be worked so as to yield
a return greater than the cost of working, it has no real value.   The controlling factor is the profit per ton or other unit of production.   The
sum any mine is worth depends upon the profit it is producing, and
will continue to produce.   When the object is revenue, not a sale,
surely the more logical way is simply to tax the profits as they are realized.
Since these cannot be accurately predicted, they cannot be accurately

capitalized; nor is there any occasion to do so, for if they prove smaller than was expected, the tax is less and no injustice has been done; should they prove greater, the tax is larger in proportion. The net-profit system automatically adjusts itself to the conditions for the time being, and takes account of all changes in expenses and returns.

Other methods than the taxation of profits have been attempted. One of these is an area tax—so much per acre, hectare, etc. This is a rough-and-ready method, and has no relation to value or ability to pay. An acre of barren rock pays the same tax as an acre of diamond-bearing earth, or of the richest quartz. A second method is a specified rate per unit of mineral, be it ton, pound, ounce, or gallon. This has the same convenience of application as the acreage tax, but takes no account of profits or expenses. A ton of coal scraped from the last workings of a mine would pay the same rate as a ton produced from the richest and most easily worked seam; an ounce of gold wrung from ore carrying $2 a ton, at a profit of ten cents, will pay as much as gold from $50 quartz; a gallon of petroleum from a well yielding a barrel a month, as much as a gallon from a gushing geyser. Such methods of taxation are easily applied, but are unscientific, and lack the essential feature of fairness.

## DISCUSSION

ALF. G. HEGGEM, Tulsa, Okla.—I have listened to the reading of these papers with much interest and feel that perhaps I can add another viewpoint to this subject of taxation based upon experience in paying income and excess profit taxes on manufacturing supplies as well as on producing oil. Dr. Arnold has said that the principles of taxation are not open to discussion. I feel that this is just the feature we should discuss for in the application of the general principles laid down I feel that the Commissioner of Internal Revenue has endeavored to use the utmost fairness with results that must appeal to all as the most satisfactory that can be obtained under the existing principles.

Excessive rate of taxation results in reduced revenue for the government. I recall one direct instance where the first well drilled on a lease of 200 acres produced over 1200 bbl. the first 24 hr.; an offer of $1,250,000 was made for the property but refused as the tax on excess profits was then fixed at 60 per cent. which would have left the owners with less than a 40 per cent. profit. It therefore seemed better to take the annual income from the property at a lower rate of tax and eventually get a much greater return. There was also a fear of an 80 per cent. tax being levied. The result was that if the 20 per cent. tax which has since been assured us had been in force the sale would have been made and the government would have received $250,000 on that transfer and the same taxes on production it has been getting. Incidentally, the value of that property has dropped about 75 per cent. so the tax would really have been levied on a surplus of capital above real value. This case is typical not in decreased value but in the effect of excessive taxation closing the avenues of revenue to the government.

In the Mid-Continent field, while development have been very active in the effort to produce a maximum of oil for war purposes, there have been practically no transfers of property during the past 12 months. The sales of property terminated about March of last year, as soon as the people began to see what the tax effect would be. Some transfers were attempted on deferred payment basis, an evasion of the tax law that did not get very far. The whole effect of high taxation has been to retard the industry and reduce the revenues. If we can keep our taxes at some point that will not interfere with the activity, the government's revenue will be at a maximum and far greater than it can be under excessively high taxes.

*Discussion continued on p. 709.*

BY R. C. ALLEN,† LANSING, MICH., AND RALPH ARNOLD,‡ LOS ANGELES, CAL.

(New York Meeting, February, 1919)

The writers have no new system and no new principle of taxation to propose. The general subject of taxation is as old as governments are and as familiar to taxpaying Americans as the general thesis on the inalienable rights of man. We believe, therefore, following the committee of the National Tax Association, that improvement in the general conditions of taxation, including mine taxation, may be brought about mainly through a better adjustment of known and tried principles and methods of taxing to modern complex forms of property and income, and that not much is to be expected from abandonment of experience for new ways and means, methods and principles, should anything new be discovered, which in our opinion is very doubtful. The committee referred to has constructed the general framework of its Model System of State and Local Taxation[1] on familiar forms of taxing, by eliminating some that have proved bad, suggesting improvements in others, and adopting the best, leaving plenty of room for the accommodation of the general plan to the varying conditions of taxation in the several states. Unfortunately, the taxation of mines is dismissed by the committee with the statement that the subject has been reassigned for further investigation and report by a special sub-committee, frankly recognizing that the peculiar difficulties of this phase of the subject require special study and investigation. Much may be hoped for from the labor of this sub-committee if the problem is approached from the viewpoint of recommendations already published.

In this paper, the writers state their opinion on some of the broader aspects of mine taxation with too little attention to the grounds on which they rest. Contrary to inclination space requirements compel them to resort to summary statements without adequate discussion for a satisfactory presentation of the whole subject. All they can do is to qualify the witnesses and record the testimony.

The senior writer was introduced to the practical aspects of the subject of this paper in 1913, when, as State Geologist of Michigan, he was

---

* Presented at the joint session with the Canadian Mining Institute.
† State Geologist, Appraiser of Mines.   ‡ Consulting Geologist and Engineer.
[1] Preliminary Report. Twelfth Annual Conference, St. Louis, Nov. 12–15, 1918.

charged with the duties of Appraiser of Mines for the Board of State Tax Commissioners. In this capacity, he assisted in developing and administering a system of mine taxation (commonly referred to as the Michigan System) under a general property tax law requiring the assessment of real and personal property for taxation at its full value. Aside from its administrative features, the main problem in this system is the determination of the values of the mines. From February to October, 1918, the writers were associated as members of the Board of Tax Reviewers, and thereafter, until February, 1919, as mine valuers in the Bureau of Internal Revenue, United States Treasury. Their duties included the development of administration and the determination of taxes under the revenue law of 1917, as applied to income derived from wasting natural assets, with main attention to the mining industry on the part of the senior writer and the oil and gas industry on the part of the junior writer. With apologies, the writers exhibit thus boldly their excuse for expressing opinions on this much-debated subject in order to claim in advance what merit they may possess.

To begin with, a line will be drawn between federal taxation and state and local taxation, for no one has suggested any means of accommodating the conditions of taxation under the federal and the state governments in a single system common to these distinct administrative spheres.

### State and Local Taxation of Mines[2]

What is the best system of state and local taxation of mines? In this form the question has no answer. Were it raised with respect to a particular state, an intelligent answer could be made eventually, but the question is meaningless when applied to all the states. It is meaningless because no one system of mining taxation, no matter how elastically framed and ingeniously devised, can accommodate the widely varying conditions of general taxation in relation to the mining industry within the various states.

With respect to a state, that system is best which most nearly meets all of the conditions of taxation within the state—economic, political, and social. Inasmuch as these determining conditions of taxation are not uniform, an ideal uniform system is an impractical concept. A tailor who cut his cloth to fit the average man would probably do little business, notwithstanding the high quality of his goods and workmanship. Likewise, the student of taxation who, considering his subject in the abstract, evolves a "best system," leaving the various jurisdictions to accommodate themselves to it as best they can, contributes little to the practical solution of tax problems. It is precisely because the deter-

---

[2] In this paper the writers use the word mine to include oil and gas wells and any operation which is based on recovering from the ground natural mineral deposits.

mining conditions of taxation in the various states are dissimilar that we shall not have, and should not have, a uniform system of mine taxation. It follows, therefore, that if one state taxes a group of mines by one method and another state taxes a precisely similar group by another method, we should not embrace the conclusion that one or the other or both methods are bad until we have familiarized ourselves with all of the determining conditions of taxation within the respective states. It may be that the method of taxation is fairly well adjusted to the general conditions in both states.

The writers insist that little is to be gained by a consideration of mine taxation purely in the abstract, which is the common refuge of the champions of uniformity. From a practical consideration of the subject, based on some knowledge of the general conditions throughout the states, nothing appears more certain to the writers than the eventual failure of any program for unifying methods of taxation of mines or even groups of similar mines throughout the states. Not long ago the American Mining Congress, by resolution, not only committed itself to the principle of taxation of all metalliferous mines in all of the states by a uniform method, but also prescribed the method.. Such a program (and every similar one) is bound to fail because it does not take into account the fact that the taxation of mines is only a part of the general problem, which has different aspects in different states, and that all classes of taxpayers and properties will have to be accommodated in a general system which will recognize the interest of government as paramount rather than that of the taxpayers in any class.

It is because the taxation of mines is so inextricably bound up with the general problems of taxation within the states that a treatment of the subject is beyond the scope of a short paper, if not also beyond the powers of the geologist-engineer who is not also an experienced political and tax economist. But if the contribution of the geologist-engineer to the general subject is necessarily limited because in its ramifications it leads him into spheres foreign to his training and experience, his aid is indispensable to improvement in the administration of all but the simplest forms of mine taxation, and it is in this direction that his services will be more and more demanded and useful.

There are three main forms of mine taxation in use in this country; namely, the specific tax on production, the ad valorem tax, and the income tax.

*Specific Taxes on Production.*—The tax on the unit of production is the simplest form of mine taxation. It is generally referred to as "gross production tax" or "tonnage tax," and is applied either as a flat rate on the unit of production or a graduated rate based on the value of the unit. It is widely used, especially in states having an important oil and gas industry. Its two main claims to preference are, it is easy to administer and difficult to evade.   The main objections to it are, it bears no relation to the ability of the taxpayer to pay it and no necessary relation to the taxes levied on other forms of property.

The specific tax cannot be unqualifiedly condemned nor unqualifiedly commended on general grounds.   For a small jurisdiction, such as a township or county, wherein mines comprise a high percentage of the total value of property and mining is almost the only industry—a condition that prevails in many localities—it is a very unsatisfactory form of taxation, because the current needs for public revenue are not closely related to the production of the mines.   Should the mines suspend operation for any cause, the public revenues are cut off and no other adequate source is available.   In periods of flush production, much more revenue is raised than necessary—a condition that has proved corrupting to civic morals.   Obviously, the small jurisdiction, dependent largely on mines for its revenue, should tax them by a method that will be independent of the vicissitudes of the mining business.   The system that best meets this condition is the ad valorem or general property tax.

The state, on the other hand, drawing its revenue from many different sources, may find that specific taxes on production are, on the whole, desirable.   The rate of taxation may be adjusted from time to time in accordance with the state of the industry and the need for public revenue, without danger of affecting too markedly the alignment of the general state finances.

The owner of mining property is generally opposed to specific taxes on production, except at merely nominal rates, in lieu of other taxes, and is almost everywhere opposed to this tax where it is combined with one or more other taxes levied by the same jurisdiction.   His opposition is well-grounded, not only on this unscientific means of measuring his tax obligations but on the violation of the principle of equality of taxation, not only with respect to different properties in this class but with respect to other classes of property.   The specific tax is a favorite measure of those who believe that mines should be taxed at a higher level than other forms of property on the theory that natural resources should be the common property of all and, therefore, inasmuch as the mines are not publicly owned the people should assert their equity in them by supertaxation.   This is the thought underlying the periodic agitation in some states, particularly in Minnesota and in Michigan, for a supertax on mines in the form of a specific tax on production in addition to the general property tax, and explains in part the determined opposition of the mine owners in those states to the imposition of this tax.   The mine owner is opposed to supertaxation as a matter of course, but even should he be compelled to submit to it he would prefer a method of taxing that would apply equitably to all the taxpayers in his class, under a system that would measure, even roughly, the amount of supertaxation in relation to the other classes of taxpayers. The mine owner has much to fear from the specific tax in the hands of a hostile or an unintelligent legislature.

Our conclusion is that the specific tax may be commended only in those jurisdictions where, for any reason, it is the public policy to tax mines at a merely nominal rate.

*Ad Valorem or General Property Tax.*—This form of taxing is common to many states. In principle, it is simple; in administration, complex and difficult, particularly in its application to mines. The taxable property is assessed for taxation at a percentage of its value. The value of the taxable property is determined each year by legally constituted assessing officers and placed on the tax roll of each local jurisdiction, usually the township or city. The total amount of taxes to be raised in the jurisdiction, including township, county, state, and special taxes, divided by the value of taxable property, determines the rate. The tax on any property is the rate multiplied by the value of property. The general property tax has its various aspects and modifications in different states, but in its application it is everywhere based on the simple principle just stated.[3]

A discussion of the general evils and merits of the general property tax may be found in all standard works on taxation and will not be repeated here. Whatever may be its defects, practically and theoretically considered—and they are serious and many—tax economists have agreed that it cannot be dispensed with, and will continue to form the basis of main taxation in local jurisdictions such as townships, cities, and counties. Those who advocate a complete separation of the sources of state and local revenue would abandon the general property tax for state purposes. In its application to non-visible personal property it has proved everywhere to be almost a complete failure.

Whatever may be the opinion of mine owners generally respecting the ad valorem tax, nothing seems more futile than a concerted attempt to replace it. As the writers have said, the taxation of mines is a problem to be solved in each state in accordance with the general determining conditions of taxation; and inasmuch as the ad valorem taxation of real property cannot be generally dispensed with, it is to be expected that mines will not generally be set apart from other forms of real property for special taxation, especially in those states where uniformity of taxation is a constitutional requirement and the classification of property for taxation is not permitted.

Of the forms of visible property, the ad valorem tax is best adapted to those that are most easily appraised. It is because mines are less easily appraised than other forms of visible property that the ad valorem

[3] In some of the western states the manner of determining the value of mining property for taxation is fixed by statute as in Idaho, Montana and Colorado.

tax is less adaptable to them than to any other form of property in this class.  In the hands of city and township officials, the appraisal of mines is uniformly unsatisfactory, because such officials are not selected with a view to the talents of a technical order possessed by the competent mine valuer; but if the mines are appraised from year to year by experts in the employ of a strong administrative state officer or body, such as now exists in some of the states, the ad valorem taxation of mines is no more unsatisfactory than that of other visible property.  In fact, this form of mine taxation, properly administered, has much to commend it. It affords, at least, a rough means of equalization as respects the various mining properties themselves and in their relation to other classes of property similarly taxed, and it does not violate, in a too serious degree, the principle of ability to pay.  To the local jurisdiction, dependent mainly on mines for support, it offers the necessary income each year as determined by the actual needs for public revenue.

In Michigan, where the administration of ad valorem taxation of mines has probably reached its highest development, the criticisms have come mainly not from mine owners but from the owners of other property in local jurisdictions where mine values constitute a large percentage of the total valuation on which taxes are levied.  In these jurisdictions, the criticisms are leveled mainly at the fluctuating values of the mines. The critics admit that mines do actually fluctuate in value with the general state of the industry and the stage of development or extinction of the mines themselves, but they insist on a recognition of something like a fixed ratio between the mine values and the values of business and residence property.  It is obvious that the mines do create value in the business and residence property in their vicinity, and the relation is particularly close where no other industry exists.  It is equally obvious that the values of the respective classes of property do not maintain a fixed ratio from year to year.  This is one of the many local aspects of taxation that have to be taken into account in good administration. Many others could be cited, but it is necessary merely to refer to them here by way of emphasizing the importance of good judgment in the successful administration of an ad valorem tax.

Most discussions of ad valorem taxation of mines consider the problem as one merely of mine valuation, another illustration of the treatment in the abstract of an intensely practical subject.  Most writers, in fact, commend or condemn this tax with respect to a group of mines on the sole basis of the ease with which values may be determined under some stated or implied definition of value.  In the earlier years of the senior writer's experience, his professional sense was shocked at times over the liberties that were taken with his conclusions respecting value by the tax commissioners in working out special and local problems of equalization, but in later years he has been an enthusiastic accomplice in bringing about equalization at the expense of a modification of his own conclusions respecting actual values arrived at through months of painstaking labor. For, after all, the main problem in taxation lies in the equalization of the tax burden, not only as respects taxpayers in a given class, but also as respects the different classes, and any administration which yields fairness and justice to the demands of a "system" of valuation merits the severest censure. It is only where there is a general lack of confidence in the ability of the appraising officials and fairness and intelligence of the administrative authorities, or the clashing of political elements with which the question of mine taxation is in issue, that a rigid systematic appraisal, based on a formula, meets support. It is in such circumstances that efforts are made to prescribe a precise method or formula of valuation by law, to avoid the greater evils that flow from the influence of politics and the blunders of prejudice and the novice.

But the ad valorem system of mine taxation is not to be condemned merely because valuations are difficult. With the coöperation of the mine owners, combined with intelligent administration and appraisals, the system may be operated with considerable satisfaction to mine owners, other taxpayers, and government alike. In Michigan, it would be difficult to displace it; those who desire supertaxation of mines have never considered the abandonment of the ad valorem tax; the mine owners defend it as against any other system whatsoever, not because they find nothing in it to criticise, but because on the whole they believe it is best adapted to the general conditions of taxation from the standpoint of both state and local jurisdictions, and that it measures their tax obligations by the same criterion that is applied to other forms of visible property. At another time, the senior writer will discuss the administration of the Michigan System in some detail, not only as a recognition of the general interest in it, but with a purpose to removing a good deal of misunderstanding prevalent among writers who have had occasion to refer to it.

In conclusion, the writers have to say that ad valorem mine taxation is neither to be commended nor condemned on general grounds. The movement in this country is toward it rather than away from it, and doubtless it will form, in the future as in the past, the most common method of state and local taxation of mines. So far as concerns the mines themselves, it may be operated satisfactorily irrespective of the nature of the properties, but that it is not generally so operated is a fact too patent for dispute. Here is a field for the geologist-engineer, not merely in the valuation of mines for taxation but in the development of better methods of tax administration, a field that has been too much neglected. So long as the ad valorem taxation of mines remains in the hands of those who have only casual knowledge of this class of property, little experience in practical tax administration, and slight interest in either, no improvement of present conditions may be hoped for.

*Income Tax.*—The successful experience of Wisconsin and Massachusetts with the income tax, the growing familiarity of the people with this form of taxation through federal administration, and the almost complete failure of the general property tax to reach invisible property, which is so large a share of the total wealth, is convincing evidence that the income tax will sooner or later be in operation in many of the states. With the general property tax, there is a disproportionately heavy burden on owners of visible property because, in its administration, it is impossible to reach a large class of able-to-pay citizens in the salaried and professional classes and those deriving income from stocks, bonds, and other forms of invisible property. These classes are reached by income taxation. It will be a long time before the ad valorem taxation of visible property is abandoned by the states, if ever, but similar taxation of invisible property will have to be abandoned in the interest of public decency; the income tax on individuals will probably be widely substituted for it. State income taxation is particularly favored by those who desire a complete separation of state and local revenues.

In speaking of the income tax, the writers refer of course to the tax on net income. The taxation of gross income, as applied to mines, is identical in principle with the specific tax on production; it may be administered with the same ease and is not more difficult to evade, but it violates in the same degree the principles of uniformity, equality, and ability to pay. Its simplicity and ease of administration commend it where, in the interest of public policy, mines may be taxed on a merely nominal basis; it has no place in a system of taxation that attempts to spread the burden uniformly and equitably over all taxpayers in each class and to bring about something like equality of taxation as respects all classes of taxpayers.

## Federal Taxation of Mines

During the war, the income tax has become the main source of federal revenue. A consideration of present and prospective federal expenditures in connection with diminishing returns from tariffs and excises on liquor, formerly main sources of revenue, with the deep-seated opposition to sales or other direct taxes on indispensable commodities of life, is evidence enough that the income tax has come to stay. In its economic, as well as its social, aspects the graduated income tax is preëminently desirable. Its basis may be broadened or contracted, the rates may be raised or lowered to meet the requirements of government and, because it is based on the principle of ability to pay, it appeals to the public sense of fairness. In direct contact with taxpayers for more than a year the writers have encountered no criticism of the simple principle of the income tax, but a good deal of criticism concerning the method of its application to mines, in law and administration.

The writers have been impressed with the good-will and cheerfulness with which the taxpayer has met his obligations to the government under this tax during the war. They have seen many a debatable point yielded by him out of sheer good-will and desire to get on with the job. In the days when government and taxpayer alike were wrestling with the interpretative and administrative difficulties of the law (war excess profits tax), hastily drawn to meet a great emergency, a good deal of accommodation was displayed from both sides; but the country is now facing a long pull ahead to liquidate the war debt and a more precise determination of the taxpayer's liability will doubtless be sought by the taxpayer as well as the government. With this in view, the writers have given some consideration to the peculiar difficulties in the administration of the tax on incomes derived from mining. They will not be able to discuss in this paper the law itself and the regulations governing its application and will, therefore, have to assume that the reader is familiar with them. Their purpose is rather to explain some suggestions for improvement in the administration of the law from the double viewpoint of government and taxpayer.

If the administration of the income tax on mines is less satisfactory than it should be, the trouble lies mainly without rather than within the Bureau of Internal Revenue. For a year, the writers have listened to suggestions from the mine operator concerning what the Bureau could do to make the situation more tolerable for him, and they have tried to digest them to discover what indeed could be done in the interest of a more intelligent consideration and efficient disposal of his tax returns. But perhaps the mine owner will be surprised to learn how often the trails of the difficulties have led squarely to his own door. It is an old and familiar rule of the courts that a plaintiff seeking relief must approach the bar of justice with clean hands. Therefore these remarks are directed first to the taxpayer, the plaintiff, that he may set his house in order before his plea is heard.

Speaking in general terms, the mine operator is a notoriously bad bookkeeper. This applies more especially to that larger class, the small operators. This is another way of stating that many mine operators do not know the true status of their own businesses. Good accounting is necessary to the return of an accurate tax statement. If the operator is content to depend on the revenue agent to set his books in order, he should be content with the findings of the agent even should additional tax liability be discovered, which is very generally the case. Defective accounting is the Pandora's box of troubles for the tax official as well as the operator. If the accounts are defective the tax return, which is based upon them, is necessarily wrong. Slipshod accounting lays the mine operator fairly open to suspicion, and he has nothing to gain by criticising the conscientious government auditor or revenue agent so long as this official has to make his way through unintelligible sets of accounts in the manner of a detective uncovering crime. A surprising number of mine operators had never made a balance sheet until the revenue agent happened along to check up his tax returns for the year 1917. Some few even did not know what a balance sheet was until confronted with one at the hands of the revenue agent. Heavy income taxation will do one unmixed good to the mining industry if it convinces the operator that careful accounting pays. Neglect of this vital governor of business not only jeopardizes investment in many a mine, but has its effect on the economics of the industry as a whole. It has been said by competent authority that in the bituminous coal fields, for instance, ignorance of the actual cost of coal on the part of great numbers of operators had a good deal to do with the demoralization of the whole industry in the years prior to the war. A mine operator who fails to keep accounts in the modern sense for tax purposes, if for no other purpose, may be likened only unto the ostrich, which buries its head in the sand to escape danger.

But good accounting is not the only essential to a correct tax return, a knowledge of the law and the regulations is equally essential. To remain willingly in ignorance of a matter that vitally affects his business is not characteristic of a good business man; but a great many mine operators seem content either to let tax matters take care of themselves or to turn them over to someone who is, all too commonly, unfamiliar with the peculiar difficulties of mine tax accounting. If the taxpayer will inform himself on the law and the regulations governing his peculiar case, he will spare the Bureau of Internal Revenue a good deal of mistaken criticism and extra labor and himself perhaps some useless embarrassment.

The true net income from a mining operation may be computed with respect to each year only after the entire operation has been terminated. There is, therefore, some fiction in almost every return of net income from a mine, even when correctly computed under the regulations of the Bureau, simply because a computation must be made with respect to a twelve-months period or fraction thereof, as the case may be. This arises mainly from the necessity of estimating, with respect to the taxable period, the intangible or investment costs of mining, which are not proportional to time. Investment costs are deductible under the items of depreciation and depletion (and also loss and obsolescence), and it is with respect to these items that the major difficulties arise in computing taxable incomes from mining. In many cases, depreciation and depletion deductions account for a large proportion of the income; and in such cases are, naturally, subject to careful scrutiny by government auditors.

The Commissioner of Internal Revenue has prescribed the rules for computing these deductions, but they are so commonly disregarded (wilfully, carelessly, or ignorantly) by mine operators, that the government auditor has developed a chronic suspicion that often leads him to cut these deductions to a figure far below a fair and reasonable estimate. Here, again, the taxpayer is in most cases to blame if injustice is done. The regulations demand specifically that the basis for depletion, particularly, and depreciation be set forth clearly in separate schedules accompanying the tax return. In the writer's experience within the Bureau, they do not recall having seen a dozen depletion schedules that were not directly solicited through correspondence with the taxpayer or presented in the report of a field audit by a revenue agent.

In cases where depletion and depreciation deductions are based on estimates of value rather than actual cost, the mine operators are especially derelict, not only as respects the computations of the deductions, but also in the establishment of the basis for them, especially the valuations. The Commissioner cannot accept an unsupported figure on the return of the taxpayer in a matter of such vital importance to the government; if he did, the government would lose millions of dollars withheld by many operators who have taken depletion deductions where none are permissible, and by many others who have taken deductions on the basis of absurdly inflated valuations, thus conveniently disposing of the taxable income. There are some mine operators, as well as other taxpayers, who seem to think that a tax return is merely an annoyance rather than a device for raising public revenue. On the other hand, there are others who have failed to take advantage of these deductions at all and some who have taken inadequate account of them.

Perhaps enough has been said to convince the reader that those who pay taxes on incomes from mining can do more to improve the administration of the law than the Commissioner of Internal Revenue and all his men working double shift, if they will give more intelligent consideration to the preparation of their tax returns in strict accordance with law and the regulations. Like many an individual afflicted with imaginary illness, the remedy for much of the trouble commonly complained of by the mine operator in his dealings with the Bureau of Internal Revenue lies in his own hands. Let him set his house in order.

The Bureau of Internal Revenue, like many of the administrative units of the government, had to expand almost over night to meet wartime demands. The Commissioner was unable to pick and choose the additional personnel on short notice. Those who were most needed in large numbers—accountants—had to be taken from the commercial field, already too thinly manned, and it was impossible to obtain the services of a sufficient number of them. Practically all of the new employees were unfamiliar with the procedure of the Bureau and knew nothing about tax administration.   They were in the position of having to learn to swim after having fallen into the water; of having to learn the job while executing it.   The organization, hastily enlarged and made over to meet the war emergency, will be perfected under peace-time conditions.

A living thing is in contact with its surroundings only through its senses.   If the senses are defective, the impulses that come from without are either not felt or not understood.   The income-tax division of the Bureau of Internal Revenue, we believe, needs to develop its senses more acutely in order to come into a better understanding of the several great industries of the country, to become more like a living thing and less like a mechanism.   It should develop more points of precise and intelligent contact with the taxpayers in the several great groups or classes.   It needs expert interpreters not only of the law in its general bearings, with which it is admirably equipped, but in its application to the peculiar circumstances of individual cases and industries.

The discretionary authority vested in the Commissioner of Internal Revenue is very large, much larger under the revenue law of 1918 than under that of 1917.   He necessarily must exercise this discretionary judgment, as applied to an enormous number of individual cases, through his agents, and it is not too much to expect that those who exercise this function on behalf of the Commissioner will be specially familiar with the matters which come under their review; the administrative officer is at a great disadvantage if he is required to pass judgment on matters foreign to his experience.   It is the fear of making mistakes of judgment when all the facts and circumstances in the case have been determined that is perhaps the most prolific source of delay and expense in the audit of tax returns.   This weakness of many officials compels them either to attempt to shift responsibility or to postpone the decision to another day.   The result is an accumulation of the more difficult cases, like a "jam" in a logging stream.   Washington has become a great gathering ground of taxpayers, each bent on the settlement of his own peculiar case.   It is a common complaint that too often the taxpayer is buffeted about from one official to another in an unsuccessful quest of a decision, returning home with nothing more than an expense account and an uncomplimentary opinion of the Bureau of Internal Revenue.

So far as concerns the mining and oil industries, of which alone the writers presume to speak, it seems that improvement of the present situation lies only through a perpetuation of the present specialized administration.   The direct contact with the taxpayer is now through men who are thoroughly conversant with the mining and oil businesses.   More men with such qualifications should be obtained to continue this work.   Such men will gladly make themselves available if they be given the same responsibilities and opportunities for the exercise of their talents as is being accorded the members of the present staff in the mining, oil and gas section of the Bureau. To the members of the Institute, the writers need only to mention that the valuation of mining and oil properties, for instance, is not the proper work of accountants, auditors, and revenue agents. To charge them with it is not only unfair to them, but also disastrous to both government and taxpayer; it irritates the taxpayer and invites from him a good deal of severe, if not contemptuous, criticism of one of the most vital arms of the government.

The geologist-engineer, in the rôle of mining taxation expert, is as yet a rather rare phenomenon. Here is a field of service almost untouched by that one class of men who are best fitted for it. It is a good deal easier, the writers think, to make a mine valuer and mine-tax administrator out of a geologist-engineer than out of an auditor or accountant. This is no disparagement of the accounting profession, for experience has shown that the technical knowledge and experience of the mining geologist should be combined with administrative capacity and sound judgment in the individual who is charged with the application of taxation to mines, whether it be under an ad valorem or income tax system. A sufficient personnel of this character has been indispensable during the war crisis, and will be, the writers believe, indispensable to the Bureau of Internal Revenue in the future, if the purpose is to make any advance over former unsatisfactory methods of determining the tax obligations of mine owners and operators. Unfortunately, the field of taxation is not only unexplored but unattractive to most mining geologists. The mature, successful man is not seeking hard labor at low pay except out of a patriotic desire to sacrifice his own interests for the public good, and the young man generally prefers to establish himself in a field that offers better reward and appreciation of initiative and energy than that afforded in a great governmental bureau under present conditions.

However, the situation is by no means hopeless if a way may be found of compensating the geologist-engineer in the employ of the government at even a fair approach to the standards of non-governmental work, and at the same time of making the positions attractive in the way of opportunities for learning and the application of individual initiative. The continuance and ample development of the special unit or division of the Bureau of Internal Revenue, manned with a specially trained and experienced personnel, in direct touch with the taxpayers, will go far in maintaining the cordial relations with the mine operators that have been built up by the war personnel of the mining and oil units.

## Industry, Democracy, and Education*

BY C. V. CORLESS,† CONISTON, ONT.

(New York Meeting, February, 1919)

WE are living at a period of the world's history in which social phenomena are on so vast a scale, are of so profoundly soul-searching a nature, and are occurring in such rapid succession in the great world drama in which we are both actors and spectators, that, in our efforts to obtain a rational point of view in relation to them, our minds may fail to discern the simple in the complex and our understanding is liable to become confused or even overwhelmed. To the social student whose scientific training has convinced him of the truth of the evolutionary law, *Natura non facit saltum*, it is neither the vast scale of the events, their deeply soul-stirring nature, nor their rapid succession that matters so much, as the discovery of the underlying principle or law in accordance with which the disintegration and reintegration, which are the two aspects of social evolution, or indeed of evolution everywhere, are occurring. He is most deeply concerned in seeking an answer to the question: Have we in the present great and perplexing upheavals in human society, whether regarded in Central Europe, in Russia, in Great Britain, or in America, a really new cause at work, or, have we, though on a very large scale, merely new manifestations of the working out of an old principle?

When we think of the appalling struggle in Europe, into which the world's greatest exemplar of democracy was finally drawn, no one has any hesitation in admitting that the great conflict was fundamentally a life-and-death struggle for self-preservation and self-propagation on the part of autocracy. The variations we meet with in the statement of the cause of the war arise mainly from differing distances of perspective; but in the last analysis all agree that the autocratic system of political government, which found its very soul and center in Prussian despotism, was consciously arrayed in a final "world-power-or-downfall" struggle against a love of freedom which was steadily widening and deepening throughout the world, and particularly throughout those parts of the world where this love of freedom and justice has found expression in democratic institutions. It is not irrelevant to our discussion to recall in passing that this most imposing structure ever conceived by the human mind as an instrument of tyranny is lying today an irretrievable mass of ruins, completely over-

---

*Presented at the joint session with the Canadian Mining Institute.
†The Mond Nickel Co., Ltd.

thrown by the eternal and irresistible power of freedom, justice and right, which it affected to despise.

If again we regard the social turmoil in Russia, the cause is at first not so clear. Hunger, deep and widespread ignorance, the misfortunes of war, and German and Bolshevistic treachery have plunged this hapless people into the abyss of anarchy. But we need carry our minds backward only a few months in order to pick up the thread which the Russian people attempted gropingly to follow but which their ignorance, their economic disorganization, and the treachery of interested factions caused them so easily to lose. For democracy is a form of government suitable not for the lowest, but only for the highest, type of civilization. It can succeed only when it is based on intelligence and when it arises from an inherent love of freedom and a deep respect for law and order.

When next we regard the widespread social, and particularly economic unrest in Great Britain and in America, which centers chiefly around the antagonism between capital and labor, we find the real cause disguised by divers names for corresponding phases of the discontent. Now the conflict centers on wages, then on hours, on holidays, on the right of dismissal, on this working condition or on that, on recognition of the labor union, or on some one of a variety of other causes of friction. But those who have experience know that to increase wages, to shorten hours, to improve now this working condition and now that, or to concede any other of the varied demands of labor, never satisfies. The discontent of labor only deepens from month to month. It is not low wages, or bad working conditions, or long hours, or want of holidays or of other concessions, or any or all of these complaints, taken singly or together, that constitute the real grievance of labor and that form the basic cause of the continual strife which frequently culminates in trials of strength or endurance. To grant any or all of these does not give any assurance of cessation of demands or of permanent improvement in the relations between capital and labor.

What then is the essence of this seemingly endless difference? What is the underlying cause of the great industrial unrest? What is the principle for which (though perhaps labor itself, or even capital, may not always recognize it clearly) the contest is in reality continually being waged? There are probably very few men who would not be willing to earn less, to work harder and for longer hours and under worse conditions, or to get on with fewer holidays if it were done for themselves in any undertaking in which they really felt a deep personal interest. But, in modern industry, they do not feel personally interested. They have no sense of ownership. They never begin and finish anything. They have little or no interest in the end product. They do not have the opportunity to think for themselves. They are generally required to perform certain definite work, or even mere mechanical movements, without consultation, or in a way that does not call for reasonable exercise or recognition of their intelligence. They have very little, if any, sense of personal responsibility. They have little or no voice in governing themselves. They are parts of an organization, cogs in an economic machine, which they do not fully understand and in which they almost lose their identity, that is, their freedom for self-development. Somebody, somewhere, shapes the organization and sets it in motion but the part of the organization they see or of the work they perform usually has little, if any, scientific or social meaning to them. Hence their spirit rebels. The human spirit, unless it has been utterly suppressed, is fortunately so constituted that it always rebels against any form of external authority in which it has no share and which it does not intelligently grasp.

This deep love of real freedom is one of the most distinguishing characteristics of the human spirit and lies at the very heart of all human progress. It is one of those basic differences that distinguish man from every other creature of the kingdom to which belongs also his long line of lowly ancestors. With almost infinite patience exercised throughout eons of time, Nature gradually fashioned an ascending line of beings with a central nervous organ, endowed with mind, which enabled them to free themselves more and more from the trammels of their physical environment. These beings, gradually becoming more clearly conscious, intelligent and human, and progressing correspondingly in their freedom, in their gradual ascent to full human status finally reached the problem of primitive man's relationships in association with his fellows. In the course of the many succeeding eras which constitute the period of human history, man has gradually evolved the democratic principle, which has arisen from his growing consciousness that the highest degree of freedom can be attained only in coöperation and which is deeply rooted in the ethics of this human association. Gradually, with many struggles and infinite discussion, with occasional set-backs but general progress, with general evolution but occasional revolution, man, where most highly civilized, has thus won for himself democratic freedom in certain aspects of his relationships to his fellows, which we group broadly as political and municipal associations. What we need now to recognize clearly is that this historic struggle for democratic freedom in political and municipal affairs, which has deep hereditary roots in the long prehistoric struggle for more purely physical freedom, is, under one disguise or another, now spreading over into this new field and is becoming a struggle for self-government in industry. If we would grasp the real inwardness and true proportions of the industrial problem, it is necessary for us to lift it out of the muck and mire of recurring squabbles over this and that petty difference between the parties in industry and to place the whole issue on a higher plane. When a great principle is at stake, a principle that has been inherent in creation since the very beginning of animal life, a principle that has divided and shaken, and is at the present moment dividing and shaking, the world of man to its very foundations, surely it is time we took the trouble to grasp it clearly, surely it is time we ceased playing and jockeying with the issues raised by it and, like men of full mental stature, frankly recognized it and deliberately so shaped our conduct as to accord with it. Evasion, under such circumstances, is senseless. What is needed is clear and fearless thought, decision, and resolution. It is not sufficient excuse that we cannot clearly see all the details of our future progress. In this matter, we may well take a lesson from your honored President in dealing with the principle of the League of Nations. We should clearly and fearlessly grasp and stick to the principle, knowing that errors in detail will be more easily corrected than failure in principle.

The sooner we recognize this single, simple principle as the root cause of the existing industrial antagonism, just so much the sooner shall we begin to work intelligently toward a permanent solution of the problem which faces us in the continual, but often clouded, struggle between capital and labor. Any step taken in clear recognition of this will be a step nearer the real solution of the problem. Other steps may be taken from the best of motives, they may be based on the most humane feelings, they may be fully justified and worthy of the highest commendation, but if they do not in some way conduce to this end, they will not assist in effecting a final solution of this great industrial problem. It is better for us to take time to see the goal clearly. Then we shall be able to judge intelligently as to our progress toward it. Society has abolished open slavery of all kinds. America did this at enormous cost in blood and treasure. The difficult problem now is to abolish gradually, and with as little economic confusion as possible, every other form and semblance of autocratic rule. Industry must now gradually come to be ruled by the intelligent consent and approval of all the interested parties. Industrial unrest will continue and will increase until society recognizes this and takes such steps as will secure suitable organization and sufficient general economic intelligence to effect it. We cannot stop half way. Democracy will not be content with mere political and municipal control. It will never leave the matters that affect most deeply and intimately our very existence and manner of life to the autocratic decisions of a few, no matter how worthy or how intelligent they may be.

England already recognizes this and, as you are all now fully aware, has, under the Whitley scheme, of which I gave a brief preliminary account before the Canadian Mining Institute last Spring, adopted the idea as a government policy. Her industries, in particular those most highly organized on the side of both capital and labor, are being organized under this scheme. Many Joint Industrial Councils are already formed. No doubt you in America will watch the organization and the working out of the functions of these with the greatest interest, as we shall in Canada. This government recognition of the principle of representation of the governed in industrial relations is a first, but an enormous, step in the right direction.

A fair and proper sharing in the control of industry among the several coöperating factors will gradually bring about that feeling of interest and responsibility and that just division of the joint product among the interested parties which will restore the incentive to do one's best and will make industry once more, as it should always have remained, the handmaid of human welfare. Industry, properly conceived, is not something to seek to run away from as long and as frequently as possible, for holidays. Work, done under proper conditions and with the right attitude of mind, can be made so mentally stimulating as to yield very real pleasures of its own. When it does thus stimulate interest and thought, it ceases to be drudgery and no longer fills the mind of the worker with envy and bitterness. When a man has reason to take deep interest and pride in his work because he, jointly with others, contributes his energy and intelligence to a common result or product in which he is interested and in which he is convinced he will have a just share, he will cease to watch the clock and will experience some of the real joy of work of which we hear so much. We need only analyze our own experiences to realize the truth of this. Any doubts we may have as to the exact truth of this probably arise from our daily experience with men whose independence and self-respect have been, in part at least, lost, owing to long years of domination by our existing industrial system. Probably the most pernicious defect of the existing system is this gradual undermining of the workmen's personal responsibility, interest, and self-respect, which are so essential to the development of personality and character.

The British machinery appears to be based on the right principle. The Joint Industrial Councils (of national scope in each industry), the District Councils, and the Shop Committees are based on equal representation of capital and labor. That experience in its operation will prove considerable modification, adaptation, and adjustment to be necessary, is to be expected, as in the case of any other new machinery; and that difficulties will arise from time to time, particularly in the less organized or wholly unorganized industries, is a practical certainty. But, if machinery is designed on the right fundamental principle, such alteration and adaptation merely indicate progress and are easy. If designed on the wrong principle, the more the machinery is perfected the deeper are we plunged into the original error.

There can be little doubt that in recognizing the representative, or democratic, principle in industry and in adopting this principle as the basis of a national industrial policy, Great Britain has started on the right track. It is peculiarly appropriate that England, the mother of parliaments as well as of the modern industrial system, should now take the lead in introducing the parliamentary idea into the organizations for controlling industrial relations, as a principle of national economic reconstruction. But let no one confuse starting in the right direction with completing the journey; creating machinery with carrying out the work the machinery was designed to accomplish. The journey is only begun and it promises to be very long and beset with many difficulties. The history of political democracy should warn us of this. The work to be performed by this new machinery will be vast in amount. Though, if it gradually removes the deep cause of the present antagonism and thus restores a high efficiency in industry, it will be the means of adding greatly to the national wealth, contentment, happiness, and prosperity; yet, because it will surely affect the distribution of wealth, it may ultimately mean serious apparent economic loss to many individuals. All the more does society need once for all to assure itself that the underlying principle it is adopting is right and just to all.

Has any one any doubt of this? Has it ever proved to be true that any people, or any section of any people, can safely entrust any aspect of their progress and well-being to any other people or class? Has it ever proved otherwise than that both parties were injured? Have you any knowledge of any instance which did not tend toward moral degeneration of both sides of such an arrangement? Do not present industrial relations further illustrate the truth of this? Have we not all some experience of the moral degeneracy that has resulted in industry from failure to recognize this principle?

Our industrial machines have been turning out two products. We have so far intently regarded the one and largely neglected the other. But the human product, which has been largely neglected, is of infinitely greater importance than the material product. Our aim has been so to beneficiate the material entering our plant that it may be of increased economic value when it leaves the plant. But, if we do this at the cost of shrinkage in value of the personality of those operating the plant, where is to be the recompense? Can this loss justly be compensated by wages, or by dividends paid to some one else? Do you know of any conditions under which human personality does not deteriorate, except those of freedom? Does the present method of organizing industry by a few, generally without consultation with or representation of the many others who are intimately concerned in it, secure this feeling of freedom? I believe that all morally just and honest men who will take the trouble to think out the fundamental issues of this industrial question, with sufficient clearness to grasp the principle involved, will readily admit the justice and rightness of the general principle on which the Whitley scheme, adopted as a labor policy by the British Government, is based.

If this opinion is right, if the democratization of industry is the principle from which all real progress in adjusting industrial relations must start, then we can hardly dwell on this matter too long or too thoroughly. It is necessary to saturate our minds with it so thoroughly that it will gradually change our habits of thought and action, that it will permeate all our views and change all our attitudes and feelings toward our fellow men in industry. When we have thus become thorough converts to this view, there will be some hope that organization to give effect to it may have a chance of success. Not otherwise. This is at bottom essentially an ethical question. If democratic organization is not based on conviction, I fear it will prove to be a form without spirit. But it is further necessary that we seek to propagate this view as widely as possible. For these principles cannot be carried out by individual firms with complete success. Economic organization is no longer local; in very few respects is it even national; more and more is it becoming cosmopolitan. Hence we should avail ourselves of every opportunity to discuss this question publicly; and it is even more necessary that, in our capacity as citizens, we should give voice to the demand that our educational systems give greater attention to these social and moral questions in the programs of our schools and colleges. As quickly as possible, these institutions should be adapted to meet society's present, urgent, social and moral needs.

For, consider for a moment what will happen in these Joint Industrial Councils if each side comes to the discussion without such sound and sane knowledge of economic questions and such ethical views as will create in the minds of both parties common standards of social justice and right. On any issue of importance the work of the representatives will end in either of two very undesirable results. Either the deadlock between the trade unions on the one side and the employers' associations on the other will have been transferred to a debating organization in which neither side can bring conviction to the other; or, even if the discussions that occur in the Industrial Councils do bring about a satisfactory compromise, such decision will be refused acceptance by the rank and file. Until both parties to the disputes are educated in economic matters up to the degree at which they can grasp clearly what is economically possible and what is socially best, and until they have clear ethical views as to the real end and purpose of all industry as social service, in other words, until they are so educated, either in early life or later, that they have the mental attitude and training practically acquired which will enable them to grip clearly and rationally these complex social, economic, and ethical questions when fully discussed with them, of what use can the mere creation of this new democratic machinery be?

With labor sharing thus in the control of industry, one of the problems that will ultimately, but inevitably, come to the front for discussion and settlement will be the principles on which must be based the equitable distribution of wealth—a fair sharing of the joint product of industry among labor, capital, management, and the community. The growth of human intelligence is making possible the production of very greatly increased wealth, that is, goods and services of all kinds necessary to supply the needs and desires of human beings. Surely the further growth of intelligence, if accompanied by a corresponding growth in moral feeling and purpose, will gradually solve the problem of preventing the partial frustration of the great purpose for which wealth is created; viz., human well-being, the highest degree of which can be brought about only by its fair and just distribution. But this is an economic and ethical question of the greatest complexity and it lies at the very core of many other economic questions, for all of which a just and practical solution must be found. Nor is it just that the solution of this difficult but important question should be left mainly in the hands of two or even three of the four interested parties. A thorough study and discussion of it by all four interested parties is the only just method. No other method would be thought of under any other than the semi-autocratic conditions prevailing in industry. We are safe in saying that very few, if any, today have given sufficient study and thought to this question, though of such basic importance, to state clearly the principles in accordance with which details of a just and right solution may be worked out in any given case. If I mention this quesion, it is not done with any intention of suggesting that I have reached clearly defined views on this vitally important matter, or of causing needless agitation by raising one of the most burning of economic questions, but rather that we may count the full cost of democratizing industry before deciding deliberately to adopt the principle. As stated, once the initial step is taken, I believe the only logical advance will lead inevitably to this problem. If, as seems inevitable, industry is to be democratized, whether agreeable to all or not, and if this is to be effected in the highest interest of all, then all sections of society should somehow be trained to think economically, socially, and ethically, so that at least a large majority will be in a position to appreciate the justice of the best solution of this and other difficult economic problems connected with industry.

The economic and ethical mental equipment for this, I believe, can best be formed by a rational educational system—a system that regards as its central aim the efficient preparation of oncoming citizens for citizenship in a commonwealth that aims to be democratic through and through, in industry as well as in matters of civil and political government. This is, in my judgment, the surest, if not the only, method by which society can avoid the economic confusion and distress, if not revolution, toward which we appear to be heading. I fear these distressing conditions may come about, if we do not clearly and frankly recognize the tendency of the present almost universal social movement and make adequate preparation, by widespread social and ethical enlightenment, to guide and assist it. Our general educational systems should be so adapted as to prepare those who will tomorrow provide the capital and perform the mental and physical labor connected with industry for a steady growth in an intelligent application of the democratic principle to industry. Industrial peace will never be attained as long as capital and management assume the right to a final say on matters intimately affecting the welfare and even the self-respecting existence of a very numerous class, whose loyal coöperation is as essential to the success of every industrial enterprise as their own.

Is it clearly recognized that we are at the beginning of a great transition period in industry? Do we realize that the autocracy of capital is coming to an end? Such periods of widespread, rapid, social change are times of peculiar danger. It is in the power of the present members of society either to recognize the principle at work and to lend intelligent assistance to the movement, or to increase the social danger by opposing it.

There is, unfortunately, here and there, a revolutionary element in labor, usually arising from a vague sense of injustice which urges the mind toward aspirations frequently impossible of attainment. Unless there exists a fair knowledge of economic facts and principles, making clear to all parties what is attainable and by what methods, and unless there exists also, in the great majority, moral purpose and determination, revolutionary schemes may have every appearance of being not only practical but legitimate and just. There is also, even among persons of fair general education who have paid but little attention to social and economic questions, too prevalent a belief that society can easily be reconstructed along the lines of some simple, socialistic formula; a belief, in other words, that a new social world can quickly be built by constitutional means. This state of mind, if widespread, is a very real danger. It may become the precursor of bolshevism and red revolution. The greatest safeguard, if not the only effective safeguard, against it is widespread social, economic, and ethical education.

He who hopes to confine the waters of a perennial stream courts disaster. The springs of social energy and change are inexhaustible. The wisest course is to recognize social aspirations and tendencies, to aim to give them intelligent direction by universal and thoroughly democratic education, and to afford them an adequate means of rational development and expression by efficient, democratic organization. This holds true in every field of social activity, inclusive of industry. It is the method that will secure the most real and rapid progress.

The democratization of industry is not more a matter of expediency than a moral necessity. The moral failure of the autocratic method in industry is shown by the hopeless divergence of view of capital and labor generally. The nation that leaped to its feet as champion of the great cause of world freedom, as soon as conscience so decreed, even though every historical tradition might be violated by the decision, will, I feel confident, never permit questions of business tradition or of economic selfishness prevent her from championing the cause of democracy in industry, when once she is convinced of its justice, its righteousness, and its practicability. Its practicability rests mainly on an intelligent adaptation of the general education of all citizens to this end. The progress of democracy is indeed almost entirely an educational problem.

Industry sustains the life of every civilized human being. It lies thus at the very foundation of every other human activity. We degrade industry when we regard it solely or mainly as a means of private gain. Participation in industry is social service of the very highest order, since no other social service is possible without it. When this view is widely and clearly grasped and when it becomes an actuating motive to industrial activity, industry will attain to the position of real dignity proper to it—the dignity that is now generally conceded only to the pursuit of pure science. It will never be possible to engender a spirit of loyal coöperation between capital and labor in industry with nothing nobler to inspire it than the low motive of private gain. A widespread unity of spirit cannot be fostered without a great and worthy ideal. If we hope to reunite society by bonds of common selfish interests, we are doomed to disappointment. Only mutual service unites; selfishness disintegrates. The honor that is supposed to hold together a band of thieves is mutual service while it lasts, but it seldom withstands for long the disintegrating selfishness of their deeper nature. The inspiring unity begotten by a noble and worthy ideal was never more convincingly shown than by the alacrity of the response of every class in America to the call for defense of world freedom. No threatened suffering, loss, or danger was a sufficient deterrent to hold men back in the presence of this great and noble ideal. Suitable organization for effecting perfect coöperation quickly resulted. But an attempted organization without the great central, unselfish purpose would have effected nothing. It is primarily not a question of form but of spirit.

This ideal, that industry is social service of the most fundamental nature and therefore of the highest order, by no means precludes the incentive of intelligent self-interest. Neither did the ideal of world freedom. If closely regarded, both will be seen rather to include than to exclude the highest self-interest. The moral universe is fortunately so constituted that the highest self-interest is always best served by considering the welfare of our fellow men. "He that saveth his life shall lose it." Selfishness is self-defeating. Unselfishness is the only efficient selfishness. The highest attainable industrial efficiency, which we are all striving after, must ultimately rest on this moral foundation of social service. This will be its ideal, and its method will be democracy and a "square deal." The nation that successfully and widely inculcates this ideal will lay a sure foundation of high individual as well as national efficiency and prosperity. The intelligent rooting in the minds of the oncoming generation of this higher economic ideal of industry as social service and the upbuilding of the less selfish character necessary to bring it increasingly into practical effect afford the greatest educational opportunity today.

In an address before the College and High School Department of the Ontario Educational Association last April, in a slightly different connection, I expressed this view in these words:

"Society already has before it a mass of unsolved social problems some of which we have broadly outlined, and of which none exists of more fundamental importance to the welfare of all than the antagonism between Capital and Labor. This problem, I believe, can be solved only by applying the democratic principle to industrial relations. The Anglo-Saxon struggle for democratic freedom has continued for more than seven centuries. It will not finally be won until the autocracy of capital is uprooted in industry. But the economic understanding, the ethical feeling and determination, the mental attitude—the indispensable psychological condition for getting together and viewing industry as a joint undertaking—must exist on the side of both Capital and Labor, before democratic organization for working amicably together can accomplish any good results. The idea must precede the expression; so must the psychological condition precede the organization. Hence we must aim to instil the truly democratic spirit of intelligent coöperation and social service in all, through our educational institutions, which citizens must rely upon to guide and quicken our social evolution in the manner outlined."

If we attach so much importance to the formation, by every educational means at our disposal, of sound economic and ethical ideas and ideals, in the minds of all oncoming citizens, why, it may be asked, do we further attach so much importance to the democratic form of organization as a means of solving the deadlock between capital and labor? Long experience, which has indeed culminated in the recent world calamity, has shown that autocratic rule in any sphere of life gives birth to ill-will, want of confidence, jealousy, and selfish ambitions. The secrecy of its methods creates the stifling and murky atmosphere in which spring up, as in a hot-bed, injustice, deception, suspicion, and distrust. This does not deny that the autocratic form of rule, in both the industrial and the political fields, has produced many results that, in outward appearance at least, were praiseworthy. But autocratic rule cannot safely be estimated at its face value. It has proved rotten at the heart. Superficially regarded, it may have appeared brilliantly successful, but it has proved itself a whited sepulchre. It talked peace but secretly plotted war. It preached "kultur" but attempted by sudden onslaught to wreck civilization and nearly succeeded in its fell purpose. Its much vaunted efficiency was only apparent. It was purely material. It failed at the very core. It failed to develop human personality and character. The world today stands aghast at the full revelation of its human, or rather, inhuman, product. Can we honestly deny that this indictment of political autocracy applies in some degree to the autocratic form of organization in industry? Does it not appear to be working similar havoc? Distrust, suspicion, and antagonism are widespread. The democratic form of organization gives the best assurance of any method which human society has so far discovered of the kind of open-and-above-board treatment and the "square deal" on which alone confidence and goodwill can rest. Thorough organization must exist in either case. The difference lies in the origin of the authority and in the consequent responsibility.

A question that deeply concerns us as engineers, managers, or superintendents of industrial enterprises is that of efficiency. In this matter, I fear, we have much blame to accept for narrowness of view. Because of our special training in the material sciences and their application to industry, we have confined our attention altogether too exclusively to machines, to processes, to arrangement of plants, and to the external forms of organization. We have paid far too little attention to the "imponderables"—to ethical standards, to psychological conditions, and to the mental attitude of those on whom real efficiency must finally depend. Surely it must be apparent to every one that there cannot be any approach to the highest attainable efficiency in production if there exists a general atmosphere of suspicion, distrust, and antagonism. Opposing forces tend to cancel one another. A high resultant can be obtained only by paralleling both the mental and the physical forces at work. Complete and efficient coöperation of the various factors in industry can be obtained only in an atmosphere of confidence and good-will. Efficient means of open discussion, knowledge of sound economic and ethical principles, and sterling character are essential prerequisites of confidence and good-will, hence also of high efficiency.

Industry has so far been under highly centralized rule. In this government, capital and management have had predominating sway. Labor has had little or no share. We are, I believe, at the threshold of self-government, or the application of the democratic principle to industry. We all need economic and ethical preparation for the more complex responsibilities this will cause to rest on us. Many old ideas and ideals regarding industry and business in general will find their way to the scrap heap. In the change, not only industry but we ourselves will rise to a new dignity. Industry will become more worthy of the best efforts of all. In a new sense it will become the servant of the community and the handmaid of human well-being, in place of being largely a source of private gain and a cause of friction and conflict between opposing classes of society. It will thus come not only efficiently to supply the physical needs but also to minister to the high moral and spiritual purposes of human life. In this new government, all interested parties, not omitting labor, the largest party, will have a voice. On this continent the entire machinery of industrial self-government has yet to be worked out but it will probably be, as in England, of such a nature as to embrace existing, even though at present antagonistic, organizations. This safe method of building brick by brick on the foundation of existing institutions and experience is characteristic of the Anglo-Saxon genius for at once securing progress and avoiding revolution. In our study of this problem during the next few years, it will well repay us to give careful attention to the working out of the Whitley scheme in Great Britain.

English-speaking peoples have been foremost in the development of democratic political government, which they have recently joined hands in defending. Let us hope they are destined now to lead the world in applying the same principle to industry. This, if carried out on a basis of high economic and ethical ideals, as indeed it must be if it is successfully to be carried out at all, will make future war impossible. With the overthrow of Prussian despotism, the fight for freedom is only half over. Political and industrial autocracy may exist peacefully side by side; but political democracy and industrial autocracy, never. We may still have a long and difficult task ahead, but the job begun in Europe must be completed at home. The seeds of social and international conflict are inherent in the present system of industrial organization, which rests on the self-disintegrating foundation of human selfishness instead of on the solid rock of social service and human welfare. The awful suffering and carnage in Europe will not have been in vain if, after it all and in some measure as a result of it, human society has successfully started reconstruction on this new industrial foundation, which, if intelligently and diligently built upon, will gradually bring conditions of permanent, social, and international peace. In the great war. the soldiers of America, Great Britain, and their allies have shown the most indomitable courage. Have we now the moral courage to look squarely at and deeply into this industrial problem and, if convinced of the soundness and justice of this method of working out the solution of it, to continue at home the courageous stand for democracy which our brave fellows have so valiantly taken in Europe?

It is the Nemesis of autocratic rule that the greater its apparent success, the greater and the more certain is its ultimate downfall. If the world war has impressed on us any one lesson more clearly than another, it is this: that no human social structure, however imposing in appearance or however brilliant in apparent success, can be permanent unless it is founded on the eternal principles of justice and right. If the principles herein discussed measure up to this standard, then, if we are really seeking a permanent solution of the great industrial problem, rather than a makeshift, we must seek to apply them. Industry will not in any case become fully democratized today or tomorrow, or even in the near future. But, if we do not mis-read the meaning of present social movements of world-wide extent, progress in this direction is as irresistible as the tides. To prolong opposition to it is to risk being overwhelmed by the flood. The wise and sensible course is for society to recognize its own movement, to undertake the most careful study of the economic and ethical conditions giving rise to it, and to arrange as quickly as may be for widespread social training of its members, in preparation for such utilization of these new powers as may result in the greatest justice and benefit to all.

One of the most encouraging signs of the times is the steadily increasing amount of time in our technical societies and of space in our technical and other more serious journals given to a discussion of the economic, social, and ethical aspects of industry. Let these discussions increase in both volume and earnestness and let this good work be backed as quickly and as strongly as possible by training the rising generation under our general educational systems, from common school to university, modified and adapted to meet the social exigencies of the times.

## DISCUSSION

S. J. Jennings, New York.—I had hoped to accentuate, and possibly emphasize from another angle, the real fundamental idea embodied in Mr. Corless' paper and trust that at some future occasion, when time is more auspicious, I will be enabled to bring before you my angle of views of the progress in the industrial relation that has been voiced by Mr. Corless. One of the things that I want you to think about and if possible bring up in some future discussion, is the idea that seems to be running through all of Mr. Corless' paper, that the world should be made "safe for democracy," that democracy is the salvation of the world. Now, that is all right if, after the world has been made safe for democracy, democracy is made safe for the world; and the only means of making democracy safe for the world is to educate the people who form the democracy. A sovereign has been defined as "A person or determinate body of persons who commands the rest of the community who are habituated to obey." Now, if democracy is all of the people, and therefore all the people are sovereigns, their command must be so based upon a clear conception of what they ought to be that they will be naturally obeyed by the community; and the only method that human beings have so far discovered of doing that is by education.

In my opinion, one of the lessons that this war has brought clearly to us is the necessity in education of some military knowledge. ⁻ An American colonel who had just returned from the front objected to this opinion, saying, "To teach a boy to be a soldier is to teach him to be a murderer. One of the main things that I taught my men in this war was how to cut the throat of an opponent with a fork. I do not want my son to be taught that thing." It seemed to me that he missed entirely the point of view, which was this: That while a soldier is an organized murderer, that is his end and aim, there are factors in his education that seem to be absolutely essential; these are obedience, discipline, and the knowledge of things rather than the knowledge of words. ⁻ Those are the points I would like you to think about and I hope that in some of our meetings in the future they will be greatly discussed, that the progress that is outlined and visualized by Mr. Corless, which in his hope and in his vision is a great step forward toward the solution of the difficulties between capital, labor, managements and the community, will be solved.

## The English-speaking Peoples*

BY T. A. RICKARD,† SAN FRANCISCO, CALIF.

· (New York Meeting, February, 1919)

WE rejoice that the world-war is ended. We are proud of the part played by the English-speaking peoples—all doing equal honor to the traditions they share in common. One of the compensations for the calamity of the past four years is the fact that the Briton and the American, striving together in the cause of human liberty, have learned to understand and to respect each other. The mother country entered the fight resolutely at the beginning, while yet unready to meet the carefully prepared onslaught of the enemy; then the sons from the overseas dominions rallied to the old battle-cry eagerly and effectively; and last, but not least, the stepsons came from across the Atlantic, speaking the same speech, playing the same game, and fighting in the same clean way.

It was a great foregathering of those that use the language of Shakespeare and idealize the principles of liberty for which the friends and associates of Shakespeare stood sponsor three centuries ago.[1] At a time like this it is pleasant to dwell upon the fact that the liberal Englishmen who organized the Virginia company were the pioneers of self-government on the American continent. The Virginia Assembly, convoked in 1619, was the first example of a domestic parliament to regulate internal affairs on this side of the Atlantic.[2] The Governor of Virginia, Sir Edwin Sandys, had been a pupil of Richard Hooker at Oxford and from that political teacher he and his friends had imbibed the idea of combining civil liberty with constitutional order. To this group of large-minded Englishmen, the American colonists owed their liberal charters and their successive triumphs over the royal prerogative. Let it be noted that the American colonists had to deal with James II and George III, the two smallest-minds in the list of British kings. Another historical note, more pleasant to record, is the connection between the two principal groups of American settlers. In 1608, when the Pilgrim fathers, William Brewster and John Robinson, led their Separatist congregation to Holland and there prepared the expedition to America, they were assisted by Sandys and the Virginia Council, who were willing to share their privileges with them. When the Pilgrims set sail in 1620 they had the promise, obtained by Sandys from King James, that they should have freedom of worship,

---

* Presented at the joint session with the Canadian Mining Institute.

† Editor, *Mining and Scientific Press.*

[1] Charles Mills Gayley: "Shakespeare and the Founders of Liberty in America," 1917.

[2] Alexander Brown: "English Politics in Early Virginia History."

equality before the law, and the right to participate in the government of themselves. Thus the men whom we may regard as the friends of Shakespeare aided the founders of New England. Together they resisted the King's arbitrary dictation. "The political principles that inspired . . . that noble company never died out of Virginia, never died out of the northern colony, called New England. These were the principles first logically developed and clearly formulated by the tutor of Sir Edwin Sandys, Richard Hooker. Disciples of Hooker, associates of Shakespeare, were the founders of the first republics in the New World."[3] These political doctrines of Hooker not only inspired the founders of the first English settlements in America but found an echo in the minds of the men who led the Revolution and subscribed to the Declaration of Independence. Hooker's ideas passed to John Locke, and through him to Benjamin Franklin, Patrick Henry, and Thomas Jefferson.

It seems worth while to make this point insistently. The old Fourth of July talk is out of date, because it is historically untrue. George III was born of German parents and married a German woman. He spoke Shakespeare's language with a guttural accent. His government failed to impose its tyrannic orders on the British colonists in America because it was not supported by the British people. Unable to conscript a British army, he hired the Hessians. It was against the forces of a reactionary German king that a great Englishman, George Washington, led his men to the winning of their independence. Lafayette and Rochambeau brought French aid to the revolutionists, but their help was prompted less by love of the colonists than by the desire to hit at England, which was then at war with France. It was the despotism of Louis XVI that sent Rochambeau and his 6000 Frenchmen "to deal England a blow where she would feel it." That was in 1780. Permit me to remind you that only 18 years afterward, in 1798, the young United States was at war with France. This is not mentioned out of ill-will, but as a historic fact of some significance. We need not belittle the romance of the Lafayette episode, even though it has been highly colored, because it is helping today to stimulate cordiality between the United States and France, but we may demur to the twisting of history in order to represent the English people as reactionary and the French people as liberal at the time of the American revolution. As one who holds that "every man has two countries, his own and France;" as one that held this view even in the days of Fashoda, I venture to say that the friendship now existing between the United States and France is all to the good, because England and France likewise are firm friends. Their *entente cordiale* joins with French-American sympathy in establishing a mutuality of good-will between the three great democracies of the world.

---

[3] Gayley: *Op. cit.*, 93.

It remains to emphasize another fact, to which allusion has been made already, namely, that the English striving for political freedom prepared the way alike for American and for French democracy. Locke, whose doctrines fed the fires of the English revolution of 1688 and those of the American revolution of 1775, derived his ideas of constitutional liberty from his fellow-countryman Hooker. It was Locke's theory that was embodied in the American Declaration of Independence in 1776 and that rationalized the French Declaration of the Rights of Man in 1789. Rousseau exercised no influence on the America of 1776; on the contrary, it was from English philosophers and from Anglo-American reformers that the French revolution derived its cue. English liberalism, disciplined by centuries of conflict, from the Magna Charta to the Bill of Rights, was the political mother of democratic institutions in the free commonwealths of the world.

Much of the prejudice against England inherited by children in the United States is due to the unfriendly tradition perpetuated in their school-books. Great Britain did treat the colonists shabbily after they had broken away from the old country, and, as the history of this nation is concerned chiefly with those early events, it is not surprising that the young American should be impressed much more by the overbearing attitude of Lord North's government, which was not so very different from that of most governments in those days, than by the shabby treatment given, in turn, to the loyalists in the American colonies. The so-called War of 1812 was bound to loom large in the school histories, because it was the first contest following the achievement of American independence; but it was a tempest in a teapot; on land it was badly fought on both sides, although at sea the young American proved himself a chip of the old block; it was a side-show started in the midst of England's great contest with Napoleonic France. At that time England had a navy many times bigger than the American, and Wellington had just returned home from his victorious Peninsular campaign at the head of 100,000 veterans—a big army in those days. In 1814 Napoleon had been defeated, yet it was Wellington that made himself responsible for an honorable peace with the United States. It must be remembered that to the United States the Revolution was the very beginning of things; to England it was only one incident in a long and eventful history. "The game was played, and she had lost. North America, in the eyes of her statesmen, was a strip of eastern seaboard; the great lakes were but dimly understood; the continent beyond the Mississippi was ignored."[4] Therefore the Revolution and the little war of 1812 left no such sting in the minds of the British as the memories that were cherished on this side of the Atlantic. It is worth noting that in 1788 the first English settlement was made in Australia,

---

[4] F. S. Oliver: "Alexander Hamilton," 115.

at Paramatta, in New South Wales. Thus Australia was added to the British dominions at the time of the American secession.

Another thought follows: looking back, it is probable that if England had been minded to recover the American colonies during the early period of their independence, she could have done so; but she had no mind to do so, and any Hanoverian king or government that had tried it would have had to face a revolution at home. Moreover, it is quite certain that the United States would have seceded sooner or later, because no government in Europe could have hoped to retain control over the growing giant of the West. I am frank to add that the idea of filling the United States with Englishmen is as unwelcome to my taste as the idea of peopling the British Isles with Americans. Variety is the spice of life.

> "Our dearest bond is this,
> Not like to like, but like in difference,
> Distinct in individualities."

Let us hope that the school-room histories in America will be rewritten in a spirit less unfriendly to the mother country. It may be more than a coincidence that the names of some of the publishers should be, for example, Kruger, Koch, and Lemp. When a New Englander or a son of Old England hears Germans, Irish, and Finns talking loudly about the time when "We licked the British in 1812," he may be pardoned for smiling. There were more Germans and Irish on the reactionary side in 1812 than on the liberal side. An acquaintance of mine, an Austrian Jew, the editor of a Jewish paper in San Francisco, began to tell me about the time when "*We* licked the British," whereupon I·called a halt and asked him *qu'allait-il faire dans cette galère?* I informed him that I had ancestors on both sides of that affair, at a time when his progenitors were wandering in the morasses of Eastern Europe without even a knowledge of the fact that the British were having a family quarrel. The foregoing story is capped by the statement appearing in a pamphlet prepared by the Sinn Feiners and intended for propaganda among the American soldiers stationed in Ireland. "We helped to win your independence," they assert.[5] The forefathers of this republic may have had some outside help, but it is a little hard on them that they should be called upon at this late date to divide the honors with such as the Sinn Feiners! The people *they* helped were the Germans, not in 1776, but in 1917.

A couple of stories will illustrate this point further. During the Spanish war of 1898, I had as neighbors a man and his wife of British birth but of American citizenship. One day the man, in the presence of his son, a boy 12 years old, remarked that the United States would defeat Spain. Whereupon the boy exclaimed; "Lick Spain, well I guess we will. Why we licked *you* twice." That boy and his two brothers have fought

---

[5] As recorded in dispatches published on January 11, 1918.

side by side with the British in France, one of them wearing the uniform of his father's native land because he could not wait until his own country entered the war.

Another boy's historical knowledge concerning the relations between the United States and Great Britain consisted of three items:

(1) Major Pitcairn spoke contemptuously of the revolutionists while he stirred his punch at Lexington.

(2) Andrew Jackson refused to black the dirty boots of a British tyrant.

(3) The Americans licked the English twice and would do it again for two cents.

Yet it is recorded that this same boy knocked down a perfectly well-behaved Bavarian in a barber-shop for expressing the opinion that England would be invaded. You could tell that boy all you pleased about the battle of Bunker Hill and the villainy of Lord North, but he would not forego his share of ownership in the Black Watch at Waterloo, of Nelson at Trafalgar, of Wolfe on the heights of Quebec, of Drake and the Golden Hind, or of the archers at Agincourt.

The tale of an ancient wrong should now be laid aside on a shelf beyond the reach of any but the most inquisitive student. The memory of Bunker Hill is overlaid by that of Manila bay. Let me recall the story as recorded by Dewey himself.

When Commodore Dewey drew the attention of Vice-Admiral von Diedrichs to the disproportion between the German naval force at Manila and the German interests in the Philippines,[6] he was met with: "I am here by order of the Kaiser, sir." The German admiral made trouble for the Americans continually, while maintaining the most cordial terms with the Spaniards; this also, it is to be presumed, by order of the Kaiser. He repeatedly ignored the blockade that Dewey had established after the battle of Manila bay, sending his warships into the harbor without allowing them to be boarded, as was necessary in order that they might be identified and assigned an anchorage. Dewey, in his autobiography, says: "Vice-Admiral von Diedrichs, in denial of the right, had notified us that he would submit the point to a conference of all the senior officers of the men-of-war in the harbor. But only one officer appeared, Captain Chichester, of the British 'Immortalité'. He informed the German commander that I was acting entirely within my right; that he had instructions from his government to comply with even more rigorous restrictions than I had laid down; and, moreover, that as senior British officer present he had passed the word that all British men-of-war upon entering the harbor should first report to me and fully satisfy any inquiries on my part before proceeding to the anchorage of the foreign fleet."[7]

---

[6] "Autobiography of George Dewey, Admiral of the Navy," 257. 1913.

[7] *Op. cit.*, 266.

It is related by General Younghusband, of the British army, who was at Manila at the time, that when Von Diedrichs asked Chichester what he intended to do, the Englishman replied: "Just what Admiral Dewey and I have agreed upon."

A more significant incident occurred later, just before the town of Manila was captured, on May 13. I quote Dewey again: "As we got under way the officers and men of the British ship 'Immortalité' crowded on the deck, her guard was paraded, and her band played 'Under the Double Eagle,' which was known to be my favorite march. Then, as we drew away from the anchorage from which for over three months we had watched the city and bay, Captain Chichester got under way also and with the 'Immortalité' and the 'Iphigenia' steamed over toward the city and took up a position which placed his vessels between ours and those of the foreign fleet."[8] Thus the British warships were differentiated from the "foreign" fleet; they stood between Dewey and the Germans.

This was no mere idiosyncrasy of Chichester; it expressed the policy of the British Government. In March, 1898, Germany asked England to join her and France in putting their fleets between Cuba and the American fleet. The British Foreign Secretary promptly refused. Great Britain was the one power that prevented the formation of a European coalition against the United States at the time of the war with Spain. It was then that the Kaiser exclaimed, "If I had a larger fleet I would take Uncle Sam by the scruff of the neck." More recently that part of his own anatomy has been in acute danger. On May 11, 1898, while the Spanish-American war was in progress, Joseph Chamberlain, the British Secretary for Foreign Affairs, said in a speech at Birmingham:

"What is our next duty? It is to establish and to maintain bonds of permanent amity with our kinsmen across the Atlantic. There is a powerful and a generous nation. They speak our language. They are bred of our race. Their laws, their literature, their standpoint upon every question are the same as ours. Their feelings, their interests in the cause of humanity and the peaceful development of the world are identical with ours. I do not know what the future has in store for us; I do not know what arrangements may be possible with us; but this I do know and feel, that the closer, the more cordial, the fuller, and the more definite these arrangements are, with the consent of both peoples, the better it will be for both and for the world—and I even go so far as to say that, terrible as war may be, even war itself would be cheaply purchased if, in a great and noble cause, the Stars and Stripes and the Union Jack should wave together over an Anglo-Saxon alliance."

Therein Chamberlain exhibited not only a brotherly spirit, but also the highest quality of statesmanship—foresight. His hope was fulfilled in 1918.

---

[8] *Op. cit.*, 277.

The echoes of Yorktown and Saratoga are smothered by the glad shouts that come from Belleau Wood, Cambrai, Lille, and other recent battlefields on which the Union Jack and Old Glory were carried forward to victory. Then was fulfilled Jefferson's hope, as expressed in a letter to President Monroe, in 1824, advising him to accept the policy, now known as the Monroe Doctrine, which had been suggested by George Canning, Secretary of Foreign Affairs for Great Britain. Thomas Jefferson wrote: "Great Britain is the one nation which can do us the most harm of any one, or all on earth; and with her on our side we need not fear the whole world. With her, then, we should most sedulously cherish a cordial friendship, and nothing would tend more to knit our affections than to be fighting once more, side by side, in the same cause." That pious wish, so like Chamberlain's, has been splendidly fulfilled. Do you remember one of the flashes of history that we found in the day's news last October? It ran like this:

Somewhere on the Western front an assault was launched at dawn under cover of a creeping curtain of shell-fire. An American division advanced shouting "Lusitania." With them went a squadron of tanks. While this attack was progressing favorably a British division on the left swam the canal and pushed forward, in the face of scores of German machine-guns, to the village of Belleglise. By nine o'clock prisoners were being sent to the rear in droves. A pause followed this first phase of the battle. The Americans, tired but elated, stood in the trenches they had captured, while an Australian regiment, moving to their support, passed over them, or leap-frogged, to form the first wave of a new advance. The storm of cheering that greeted this manoeuvre rose high above the din of battle.

We echo those cheers today. The word "Lusitania" made those English-speaking soldiers a unit against

"A people with the heart of beasts
Made wise concerning men."

One fateful consequence of the war is the suicide of the German tradition. Before 1914 the Germans had a growing hold upon American business, they were grafting their kultur upon the American people, chiefly through the scheme of exchange-professors, whereby German propagandists were given a free hand at American universities, and, what was worse, sundry American professors went to Berlin, where they succumbed to the hospitality of the Kaiser and became sycophants to his purpose. The Germans were even obtaining success in imposing their language upon a large number of native-born Americans; and in doing this, they were undermining the English tradition, inherited legitimately from the founders of the United States. They were assisted in their propaganda by the fact that many Germans of the highest character migrated to the United States at the time of the War of Liberation in 1848. These proved excellent American citizens because they came mostly from the South German States and brought with them none of the Prussian idea. They were followed in later years by other Germans, not so liberal-minded, but of undoubted capacity in business. Clannish always, they coöperated, they became pioneers of the German idea, which had made considerable headway when William of Hohenzollern and his military caste, supported by the German people, began their onslaught upon Western civilization. During the time the German tradition waxed in the United States that of the mother country waned; for many reasons, some large, others small. The American alienation from the people of "the sceptred isle" has been due in part to sympathy for those of "the emerald isle." Undoubtedly the blundering policy of the British Government in handling the Irish question has tended to perpetuate the prejudice against England; the Irish are born politicians; in the big cities of this country they exert an influence far out of proportion to their numbers or their character; they have played into the hands of the Germans and together they have fostered a sentiment that has tended continually to hinder the development of good-will between our peoples. By "our peoples" I mean those represented by the Canadian Mining Institute and the American Institute of Mining Engineers. It has been the popular thing for generations in the United States "to twist the lion's tail;" it pleased both Irish and American prejudice; it was the regular stock-in-trade of frothy orators and jingo editors.

The English-speaking peoples have so many proud and happy memories in common that it is about time to balance the account. England did treat the colonists arbitrarily and they treated the loyalists shabbily; the young United States soon after achieving independence did have a further fuss with the mother country, which withdrew from the quarrel voluntarily. So much for that. From her independent sons in America England learned a lesson she has never forgotten, as is proved by the record of her relations with Canada, Australia, India, and Egypt, and her other territories, particularly her treatment of the Boers. She holds her overseas dominions by the silken thread of good-will, by that and nothing more. The American people share with the English people the glorious traditions derived from the men that helped to develop constitutional liberty before the Declaration of Independence, which was a logical sequel to the Magna Charta and the Bill of Rights. Many Americans, even those of British descent, may choose to forego the privilege of sharing those ancient glories, but they will not refuse to claim the inheritance of Chaucer and Spenser, of Shakespeare and Milton, of the King James version of the Bible and the Book of Common Prayer, of the English common law and the unwritten rules whereby both alike "play the game" in war and peace. For those traditions we are joint trustees. To them we add now the vivid, the searing, the proud memories of the Great War, in which at last our men stood shoulder to shoulder to assert the principles of freedom on earth. Abraham Lincoln closed his first inaugural address by an appeal for reconciliation with the South; "The mystic chords of memory, stretching from every battlefield and patriot grave to every living heart and hearthstone all over this broad land, will yet swell the chorus of the Union, when again touched, as they surely will be, by the better angels of our nature." Does not this find an echo in our hearts today; do not the chords of our memory vibrate to the stories that have come from the battlefields in France, and will not the better angels of our nature play on those mystic chords a song to which we can pull together in unison for liberty, justice, and peace?

The visit of President Wilson to England, following our comradeship in arms during the war, is one of the great events of history. The intensely cordial greeting that he received during his visit, not only from King George but from the crowd, augurs well for the friendship between the English-speaking peoples. One of our miserable San Francisco papers spoke of the reception accorded to the President as the most enthusiastic ever given to "a foreign citizen," as if a man who found himself in a country where his native tongue is spoken, where his mother was born, and from which his paternal grandfather came, could feel himself a "foreigner." Legally he may be, but setting aside the legal fiction, Mr. Wilson found himself among his own kinsmen. There are three kinds of people in the world: Americans, Britons, and foreigners. Does any one of us feel like a foreigner when he is either in Canada, England, or the United States? I trow not. However, I have lived so long in the United States that I venture to warn Britons against over-playing the "kinsman" note. Mr. Wilson's ancestry brings him within the category, but most Americans do not like to be dubbed "kinsmen" or "cousins," because they are strongly assertive of their nationality, and of their own identity as a people; moreover, the influx of alien blood from the other countries of Europe is so considerable that it is incorrect to regard the American as a cousin of the Briton. Indeed, this assumption was at the bottom of much of the chagrin felt in Great Britain and in Canada when the Government of the United States deliberately adopted an attitude of neutrality during the early part of the war. Impatient as most of us may have been at the aloofness of the United States during that period, we should have reminded ourselves that the American people includes a large proportion of citizens of other than British descent, about one-half, of whom ten millions were born in Germany or born of German parents, and perhaps twice as many more have German ancestry. We ought to have reminded ourselves that even those who are of British descent feel their separateness strongly, partly on account of old revolutionary prejudice, and partly because there is that constant urge to emphasize the individuality of the American nation. That is why I deem it more tactful, and also more in accord with the facts, to lay stress on our common notions of fair play and our common insistence on the right to live and let live—the right that the Prussian and his cohorts undertook to suppress. The Briton must accept the fact that the American, especially those Americans who have no English blood in their veins, dislikes an excess of emphasis on kinship. For instance, Theodore Roosevelt objected to it and he was a typical American if ever there was one. He had Dutch blood, English blood, French blood, even German blood, in his veins, but there is no mistaking the fact that in him it was no mere mechanical mixture but an ethnical compound, called "American," because it is entirely different from any of the ingredients of which it is composed. As he himself said: "We are a new and distinct nationality. We are developing our own distinctive culture and civilization, and the worth of this civilization will largely depend upon our determination to keep it distinctively our own." That undoubtedly is the voice of young America, the expression of the virile nationalism that Roosevelt typified so splendidly. If the call of the blood were to be taken literally, the cries from America to Europe would be as confused as those that arose from the Tower of Babel. President Poincaré at the opening of the Peace Conference said appropriately: "America, the daughter of Europe, crossed the ocean to rescue her mother from thraldom and to save civilization." Therefore, it is wiser to base our international friendship upon the other common factors: language, literature, law, sports, ideas, and ideals, themselves largely a consequence of our common ancestry.

Our people are different in their traditions and in their outlook, or, to be more nearly correct, the Briton cares more for tradition than the American, who, on the other hand, cares more for outlook. The one feels his background, the other his foreground. The Englishman accepts his social environment; he is proud that his father and grandfather did as he is doing; he loves the continuity of custom. The aim of the American is social extrication; he sees no reason for following in the footsteps of his forebears; he blazes a fresh trail, and rejoices in breaking into a new environment. Britain is politically a democracy, but socially she still preserves many of the traditions of feudal days. These make for the amenities of life, but they, and the social manners derived from them, are distasteful to the unconventional men and women of a country that has broken definitely with all that such customs imply. To us, "the rank is but the guinea-stamp; the man's the gold for all that." In freeing ourselves from such trappings we may have gone to the other extreme; the lack of respect for authority is not the most desirable trait of the democrat.

Life is full of compensations; every loss has some gain. The engaging frankness of the American contrasts with the starving of emotion in the Englishman. He thinks it good form to suppress any expression of enthusiasm to the point of making himself appear cold or supercilious. It is an unlovely trait and destroys the natural grace of an intelligent human being. In England the religion of "good form" is a disease among well-bred men, making spontaneity a mark of the socially uneducated. An American officer says to a British officer: "Well, I guess we'll have to clean up the Boches together!" The English officer, adjusting his eye-glass, says, "Really." That reply was not meant to be insulting, but it chilled any *rapprochement*. We are reminded thereby of "a certain condescension among foreigners," on which an American essayist expatiated. May I suggest that some of that British superciliousness is due to shyness, not to impertinence; it is what our French friends call *mauvaise honte;* it springs from the Englishman's inbred fear of making himself ridiculous. On the other hand, the autobiographical garrulity of an opposite kind of American and the boyish inquisitiveness with which he will dive into the affairs of a comparative stranger bear the marks of a crudity that may sometimes be repellant. Again, an Englishman will be severely critical of his own country, because to him to speak well of her is like speaking well of himself, which is taboo, whereas the American will sail in boyishly to praise his native country and to assert how superior it is to every other. Such small national differences should be taken with good humor. The social code of a small island 3000 miles away does not fit the less formal, more spontaneous, life of a younger people sprawling across a continent more than 3000 miles wide. The American ought to understand the Briton if anybody is to do so, and, conversely, the Briton ought to meet the American half-way quicker than anybody else. Our mannerisms may be different, but our ideals are much the same. I am reminded of the story told of the judges who were preparing a congratulatory address for presentation to Queen Victoria, on the occasion of her jubilee. They were discussing the phrase "conscious as we are of our many infirmities," whereupon Bowen, the Master of the Rolls, suggested that the wording should be changed to "conscious as we are of each other's many infirmities." That, I regret to think, is what we do internationally. We are too much like the old man who said to his wife: "Everybody is queer except thee and me, and sometimes I think thee's a bit off." We need more tolerance— a more tolerant humor—remembering that a friend is a man whom you know well and still like. Britons and Americans can risk the closer acquaintance that leads to friendship, because they have fewer divergencies than common aims; and their friendship will be on a safer footing if they take each other as they are and determine to make the best of their interesting differences.

Let us look forward instead of backward. Whatever our differences in the past, let us realize the similarity of our aims, the identity of our political ideals, and endeavor so to act and speak that the harvest of this calamitous war shall be not the barren thistle of discord but the wholesome wheat of good-will.   The great sacrifices entailed by the war will be inadequately compensated if the result is not to establish closer relations of friendship between Britain and America.   Indeed, it would be an immeasurable loss if the peace settlement should provoke any discord between the two English-speaking peoples.   If we cannot agree to keep the peace, nay more, to work for human progress together, then no league of nations is conceivable that will do so.   On the contrary, if there be any hope that mankind will advance not only from a jungle existence but from the organized vendetta of a semi-civilized state of society, if there be any hope of improvement in national relations, then that hope lies in one fact —a fact of which we have common reason for being intensely proud— and that is the 3500 miles of unfortified frontier between Canada and the United States.   If we can live on such terms, with a willingness not only to arbitrate international differences but with a constant desire not to provoke them, then other nations—in time, all the nations of the world— will find it desirable, will find it imperative, to do the same.   Let that physical frontier, without the menace of a fort, without the provocation of a single cannon, let it be the symbol of unaggressive neighborliness, of a mutual goodwill, of a promise of that "far off divine event"—of universal peace and amity—"to which the whole creation moves."

## International Coöperation in Mining in North América*

BY A. R. LEDOUX,† NEW YORK

(New York Meeting, February, 1919)

I WAS wondering whether we were going to adhere to our text. It seems to me that we are having a very remarkable meeting of mining engineers this year, because no matter what the texts may be that are assigned to us, we get back toward something higher; we turn to the consideration of questions of ethics, and of applied Christianity. Yesterday we listened to the most eloquent and remarkable address by Mr. Rickard in memory of Dr. Raymond, and it was told us that he was to speak about Raymond's achievements as an engineer. He did say considerable on that subject, but he was himself carried away with the spirit of the man and moved us all by his portrayal of Dr. Raymond's continued and fruitful labors for the betterment of mankind—that is to say, his religion.

Our President asked me if I would say something this afternoon and I find in the bulletin that the text of the discourses which we are to have here this afternoon is "International Coöperation in Mining in North America." But again I find that the eloquent speaker who has just preceded me has dwelt largely on ethical and spiritual coöperation rather than upon coöperation in mining. That is, coöperation as men, rather than solely as engineers.

The imaginary line dividing Canada from the United States, unmarked by fort or military post, was mentioned by a previous speaker. There are going to be in the future, I am sure, fewer entanglements of the barbed wire of tariff discrimination along our border—to that end let us coöperate. But there has always been a reciprocity in men. A Canadian conceived and created our great transcontinental line, the Great Northern Railway; an American made the Canadian Pacific what it is.

In mining and metallurgy, American engineers, backed by American capital, have built up great Canadian industries while Canadian experts have developed some of the resources on this side of the line. They have done much more than develop our resources: some of them have been our leaders in sympathy for labor. It is therefore not a question of creating a spirit of coöperation, but of giving it freer scope and a fuller realization of fraternity—or, to use the word now more common and of deeper significance—brotherhood.

These are days of combination. Corporate interests in this country are now compelled, or at least encouraged, to combine in ways that

---

* Presented at the joint session with the Canadian Mining Institute.

† Mining Engineer and Assayer.

would have landed their directors in jail, if attempted a few years ago. The Sherman law is still on our statute books, but so are some of the "blue laws" of New England—more honored in their breach than in their observance.

The friendly relations, growing even closer, between the two great Mining Institutes, point the way to increasing coöperation. We have much to learn from Canada. I venture to suggest that if some of the distinguished Canadians, our guests today, could be given an·opportunity to tell the solons in Washington wherein Canadian mining laws are an improvement on those still prevailing on this side of the boundary, it would be of real service, for they would be disinterested witnesses. There are precedents for this: Some of us have been asked to appear before committees of the Dominion and Provincial Parliaments of Canada to give evidence and advice on laws proposed for the control or assistance of mining and metallurgical industries.

It is for those more actively engaged than I in mining to tell how best we may advance the interests of the industry, which is alone surpassed in importance by that of agriculture in Canada and the United States. Can we not appoint in each Institute permanent committees on coöperation that will hold frequent joint sessions, and make suggestions to their respective societies leading to action that may benefit us all?

I suppose "North America" in the title was intended to include Mexico. What chance is there to coöperate in that land of chaos and revolution? Personally, I think there will be such opportunity in the near future. The Mexican is a proud man; his complaint that foreigners alone have exploited and profited from the resources of his country is to some extent justified. But some of our mining companies operating in Mexico have employed and still employ Mexicans in positions of trust and authority, with satisfaction. When the time comes for the United States to consider and to act upon questions of rehabilitation of railroads, mining and other industries, in which so much foreign money has been lost, and to propose safeguards for the future, we shall have to decide, as in the Philippines, whether it will be easier to "civilize 'em with a Krag" as militarists proposed, or to suggest a Joint Commission of foreign and Mexican composition, as in our far-eastern dependency. This is a question—this one of policy—that is soon to come to the front; it is one on which there are differences of opinion.

Mr. Chairman, I have purposely injected this Mexican problem into this meeting, that if there be time it may be discussed, and that our two Institutes may perhaps become pioneers in influencing public opinion. To this task, our Canadian brethren can bring more open minds; they are farther away from the disturbance and can better understand both sides.

## DISCUSSION

S. J. Jennings, New York.—I am sure we all appreciate Dr. Ledoux's suggestion that a permanent committee should be appointed by both the American Institute of Mining and Metallurgical Engineers and the Canadian Mining Institute, in order to suggest closer and more constant coöperation. This suggestion will undoubtedly be brought before the Directors of the American Institute of Mining and Metallurgical Engineers and I trust that the Canadian Institute will also take it into account and have a committee appointed, that will serve with that spirit of closer coöperation, that idea of common ancestry, which has so eloquently been placed before us this afternoon.

Dr. Ledoux's suggestion of coöperation with Mexico is one that is extremely close to my mind. I am planning to go to Mexico the end of this week and hope to meet with a local session in the City of Mexico soon after my arrival there. We may then initiate some discussion of the problems that confront the American engineer, or the members of the American Institute of Mining Engineers located in America, whether he be an American, or Mexican, and ventilate those problems so that we will appreciate his ideas and help solve them.

## Uniform Mining Law for North America*

BY T. E. GODSON,† OTTAWA, ONT.

(New York Meeting, February, 1919)

As this is the age of reform, a uniform mining law for North America is a moot subject for discussion at this meeting of the Institute. The question is one of peculiarly technical and, in many respects, local character. We all appreciate the value of the metal-producing industries and recognize the necessity and importance of a well-considered and evenly balanced mining law as a branch of our jurisprudence either for the Provinces or the Dominion as a whole.

Mining has taken its place as a foremost industry and legislation that seeks to control and govern it is the concern of not only the active miner but the citizens of the state. A law that, through its charitable measures brings into being the prospector, the pathfinder and superstructure of all mining activity and industry, protects his interests, safeguards invested capital, and encourages and enforces development, possesses at least the salient requirements of a businesslike Act. The manner of doing so is, I think, one of local and internal concern.

May I be permitted to say without presumption that of the five mining codes, or Acts, that we have in the Dominion of Canada born and bred within the Provinces of British Columbia, Ontario, Quebec, New Brunswick, and Nova Scotia, some of them could be materially improved. In many respects they do not measure up to the times; they were no doubt creations of merit at the date of their birth but they have not been clothed in seasonable raiment as the years passed by.

It is a stern fundamental principle that all natural resources are the inalienable property of the crown. If it is mineral land, then it follows for the protection of the state and the direct advancement of the occupant that the land should be developed as such. The method or machinery by which this must be accomplished is a matter of detail, not of principle. If the right to possession as a licensee depends on discovery, it is not inconsistent with the duty the crown owes to the state that a grant in fee should pass to the discoverer if there is imposed, after patent issues, a requirement of continuous work or the alternative of an acreage tax sufficiently burdensome to make it unprofitable that the land should be held in an undeveloped state. If this requirement is recognized, the mental distinction between a grant and lease ceases to exist. A

---

* Presented at the joint session with the Canadian Mining Institute.
† K. C., Mining Commissioner.

grant contingent upon work or payment of taxes is not dissimilar in effect to a lease upon terms. In my opinion it is a matter of indifference which method is adopted provided title is dependent on development. Capital will follow mineral and should be willing to accept a leasehold tenure, provided the conditions of the lease are fixed and unalterable and the term of occupation is sufficiently long to permit proper development and a reasonable return for money invested.

From time to time spasmodic attempts are made in Canada to assimilate various Provincial laws. We have demanded a Dominion Commercial Code, a common Bankruptcy Law, and a uniform Companies Act, but they are yet to come into being. Nearly all the Provinces of Canada have a Workmens' Compensation Act, but these laws differ in essentials if not, in some respects, in principle. Public sentiment favors a systematized, assimilated, and uniform law upon a given subject when possible. Then why the necessity for a uniform mining law for North America? I prefer to discuss it from its applicability to Canada. I do not know what suggested the discussion. Is it to be said a real necessity has arisen; if so what is the foundation of the complaint? Is it not rather that discussion has been invited upon this broad and comprehensive subject in order that the attention of the Provincial Legislatures within Canada and its Federal Parliament and the Federal authorities of the United States of America may awaken to the knocking hand of opportunity and so frame their laws that the opportunity is not allowed to pass by unheeded? Each Province should know best its own requirements. A fundamental change in a law recognized, adopted, and followed for years should not, in my opinion, be made unless imperative; but the foundation of a particular law or act, if sound, should be built upon and extended to meet all economic changes, and it is just there our respective mining laws need attention.

If one is to be guided by the silence of comment, the Mining Act of the Province of Ontario might be said to be satisfactory; but that it can be improved is admitted. We have progressed and prospered under a discovery clause, harsh and stern in its requirements, it will be said; that the law has not been observed, probably not since inspection has been done away with, but the clause has served a good and useful purpose.

The principle of discovery I believe to be absolutely sound, the difficulty is its application; but that opens up too broad a field to discuss at this time. If you remove the prerequisite of discovery you must tighten the strings of control in the crown's hands after patent.

Development is a condition of possession and should be a fundamental principle underlying all mining law. The right to a claim should only exist as long as it is worked or taxed to insure work. Once depart from that sound maxim, there results improper control of the natural resources of the country, which primarily belong to the people. A requisite of discovery at some period prior to a grant with the imposition of an acreage tax after patent of an amount that will prevent inaction is sufficient justification for a crown grant as adopted in Ontario.

It is the retention of public control, not necessarily revenue, that should govern in fixing the amount of the tax. The tax must be consistent with the rights of the prospector, who is entitled to the fruits of his labor, and must not offend capital, which is sensitive.  .

The point of view of the Provinces of Manitoba, Saskatchewan, and Alberta when they obtain their natural resources is problematical. Are they to be guided by our own experience in the adoption of an Act, or by what appears to be essential to the building up of a mining industry . within their Provinces? Prospectors must be encouraged; capital must be coaxed and flirted with. How best can this be done is their problem, not ours; and this suggests one of the many difficulties of a common mining law for the whole Dominion.

We feel that we have an Act which meets and encourages the prospector, welcomes the capitalists, and promotes the development of our mineral resources. What has been and is being done is evidence of this fact. Each year, it is sought, by material changes in the Act, to meet conditions as they arise. We have yet to sit in final judgment upon the present use and future necessity of a discovery clause, but that will be dealt with when the question has been thoroughly canvassed.

Under our Act, if certain requirements are not performed within the time required by the Act, an *ipso facto* forfeiture occurs; but that forfeiture may be relieved against by an application to the Mining Commissioner within 3 months from default. If no adverse interests have arisen, usually the application is granted. If new interests have appeared, it is a question for compensation if the application is to be allowed. There is a further provision allowing an application, after 3 months from default, to go to the Lieutenant-Governor in Council upon the recommendation of the Minister and report of the Commissioners; this salient provision is not extensively used but it has met many cases of merit where a serious loss would have occurred if there had been no relieving section. It will be observed that the Minister is protected by the condition requiring a report to be filed- by the Commissioner based upon evidence heard *viva voce*, or otherwise, as is deemed expedient.

The office of Mining Commissioner is a creation of the Mining Act of Ontario and provides machinery not to be found in other Acts. It is judicial in its nature, with right of appeal from the decision of the Commissioner to the Appellate Court in the same manner an appeal is carried from a trial judge to that Court. Disputes, notices of claim of interests prior to patent, applications for relief from forfeiture, and all litigious matters come before the Commissioner. The procedure is simple and unencumbered by pleadings with the happy result that litigation is cheapened and expedited and there is no undue delay in determining the issue such as necessarily occurs in the congestion of our civil courts. As it is a moving court, the Commissioner has the advantage of hearing the point of view of many of the prospectors with regard to the working of the Act. This close contact has the natural tendency of keeping the Act in step with public opinion. Another salutary provision of our Act is the right given to the Commissioner, where required in connection with the proper working of a mine, mill for treating ore, or quarry, after hearing the interested parties to vest in the owner, lessee or holder, the right to discharge water upon any land, to draw off or divert, collect or dam back water and to take the same, rights of way over other lands, to enter upon and use specified areas and to deposit tailings, slimes, etc., all of which power must not be exercised unless any wrong or damage caused to any other person can be adequately compensated for nor unless it is reasonable and fitting to grant the same. This section, if used advisedly, greatly promotes mining and prevents selfish and antagonistic interests from defeating a legitimate mining enterprise.

I have touched upon some of the features of our Act for the purpose of showing that they are outstanding and are a step in advance of any other Act I have read and if there is to be (and I would welcome it) a uniform Act for Canada or North America, I say, from experience and with due modesty, that our Act in many respects might be used as a pattern.

Let there be a general awakening to our necessities. If nothing more is accomplished than a remodeling of the different mining laws of the respective Provinces, time and discussion will have been well spent. If each Province is to retain its own creation, it should be put in concise readable form so that "he who runs may read." At present, it strains the ingenuity of a lawyer to grasp the requirements of some of the Acts. This is the result of inattention and want of due consideration in their formation.

*Discussion on pp.* 700 *and* 712.

## Uniform Mining Law for North America*

BY H. V. WINCHELL,† MINNEAPOLIS, MINN.

(New York Meeting February, 1919)

IT seems to me that uniformity of mining law in North America must of necessity start with uniformity in each country. I understand that the Provinces of Canada have the right to make their own individual laws and conditions, to prescribe the terms under which the prospector or the applicant for mineral titles may secure the same. That is not true of the states in the United States. In this country, the Federal Mining Law was made applicable to the public domain, which, at the time of the. adoption of this Act in 1872, was chiefly west of the Missouri River; it is impossible to locate a mining claim in most of our eastern states. It is true, however, that there are certain holdover rights, mostly obsolete, arising from the fact that at the time of the formation of the Confederation of States, the land that belonged to each state became the property of the Federal Government. I am not quite correct in that, it was sometime after the adoption of our Constitution, I think it was 1778, but those rights were largely, by agreement, surrendered to the Federal Government; but it is a fact that the great acquisitions of territory in the West, with the single exception of Texas, were of domain that came into the possession of the Federal Government and our Federal Mining Law applies to that territory. No state has the right to (by legislative enactment) enlarge upon the rights given by the Federal Mining Law, governing the acquisition of mining claims in its territory, within its boundaries. Each state, however, may, and some of them do, limit somewhat the rights given to the prospector. For instance, the Federal Mining Law provides that a claim may be taken, staked upon the public domain, of 600 ft. in width and 1500 ft. in length. Some of the states, like South Dakota and Colorado, have at times said "that is an excessive amount of land for a single mining claim; we will therefore limit it to 300 ft. or 200 ft. in width." There is, therefore, not any direct and immediate need for uniformity of mining laws in the United States, such as there might possibly be if it were considered at all desirable upon the whole in Canada.

The initiation of mining rights under the Federal Mining Act is the act of discovery of valuable mineral in place, referring now to the lode mining claims, but the same is true of the placer claim. It is true that the emphasis has never been laid upon the word valuable and that

---

* Presented at the joint session with the Canadian Mining Institute.
† President, American Institute of Mining and Metallurgical Engineers for 1919–20.

for a long time there was such a vast surplus, or an excessive amount of unwanted, undesirable land in the West that the land offices did not scrutinize the claim sought to be located. They passed upon it perfunctorily and granted the patents upon affidavits of the owner. As a result it became customary to allow a prospector to locate a claim or any number of claims, and to do a certain amount of work each year; then after filing his affidavit for assessment work and going through the preliminaries and paying the required purchase price he would secure his patent. It is a fact, however, that mineral must be discovered. The right to the enjoyment of possession is based upon that fundamental fact; if a contest arises, before patent, and it is shown by the contesting applicant that there was a defect or an untruth in the statement of the discovery of minreal, such a claim may be canceled.

We recognize very many points of excellence in the Canadian mining laws and particularly in that of Ontario. We have for a long time felt that our law is woefully defective in some respects. There is, however, apparently a fundamental difference of principle as to the character of title to mineral land between that of the United States and that of Ontario. The United States law grants the title to a piece of ground together with all veins or lodes whose outcrops or apices lie within the exterior boundaries of the claim granted, although upon their descent into the earth, such veins depart from a perpendicular and pass beneath the surface of the property of an adjoining claim.

It is probably true that there is still a vestige of the regalian right in our mining law; it is perhaps true that by no act of Congress nor any specific general rule has the fundamental and inherent sovereignty, sovereign ownership, of the right to minerals, the ownership of minerals ever been alienated by the United States. The situation is somewhat similar to that in some of the British Provinces, where the first or some early charter conveyed or authorized the conveyance of titles to all of the land with the underlying minerals. Such was the case in South Australia where, for seventeen years, grants were made of lands, together with the minerals, under the full authority of the Acts of Parliament. When minerals were discovered, the then Governor-General said, "Minerals belong to the crown, we will take possession and charge a royalty, and so on." The owners of lands protested and said, "But you cannot do that, it has all been threshed out, we have our patents, we have them under authority of the Acts of Parliament." The Governor-General replied, "Nevertheless, minerals belong to the crown." Upon appeal to the higher authority, it was decided, I understand, that no Governor-General and no sovereign could alienate from himself the inherent right and title to those minerals.

The Government, therefore, resumed possession and occupation; but in the true English, generous, equitable fashion said to the owners of this mining grant, "Now we have given you that land and taken away your minerals, but we will return your royalty minus expenses." We have never attempted, so far as I know, to thoroughly test out that question. The courts have held that if in the government grant there is no reservation of mineral (not referring now to a mining claim but to a homestead, a ranch, anything of that sort) the grant conveys also subterranean rights to everything beneath the area of the surface conveyed. Recently, there has arisen litigation over the construction of railroad grants in the West and an effort has been made in some cases to get the government to reclaim the oil and other minerals beneath the surface of lands that were conveyed without a mineral reservation. It is probable that in the United States, therefore, under a long line of decisions, that the title to the minerals beneath the surface, where the land has been conveyed in some other form than as mining land, will not be attacked.

The custom of locating square claims with cardinal boundaries, on the latitude and longitude parallels and meridians, is one that appeals to a good many of us in this country. We should be very glad, indeed, I think, to see a modification in our laws to permit some such arrangement as that. The mass of mining locations that are plastered over some of the ground in our mining camps in the West is simply worse than any Chinese puzzle; and it leads to endless confusion and litigation.

Reverting to that question of discovery, it is my opinion (I am not in accord with all of the mining men of this country, as I well know), that under the new conditions with the new kinds of prospecting made possible by diamond drilling and other methods of exploration and the new developments of metallurgy, in which ore may be a rock containing an infinitesimal amount of value, the requirement for discovery is putting the cart before the horse. You say to the prospector, "You find the mine and we will let you prospect for it.' Now, why not say, "Go out and stake a claim, and as long as you spend your money in good faith looking for minerals, you may hold it." Or, if you prefer, say, "Hunt for five years spending so much per acre, per annum, and if you do not find it, let somebody else have a chance." But the idea of requiring a mine to be discovered before the man has a mine seems, to me, ridiculous.

Of course, we have that very ingenious aid to mine litigation, the apex provision in our law, and it is not necessary to dwell upon that here. I certainly would not recommend its adoption in any uniform law to cover territory in Canada. I believe you had a little experience with it in British Columbia. We are the only people, with the single possible exception of Rhodesia, in Africa, where that extralateral right provision still lingers. We have found it difficult to convince our Congressmen of our sincerity in advocating its abolition, and the mining men have encountered considerable opposition on the part of Western Bar Associations, when they attempted to minimize the amount of litigation called for.

One thing that we have admired in the Ontario laws in general is the fair treatment given to the owner, prospector, and the man endeavoring to develop mines. We have sometimes chafed a little under the regulations, perhaps, simply because they were a little different; we have not realized the importance of the limitation of the number of claims that a man can buy after the prospector has found something; and if a patent is issued and five cents an acre tax paid, if some person wishes to buy the claims that have been located by one hundred prospectors he can do it. The purpose of prospecting and mining laws is development of mineral resources. It seems to me the requirement for continuous development and work is the vital point and not limiting the number of claims that a man might take. One man might be able to work only three claims, another might work one hundred, but if the work went on and the mines were developed, that is the result desired.

I have always been impressed by the liberality and fairness of treatment received by the prospector and miner in Ontario and I am glad to have this occasion of saying so.

I should like to ask one question, I am not quite clear on that point: Do the laws of Ontario limit these prospector's licenses to citizens?

T. W. GIBSON.—A miner's license can be had by anybody. They are limited at the present time to those nations who were allied with us in the war. We would not give a German a license, of course.

MR. WINCHELL.—That is a more liberal provision than is contained in the United States mining law. An applicant for a mining claim in the United States must be a citizen of the United States or one who has declared his intention of becoming such. Of course that simply prevents some prospectors; it does not hinder a German or anybody else from buying a claim after it has been located, but there is that restriction. That fact was pointed out to me by an eminent attorney in Argentina, who said, "Down in Argentina anybody can take a mining claim, the United States has shut everybody out unless they are citizens, that is why our laws are better than yours." Well, it is true that provision is contained in one of the few opening paragraphs of their code, but it is not followed by the same liberality and simplicity of provision.

The Mexican mining law has been, I cannot speak from actual knowledge of what it is right now, a very good mining law. It has been possible to acquire a title to property, to operate with few restrictions, to pay a return to the government upon the material produced. In general, the mining law has been admirable. Of course, it was based upon the experience of a long time; it was in some respects built quite similar to the old

Spanish code. There, however, again the government does not part with the title to the minerals. There is a little different system.

It seems to me, therefore, when we consider the question of uniformity throughout North America, that uniformity would not be in every respect advisable. I believe, as Mr. Godson said, $conditions$ are so different, the character of the minerals in different countries are so different, the habits and customs and wants of the people are so different, that it is not altogether desirable. What is desirable is a closer appreciation of each other's needs, and of the needs of the industry as a whole, better coöperation in development, in the exchange of ideas, in the exchange of commodities, and in freedom of access to information, and, as Dr. Ledoux said, possibly the exchange of materials, machinery, products, without tariff limitations.

## DISCUSSION*

H. H. Rowatt,† Ottawa, Canada.—One of the previous speakers spoke very fully about the Ontario mining law and the difference which exists between the law and the mining law of some of the other Provinces. The Federal Government administers the minerals in a portion of the Dominion of Canada; that is, the Government administers the minerals in those Provinces that were not originally included in the federation and which have been brought into confederation more recently. Manitoba, Saskatchewan, Alberta, and the unorganized northern territories are administered by the Federal Government; the minerals in these provinces and territories are also administered by that Government. Up to the year 1887, it was not thought that these western territories, then known as the Northwest Territories, contained any minerals. But coal has been discovered in several parts of the western territories. The precious metals have always been the property of the crown, but the baser were conveyed in fee-simple, issued for such rights in the form of homesteads and purchases. On Nov. 1, 1887, all mines and minerals were reserved to the crown, and all titles that were earned subsequent to that date contained a reservation of the minerals. Titles earned prior to that date reserved only gold and silver.

In the Western Provinces, there are minerals that are not found in Ontario and the mining law, as a result, is of a broader character, but this afternoon regulations that were particularly discussed were the lode regulations and in these the Federal law differs only very slightly from the laws of the Province of Ontario. Up to the year 1914, these laws were practically similar; but by amendment to the Dominion's Land Act at that time the title was changed from fee-simple to lease, and all titles

---

* This discussion includes also the paper of T. E. Godson, p. 692.

† Controller of Mining Lands, Dept. of Interior, Canada.

earned subsequent to that year are in the form of leases. But this lease is practically a perpetual lease; that is, it is issued for a period of 21 years and is renewable for an additional period. In nearly every respect, it is quite as good as the fee-simple.

I think that so far as regulations dealing with lode mining are concerned, there would be very little difficulty indeed, so far as the Dominion of Canada is concerned, to have a uniform mining law. There are only a few particulars in which the Federal mining law, in respect to lode mining, differs from the laws of the Provinces. There is, perhaps, a little more liberality in the Federal law, in that a prospector for minerals may stake claims and may acquire the final form of title. It is not necessary to be a British subject, it is only necessary to be a subject of a country that is not an alien-enemy country, in order to acquire all rights to minerals under Federal lands. Citizens of allied countries are entitled to precisely the same privileges as citizens of the British Empire, and I think that in nearly all of the Provinces of the Dominion, this law is the same; that is, that only persons of alien-enemy nationality are barred from acquiring licenses. That is the only regulation that has been discussed, and therefore I have confined my remarks to that regulation and from my knowledge of the American mining law, dealing with lode mining, there is a very great similarity in the laws of each country.

PRESIDENT JENNINGS.—I would like to call the attention of some of the gentlemen, who may not be familiar with it, to the hardship caused by the necessity for discovery before the issuance of the right to prospect. In the state of Nicaragua, where the necessity for discovery prior to the issuance of a title obtains, there is a lode which outcrops for several thousand feet; it starts in with a dip of something like 60° and as it goes down, it changes its dip to about 40° and at a depth of less than 800 ft. dips out of the property which was originally available to locate the crop. The hanging wall of that lode is a siliceous material which, so far, has absolutely failed to reveal any other valuable mineral except that one vein, so you would have to sink a shaft at least 800 ft. deep through the hanging wall of that vein in order to get the deep levels of the property. While you know that valuable deposits exist there, you cannot discover it in the surface area which would be covered by the deep levels of that claim and therefore you cannot get them. That was brought to the attention of the Nicaraguan Government and an attempt was made to persuade it to change the laws. The last I heard these laws had not been changed.

THOS. W. GIBSON, Toronto, Ont.—Discovery has been referred to as a matter of importance. That was and is a requirement of the Ontario law, and on one occasion a vigorous effort was made to enforce this requirement. In 1903, the first discoveries were made in what speedily proved to be one of the most valuable silver fields ever found—the Cobalt camp.   Proof of discovery under the regulations was met by filing the affidavit of the discoverer.   In many cases, especially near a known and valuable deposit, doubts were freely cast on the genuineness of the alleged discovery, and it was felt that an affidavit was not sufficient proof.   Accordingly, a demand arose from the prospectors that all discoveries on the strength of which claims were staked should be inspected by the government, and passed on officially.   Inspectors of known skill and probity were appointed for this purpose, and "passed" or canceled claims according to whether or not in their judgment, a real find had been made.   The practice was to notify the locator, and have him accompany the inspector while the latter made his examination.

These silver veins were unusually valuable, containing up to 10,000 oz. of silver per ton of ore.   Carloads of 25 or 30 tons were taken out and shipped, bringing as much as $100,000 or $120,000 in net returns, with silver at 60 c. per oz. or less.   Possession of ground so rich was a prize worth striving for; consequently it is doubtful whether any other area of ground of equal size anywhere was ever more closely or minutely prospected than the Cobalt silver area.   A prospector simply had to make a real discovery in order to obtain a valid claim.   This plan was followed for several years, until in fact most of the outcropping veins has been located and staked.   The sentiment of the prospectors changed; they said: If a man is willing to spend his time and money in trying to make a mine out of any piece of ground, why not let him do so and give him the claim?   The inspectors were accordingly withdrawn, and the system of inspecting discoveries discontinued.

I wish also to speak of the mineral rights as apart from the surface rights.   It used to be the case that in granting a piece of land for agricultural purposes, the mineral rights were reserved to the crown, and prospectors were permitted to go on privately owned land and look for minerals.   If they made a find, they could secure the mineral rights, on due compensation being made to the land owner for damages to the surface.   In case of dispute the matter was referred to the Mining Commissioner.   That system was brought to an end about 11 years ago. It would not become me to say anything disrespectful of the law, but in the interest of mining it is doubtful whether the change was an improvement.   Much of the land was taken up by farmers who are not prospectors and have little knowledge or interest in minerals, perhaps would not recognize valuable mineral if they saw it.   The consequence is the land so granted has gone unprospected, because no prospector will spend his time looking for minerals on land if he cannot get the benefit of what he may find.   There is a possibility of mineral deposits lying concealed on agricultural grants, which if located and worked would strengthen and expand the mining industry.

H. V. WINCHELL.—I have just one or two words appropriate to Mr. Gibson's remarks. I have called attention to the situation in writing. A few years hence our public domain will be no more. Where then is our prospector going to find scope for his activities? There is no provision in our law by which a prospector may go upon privately owned property and by compensating the owner for damage to surface, or in any other way have the right to prospect. Some day we are going to be face to face with the necessity for an amendment to the Federal Constitution, giving that right again to the government to say to the prospector, "You may go where you will and sink diamond-drill holes, or otherwise, upon lands that heretofore have been conveyed to private owners." It will be an absolute necessity to provide some such provision as that.

Now I wish to say further that our Land Department has within the last few years attempted to inspect the validity of discoveries to inquire into mining claims for which applications have been made. We have had a class of mineral examiners who have gone through the land looking at these things, not always in company with the applicant or the claimant, and reporting upon them. Some of them have been good men and some have not, but what difference did that make? How could a man go to Tonopah and decide upon the prospect of finding ore on a claim where the lava was 2000 ft. deep over the outcrop of the vein? How could he go to Utah copper mine and decide as to that whole mountain of material which occupies the surface, which is highly discolored in places, underneath which lies the largest copper-ore deposit being worked in the United States today? How could he go to Butte and look at the outcrop or look at the first 450 ft. of development work on the Anaconda Copper Mine and its veins, which in many cases do not carry one trace of copper, and in some cases only a trace of gold, and say it is going to be a copper mine? The opinion of no one is sufficient to say a prospector has a claim that is going to develop into a mine. He must have time to prospect it and develop it, and that is the only thing.

PRESIDENT JENNINGS.—I would like to just add one word to this discussion on the point brought out by Mr. Gibson, and that is, when, as I understand it, land has once been allocated to a farmer, the present law in Ontario is that the minerals underneath that land go with it and, therefore, also belong to the farmer. The question arises in his mind, and in that of Mr. Winchell also, how are you going to develop mines under these circumstances? That is what happened in the Transvaal. When a land passed to a farmer, a certain amount of the mineral rights passed with it, but while the state nominally reserved to itself the right of minerals, it granted at the same time the right to the farmer to say whether or not his land should be prospected. If he did not want it prospected, it was not, but if he did want it prospected and valuable minerals were found, he had for his own right a certain fraction of that farm, which was called the mynpact. A similar way of handling the situation, it seems to me, can easily enough arise in the United States. If I think that on any given man's farm or piece of ground there is a chance to discover mineral rights, I can go and agree with him and say: "If you will allow me to prospect on your land, I will do so under such and such terms. I will spend my money and my time in the chance of finding the mineral, and if I do find mineral, we will share in that mineral in any given proportion." This method was adopted in the Transvaal, and certainly that is one of the largest fields of mineral discoveries of the world. It seems to me that to inject a new principle in the ownership of land is not necessary. An agreement with the owner is always possible, and if not with that particular owner, when he dies, with his heir.

T. C. DENIS,* Quebec, Canada.—With reference to one of Mr. Gibson's arguments, I may say that the Province of Quebec adopted, in 1880, the principle of the separation of the mineral rights from the surface rights. On all crown lands granted to colonists for agricultural purposes, since 1880, the legislature reserves the mineral rights, and those holding prospecting licenses may go and prospect those lands for minerals. When the owner of the land objects to the prospector going on the land the prospector comes to our office and makes certain deposits, determined by the Mining Bureau, to guarantee against damages.

In the province of Quebec on nearly all lands granted by the King of France before the conquest by England, the mineral rights were reserved and in the old settled parts of Quebec the government owns the mineral rights. We often have granted mineral claims on lands that were ceded to French seigniors back in the 17th century, and we really have had no trouble. Of course we have not had intensive mineral activities, our annual mineral production being around $16,000,000, but I do not think any trouble would arise from this condition.

A. R. CHAMBERS,† New Glasgow, N. S.—In the old colony of Newfoundland, the rights to all minerals are vested in the crown, and a mining man or prospector having placed his application with the crown, or more correctly, with the Mining Department, proceeds to prosecute either the prospecting or the mining, as the occasion demands, of any or all minerals that may be found thereon, and so he has complete authority to exercise all his ability and interest capital as well during the term of his mining right. There is here, possibly, a tendency to tie up a lot of mining lands without work being energetically carried on. But, in listening to this discussion, I have been wondering whether or not some compromise between these two ways would not be better. That is to say, if we have

---

* Supt. of Mines, Province of Quebec.     † Nova Scotia Steel & Coal Co.

a claim in which we can get full right and title to all mineral, without the burden of previously proving the existence of the same, let it be so arranged that full advantage can be taken of this, with proper precautions, of course, to avoid non-working.

In Nova Scotia the situation is much more complicated. The crown has been giving surface grants for many years and some of the older grants gave nearly all minerals to the surface owners. But, as time passed, the crown reserved more and more minerals, until in 1858, practically all mining rights were reserved to the crown. In this field we have, as I say, a very complicated situation and, while I am not prepared to make any comment on it, the information given here this afternoon will, I know, help us in that country, for I fear we will have to do something in the near future to revise these laws.

(*Discussion continued on p.* 712.)

## A Study of Shoveling as Applied to Mining

Discussion of the paper of G. TOWNSEND HARLEY, continued from p. 187.

GERALD SHERMAN, Bisbee, Ariz. (written discussion*).—The fact that, after a change in the hours of labor, as much work is found to be done in 8 hr. as was accomplished in 10, probably results from the more rigid examination of operating methods that is made when the management faces new and more difficult conditions. Although a better working pace will prevail, most of the gain will result from economies of plan and only a part of it from the improved efficiency of labor. The labor of shoveling is so severe that it may be easily believed that a working period of less than 7 hr. will give maximum results, but there are many other occupations about a mine in which this is not true, and in which the rate of accomplishment is in proportion to the time employed, up to 9 or 10 hours.

Mr. Harley's study of shoveling has been thorough, and demonstrates the benefit of investigation and education even in such simple work, but his reference to methods of payment opens a subject even more important in mine management. The efficiency of labor may be excellent, but unless a method of equitable payment accompanies it, the result will be only temporary.

With strict but sympathetic supervision, a good average day's work can be obtained; but with the exception of those unusual men who cannot be restrained from working to their limit, an average daily wage does not bring out the remarkable performance of which many men are capable. There is also a considerable variation in the average amount accomplished, dependent on the supply of labor. This can readily be shown by any labor record that covers several periods of shortage or superabundance of labor. An increase in the daily wage in the time of labor shortage does not necessarily bring out an increased day's work; there is usually a decrease in labor efficiency at such times regardless of the wage, whenever a daily rate is used.

Rewards to individuals for exceptional performance appearing as an increased rate of pay cannot be made in a large organization without bad feeling. In a small organization, an individual's ability may be so well recognized that different rates of pay can be applied.

Exceptional work can only be obtained by a method of payment based on the valuable results accomplished. As a basis for such payment, an

---

* Received July 22, 1919.

accurate measurement of the work is essential, which for many kinds of underground work is impossible. Unless an unquestioned measurement can be made, payment on the piece-work or bonus system must be dismissed from consideration, and the judgment of the foremen relied upon to determine the minimum day's work that will be accepted. The advantage of the piece-work, contract, or bonus method of payment is to make the reward correspond with the work done, thus reducing supervision merely to directing the work or an educational function. It also allows the exceptional man to receive a larger daily wage than can be paid if the scale is the same for all, for in the latter case, good workmen have to carry the incompetent. The contract or piece-work system is not entirely satisfactory in its operation. An incompetent or lazy workman who earns a small wage may create a strong adverse sentiment among sympathetic outsiders who know nothing of the circumstances, by complaining of his meager daily earnings. It is probably better to pay such a man a daily wage until he is dropped from the payroll for incompetence.

The task and bonus system based on a minimum has, therefore, great practical advantages as a method of payment for such work as can be measured. If the bonus payments are based on a proper foundation, the better workman will be paid according to his accomplishment. A small proportion of the workmen will not earn a bonus—in fact will not earn the guaranteed standard wage—but so long as the labor supply is limited, must be carried. The standard performance beyond which bonus payment is made must be determined by experiment, and will conform to the general rates prevailing in the district for the kind of work done, but it can only be applied with confidence after a thorough and continued investigation, as the standard can only be raised afterward in case of some undisputed improvement of method or equipment that is introduced by the mine management by which more work can be accomplished with the same effort. After it is fixed, a standard must not be raised to cut down earnings that may seem excessive, if they are the result of the effort of the workman. The rate of bonus payments should be increased or reduced to correspond with changes in the basic wage scale.

Mr. F. W. Taylor increased the capacity of iron-ore shovelers from 16 to 59 tons per day, raising wages 63.5 per cent., but the operating cost was reduced 54.2 per cent. It may be asked on what the rate of bonus payment was fixed, and if a bonus payment at less than the unit rate prevailing for a standard day's work is equitable.

In general, the standard rate of work before investigations of efficiency are made is below what should be obtained. It is advisable, therefore, to introduce improved methods before setting a standard. To introduce the plan, it may be good policy to set the standard slightly lower than may be expected from the average performance of the average workman. The average workman would therefore receive a small bonus.

The rate of pay per unit of work is therefore raised slightly before the bonus rate is applied.

In a manufacturing business, there is a greater output of the plant and a reduction in fixed charges if the bonus system raises the working pace. Increased pay up to the average unit cost will still leave a greater profit to the manufacturer. In mining, this is not necessarily the case. The maximum production of the mine does not always depend on the mine or the workman. It may be determined by the capacity of the reducing plant, concentrator, or smelter that treats its product. In a manufacturing business, an increased daily production means cheaper goods, more orders and an increased and perhaps permanent life. In a mine, it may mean some profit in taking advantage of a better market, but the total production is limited by the quantity of ore in the ground and can only be increased by decreasing costs.

Is a reduction in the rate for the amount of work done above the standard permissible? It may be justified by the fact that a miner has an overhead expense as well as a mine and that the majority of the improvements are introduced by the employer. The addition of a dollar or more per day to the standard wage, which in these days is assumed to be based on living conditions, is worth more to the miner than the first dollar in the standard rate, as it is a clear surplus. Some credit may also be claimed for the employer who introduces a system that has the effect of raising wages.

The object of this contribution is to bring up the principles of payment for discussion. The development of a general principle on which increased and equitable wages can be paid and which will depend on the intelligent interest of the workmen in the work is the most important subject now before the engineer.

The objection of most labor organizations to the bonus system or to so-called efficiency methods is that the workman may injure his health. If this is likely to be the result, the system should not be encouraged. It is a shortsighted policy to permit a good workman to injure himself, and this fundamental principle must be accepted. In most mines, the hours of labor are limited and the danger of injury to health, if working conditions are proper, is not great.

Discussion of the paper of T. W. GIBSON, continued from p. 648.

R. B. BRINSMADE, Ixmiquilpan, Hgo., Mexico (written discussion*).—While agreeing with Mr. Gibson that his net-profit tax is the fairest and best for mines, I believe that the tax he describes can be more accurately defined by another name. In accounting, net profits are usually considered to be the residual of the gross income after operating expenses and the annual charge for upkeep of plant have been deducted. Even the necessary interest on bonds or mortgages should not be subtracted in this calculation, for this interest represents merely that fixed portion of the net profits going as a return to that fraction of the total property value owned by the bondholders. Applying this method of calculation to the gross income of a mining property, we should only deduct operation expenses and upkeep of plant to get the net profits on its whole property value. So Mr. Gibson in deducting a further sum, to cover the interest and amortization of the capital expended in mine development and equipment, really arrives at something quite different from net profits; which can properly only be defined as the annual value of the mineral in the ground, or the royalty value in miners' parlance.

From one viewpoint, Mr. Gibson's first principle, which assesses taxes, "proportionate to the benefits received by the person or property taxed," is identical with his third principle which assesses "with reference to the extent and value of the natural resources enjoyed." With one exception, everything of use around a mine represents a cash expenditure, or its equivalent in human labor, and state action has not added an iota to its value. The one exception is what makes the enterprise possible and it includes the mineral deposits and their associated surface; water and other landed rights. These natural resources are originally the gift of the people acting through their agent, the state, and it is evident that a tax laid on whatever profits may arise from their enjoyment will be exactly "proportionate to the benefits received by the person or property taxed."

Mr. Gibson is mistaken in his belief that the general property tax of the United States has the same incidence as his system of taxing profits, because the general property tax falls not only on land[1] value—as does

---

* Received July 30, 1919.

[1] *Land* is defined by J. E. Syme ("Political Economy," 5) as "Such material gifts of nature as can be monopolized."

Mr. Gibson's tax exclusively—but also on capital[2] for it assesses buildings, machinery, etc. as well as mineral deposits. It thus goes directly contrary to Mr. Gibson when he says (p. 640): "Any tax upon legitimate industry which diminishes the amount of capital available or necessary for that industry is objectionable for the reason that its effect must be to cripple, or in the end even to extinguish such industry." I demonstrated this same conclusion nearly four years ago.[3] As early as 1913[4] I voiced Mr. Gibson's opinion that the general property tax was inequitable for mines even when buttressed by Mr. Finlay's expert valuations. However accurate the latter were, they had two incurable faults; they taxed capital, and in their taxation of all explored ore reserves they tended to penalize the very operators who planned their exploitation most scientifically. Besides, numerous Michigan iron mines were worked as leaseholds on contracts that required the operators to pay all taxation in addition to the fee-owner's royalty. Thus the landowners, the very persons who under Mr. Gibson's system would have to pay all the taxes, in Michigan escaped taxation altogether and shifted the burden of Mr. Finlay's increased valuations upon the backs of the already distressed operators. Such leasing contracts, which exempt fee owners from taxation, are clearly void in equity and social morals and should be also declared invalid in law because contrary to the public policy of a democratic people.

In Mexico, metal mines are held only under leases from the Federation and, besides a light state tax on surface real estate, have to pay two taxes to the nation. The first is an output tax on the metal produced, which is practically an export tax, and is unfair to operators of lean ores, for they have to pay just as much on their bullion as the bonanza miners, as Mr. Gibson points out. The second tax is an areal tax and at present is graduated from an annual rate of 6 pesos the hectare for claims under 5 hectares to 16 pesos for claims above 100 hectares in extent. In spite of Mr. Gibson's apparent dislike of the areal tax, it fulfills the highly useful function here of acting as a deterrent to that curse of United States mining camps—the forestaller. In fact, there is but one alternative remedy to the areal tax for this class of parasite and that is obligatory continuous work, as in West Australia.[5] The present high rate of areal tax here, being now supplemented by a requirement for continuous obligatory work (as in the older Mexican codes), will henceforth make hard the way of the would be mineral-land monopolist who, under President

---

[2] *Capital* is defined by Symes as "Such material products of labor as are devoted to the production of other objects of desire." (*Ibid.*)

[3] Natural Taxation of Mineral Land, *Min. & Sci. Pr.* (Oct. 29, 1915).

[4] Valuation of Iron Mines. *Trans.* (1913) **45**, 322; (1914) **50**, 194.

[5] A. C. Veatch: Mining Laws of Australasia. U. S. Geol. Survey *Bull.* 505.

Diaz, had everything his own way, for then there was no obligatory work at all and the annual areal tax was only 6 pesos the hectare up to 25 hectares and 3 pesos for all larger tracts.

The New Mexico petroleum decrees, which have aroused so many foreign protests, were issued in 1910 to put in force the nationalization of combustible minerals, as prescribed by the national constitution of 1917. Though the Mexican mining codes of 1884 and 1892 had reversed the ancient national policy and had given away all the non-metallic mineral deposits to the surface landowner, this gift, being unconstitutional and a gross abuse of legislative power, was wisely abrogated by the constituent assembly of Queretaro. The two chief fiscal features of the 1918 decrees are an areal tax of 3 pesos the hectare on petroleum land, and a tax of 5 per cent. on the gross output; the latter is practically a royalty tax of 50 per cent. as the landowners' royalty has usually been 10 per cent. of the production.

The new areal tax will tend to give the independent operators a chance again by breaking up the land monopoly, existing in 1918, by which four big interests had gobbled up $3\frac{1}{3}$ million acres of the likely petroleum land while they had only bored a few score wells.[6] This artificial land famine had caused the price of the little good land still available to outsiders for exploration in recent years to reach skyscraper heights, and annual leasing premiums of 3675 pesos per hectare, in addition to a royalty of 25 per cent. of the production to the landlord, had been paid. Both the areal and the royalty taxes were suggested by me, in 1916[7] as the only effective cure for the stranglehold of landlordism on the Mexican petroleum industry. But in my book I advocated that the renationalization of non-metallic minerals should be carried out gradually, by means of the government absorption of all such land rights (for which no private bidders should appear), at the annual tax sales under the new schedule of rates.

---

[6] Projecto de Ley Dept. de Petroleo per los Ings. Schiaffino, Santaella Landiguri y Lengaxicon. *Sria. de Industria*, Mexico City, 1918.

[7] El Latifundismo Mexicano, su Origen y su Remedio, Chap. 12. Published by Federal *Sria. de Fomento*, Mexico City, 1916,

## Uniform Mining Law for North America

Discussion of the paper of T. E. GODSON, continued from p. 705.

R. B. BRINSMADE, Ixmiquilpan, Hgo., Mexico (written discussion[*]).—How far ahead of the United States are the self-governing British colonies in their application of economic democracy is well illustrated by Mr. Godson's paper. To prove this, I need quote only two of his assertions, the italics being mine: "It is a stern fundamental principle that all natural resources are the *inalienable property of the crown.*" "The right to a claim should only exist as long as it is worked or taxed to insure work. Once depart from that sound maxim, there results *improper control* of the natural resources of the country, *which primarily belong to the people.*" These assertions formed the basis of my criticism[1] of Dr. Raymond's paper in 1913, and they should be inscribed in letters of gold above the entrance to the office of every county recorder in the United States.

I venture to assert that the existing United States mineral land laws are essentially unfair to the nation and a hindrance to the legitimate miner, because they fail to embody the above quoted principles; and this applies both to state laws east of the Mississippi and to federal laws in the West. That these laws favor the forestaller and the monopolist, defraud the government of its fair share of the profits of mineral production, and tend to industrial feudalism must be evident to any competent unbiassed observer. The fact that no serious political movement arose to change them, until the conservation agitation of the last decade, may be largely explained by two salient conditions. First, as stated by President Winchell, "for a long time there was such a vast surplus, or excessive amount of unwanted, undesirable land in the West that the land offices did not scrutinize the claim sought to be located;" second, that the millions of individual farm owners throughout the country could easily be aroused, by the special pleaders hired by the great land speculators and monopolists, to use their vast political power for the suppression of any statesmen striving to incorporate in our semifeudal land laws any such principles as those quoted from Mr. Godson.

While agreeing with Mr. Godson that a freehold or fee-simple title, if coupled with obligatory continuous work, may have the same practical effect as a lease, it seems to be the experience of the most active and

---

[*]Received July 30, 1919.     [1] *Trans.* (1912) **44,** 633.

oldest mining countries that the lease title is preferable. Thus the Australasian colonies which started with the freehold system have all, except one, changed to leasing;[2] while of the many Latin countries, as far as I know, Brazil is the only one allowing freehold mineral titles and she is probably the least developed of all in proportion to her natural resources.

The right of eminent domain for the miner, as described by Mr. Godson for the Ontario law, is usually in the Latin mineral laws based on the Napoleonic code, which distinctly separates surface and subterranean property and reserves the latter to the state. Where subterranean and surface property is titled separately, the only good excuse ever given for the continuance of the federal "law of the apex" becomes invalid, as a recent defender acknowledges.[3]

With mineral property distinct from surface rights, the requisite of discovery before getting title, which (as stated by Mr. Jennings for Nicaragua) often hinders development, at once becomes unnecessary. And with obligatory continuous work assured, by provisions in the leasing contract or by heavy areal taxation, it matters little to the nation whether one man leases one claim or a hundred; because, as President Winchell said, "The purpose of prospecting and mining laws is development of mineral resources."

While the Transvaal law, giving the farm owner control of the mineral deposits under his land, evidently achieved its object, of enriching the native Dutch landholders at the expense of the foreign mining investors, it can scarcely be commended as a democratic or equitable policy. It is certainly completely at variance with Mr. Godson's quoted principles and may be said to duplicate the absurd land laws of the United States, which allow a West Virginian farmer, who never discovered a seam or dug a ton of coal, to extract a large "royalty" from the company operating under his fields, and which permit an Oklahoma Indian to grow rich from tolls on petroleum extracted from under his ranch, which he could never have found or developed by himself. Mr. Jennings may enjoy supporting such industrial parasites rather than "to inject a new principle into land ownership" but I prefer the latter. Indeed, the principle that "the land belongs to the whole people and that its exclusive possession shall only be given to individuals able to develop it and pay a fair share of its rent to the state for this privilege" is as old as the Anglo-Saxon democracy. Any other system which allows *absolute* private property in land means the allodial tenure of republican Rome and can only end, as that one did in the first century B.C., in the general economic

---

[2] A. C. Veatch: Mining Laws of Australasia. U. S. Geol. Survey *Bull.* 505.
[3] W. E. Colby: Extralateral Right. *Calif. Law Review* (1917), 324.

enslavement of the civilized world by a few hundred patrician land monopolists.[4]

The Great War has shown us that some of our most rooted beliefs as to individual rights had to be upset before the allied nations could become sufficiently coherent to resist the menace of autocratic militarism. Just so must much of our awe of ancestral codes and court rulings be overcome before we can hope to democratize effectively our land laws. The idea that one generation has the right to dispose of the peoples' heritage, the national territory, for any period beyond its own short life (50 years) is one that must be relegated to the limbo of exploded superstitions. We must either moralize industry by cutting off its parasites— chief of whom are idle landholders and their allied classes of private monopolists—or our brilliant mechanical civilization will be overwhelmed by a flood from its own moral cesspools. The inevitable rise of such a destructive movement as that of the I. W. W. and the Bolsheviki was accurately predicted as early as 1879, in an epoch-making work[5] whose studious perusal no engineer can afford to neglect.

---

[4] Gibbon: "Decline and Fall of the Roman Empire," 1.
[5] Henry George: "Progress and Poverty."

# BIOGRAPHICAL NOTICES

## Biographical Notice of Frank Firmstone

Frank Firmstone was born Aug. 29, 1846, at Glendon Iron Works, near Easton, Pa., the residence of his father, William Firmstone, one of the pioneers of the anthracite-iron business. After preliminary school education at Easton, Pa., he attended the Saunders Military Academy in Philadelphia and subsequently entered the Polytechnic College of Pennsylvania, where he was graduated as Mining Engineer in 1865, at the age of nineteen. In 1865 and 1866, he was employed in railroad surveying. In January, 1867, he became draftsman and in 1869, assistant manager of the Glendon Iron Works, of which his father was manager until his death in 1875, when the son practically succeeded to that position, becoming formally in January, 1877, and remaining until February, 1887, the manager of the works and business agent of the Glendon Iron Company. In February, 1889, he became President of the Cranberry Iron and Coal Company, and held that position until November, 1897.

He was a director of the Longdale Iron Company, Longdale, Va., from its organization in 1870 until its dissolution in 1911.

With the termination of this official responsibility, Mr. Firmstone retired from active business, although he continued to maintain an intelligent interest, and to play a helpful part, in the scientific progress of the iron industry, in which he had been a leader.

PAPERS OF FRANK FIRMSTONE IN THE *Transactions.*

| Title | Vol. | Page | Year |
|---|---|---|---|
| A Comparison between Certain English and American Blast Furnaces as to their Capacity by Measurement and their Capacity by Weight | i | 314 | 1872 |
| A Modification of Coingt's Charger | ii | 103 | 1873 |
| Repairing the Upper Part of a Furnace-Lining without Blowing Out | iv | 29 | 1875 |
| Results from Open-Topped and Close-Topped Furnaces. | iv | 128 | 1875 |
| Indicator Cards from a Water-Pressure Blowing Engine. | vii | 339 | 1879 |
| A New Charging Bell | xiii | 520 | 1875 |
| Form of Crater Produced by Exploding Gun-powder in a Homogeneous Solid | xviii | 370 | 1889 |
| Magnesia and Sulphur in Blast-Furnace Cinder | xxiv | 498 | 1894 |
| An Example of the Alteration of Fire-Brick by Furnace Gases | xxxiv | 427 | 1903 |
| An Old Specimen of American Spiegeleisen | xxxvii | 198 | 1906 |
| An Early Instance of Blowing-In without "Scaffolding Down" | xxxviii | 124 | 1907 |
| An Unusual Blast-Furnace Product; and Nickel in Some Virginia Iron Ores | xxxix | 547 | 1908 |
| Development in the Size and Shape of Blast Furnaces in the Lehigh Valley, as shown by the Furnaces at the Glendon Iron Works | xl | 459 | 1909 |
| First-aid and Rescue Work in the Lake Superior Iron Region | l | 760 | 1914 |

As Mr. Firmstone's father had been a pioneer in anthracite-iron practice, he himself was a pioneer in the application of scientific theory to that practice, bringing to bear upon it his technical education and his habit of earnest and wide study of technical literature. Although he was not present at the Wilkes-Barre meeting of 1871, when this Institute was organized, he was quick to see the promise of professional progress involved in such a movement, and became a member in that first year. His interest was active and eager from the first. In 1873–1875, 1884–1886, and 1889–1891, he was a manager, and in 1875–76, a vice-president. In those days, it was customary to elect as officers of the Institute men who had shown their sympathy with its purpose by valuable contributions to its *Transactions;* and the abundant claim of Mr. Firmstone to this recognition is evident from the above catalog of his contributions.

In addition to these papers, Mr. Firmstone contributed discussion to numerous papers on sundry branches of the iron and steel industry, his remarks having been printed in no fewer than 34 different places in the *Transactions.*

A simple inspection of the titles of his papers indicates, what the respective contributions confirm, that Mr. Firmstone was a vigilant student of current technical literature, and also of contemporary practice; that he understood and employed scientific tests of apparatus and methods, and that his position, on the whole, was that of an appreciative, but conservative, critic. In a period of rapid progress like that of the last forty years, such a man, though he may not accept novelties with eagerness, performs an important function in "proving all things," that the pioneers may "hold fast that which is good."

Mr. J. E. Johnson, Jr., a son of Capt. J. E. Johnson, for many years the skillful manager of the Longdale furnace, writes in the *Iron Age:*

Mr. Firmstone learned the blast furnace under the hard conditions prevailing when anthracite fuel with its slow rate of combustion was used to smelt the refractory magnetite ores of New Jersey, an operation which could only be carried on at a very slow speed. He was out of sympathy with the modern practice in hard-driving which has developed from the use of coke fuel, high blast pressures and soft ores, so he retired from active service rather than adapt himself to conditions which seemed wrong to him.

Mr. Firmstone was a profound student with a brilliant mind and a retentive memory. His mind was a treasure house of information and one associated with him for many years has said that he had never known any other man to possess a mass of information so vast and at the same time so accurate, and that if Frank Firmstone made a statement it need not be checked up but could be accepted as a fact. This might well serve as the epitaph of this iron master, whose personality was so little known during his later life to the modern school of furnacemen.

Since his death, the greater part of his splendid scientific and engineering library has been given to Lafayette College, Easton, Pa.

Besides his connection with this Institute, through which he made most of his contributions to technical literature, Mr. Firmstone was also a member of the American Society for Testing Materials, the American Society of Civil Engineers, the American Society of Mechanical Engineers, and the Engineers' Club of New York.

He died June 27, 1917, at Glendon, his birthplace, leaving behind him a stainless reputation for honor, courage, kindliness, genial fellowship and loyal friendship, as well as for keen practical judgment, wide scholarship, and unwearied activity.

ROSSITER W. RAYMOND.

## Biographical Notice of James W. Malcolmson*

James W. Malcolmson died suddenly on Dec. 26, 1917, at Kansas City, Mo., where he had made his home for the past ten years. He was born at Dover, Kent, England, on Oct. 6, 1866.

He graduated from the Royal School of Science, London, England, in 1889, and almost immediately after graduation came to Mexico, accepting a position with the Michoacan Railway and Mining Co., as mining and mechanical engineer. His work in the mining profession was carried on principally in Mexico and the United States.

In 1888, he married Katherine Haden Krause, of Woolwich, Kent, England, and leaves, besides his wife, five sons and one daughter.

During Mr. Malcolmson's long and successful career as a mining engineer, he was connected with the following companies: from 1889 to 1893, with the Michoacan Railway and Mining Co., as mining and mechanical engineer; from 1893 to 1898, in various capacities with the Kansas City Smelting & Refining Co.; after the consolidation of the Kansas City Smelting & Refining Co. with the American Smelting &'Refining Co., he remained as manager of the mining department in Mexico of the American Smelting & Refining Co. until 1902. In 1902, he located in El Paso; and took up the practice of his profession as a consulting engineer. While there, he was consulting engineer for the famous and rich Pedrazzini mines at Chispas, in the State of Sonora. He was instrumental in organizing the La Republica mine in Chihuahua, and for a time acted as consulting engineer for the property. He remained in El Paso until 1907 and then moved to Kansas City and became associated with the United States & Mexican Trust Co., as consulting engineer. In 1909, he became connected with The Lucky Tiger Combination Gold Mining Co., as its consulting engineer, and continued in this connection and in the general practice of his profession as consulting engineer until his death. His general policy in connection with the Lucky Tiger company was always a progressive one and the success of the enterprise can be largely attributed to him and his good judgment.

Mr. Malcolmson contributed several articles to the Institute at different times, one of them being on the ore deposits of the Sierra Mojada Mining District, in 1901. Much of his work was in Mexico, and during his long residence in that country he acquired a very thorough knowledge of the Mexican people and their customs. During the revolution, from 1910 to 1916, this knowledge was of great value in connection with his work for the Lucky Tiger company, and often assisted greatly in keeping the property going.

Mr. Malcolmson was one of the very widely known men in his profession; he was everywhere held in the very highest regard, and his position among men in the mining profession was an enviable and honorable one. At the time of his death, besides being.consulting engineer for the Lucky Tiger company, he was interested in mines in the Miami zinc fields and was acting as consulting engineer for two of the operating properties there. He invested to quite an extent in real estate when he moved to Kansas City in 1907 and his investments in this line proved successful.

Mr. Malcolmson was a member of the Institute of Mining Engineers, associate member of the Institute of Civil Engineers of London, and a member of the Mining and Metallurgical Society of America. He was also a member of several clubs in Kansas City, including the University Club and the Engineers' Club. He served in the capacity of vice-president of the University Club,. and was the first president of the Engineers' Club. He was very widely read and well informed and much interested in subjects of general interest. He was vitally interested in the war, and during the past year served the Government on a mission gathering data on sulfur deposits and production of sulfur in the south. He was one of the commission sent from southwest Missouri and Oklahoma by the zinc-mine operators to Washington in December, 1917, in connection with the regulation of the price of zinc by the Government and the adjustment of the excess-profit taxes to be levied on the zinc-mining companies of that section.

By his friends and associates he was held in the highest esteem, and his honorable dealings with all those with whom he came in contact placed him in a position to command the highest respect from all who had dealings with him of a personal or professional nature.

He was always willing to take more than half the blame and willing to give more than half the credit in dealings with his close associates, and was a very loyal friend. His death was a great loss to his many personal friends as well as to the profession.

L. R. BUDROW.

### Biographical Notice of Joseph Hartshorne

Joseph Hartshorne was born in Philadelphia in 1852. He died Aug. 23, 1918. After graduating from Haverford College, he took a special course in chemistry at the Massachusetts Institute of Technology, and later at the University of Pennsylvania. He started with The Pennsylvania Steel Co. as an apprentice in 1873, working in the laboratory and mills for two years, and then one year as foreman in the Bessemer Department. Next, he was assistant chemist in the second Geological Survey of Pennsylvania. In 1878, he went to France to investigate the Pernot open-hearth furnace for the Cambria Iron Co. Upon his return, he assisted in designing and erecting the plant in Johnstown and was foreman of a turn for one year, and superintendent for the next. In 1881, he became superintendent of the Bessemer and rolling-mill departments. In 1884, he went to Pottstown and also made a trip to Europe to investigate the Bessemer basic process. Upon his return he designed and erected the basic Bessemer plant for the Pottstown Iron Co., and was superintendent there until 1893, when the company failed. Since then he has been a consulting metallurgical engineer, and expert in patent cases. The most important of the latter was that of Krupp versus Midvale on the armor-plate patent case, in which he was leading expert for Midvale. He was a good linguist, having command of French and German, and a fair knowledge of Italian and Spanish. In all, he made eight professional trips to Europe.

## Biographical Notice of Frederick William Matthiessen

F. W. Matthiessen, who, with E. C. Hegeler, of La Salle, was one of the creators of the zinc industry in the United States, was born in Altona, Schleswig-Holstein, Germany, Mar. 5, 1835. He was one of a family of remarkable brothers, two of whom were long associated in this country with the sugar and glucose industry. Mr. Matthiessen was educated at Freiberg, and came to this country in 1857 with Mr. E. C. Hegeler, after having, in company with that gentleman, visited the mining districts of Germany, Belgium, and England. They first went to Friedensville, near Bethlehem, Lehigh County, Penn., where attempts had already been made to produce zinc from the ore deposits found in that place. The ore was a fine silicate but all attempts to extract zinc from it had failed. Mr. Matthiessen and Mr. Hegeler were successful, but declined to invest their money in the zinc industry at Bethlehem because oi certain local conditions.

Hearing of the discovery of zinc ore in the West, they visited both Wisconsin and Missouri and thought of settling in the coal region of East St. Louis. Finally La Salle, as being near to the Wisconsin fields and having much coal, was selected as the site of the new smelting company, and on Dec. 24, 1858, the first shovelful of dirt was turned for the buildings of the new industry. The Civil War at first interfered with the manufacture of zinc, but afterward a lively demand for the metal arose and the future success of the institution was established. A large rolling mill was added in 1866, and in 1881 the manufacture of sulfuric acid as a byproduct was begun.

Mr. Matthiessen also developed the Western Clock Manufacturing Co. of La Salle, which manufactures the Big Ben alarm clock, from a small plant having 25 men to the present large institution which employs some 2000 workers. Mr. Matthiessen's affiliations with other metal industries have been numerous.

But the contributions of the great zinc manufacturer to the causes of philanthropy and education have made his name known throughout the country, beyond industrial circles. He has given to the city of La Salle probably a sum total of $500,000 in various benefactions. Among the objects of his philanthropy were the La Salle-Peru Township High School, which he has made a model institution of secondary education for the entire United States; also the Tri-City Hygienic Institute of La Salle, Peru, and Oglesby, Ill., which he endowed for $200,000. This Hygienic Institute, which is unique in the United States, has a trained staff of health officers, bacteriological laboratory, medical library, infant welfare station, milk station, free dental clinic, school nurse, and isolation hospital.

Mr. Matthiessen was elected mayor of the City of La Salle three times and gave generously to the municipality of which he was the head. He also saved for the people of Illinois one of the great scenic districts of the state, Deer Park, with its cañons and natural beauties, which he converted into a model scenic resort, and the proceeds of which he gave to the charities of the Tri-Cities in which he lived. He died on Feb. 11, 1918.

## Biographical Notice of Arthur Brice deSaulles

In the death of Major A. B. deSaulles at South Bethlehem, Pa., on Dec. 24, 1917, the Institute lost a valued and esteemed member, one of the last few of those who, in May, 1871, at Wilkes-Barre, attended the first meeting of the Institute, in response to a call that had been issued by R. P. Rothwell, Eckley B. Coxe, and Martin Coryell.

Major deSaulles was born Jan. 8, 1840, at New Orleans, and was the son of Louis and Armide Longer deSaulles, each of French descent. As a boy he was privately tutored in the South, and then schooled in New England to enter the Rensselaer Polytechnic Institute, where he was graduated with honors in the class of '59. Following his graduation he studied in France and Germany for two years, returning to New Orleans in 1861 to enter the Confederate Army, in which he served with distinction as an engineer throughout the Civil War, rising to the rank of Major and the command of his Corps. After the close of the war he pursued his studies abroad for another year, returning to New York in April, 1866, when he became engineer for the New York & Schuylkill Coal Company, and remained at its plant near Wilkes-Barre until October, 1871, when it was sold to the Philadelphia & Reading Coal and Iron Company. He then returned to New York where for several years he was engaged in the practice of his profession.

In 1869, he married Catharine M. Heckscher, daughter of Charles Heckscher of New York City. Mrs. deSaulles survives him.

In 1876, Major deSaulles became superintendent of the Dunbar Furnace Company in Fayette County, Pa., with which company he remained until 1883. As a feature of the Pittsburgh meeting in May, 1879, Major and Mrs. deSaulles entertained the members of the Institute at dinner at Dunbar on the occasion of the Institute's visit to Dunbar furnace. In 1883, he took up the formation of the Oliphant Furnace Company in that county, of which he became and remained the head for five years. He was then called to the superintendency of the New Jersey Zinc Company at South Bethlehem, Pa., in which position he remained until his retirement from active service in 1911, maintaining, however, his keen interest in the metallurgy and smelting of zinc, in collaboration with his son Charles, a member of the Institute and a Director of the American Smelting and Refining Company. From 1888, until his death, he resided in South Bethlehem, a prominent and highly respected citizen of that progressive and stirring community.

Major deSaulles was a member of the Protestant Espicopal Church, and at the time of his death was a vestryman of the Church of the Nativity at South Bethlehem. He was president of the Men's Club of the Church of the Nativity. His kindly nature led him to take an active interest in the promotion of healthful sports among young men. He was ever active in all things pertaining to the general welfare and uplift of his fellowmen. In his death the profession has lost a distinguished engineer and metallurgist and our country a patriotic able citizen and man of affairs—one greatly beloved by those whom he honored with his friendship, and respected and looked up to by all who were privileged to know him.

## Biographical Notice of Charles Richard Van Hise

The sudden and untimely death of Dr. Charles R. Van Hise, late president of the University of Wisconsin, was one of the greatest losses, not only to the educational world and science of geology, for which he was a great leader and pathfinder, but also to the world of mining and allied interests, for which he was an adviser and helper who blazed trails and pointed out paths of development, which the practical men of engineering and industry will follow long after his death.

To the general public, Dr. Van Hise stood as a great educator who conceived and wrought out a new idea in the people's university of a commonwealth and as a great active mind that could not be held within the bounds of education and science but, by its own bigness and broadness, was forced into the contemplation of the larger affairs of the nation. To the engineering profession, Dr. Van Hise will be remembered as a tireless worker who solved some of the most complex problems in geology and the development of mineral resources and recorded these findings in a form that is of lasting value to those who follow him in similar labor. In this combination of thinker and worker, educator and economic student, scientist and practical engineer, lies Dr. Van Hise's contribution to his generation.

As an educator, Dr. Van Hise attained perhaps his widest recognition, for in the University of which he was president he evolved and accomplished a new idea, which caused it to be called, in 1908, by President Eliot of Harvard, "the leading State University." At the time of his death, Nov. 19, 1918, he was just completing fourteen years as president and forty-three years as student and teacher in the University. His entire life of 61 years had been devoted to his State and its University. He was the first alumnus of the University and the first Wisconsin-born citizen to become its president—and he enjoyed the longest term as president of the institution. Since his graduation from its College of Engineering in 1879, he had been constantly a member of its faculty. As Chief Justice John B. Winslow, of the Wisconsin Supreme Court said at the time of President Van Hise's death, "Wisconsin has had many able sons, men who have served their Country and their State with distinguished honor in various fields of effort, but among them all none, I believe, has rendered greater service in his time than President Van Hise. The University will be his true monument, for to him, more than to any one person, we owe the present commanding position of that great institution."

What his "new ideal of a State University" was may be best expressed in the words which he used to outline it in his inaugural address in 1904. "I shall never be content until the beneficent influence of the University of Wisconsin reaches every family of the State. This is my ideal of a State University. When the University of Wisconsin attains this ideal, it will be the first perfect State University. . . . The University of Wisconsin desires to prevent that greatest of all economic losses to the State, the loss of talent. To prevent this loss of talent, the University must not only provide for those who come to Madison for instruction, but must go out to the people of the State with the knowledge which they desire and need." Hence it was said early in his administration that "the boundaries of the State are the campus fence."

To engineers and scientists among his colleagues and associates, it has always seemed that his achievements in all lines—education, geology, public thinking—have shown the influence of his engineering training and point of view. He was educated as an engineer, his first degree was in engineering, his first teaching work was in metallurgy, and it was not until later that he turned to mineralogy and geology. Throughout his life's work he kept the engineer's point of view, as could be readily seen in his conduct of the University, his geological researches, his writing on public questions, and his attack on concrete problems. His continued and constant interest in engineering matters was evidenced strikingly by his appointment, in 1915, as chairman of the committee of the National Academy of Science to investigate causes and suggest remedies for the great slides of the Panama Canal.

His first college teaching was as instructor in metallurgy, a work which he carried on from 1879 to 1883. About that time his interest grew over toward the field of geology and, in 1888, he was holding the chair of professor of mineralogy. Two years later, he was professor of archean and applied geology; and in 1892, he was also lecturing as non-resident professor of structural geology at the University of Chicago while continuing to occupy the chair of geology at Wisconsin. This change of interest from engineering to geological and mining problems grew out of his close association with Professor R. D. Irving, who was then chairman of the Department of Geology and was carrying on investigations for the United States Geological Survey in the Lake Superior region. His association with Dr. Irving led him into the investigation of this region, and, at Professor Irving's untimely death at the age of 39, Dr. Van Hise took up his work and carried it on toward completion.

In the solving of the problems of the Lake Superior region, Dr. Van Hise showed at its best his remarkable mental grasp and his tireless energy. The Lake Superior region is one of great complexity from the geological standpoint. Being a region of great economic importance, because of its ores of iron and copper, it was important that the geology be thoroughly mapped and studied. The problems and conditions met by Dr. Van Hise in this work were largely new ones and, with his tireless love of hard work, his boundless fund of physical energy, and his wonderful resourcefulness and vision, he met and solved these difficult problems to the great benefit of the mining industry, and with the spirit of the true student of science put together the results of his studies to the lasting benefit of the science of geology. The Lake Superior rocks as a whole are very ancient and have been contorted and deformed and altered to such an extent as to present many difficult problems. In working out the complicated structure of the region, Dr. Van Hise had to blaze new trails and develop new principles for the deciphering of complicated structures. This work has made him a leader in the field of structural and dynamic geology. Similarly, the great variety of the rock alteration that he found in this interesting region led to the development of fundamental principles, which he has summarized in one of his most important publications, "A Treatise on Metamorphism," which is recognized as the classical work on that subject. The interest in pre-Cambrian geology that grew out of his work in the Lake Superior region led to studies of other areas of pre-Cambrian geology and for many years Dr. Van Hise was in charge of the division of pre-Cambrian geology for the United States Geological Survey.

Dr. Van Hise was one of the small group instrumental in the creation of the Geological and Natural History Survey of Wisconsin and was closely identified with the Survey until his death. From 1903 until 1918, he was president of its board of commissioners and interested himself in all phases of geological work in the state; not only that leading to the development of mineral resources, but just as keenly in the study of the way in which the surface of the state has been carved out through the ages. It was on his suggestion that the State Geological Survey started the present state highway work and carried it on until the State Highway Commission was created.

In his later years, his interest in mineral resources became more broadly philosophical as his widening experience taught him the tremendous part played by these resources in the development of civilization. He saw keenly the limited nature of many of these resources and that the welfare of future generations was dependent on their possession. This led him to bend his full energies to the propaganda for wise use and curtailment of waste by the present generation—to the development of conservation in state and nation. Out of this interest grew many other public services relating to government problems, so that at his death the nation found him writing, not on scientific problems or upon conservation, but upon regulation, the trust problems, and other great economic questions.

At the moment of his death, the problems of the war were his greatest interest. He had devoted all the time that he could spare for the better part of a year to the work of the food administration and other war boards and had just completed a volume on conservation and regulation during the war. A week before his death he had returned from a trip to England and to the battle fronts in France and Belgium and brought back with him enthusiastic plans for comprehensive after-war work, both through the University and through energetic advocacy of the plan of a League of Free Nations to prevent future wars. Among his associates it is well-known that he realized the danger of German aggression long before it was realized by the public of his section of the country and was eager to see the overthrow of the German military autocracy. When America entered the war, he was full of enthusiasm over what he considered the great, and almost holy, enterprise that the nation had undertaken. And under his guidance the University threw all of its energies and endeavors into a wide range of war activities.

But, now that he is gone, the memory that remains with his colleagues is not alone the recollection of his works, but a simple and affectionate memory for the man himself—an intimate personal regard, shared by all who knew him—students, academic colleagues, field associates—such as is left by few public men. The loss to his friends was not so much a public loss as a personal loss to every one who has ever known or worked with President Van Hise. The tributes written by his colleagues and the great men of his state were not alone eulogies of his great works but expressions of personal sorrow. Among his former students, many of whom are now leading geological or mining men, the inspirational quality of his teaching developed a personal affection that has bound them to him since they left the college halls.

## DIED IN SERVICE

Undoubtedly other members have given their lives in the Service of the United States and the Allies during the past four years, but the following biographical notices are all that have reached us as yet.

### Lewis Newton Bailey

Lewis Newton Bailey died of pneumonia at Camp Merritt, N. J., Apr. 30, 1918, at the age of 34 years. He was Master Engineer, Senior Grade, in the Fourth Regiment, U. S. Engineers, and although he was offered a commission as First Lieutenant shortly after enlisting, he preferred not to accept it until he had reached France. By those who knew him well, this lack of eagerness for promotion will be recognized as consistent with his whole character.

LEWIS NEWTON BAILEY.

Mr. Bailey was born at San Diego, Cal., May 28, 1884, and graduated from the University of California, College of Mines, in 1906. He joined the American Institute of Mining Engineers in 1916. During 1904 he was employed at the cyanide plant of the Buckeye mines, Forest Hills, Cal. In the autumn of 1909 he went to the Coeur d'Alene District, Idaho, and was first employed on construction work of the Bunker Hill & Sullivan Mining & Concentrating Co., with a short intermission at the Hercules mill, after which he was sampler and assayer for the Bunker Hill & Sullivan Co. until May, 1910, when he was obliged to return to California to attend to family interests.

Returning to the Coeur d'Alene district in November, 1911, he was given charge of remodelling and operating the mill of the Coeur d'Alene Development Co., treating the output of the Ontario mine; this work he performed with credit and complete success. In December, 1914, he returned to the Bunker Hill & Sullivan Co., as assistant metallurgist and mill superintendent, his work consisting mainly in the testing and improvement of concentrator practice, which he continued with great credit until August, 1915, when family considerations required that he again return to the home of his parents at San Diego.

Mr. Bailey was characterized by energy, both physical and mental, by devotion to his friends, associates and employer. He was particularly successful in his relations with the men under him and, in fact, was thoroughly well liked by everybody with whom he came in contact.

He took a great interest in technical matters and his inclination was strongly toward research investigations. At the same time he was effective in manual and mechanical work and was generally found working side by side with the men under his direction. His nature was unselfish and he was extremely popular socially. He was a member of the Coeur d'Alene Commandery, Knights Templar. STANLY A. EASTON.

## Lieutenant Louis Baird

Louis Baird, a Lieutenant in the Royal Field Artillery of the British Army, died on the battlefield in 1915.

Lieutenant Baird was born July 28, 1880, at Stirling, Scotland. His technical education was obtained at King's College, Melbourne, Australia, in 1893 to 1895, and at the Technical College at Glasgow, Scotland, in 1897 to 1900. After an apprenticeship with the Mexican J. & S. Rec. Co. of Mexico, lasting from 1900 to 1903, he became mill superintendent for the Cia. Minera Benito, at Juarez, Zac. During the year 1905–6 he was assistant manager of the Candelaria y Anexas, at Pinos, Zac., and from 1906 to 1907 was engineer with the Cia. Minera Jalisco, San Sebastian, Jalisco.

At the time of his admission to the Institute, in 1908, he was practising as a mining engineer at Etzatlan, Jalisco, Mexico. For the next four years he practised his profession in the City of Guadalajara, Jaliseo, removing, in 1913, to Ixtlan del Rio, in the State of Tepic. It was from here that he returned to England to put his services at the disposal of the British Government. The Institute is not definitely informed as to how, where, or when Lieutenant Baird met his death.

CAPTAIN JOHN H. BALLAMY.

## [Captain John H. Ballamy

John H. Ballamy, Captain on the Regimental Staff of the 103d Engineers, was killed near Fismes, on Aug. 9, 1918.

Captain Ballamy was born at Plymouth, Pa., in 1886 and graduated from the High School of that town in 1903. In the following year he began work as a coal miner, and at the same time began the study of mining engineering in the evening and any other spare time. He continued this diligent and energetic practice for many years. For sixteen years he continued in the employment of the Delaware, Lackawanna & Western Railroad Company's coal mining department, first as a practical miner, next as rodman on the survey corps, of which he soon became chief, then draftsman and mining engineer in the company's office.   In January, 1916, he was promoted to the position of district engineer in charge of mining engineering work.   He became a member of this Institute in September, 1917.

Captain Ballamy's military experience began at the same time as his engineering practice.   He was a charter member of Company A, Engineers, N. G. P., which was organized ten years ago.   He saw service on the Mexican border as First Lieutenant of this Company, and remained in that position until November, 1916.   In July, 1917, he was again called into service and was sent to Camp Hancock, Ga., with the 103d Regiment of Engineers.   Before joining this regiment he attended a school for officers at Fort Sill, Okla., where he received special training in field fortifications.   In December, 1917, he was commissioned Topographical Captain on the Regimental Staff, and in April, 1918, preceded his regiment by about six weeks to France, where he again attended an officers' training school.   He rejoined his regiment on July 15, near Chateau Thierry and was killed near Fismes on Aug. 9, 1918.

--------

### Lieutenant Martin F. Bowles

Martin F. Bowles, born Apr. 25, 1893, at Bonne Terre, Mo., and graduated from the Neodesha, Kans., High School, had finished all but

LIEUTENANT MARTIN F. BOWLES.

one month of a four-year course in metallurgical engineering at the Missouri School of Mines, Rolla, Mo., when on May 12, 1917, he entered the first Officers' Training Camp at Fort Riley, Kans. At the close of the training period, he was commissioned a Second Lieutenant and assigned to Co. B, 355th Infantry, 89th Division, and stationed at Camp Funston.   He left this camp with his company on May 21, 1918, and sailed from an eastern port on June 4, arriving in England on June 15, 1918.   After a short stay at a rest camp there, he proceeded to France.   On August 14, he wrote that he was with Battalion Headquarters as Scout Officer, 1st Battalion, 355th Infantry, 89th Division. The last letter received from him, dated September 2, was written in his dugout and contained the following: "Here it is 2.45 A. M.   Sitting in a dugout waiting for some patrols to report back in, I am writing this by the light of two candles."   He was killed in action on the night of Sept. 3, 1918.   The letter of notification received from Brigadier General Thos. G. Hanson said: "It is my painful duty to communicate to you the fact of the loss of your son, Martin F. Bowles. About 11 o'clock on the night of September 3, Lieutenant Bowles, with Lieutenant Joseph B. Keckler, 355th Infantry, in command of a reconnoitering patrol of our troops, encountered the enemy. In the ensuing engagement your son received a rifle bullet through his heart. His death was instantaneous and painless. His remains were interred with full military honors on the 4th of September." He was promoted to First Lieutenant, but his commission did not reach him before his death. He was president of the Missouri Mining Association, Missouri School of Mines, for the year 1916–17.

## William Morley Cobeldick

William Morley Cobeldick, one of the British Royal Engineers, died from gas poisoning on Oct. 7, 1915.

Mr. Cobeldick was born Mar. 21, 1882, in London, England, where his early education was obtained at the Finsbury Technical College. In 1898 he was employed by the British Columbia Exploring Syndicate as electrical engineer with their gold dredging plant on the Fraser River, British Columbia, his principal work being the erection and operation of the power plant, to which were added certain assaying and metallurgical duties. The Government of British Columbia awarded him a diploma for efficiency in work of this character.

In 1901 he became works manager and chemist to the Metal Trust, Ltd., of London, and for four years was instrumental in developing the Swinbourne-Ashcrofts chlorine process for treating lead-zinc sulfides. The years 1905 to 1907 he spent at the Royal School of Mines, London, at the conclusion of which he was awarded the Bessemer Medal and prize in the Department of Metallurgy. The next six months he spent in inspecting metallurgical plants in Great Britain and Europe.

Beginning in 1907, he spent the next two years in the development of a process for treating tin-copper sulfides from the Oonah Mines, Ltd., of Tasmania, which involved the erection and operation of an experimental reverberatory plant and leaching appliances for the treatment of this ore at Swansea; in this work he was associated with Messrs. A. Hill and Stewart of London. The first part of 1910, he spent in making a special study of the treatment of copper-tin mattes and alloys and was then appointed chemist to the Wallaroo & Moonta Mining & Smelting Co., at Wallaroo, South Australia, which position he occupied at the time of his admission to the Institute, in 1913. The following year he returned to England and at once entered upon his military service.

## Ralph Dougall

Ralph Dougall enlisted as a private in the Fourth University Company of the Princess Patricia Regiment, declining the solicitation of his friends that he should apply for a commission. He was killed in action early in the war and only shortly after having become a member of the Institute.

He was born at Montreal, Canada, in 1875. His earliest education was received in Montreal but was continued from 1885 to 1892 at schools

RALPH DOUGALL.

and academies in Brooklyn, Whitestone, and Flushing, L. I. For his technical education he returned to McGill University, where he graduated in 1897.

His first professional occupation was as transitman engaged in the running of township lines in the Rainy River District of Ontario. In 1899, he was made assistant chemist at the Guggenheim smelter at Aguascalientes, Mexico. From 1900 to 1903, he was chief chemist for La Compañia Minera de Peñoles, at Mapimi, Durango. We have no information as to his pursuits during the next eight years, but from 1911 until 1914, when he became a member of the Institute, he was chief engineer for the Bankhead mines in Alberta, Canada.

## Lieutenant-Colonel Alfred Winter Evans

Lieutenant-Colonel Evans, who was in command of the Third Battalion of the New Zealand Rifle Brigade, and had been cited in Distinguished Service Orders and received the Distinguished Conduct Medal, was killed in action on Oct. 12, 1917.

Alfred Winter Evans was born in Natal, South Africa, in 1881, and received his preliminary education in the Durban High School. For the next five years he attended St. George's School at Harpenden Heights, England, until 1898. Returning to South Africa, he worked for three months of 1898 as underground sampler for the Crown Deep Gold Mining Co., at Johannesburg, and for six months of the following year he was shift-boss for the Crown Deep, Ltd. At the outbreak of the Boer War he took an active part and was engaged on military service from 1899 to 1902. At the close of the war, he came to New York and attended the Columbia School of Mines, where he graduated with the degree of E. M. in 1906. A part of that year he spent as assayer to a mining company at Poland, Arizona.

In 1907, he returned to South Africa and engaged as underground contractor in charge of mining and development at a number of the South African mines, notably the French-Rand, the Village Main Reef, the City Deep, and others. In 1908, he acted as shift-boss for the Ferrerra Gold Mining Co. at Johannesburg. In 1909, he was appointed acting general manager for the Consolidated Goldfields of New Zealand, but, returning to South Africa, for the next two years he was assistant general manager of the Simmer Deep Gold Mining Co., at Johannesburg. In June, 1911, he returned to New Zealand as general manager and consulting engineer for the Consolidated Goldfields of New Zealand, at Reefton, which position he held at the time of his admission to the Institute in 1914. Early in September, 1915, he left New Zealand for active service in France.

### Corporal Sheppard B. Gordy

Sheppard B. Gordy, born in Ansonia, Conn., in 1889, graduated from the Sheffield Scientific School in 1910 and with the degree of E. M. in 1912. He entered the employ of the Braden Copper Co. immediately on his graduation and remained with this company four years, working up through the various grades to assistant foreman and foreman to one of the individual mines. The last year he was in its employ, he acted as general mine foreman. He was for a short time with the Chile Exploration Co. and then went with the Andes Exploration Co. (Anaconda) and remained with the latter, doing mine exploration and development work, until he sailed for home in June, 1918, to give a more personal and active service to his country in her need.

CORPORAL SHEPPARD B. GORDY.

He declined a tender of induction into an officers' training camp, at the urgent request of Professor McClelland supplemented by Mr. Potter, because they demonstrated to his satisfaction that he could give more valuable service to his country in helping out with the aircraft production. Being within the draft age, in order to carry out this plan, it was necessary for him to be called by his Local Board and sent to camp. This call came Aug. 26. Two weeks before that time he had gone to Dayton, Ohio, to begin the study of aircraft production, and was sent by the Dayton Board to Camp Sherman, Chillicothe, Ohio, where he died of pneumonia on Oct. 9.

### Lieutenant Thomas Clarence Gorman

Thomas C. Gorman, a lieutenant in the Second Tunnelling Company of the Canadian Engineers, was killed in France on Mar. 18, 1918. He was resting in his sleeping hut and was in the act of writing a letter when a bursting shell killed him.

Lieutenant Gorman was born in Ottawa, Canada, in 1888, and after preliminary education at Ottawa University and the Ottawa Collegiate Institute, he graduated from McGill University in 1913, as a mining engineer.

LIEUTENANT T. CLARENCE GORMAN.

After graduation, Mr. Gorman spent several months in a tour of Europe.

The year 1909 was occupied at one of the Canadian graphite mines, and the summer of 1911 he spent with the Granby Mining, Smelting & Power Co., Ltd. At the time of his admission to the Institute, in 1914, he was sampler with the Dome mines at South Porcupine, Ontario. His next engagement was at the Creighton mine, Ontario, but in 1916 he joined the Canadian Expeditionary Force and went to England.

A letter written by Major Ritchie, commanding the Second Tunnelling Company, to Lieut. Gorman's mother, contains the following words.

Tommy was killed at our headquarters by a nine-inch shell which struck the officers' quarters. The Hun had been shelling a small village close to the camp with a long-range gun and a shot fell into our camp. Tom was killed instantly by the concussion. He was buried this morning with Military Honours in the military cemetery close to our camp. Tommy was an efficient officer and a good soldier and his loss is very keenly felt by the officers and men of this unit.

---

### Lieutenant William Hague

William Hague, First Lieutenant Co. F, 116th Regiment of Engineers, and a member of the Institute since 1906, died of pneumonia in France, on Jan. 1, 1918.

Lieutenant Hague was born in Orange, N. J., Mar. 31, 1882, the son of the late James D. Hague, a distinguished mining engineer and a life member of the Institute, and Mary Ward (Foote) Hague, of Guilford, Conn. He attended Milton Academy, Milton, Mass., and was graduated from Harvard in the class of 1904. He was a nephew of the late Arnold Hague, of the United States Geological Survey. Seven years ago he married Elizabeth Stone, of Milton, who with their son, James D. Hague, six years old, survives him.

Lieutenant Hague's mining career began immediately after his graduation from Harvard, when he went to Bisbee, Ariz., to be a surveyor's helper in the mines of the Copper Queen Consolidated Mining Co. In 1905, he became an instrument man, being engaged on the construction work of the Copper Queen smelting plant at Douglas, Ariz. In the latter part of the same year he was transferred to the geological department of the company, being occupied in that work until May, 1906. The summer of 1906 was spent in prospecting in Michigan; but in the autumn he returned to Arizona as assistant in construction of the Copper Queen plant at Douglas, and remained on that work until October, 1907. The autumn and winter of 1907–08 he spent in traveling in the United States and Mexico, his purpose being to broaden his experience; wherefore he proceeded leisurely, occasionally taking a position for a short time. Thus, for two months he was employed as a shift boss in the cyanide plant of the Guanajuato Consolidated and Milling Company.

A serious illness that befell him in 1908 kept him from work during the major part of that year, but upon his recovery, in December, he was appointed managing director of the North Star Mines Co., a famous and successful gold-mining enter-prise in California, with which his distinguished father had been identified for a great many years.

However, William Hague could not keep away from purely professional activities in directions wherein he was intensely interested, and dur-ing a considerable part of 1909 and 1910 he was engaged in geological work at Bisbee, Ariz., for the Copper Queen Consolidated Mining Co., making occasional trips to Grass Valley, Cal.

LIEUTENANT WILLIAM HAGUE.

From June, 1910, up to the time when he entered the United States Army, Mr. Hague resided at Grass Valley as managing director of the North Star Mines Co.; but during 1911 he joined J. R. Finlay as assistant in the ap-praisal of copper mines for the State of Michigan.

Lieutenant Hague entered the Officers' Training Camp at Plattsburg in September, 1916. After receiving his commission, he was called to the Engineers' Training Camp at Vancouver Barracks, Ore. Later he was transferred to the camp at American Lakes, near Tacoma, and from there to Charlotteville, N. C. He was ordered with his regiment to Mineola in November, and soon afterward left for France. His family received news of his safe arrival abroad on Dec. 15. A cable of Christmas greetings to his family was the only other word received from him.

In the *New York Evening Post* a few days after news of his death had been received, there was a tribute from an anonymous friend which may well be repeated:

A few short weeks ago there was the bustle of camps; then a great silent flitting of our boys going "over there," and now there are commencing the first brief lists of those who are to lie in the torn fields of France. Today we read of Lieut. William Hague, whom we said good-by to hardly more than a month ago—so clean, so young, so strong—who, abandoning the professional career in which he had won such commendation and which held for him such promise, leaving his wife and his little boy to whom he was so dear, answered at once the call for men of his training, and is now dead "in the service of his country."

There are many friends of that courtly and dignified gentleman James D. Hague who recall, both here and in Stockbridge, the parental pride in the promising lad of such a little time ago—the eager schoolboy at Milton, the rather grave youth at Harvard, his entry into new experiences in the Western mining world, and who, seeing him during his stay at Camp Upton, realized that the old Puritan stock was still sound and true—and now with him the struggle is over and the sacrifice made.

## Captain William Teasdale Hall

Captain Hall, who was admitted to Junior Membership in the Institute in 1915, while still a student of mining at the University of Toronto, was killed in action in France on May 19, 1917.

Captain Hall was born in Toronto in 1893, and received his academic training at Harbord College, from 1908 to 1911, when he entered the

CAPTAIN WILLIAM T. HALL.

University of Toronto, School of Practical Science. During the three summer vacations of his university course, he was employed by the O'Brien mine at Gowganda, next by the McIntyre Mining Co., at Porcupine, and finally by the Mines Branch of the Geological Survey, which was conducting magnetic surveys to the north of Port Arthur.

When war was first declared, Mr. Hall endeavored to join the artillery as an officer, and for the purpose secured a provisional lieutenancy with a Hamilton battery, but he did not receive an overseas appointment, and on being graduated as a mining engineer in 1915, he was offered a position in Chile, which he accepted. He left Toronto on May 27, 1915, and was engaged in Chile as a mining engineer for nearly a year. In the latter part of April, 1916, he decided to offer himself again for service at the front. He crossed the Andes and sailed for Liverpool, where he landed on May 13, 1916.

On May 27, 1916, Captain Hall was given a commission as a lieutenant in the Royal Flying Corps, 21st Squadron. He went to the front on Sept. 1, 1916, and remained there continuously. His record with the Royal Flying Corps was considered a remarkable one, for he was at the front within two months of the date on which he was granted a commission, during which interval he took the prescribed technical course at Oxford, and aviation training at Netheravon and Bristol.

On Sept. 14, 1916, he took his first flight over the German lines, when his engine stalled and he had to volplane down, fortunately landing on the French side of the lines. On Sept. 16, he went over the lines with six others, and not returning immediately with the others, was lost. He tried to make for the sea and come back following the course of a river, but when he did descend he found that he was 60 miles west of Paris and 100 miles south of his lines. He flew over to Paris and stayed there for the night, going back to his squadron the next morning. Then on Oct. 4, he was sent up at night for the first time, and in making a landing he passed the flares just as he struck the ground, which put him in total darkness. He struck something and his machine turned a complete somersault, which accident put him in the hospital for ten days.

In December, 1916, he left the 21st Squadron and was transferred to the 24th, as the 21st was made an Artillery Squadron and he preferred to remain a fighting scout. In the same month, he was made Acting Flight Commander for about 6 weeks, and about April 3, 1917, his rank of Flight Commander was confirmed. The beginning of May he started to introduce a number of new machines to his squadron, these being of an entirely new type. On May 19, he was up with one of the machines in the evening, practising sharp turns and dives, when the wings suddenly crumpled up and he fell 700 ft., being killed instantly. He was buried in a French Cemetry at Monchy Lagache about 8 miles southeast of Perronne, and about 25 miles east of Amiens.

### Sergeant Herbert Moore Harbach

Herbert Moore Harbach was born in Lebanon, Pa., Apr. 4, 1891, graduated from the Lebanon High School in 1911, and entered State College from which he received his degree of Bachelor of Science in Metallurgical Engineering in 1915. He was employed by the Lackawanna Iron Co., at Buffalo, N. Y., and was later transferred to their Lebanon plant to start up the newly erected benzol plant. He was later made foreman of the coke ovens, which position he held after it was bought by the Bethlehem Steel Co., and until he enlisted in the service of his country in September, 1917. He was stationed at Camp Meade, Md., until the spring of 1918, when he was transferred to the C. W.

SERGEANT HERBERT MOORE HARBACH.

S., at Niagara Falls, as an expert chemist. There he was engaged in perfecting the various kinds of gases, and while thus employed was gassed several times, which so affected his heart and lungs that he was unable to withstand an attack of the influenza which developed into pneumonia, from which he died at the end of 24 hours.

He was a member of St. Luke's Episcopal Church, Lebanon, and belonged to the Ancient Accepted Scottish Rite of Freemasonry, Lebanon Lodge 121, I. O. O. F. and Camp 65, P. O. S. of A.

---

### Lieutenant Bernhardt Edward Heine

Bernhardt Edward Heine, a lieutenant in the Aviation Service of the U. S. Army, died as the result of a fall in an aeroplane at Fort Sill,

LIEUTENANT BERNHARDT E. HEINE.

Okla., on Aug. 10, 1918. The accident occurred on Aug. 2, when the machine in which he was flying with Lieutenant Carsons fell from a height of 1500 ft.; Lieutenant Carsons was instantly killed, but Lieutenant Heine died from his injuries a week later.

Lieutenant Heine was born at Mount Clemens, Mich., in 1895. After attending the local High School he entered the Michigan College of Mines, and became a Junior Member of the Institute while still a student. He graduated in 1916, with the degree of B. S.

In 1914 he served as a transitman for the engineering department of the City of Hancock, Mich.

Lieutenant Heine had been in the army since 1916, having been with the troops at the Mexican border. He had been in the Aviation Section since August, 1917.

## Captain John Duer Irving

John Duer Irving, who left his post as Professor of Economic Geology at the Sheffield Scientific School, New Haven, Conn., to join the Eleventh Regiment of Engineers shortly after the declaration of war, died in France, July 26, 1918, from an attack of pneumonia. Through the initiative of Mr. Benjamin B. Lawrence, a memorial service to Captain Irving was held at St. Paul's Chapel, Columbia University, on Sunday afternoon, August 4, which was attended by members of Captain Irving's family, and about four hundred of his friends from Columbia, Yale, the American Institute of Mining Engineers and the Association of the 11th Regiment of Engineers. Professor James F. Kemp delivered the memorial address, and has prepared for the *Institute* the following account of Captain Irving's life, Professor Kemp and Captain Irving having been associated on the most intimate terms for a number of years.

The war has brought home to friends and kin in the western shores of the Atlantic many losses whose suddenness has the shock of a blow. Many more will follow, but none can leave a sharper sense of regret than the news that the keen and productive mind; the inspiring teacher; the successful and considerate editor, the fine, true man and friend had all passed away when John Duer Irving made the final great gift to the service of his country, in the cause of decency and right. In the closing days of July, the cable brought the tidings that toward midnight on the twentieth of the month pneumonia had proved too great a tax upon one already worn by excessive labors.

John Duer Irving was born in Madison, Wis., August 18, 1874. His father, Roland Duer Irving, was at the time professor of geology, mineralogy and metallurgy in the State University of Wisconsin. Roland Irving, the father, had entered Columbia College in 1863, but trouble with his eyes compelled him to suspend his studies in the classical course in his sophomore year, and to replace study with some months of life and travel in England. Ultimately his eyes became stronger, and he entered the School of Mines, completing the course for the degree of Engineer of Mines in 1869, with the third class graduated. The class of eleven contained other future geologists. It numbered on its roll Henry Newton and Walter P. Jenney of the early survey of the Black Hills, where John, future son of Roland, was in the course of years to make his doctor's dissertation, and with its publication his really serious entrance into the profession.

Roland Irving was a favorite student of Professor J. S. Newberry, affectionately known to all of his students as "Uncle John," and after some experience in mining and smelting in Pennsylvania and New Jersey, joined the Ohio Survey under Dr. Newberry. The work in Ohio was largely performed during the vacations of the University of Wisconsin, to whose chair Roland Irving was called in 1870. Later, his residence in Wisconsin led to two events of prime importance in the history of American geology. He began the study of the Gogebic iron range and caught the clue to the origin and stratigraphy of the iron ores of all the Lake Superior ranges; and he later undertook the mapping and description of the copper-bearing rocks of the Lake Superior basin. The formative years of the son, John, were passed in a home where the father,

Roland, was preparing the now famous monographs on these two subjects.   Not alone did John's father work upon their preparation, but the
skilful hand and fine sense of color of his mother were placing on paper
the beautiful illustrations of the microscopic mineralogy of the copper-
bearing rocks, and all in the very early days of microscopic rock study in
America.

CAPTAIN JOHN DUER IRVING.

In John's fourteenth year, illness, preying on a. constitution never
over-strong, deprived him of his father, and left him the almost sacred
duty of following in the footsteps and continuing the work of one interrupted at thirty-nine in mid-career.   Mrs. Irving removed to the East
among her kinsfolk and John was prepared for Columbia College to represent the fourth generation of his name upon the alumni rolls.   His
boyhood period passed and his college and university life began.

He entered in 1892 and took the prescribed courses of the time. The vacation of 1895, following his junior year, was passed with one of Professor Osborn's parties in searching for Tertiary vertebrates in the Browns Park beds of northeastern Utah. On the observations there obtained was based the first contribution of John Irving to geological literature. Following his senior year, he went with the writer to the Adirondacks and had a month or two of experience with old-time crystalline rocks in the mountains around Elizabethtown. In the vacation after his first year of graduate study a place was found for him in the field-party of Dr. Whitman Cross in the steep and rugged mountains of the San Juan region of southwestern Colorado.

About this time the writer made a visit to the northern Black Hills of South Dakota and became impressed with the extremely interesting problems presented by the local geology, which, as earlier stated, had been covered by the pioneer work of Roland Irving's old classmates, Henry Newton and Walter P. Jenney. A large-scale map had been prepared by Professor Frank C. Smith and Dr. MacGillicuddy of the Dakota School of Mines, of Rapid City, and by the generous coöperation of these two friends, John Irving was suitably started upon his four months of field-work for his doctor's dissertation. With the completion and publication of his results and the taking of his doctor's degree, the second period of his life merged into the third—that of active work in the outside world.

Combined with a most creditable passing of the civil service examinations, his work in the Black Hills brought to its author an appointment in the ranks of the U. S. Geological Survey, and assignment to the party then undertaking the investigation of the mineral resources of this area. The study of the mines placed young Dr. Irving in close association with the most respected of American mining geologists, that fine, true, Nature's nobleman, Samuel Franklin Emmons, and in the end, made of John Irving Dr. Emmons' closest helper. Years afterward, when Dr. Emmons passed away, leaving in fragmentary and uncompleted state the great new monograph on Leadville, it was John Irving who finished the manuscript and forwarded it to the Survey, ready for the printer.

Following the field experience in the Black Hills, John Irving was busied in association with F. L. Ransome in the Globe district of Arizona; with J. M. Boutwell in the Park City district of Utah; with W. H. Emmons in the Needle Mts. quadrangle of Colorado; and with Howland Bancroft at Lake City in the same state. He also did mapping of coal-bearing quadrangles in Indiana under M. L. Fuller, and in Pennsylvania, under M. R. Campbell. Most important of all, he became in time the right-hand man of Dr. Emmons in the revision of the Leadville monograph. While on the Survey and spending the winters in Washington, John Irving entered heartily into the life of the younger men, and was active in the so-called Association of Aspiring Assistants of the Survey; whose initials were a sort of caricature of the American Association for the Advancement of Science.

But the desire to teach, partly inherited, partly acquired, was very strong, too strong to be longer suppressed; and with its gratification in 1903 John Duer Irving took up, after ripe preparation, the teacher's part of his career. The opportunity came to the writer to suggest to an old friend, Professor Wilbur C. Knight, of the University of Wyoming, one who might substitute for him during a year's leave of absence.

The offer proved agreeable to Dr. Irving, and he undertook the work, retaining his connection with the Survey for the free time of the vacations. Some of the Survey work outlined above, was indeed done after he had begun to teach. In 1904 he was called to be assistant professor of geology at Lehigh University, and was promoted to the full chair in 1906. In 1907 he was called to be professor of economic geology at the Sheffield Scientific School of Yale, where he shared in developing the mining course made possible by the gift of the Hammond Laboratory. This professorship Dr. Irving held at the time of his death.

In 1905, while John Irving was at work at Lehigh University, the plan **was** developed of establishing a magazine which might be the special means of expression and record for the vigorous young school of American students of ore deposits and applied geology, which had then become a marked feature of our scientific life. We gathered a little band of pioneers, willing to risk a part of their not over-abundant worldly possessions in the venture with no thought of return. John Irving was our choice for managing editor. To his untiring efforts, ably aided by the unselfish work of W. S. Bayley, as business manager, we chiefly owe the thirteen volumes of this most valuable and interesting journal, *Economic Geology*. As we turn its pages we see the subjects which appealed from time to time to its editor. His work in the mining districts emphasized the importance of ore shoots and localizations of values in veins, and he seeks to classify and systematize their causes. He is again impressed with the importance of a comprehensive study of special problems, wherever one particular case of them may be illustrated, as contrasted with the generally localized investigations carried on by one individual. The committee which has been studying for several years the problem of secondary enrichment, under L. C. Graton, carries out exactly this idea. John Irving's extended Leadville experience, in the mines where S. F. Emmons first formulated the ideas of replacement of rock by ore, leads him to seek to establish criteria whereby replacement bodies may be identified. Thus as we turn the pages of the magazine we see how, year by year, an active and thoughtful mind was philosophically pondering, now this, now that important phase of his special branch of science.

And then, while in the full exercise of his many useful activities, rose above the horizon the dark cloud of the Robber Barons' war. He, as all of us, awoke to the growing conviction that the world in which we lived was not what we had thought it; that our ideals of government, our rules of openness, loyalty and truth in the relations of life, the very conditions of our existence were in danger and would need the supreme effort to defend them. In the spring and early summer of 1916, the writer knows from intimate talks that John Irving felt the danger menacing our country and others like it, from the growing threat of the worst features of medi-. ævalism. Being unmarried, he believed that even though he was past forty years of age, it was his duty to go to Plattsburg and enter the officers' training camp. He took up the routine earnestly and seriously and wrote with great pride of his promotion to be a non-commissioned officer. When the camp closed he entered his name as one available for service, if conditions should call for him. In the spring of 1917 these conditions materialized. He took his officer's examinations, was appointed captain in the 11th Regiment of Engineers, and was granted leave of absence by the authorities of Yale. At first he expected to serve as topographic engineer, or as an interpreter of air-plane maps. He jestingly referred to himself as "skyographer." Actually he became recruiting officer for his regiment, in whose ranks are so many of the Institute's members. With them he sailed for France in July, 1917. Of his special field of work we know only what the censorship has permitted us to learn from his letters. He was early engaged in railway construction and worked just as long hours as anyone could in building the arteries of supply and nourishment of the Army. His duties last Fall we know brought him under shell-fire and he learned to keep his nerve amid these trying conditions. Like so many of our boys he writes jestingly of them, taking the dangers in the light-hearted way of our countrymen, as being all in the day's work.

Later, as tunneling and the exploding of mines beneath the enemy's works, and all sorts of excavation became so important, he was detailed to the engineers' school at headquarters and was busy with a seemingly endless procession of classes to be instructed in the rudiments of mining engineering. His letters show that the calls were hard, exacting and exhausting. He speaks of never having worked so hard in all his life. Probably his vitality ran low and his powers of resistance and recuperation were exhausted. We only know that he returned to the front, became a victim of the so-called Spanish grippe, which, despite every effort on the part of the medical staff, developed into pneumonia, and that one more name was added to the Roll of Honor.

As we look back over this brief sketch of a busy and useful life, we see that it was filled to the very limit. Those of us who knew John Irving well know that the duties and calls were met faithfully and with a high sense of responsibility. Many of them were essentially unselfish. The magazine *Economic Geology* was a labor of love. The time of Professor Irving was never so valuable but that a student could command advice and guidance. The full strength and more of Captain Irving were given to his country. And yet we know that the calls often bore heavily on a physique none too robust. Although tall and broad-shouldered of stature, our friend had some delicacy of constitution which asserted itself at times of special stress and which gave warning that certain limits must not be passed.

Although I have written of his scientific work, I have not mentioned that many talks while we have been off on trips together revealed to me ambitions in literary expression and composition which were in keeping with the high traditions of his family, bound up, as they are, with one of the very first of our really great men of letters. Insistent calls prevented John Irving from realizing all that was in his mind; but the ambition certainly gave lucidity and clearness to his scientific writings.

It is, however, as friend that we like best to recall him, and in these respects we know well, that while his work lives after him, the fine sweet inspiration of his life will endure still longer. We live in the most trying times with which we have all been confronted for nearly three generations. Faith is called for, as almost never before. We must have the grim and unshaken holding on to calm belief in the triumph of right and in the subordination of cruelty and unbridled selfishness to the control of law and justice. A decent consideration for the rights of others and observance of these must not perish from the earth, nor must the rule of abominable materialism be permitted to exercise its sway over victims unable to protect themselves. For these high causes John Duer

Irving gave his all.   Useful and inspiring as he was in life, he rises to yet greater heights in his supreme sacrifice, and to us who yet remain he is a sermon whose moving appeal causes to sound aloud the deepest chords of our nature.

JAMES F. KEMP.

A copy of the following letter written by Major Evarts Tracy to President Hadley, of Yale University, has been handed us for publication through the courtesy of Mr. B. B. Lawrence.

July 27, 1918.

Dear President Hadley:

Before this reaches you, you may have heard of the death of Captain Irving, which occurred last night.

I only wish to tell you what a loss it has been to the Army.   It is always sad to have anyone die here otherwise than in action.   In the latter event one always feels that it is a proper and glorious death.

Captain Irving died as gloriously as any man in the service ever died.   He gave all he had.   The amount of work he accomplished here in the design and adoption of his methods in mining and shelter dugouts, which are the only life savers when batteries are registered by the enemy, was beyond calculation.

He worked himself to death, and in the face of opposition proved that he was right, time after time.   We all remonstrated with him at his hours, but his devotion to duty, as he conceived it, lowered his vitality, and pneumonia, following a bad attack of the so-called Spanish grippe, cut him down.

When I tell you that since he was taken ill, the personnel department has been over the records of over fifty men, trying to find someone to take his place, without success, you can appreciate his value to the service.

If you can let any of his friends know what we, his close associates here feel, about his loss we will appreciate it.   No one here has done more for the United States than he has.

BIBLIOGRAPHY OF THE WORKS OF JOHN DUER IRVING

1896.—The Stratigraphical Relations of the Browns Park Beds of Utah.   *Transactions*, New York Academy of Sciences (1896), **15**, 252–259.

1898.—Contact-Metamorphism of the Palisades Diabase.   Abstract.   *American Geologist* (1898), **21**, 398.

1899.—A Contribution to the Geology of the Northern Black Hills.   *Annals*, New York Academy of Sciences (1899), **12**, 187–340.

Contact Phenomena of the Palisade Diabase (New Jersey).   *School of Mines Quarterly* (1899), **20**, 213–223.

1901.—Some Recently Exploited Deposits of Wolframite in the Black Hills of South Dakota.   *Trans.* A. I. M. E. (1901), **31**, 683–695.

1903.—Ore Deposits of the Northern Black Hills.   *Mining and Scientific Press* (1903), **87**, 166–167, 187–188, 221–222.

1904.—Ore Deposits of the Northern Black Hills.   *Bulletin* **225**, U. S. Geological Survey (1904), 123–140.

The Ore Deposits of the Northern Black Hills.   *Mining Reporter* (1904), **50**, 430–431.

JOHN DUER IRVING AND S. F. EMMONS.   Economic Resources of the Northern Black Hills.   Part II.   Mining Geology.   *Professional Paper* **26**, U. S. Geological Survey (1904), 43-222.

1905.—JOHN DUER IRVING AND W. H. EMMONS.   Economic Geology of the Needle Mountains Quadrangle (Colorado).   *Geologic Atlas*, U. S. Geological Survey (1905), **131**.

Microscopic Structure and Origin of Certain Stylolithic Structures in Limestone.   Abstract.   *Annals*, New York Academy of Sciences (1905), **16**, 305–306.   Same.   *American Geologist* (1904), **33**, 266–267.

[Ore Deposits in the Vicinity of Lake City, Colorado.   *Bulletin* **260**, U. S. Geological Survey (1905), **260**, 78–84.

1905,—Ore Deposits of the Ouray District, Colorado.   *Bulletin* **260**, U. S. Geological Survey (1905), 50–77.

University Training of Engineers in Economic Geology. *Economic Geology* (1905), **1**, 77–82.

1906.—Review of "The Geological Map of Illinois," by Stuart Weller (Ill. State Geol. Survey, *Bull.* No. 1). *Economic Geology* (1906), **8**, 816–818.

1907.—JOHN DUER IRVING, WHITMAN CROSS AND ERNEST HOWE. Description of the Ouray Quadrangle (Colorado). *Geologic Atlas*, U. S. Geological Survey (1907), **153**, 20 pp.

JOHN DUER IRVING AND S. F. EMMONS. The Downtown District of Leadville, Colorado. *Bulletin* **320**, U. S. Geological Survey (1907), 75 pp.

1908.—The Localization of Values or Occurrence of Shoots in Metalliferous Deposits. *Economic Geology* (1908), **3**, 143–154.

1910.—Special Problems and Their Study in Economic Geology. *Economic Geology* (1910), **5**, 670–677.

1911.—JOHN DUER IRVING AND HOWLAND BANCROFT: Geology and Ore Deposits near Lake City, Colorado. *Bulletin* **478**, U. S. Geological Survey (1911).

Replacement Ore-Bodies and the Criteria for Their Recognition. *Economic Geology* (1911), .**6**, 527–561, 619–669.

JOHN DUER IRVING AND WALDEMAR LINDGREN. The Origin of the Rammelsberg Ore Deposit. *Economic Geology* (1911), **6**, 303–313.

1912.—Discussion on a Concise Method of Showing Ore Reserves. *Trans.* (1912), **43**, 714.

Geological Diagnosis (Editorial). *Economic Geology* (1912), **7**, 83–86.

Some Features of Replacement Ore Bodies and the Criteria by Which They May be Recognized. *Journal,* Canadian Mining Institute (1912), **14**, 395–471.

(Geological Field Methods. Editorial.) *Economic Geology* (1913), **8**, 64–65.

JOHN DUER IRVING, H. D. SMITH AND H. G. FERGUSON. A Selected List of the More Important Contributions to the Investigation of the Origin of Metalliferous Ore Deposits. In S. F. EMMONS' Volume on Ore Deposits. A. I. M. E. (1913), 837–846.

1913.—The Substructure of Geological Reports. *Economic Geology* (1913), **8**, 66–96.

1915.—Discussion on Recrystallization of Limestone at Igneous Contacts. *Trans.* (1915), **48**, 214.

Discussion on To What Extent is Chalcocite a Primary, and to What Extent a Secondary Mineral in Ore Deposits. *Trans.* A. I. M. E. (1915), **48**, 194.

1916.—Discussion on the Iron Deposits of Daiquiri, Cuba. *Trans.* A. I. M. E. (1916), **53**, 65.

Discussion on the Iron Ore Deposits of the Firmeza District, Cuba. *Trans.* A. I. M. E. (1917), **56**, 135.

1918.—Chapter on "Geological Data for Prospecting and Exploration," in *Mining Engineers' Handbook* (Peele), 392–408.

---

## Lieutenant S. A. Lang, C. E. B. A. SC.

From "Knots and Lashings," of Oct. 10, 1918, the military publication of Canadian Engineers' Training Depot, St. Johns, Que.

During the early hours of Sunday morning last (the 6th inst.) there passed from our midst one of the most affable and estimable officers who, during our brief connection with the Depot, it has been our fortune to meet.

Sidney A. Lang was born in Ontario in the year 1884 and lived in Toronto for a number of years. He attended Toronto University, where he displayed remarkable ability, and after graduating followed the profession of mining engineer. Being young and endowed with the spirit of adventure, he followed his profession in the mines of Chile and Peru; he could speak Spanish fluently. Always a reserved man, of refined tastes, it was difficult to induce him to relate some of his adventures in those countries so shrouded in mystery and romance. He joined the Canadian Engineers in Toronto as a sapper and shortly after came to this Depot.

We remember him as a messmate, as a roommate, and when we rubbed shoulders with him on the square, for his kindly thought and unassuming manners.

The Spanish grippe has claimed many from our depot. As we read down the names we pause here and there to recall some incident to establish permanently in our memory as a link to bind us, the living, with the dead.

> "And there's one name we cannot pass
> So we pause, and ponder awhile.
> And recall all his cheery words,
> And his smile—his perpetual smile."

Men detained in the Quarantine Camp will have cause to remember him too, for he displayed his gentlemanly bearing and courtesy to all he came in contact with.

----

### Lieutenant Norman Lloyd Ohnsorg

First Lieutenant Norman Lloyd Ohnsorg died in Nashville, Tenn., on Oct. 11, 1918. He was born in St. Louis, Mo., Jan. 10, 1889, and lived in that city until he was 11 years of age when his parents removed to Iron Mountain, Mo., where he spent the next 6 years. On Jan. 10, 1906, he entered the Missouri School of Mines and Metallurgy at Rolla, Mo., and graduated from that institution in 1910, receiving both the Bachelor of Science in Mining Engineering and the Engineer of Mines degrees. In 1912, he received the degree of Bachelor of Science in Metallurgy, and in 1916, the degree of Metallurgical Engineer. After leaving college he was connected with the following companies: Noble Electric

LIEUT. NORMAN LLOYD OHNSORG.

Steel Works, Heroult, Shasta Co., Cal.; Mammoth Copper Mining Co., Kennett, Cal.; St. Joseph Lead Co., Herculaneum, Mo.; Granby Smelting & Mining Co., Neodesha, Kans.; The Phosphate Mining Co., Nichols. Fla. While with the last company, he volunteered, in April, 1917, for service in the engineering corps of the U. S. Army. He passed an excellent examination, was accepted and called to Washington and placed in Division T. of the Ordnance Department where he was engaged in research work pertaining to nitrogen. He was then ordered to Sheffield, Ala., where he was second in command of the construction of the Government nitrogen plants at Muscle Shoals.

After the completion of the plants, he was ordered to Buffalo, N. Y., on Sept. 28, 1918, to inspect and report on a government plant. After completing this work, he was called back to Sheffield. He left Buffalo on Oct. 6, but delayed trains caused him to miss connections in Cincinnati where he had to wait 12 hours. He had not been feeling well for several days before leaving Buffalo; influenza developed, and when his train reached Nashville, Tenn., he was too ill to continue his journey and was taken to the Kissam Hall Hospital, Vanderbilt University, where he died of pneumonia.

He was a young man of unusual ability and high moral character, always bright, happy and kind, faithful and honorable in all things; loved by both his associates and those under his authority. His life and character cannot be better described than by the following quotation from one of his superior officers: "Lieutenant Ohnsorg was a man loved by all of us, always cheerful, just and kind, with unusual executive ability. We can never expect to find a man who has his ability and fine temperament and his position can never be filled as he filled it."

He was happily married and leaves to mourn his loss a wife, Constance E. (Rogers) Ohnsorg and a baby girl born since his death, Nov. 29, 1918, a father, W. H. Ohnsorg, mother, Ida May Ohnsorg, and a sister, Mrs. Stuart Strathy McNair.

He was a member of the American Institute of Mining Engineers and of the Florida Phosphate Miners' Association.

---

### Lieutenant Edward Hale Perry

At the height of the first great German offensive of the Spring of 1918, Edward Hale Perry, of Boston, First Lieutenant, Company D, Sixth Regiment Engineers, U. S. Army, was killed on March 30, near Warfusee—Abancourt, Picardy, France, while defending the Bois des Tailloux against the terrific plunge aimed at Amiens.

Lieutenant Perry was born in Boston, Jan. 23, 1887, the son of Georgianna W. and the late Charles F. Perry. After completion of his college preparatory course, he travelled for a year in South America and Europe before entering Harvard with the class of 1910. It might have been regarded as the natural thing for Perry, upon graduation, to choose a path that would lead to a business or professional career at home, but there were in his character a solidity, a horror of sham, a contempt for the "soft"

LIEUTENANT EDWARD HALE PERRY.

things, and a love of the open which caused him to be attracted to a life of stern and sturdy reality. Accordingly, he entered the graduate mining school at Harvard, and received the degree of Mining Engineer in 1913. In the meantime, two summers spent in Western mining camps had attracted him particularly toward the geological aspects of mining so that the latter part of his course was directed definitely toward mining geology.

Because of his evident aptitude for geological problems, his mental and moral integrity, and his boundless enthusiasm, Perry was asked upon his graduation from the mining school to join the staff of the Secondary Enrichment Investigation. This he did, giving his services without compensation, though relinquishing in consequence an attractive opening in the geological department of one of the large mining companies of the Southwest. For two years he was thus engaged in intensive geological study of the principal copper mines of the country. During this period, his scientific development and his growth in judgment and poise made a profound impression on those most closely associated with him. And the value of his efforts and his spirit in the work of the organization is beyond measure or recompense.

At the conclusion of the field work of this investigation in 1915, Perry joined Dr. Augustus Locke, who had been associated in the same research, and took up professional practice in mining geology. In this Perry met with instant and conspicuous success, winning as much by his personal force, his ready grasp of every phase of a situation, and his ability to bring men to his point of view, as by his conscientious study and keen understanding of the conditions of ore occurrence and his sanity in interpretation and recommendation.

Notwithstanding his unusual success in commercial work, Perry maintained with keen relish and devotion his interest in the scientific aspects of geology. With Dr. Locke, he contributed a paper[1] on "The Interpretation of Assay Curves for Drill Holes." He sacrificed time and income in order to spend two or three months each year in continuing his special research upon the relations of rock alteration to ore deposition. His last days at home, even to his last hour before going to Plattsburg, were spent completing in outline the record of four years of study upon this subject, which Dr. Locke and the writer of this inadequate tribute to his memory will enjoy putting into final shape for publication, and which is certain to prove a noteworthy and valuable contribution to the science.

While Perry was in the midst of a professional engagement in Arizona, our country entered into the European War. He immediately advised his closest associates of his intention to enlist, and as soon as he could, with added help, complete the work then in hand, he came East and entered the officers' training school at Plattsburg, in May, 1917. In June, because of his technical training and experience, but particularly because of his application and ability, he was transferred to the Engineer Officers' Camp at Washington, and soon thereafter was commissioned First Lieutenant in the 6th Regiment Engineers, as reserve officer in charge of mining, sapping, and demolition.

Perry's work of instruction with his men won quick recognition and commendation. He was offered positions as instructor in this country, carrying with them higher rank than he could hope to reach in the regular army, but believing that his duty lay at the Front, he declined to consider them. He sailed for Europe in December, 1917. In January, Companies D and B were detached from the rest of the Regiment and, because of the ability of their officers, were brigaded with the 5th British

---

[1] *Trans.* (1916) **54**, 93.

Army and sent to Peronne to build heavy steel bridges over the Somme. While this work was going on, the Germans launched their great drive on March 21. For the ensuing few days it was the duty of the Engineers to stand by their bridges until the retiring British Army had crossed, and then demolish them. This they did, Perry and his platoon being the last to leave after the British Artillery had all passed. Then, on the 27th, these two companies joined that motley but determined and immortal band which General Carey, realizing the imminence of disaster to the entire Allied forces due to the crumbling and withdrawal of part of the British line, picked up and threw in to close the fast-widening breach. Lieutenant Perry had command of a section of the front line trench between Hamel and Villers-Bretonneux near the middle of this gap.

The energy and devotion which Perry put into his work as a soldier, and the spirit and fine courage with which he faced and paid the Great Price, may best be revealed by extracts from letters written to relatives by his associate officers since his death.

His fellow-Lieutenant of Company B wrote:

I have never worked with a man who put as much spirit and energy into his work, and who inspired men under him, causing them to exert their best efforts to help a common cause.

The officers and men who were privileged to know Edward feel that they have lost a true friend, and the men under him knew they possessed a leader of remarkable qualities, one who knew their wants and who cared for them before thinking about himself and his own comforts.

### Perry's Captain said in part:

During the previous months he was a tireless worker, never satisfied unless he was doing his own job and most of his neighbor's. In the early part of March, when we were on heavy bridging operations, he used to leave camp at 5 a. m. and return at 8 p. m. while two shifts of men worked under him; then he would spend a good part of the night on plans and lists of material.

No officer in the regiment was so trusted and looked up to by the men; they gave him their money to keep for them, asked his advice on all sorts of affairs, and besieged me with requests to transfer to his platoon. In his ability to get work done by leading instead of driving, he had no equal. And as a friend and brother officer, he leaves an unfillable gap that is brought to our attention every day. He had been recommended for promotion not long before his death.

He died as he had lived, helping others. It was Saturday, March 30. We underwent a good preliminary bombardment followed by the infantry attacks, supported by heavy barrages. Our trenches were pretty poor, as we had to get underground at the same time that we were keeping Fritz out of the way, and the artillery smashed a good deal of our defenses. A shell had demolished a traverse in Perry's section of trench, killing four men. He was working in the gap repairing the damage with his own hands, when a bullet, probably from a machine gun in an enemy aeroplane which was raking the trenches, penetrated his skull.

We all feel that his place in this organization, which he helped to build up, will never really be filled, but we draw what satisfaction we can from the circumstances of his death; as we must all go sometime, I know of no straighter, cleaner way than his.

### Colonel J. M. Hodges, his regimental commander, has written:

At a critical time during the German offensive in March, this organization was given a section of the front-line trench which was essential to the scheme of defense, and orders had been received that it was to be held at all costs. Lieutenant Perry was commanding a platoon of his company in the front line. He was killed instantly by a bullet through the forehead. At the time of his death, he was engaged in reconsolidating a section of trench that had been demolished by a previous bombardment and in arranging for the burial of his men who had been killed.

Lieutenant Perry was an excellent soldier and an exemplary officer. I had always considered him as one of the best, if not the best, of the young officers of the Regiment. He had real ability and could be counted on for results. At the critical time he did not weaken; I saw him shortly before he was killed; his conduct under fire was splendid and an inspiration to his men. His loss is felt deeply by all ranks. Thanks to him and to others, who like him, paid the full measure of devotion to their country, our line was held until the critical situation in that vicinity was at an end. He died the true death of a soldier, with his face to the enemy.

As we now look back, it is easy to believe that this holding of the line of defense intact by General Carey and his men was a determining factor in the outcome of the war. To have played so important and noble a part in this vital effort as that taken by Perry is assuredly the privilege of few. Our lives, it seems, are like capital entrusted to us to be expended as wisely and effectively as we may. With them we purchase whatever of accomplishment the stuff that is in us permits. It is impossible to escape profound regret that a career so full of the highest promise, and a personality so overflowing with all that is fine and lovable, should have been cut short at the age of thirty-one. Yet who can doubt that in a few months Perry bought with his life the fullest achievement of a life-time—a glorious part in the salvation of Liberty and Justice and Decency, indeed of Civilization itself!          L. C. Graton.

### Lieutenant Frank Remington Pretyman

Frank Remington Pretyman, Second Lieutenant with the Royal Engineers, was killed in action on June 17, 1916.          ·

Lieutenant Pretyman was born at Chicago, Ill., in 1890. He received his academic education at Marlborough College, Wiltshire, England, and in 1908 entered the Royal School of Mines, London, where he remained for two years. From 1910 to 1911, he was engaged as mine surveyor for the Mazapil Copper Co. in Zacatecas, Mexico, and for the next year with the Foundation Co. of New York, which was engaged in sinking a concrete drop-shaft at St. Albert, Alberta, Canada. At the conclusion of this engagement, Mr. Pretyman returned to the Royal School of Mines and finished his course there in the next two years, receiving the degree, Associate of the Royal School of Mines, in 1913. In the same year he became a Fellow of the Geological Society of London. At the time of his admission to the Institute, in May, 1914, he was taking a post-graduate course in geology at Columbia University, New York.

### Captain Frederick Bennett Reece

Frederick B. Reece, who was a Captain in the Royal Engineers, 232d Army Troops Company, of the British Expeditionary Forces, was killed in action. ·

He was born in Liverpool, England, in 1877, and received his academic education in England. The year 1899 he spent in the shops of Lingford Gardiner & Co., colliery engineers, at Durham. In 1903, he entered the College of Mines of the University of California, at Berkeley, and graduated in 1906. His summer vacation of 1905 was spent in timbering and machine drilling at the App mine, Tuolumne Co., California.

His first employment after graduating from the University of California was in El Oro, Mexico, where he spent two years with the Hacienda Vieja. In 1908, he was practising mining engineering at Lead, So. Dak., and in 1909 was engineer with the Socorro mine, at Mogollon, New Mexico.

In 1910, he was with the Hidalgo Guadalupe, Pachuca, Mexico, and in 1911 was engineer with the Gualcola Mines Co., at Tuquerres, Colombia. In 1913, he returned to the Southwest, and was employed fᴝr that year by the Inspiration Cons. Copper Co. at Miami, and in the middle of the following year he became engineer with the Detroit Copper Co., at Morenci.

In the autumn of 1914, Mr. Reece, stirred by his English blood, resolved to return to England to take part in the war. At this time he wrote "I think it will be better for me to resign my membership. If I come through all right, I will apply for re-election." The Institute, however, instead of accepting Mr. Reece's resignation, voted him an indefinite leave of absence, and this precedent soon crystallized into the regular procedure, now in force, of remitting the dues of members in active service. On Dec. 17, 1914, he wrote from Montreal that he fully expected to go over shortly in the second contingent from Canada, with Borden's Armored Brigade, operating with armored motor cars. As he said "this will be active enough service to satisfy any enthusiast, as they are used generally for reconnaissance duty."

---

### Soren Ringlund

Soren Ringlund died suddenly on July 24, 1918, at Fort Logan, Colo., where he was engaged in service with the Medical Department of the U. S. Army.

SOREN RINGLUND.

Mr. Ringlund was born in Denmark in 1875. He came to the United States in 1902, and became a citizen in 1908. He graduated as

B. S. and E. M. from the School of Mines at Socorro, New Mexico, in May, 1912.   Immediately after graduating, he began as a practical miner in the U. S. mines at Bingham, Utah, but in November, 1912, he entered the operating department of the Chino Copper Co.'s concentrator at Hurley, New Mexico, where he stayed until July, 1913.   During the latter half of 1913, he was employed as chemist in the laboratory of the El Paso smelter, at El Paso, Texas.   In December, 1913, he became engineer and geologist with the Empire Zinc Co., at Socorro, New Mexico, which position he held at the time of his admission to the Institute in 1914, and until he entered the army in June, 1918.   Before beginning the study of mining, Mr. Ringlund had completed a course in pharmacy, which accounts for his enlisting in the Medical Corps.

### Lieutenant George Roper, Jr.

George Roper, Jr., a Junior Member of the Institute, was killed in an aeroplane accident near Shotwich, England, on May 25, 1918.

He was born at Steubenville, Ohio, in 1893, and at the time of his admission to the Institute in 1916 was a student in mining and metallurgy

LIEUTENANT GEORGE ROPER, JR.

at the Massachusetts Institute of Technology, Boston, Mass., from which he received the degree of S. B. in June, 1917.

In August, 1917, he enlisted in the Royal Flying Corps, and after preliminary training on this side of the water, was sent to England. He was making the final cross-country flight of his course of training when the accident occurred.

## Raymond Weir Smyth

Raymond Weir Smyth, born Nov. 3, 1888, was the son of Herbert Weir Smyth, professor of Greek Literature at Harvard University. He graduated (A. B.) from Harvard in 1909 and later pursued advanced studies there in chemistry and in metallurgy. He held the position of metallurgist for the American Steel and Wire Co. at Worcester, Mass., and also for the Steel and Tube Co. at Youngstown, Ohio. For a time he was inspector of munitions for the British Government and stationed at the Railway Steel and Spring Co. at Latrobe, Pa. For a short time, in 1917, he was Inspector of Ordnance for the U. S. Government, and the following year, having entered the U. S. Naval Reserve Force, he

RAYMOND WEIR SMYTH.

was ordered in a similar capacity to the Midvale Steel Co. In September, he removed to the Navy Yard, League Island, Philadelphia, to train for the obligatory minimum of six months' sea duty. There he contracted influenza, which was followed by pneumonia, from which he died, September 27, at the Navy Yard Hospital.

## Captain Braxton Bigelow*

Captain Braxton Bigelow was the son of Major John B. Bigelow, U. S. A., retired, and grandson of the late John Bigelow, author, and Minister to France under President Grant. Captain Bigelow graduated

* Material for this notice was received just as this Volume was going to press.

from the Massachusetts Institute of Technology in 1910, and became a member of the A. I. M. E. in 1914.

He was in South America at the outbreak of the war, and left at once for Europe, where he served at first in the ambulance service. During the time of this service, he was decorated for bravery. Later on he obtained a commission in the British army as Lieutenant in the Royal Artillery, and at the time of his disappearance was Captain in the 170th Tunnelling Company, Royal Engineers, B. E. F. On the night of July 23, 1917, near Lens, he volunteered to lead a small party of sappers

CAPTAIN BRAXTON BIGELOW.

to investigate some mining work which the enemy was suspected of conducting. He "went over the top" with eight men. After bombing several dug-outs, all but two of his men were either disabled or killed. With these two he entered a mine. Leaving one man at the entrance and the other at a crossing, he went on with a man he had taken prisoner. The prisoner walked ahead of him. In a short time, the man at the crossing heard the Captain's voice, then a shot, and, running ahead, he found the prisoner alone. The latter, on being asked about the Captain, pointed up a mine shaft. Captain Bigelow never returned, and no trace of him has been discovered.

# INDEX